Functional Materials from Colloidal Self-assembly

Functional Materials from Colloidal Self-assembly

Edited by Qingfeng Yan and George Zhao

WILEY-VCH

Editors

Prof. Qingfeng Yan
Tsinghua University
Department of Chemistry
Haidian District
100084 Beijing
China

Prof. George Zhao
The University of Queensland
School of Chemical Engineering
Don Nicklin Bldg
4072 Brisbane St. Lucia
Australia

Cover Image: © oxygen/Getty Images

Library of Congress Card No.: applied for

British Library Cataloguing-in-Publication Data
A catalogue record for this book is available from the British Library.

Bibliographic information published by the Deutsche Nationalbibliothek
The Deutsche Nationalbibliothek lists this publication in the Deutsche Nationalbibliografie; detailed bibliographic data are available on the Internet at <http://dnb.d-nb.de>.

Print ISBN: 978-3-527-82795-4
ePDF ISBN: 978-3-527-82871-5
ePub ISBN: 978-3-527-82873-9
oBook ISBN: 978-3-527-82872-2

Typesetting Straive, Chennai, India
Printing and Binding CPI Group (UK) Ltd, Croydon, CR0 4YY

Printed on acid-free paper

C106593_110122

Contents

Preface *xv*

1 **Colloidal Molecules and Colloidal Polymers** *1*
Etienne Duguet, Etienne Ducrot, and Serge Ravaine
1.1 Introduction *1*
1.2 Colloidal Molecules: Mimicking Organic and Inorganic Molecules *3*
1.2.1 Clustering of Isotropic Colloids *3*
1.2.1.1 Clustering Assisted by External Fields *3*
1.2.1.2 Clustering Assisted by Geometrical Confinement *4*
1.2.1.3 Clustering Assisted by Physical or Chemical Interactions *6*
1.2.2 Clustering of Patchy Particles *8*
1.2.2.1 Self-aggregation of One-Patch Particles, i.e. Janus Particles *8*
1.2.2.2 Stoichiometric Attachment of Colloidal Satellites Around Patchy Particles *10*
1.3 Colloidal Polymers: Mimicking Organic Macromolecules *13*
1.3.1 Dipole-Directed Formation of Colloidal Polymers *14*
1.3.1.1 Electric Dipoles *14*
1.3.1.2 Magnetic Dipoles *14*
1.3.2 Formation of Colloidal Polymers by Nanowelding *15*
1.3.3 Formation of Colloidal Polymers Under Physical or Chemical Confinement *15*
1.3.4 Field-Directed Formation of Colloidal Polymers *17*
1.3.4.1 Electric Fields *17*
1.3.4.2 Magnetic Fields *19*
1.3.5 Ligand-Directed Formation of Colloidal Polymers *22*
1.3.5.1 Electrostatic Interactions *22*
1.3.5.2 Covalent Bonding *23*
1.3.5.3 Noncovalent Attachment *25*
1.4 Conclusion and Outlook *30*
References *31*

2 Self-assembly of Anisotropic Colloids in Solutions *37*
Yiwu Zong, Huaqing Liu, and Kun Zhao
2.1 Introduction *37*
2.2 Fabrication of Anisotropic Colloids *38*
2.2.1 Bottom-Up Routes *38*
2.2.1.1 Template(Seed)-Assisted Synthesis *38*
2.2.1.2 Cluster-Based Methods *41*
2.2.1.3 DNA-Based Assembly Methods *41*
2.2.1.4 Surface Modification Methods *42*
2.2.2 Top-Down Routes *42*
2.2.2.1 Photolithographic Fabrication Methods *42*
2.2.2.2 Physical Methods *43*
2.2.3 Anisotropic Colloids from Natural Materials *44*
2.3 Self-assembly Mechanisms of Anisotropic Colloids *44*
2.3.1 Self-assembly Through Specific Interactions *44*
2.3.1.1 Electrostatic Interactions *44*
2.3.1.2 Hydrophobic Interactions *45*
2.3.1.3 Entropic Depletion Interactions *47*
2.3.1.4 DNA-Mediated Interactions *52*
2.3.2 Assembly in External Fields *54*
2.3.2.1 Electric Field Assisted *54*
2.3.2.2 Magnetic Field Assisted *56*
2.3.2.3 Self-propelled Colloidal Motors *58*
2.4 Applications of Self-assembly of Anisotropic Colloids *60*
2.4.1 Liquid Crystals *60*
2.4.2 Photonic Crystals *61*
2.4.3 Sensors *63*
2.4.3.1 Metal Ions *63*
2.4.3.2 Biomolecules *65*
2.4.3.3 Gases *66*
2.4.3.4 Other Environmental Cues *68*
2.4.4 Electrode Materials *68*
2.4.5 Other Applications in Making Functional Materials *71*
2.5 Summary and Outlook *74*
Acknowledgment *74*
References *74*

3 Self-assembly Enabling Materials Nanoarchitectonics *87*
Katsuhiko Ariga
3.1 Introduction *87*
3.2 Fullerene Nanoarchitectonics *91*
3.3 Layer-by-Layer Nanoarchitectonics *97*
3.4 Conclusions *104*
Acknowledgments *105*
References *105*

4 Self-assembly of Colloidal Crystals: Strategies *109*
Junchao Liu, Jingxia Wang, and Lei Jiang
4.1 Introduction *109*
4.2 Assembly Mechanism of Latex Particles *110*
4.2.1 General Assembly Process *110*
4.2.2 Influence of Substrate Wettability on the Assembly Process *111*
4.2.3 Influence of Magnetic/Electric Field on the Assembly Process *113*
4.3 Assembly Strategies of Colloidal Crystal *115*
4.3.1 Large-Area Colloidal PC *115*
4.3.1.1 Vertical Deposition *115*
4.3.1.2 LB Approach *116*
4.3.1.3 Doctor Blade *118*
4.3.1.4 Spin Coating *118*
4.3.1.5 Spray Coating *120*
4.3.1.6 Other *121*
4.3.2 Patterned Colloidal PC *123*
4.3.2.1 Physical Confinement: Geometric Defect *123*
4.3.2.2 Wettability-Induced Template Assembly *124*
4.3.2.3 Direct Dropping *125*
4.3.2.4 Direct Writing *126*
4.3.2.5 Inkjet Printing *127*
4.3.3 Specific Structure Colloidal PC *128*
4.3.3.1 Cylinder Assembly from Capillary Tube *128*
4.3.3.2 Electrospun *130*
4.3.3.3 Microfluidics *131*
4.4 Conclusions *133*
References *133*

5 2D and (2+1)D Colloidal Photonic Crystal *139*
Lijing Zhang, Jiaqi Han, and Bofan Liu
5.1 Colloidal Photonic Crystals *139*
5.2 2D Colloidal Photonic Crystal *140*
5.2.1 Preparation Methods *140*
5.2.1.1 Solvent Evaporation-Induced Self-assembly *141*
5.2.1.2 Spin Coating *142*
5.2.1.3 Electrophoretic Deposition *142*
5.2.1.4 Air/Liquid Interface Self-assembly *142*
5.2.1.5 Other Strategies *146*
5.2.2 Optical Properties *150*
5.2.2.1 Diffraction *150*
5.2.2.2 Debye Diffraction Ring *151*
5.2.3 Application *152*
5.2.3.1 Templates *152*
5.2.3.2 Detection and Sensing *157*
5.3 (2+1)D Colloidal Photonic Crystal *160*

5.3.1 Preparation Method *161*
5.3.1.1 The Langmuir–Blodgett Method *162*
5.3.1.2 PDMS Sheet-Assisted Layer-by-Layer Transfer Technique *164*
5.3.1.3 Layer-by-Layer Scooping Transfer Technique *164*
5.3.1.4 *In Situ* Annealing Combined Layer-by-Layer Transfer Technique *164*
5.3.2 Optical Properties *167*
5.3.2.1 Comparison Between (2+1)D and 3D *167*
5.3.2.2 Bandgap Engineering *171*
5.3.3 Application *173*
5.3.3.1 Engineering of Complex Macroporous Materials *174*
5.3.3.2 High-Performance Back Reflector *174*
5.3.3.3 FL Enhancement Substrate *174*
5.4 Outlook *179*
References *180*

6 Structural Color due to Self-assembly *183*
Yong Qi and Shufen Zhang
6.1 Structural Color in Nature *183*
6.2 The Type of the Structural Color and Its Formation Mechanism *186*
6.2.1 Structural Color due to Interference *186*
6.2.2 Structural Color due to Scattering *187*
6.2.3 Structural Color due to Diffraction *188*
6.3 The Assembly Methods of the SCMs *191*
6.3.1 Evaporation Self-assembly *191*
6.3.2 Membrane Separation-assisted Assembly *194*
6.3.3 Air–Liquid Interface Self-assembly *195*
6.3.4 Oil–Oil Interface Self-assembly *200*
6.3.5 Oil–Water Interface Self-assembly *200*
6.3.6 Controlled Micellization Self-assembly *202*
6.3.7 Layered Hydrogels Self-assembly *205*
6.3.8 Spray Coating Self-assembly *207*
6.3.9 Unidirectional Rubbing Self-assembly *210*
6.3.10 Edge-Induced Rotational Shearing Self-assembly *210*
6.3.11 Screen Printing Self-assembly *213*
6.3.12 Magnetic-Induced Self-assembly *215*
6.3.13 Photoinduced Self-assembly *218*
6.3.14 Atomic Layer Deposition Self-assembly *220*
6.3.15 Physical Vapor Deposition Self-assembly *220*
6.3.16 Surface Wrinkling *223*
6.4 Typical Applications of Structural Color *226*
6.5 Conclusions *228*
Acknowledgments *230*
References *230*

7 Colloidal Photonic Crystal Sensors *237*
Zhongyu Cai, Xiaoying Xu, Zihui Meng, Bushra Rafique, and Ruixiang Liu
7.1 Introduction *237*
7.2 Fundamentals of Colloidal Photonic Crystal Sensors *239*
7.3 Responsive Materials and Novel Photonic Structures *242*
7.4 Colloidal Photonic Crystal Sensors Responsive to Physical Stimuli *243*
7.4.1 Colloidal Photonic Crystals Responsive to Humidity *243*
7.4.2 Mechanically Responsive Colloidal Photonic Crystals *245*
7.5 Colloidal Photonic Crystal Sensors Responsive to Chemicals *247*
7.5.1 Colloidal Photonic Crystals Responsive to Solvents *247*
7.5.2 Colloidal Photonic Crystals Responsive to Vapors (VOCs) *249*
7.5.3 Colloidal Photonic Crystals Responsive to Ions *250*
7.5.4 Colloidal Photonic Crystals Responsive to Organophosphates *253*
7.5.5 Colloidal Photonic Crystals Responsive to Surfactants *256*
7.6 Colloidal Photonic Crystal Sensors Responsive to Biological Species *256*
7.6.1 Colloidal Photonic Crystals Responsive to Carbohydrates *257*
7.6.2 Colloidal Photonic Crystals Responsive to Proteins (Enzyme, Antibodies) *260*
7.6.3 Colloidal Photonic Crystals Responsive to Amino Acids *264*
7.6.4 Colloidal Photonic Crystals Responsive to Biomarkers *265*
7.6.5 Colloidal Photonic Crystal Sensors for Microorganisms Detection *267*
7.7 Summary and Outlooks *269*
Acknowledgment *269*
References *270*

8 Self-assembled Graphene Architectures for Electrochemical Energy Storage *277*
Pei Li, Junwei Han, and Quan-Hong Yang
8.1 Introduction *277*
8.2 Self-assembly Strategies *278*
8.2.1 Langmuir–Blodgett (LB) Technique *281*
8.2.2 Layer-by-Layer (LbL) Assembly Approach *281*
8.2.3 Flow-Directed Self-assembly *282*
8.2.4 Interface-Induced Self-assembly *283*
8.2.5 Template-Directed Self-assembly and Hydrothermal Process *284*
8.2.6 Spinning and Space Confinement Self-assembly *286*
8.3 Methods for Tailoring the Assemblies *286*
8.3.1 Structural Tuning and Surface Modification *287*
8.3.1.1 Compact Assembly via Capillary and Mechanical Compression *287*
8.3.1.2 Tunable Porous Structure and Surface Chemistry *289*
8.3.1.3 Assembly in the Electrostatic and Magnetic Field *291*
8.3.2 Composite Materials Prepared by Self-assembly *291*
8.3.2.1 Graphene/Carbon Material Assembly *291*

8.3.2.2 Graphene/Polymer Assembly *292*
8.3.2.3 Graphene/Metal or Metal Oxide Assembly *292*
8.4 Applications of Self-assembled Graphene Architecture *293*
8.4.1 SCs *293*
8.4.2 LIBs *296*
8.4.3 Li–S Batteries *297*
8.4.4 Challenges for Practical EES Applications *299*
8.5 Conclusions *300*
Acknowledgments *300*
References *300*

9 Patterning Assembly of Colloidal Particles *305*
Minxuan Kuang and Yanlin Song
9.1 Introduction *305*
9.2 Strategies of Patterning Assembly *306*
9.2.1 Inkjet Printing of Assembly Patterns *306*
9.2.1.1 Ink Formulations *306*
9.2.1.2 Surface Wettability *310*
9.2.2 Patterned Substrate-Induced Assembly *314*
9.2.3 External Stimuli-Induced Assembly *316*
9.3 Applications of PC Patterns *316*
9.3.1 Displays *316*
9.3.2 Sensors *319*
9.3.3 Anticounterfeiting *324*
9.4 Summary *326*
References *326*

10 Light Extraction Efficiency Enhancement in GaN-Based LEDs by Colloidal Self-assembly *331*
Tongbo Wei
10.1 Introduction *331*
10.2 Nanostructure Fabrication by MCC *333*
10.3 Applications of Nanostructures to LEDs *335*
10.3.1 Surface Texturing *335*
10.3.2 Submicron Lenses *339*
10.3.3 Photonic Crystals (PhCs) *341*
10.3.3.1 Surface PhCs *342*
10.3.3.2 Bottom PhCs *344*
10.3.4 Localized Surface Plasmon (LSP) *346*
10.3.5 Nano-patterned Sapphire Substrates (NPSS) *348*
10.3.6 Optical Reflector *352*
10.4 New Applications for Optoelectronic Devices *354*
10.4.1 Nanorod LEDs *354*
10.4.1.1 Top-Down Nanorod LEDs *355*
10.4.1.2 Bottom-Up Nanorod LEDs *357*

10.4.2 Microdisk Lasers *357*
10.5 Conclusions and Perspectives *359*
References *361*

11 Self-assembled Photonic Crystals for Solar Cells *365*
Likui Wang
11.1 Introduction *365*
11.2 The Application of Self-assembled PCs in DSSCs *368*
11.2.1 Application of Dye-Sensitized IOs as Photoanodes of DSSC *369*
11.2.2 Self-assembled Photonic Crystals for Light Harvesting Enhancement in DSSCs *375*
11.3 Self-assembled Photonic Crystals for Perovskite Solar Cells *384*
11.4 The Application of Self-assembled PCs in Silicon-Based Solar Cells *389*
11.5 Summary and Outlook *390*
References *391*

12 Mesoporous Zeolites: Synthesis and Catalytic Applications *397*
Feng Yu and Feng-Shou Xiao
12.1 Introduction *397*
12.2 Synthesis of Mesoporous Zeolite *398*
12.2.1 Bottom-up Zeolite Synthetic Strategies *399*
12.2.1.1 Hard Templating Methods *399*
12.2.1.2 Soft Templating Methods *404*
12.2.1.3 Non-templating Methods *411*
12.2.2 Top-down Approaches via Demetallization *414*
12.2.2.1 Dealumination *414*
12.2.2.2 Desilication *415*
12.2.3 Mixed Synthetic Approaches *417*
12.3 Catalytic Application of Mesoporous Zeolites *419*
12.3.1 Fuel Chemistry *419*
12.3.1.1 Catalytic Cracking *419*
12.3.1.2 Fischer–Tropsch Reaction *422*
12.3.1.3 Methanol to Gasoline or Olefins *423*
12.3.2 Selective Organic Reactions *425*
12.3.2.1 Hydrogenation *427*
12.3.2.2 Oxidation *428*
12.3.2.3 Friedel–Crafts Alkylation *429*
12.3.2.4 Isomerization *430*
12.3.2.5 HDO Phenolics *432*
12.3.3 Catalytic Combustion *432*
12.3.4 Biomass Valorization via Catalytic Fast Pyrolysis *433*
12.4 Conclusions and Perspectives *435*
References *436*

13 Colloidal Self-assembly of Block Copolymers for Drug Loading and Controlled Release *455*
Guangze Yang, Yun Liu, and Chun-Xia Zhao
13.1 Introduction *455*
13.2 Block Copolymers *456*
13.2.1 Diblock Copolymers *456*
13.2.2 Triblock Copolymers *458*
13.2.3 Self-assembly Mechanism *459*
13.3 Nanoscale Structures *460*
13.3.1 Micelles *460*
13.3.2 Nanoparticles *461*
13.3.3 Vesicles or Polymersomes *462*
13.4 Drug Loading *463*
13.4.1 Single Drug Encapsulation *463*
13.4.2 Dual Drug Encapsulation *464*
13.4.2.1 Similar Hydrophobicity *464*
13.4.2.2 Opposite Hydrophobicity *465*
13.4.3 Multiple Drugs *466*
13.4.4 Drug and Imaging Reagent *468*
13.5 Drug Release *470*
13.5.1 Drug Release Mechanism *470*
13.5.2 pH-Triggered Release *471*
13.5.3 Thermo-Triggered Release *472*
13.5.4 Redox-Triggered Release *472*
13.5.5 Other Triggered Release *473*
13.5.6 Multiple-Triggered Release *475*
13.6 Conclusions *475*
References *477*

14 Heat Management by Colloidal Self-assembly *493*
Qimeng Song, Marius Schöttle, Pia Ruckdeschel, Fabian Nutz, and Markus Retsch
14.1 Introduction *493*
14.2 Fabrication Methods *494*
14.2.1 Categories of Colloidal Assemblies *494*
14.2.2 Self-assembly Strategies *495*
14.3 Fundamentals of Thermal Transport in Dielectric Materials *497*
14.3.1 Thermal Transport in Crystalline Matter *497*
14.3.2 Thermal Transport in Amorphous Matter *498*
14.3.3 Thermal Transport in Composite and Porous Materials *499*
14.4 Characterization Methods *501*
14.4.1 Laser Flash Analysis (LFA)/Xenon Flash Analysis (XFA) *501*
14.4.2 Transient Plane Source (TPS)/Modified Plane Source (MPS) Technique *503*

14.4.3 Time-Domain Thermoreflectance (TDTR) and Frequency-Domain Thermoreflectance (FDTR) Methods *504*
14.4.4 Guarded Hot Plate (GHP) and Heat Flow Meter (HFM) *506*
14.4.5 Photoacoustic (PA) Method *509*
14.4.6 Lock-In Infrared Thermography (LIT) *509*
14.5 Heat Transport in Colloidal Structures *512*
14.5.1 Porous Structures *512*
14.5.2 Dense Structures *520*
14.6 Heat Management Applications *527*
14.6.1 Thermal Recorders *527*
14.6.2 Thermal Switches *527*
14.6.3 Thermal Rectification *530*
14.6.4 Thermoelectrics *531*
14.6.5 Passive Cooling *532*
14.7 Conclusion and Summary *532*
References *533*

Index *539*

Preface

Functional materials represent a fast-growing research field in materials science. Functional materials generally refer to those that possess particularly intrinsic properties and functions of their own. For example, ferroelectric materials, piezoelectric materials, thermoelectric materials, magnetic materials, semiconductor materials, optoelectronic materials, and many others. However, there is a special type of functional materials, i.e. metamaterials, the function of which comes from their designed structures. The metamaterial is usually an artificial material that exhibits unique properties not found in nature, such as a negative index of refraction, in its interaction with electromagnetic radiation, sound, or other wave phenomena. Such properties are a consequence of the metamaterial's structure at the microscopic or macroscopic level rather than of the intrinsic properties of its components.

As a representative optical metamaterial, photonic crystals have some extraordinary physical properties, such as effective negative parameters, bandgaps, negative refraction, etc., and have attracted extensive attention due to their promising application potentials. Photonic crystals exist in nature and can be artificially created by controlling structural periodicity of dielectric materials. When their feature size goes to micro- or nanometer and their periodicity extends to three dimensions, the fabrication of photonic crystals with a full bandgap has been a grand challenge.

In this regard, nature has been a great mentor, teaching humans how to construct a substance having a complex structure. The beauty of gemstone opal originates from a densely packed, highly ordered arrangement of silica spheres with a diameter of several hundred nanometers. Silica spheres themselves have no intrinsic color, while a periodic arrangement endows their entirety an amazing color. Such ordered nanostructure is a typical example of natural photonic crystals, which can be formed by spontaneous sedimentation of silica spheres in nature or by a colloidal self-assembly process.

Colloidal particles normally have a diameter in the range from a few nanometers to a few microns – large enough to be visualized under optical microscope, yet small enough to sustain Brownian motion when dispersed in a fluid medium. Self-assembly of colloidal particles provides an efficient approach for the biomimetic construction of functional materials for manipulation of the flow of light, energy storage and conservation, anticounterfeiting, green printing, dynamic color display, environmental sensing, fabric coloring, heat management, and so on.

The past several decades have witnessed a number of encouraging advancements in colloidal self-assembly for functional materials. This book intends to provide an overview of the state-of-the-art colloidal self-assembly for functional materials ranging from emerging colloidal building blocks and self-assembly mechanisms and strategies to applications in structural color, light emitting diodes (LEDs), solar cells, catalysts, drug load and release, and heat management. In addition to providing accounts of recent developments in the respective topics, the authors also note the challenges and present their perspectives on the individual directions in the future.

Traditional colloidal particles have been spherical in shape, being viewed as "big atoms" suitable for studying various physical phenomena in atomic crystals, such as crystallization, glass transition, and defect evolution. Advances in the areas of functional materials demand for the development of complex colloidal particles, which are rich in component, shape, and surface chemistry. Benefiting from recent remarkable research breakthroughs, in advanced synthesis methods, and fabrication techniques, new colloidal building blocks have emerged and expanded the boundaries of conventional colloidal self-assembly.

In Chapter 1, Ravaine and coworkers discuss recent advances in the synthesis of clusters of isotropic colloids and patchy particles, as well as the self-assembly of these complex colloidal particles toward the so-called colloidal molecules and colloidal polymers. Compared with their isotropic spherical counterparts, anisotropic colloids show unique physiochemical properties and much richer phase behavior. In Chapter 2, Zhao and coworkers summarize the latest advances in both fabrication techniques and self-assembly mechanisms of anisotropic colloids, as well as the progress in the applications of assembled structures. Due to the extraordinary characteristics either from their unique shapes or well-designed particle interactions, they show a much broader range of assembled structures than isotropic colloids. With such a variety of accessible structures that could possess special properties, self-assembly of anisotropic colloids show great potential for applications in various fields. In Chapter 3, Ariga introduces the concept of nanoarchitectonics through self-assembly of different colloidal particles, focusing on fullerene assemblies at gas/liquid interfaces and assemblies based on layer-by-layer (LbL) technique. Due to the abundant surface and interface properties of graphene-based colloids, especially chemically modified graphene, various self-assembly strategies are developed to construct hierarchical structures and composite materials. In Chapter 8, Yang and coworkers review the recent advances in the self-assembly and tailoring methods of graphene-based colloids, highlighting important applications and perspectives on self-assembled graphene structures. In Chapter 12, Xiao and Yu present an overview of diverse synthesis approaches to the fabrication of mesoporous zeolites and their catalytic applications. In Chapter 13, Zhao and coworkers provide an overview of the colloidal self-assembly of diblock and triblock copolymers for drug loading and controlled release. The advantages and disadvantages of block copolymers as well as future directions for various drug delivery applications are also discussed.

In Chapter 4, Wang and coworkers discuss research progress on self-assembling strategies of colloidal photonic crystals, including assembly mechanisms and approaches. Researches on 3D colloidal photonic crystals have been very extensive and in-depth, but less attention was paid to 2D colloidal photonic crystals and (2 + 1)D photonic crystals. In Chapter 5, Zhang and coworkers discuss the preparation, optical properties, and applications of 2D and (2 + 1)D colloidal photonic crystals. Furthermore, the current limitations and perspectives for their future developments are also discussed.

Structural colors are all kinds of colors produced by the microscopic structures corresponding to the light wavelength through interference, diffraction, scattering, and other ways to reflect the incident light at a specific wavelength and in a specific direction. Self-assembly of nanoparticles provides effective approaches for the fabrication of functional structural color materials. In Chapter 6 by Zhang and Qi, assembly mechanism and coloring mechanism during constructing functional structural color materials are discussed in detail.

A colloidal photonic crystal sensor is usually implemented with responsive materials or functionalized with recognition elements or created with defects within the colloidal photonic crystal for the detection of target analytes. In Chapter 7, Cai and coworkers highlight the significant progress achieved in the development of a variety of colloidal photonic crystal sensors, such as physical sensors, chemical sensors, and biosensors.

Device applications typically involve patterning. In Chapter 9, Song and Kuang present an overview of the strategies for fabricating patterned colloidal photonic crystals. The applications of patterned colloidal photonic crystal devices are also discussed, showing great potential in the area of display, sensor, and anticounterfeiting.

One of the important properties of photonic crystals is their photonic bandgap, with which photonic crystals can control the flow of light in a manner similar to what semiconductors do in controlling the flow of electrons. Such a property has found wide applications in photon management. In Chapter 10, Wei illustrates how to enhance the light extraction efficiency in GaN-based LEDs by using colloidal photonic crystals and colloidal monolayer-derived nanostructures. While in Chapter 11, Wang demonstrates the latest advancement on the enhancement of power conversion efficiency in dye-sensitized solar cells, perovskite solar cells, and silicon solar cells by integration of colloidal photonic crystals.

In the field of heat management, colloidal ensembles provide a versatile pathway to study thermal transport on a small length scale and reveal the governing structure–property relationships. In Chapter 14, Retsch and coworkers review the recent progress of advanced heat management by using colloidal self-assembly strategies.

Colloidal self-assembly holds a great promise for the creation of functional materials. The past decades have witnessed a number of encouraging advancements in this burgeoning area. A vast toolbox of appropriate colloidal building blocks has been established while more and more new building blocks are emerging and enriching the toolbox. Consequently, abundant functional materials can be accessed once these colloidal building blocks are able to self-assemble via a proper strategy. While the present book cannot provide a comprehensive collection of the current state-of-the-art colloidal self-assembly for functional materials, we do expect that the chapters included in this book cover the most exciting developments in this prosperous field.

We would like to express our deep appreciations to all the contributing authors. Without their contributions and strong supports to this endeavor, it would not be possible to reach to the moment of having this book published, especially during this unprecedented challenging time due to the COVID-19.

October, 2021

Qingfeng Yan
George Zhao

1

Colloidal Molecules and Colloidal Polymers

Etienne Duguet[1], Etienne Ducrot[2], and Serge Ravaine[2]

[1] *Université de Bordeaux, Institut de Chimie de la Matière Condensée de Bordeaux (ICMCB, UMR 5026), CNRS, Bordeaux INP, 87 avenue du Dr Albert Schweitzer, Pessac, 33600, France*
[2] *Université de Bordeaux, Centre de Recherche Paul Pascal (CRPP, UMR 5031), CNRS, 115 avenue du Dr Albert Schweitzer, Pessac, 33600, France*

1.1 Introduction

The self-assembly of colloidal objects, inspired by self-organization phenomena observed in nature, is a relevant alternative to chemical synthesis route and top-down fabrication techniques for building increasingly complex structures and materials. It is also an effective tool to understand physicochemical processes that drive self-assembly phenomena as colloidal units are typically three orders of magnitude larger than atoms and molecules and thus can be tracked by optical microscopy techniques [1, 2]. For instance, isotropic spherical particles, often called colloidal atoms (CAs), helped to clarify crystallization and phase transition mechanisms and were observed as they assembled in different phases such as gels, glasses, and several crystal phases [3, 4]. It was shown that interactions between CAs can be tuned by varying the composition of the solvent, by applying external fields and/or concentration gradients, and by modulating the surface chemistry of the particles [5]. Driven by the desire to elaborate colloidal materials possessing new, interesting properties, the field of colloidal self-assembly has made a leap over in the last decades by synthesizing monodisperse particles of various shapes and sizes and by focusing on their assembly through a wide range of interactions [6–8]. In particular, nonspherical colloids stemming from the aggregation of a small number of particles, named "colloidal molecules" (CMs) by van Blaaderen [9], have attracted a lot of attentions these recent years because they are expected to show complex behavior (like low-molecular-weight compounds) dictated not only by the shape of the clusters but also by the variety of the interactions they could generate [10–13]. We have recently proposed a classification of CMs using and extending the well-known formalism of Gillespie derived from the valence shell electron pair repulsion (VSEPR) model (Figure 1.1) [14], which facilitates the reading of this chapter.

Functional Materials from Colloidal Self-assembly, First Edition. Edited by Qingfeng Yan and George Zhao.

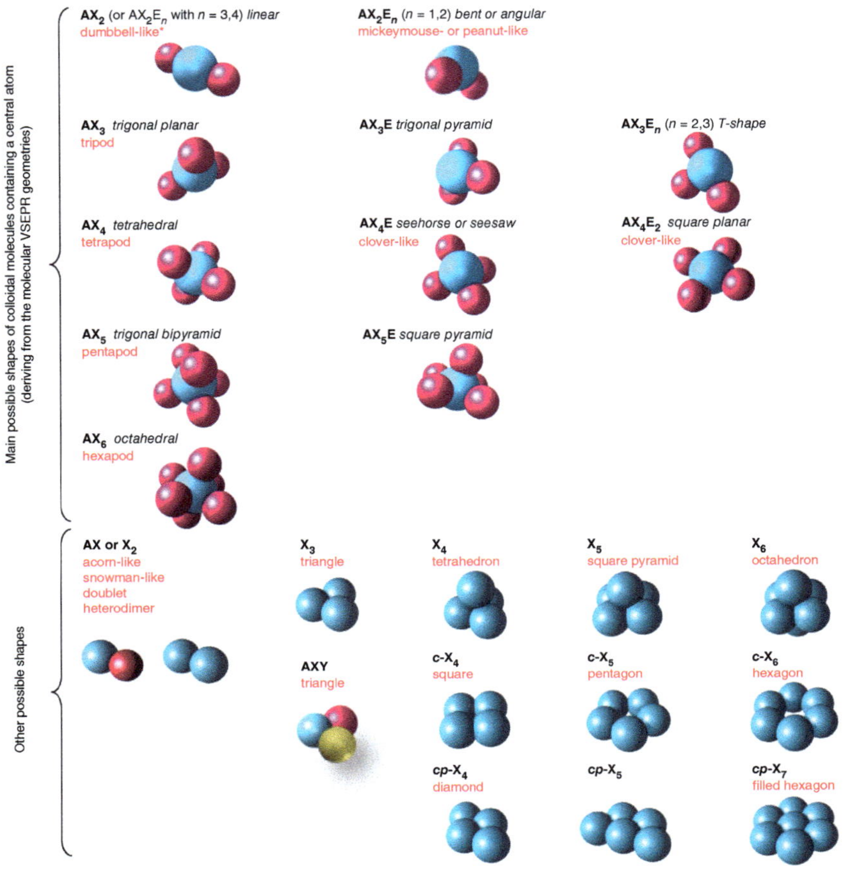

Figure 1.1 Proposition of classification for colloidal molecules based on spheres and mimicking space-filling models of simple molecules. Source: Duguet et al. [14] / with permission from the Royal Society of Chemistry.

The concept of using preformed particles as colloidal monomers has also received recent attention for the formation of one-dimensional (1D) structures, also called "colloidal polymers" (CPs) [15, 16]. A big challenge in the formation of CPs is that it requires strong anisotropic interparticle interactions to minimize side reactions, which is possible to achieve through the development of synthetic methods that allow one to embed anisotropic character to colloidal monomers.

We aim here to highlight and classify the major strategies hitherto reported to fabricate CMs and CPs by self-assembly of preformed particles. We first describe the synthesis pathways to CMs, emphasizing in particular the use of patchy particles as building units. We finally present the different inventive routes that have been developed to create CPs. We restrict the discussion to the synthesis routes and the morphology control, whatever the physical, biological, or chemical properties and potential applications of the CMs and CPs may be.

1.2 Colloidal Molecules: Mimicking Organic and Inorganic Molecules

From the point of view of the synthetic strategy, there exist two main ways to access CMs [14, 17]. The first one concerns essentially AX_nE_m-type CMs and starts from soft or hard preformed CAs decorated by satellite colloids directly generated at their surface, thanks to controlled phase separation or nucleation/growth phenomena, respectively. The second pathway consists in the controlled clustering of preformed CAs through physical routes, chemical routes, or 2D/3D geometrical confinement. The readers who want to embrace the full story have to refer to our previous reviews [14, 17] and/or to read excellent recent reviews written by others and addressing the field from other angles [10–12, 18–25]. In the present chapter, we propose to focus our description of the work in the literature on this second strategy, which relates to the general assembly process.

1.2.1 Clustering of Isotropic Colloids

The following pathways use isotropic spherical CAs, i.e. particles, which surface chemistry and topology are not patterned, meaning they are not spontaneously able to establish interactions in specific directions with other particles.

1.2.1.1 Clustering Assisted by External Fields

These strategies are usually implemented with microparticles on solid surfaces under optical microscope and generate essentially quasi-planar CMs. The strength of the used fields shall also be tuned to overcome the gravity and surface adhesion forces. In this context, Wu and coworkers have reported that applying external alternating current (AC) electric fields is an efficient and simple way to generate anisotropic dipolar and dielectrophoretic interactions between negatively charged polystyrene (PS) microparticles [26]. The morphology of the as-obtained CMs may be quite complex and varied by changing the field frequency via different assembly pathways that resemble chemical reactions of real molecules. Grzybowski and colleagues have shown that magnetic field microgradients established in a paramagnetic fluid can be used as virtual molds to serve as templates for the production of a variety of CMs, including A_4, A_5, A_6, AX_2, AX_3, and AX_4 types (Figure 1.2) [27]. The assembly of CAs into CMs has also been driven by a magneto-acoustic method by Yellen and coworkers. The CM morphology can be tuned by varying the magnetic and acoustic fields that allow to control the interparticle interactions and the local density of particles by attracting them to the pressure nodes of the standing wave, respectively [28]. Although attractive for their ease of use and high morphology yield, these synthesis routes assisted by external fields suffer from very low production yield and sometimes from lacking fixation strategy for making the CM structure permanent (after suppression of the fields). However Grzybowski and colleagues have shown that this last issue may be solved by attaching the CMs to the substrate with carbamide bonds, making them permanent by silica deposition before their release into a liquid by sonication [27].

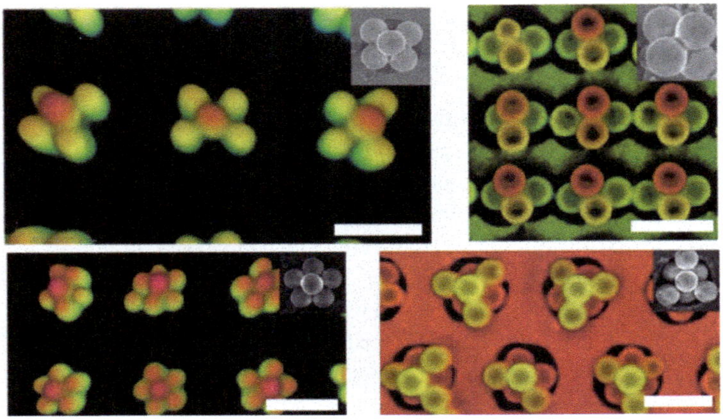

Figure 1.2 CMs of diamagnetic particles with X_4, X_5, X_6, and X_7 geometries obtained under magnetic field microgradients established in a paramagnetic fluid. Scale bars: 2 μm. Source: Demirörs et al. [27] / with permission of Springer Nature.

1.2.1.2 Clustering Assisted by Geometrical Confinement

The patterning of 2D substrates with hydrophilic patches [29] or traps [30] allows the controlled sequential deposition of microparticles, resulting in the creation of a wide library of CMs, including 3D chiral objects. Nevertheless, the production yields remain low. Another strategy consists in the adsorption of particles onto emulsion droplets used as 3D templates, i.e. taking advantage of the Pickering effect, followed by their aggregation through the droplet phase evaporation. This is the pioneering pathway opened by Manoharan and colleagues [31] which has inspired van Blaaderen to develop the concept of CMs [9]. When during evaporation the CAs touch another on the surface of the droplets, removing more liquid causes the droplet to deform and generates capillary forces that collapse the CAs into a CM [31]. The structures of these CMs include familiar polyhedra, e.g. X_4 and X_6, and ones that are more unusual. The selection of a unique packing arises almost entirely from geometrical constraints [32]. Nevertheless, the size polydispersity of the initial emulsion droplet leads to complex mixtures of different CMs that can be finally sorted by density gradient centrifugation. Using narrow droplet size distribution through ultrasonication, Witteman and coworkers have obtained similar CMs from 150-nm particles (Figure 1.3) [33]. Later, Monte Carlo simulations have allowed a full description of the dynamics correlated to experiments [34]. Crassous and colleagues have recently investigated again this route with thermosensitive particles, e.g. poly(*N*-isopropylacrylamide) (PNIPAM) microgels, and achieved CMs with externally tunable interaction sites by playing with temperature or ionic strength [35].

Another method of interest for the preparation of CMs is the crystal templating strategy, which relies on the use of binary colloidal crystals (CCs) as templates and leads exclusively to AX_{12} CMs. Zhang and coworkers have applied this strategy to soft PNIPAM microgel submicron CAs [36]. In *fcc*- or *hcp*-like CCs prepared by sedimentation from a 40 : 1 mixture of PNIPAM microspheres with surface thiol groups and surface vinyl groups, the latter are statistically in close contact with 12 CAs bearing

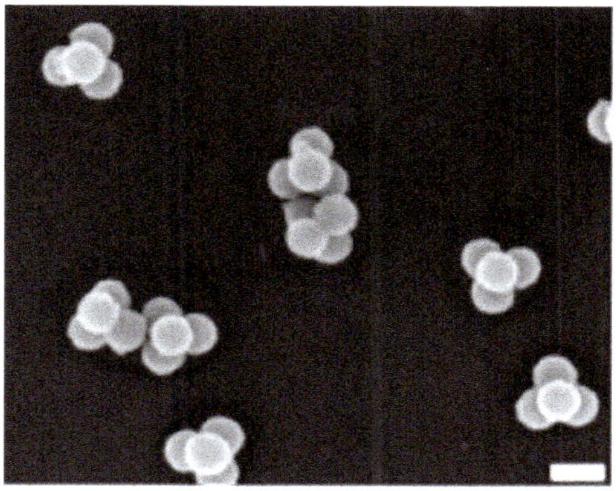

Figure 1.3 FESEM micrograph of tetrahedral CMs. Scale bar: 200 nm. Source: Wagner et al. [33] / with permission of American Chemical Society.

thiol groups. Under UV light irradiation, the thiol–ene reaction allows the covalent binding of the 13 microspheres, which are recovered as independent CMs after the destruction of the CCs. By-products with a higher number of microspheres have been found in the same batch but easily removed by density gradient centrifugation. The same strategy has been extended to AX_4, AX_5, and AX_6 CMs using vinyl-PNIPAM microspheres smaller than the thiol-PNIPAM ones in order to accommodate them in the interstitial sites of different close-packed lattices [37]. An enhanced crystal templating method has been investigated by Crocker and colleagues using PS microspheres coated with DNA strands (Figure 1.4) [38, 39]. They have demonstrated that interactions between these colloids can be reprogrammed using enzymes or reinforcing DNA strands to prepare CMs (AX_n with $n = 4, 6, 8, 10$, or 12). In the enzymatic approach, two populations of DNA-coated CAs are mixed, with a large excess of the larger ones, and forced into close-packed crystal structures via centrifugation at low ionic strength [38]. At higher ionic strength, sequence-specific enzymatic ligation selectively creates a covalent bond between colloids of different natures. A final

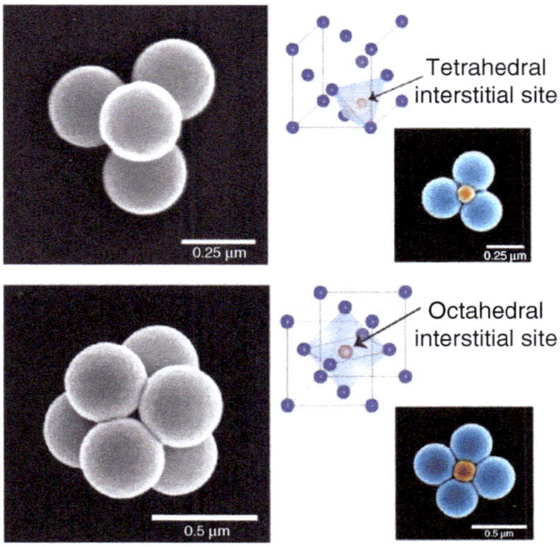

Figure 1.4 SEM images of tetrahedral and octahedral CMs produced using a colloidal crystal as a template (scale bars: 250 nm (top) and 500 nm (bottom). Source: McGinley et al. [38] / with permission of American Chemical Society.

washing step at low ionic strength destroys the crystals, thereby releasing the covalently bound CMs, which type depends on the size ratio between the two populations of colloids, i.e. the valence of the interstitial site occupied by the minority colloid.

According to a similar clustering strategy in 3D confinement, Tabeling and coworkers have reported a strategy to obtain CMs (X_n with n = 3–6) in microfluidic channels, thanks to the coupling of hydrodynamic dipolar interactions and depletion forces [40].

1.2.1.3 Clustering Assisted by Physical or Chemical Interactions

Clustering into AX_n CMs can also be quite simply achieved, thanks to the single sticky protrusion emerging from the surface of cross-linked polymer CAs when they are swollen with monomer and temperature is raised. In these conditions, the central CA is generated *in situ* and made of the merged protrusions. In this way, Kraft et al. have obtained CMs with AX_2–AX_9 morphologies from PS nanoparticles previously coated with vinyl acetate by delaying the polymerization of the protrusions for promoting coalescence upon collision before solidification upon polymerization [41, 42]. Interestingly, the swelling liquid can also be an apolar solvent allowing to reconfigure any CM into regular ones through minimization of the interfacial energy, thanks to a diffusion-limited aggregation process. The final morphology can be tuned by varying the surfactant concentration and swelling conditions, i.e. time, solvent nature, and ratio.

Close-packed planar CMs obtained under depletion conditions in a 2D cell have been achieved by Manoharan and coworkers [43]. Using this strategy it was possible to form CMs with AX_2, AX_3, AX_4, and AX_6 morphologies with high yield, simply by choosing the appropriate size ratio between the spherical building units [44]. They have shown that a yield as high as 90% is achievable for the fabrication of AX_4 morphologies in the absence of confinement just by adjusting the size ratio between two size-monodisperse populations of colloids to 2.41. This strategy seems efficient whatever the driving force used for assembly, e.g. DNA hybridization or electrostatic interactions.

Thanks to the pioneering works of Alivisatos', Mirkin's, and Sleiman's groups, DNA hybridization anchored at the surface of CAs provides a flexible tool to control in solution the formation of CMs and in particular from CAs of different sizes or chemical compositions [45–48]. DNA origamis provide a new opportunity for CM design. Gang and coworkers have shown that they can guide directional and specific interactions between isotropic gold nanoparticles to achieve X_6, planar c-X_4, and planar AX_4E_2 CMs in solution [49, 50]. Recently, Fan and coworkers developed a strategy to assemble gold nanoparticles into a large variety of CMs relying on a long single-stranded DNA chain containing polyadenine domains and encoding segments [51]. Once adsorbed at the surface of gold nanoparticles, thanks to the polyadenine domains, the DNA strands guide the assembly of satellite particles by hybridization of the encoding domains. They show fine control of the valence of the CMs as well as the size of each bound particles. Furthermore, chiral CMs have been realized by the introduction of three different DNA strands at chosen vertices of the origami frame combined with the use of three pairs of gold nanoparticles with

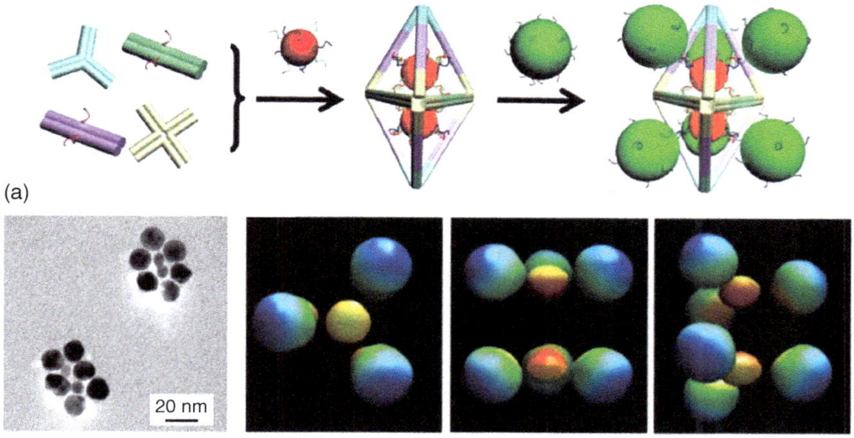

(a)
(b) (c)

Figure 1.5 DNA frame guided-assembly of ethane-like CMs: (a) schematic representation of the assembly guided by a DNA cage, (b) corresponding TEM image, and (c) different views of the 3D reconstruction obtained by cryo-electron microscopy. Source: Li et al. [52] / with permission of American Chemical Society.

different sizes and complementary DNA shells [49]. Mao and coworkers have used DNA cages, i.e. smaller DNA constructs than origami that consist only of synthetic strands for trapping gold nanoparticles and guiding their assembly into AX_4 CMs with yields as high as 67%. This grants control over the geometry of AX_6 CM between octahedral and trigonal prismatic, as in SF_6 and $W(CH_3)_6$, respectively. Even more complex CMs such as dual-core ethane-like CMs frozen in an eclipsed conformation have been also achieved (Figure 1.5) [52].

To overcome the recurring problem of the too flexible DNA architectures that fail to maintain in the desired geometry the CMs made of bulky and heavy CAs, Chaikin and coworkers have designed a DNA origami belt that binds to the surface of 700-nm CAs and serves as guide for the later assembly of other DNA-coated CAs. This strategy not only yields X_4-derived chiral CMs but also allows controlling the dihedral angle between the particles (Figure 1.6) [53].

In an opposite approach, it may be relevant to fabricate CMs flexible at a will to mimic, to a certain extent, scissoring, rocking, twisting, or wagging vibrations of conventional molecules. The recent emergence of surface mobile DNA linkers provides new types of AX_n CMs ($n = 2$, 3, and 4) with flexible joints, where the satellite CAs remain free to move around the central one [54]. These central CAs forming these structures can be silica particles surrounded by a lipid-bilayer membrane into which DNA strands functionalized at their ends by cholesterol or stearyl moieties are anchored. As long as the lipid bilayers remain fluid, the DNA strands are free to move around the particles and so are the joints formed between CAs functionalized with complementary strands. These flexible CMs are physically stable but reorganize in permanence at high speed.

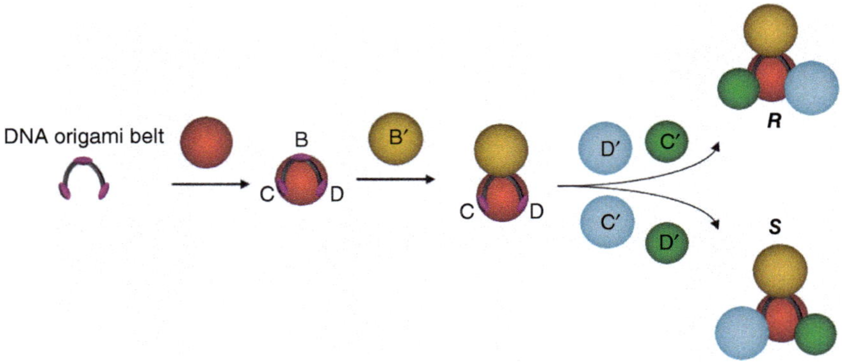

Figure 1.6 Schematic representation of the wrapping of a DNA origami belt around a first CA that guides the assembly of three other CAs into a chiral CM. Source: Modified from Ben Zion et al. [53].

1.2.2 Clustering of Patchy Particles

To mimic molecular geometry as best as possible, it is important to control not only the valence of the CAs but also the bond angles between them. We have shown in the previous section that this control can be exercised in an intrinsic way with more or less success by steric hindrance, electrostatic repulsions, or even the functionality of a DNA belt or origami. An increasingly explored alternative consists in programming this valence and directionality of interactions within CAs just as they are in conventional atoms, thanks to valence electrons. This is the concept of valence-endowed CAs, or patchy particles, where a patch is a surface discontinuity from the viewpoint of the chemistry or topography. They are promising colloidal building blocks not only for engineering CMs but also for open frameworks, if the number, relative location, size, and geometry of the patches can be precisely controlled [55]. The readers interested in a comprehensive description of the variety and synthesis pathways of patchy colloids is invited to refer to several recent reviews [17, 55, 56]. We focus here on their assembly capability to provide CMs. The patch-to-particle size ratio is one of the critical parameters to control the valence; roughly, the larger the patch, the higher the number of establishable bonds, especially as the patch number is low. Thus for getting CMs from patchy particles, there is mainly two routes: some X_n CMs may be more readily achieved from the clustering of one-patch particles, while AX_n ones can be generated from a central CA with the desired valence and the attachment of a single conventional CA to each of its patch.

1.2.2.1 Self-aggregation of One-Patch Particles, i.e. Janus Particles

As a subclass of patchy particles, Janus particles, named after the ancient roman god Janus [57], have extensively served as building blocks for CMs in the last few years, and the field has been recently reviewed [58–60]. Published examples concern Janus particles in the size range of 50-nm to 200-μm clustering into complex mixtures of CMs with aggregation numbers varying in the 2–20 range (Table 1.1).

Table 1.1 Representative examples of X_n CMs obtained from one-patch particles.

One-patch particles	Assembly driving force	Aggregation number	References
1-μm carboxylate-coated PS spheres then half-coated with gold treated with (11-mercaptoundecyl) ammonium chloride	Electrostatic attraction	2–12	[61]
1-μm carboxylate-modified PS spheres half-coated with gold treated with alkanethiol	Solvophobic effect helped with electrostatic screening	2–9	[62]
1-μm sulfate PS spheres half-coated with gold treated with alkanethiol	Hydrophobic attraction and electrostatic repulsion	2–7	[63]
4-μm snowman-like PS particles with smooth body and rough head	Roughness-controlled depletion attraction	2–12	[64]
2-μm organosilica particles with a partially embedded 1-μm hematite cube	Magnetostatic binding forces	2–3	[65]
3-μm gold half-coated silica spheres	At tipping points of thermally switched wetting	4–~20	[66]
4-μm iron oxide half-coated latex spheres	Lipid-induced capillary bridging	2–5	[67]
120-nm organosilica sphere with one 44-nm gold head	van der Waals force and electrostatic force	2–3	[68]
50-nm silica sphere with one 45-nm gold head coated with alkanethiols	Solvophobic effect	4–7	[69]
1-μm organosilica sphere with a PS patch (patch ratio of 0.13) that is functionalized with DNA having self-complementary sticky ends	DNA-mediated attraction varied by varying the temperature	2–4	[70]

The nature of the driving forces is generally physical, e.g. electrostatic, hydrophobic, or solvophobic interactions, roughness-controlled depletion, capillary bridging, or temperature-mediated DNA hybridization, meaning that bonding is quite weak or easily reversed. Therefore, the as-obtained CMs are fragile, making their purification difficult to impossible to succeed [71]. As far as we know, their strengthening by a chemical or physicochemical process has not yet been reported.

Nevertheless, three studies have been reported about the preparation of robust AX_n CMs. Bon and coworkers have synthesized asymmetric dumbbells made of a PS microsphere and a poly(*n*-butylacrylate) lobe and showed their ability to self-aggregate in aqueous media [72]. The desorption of the poly(vinylpyrrolidone)

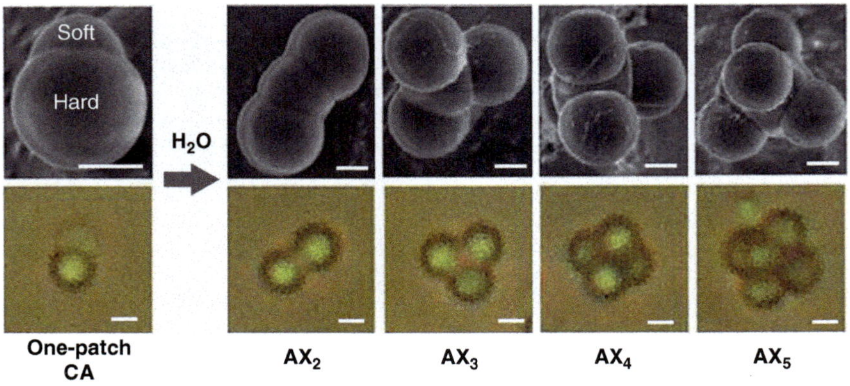

Figure 1.7 High resolution cryo-SEM (top row) and optical (bottom row) images of AX_n CMs in suspension obtained from polystyrene hard microspheres with a soft poly (*n*-butylacrylate) lobe (first column). Scale bars: 1 μm. Source: Skelhon et al. [72] / with permission of Royal Society of Chemistry.

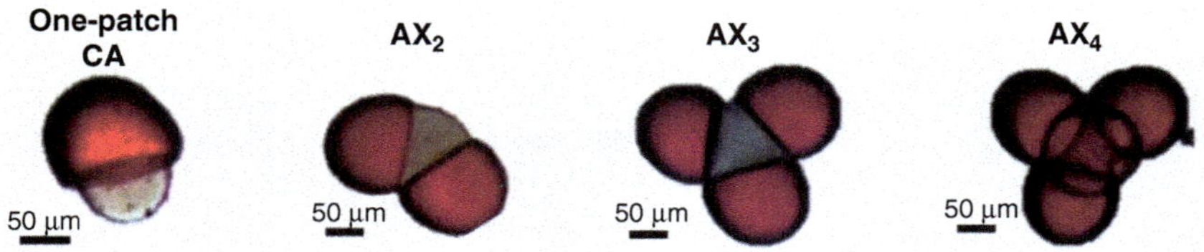

Figure 1.8 Optical microscopy images of CMs obtained from hydrophilic poly(ethylene glycol) diacrylate sphere with ethoxylated trimethylolpropane tri-acrylate lobe. Source: Ge et al. [73] / with permission of Wiley-VCH GmbH.

stabilizer from the particles trigger the assembly process through collision and merging of the soft poly(*n*-butylacrylate) lobes upon contact (Figure 1.7). Xu and colleagues have exploited the drying of aqueous Pickering emulsions made from 200-μm hydrophilic poly(ethylene glycol) diacrylate sphere with 160-μm ethoxylated trimethylolpropane tri-acrylate lobe prepared by microfluidics [73]. They have obtained quite regular CMs with aggregation numbers up to 8 (Figure 1.8). Lastly, we have recently reported the fabrication of CMs from silica/PS asymmetric dumbbell-like nanoparticles (Figure 1.9) [74]. In ethanol/DMF mixtures, the one-patch particles stick together by their PS lobe after swelling/plasticization by DMF. The as-obtained CMs present aggregation numbers from 2 to more than 6.

1.2.2.2 Stoichiometric Attachment of Colloidal Satellites Around Patchy Particles

Surface dimples or cavities on patchy particles are also named entropic patches because they can be exploited to attract and bind spheres with similar curvature radii through depletion forces [75]. Already exploited for the formation of CCs [76], the depletion technique has been extended to fabricate CMs from microparticles

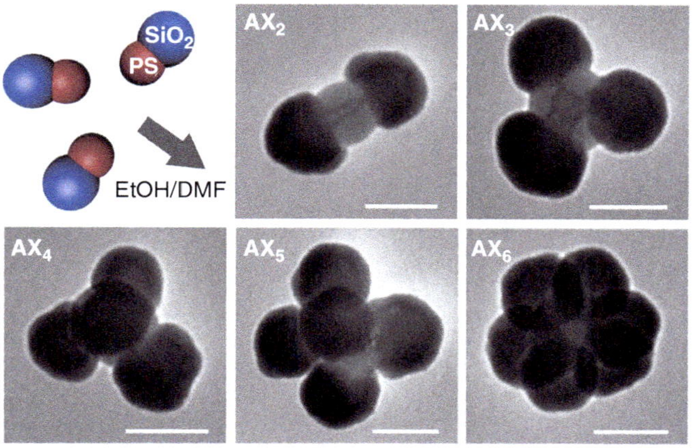

Figure 1.9 TEM images of CMs made of silica satellites and a PS central core obtained in ethanol/DMF mixture with a DMF fraction of 30 vol.%. Scale bars: 100 nm. Source: Li et al. [74] / with permission of Elsevier.

according to the lock-and-key principle proposed by Pine and coworkers [77–79]. They used silica microparticles with a well-defined number of cavities combined to conventional spheres, which diameter fits that of the dimples in the presence of poly(ethylene oxide) as depletant [79]. AX, AX_2, AX_3, AX_4, and AX_5 CMs were stoichiometrically achieved when increasing the valence of the patchy microspheres (Figure 1.10). This multivalent lock-and-key binding is reversible, i.e. disassembly occurring when the depletant concentration is lowered, and flexible as the spheres are rotating freely within the cavities. The authors have not reported a pathway to make these CMs permanent.

The same group has reported that DNA hybridization is another way to assemble one-patch particles around a multipatch one [80]. They have obtained new CMs from AX to AX_4 morphology (Figure 1.11). For high patch-to-particle size ratio, they have observed original CMs, which mimic ethylene molecules if the single oblong central particle is considered as embedding both carbon atoms. Stepwise self-assembly has been demonstrated with similar particles bearing two types of DNA strands, i.e. one type on the patches and another one on the remaining particle surface, taking advantage of the 10 °C difference between their melting temperature [81].

Lastly, our group has successfully implemented a covalent strategy and obtained silica CMs from dimpled nanoparticles by aminating PS macromolecules anchored at the bottom of the dimples [82]. Then, we have incubated them in DMF with silica nanospheres, which surface ester groups are activated in order to lock them within the dimples via amide bonds. TEM pictures show that the satellite particles occupy most of the dimples in a robust way, making centrifugation sorting possible (Figure 1.12). AX_2, AX_3, AX_4, AX_nE_m, and AX_nY_m CMs have been obtained from di-, tri-, and tetravalent particles, respectively, by playing with the stoichiometry and number of different types of silica satellites [83]. Chiral CMs have been also randomly observed by using four differently sized satellites [82].

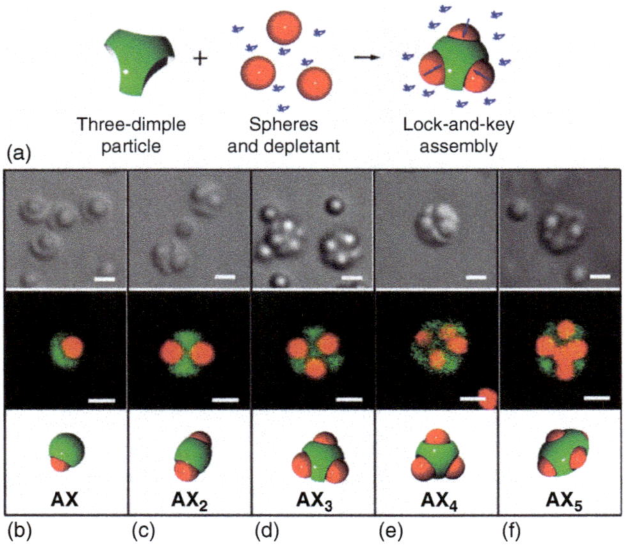

Figure 1.10 (a) Schematic illustration showing a three-patch particle with three assembled spheres using depletion interaction. The depletant (blue coil) causes osmotic pressure (arrows) between adjacent colloids, which is maximized when a sphere assembles into a cavity. (b–f) Bright field micrographs (top panel), confocal micrographs (middle panel), and cartoons (bottom panel) showing multivalent lock particles with (b) one, (c) two, (d) three, (e) four, and (f) five cavities binding to red fluorescent spheres stoichiometrically. Scale bars: 1 μm. Source: Wang et al. [79] / with permission of American Chemical Society.

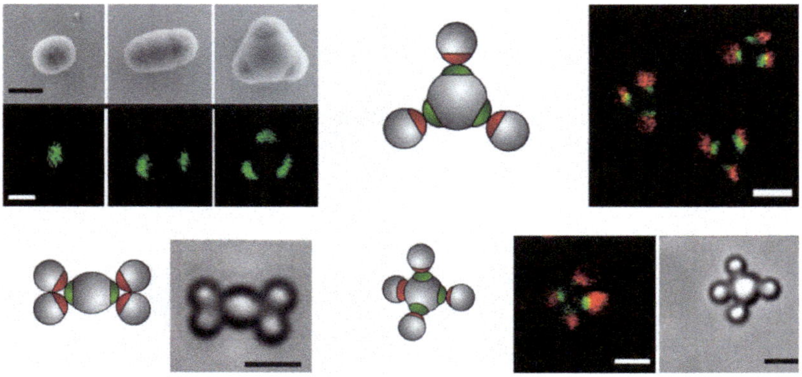

Figure 1.11 SEM and confocal fluorescent images of patchy particles used as CAs (scale bars: 500 nm) and bright-field confocal fluorescent images and schematics of CMs assembled from them through DNA hybridization (scale bars: 2 μm). Source: Wang et al. [80] / with permission of Springer Nature.

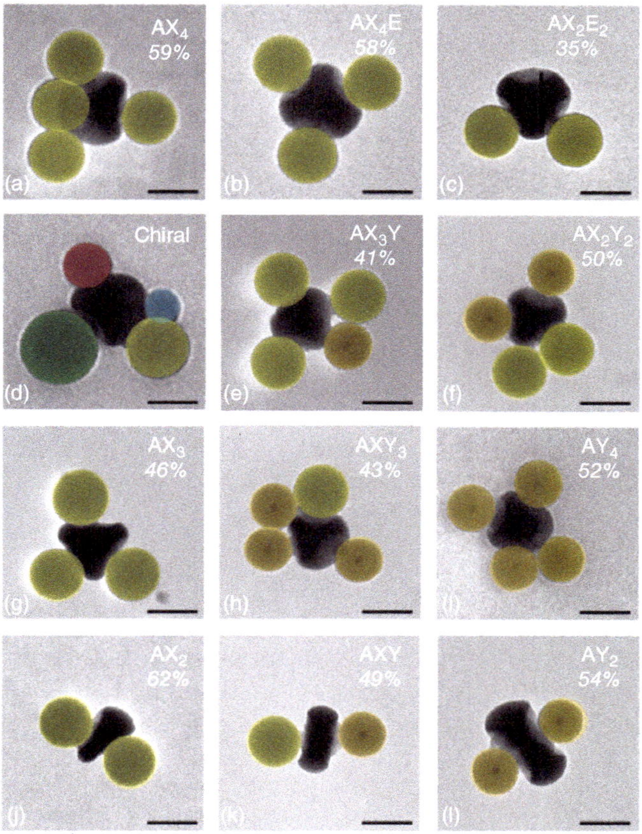

Figure 1.12 TEM images of silica CMs obtained by directional covalent bonding of patchy particles with four dimples (a–f; h–i), three dimples (g) and two dimples (j–l) and satellites of pure silica and different sizes labeled in yellow (100 nm), red (80 nm), blue (60 nm), and green (130 nm) or 90-nm silica satellites including a central gold dot (labeled in orange). Scale bars: 100 nm. The percentage value indicates the prevalence of the shown morphology knowing that exact stoichiometry of nanoparticles was used to obtain it. Scale bars: 100 nm. Source: Modified from Rouet et al. [82] and from Rouet et al. [83] / with permission of Wiley-VCH GmbH.

1.3 Colloidal Polymers: Mimicking Organic Macromolecules

The concept of using nano- or microparticles as "colloidal monomers" has received recent attention for the formation of 1D mesostructures, or "CPs." [15, 84] These colloidal monomers form linear assemblies through directional, attractive, interparticle interactions, which are similar to covalent or supramolecular interactions in classical polymer science. However, in contrast to the high degree of structural control available in the synthesis of classical molecular polymers, methods to control fundamental structural features such as chain length or degree of polymerization

(DP), composition (copolymers), and architecture (linear, branched, etc.) are still being developed for CP systems. In this section, we provide a description of the different strategies that have been developed for the formation of CPs with examples chosen to evidence control of structure and composition. The use of patchy particles as monomers will be particularly emphasized given the growing number of studies on this subject in the recent years.

1.3.1 Dipole-Directed Formation of Colloidal Polymers

Dipole-directed colloidal polymerization refers to the formation of linear (or lightly branched) assemblies arising from coupling through space of charge or spin dipoles inherent to the inorganic cores of colloidal monomers. The colloidal polymerization is sensitive to changes in the composition and crystal phase of the material, where the attractive interparticle dipolar forces can be mitigated by ligand-induced repulsive interactions or increased temperature. Moreover, the architecture of the resulting CPs is dependent upon the balance between dipolar and van der Waals interactions between the constituent colloidal monomers.

1.3.1.1 Electric Dipoles

Tang et al. reported the spontaneous chaining of CdTe nanocrystals of different diameters stabilized by thioglycolic acid when they were allowed to age in the dark at room temperature for up to 48 hours [85]. The formation of chains of nanocrystals was attributed to the existence of strong dipole–dipole attraction, the energy of nanoparticle dipole attraction being estimated to 8.8 kJ/mole for a nanoparticle diameter of 3.4 nm and a center-to-center interdipolar separation of 4.4 nm. A face-to-face orientation of the CdTe nanocrystals in the CPs was later suggested by Monte Carlo simulations [86]. The same team also showed that the electric dipole induced formation of CPs of CdTe nanowires [87] and of "nano-centipedes," which resembled classical polymer brushes when silica-coated CdTe nanowires were used as monomers [88]. Yi et al. recently reported an elegant paradigm for the dipole-directed copolymerization of nanoparticles grafted with reactive block copolymers ligands into linear chains with periodic sequence [89]. Upon mixing in the presence of acid catalyst, the neutralization between colloidal monomers of two different types, A and B, generates electric dipole-like AB dimers that further assemble into alternating copolymers composed of tens of sequentially positioned monomers (Figure 1.13a).

1.3.1.2 Magnetic Dipoles

In the absence of any external magnetic field, strong interparticle interactions of the magnetic spin dipoles of the colloidal monomers induce the formation of linear chains. Thomas showed in a pioneering work that polymer-stabilized 20-nm cobalt nanoparticles spontaneously form chains [92]. Wei and coworkers synthesized dipolar cobalt nanoparticles that formed bracelets under zero-field conditions [93–96], and Puyn and coworkers developed the synthesis of dipolar cobalt nanoparticles functionalized with end-functionalized PS ligands, which spontaneously formed linear CPs [90, 97–99]. Interestingly, a nematic-type liquid crystalline assembly was

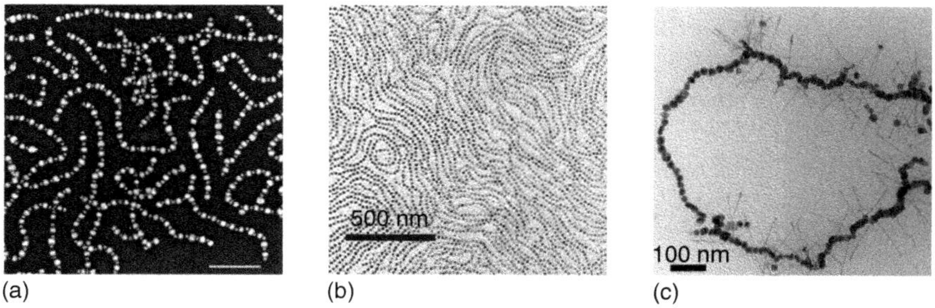

Figure 1.13 (a) SEM image of alternating copolymers made of inorganic nanoparticles of two different sizes (32 and 18 nm). Scale bar: 300 nm. Source: Yi et al. [89] / with permission of American Chemical Society. (b) TEM image of 1D assemblies of PS-coated cobalt nanoparticles. Source: Benkoski et al. [90] / with permission of American Chemical Society. (c) TEM image of cobalt-tipped semiconductor nanorods blended with free cobalt nanoparticles. Source: Hill et al. [91] / with permission of American Chemical Society

observed, where both strong north–south dipolar and weak antiferromagnetic coupling between nanoparticles dipoles was found, resulting in intermittent folding of the colloidal chains (Figure 1.13b). The same team prepared core–shell Au@Co nanoparticles by deposition of cobalt shells onto preformed gold seed nanoparticles [100]. These core–shell nanoparticles were capable to form linear assemblies due to spin dipolar interactions of the ferromagnetic cobalt shells. More complex polymer architectures were prepared from heterostructured monomers consisting in cobalt nanoparticles deposited onto semiconductor nanorods. Linear assemblies of these "matchstick" monomers were obtained due to dipolar magnetic coupling of the cobalt tips [91]. The preparation of block-type colloidal copolymers was also demonstrated by blending dipolar "bare" cobalt nanoparticles with cobalt-tipped semiconductor nanorods possessing cobalt nanoparticles of similar size and magnetization (Figure 1.13c).

1.3.2 Formation of Colloidal Polymers by Nanowelding

CPs composed of gold nanoparticles-tipped CdSe nanorods and consisted of several tens of monomers were fabricated through the destabilization of surface ligands on the gold domains by the addition of molecular iodine and the resulting coalescence of the exposed gold domains [101]. These coalesced gold tips were observed to be polycrystalline after nanowelding.

1.3.3 Formation of Colloidal Polymers Under Physical or Chemical Confinement

Physical confinement and attractive capillary forces have been successfully combined to organize few monodisperse spherical particles into a variety of colloidal chains with a predetermined geometric structure.

In order to fabricate chains of colloids ranging in size from several microns to a few nanometers, Xia et al. have used a fluidic cell with a lithographically patterned bottom surface, which is tilted at an angle and filled with a colloidal solution [102–104]. As the liquid dewets across the cell during template-assisted self-assembly (TASA), the capillary force dominates the assembly process if the particle density is chosen similar to that of the assembling liquid. Colloidal chains were also produced by using lithographically patterned surfaces to trap PS and silica beads into grooves. The length of the chain-type aggregates was determined by the longitudinal dimension of the templates, and the internal structure (linear versus zigzag) was defined by the relative ratio between lateral dimension of the templates and the diameter of the spherical colloids (Figure 1.14a–d). Next, the system is heated slightly above the glass transition temperature of the colloidal material to permanently bond together the spherical particles. Finally, the patterned substrate is dissolved to release the resulting colloidal chains. More complex heterogeneous linear chains were obtained by Wolf and coworkers, applying consecutive depositions with particles of different sizes or composition [30]. Lee et al. have developed an alternative method, called the polymorphic meniscus convergence (PMC) method in order to overcome the main limitation of TASA, which is the achievable linewidth of lithography needed to pattern the bottom surface of the fluidic cell, thereby limiting the technique to colloids with diameters of 50 nm and larger [105]. They used a flow cell made of a hydrophobic auxiliary substrate with templated features and a hydrophilic support arranged such that the colloidal dispersion can only infiltrate specific regions of the cell. As the colloidal dispersion dries in those regions, the menisci converge through the action of lateral capillary

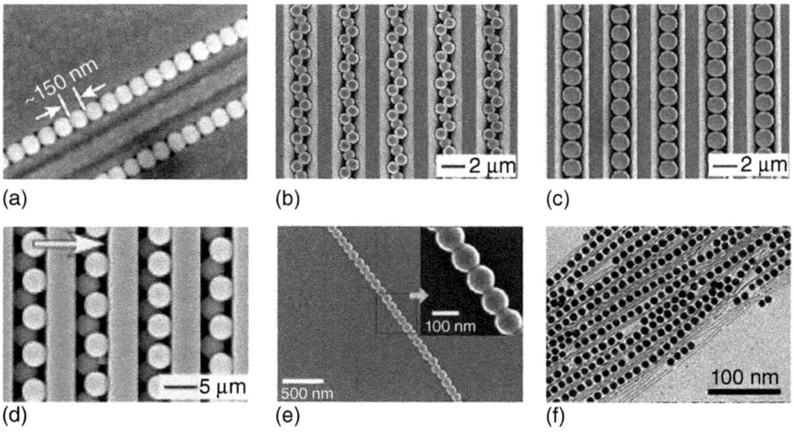

Figure 1.14 (a) SEM image of two linear chains of 150-nm PS beads. Source: Rycenga et al. [104] / with permission of Royal Society of Chemistry. (b–c) SEM images of two chainlike structures of PS beads assembled in 2D arrays of V-grooves. Source: Rycenga et al. [104] / with permission of Royal Society of Chemistry. (d) SEM image of double-layered zigzag chains of 4.3-µm PS beads assembled in an array of channels whose cross-sections were 5 µm in width and 5.5 µm in height. Source: Xia et al. [103] / with permission of Wiley-VCH GmbH. (e) SEM image of 1D chain of 97-nm particles. Source: Lee et al. [105] / with permission of Wiley-VCH GmbH. (f) TEM image of 1D chains of gold nanospheres driven by gold nanowires. Source: Sánchez-Iglesias et al. [106] / with permission of Wiley-VCH GmbH.

forces, and the colloidal particles align with the features of the auxiliary substrate (Figure 1.14e). The colloidal chains structure is determined by the height of the cell gap between the two surfaces as well as the width and depth of the template pattern. Nanoporous alumina membranes were also used by Sawitowski et al. to arrange gold nanoparticles in the form of chains [107]. Pre-synthesized nanoparticles were filled into the porous system by vacuum induction. By adding different amounts of gold colloid, the number as well as the length of the chains can be increased. Liz-Marzan and coworkers used highly anisotropic gold nanowires that drive the oriented assembly of spherical and rodlike gold nanoparticles into 1D chains (Figure 1.14f) [106].

A 2D template made of four different DNA tiles was employed to align gold nanoparticles into chains with nanometer-scale precision [108]. One of the tile types contained a long polyadenosine sequence designed to be unhybridized upon assembly of the 2D array, thereby allowing it to be used as an anchor for particle binding. Six nanometer gold nanoparticles were then functionalized with multiple strands of 3′-thiolated polythymine DNA, which allowed the NPs to bind to the polyadenosine sequences on the 2D DNA tile arrays. The same approach allowed the same research group to produce alternating parallel chains of 5-nm and 10-nm gold nanoparticles [109].

1.3.4 Field-Directed Formation of Colloidal Polymers

Applying external fields offers several advantages for the assembly of colloids. First, the external field can be turned on, off, or programmed with different patterns conveniently, which allows the annealing of assembled colloidal structures due to a temporary release of the colloids from the directing force. Second, the external field does not contaminate the sample, and, third, the strength, type, and effective length of field-induced interparticle can be tuned over a much wider range than that in self-assembly.

1.3.4.1 Electric Fields

Electric field is widely employed to direct the assembly of colloidal particles because most of them are responsive to electric field. Indeed, colloidal particles that have dielectric constant values that are different from that of the solvent acquire a dipole moment that is parallel to the external field. The use of electric fields in the directed assembly of spherical colloids has been nicely reviewed by Velev and Bhatt [110]. Electric field assembly can be performed in direct current (DC) and alternating current (AC) fields. Controllable parameters such as amplitude, frequency, wave shape/symmetry, and phase allow the precise adjustment of driving forces important during the assembly process. By applying an alternating electric field to latex spheres entrapped between planar electrodes, an alignment of the particles into chains due to the attractive interactions between the induced dipoles was observed after 2 s [111]. The chains are parallel to the direction of the applied field and thus are perpendicular to the gap between the two electrodes. Chaining of charged and sterically stabilized polymethyl methacrylate (PMMA) spheres in a density- and

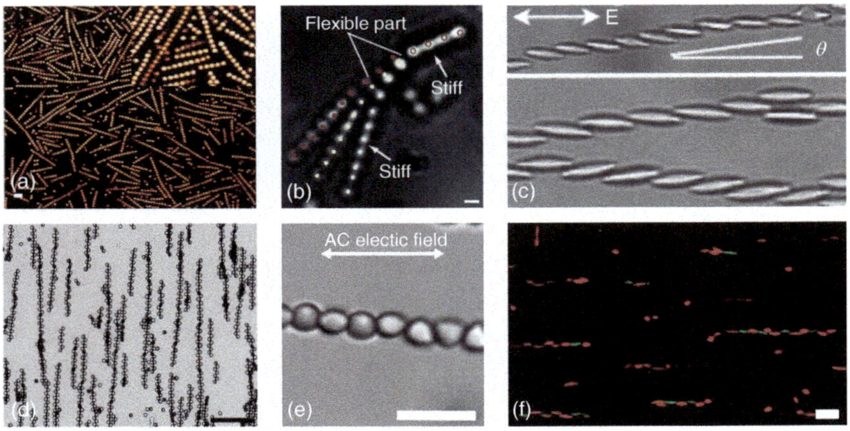

Figure 1.15 (a) Confocal micrographs of permanent PMMA chains in cyclohexyl bromide; the upper inset is a magnified view of the bead chains. Scale bar: 5 μm. (b) Color-coded overlay of optical micrographs of a triblock copolymer-like chain of PS beads taken at different times; the overlay was constructed by placing the rigid end in the same position and orientation. Scale bar: 2 μm. Source: Vutukuri et al. [114] / with permission of Wiley-VCH GmbH. (c) Optical micrographs of PS latex ellipsoids into chains orienting at an angle with respect to the field. Source: Singh et al. [115] / with permission of American Physical Society. (d) Optical micrographs of staggered chains formed from Janus particles in AC field of 56 V/cm at 40 kHz. Scale bar: 70 μm. Source: Gangwal et al. [116] / with permission of American Chemical Society. (e) Optical microscopy image of a chain of three-patch particles. Scale bar: 5 μm. Source: Song et al. [117] / with permission of American Chemical Society. (f) Confocal laser-scanning microscopy image of staggered chains of Janus ellipsoids on application of an AC electric field. Scale bar: 5 μm. Source: Shah et al. [118] / with permission of Springer Nature.

refractive-index-matched solvent mixture [112] and of PMMA particles in tetrahydrofuran [113] or cyclohexyl bromide (Figure 1.15a) [114] was also observed under AC electric field. In the latter case, the timescale for chain formation is on the order of a few seconds and the average chain length could be increased by increasing the applied field strength. The same approach was followed to make rigid chains consisting of PS or silica@PMMA core–shell particles. By mixing flexible and rigid chains together and by subjecting the mixture to the same protocol that was used in making the constituent bead chains, triblock copolymer-like chains were obtained (Figure 1.15b) [114].

AC electric fields have also been used to assemble anisotropic particles such as PS latex ellipsoids with aspect ratios 3.0, 4.3, and 7.6 into chains orienting at an angle with respect to the field [115]. This angle decreases as the ellipsoid aspect ratio increases (Figure 1.15c). Chains of asymmetric colloidal dimers have been created by inducing anisotropic interactions among them under AC electric field when the particle density is ~15% [119]. The 1D chains are uniquely formed by alternating association between dimers with opposite orientations. Such a pattern was theoretically attributed to an exquisite balance between electrostatic (primarily dipolar) and electrohydrodynamic interactions. Patchy metallodielectric particles have been extensively aligned under DC or AC electric fields because of the ease to directionally deposit a metallic cap on dielectric spheres [120, 121]. Janus particles prepared from silica microspheres with one hemisphere coated with metal, covered with a thin silica protective layer, have been subjected to perpendicular AC electric field, and chains were observed when the electric field frequency reached the megahertz range [122]. When Janus particles consisting of PS particles covered by a gold cap are exposed to an AC electric field at low frequencies (<10 kHz) and medium field strengths (40–100 V/cm), the formation of chains that resemble to those obtained from plain dielectric particles is observed, whereas staggered chains form when frequency is increased [116]. In the staggered chains (Figure 1.15d), the gold caps touch each other to minimize the electrostatic energy. An expansion of this work to patchy particles with two patches on opposite poles reveals how the particles assemble into regular, straight chains parallel to the electric field lines at frequencies lower than ~50 kHz, no distinguishable orientation of the patches being observed at such low frequencies [123]. At higher frequencies (> ~50 kHz) of the field, however, the two-pole microspheres form staggered chains and unanticipated diagonal chains oriented at ~45° to the field direction. Colloidal particles with two or three negatively charged patches fabricated by the cluster encapsulation method form CPs when polarized by an AC electric field [117]. Interestingly, the chains exhibit segments with a 2_1 screw axis symmetry due to a 180° rotation of successive particles along the chain, which creates favorable dipole–dipole interactions between polarized patches according to solid models (Figure 1.15e). Shah et al. showed that Janus ellipsoids that have been synthesized by the sequential deposition of 7.5 nm of chrome and 15 nm of gold on a monolayer of ellipsoidal particles assemble into staggered chains at high frequencies (Figure 1.15f) [118]. Chaining corresponds with a minimization of electrostatic energy, as the electric dipole moment of the Janus ellipsoids moves away from the ellipsoid's center of mass toward the extremities of the gold half of the particles. Shields et al. synthesized a wide range of anisotropic, patchy microparticles combining photolithography and metal deposition [124]. They showed that particles were attracted toward each other dielectrophoretically and they reoriented and aligned so that their edges were parallel and orthogonal to the field direction such that the particles organized into chains, usually with the metallic patch aligned with the field direction. It was also shown that composite dumbbells incorporating an asymmetrically placed titania sphere formed colloidal chains, in which they contacted their core–shell parts and oriented perpendicularly to a low-frequency (kHz) field, whereas they oriented parallel to a high-frequency (MHz) field [125].

1.3.4.2 Magnetic Fields

Colloids with magnetic properties in magnetic fields behave very similar to polarized particles in electric fields. Magnetic fields have also been extensively employed to direct the assembly of magnetic colloidal particles, with key factors that determine the final assembly equilibrium being the magnitude of magnetic interactions as well as the local particle concentration [126]. The strength of external magnetic fields directly determines both factors and enables full control of the assembly of particles into desired structures. Furst et al. created colloidal chains by applying a magnetic field to a suspension of paramagnetic PS spheres containing magnetite [127]. Chains were made permanent by covalent linking of the beads using glutaraldehyde. A similar strategy using streptavidin–biotin binding was employed by Biswal and Gast to produce flexible magnetoresponsive chains, which mechanical properties were studied [128]. Zhang et al. produced chains of 150-nm Fe_3O_4 particles under magnetic field [129]. The length of the chains increased with increasing field intensity, while the diameter of the particle in the chains remained nearly constant. The average length of the chains increased from 1.5 to 4 μm with the synthesizing field varying from 1000 to 3500 Oe. A convenient and flexible approach for the fabrication of colloidal chains exhibiting photonic properties was proposed by Yin and colleagues [130–132]. The colloidal monomers were Fe_3O_4 colloidal nanocrystal clusters, which were assembled into chains by applying a magnetic field. All the chains aligned along the field direction so that the dispersion diffracts light and shows brilliant colors. In response to the change of the strength of the external field, the periodicity of the chainlike structures alters. The diffraction wavelength is therefore responsive to external magnetic fields. After carefully optimizing the sizes of nanocrystal clusters, their diffraction wavelength or color can be effectively tuned within the visible light spectrum, from blue to green and to red. The color change occurs instantly upon the change in field strength. A similar study was reported by Hu et al. in which the Fe_3O_4 colloidal nanocrystal clusters were coated with a thin layer of silica and the chains were overcoated with an additional layer of silica to stabilize their structure (Figure 1.16a) [133]. Recently, Yellen and coworkers have chained superparamagnetic particles by the magneto-acoustic method, already described earlier [28]. At low particle concentration, they observed the formation of CPs when the tilt angle of the applied magnetic field was higher than 60°, whereas chaining occurs as soon as the tilt angle is >0° for intermediate particle concentrations. Bannwarth et al. obtained homopolymers under magnetic field from monodisperse superparamagnetic PS nanoparticles [134]. The chain length can be tuned either by the concentration of the nanoparticles dispersion or by the growth time of the chains. When dispersions with higher polydispersity were used, a sequencing analogous to that known from polymer chemistry can be observed. Indeed, block- or statistical copolymer-like sequencing of nanoparticles with different sizes was observed depending on the assembly conditions. This is because a particle's dipole is proportional to its volume. At a given field strength, stronger dipolar forces between larger particles always assemble them first into linear chains. Chains made of smaller particles form later and attach to chains of larger particles as the field strength is further increased. If the field is ramped up instantaneously, such a sequential assembly will be suppressed, and the resulting chains exhibit statistically random sequences. It was also observed that the size ratio of two neighboring particles can be used to introduce junction points into the linear arrangement of the hybrid nanoparticles induced by the magnetic field, giving rise to branched polymer networks (Figure 1.16b). More precisely, the combination of a large particle with several small binding partners provides the basis for a junction. The larger the size ratio of two neighboring particles, the more small particles "fit" next to a large particle.

Anisotropic magnetic particles have also been used as precursors of CPs when exposed to magnetic fields. Lee and Liddell reported the synthesis of peanut-shaped hematite particles coated by thin layer of silica and their assembly to form kinked zigzag chains under a DC field [138]. Chaining was also observed when a magnetic field was applied to monodisperse hematite ellipsoids [139]. Zerrouki et al. have shown that silica particles with a magnetic cap self-assemble into chains under an external magnetic field by ordering alternatively up and down along the chain direction [135]. This specific ordering results from the capped particles that have a magnetic dipole interaction, which is maximized when the circular bases of the caps are parallel to each other, parallel to the field, and as close together as possible (Figure 1.16c). In the same paper, they reported that symmetric dumbbells made of two silica particles with a solid magnetic ring located around the contact point between the two silica spheres also formed chains in which they rotate by 90° relative to their neighbors along the chain and field direction. Sacanna et al. synthesized magnetic particles made of an organosilica polymer sphere with a single hematite micromagnet embedded and their arrangement into long linear structures when an external magnetic field is imposed [65].

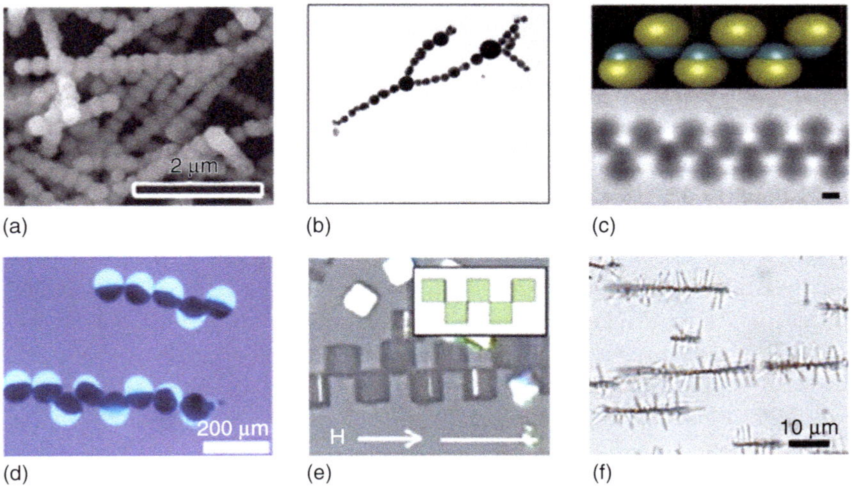

Figure 1.16 (a) SEM image of CPs made of Fe_3O_4 nanoparticles. Source: Hu et al. [133] / with permission of Wiley-VCH GmbH. (b) TEM image of a branched polymer network where the larger nanoparticles act as junction point. Source: Bannwarth et al. [134] / with permission of American Chemical Society (c) Optical microscope image and its corresponding schematic representation of chains of silica particles capped with a magnetic cap. Scale bar: 1 μm. Source: Zerrouki et al. [135] / with permission of Springer Nature. (d) Microscopic image of single pearl chains of Janus hydrogel microparticles under precessional magnetic field. Source: Yoshida et al. [136] / with permission of American Institute of Physics. (e) Optical micrograph of zigzagged chains of top-coated cubes. Scale bar: 25 μm. Source: Wyatt Shields Iv et al. [124] / with permission of Royal Society of Chemistry. (f) Optical microscopic image showing the directed assembly of Janus magnetic rods into chains under an external magnetic field. Source: Zhao et al. [137] / with permission of Royal Society of Chemistry.

The chaining behavior of magnetic Janus particles under an external magnetic field has also been extensively studied. Compartmental magnetic Janus spheres produced by phase separation approaches form staggered chains parallel to the field lines with the magnetic hemispheres pointing inward [140, 141], whereas the application of a precessional magnetic field to compartmental magnetic Janus hydrogel particles led to the formation of single or double pearl chains (Figure 1.16d) [136]. Staggered chains were also produced by applying a magnetic field to Janus particles prepared by partially coating one hemisphere of PS microspheres with a 34-nm layer of iron [124] or iron oxide [142]. When the thickness of the iron layer was smaller (8 nm), Janus particles form staggered or double chains in a magnetic field of 0.15 T, both oriented along the magnetic field direction. Recently Erbe and coworkers developed Janus magnetic particles prepared by coating a PS sphere with a multilayer stack of cobalt, palladium, and tantalum [143]. Under no applied field and guided by their long-range magnetic interaction, the particles assembled into a network of chains, connected by branching points. The authors showed that an applied oscillating magnetic field in plane or out of plane lead to the formation of chains with the amplitude of the field as a knob to tune the chain's conformation: staggered or linear. Shields et al. have fabricated microcubes and microcylinders in SU-8 coated with a thin layer of nickel [124]. With the application of a magnetic field parallel to the plane of a chamber containing an aqueous dispersion of top-coated microcubes, the authors observed the formation of zigzagged chains, whereas linear chains were fabricated when the particles were dispersed in ferrofluid due to the efficient formation of magnetic quadripoles resulting from the presence of the diamagnetic SU-8 component with respect to the ferrofluid magnetization (Figure 1.16e). Top-coated cylinders in water and in ferrofluid exhibited a similar behavior than top-coated cubes. Another interesting class of anisotropic magnetic precursors consists in magnetic silica rods. Yan et al. produced such objects by coating silica rods with a thin hemicylindrical magnetic layer of nickel [144]. When exposed to a magnetic field, the particles stack into kinked zigzag chains. Another synthetic approach based on the anisotropic growth of silica on the surface of Fe_3O_4 microspheres was developed by Zhao et al. to fabricate magnetic silica rods [137]. Under a DC magnetic field, the rods assemble into chains (Figure 1.16f).

1.3.5 Ligand-Directed Formation of Colloidal Polymers

The chemical interactions of surface ligands between neighboring particles have been widely used to dictate their 1D assembly into CPs and can be classified in several categories: electrostatic, covalent, or noncovalent interactions.

1.3.5.1 Electrostatic Interactions

Lin et al. reported the preparation of discrete chains, bifurcated and looped chains, or interconnected chain networks of negatively charged gold nanoparticles using a facile method based on the controlled replacement of citrate ions adsorbed onto the surface of the nanoparticles with covalently bound neutral mercaptoethyl alcohol molecules [145]. In so doing, electrostatic repulsion between the gold nanoparticles is progressively reduced and the stability of the electric dipole associated with charge separation on the nanoparticle surface is potentially enhanced by spatial partitioning of the mercaptoethyl alcohol molecules and citrate capping ligands. Many groups have also reported that colloidal chains formed during the addition of ethanol and other polar organic solvents to aqueous gold nanoparticle solutions [146–149]. In order to clarify the assembly mechanism of gold nanoparticles in such solvents, Yin and coworkers carried out systematic studies indicating that residual salt in conjunction with ethanol, instead of ethanol itself, induces the assembly of gold nanoparticles in ethanol [150]. It was in particular shown that the CP length increased with increasing salt concentration. Xia and coworkers induced the assembly of gold nanoparticles into chainlike structures with tunable lengths and interparticle separations by adding $HS(CH_2)_nCOOH$ ($n = 2$, 10, and 15) into a suspension of gold nanoparticles in a mixture of ethanol and water [151]. The number of gold nanoparticles in the chainlike assemblies could be altered by varying the concentration of the thiol, while the interparticle distance between gold nanoparticles in the chain could be adjusted by using thiols of different chain lengths. The formation of linear chains induced by the addition of salt into a suspension in DMF of gold nanoparticles capped with a monolayer of 2-naphtalenethiol was reported by Yang et al. [152]. To prevent the dissociation or the aggregation of the chains upon drying during the preparation of the substrate for SEM or TEM imaging, the 1D assemblies were encapsulated in shells of polystyrene-block-poly(acrylic acid) (Figures 1.17a–b). The authors proposed that the linear chain formation was kinetically controlled and that charge repulsion among gold nanoparticles was the key factor directing linear aggregation. Block copolymers consisting in segments of gold nanoparticles of different diameters were also fabricated, taking benefit on the strong dependence of the repulsive forces between nanoparticles on their respective sizes. Linear composite chains with configurations reminiscent of those of di- or triblock copolymers have also been prepared from pH-sensitive gold nanoparticles by blending chains of 25-nm and 36-nm nanoparticles and by decreasing the pH value from 3.0 to 2.5 [155].

1.3.5.2 Covalent Bonding

Stellacci and coworkers demonstrated the ability to position two individual molecules at opposite poles of a spherical gold nanoparticle and use these molecules to direct the formation of linear nanoparticle chains [156]. These bifunctionalized nanoparticles are created by first forming a mixed monolayer on the surface of nanoparticles consisting of 1-nonanethiol and 4-methylbenzenethiol. These two molecules were reported to form ordered rings of alternating phases on the surface of the nanoparticles, creating two diametrically opposed defect points on the particle surface. The small molecules at the defect points were then replaced by 1-mercaptoundecanoic acid activated by *N*-hydroxysuccinimide in order to induce the formation of a large number of linear chains, containing between 3 and 20 gold nanoparticles. The interparticle distances along the chains were changed by performing the synthesis with one of two divalent linking molecules of different lengths. The same group adopted a similar approach to form chains

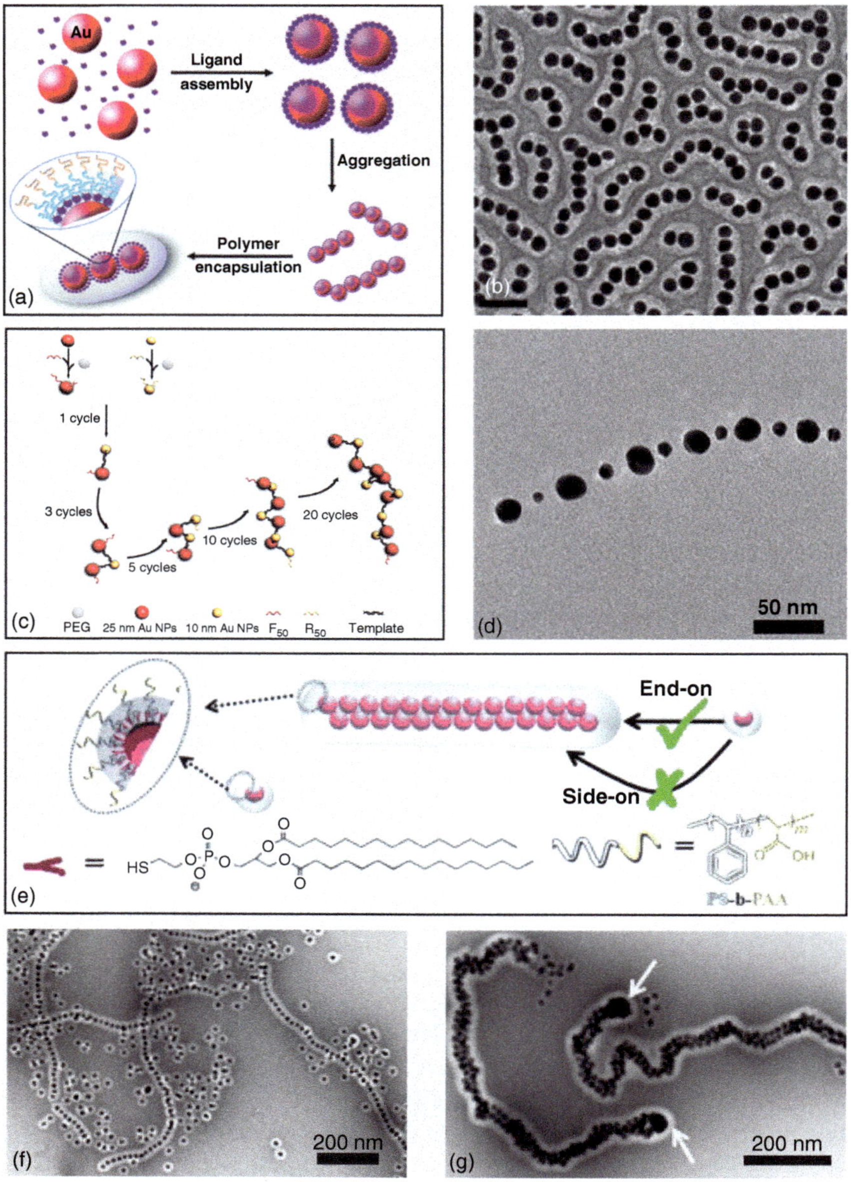

Figure 1.17 (a) Schematic illustration of the chains formation. (b) TEM image of chains of gold nanoparticles. Scale bar: 50 nm. Source: Yang et al. [152] / with permission from the Royal Society of Chemistry. (c) Schematic illustration of heterochains assembled based on PCR. (d) TEM image of heterochains of gold nanoparticles obtained by PCR. Source: Zhao et al. [153] / with permission of American Chemical Society. (e) Schematics illustrating the 1D assembly of gold nanoparticles encapsulated in Ps-b-PAA. (f) TEM image of single chains of AuNP@PSPAA. (g) TEM image of double-line chains of AuNP@PSPAA. The white arrows indicate the larger core–shell gold NPs used as seeds. Source: Wang et al. [154] / with permission of Wiley-VCH GmbH.

of superparamagnetic 13-nm iron oxide nanoparticles coated with a mixture of nonanoic acid and 4-phenylbutyric acid, which have "rippled" ligand shells with reactive polar point defects [157]. The chains showed ferromagnetic interactions between the nanoparticles that compose them, which raise the blocking temperature and hence the ferromagnetic phase of 40 K. α,ω-Dithiol molecules were also employed as linkers between gold nanorods [158] and CdSe nanorods possessing gold tips [159] to promote the formation of linear chains. In the latter system, the thiol groups bind strongly onto the Au tips, and the bifunctional molecules allow the preferential organization of CdSe nanorods as nanochains.

1.3.5.3 Noncovalent Attachment

Using Isotropically Capped Particles as Monomers Zhao et al. used polymerase chain reaction (PCR) to fabricate chains consisting of gold nanoparticles of different sizes, selectively capped with two oligonucleotide primers [153]. Alternating big and small chains were prepared by the DNA polymerization reaction directly on the surface of gold nanoparticles (Figures 1.17c–d). With an increasing number of PCR cycles, the number of nanoparticles in the assembled chains was found to increase from 2 up to 12. Taton and coworkers nicely showed that gold nanoparticles encapsulated within polystyrene-block-poly(acrylic acid) (PS-b-PAA) micelles assemble into regular 1D chains when they are exposed to solvent conditions that relax interfacial curvature in the micellar shell [160]. Nanoparticle chaining was induced by adding salt, acid, or cationic carbodiimide to the nanoparticles suspension. A similar approach was adopted by Chen and coworkers to fabricate linear assemblies (Figure 1.17e) [154]. The formation of "double-line chains" was observed through the addition of hydrochloric to a dispersion of the (PS-b-PAA) core–shell nanoparticles in a DMF/water (7 : 3, v/v) mixture at 70 °C and was attributed to the aggregation of the core–shell nanoparticles and subsequent reorganization of the polymer shells to form cylindrical micelles. When the DMF/water ratio was increased to 6 : 1, ultralong single chains were obtained (Figure 1.17f). In another set of experiments, the authors used larger core–shell gold nanoparticles of core–shell silver nanocubes as seeds to grow chains. The concentration of the seeds was too low that they could not aggregate upon destabilization under the acidic conditions. Upon addition of a large amount of small core–shell gold nanoparticles, "double-line chains" with large seeds at their end were obtained (Figure 1.17g). Choueiri et al. used gold nanospheres or silver nanocubes capped with PS ligands of different molecular weights as monomers and induced their assembly into chains by reducing the solvent quality for PS ligands [161]. Tiopronin-coated silver nanoparticles were also assembled under sonication into linear chains, which were held together by a combination of van der Waals forces, argentophilic interactions, and hydrogen bonding [162]. Periodically spaced and highly aligned 1D chains of hydrophilic polymer-grafted silver nanocubes were produced by introducing the metal blocks into a hydrophobic polymer matrix, which was annealed using thermal or solvent vapor treatment [163]. Chen and coworkers studied the self-assembly behavior of ultra-anisometric silver nanoplates capped with carboxylate-thiols [164]. Nanoplates behaved as bifunctional monomers and polymerized into chains upon the addition of sodium chloride. The assembly was found to proceed by a mechanism characteristic of molecular step-growth polymerization, since the number-average degree of polymerization grows linearly with time.

The precise control of the monomers' valence is crucial in order to self-assemble CPs. A valence 2 is indeed required to prevent branching or chain ends. Brujic and coworkers explored the use of emulsion droplets, decorated with DNA strands on the surface to assemble CPs [165]. Attached to the surface by lipids, the surface strands are still mobile. When two droplets bind through DNA hybridization, the authors showed that DNA strands were recruited at the contact points, forming a patch of high density while the rest of the surface was depleted of chains. A fine-tuning of the DNA coverage was shown to lead to a valence control when the amount of remaining DNA strands on the surface is too small to allow one more particle to bind. The same group showed a control of the monomer sequence by a careful design of the DNA sticky sequences [166]. They also recently showed that the polymers formed are dynamic and were models of freely joined polymers, at the micrometer scale [167].

Using Patchy Particles as Monomers The differential binding of ligands to the long sides or to the edges of gold nanorods was largely used to promote their assembly into CPs. Caswell et al. synthesized gold nanorods that were coated with cetyl trimethylammonium bromide (CTAB) and postulated that the molecular ligands preferentially bind along the long lateral facets of the nanorods [168]. They further bound biotin ligands onto the nanorod tips and induced chaining through the addition of streptavidin, since biotin binds tightly to streptavidin, with four biotins binding per protein. CPs containing at least ten nanorods were obtained. The same group reported a little bit later the formation of linear assemblies of CTAB-coated gold nanorods, nanospheres, and nanobipyramids, and attributed this assembly behavior to the lower stability of CTAB bilayers on nanoparticle surfaces with high curvature [169]. The regioselective binding of CTAB molecules along the lateral facets of gold nanorods and bipyramids was also exploited by Yan and coworkers to promote the formation of linear chains, branched chains, and necklace structures by the preferential attachment of glutathione or cysteine onto the terminal facets of the nanoparticles [170]. In a similar way, Kotov and colleagues have utilized the chromonic material disodium cromoglycate (DSCG) to end-to-end assemble negatively charged poly(acrylic acid) (PAA)-coated gold nanorods, since PAA was able to bind the positively charged CTAB-coated sides of the nanorods but not the tips, which were left to interact with DSCG via electrostatic or van der Waals forces [171]. By specifically grafting thiolated oligonucleotides on the tips of CTAB-coated gold nanorods, Pan et al. reported that metal blocks can be assembled preferentially in an end-to-end fashion upon the addition of oligonucleotides with complementary sequences [172]. Kumacheva and coworkers grafted thiol-terminated PS oligomers onto the tips of laterally CTAB-capped gold nanorods [173]. By adding water (20 wt%) (a bad solvent for PS) to a suspension of nanorods in dimethylformamide (a good solvent for both CTAB and PS), they reduced the solubility of PS, thereby triggering nanorod assembly in chains. The spacing between the adjacent nanorods in the chains was found to increase with increasing water content, due to a relocation of the PS tails from the long faces of the nanorods to the gaps between the metal blocks. The authors further investigated the evolution of the self-assembled chains, following the gradual change in the selectivity of solvent for the central CTAB-coated metal block and PS molecules [174]. They also performed a quantitative study of the assembly process in order to predict the chain topology and the kinetics of the chain growth [175]. The colloidal polymerization was shown to follow a step-growth polymerization model, the nanorods acting as multifunctional monomers and forming reversible, noncovalent bonds at specific bond angles. A similar conclusion was made when the polymerization was induced by the addition of salts, which was found to increase the assembly rate due to a salt-mediated reduction in the solubility of the PS ligands [176]. In order to provide insight into the polymerization kinetics, Au–Fe_3O_4 heterodimers were blended with PS-terminated gold nanorods in DMF/water mixtures, the heterodimers being utilized as "chain stopper." [177] The replacement of PS oligomers by poly(*N*-isopropylacrylamide) or random poly(styrene-*co*-isoprene) copolymers allowed the same group to photothermally trigger the assembly of gold nanorods in 1D structures [178] and covalently fix the bond angles in the CPs [179], respectively. Finally, the co-assembly of gold nanorods with different dimensions into random and block copolymers and of gold and palladium nanorods into random copolymers was also demonstrated [180] (Figure 1.18a).

Monodisperse and well-defined patchy micelles prepared by the solution self-assembly of amphiphilic block copolymers represent other very popular building blocks for the preparation of CPs. Pioneering work by Li et al. in 2004 [184] and by Cui et al. in 2007 [185] showed that patchy micelles prepared from triblock copolymers undergo further self-assembly to afford long linear chains, whereas branched assemblies could also be observed, probably due to polydispersity in size and shape of assembling micelle units [185]. By immersing assembled 1D structures into primary amine-coated gold nanoparticle aqueous suspension for several minutes, it was possible to regioselectively decorate the negatively charged zones of the CPs, giving rise to very interesting hybrid CPs [185]. Gröschel et al. synthesized a series of patchy micelles using triblock copolymers as precursors and induced their assembly in linear and branched chains by the introduction of a poor solvent for corona patches of the micelles [186]. Adding monovalent micelles in specific ratios to a solution of divalent micelles allowed the authors to control the length of the polymer chains, while multiblock co-assemblies were obtained by combining two types of divalent micelles (Figure 1.18b) [181]. In the same vein, Winnik and coworkers produced colloidal chains by placing cylindrical amphiphilic triblock micelles in nonsolvents for the central or terminal segments [187]. Sohn and coworkers fabricated CP chains by adding water in a DMF solution of two-patch micelles consisting in a central poly(4-vinyl-pyridine) core surrounded by two separated PS patches [188]. Random and block copolymer chains were also obtained by mixing two types of patchy micelles or two pre-polymerized chains, respectively. In remarkable later studies, the same group reported controlled branching of the CP chains by introducing well-defined trifunctional patchy micelles (Figure 1.18c) [182], the production of fluorescent chains from micelles containing fluorophores [189] and of chains containing gold or silver nanoparticles [190].

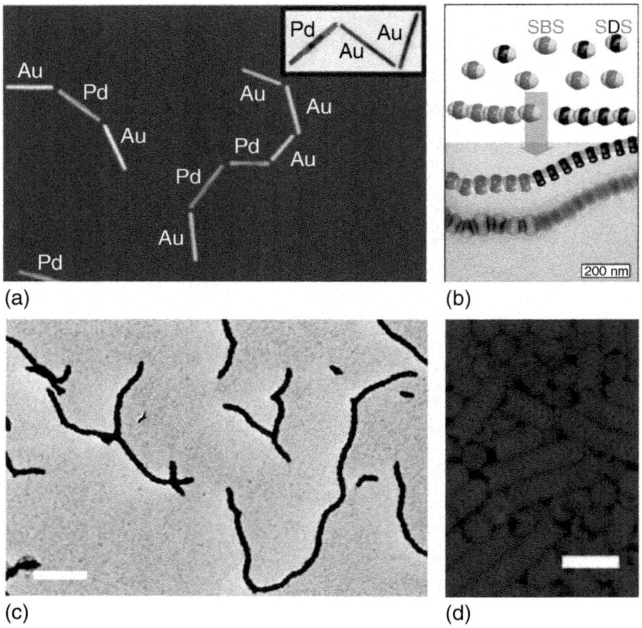

Figure 1.18 (a) STEM image of copolymers of palladium and gold nanorods. Scale bar: 500 nm. Source: Liu et al. [180] / with permission of Wiley-VCH GmbH. (b) Scheme and TEM image of the formation of colloidal copolymers by combining two types of divalent micelles. Source: Gröschel et al. [181] / with permission of Springer Nature. (c) TEM image of branched chains obtained by adding trifunctional micelles. Scale bar: 300 nm. Source: Lee et al. [182] / with permission of Royal Society of Chemistry. (d) SEM image of CPs made of cone-shaped particles. Scale bar: 10 μm. Source: Tigges and Walther [183] / with permission of Wiley-VCH GmbH.

Dielectric patchy particles were also employed as building units to fabricate CPs. Onoe and coworkers described sequential assembly of silicon microparts with two different binding sites in the form of 1D chains, thanks to hydrophobic and van der Walls interactions [191]. Depletion forces were also nicely exploited by Pine and coworkers [78] and by Walther and coworkers [183] to assemble colloidal particles with complementary geometrical forms into colloidal chains (Figure 1.18d). The co-assembly of cone-shaped particles with double-cone particles, which correspond to a homobifunctional monomer, nicely indicated a similar reactivity of both particles and a smooth "copolymerization." [183]

The formation of colloidal chains triggered by hydrogen bonding interaction [192] or biorecognition between complementary DNA strands [70] was also reported. In the latter case, the influence of the Janus balance of the chain structure was nicely investigated, and it was shown that dimer and trimer chains were obtained for intermediate and high patch ratio values (Figure 1.19a), respectively. Particles with two patches were also successfully employed as colloidal monomers. Combining depletion forces and triblock particles, Weck and coworkers recently demonstrated the selective formation of various chains (cross-chains, ladder, and tilted ladder

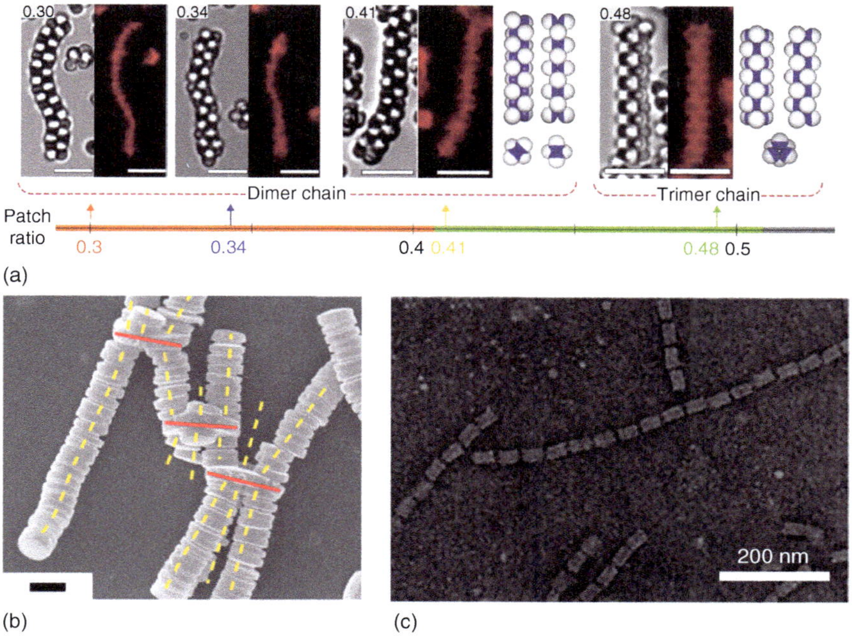

Figure 1.19 (a) Bright-field and fluorescent images of CPs made of DNA-coated Janus particles with patch ratios varying from 0.3 to 0.6. Scale bars: 10 μm. Source: Oh et al. [70] / with permission of Springer Nature. (b) SEM image of branch-like CPs resulted from the introduction of larger disks (represented by the red lines). Scale bar: 2 μm. Source: Zhao et al. [193] / with permission of American Chemical Society (c) TEM image showing polymeric chains of patchy DNA origami cuboids. Source: Tigges et al. [194] / with permission of American Chemical Society.

chains) of patchy particles [195]. They exploited the fact that the monomers formed of patches of PS with a core of polymerized 3-(trimethoxysilyl)propyl methacrylate (TPM) exhibit a directional and selective depletion interaction. In the presence of micelles of Pluronic F127 or Superonic F108, the particles bind selectively with a patch/patch or core/core interaction forming cross-chains or ladder chains depending on the overall shape of the monomer. The contrast of material giving rise to a hydrophobic contrast was also used by Shall and coworkers to assemble two-patch CAs into polymers using Casimir forces [196]. The monomers used were consist of hydrophobic patches of poly(methylmethacrylate) with an hydrophilic core of poly(methylmethacrylate-*co*-methacrylic acid). Once suspended in a mixture of heavy water and 3-methylpyridine (3MP) near the critical temperature, the particles interact through Casimir forces with temperature as the adjustment parameter, allowing the tuning of the bending stiffness between monomers as well as the interaction range. The authors showed that starting from a water-rich solvent gave rise to an attractive force between the hydrophilic cores of the particles forming a ladder-type polymer. On the contrary, a 3MP-rich solvent promoted the hydrophobic interaction between the patches, forming linear chains.

Liu and coworkers recently reported that capillary force can be used for directional 1D assembly of colloidal disks with two liquid patches [193]. By incorporating disks with larger diameters, which served as cross-linkers, the authors formed branch-like CPs (Figure 1.19b). We have produced colloidal chains from two-patch silica nanoparticles with PS chains at the bottom of their two cavities through reduction of the solvent quality for the PS chains [197]. We also showed that chain networks can be formed through the addition of three-patch nanoparticles, which play the role of branching points. Furthermore, block copolymers were also achieved by co-assembling preformed homopolymers composed of patchy NPs of different sizes or surface chemical groups. Pine and coworkers engineered PS-based particles with two patches functionalized with palladated pincer receptors and initiated the assembly by the addition of $AgBF_4$, which frees a coordination site for the pyridine-functionalized particles, thus allowing for the formation of linear and branched CPs [198]. The same research team took benefit of the inherent selectivity of host–guest assemblies to assemble particles with two patches functionalized with cucurbit[7]uril (CB[7]) into chains by the addition of diphenyl viologen (DPV) [199]. The resulting chains were sustained through the formation of CB[7]-DPV-CB[7] bridges and could be reversibly disassembled/reassembled by one-electron reduction/oxidation of the viologen. The group of New York University also fabricated alternating copolymers or homopolymers from two-patch particles site-selectively functionalized with complementary DNA strands [81, 200]. Programmable DNA hybridization was also employed to trigger the assembly of patchy, divalent DNA origami nanocuboids into chains (Figure 1.19c) [194]. Heating of the self-assembled chains led to disassembly into monomers, whereas subsequent cooling induced their reassembly.

1.4 Conclusion and Outlook

In this chapter, we have discussed the synthesis of CMs and CPs through the assembly of preformed particles. Indeed, a tremendous progress has been made in the synthesis of colloidal particles with a good size and shape selectivity over the last two decades, which are capable of exhibiting binding specificity via directional interactions and thus emerged as attractive building blocks [201]. Nevertheless, one can be sure that numerous advances in this field will be made in the near future, which would allow one to envision the preparation of a wider range of novel CMs and CPs, which will further open new avenues of research in the area of materials chemistry. For example, one can introduce orthogonal surface chemistries in designated patches [81] to guide the assembly of patchy particles into novel CMs. The orthogonal surface functionalization of colloidal monomers would also allow one to encode the required instructions to promote the folding of the resulting CPs into a 3D target structure [84, 202, 203]. Robust synthetic methods to control CPs architecture, regiochemistry, and composition also still need to be further developed, in order to afford useful, synergistic properties. For instance, the use of a colloidal analog of solid phase peptide synthesis is highly promising in order to control the sequence of colloidal monomers along the chains. Hence, we are convinced that the synthesis of CMs and CPs will remain a great source of inspiration for colloidal chemists, physical chemists, and physicists in a close future.

References

1 Grzybowski, B.A., Stone, H.A., and Whitesides, G.M. (2000). *Nature* 405: 1033.
2 Park, S., Lim, J.-H., Chung, S.-W., and Mirkin, C.A. (2004). *Science* 303: 348.
3 Pusey, P.N. and van Megen, W. (1986). *Nature* 320: 340.
4 Lu, P.J. and Weitz, D.A. (2013). *Annu. Rev. Condens. Matter Phys.* 4: 217.
5 Palberg, T. (2014). *J. Phys. Condens. Matter* 26: 333101.
6 van Blaaderen, A. (2006). *Nature* 439: 545.
7 Glotzer, S.C. and Solomon, M.J. (2007). *Nat. Mater.* 6: 557.
8 Sacanna, S. and Pine, D.J. (2011). *Curr. Opin. Colloid Interface Sci.* 16: 96.
9 van Blaaderen, A. (2003). *Science* 301: 470.
10 Plüisch, C.S. and Wittemann, A. (2013). *Macromol. Rapid Commun.* 34: 1798.
11 Morphew, D. and Chakrabarti, D. (2017). *Curr. Opin. Colloid Interface Sci.* 30: 70.
12 Elacqua, E., Zheng, X., Shillingford, C. et al. (2017). *Acc. Chem. Res.* 50: 2756.
13 Yu, B., Cong, H., Peng, Q. et al. (2018). *Adv. Colloid Interf. Sci.* 256: 126.
14 Duguet, E., Désert, A., Perro, A., and Ravaine, S. (2011). *Chem. Soc. Rev.* 40: 941.
15 Hill, L.J., Pinna, N., Char, K., and Pyun, J. (2015). *Prog. Polym. Sci.* 40: 85.
16 van Oostrum, P. (2017). *Design of Self-Assembling Materials* (ed. I. Coluzza), 91–106. Cham: Springer International Publishing.
17 Li, W., Palis, H., Mérindol, R. et al. (2020). *Chem. Soc. Rev.* 49: 1955.
18 Ning, H., Zhang, Y., Zhu, H. et al. (2018). *Micromachines* 9: 1.
19 Li, F., Josephson, D.P., and Stein, A. (2011). *Angew. Chem. Int. Ed.* 50: 360.
20 Rogers, W.B., Shih, W.M., and Manoharan, V.N. (2016). *Nat. Rev. Mater.* 1: 16008.
21 Scarfiello, R., Nobile, C., and Cozzoli, P.D. (2016). *Front. Mater.* 3: 1.
22 Plüisch, C.S. and Wittemann, A. (2016). Assembly of nanoparticles into "Colloidal Molecules": Toward complex and yet defined colloids with exciting perspectives. In: *Advances in Colloid Science* (eds. M.M. Rahman and A.M. Asiri), 237–264. InTech.
23 Löwen, H. (2018). *Europhys. Lett.* 121: 58001.
24 Wintzheimer, S., Granath, T., Oppmann, M. et al. (2018). *ACS Nano* 12: 5093.
25 Genix, A.-C. and Oberdisse, J. (2018). *Soft Matter* 14: 5161.
26 Ma, F., Wu, D.T., and Wu, N. (2013). *J. Am. Chem. Soc.* 135: 7839.
27 Demirörs, A.F., Pillai, P.P., Kowalczyk, B., and Grzybowski, B.A. (2013). *Nature* 503: 99.
28 Yang, Y., Pham, A.T., Cruz, D. et al. (2015). *Adv. Mater.* 27: 4725.
29 Wirth, C.L., De Volder, M., and Vermant, J. (2015). *Langmuir* 31: 1632.
30 Ni, S., Leemann, J., Buttinoni, I. et al. (2016). *Sci. Adv.* 2: e1501779.

31 Manoharan, V.N., Elsesser, M.T., and Pine, D.J. (2003). *Science* 301: 483.
32 Lauga, E. and Brenner, M.P. (2004). *Phys. Rev. Lett.* 93: 238301.
33 Wagner, C.S., Lu, Y., and Wittemann, A. (2008). *Langmuir* 24: 12126.
34 Schwarz, I., Fortini, A., Wagner, C.S. et al. (2011). *J. Chem. Phys.* 135: 244501.
35 Månsson, L.K., Immink, J.N., Mihut, A.M. et al. (2015). *Faraday Discuss.* 181: 49.
36 Yuan, Q., Gu, J., Zhao, Y. et al. (2016). *ACS Macro Lett.* 5: 565.
37 Yao, L., Li, Q., Guan, Y. et al. (2018). *ACS Macro Lett.* 7: 80.
38 McGinley, J.T., Wang, Y., Jenkins, I.C. et al. (2015). *ACS Nano* 9: 10817.
39 McGinley, J.T., Jenkins, I., Sinno, T., and Crocker, J.C. (2013). *Soft Matter* 9: 9119.
40 Shen, B., Ricouvier, J., Malloggi, F., and Tabeling, P. (2015). *Adv. Sci.* 3: 1.
41 Kraft, D.J., Vlug, W.S., van Kats, C.M. et al. (2009). *J. Am. Chem. Soc.* 131: 1182.
42 Meester, V., Verweij, R.W., van der Wel, C., and Kraft, D.J. (2016). *ACS Nano* 10: 4322.
43 Perry, R.W. and Manoharan, V.N. (2016). *Soft Matter* 12: 2868.
44 Schade, N.B., Holmes-Cerfon, M.C., Chen, E.R. et al. (2013). *Phys. Rev. Lett.* 110: 148303.
45 Alivisatos, A.P., Johnsson, K.P., Peng, X. et al. (1996). *Nature* 382: 609.
46 Aldaye, F.A. and Sleiman, H.F. (2007). *J. Am. Chem. Soc.* 129: 4130.
47 Mastroianni, A.J., Claridge, S.A., and Alivisatos, A.P. (2009). *J. Am. Chem. Soc.* 131: 8455.
48 Xu, X., Rosi, N.L., Wang, Y. et al. (2006). *J. Am. Chem. Soc.* 128: 9286.
49 Tian, Y., Wang, T., Liu, W. et al. (2015). *Nat. Nanotechnol.* 10: 637.
50 Liu, W., Halverson, J., Tian, Y. et al. (2016). *Nat. Chem.* 8: 867.
51 Yao, G., Li, J., Li, Q. et al. (2020). *Nat. Mater.* 19: 781.
52 Li, Y., Liu, Z., Yu, G. et al. (2015). *J. Am. Chem. Soc.* 137: 4320.
53 Ben Zion, M.Y., He, X., Maass, C.C. et al. (2017). *Science* 358: 633.
54 Chakraborty, I., Meester, V., van der Wel, C., and Kraft, D.J. Colloidal joints with designed motion range and tunable joint flexibility. *Nanoscale* https://doi.org/10.1039/C6NR08069C.
55 Ravaine, S. and Duguet, E. (2017). *Curr. Opin. Colloid Interface Sci.* 30: 45.
56 Pawar, A.B. and Kretzschmar, I. (2010). *Macromol. Rapid Commun.* 31: 150.
57 Perro, A., Reculusa, S., Ravaine, S. et al. (2005). *J. Mater. Chem.* 15: 3745.
58 Walther, A. and Müller, A.H.E. (2013). *Chem. Rev.* 113: 5194.
59 Zhang, J., Luijten, E., and Granick, S. (2015). *Annu. Rev. Phys. Chem.* 66: 581.
60 Zhang, J., Grzybowski, B.A., and Granick, S. (2017). *Langmuir* 33: 6964.
61 Hong, L., Cacciuto, A., Luijten, E., and Granick, S. (2006). *Nano Lett.* 6: 2510.
62 Hong, L., Cacciuto, A., Luijten, E., and Granick, S. (2008). *Langmuir* 24: 621.
63 Chen, Q., Whitmer, J.K., Jiang, S. et al. (2011). *Science* 331: 199.
64 Kraft, D.J., Ni, R., Smallenburg, F. et al. (2012). *Proc. Natl. Acad. Sci. U. S. A.* 109: 10787.
65 Sacanna, S., Rossi, L., and Pine, D.J. (2012). *J. Am. Chem. Soc.* 134: 6112.
66 Yu, C., Zhang, J., and Granick, S. (2014). *Angew. Chem. Int. Ed.* 53: 4364.
67 Bharti, B., Rutkowski, D., Han, K. et al. (2016). *J. Am. Chem. Soc.* 138: 14948.

68 Hu, H., Ji, F., Xu, Y. et al. (2016). *ACS Nano* 10: 7323.
69 Castro, N., Constantin, D., Davidson, P., and Abécassis, B. (2016). *Soft Matter* 12: 9666.
70 Oh, J.S., Lee, S., Glotzer, S.C. et al. (2019). *Nat. Commun.* 10: 3936.
71 Ben Zion, M.Y., Caba, Y., Sha, R. et al. (2020). *Soft Matter* 16: 4358.
72 Skelhon, T.S., Chen, Y., and Bon, S.A.F. (2014). *Soft Matter* 10: 7730.
73 Ge, X.-H., Geng, Y.-H., Chen, J., and Xu, J.-H. (2018). *ChemPhysChem* 19: 2009.
74 Li, W., Ravaine, S., and Duguet, E. (2020). *J. Colloid Interface Sci.* 560: 639.
75 Asakura, S. and Oosawa, F. (1954). *J. Chem. Phys.* 22: 1255.
76 Dinsmore, A.D., Yodh, A.G., and Pine, D.J. (1995). *Phys. Rev. E* 52: 4045.
77 Kim, S.H., Hollingsworth, A.D., Sacanna, S. et al. (2012). *J. Am. Chem. Soc.* 134: 16115.
78 Sacanna, S., Irvine, W.T.M., Chaikin, P.M., and Pine, D.J. (2010). *Nature* 464: 575.
79 Wang, Y., Wang, Y., Zheng, X. et al. (2014). *J. Am. Chem. Soc.* 136: 6866.
80 Wang, Y., Wang, Y., Breed, D.R. et al. (2012). *Nature* 491: 51.
81 Zheng, X., Wang, Y., Wang, Y. et al. (2016). *Chem. Mater.* 28: 3984.
82 Rouet, P.-E., Chomette, C., Duguet, E., and Ravaine, S. (2018). *Angew. Chem. Int. Ed.* 57: 15754.
83 Rouet, P.-E., Chomette, C., Adumeau, L. et al. (2018). *Beilstein J. Nanotechnol.* 9: 2989.
84 Cademartiri, L. and Bishop, K.J.M. (2015). *Nat. Mater.* 14: 2.
85 Tang, Z., Kotov, N.A., and Giersig, M. (2002). *Science* 297: 237.
86 Zhang, Z., Tang, Z., Kotov, N.A., and Glotzer, S.C. (2007). *Nano Lett.* 7: 1670.
87 Wu, L., Shi, C., Tian, L., and Zhu, J. (2008). *J. Phys. Chem. C* 112: 319.
88 Wang, Y., Tang, Z., Liang, X. et al. (2004). *Nano Lett.* 4: 225.
89 Yi, C., Yang, Y., and Nie, Z. (2019). *J. Am. Chem. Soc.* 141: 7917.
90 Benkoski, J.J., Bowles, S.E., Korth, B.D. et al. (2007). *J. Am. Chem. Soc.* 129: 6291.
91 Hill, L.J., Richey, N.E., Sung, Y. et al. (2014). *ACS Nano* 8: 3272.
92 Thomas, J.R. (1966). *J. Appl. Phys.* 37: 2914.
93 Tripp, S.L., Pusztay, S.V., Ribbe, A.E., and Wei, A. (2002). *J. Am. Chem. Soc.* 124: 7914.
94 Wei, A., Tripp, S.L., Liu, J. et al. (2009). *Supramol. Chem.* 21: 189.
95 Wei, A., Kasama, T., and Dunin-Borkowski, R.E. (2011). *J. Mater. Chem.* 21: 16686.
96 Wei, A. (2006). *Chem. Commun.*: 1581.
97 Keng, P.Y., Shim, I., Korth, B.D. et al. (2007). *ACS Nano* 1: 279.
98 Bowles, S.E., Wu, W., Kowalewski, T. et al. (2007). *J. Am. Chem. Soc.* 129: 8694.
99 Korth, B.D., Keng, P., Shim, I. et al. (2006). *J. Am. Chem. Soc.* 128: 6562.
100 Kim, B.Y., Shim, I.B., Araci, Z.O. et al. (2010). *J. Am. Chem. Soc.* 132: 3234.
101 Figuerola, A., Franchini, I.R., Fiore, A. et al. (2009). *Adv. Mater.* 21: 550.
102 Yin, Y., Lu, Y., Gates, B., and Xia, Y. (2001). *J. Am. Chem. Soc.* 123: 8718.
103 Xia, Y., Yin, Y., Lu, Y., and McLellan, J. (2003). *Adv. Funct. Mater.* 13: 907.
104 Rycenga, M., Camargo, P.H.C., and Xia, Y. (2009). *Soft Matter* 5: 1129.

105 Lee, S.W., Park, S.C., Lim, Y. et al. (2010). *Adv. Mater.* 22: 4172.
106 Sánchez-Iglesias, A., Grzelczak, M., Pérez-Juste, J., and Liz-Marzán, L.M. (2010). *Angew. Chem. Int. Ed.* 49: 9985.
107 Sawitowski, T., Miquel, Y., Heilmann, A., and Schmid, G. (2001). *Adv. Funct. Mater.* 11: 435.
108 Le, J.D., Pinto, Y., Seeman, N.C. et al. (2004). *Nano Lett.* 4: 2343.
109 Pinto, Y.Y., Le, J.D., Seeman, N.C. et al. (2005). *Nano Lett.* 5: 2399.
110 Velev, O.D. and Bhatt, K.H. (2006). *Soft Matter* 2: 738.
111 Lumsdon, S.O., Kaler, E.W., and Velev, O.D. (2004). *Langmuir* 20: 2108.
112 Yethiraj, A. and Van Blaaderen, A. (2003). *Nature* 421: 513.
113 Peng, B., van der Wee, E., Imhof, A., and van Blaaderen, A. (2012). *Langmuir* 28: 6776.
114 Vutukuri, H.R., Demirörs, A.F., Peng, B. et al. (2012). *Angew. Chem. Int. Ed.* 51: 11249.
115 Singh, J.P., Lele, P.P., Nettesheim, F. et al. (2009). *Phys. Rev. E.* 79: 050401.
116 Gangwal, S., Cayre, O.J., and Velev, O.D. (2008). *Langmuir* 24: 13312.
117 Song, P., Wang, Y., Wang, Y. et al. (2015). *J. Am. Chem. Soc.* 137: 3069.
118 Shah, A.A., Schultz, B., Zhang, W. et al. (2015). *Nat. Mater.* 14: 117.
119 Ma, F., Wang, S., Zhao, H. et al. (2014). *Soft Matter* 10: 8349.
120 Pawar, A.B. and Kretzschmar, I. (2008). *Langmuir* 24: 9057.
121 Zhang, G., Wang, D., and Möhwald, H. (2005). *Angew. Chem. Int. Ed.* 44: 7767.
122 Yan, J., Han, M., Zhang, J. et al. (2016). *Nat. Mater.* 15: 1095.
123 Gangwal, S., Pawar, A., Kretzschmar, I., and Velev, O.D. (2010). *Soft Matter* 6: 1413.
124 Wyatt Shields Iv, C., Zhu, S., Yang, Y. et al. (2013). *Soft Matter* 9: 9219.
125 Nagao, D., Sugimoto, M., Okada, A. et al. (2012). *Langmuir* 28: 6546.
126 Melle, S., Calderón, O.G., Rubio, M.A., and Fuller, G.G. (2002). *Int. J. Mod. Phys. B.* 16: 2293.
127 Furst, E.M., Suzuki, C., Fermigier, M., and Gast, A.P. (1998). *Langmuir* 14: 7334.
128 Biswal, S.L. and Gast, A.P. (2003). *Phys. Rev. E.* 68: 021402.
129 Zhang, Y., Sun, L., Fu, Y. et al. (2009). *J. Phys. Chem. C* 113: 8152.
130 Ge, J., Hu, Y., Zhang, T. et al. (2008). *Langmuir* 24: 3671.
131 Ge, J., Hu, Y., and Yin, Y. (2007). *Angew. Chem. Int. Ed.* 46: 7428.
132 Wang, M. and Yin, Y. (2016). *J. Am. Chem. Soc.* 138: 6315.
133 Hu, Y., He, L., and Yin, Y. (2011). *Angew. Chem. Int. Ed.* 50: 3747.
134 Bannwarth, M.B., Utech, S., Ebert, S. et al. (2015). *ACS Nano* 9: 2720.
135 Zerrouki, D., Baudry, J., Pine, D.J. et al. (2008). *Nature* 455: 380.
136 Yoshida, S., Takinoue, M., Iwase, E., and Onoe, H. (2016). *J. Appl. Phys.* 120: 084905.
137 Zhao, B., Zhou, H., Liu, C. et al. (2016). *New J. Chem.* 40: 6541.
138 Lee, S.H. and Liddell, C.M. (2009). *Small* 5: 1957.
139 Martinez-Pedrero, F., Cebers, A., and Tierno, P. (2016). *Phys. Rev. Appl.* 6: 034002.

140 Yuet, K.P., Hwang, D.K., Haghgooie, R., and Doyle, P.S. (2010). *Langmuir* 26: 4281.

141 Dyab, A.K.F., Ozmen, M., Ersoz, M., and Paunov, V.N. (2009). *J. Mater. Chem.* 19: 3475.

142 Ren, B., Ruditskiy, A., Song, J.H., and Kretzschmar, I. (2012). *Langmuir* 28: 1149.

143 Steinbach, G., Schreiber, M., Nissen, D. et al. (2019). *Phys. Rev. E* 100: 012608.

144 Yan, J., Chaudhary, K., Chul, B.S. et al. (2013). *Nat. Commun.* 4: 1516.

145 Lin, S., Li, M., Dujardin, E. et al. (2005). *Adv. Mater.* 17: 2553.

146 Liao, J.H., Chen, K.J., Xu, L.N. et al. (2003). *Appl. Phys. A Mater. Sci. Process.* 76: 541.

147 Zhang, H. and Wang, D. (2008). *Angew. Chem. Int. Ed.* 47: 3984.

148 Liao, J., Zhang, Y., Yu, W. et al. (2003). *Colloids Surf. A Physicochem. Eng. Asp.* 223: 177.

149 Zhang, H., Fung, K.-H., Hartmann, J. et al. (2008). *J. Phys. Chem. C* 112: 16830.

150 Han, X., Goebl, J., Lu, Z., and Yin, Y. (2011). *Langmuir* 27: 5282.

151 Cho, E.C., Choi, S.-W., Camargo, P.H.C., and Xia, Y. (2010). *Langmuir* 26: 10005.

152 Yang, M., Chen, G., Zhao, Y. et al. (2010). *Phys. Chem. Chem. Phys.* 12: 11850.

153 Zhao, Y., Xu, L., Liz-Marzán, L.M. et al. (2013). *J. Phys. Chem. Lett.* 4: 641.

154 Wang, H., Chen, L., Shen, X. et al. (2012). *Angew. Chem. Int. Ed.* 51: 8021.

155 Xia, H., Su, G., and Wang, D. (2013). *Angew. Chem. Int. Ed.* 52: 3726.

156 DeVries, G.A., Brunnbauer, M., Hu, Y. et al. (2007). *Science* 315: 358.

157 Nakata, K., Hu, Y., Uzun, O. et al. (2008). *Adv. Mater.* 20: 4294.

158 Shibu Joseph, S.T., Ipe, B.I., Pramod, P., and George, T.K. (2006). *J. Phys. Chem. B* 110: 150.

159 Mokari, T., Rothenberg, E., Popov, I. et al. (2004). *Science* 304: 1787.

160 Kang, Y., Erickson, K.J., and Taton, T.A. (2005). *J. Am. Chem. Soc.* 127: 13800.

161 Choueiri, R.M., Galati, E., Klinkova, A. et al. (2016). *Faraday Discuss.* 191: 189.

162 Shiers, M.J., Leech, R., Carmalt, C.J. et al. (2012). *Adv. Mater.* 24: 5227.

163 Gao, B., Arya, G., and Tao, A.R. (2012). *Nat. Nanotechnol.* 7: 433.

164 Luo, B., Smith, J.W., Wu, Z. et al. (2017). *ACS Nano* 11: 7626.

165 Feng, L., Pontani, L.-L., Dreyfus, R. et al. (2013). *Soft Matter* 9: 9816.

166 Zhang, Y., McMullen, A., Pontani, L.-L. et al. (2017). *Nat. Commun.* 8: 21.

167 McMullen, A., Holmes-Cerfon, M., Sciortino, F. et al. (2018). *Phys. Rev. Lett.* 121: 138002.

168 Caswell, K.K., Wilson, J.N., Bunz, U.H.F., and Murphy, C.J. (2003). *J. Am. Chem. Soc.* 125: 13914.

169 Sau, T.K. and Murphy, C.J. (2005). *Langmuir* 21: 2923.

170 Zhang, S., Kou, X., Yang, Z. et al. (2007). *Chem. Commun.*: 1816.

171 Park, H.S., Agarwal, A., Kotov, N.A., and Lavrentovich, O.D. (2008). *Langmuir* 24: 13833.

172 Pan, B., Ao, L., Gao, F. et al. (2005). *Nanotechnology* 16: 1776.

173 Nie, Z., Fava, D., Kumacheva, E. et al. (2007). *Nat. Mater.* 6: 609.

174 Fava, D., Nie, Z., Winnik, M.A., and Kumacheva, E. (2008). *Adv. Mater.* 20: 4318.
175 Liu, K., Nie, Z., Zhao, N. et al. (2010). *Science* 329: 197.
176 Liu, K., Resetco, C., and Kumacheva, E. (2012). *Nanoscale* 4: 6574.
177 Klinkova, A., Therien-Aubin, H., Choueiri, R.M. et al. (2013). *Proc. Natl. Acad. Sci.* 110: 18775.
178 Fava, D., Winnik, M.A., and Kumacheva, E. (2009). *Chem. Commun.*: 2571.
179 Lukach, A., Liu, K., Therien-Aubin, H., and Kumacheva, E. (2012). *J. Am. Chem. Soc.* 134: 18853.
180 Liu, K., Lukach, A., Sugikawa, K. et al. (2014). *Angew. Chem. Int. Ed.* 53: 2648.
181 Gröschel, A.H., Walther, A., Löbling, T.I. et al. (2013). *Nature* 503: 247.
182 Lee, S., Jang, S., Kim, K. et al. (2016). *Chem. Commun.* 52: 9430.
183 Tigges, T. and Walther, A. (2016). *Angew. Chem. Int. Ed.* 55: 11261.
184 Li, Z., Kesselman, E., Talmon, Y. et al. (2004). *Science* 306: 98.
185 Cui, H., Chen, Z., Zhong, S. et al. (2007). *Science* 317: 647.
186 Gröschel, A.H., Schacher, F.H., Schmalz, H. et al. (2012). *Nat. Commun.* 3: 710.
187 Qiu, H., Hudson, Z.M., Winnik, M.A., and Manners, I. (2015). *Science* 347: 1329.
188 Kim, J.-H., Kwon, W.J., and Sohn, B.-H. (2015). *Chem. Commun.* 51: 3324.
189 Kim, K., Jang, S., Jeon, J. et al. (2018). *Langmuir* 34: 4634.
190 Jang, S., Kim, K., Jeon, J. et al. (2017). *Soft Matter* 13: 6756.
191 Onoe, H., Matsumoto, K., and Shimoyama, I. (2007). *Small* 3: 1383.
192 Onishi, S., Tokuda, M., Suzuki, T., and Minami, H. (2015). *Langmuir* 31: 674.
193 Zhao, S., Wu, Y., Lu, W., and Liu, B. (2019). *ACS Macro Lett.* 8: 363.
194 Tigges, T., Heuser, T., Tiwari, R., and Walther, A. (2016). *Nano Lett.* 16: 7870.
195 Liu, M., Zheng, X., Grebe, V. et al. (2020). *Nat. Mater.* 19: 1354–1361.
196 Nguyen, T.A., Newton, A., Veen, S.J. et al. (2017). *Adv. Mater.* 29: 1700819.
197 Li, W., Liu, B., Hubert, C. et al. (2020). *Nano Res.* https://doi.org/10.1007/s12274.
198 Wang, Y., Hollingsworth, A.D., Yang, S.K. et al. (2013). *J. Am. Chem. Soc.* 135: 14064.
199 Benyettou, F., Zheng, X., Elacqua, E. et al. (2016). *Langmuir* 32: 7144.
200 Wang, Y., Wang, Y., Breed, D.R. et al. (2012). *Nature* 490: 51.
201 He, M., Gales, J.P., Ducrot, É. et al. (2020). *Nature* 585: 524.
202 Coluzza, I., van Oostrum, P.D.J., Capone, B. et al. (2013). *Phys. Rev. Lett.* 110: 075501.
203 Coluzza, I., van Oostrum, P.D.J., Capone, B. et al. (2013). *Soft Matter* 9: 938.

2

Self-assembly of Anisotropic Colloids in Solutions

Yiwu Zong, Huaqing Liu, and Kun Zhao

Tianjin University, Department of Pharmaceutical Engineering, Key Laboratory of Systems Bioengineering (Ministry of Education), School of Chemical Engineering and Technology, No. 92 Weijin Road, Tianjin 300072, PR China

2.1 Introduction

New functional materials are in urgent demand due to their tremendous potential in coping with critical issues emergent in the fields of environment, energy, and health. The self-assembly of colloids has been viewed as a promising bottom-up way to fabricate the functional materials with desired properties from custom-designed building blocks [1, 2]. However, isotropic spherical particles generally have restricted accessible structures through self-assembly such as face-centered cubic (FCC), body-centered cubic, or variant configurations [3–5]. By contrast, anisotropic particles can form much richer structures due to their anisometric shapes and/or anisotropic interactions [6]. For instance, nematic liquid crystal (LC) structures can be formed by classical Onsager rods [7]. More exotic phases/structures such as lamellar structures [8], fibrillar triple helices [9], and kagome lattice [10] have also been successfully realized using anisotropic building blocks. This makes anisotropic particles highly desirable for making new functional materials applicable in a broad range of fields such as photonic crystals [11], optoelectronic materials [12], biosensors [13] as well as environmental cue sensors [14], biomedical materials [15], energy storage materials [16], etc.

In the last three decades or so, with the advances in both fabrication techniques and our understanding of various physical phenomena, we have witnessed a rapid progress in the self-assembly of anisotropic particles, including continuously emerging new synthesis techniques, new assembled structures, and new applications of anisotropic particles.

In this chapter, we will review the recent progress on the self-assembly of anisotropic colloids in solutions, including both anisometric-shaped colloids and colloids with anisotropic interactions. Although most linear, block, graft, or other non-particulate polymer molecules are anisotropic, their self-assembly is a vast

Yiwu Zong and Huaqing Liu contributed equally to this work.

Functional Materials from Colloidal Self-assembly, First Edition. Edited by Qingfeng Yan and George Zhao.

research area and deserved a special treatment to fully appreciate the achievements in the corresponding field. So in this review we do not include the self-assembly of anisotropic non-particulate polymer molecules.

The chapter is constructed as the following. In Section 2.2, an overview of the fabrication methods for anisotropic colloids is presented. In Section 2.3, we categorize the mechanisms governing the self-assembly of anisotropic colloids and show examples of assembled structures. In Section 2.4, the progress in the applications of assembled structures of anisotropic colloids is reported. Lastly, in Section 2.5, we conclude this chapter by discussing the perspectives in this field.

2.2 Fabrication of Anisotropic Colloids

Particles can be anisotropic either in shape, in particle–particle interaction, or in both. With the advances in technology, new ways to fabricate anisotropic colloids continually emerge, and the library of anisotropic colloids has been greatly expanded. In general, the fabrication methods of anisotropic colloids can be classified into three categories: bottom-up routes, top-down routes, and those harvested directly from natural materials. Under each category, different types of methods have been developed and have displayed their capabilities in obtaining anisotropic particles. Table 2.1 lists typical methods under each category, together with their pros and cons.

2.2.1 Bottom-Up Routes

Bottom-up routes have been a common way to synthesize colloids in general. One classical bottom-up method to make colloids is the Stöber process [31]. But such method typically produces spherical colloids driven by the minimization of the interfacial free energy. To synthesize anisotropic colloids, a variety of different bottom-up methods have been developed.

2.2.1.1 Template(Seed)-Assisted Synthesis

One widely used method to fabricate anisotropic-shaped colloids is the template-assisted or seed-assisted synthesis. For example, by growing silica shells on anisotropic hematite spindles templates, hematite–silica core–shell ellipsoids with controlled aspect ratios have been fabricated [32]. If the hematite cores are then dissolved, hollow silica ellipsoids are obtained, which are suitable for optical applications. Similarly, by changing the cores to be peanut-shaped or cube-shaped hematite, corresponding shaped silica particles can be synthesized [17, 33]. In addition, by using rodlike tobacco mosaic virus (TMV) as templates, one-dimensional (1D)-conducting polyaniline and polypyrrole nanowires have been fabricated [34], showing a wide range of applicability of this method.

Besides nonspherical templates, Sacanna et al. [18] also developed a method to fabricate nonspherical colloids using spherical templates. To do this, silicon oil emulsion droplets of 3-methacryloxypropyl trimethoxysilane (TPM) were first formed through a polymerization step, then these particles were encapsulated

Table 2.1 Typical examples used for fabricating anisotropic colloids.

		Example methods	Pros and cons	Ref*
Bottom-up routes	"Lock" particles Dumbells	Template(seed)-assisted synthesis 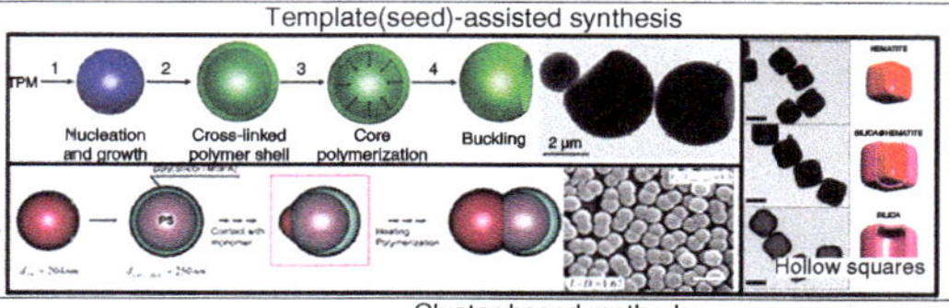	Pros: Easily applied; large yield suitable for industrial applications; moderate size polydispersity. Cons: Limited shape choice; extra sorting process needed for some methods to get uniform samples.	[17, 18] [19]
Bottom-up routes	"Colloidal molecules" formed through emulsion confinement Template-assisted self-assembly approach (TASA)	Cluster-based methods 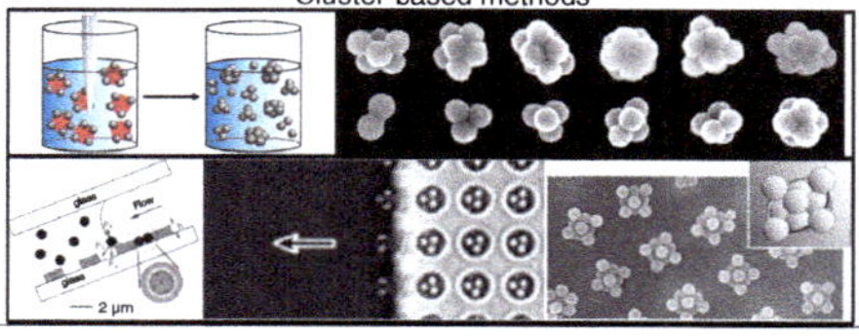	Pros: Easily implemented; large quantity can be obtained; small polydispersity both in size and in shape. Cons: Limited shape choice; limited type of materials that can be used; sorting process needed for some methods to get uniform samples.	[20] [21]
Bottom-up routes	DNA origami	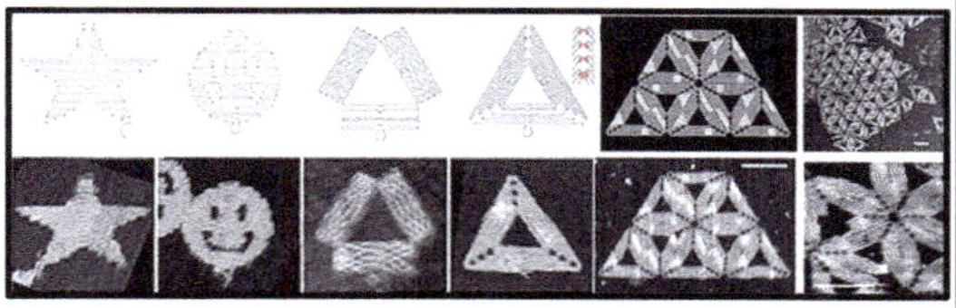	Pros: Excellent ability to fabricate particles with designed complex structures having nanoscale resolution. Cons: Low yield; relative high cost of materials; sorting process needed to purify samples; limited range of application conditions for the DNA containing products.	[22]
Bottom-up routes	Triblock Janus particles	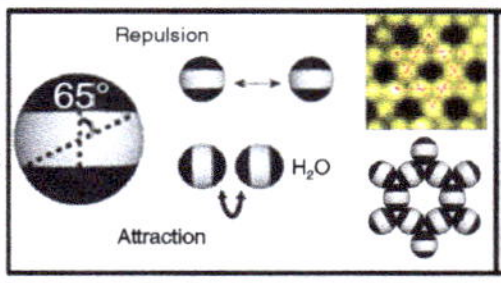 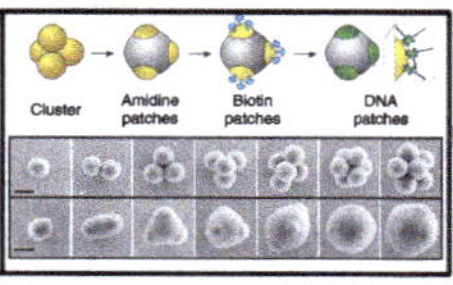 Patchy particles	Pros: Tunable anisotropic interactions independent of particle shape. Cons: Challenging to precisely control the distribution of patchiness; relatively complicated modification process.	[10] [23]
Top-down routes		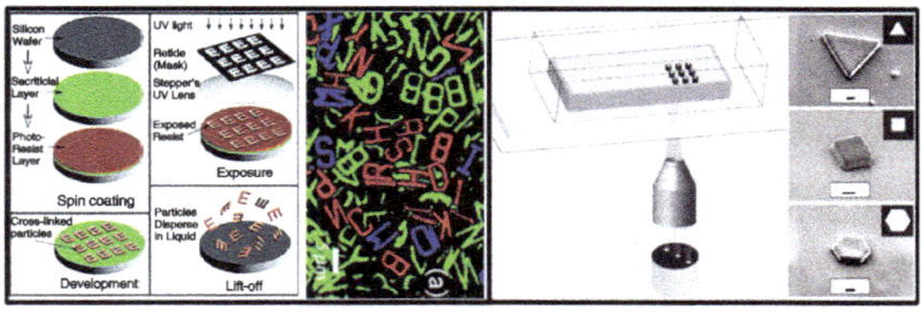 Photolithographic fabrication — Continuous-flow lithography	Pros: Excellent ability to fabricate particles with arbitrary 2D shape; good scalability for some method; small polydispersity both in size and in shape. Cons: Relative expensive photolithography facilities needed; limited type of materials; limited 3D shape.	[24] [25]
Top-down routes		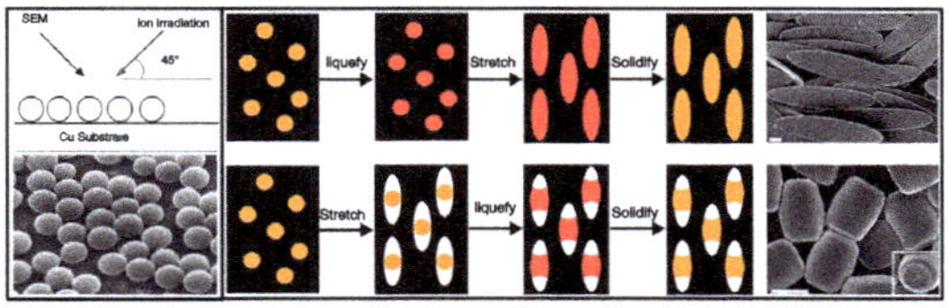 Ion-irradiation — Heating and/or mechanical stretching	Pros: Easily implemented; relatively good scalability for some method; small polydispersity in both size and shape. Cons: Limited type of materials; requiring specific facilities to generate high energy ions for ion-irradiation method.	[26] [27]
		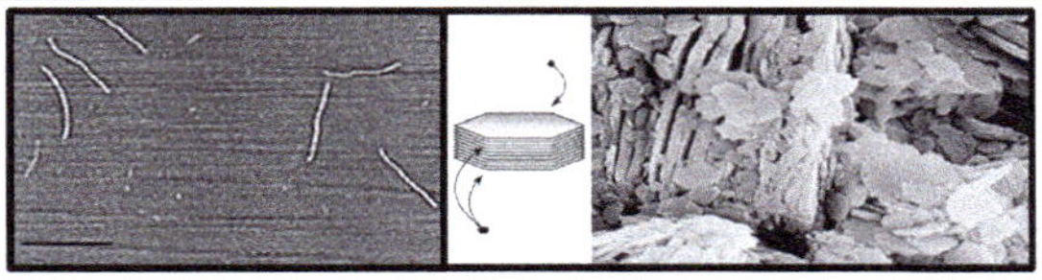		

(*Continued*)

Table 2.1 (Continued)

	Example methods	Pros and cons	Ref*
Natural materials	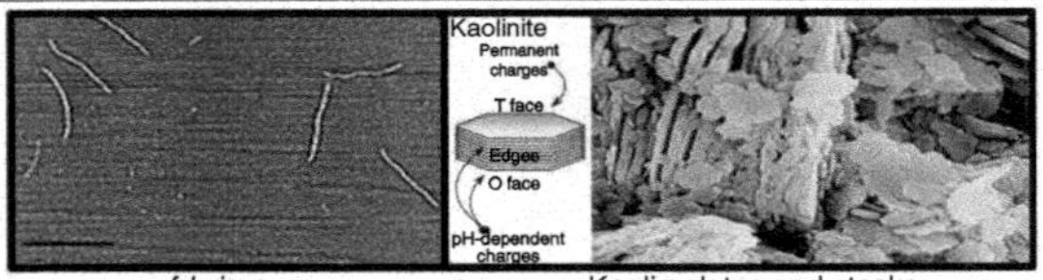 *fd* virus — Kaolin plates and stacks	Pros: Low cost; large yield. Cons: Limited shape choice; extra purification and sorting processes needed.	[28] [29, 30]

Examples are from Sources: Adapted from Chen et al. [10] / with permission of Springer Nature; Adapted from Rossi et al. [17] / with permission of Royal Society of Chemistry; Adapted from Sacanna et al. [18] / with permission of Springer Nature; Adapted from Park et al. [19] / with permission of American Chemical Society; Adapted from Wagner et al. [20] / with permission of Springer Nature; Adapted from Xia et al. [21] / with permission of Wiley-VCH GmbH; Adapted from Rothemund [22] / with permission of Springer Nature; Adapted from Wang et al. [23] / with permission of Springer Nature; Adapted from Hernandez and Mason [24] / with permission of American Chemical Society; Adapted from Dendukuri et al. [25] / with permission of Springer Nature; Adapted from Champion et al. [26] / with permission of The National Academy of Sciences of the USA; Adapted from Liu et al. [27] / with permission of Elsevier; Adapted from McMullen et al. [28] / with permission of Springer Nature; Adapted from Tombácz and Szekeres [29] / with permission of Elsevier; Adapted from Murray [30] / with permission of Elsevier.

into a core–shell structure through a second polymerization step, during which nonspherical TPM particles with a morphology containing a spherical cavity were finally obtained due to the shell buckling induced by the contraction of core materials [18].

The aforementioned template-assisted method is easily implemented and has excellent scalability, which can satisfy the large yield requirement for industrial applications. Particles made by this method have a reasonable polydispersity in size, but their shape is limited by available templates.

There is also a modified version of template-assisted method, the seeded polymerization-based method, by which anisotropic particles are synthesized by taking advantage of the phase separation happened during the later steps of seeded polymerization process [19, 35, 36]. For instance, to fabricate dumbbell-shaped colloids, starting with polystyrene (PS) spheres, Park et al. [19] first synthesized core–shell particles with a shell consisting of random copolymer of styrene and trimethoxysilyl propyl acrylate through seeded emulsion polymerization. Then these core–shell particles were swollen with styrene. Upon heating and polymerization, monomers induced the PS core to locally squeeze past the shell and preferentially absorb all the additional monomers, which finally led to the formation of dumbbell-shaped particles [19]. In another work, by delaying the polymerization of the liquid protrusions formed due to the heating of monomer-swollen PS spheres, which then led to coalescence of the liquid protrusions when PS spheres collided, Kraft et al. [36] fabricated a variety of anisotropic particles such as triangles and tetrahedra. By incorporating multiple chemical species into the synthesized particles, the chemical composition of particles can also be controlled. So not only shape-anisotropic but also chemically anisotropic particles such as Janus particles can be obtained by this way. This has been clearly illustrated in a recent work done by Liu et al. [37], where using this method, biphasic triblock particles with polymerized TPM in the center part and PS along the two poles were successfully fabricated. This seeded polymerization-based method is facile and has a high yield. Depending on the preparation method, particles made by without delaying polymerization upon heating are highly uniform both in size and in shape, but the range of shape is very limited, whereas particles fabricated through coalescence of liquid protrusions by delaying polymerization are typically a mixture of different shaped particles and need a sorting process to obtain uniform samples.

2.2.1.2 Cluster-Based Methods

Clusters formed by aggregation of several spheres, which often have anisotropic envelopes, can be used as anisotropic building blocks for self-assembly. One specific type of cluster is the so-called "colloidal molecules," which mimic the molecules formed by atoms and have drawn a lot of attention due to their great potential in making complex materials.

"Colloidal molecules" consisting of silica (seed) and PS (nodules) can be fabricated by controlling the reaction sites on seed particles [38]. A more efficient way to fabricate "colloidal molecules" is through emulsion confinement [20, 39, 40]. In a work done by Manoharan et al. [39], emulsions of toluene in water were first obtained, in which PS spheres were constrained to the oil–water interface by surface tension. Then as the toluene was gradually evaporated, the droplets shrunk, inducing a capillary force to pack the PS spheres. When the toluene completely evaporated, a cluster of PS spheres were obtained whose configuration is determined by the minimization of the second moment of the mass distribution [39].

The products made by this method often contain a mixture of clusters due to the broad number distribution of particles confined on each droplet. By optimizing the emulsification conditions, Zerrouki et al. [40] narrowed the number distribution of particles on the droplets and increased the yield for doublets, triplets, and quadruplets. However, to get a single type of cluster through this method is still quite challenging. To overcome this problem, Xia et al. [21] developed a template-assisted self-assembly (TASA) approach using microfluidics, which can produce a single type of cluster with well-controlled size and shape. The template is a two-dimensional (2D) array of relief structures (typically trenches) formed by photolithography. As the liquid slug slowly dewets across the relief structures, colloidal particles are trapped in them. Then after the solvent evaporates, particle clusters with closely packed structures are formed, which can then be welded by heating and released from the relief structures [21]. Using this method, not only closely packed clusters consisting of same spheres but also complex clusters consisting of different materials or with specific structures such as helical structures can be fabricated. However, particles fabricated by the cluster-based methods are limited by the possible cluster structures and thus have a narrow range of possible shapes to choose.

2.2.1.3 DNA-Based Assembly Methods

DNA molecules, due to their unique Watson–Crick base-pairing interactions, have been used to make particles with complex architectures through a programmable self-assembly fashion [41]. Two different approaches have been developed for realization of the DNA-mediated programmable assembly. One way is through hybridization-based DNA bonds, first developed by Seeman and coworkers [42], in which the rigidity of assembled final structure is mainly obtained through subunits called multiple strand crossover structures. Based on this approach, Rothemund [22] developed a scaffolded DNA origami technique, which can essentially fold long, single-stranded DNA molecules into arbitrary 2D shapes. Using this technique, single shapes such as squares, triangles, or stars, as well as hierarchical complex shapes like extended periodic lattices and a hexamer of triangles, have been successfully fabricated [22], which demonstrates the excellent capability of the DNA origami technique in fabricating particles with specific designed complex nanostructures. But due to the mismatch or mis-binding between DNA molecules during the thermodynamic assembly process, the yield of the particles with exactly designed structures is low. The final products typically contain a mixture of particles with varied structures and thus require separation processes in order to get homogeneous pure samples. In practice, the limited yield makes the DNA origami technique not suitable for applications where large quantities of anisotropic particles are needed.

The other way is through nanoparticle-templated DNA bonds, first developed by Mirkin and coworkers [43]. In this approach, nanoparticles functionalized with appropriate DNA strands act as building blocks, and a variety of superlattice structures can be assembled from them. Different from the former approach, where the obtained particles are made of DNA molecules, the composite particles obtained by this approach are mainly consisting of nanoparticles, and DNA molecules act as bonding agents between them.

2.2.1.4 Surface Modification Methods

Beyond the aforementioned anisotropic colloids grown from single monomers/molecules through chemical synthesis, other types of anisotropic particles that have anisotropic interactions such as Janus particles or patchy particles can also be obtained by surface modification of existing colloids (often spherical ones). Using glancing angle deposition of metal films followed by the deposition of self-assembled monolayers of *n*-octadecanethiol, Chen et al. [10] fabricated triblock Janus particles with electrostatic repulsion in the middle and hydrophobic attraction at the poles. Due to their anisotropic interactions, particles can self-assemble into an interesting kargome structure under proper conditions. Similarly, to form patchy particles that have directional interactions, surfaces of colloidal spheres can be asymmetrically functionalized using biopolymers that have specific binding interactions such as DNA base-pairing or biotin–streptavidin interactions [23, 44].

2.2.2 Top-Down Routes

Beyond bottom-up synthetic methods, there are also top-down fabrication methods developed to make anisotropic colloids.

2.2.2.1 Photolithographic Fabrication Methods

Borrowing from the field of integrated circuit industries, photolithographic technique has been applied to fabricate anisotropic particles. To do this, UV light shines on a substrate (often silicon wafer), which are coated with appropriate photoresist material, through a mask containing an array of shape-designed patterns. The UV illumination followed by heating causes the photoresist to react (e.g. cross-link). Then, after developing the patterns and releasing them from the substrate, anisotropic particles can be obtained. The shape of particle made in this way is determined by the mask patterns. So particles with an arbitrary 2D shape can be made by this method. For instance, Hernandez and Mason [24] have used this method to fabricate an alphabet soup that contains micron-sized all 26 English letters. Other shapes like disks and polygonal platelets have also been made [45–48].

Benefiting from the high precision and uniformness control of photolithographic technique, particles obtained by this method have a very narrow distribution both in size and shape, which are ideal as model systems for fundamental studies. However, the yield is limited due to the constrains like wafer size. To alleviate this limitation, Dendukuri et al. [25] developed a continuous-flow lithography by combining photolithography with microfluidics, which enabled a high-throughput production of anisotropic colloids. Another limitation of the photolithographic method is in fabricating anisotropic particles with three-dimensional (3D) shapes. Although in principle 3D-shaped particles can be made by multiple exposures with different patterned masks, in practice, the difficulty in controlling the alignment between multiple exposures and the limited range of shape that can be chosen have constrained their applications.

2.2.2.2 Physical Methods

Beyond chemical synthesis, anisotropic polymeric colloids can also be made by physical methods through heating and/or mechanical stretching [26, 49]. To do this, suspensions of polymeric spheres in polyvinyl alcohol (PVA) solutions are first cast into a film, and then the film is stretched either before or after the liquefacation of polymeric particles induced by heating (or adding solvent). Next, after the re-solidification of polymer by cooling (or removing solvent), particles with new shapes can be obtained. The final shape of polymeric particles is subject to several factors, including the stretching parameters, the material properties of the film and polymeric particles, and the film–particle interactions. Using this method, Champion et al. [26] have shown that as many as over 20 different shaped PS colloids can be fabricated. This method is easily implementable and has a good scalability as the yield can be largely raised, for instance, by increasing the loading of particles in films and/or employing parallel processing. However, the materials that can be used in this method is limited, thus stretching inorganic material is difficult.

To make anisotropic inorganic particles through physical methods, ion irradiation can be used. By bombing silica spheres with high-energy heavy ions, ellipsoidal particles have been made [50]. In addition, this method has also been shown to be applicable for polymeric materials [27], indicating its generality in the range of applicable materials although the shape of fabricated particles is very limited. Another drawback of this method is that it requires facilities that can generate high-energy ions, which are not easily accessed for normal laboratories.

2.2.3 Anisotropic Colloids from Natural Materials

In earlier days when the aforementioned bottom-up and top-down routes are not available, natural materials are the main resources from which anisotropic colloids can be obtained.

One of the widely used model systems from nature for studying phase behavior of rodlike colloids is the virus system such as TMV [51] and *fd* virus [28, 52]. Such virus particles can be obtained relatively easily in a large quantity and have a narrow polydispersity in both size and morphology that can meet the requirement for fundamental studies in colloid field. One disadvantage is that virus surfaces are complex and not easily modified to control the interactions between them.

Clay particles are another example of colloids obtained directly from nature. They are the major solid constitutes of soil, some sediments, and fine-grained natural rocks. Clay particles typically show nonhomogeneous distribution of charges on surfaces, which also vary depending on the environmental conditions. For example, kaolinite particles have one-face surface (the outmost tetrahedral sheet) permanently negative-charged, but their edge surfaces have a pH-dependent charge due to broken bonds and are positive charged under acidic conditions [29]. Because of the complexity in surface charges, as well as their irregular shapes and flexible layers, typical suspensions of clay particles contain both single particles and aggregates, and have a large distribution in both particle size and shape [30, 53]. Thus, further purification and modification processes are required for their applications in controlled assembly as model colloidal systems.

2.3 Self-assembly Mechanisms of Anisotropic Colloids

Anisotropic particles could assemble into rich phases as either the intrinsic properties of particles such as nonspherical shapes and/or anisotropic interactions, or the special induced character by external fields like electric fields would help them organize into clusters with certain structures. In this section we will introduce the mechanisms of the self-assembly of anisotropic colloids through different interactions, including electrostatic interactions, hydrophobic interactions, entropic depletion interactions, DNA-mediated interactions, and external field-assisted interactions.

2.3.1 Self-assembly Through Specific Interactions

2.3.1.1 Electrostatic Interactions

An inverse patchy colloid (IPC) refers to a colloidal particle with its surface divided into positive- and negative-charged regions. As the like charge parts between two particles are repulsive and the opposite charge parts are attractive, the competitive electrostatic interaction will align IPC particles and organize them into certain structures that depend on the relative coverage of the charges on the IPCs.

The simplest type of IPC is single-patched colloids. Hong et al. [54] synthesized spherical particles with opposite electric charges on both hemispheres, which then can assemble into clusters (Figure 2.1a). These experimental observations were also confirmed by Monte Carlo (MC) simulations. In this work, the proportions of oppositely charged areas are fixed to be equal [54]. To further understand the patch size effect on the assembled structures, Dempster and de la Cruz [55] studied the phase behavior of single charge-patched IPCs with varied patch sizes using MC simulations. They found that particles with large patch sizes tend to form ferroelectric crystals, whereas those with small patch sizes tend to form cross-linked gels (Figure 2.1b) [55]. Similar results were also obtained using Brownian dynamics simulations [56]. Compared with single-patch IPCs, double-patch IPCs show even richer phase behavior. Noya et al. [8] studied the phase behavior of IPCs with two charged patches at the poles and an oppositely charged equatorial belt by MC simulations. The obtained phase diagram (Figure 2.1c) shows a broad region of laminar structures composing of parallel colloidal monolayers arranged in a close-packed triangular lattice. They further showed that the region of stability of the layered solid phase can be expanded by increasing the charge imbalance and/or by reducing the interaction range [57]. On the other hand, Ferrari et al. [58] studied different lamellar phases spontaneously assembled by this kind of double-patch IPCs via molecular dynamics (MD) simulations . They found that the remarkable capacity to maintain the assembled structures' stability is related to a characteristic bonding mechanism: stable intra-layer bonds guarantee the formation of planar aggregates, while strong inter-layer bonds favor the stacking of the emerging planar assemblies. These two types of bonds together promote the self-healing processes during the spontaneous assembly [58]. IPCs with nonspherical shapes such as disks with a central charge and having their surfaces decorated with oppositely point-like charged patches have also been investigated, and rich phases and reentrant percolation behavior of these IPC disks were reported [59]. Since IPC particles interact through electrostatic interactions, their assembly behavior can be finely tuned by means of external parameters, for instance, the pH and salt concentration [60–62], which affect the Debye screening length [56], or by introducing unpatched charged colloids [55, 61], either before or after the assembly process.

Although interesting ordered assemblies of IPCs have been shown in simulations, experimental realizations of such systems typically end with a mixture of various structures [62], primarily due to the polydispersity in the distribution of the charged areas of IPC. So, to make it practical to fabricate ordered materials through the assembly of IPCs, how to synthesis monodisperse IPCs with the same configuration of charged area pattern on each particle is the key for success.

2.3.1.2 Hydrophobic Interactions

In aqueous dispersions, bare hydrophobic particles tend to aggregate to minimize the interfacial free energy. Thus, by controlling the hydrophobic properties of anisotropic colloids, various structures can be assembled. Hong et al. [63] prepared Janus particles with a hydrophobic hemisphere on one side and a hydrophilic (charged) one on the other side. As the interaction is attractive between the hydrophobic parts and electrostatic repulsive between the charged parts, by tuning the relative strength of these two interactions, different structures can be assembled

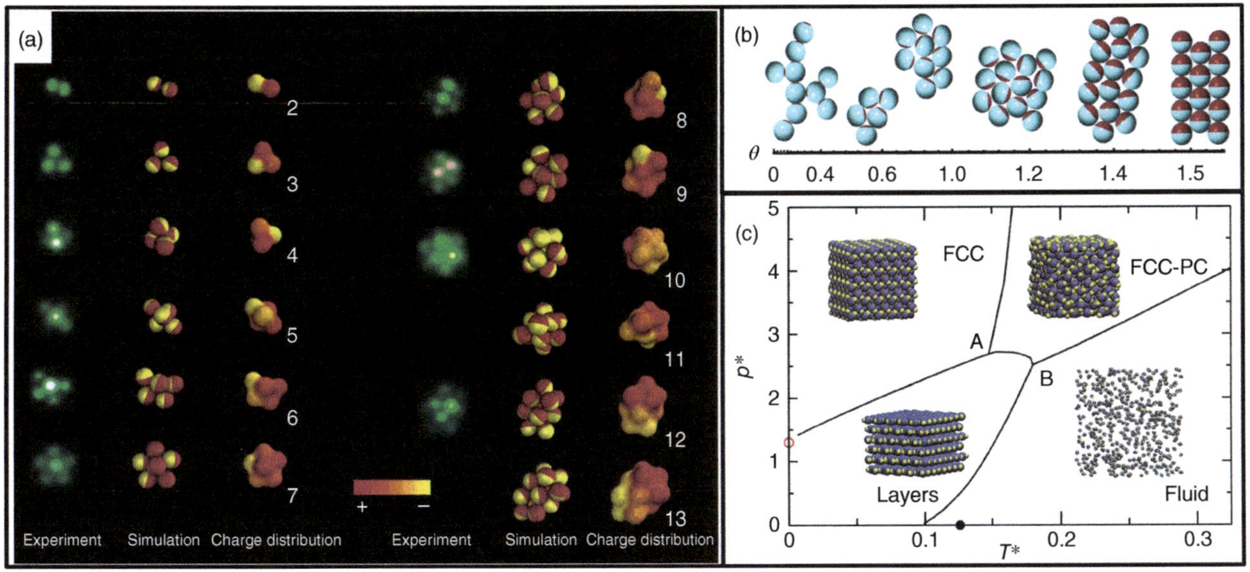

Figure 2.1 (a) Clusters assembled by particles with near-equal positive and negative charges on the two hemispheres, obtained both from experimental results and simulations. Source: Hong et al. [54]/with permission of American Chemical Society. (b) Primitive figure showing the transition from cross-linked gels at very small patch sizes to polarized crystals at opening angle $\theta \geq 1.5$. Source: Dempster and de la Cruz [55]. Reproduced with permission of American Chemical Society. (c) Phase diagram in the p-T representation of the IPCs with two charged patches at the poles and an oppositely charged equatorial belt. Source: Noya et al. [8]. Reproduced with permission of Royal Society of Chemistry.

by these Janus particles. For instance, at low ionic strengths where the electrostatic repulsion is strong, oligomers of particles are observed, whereas at high ionic strengths, which reduce the repulsion due to the charge screening effect, wormlike long strings can be formed (Figure 2.2a) [63]. On the other hand, Chen et al. [9] studied the supracolloidal reaction kinetics of this kind of Janus particles as a function of ionic strength. They showed that through a coordinated control over the reaction kinetics with the chemical anisotropy, fibrillar triple helices with at most six nearest neighbors per particle can be formed by fusing small, kinetically favored isomers timely before they equilibrate [9].

Beyond the ionic strength, the number and distribution of hydrophobic patches can also be varied to control the assembly structures. Chen et al. [10] fabricated "triblock Janus" particles with hydrophobic patches located at two poles and hydrophilic (charged) face in the middle. With the as-prepared triblock Janus particles, a kagome lattice can be assembled. Further maximizing the hydrophobic node by stacking kagome lattices could result in a bilayer structure containing six hydrophobic poles in each octahedron [10]. More complicated "triblock Janus" particles, such as the so-called X, Y, and K-shaped ones, can be made by specifically controlling the distribution pattern of hydrophobic patches on particles. They can form exotic structures like the truncated hexagonal lattice with dodecagonal pore assembled from Y-shaped triblock Janus particles (Figure 2.2b) [64]. Similarly, nonspherical colloids can also be modified to have a Janus type of surface including both hydrophobic and hydrophilic parts. Repula et al. [65] grafted fluorescent dyes onto the tips of filamentous viruses to form a small hydrophobic patch (Figure 2.2c). The attraction between the filamentous viruses can be tuned by varying the number of bound dye molecules so that the structure of their assembles can be controlled: as the single tip attraction increases, the smectic phase is stable at an expense of the nematic phase; when the tip attraction is strong enough, the nematic state would be suppressed completely to get a direct isotropic-liquid-to-smectic phase transition.

2.3.1.3 Entropic Depletion Interactions

Depletion force driven by entropy is often used in the self-assembly of anisotropic colloidal particles. It is an effective short-range attractive force that arises between large colloidal particles when they are suspended in a solution of non-absorbable depletants. The depletion attraction between two large colloids is proportional to the overlapped excluded volume between them. Depending on the way to control the overlapped excluded volume, there have been two methods developed to manipulate the self-assembly by depletion attractions.

One method is through shape-complementary matching. A typical example is the so-called "lock-and-key" assembly, in which only when locks and keys are shape-matched, the overlapped excluded volume between them are maximized, leading to stable lock–key assemblies under an appropriate concentration of depletants. For instance, by mixing one kind of particles with a cavity on their surface (as locks), and spherical particles whose sizes match with that of cavity (as keys), Sacanna et al. [18] produced stable lock–key assemblies (i.e. clusters) when small depletants were present. A unique feature of as-obtained lock-and-key assemblies

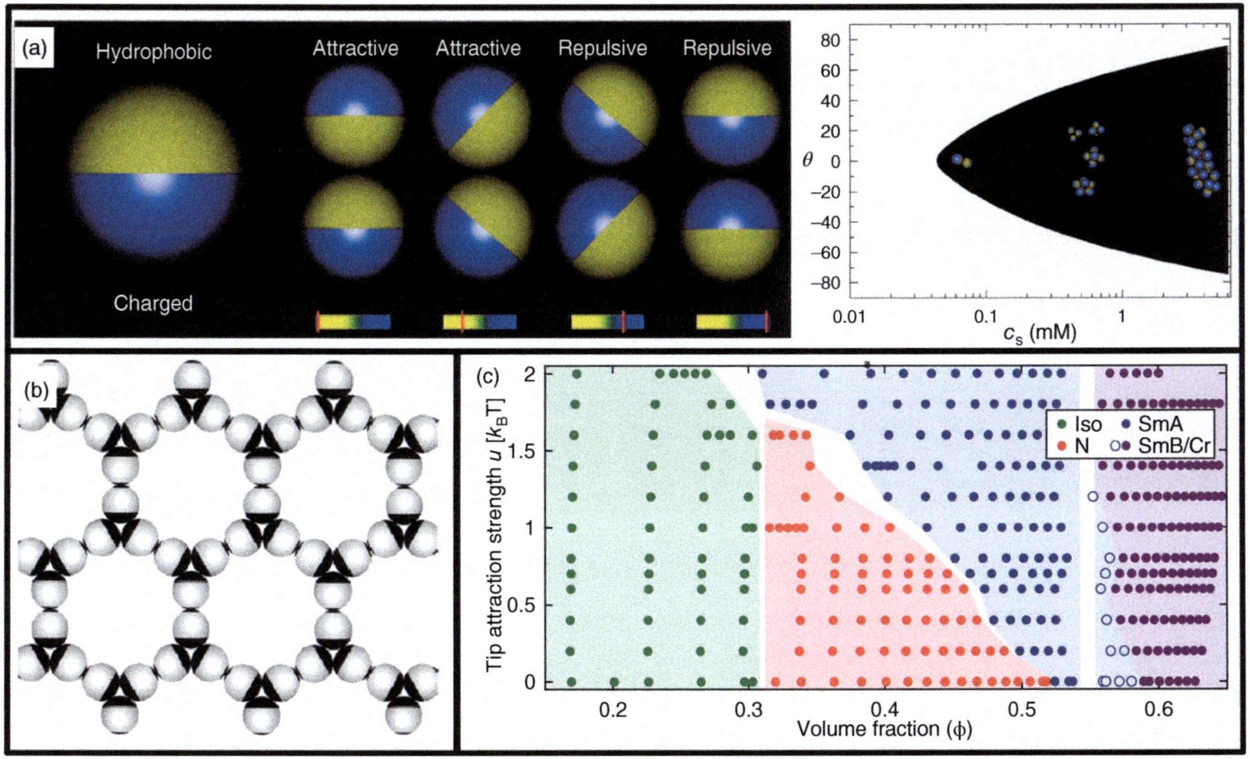

Figure 2.2 (a) Left: the interaction potential between Janus pairs switches from attractive (left) to repulsive (right) in water. Right: region of permitted tilt angles and corresponding assembled structure as a function of the concentration of monovalent salt. Source: Hong et al. [63]/with permission of American Chemical Society. (b) Dodecagonal pore of a truncated hexagonal lattice formed from triblock silica spheres with Y-shaped bonding geometry. Source: Chen et al. [64]/with permission of American Chemical Society. (c) Calculated phase diagram in terms of the attraction strength u between the end groups of the semiflexible rodlike particles as a function of their volume fraction ϕ. Source: Repula et al. [65]. Reproduced with permission of American Physical Society.

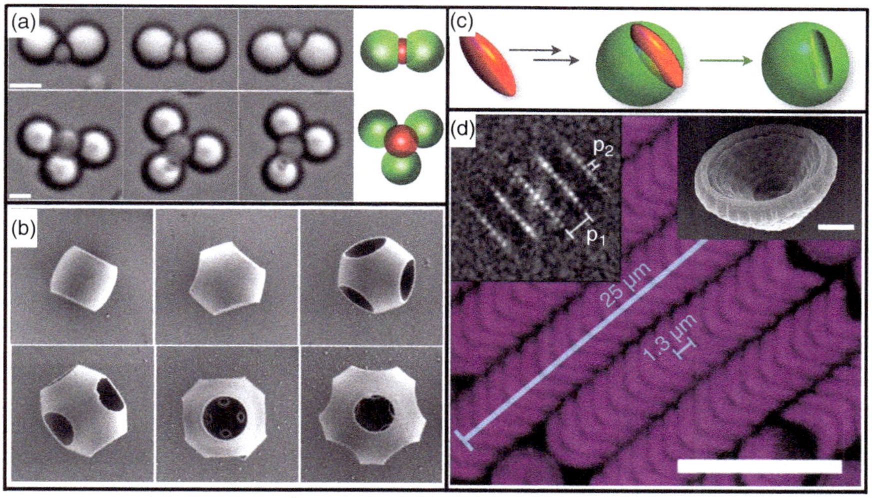

Figure 2.3 (a) The clusters formed by lock-and-key assemblies by changing the size of key particles. Source: Sacanna et al. [18]/with permission of Springer Nature. (b) Electron micrographs of well-designed multicavity particles. Source: Wang et al. [66]/with permission of American Chemical Society. (c) Schematic showing a lock particle with a well-defined elongated cavity to capture a shape-complementary ellipsoid-like key particle. Source: Sacanna et al. [67]. Reproduced with permission of Springer Nature. (d) A confocal laser scanning microscopy image of self-assembled supra-colloidal fibrils by cone-shaped particles (top right shows SEM image). Source: Tigges et al. [68]/with permission of John Wiley & Sons, Inc.

is that the bonds between particles are flexible and reversible. Depending on the size ratio of the key particle and the cavity, different assembled clusters like dimers, trimers, and tetramers can be fabricated (see examples in Figure 2.3a) [18]. Another easier and more precise way to make multimers through lock-and-key assembly is to use lock particles with multicavities. Wang et al. [66] develop a method to synthesize colloids having well-defined multicavities with geometry spanning a wide range of shapes including spherical, linear, triangular, tetrahedral, trigonal dipyramidal, octahedral, and pentagonal dipyramidal (Figure 2.3b). Using these lock colloids, assemblies with well-defined structures could be produced. Similarly, lock particles with cavities having varied shapes have also been fabricated (Figure 2.3c) [67]. Since the lock-and-key assembly through depletion attractions depends on the shape-complementary, such locks with shape-varying cavities will only react to key particles that have complementary shapes to form stable lock–key assemblies and thus can be used to separate different shaped colloids.

Beyond modifying spherical colloids with cavities, lock particles can also be made directly from nonspherical colloids having concave components. Using 3D laser writing, Tigges and Walther [68] fabricated anisotropic cone-shaped particles (Figure 2.3d, right top inset), which can act as both locks and keys. When small depletants are added, these cone-shaped colloids self-assemble into nematic liquid crystalline through shape recognition driven by depletion attractions (Figure 2.3d).

Compared with typical nematic phases applied in electronic displays, this nematic liquid crystalline structure is realized without applying an electric field. The aforementioned lock and key particles have fixed sizes once they have been made, which may limit their applications. To make lock and/or key particles with tunable sizes, Mihut et al. [69] constructed charged lock and key microgel particles, whose hydrodynamic radius and electrophoretic mobility both change with temperature and the ion concentration in solutions. Thus, the self-assembly process in such temperature-responsive systems can be finely controlled through an external temperature field.

The other method to realize controllable depletion attractions for the self-assembly is through the "roughness-controlled depletion attractions" (RCDAs), in which the overlapped excluded volume between large colloids can be tuned by the size ratio of the surface roughness of large colloids and the size of depletants [70, 71]. For instance, taking platelets as an example (Figure 2.4a, top), if the size of depletants is smaller than the surface roughness, even when the planar surfaces of the two platelets contact with each other, the overlapped excluded volume between them is still small due to the relative large surface roughness, and thus the depletion attraction is not strong enough to form a thermally stable dimer; by contrast, if the size of depletants is increased to be bigger than the surface roughness, then these platelets will behave like smooth particles, and the depletion attraction between them can form stable columns. Interestingly, the depletion attraction can also be enhanced if rough patterns are shape-complementary such as regular concave–convex ridges.

Using RCDA, Zhao and Mason [70] successfully fabricated a dimer-only phase by employing a Janus type of pentagonal platelets whose two flat surfaces have different surface roughness (Figure 2.4 a bottom), together with appropriately choosing the size of depletants. By taking advantage of the different surface roughness between the flat surfaces and the edge surfaces of the lithographically fabricated particles, Zhao et al. further explored the applications of RCDA in making 2D monolayers of a variety of platelet-like particles including triangles, squares, pentagons, hexagons and kites, etc. Different assembled structures in these 2D systems including LC, crystal, and even glass have been reported (Figure 2.4b) [46–48, 72, 73, 75, 76].

More delicate hierarchically organized structures can be obtained through roughness-controlled regioselective depletion attractions, induced by the site-specific surface roughness patterns on otherwise uniform surfaces of colloids. Kraft et al. [74] synthesized patchy particles with curved, smooth patches on rough colloids as shown in Figure 2.4c. When appropriately sized depletants are added into the dispersions of such patchy particles, the depletion attraction between the smooth parts of the particles is larger than that between the rough parts, which then direct them into structures of complex colloidal micelles [74]. Similarly, the "Micky Mouse"-shaped colloids consisting of one central smooth lobe connected to two rough lobes on either side at an angle of 90° have also been produced, and they can self-assemble into tubelike structures at sufficiently strong depletion attractions [77]. Recently, a new assembly method by regioselective depletion attractions based on material intrinsic properties rather than surface roughness has been demonstrated by Liu et al. [37]. Using PS and TPM materials, they fabricated smooth

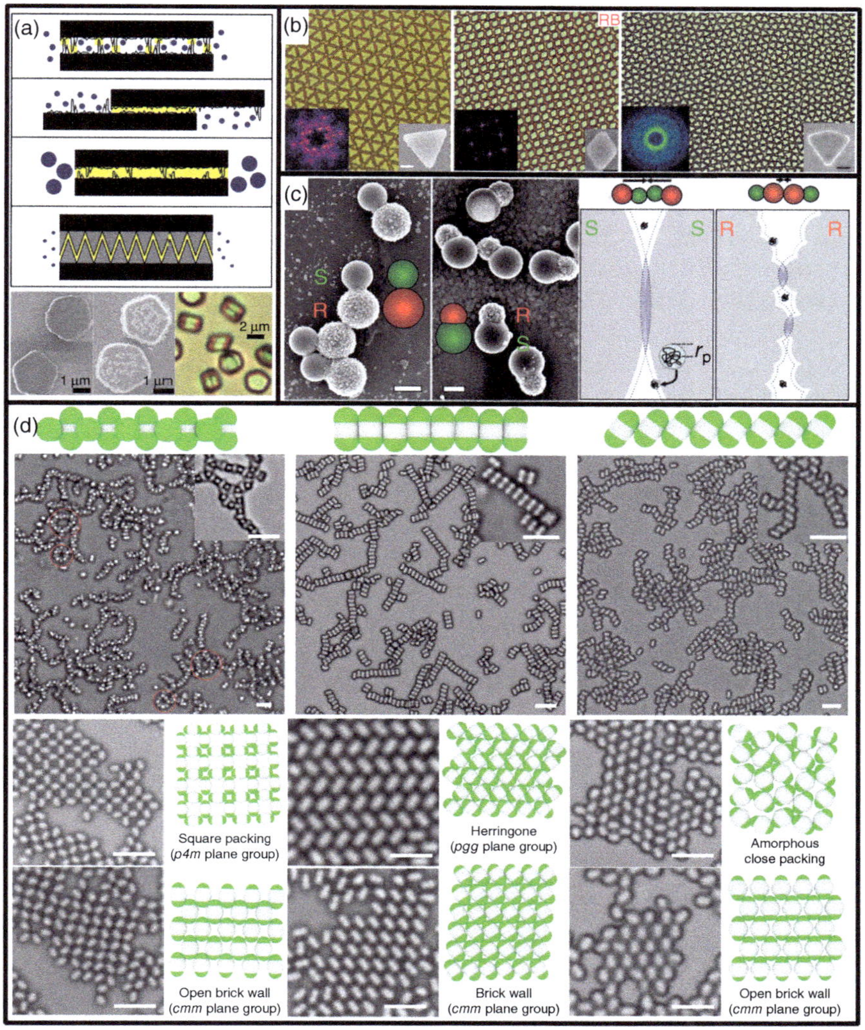

Figure 2.4 (a) Top: a schematic illustration of roughness-controlled depletion attractions (RCDAs) between platelets. Bottom: from left to right, the untreated smooth surface of pentagons, the opposite silica-attached surface of pentagons, and dimers assembled by these Janus pentagons Source: Zhao et al. [70]/with permission of American Physical Society. (b) Example structures (liquid crystal, rhombic crystal, and glasses, from left to right) assembled by certain platelets (bottom right are their SEM images). Source: Refs. [48, 72, 73]/with permission of Springer Nature, with permission of American Chemical Society, with permission of National Academy of Sciences. (c) Left: colloid particles consisting of one smooth sphere and one rough sphere. Right: a schematic illustration showing that the overlapped excluded volume between two smooth surfaces is significantly larger than that between two rough surfaces Source: Kraft et al. [74]/with permission of National Academy of Sciences. (d) Schematic illustration and optical microscopic images of 1D (the first row) and 2D (the second row) structures assembled from triblock biphasic colloids with different aspect ratios. Source: Liu et al. [37]/with permission of Springer Nature.

PS-TPM-PS triblock biphasic colloids with PS at the poles and TPM at the equator. Pure PS and TPM particles show different material-dependent assembly behavior under the same depletant and salt conditions, presumably due to different surface zeta potentials and/or different depletant-particle interactions of two materials. Thus, for the PS-TPM-PS triblock colloids, by tuning the pole-to-pole and/or center-to-center interactions, a variety of superstructures and superlattices ranging from 1D cross-chains, ladder-like chains, and tilted ladder-like chains (Figure 2.4d, top) to 2D wall-like structures (Figure 2.4d, bottom) have been fabricated [37]. Compared with the shape-complementary and RCDAs, this method does not need shape recognition elements or surface functionalization for colloids. But the type of materials that can be used in this method is limited. In the future, toward a better self-assembly process guided by depletion attractions, a combination of all these methods is likely needed, which can enable us to have a more flexible and delicate control on the depletion attractions.

2.3.1.4 DNA-Mediated Interactions

Colloidal particles grafted with DNA chains can self-assemble into various structures, including crystalline structures [78–81] and certain gelation clusters [82], through specific base-pairing interactions between DNA chains. Attaching DNA chains to certain part of colloidal particles, one could get finely designed stereotype colloidal assembles. For instance, Ben Zion et al. [83] first bound DNA origami L belts with polymethyl methacrylate (PMMA) colloids using four sticky DNA ends, which were then subjected to annealing to fix the relative position and angle of the sticky ends. Next, they added other kinds of colloids with sticky DNA ends that can selectively bind to certain ends of the L belt, and finally stereotype colloids with designed configurations were successfully assembled.

Similar to the previous discussed Janus or patchy colloid particles in the context of surface charges, hydrophobicity, materials or surface roughness, Janus or patchy particles in the context of DNA-functionalized surfaces have also been explored. Oh et al. [84] synthesized Janus particles with a varied patch ratio (χ) that is DNA functionalized. They found that when χ is small, only small aggregates can be formed, while when χ reaches 0.3, chains or multilayered crystal structures can be obtained (Figure 2.5a). Besides the specificity given by base-paring, another unique property of the DNA-mediated interaction is thermally reversible. Thus, by combining the temperature control with the specific design of DNA sequences as well as controlling the distribution of DNA-functionalized patches, hierarchical multistage self-assembly can be realized [85, 86]. For example, using "chromatic" patchy particles that have distinguishable ("colored") pairwise interactions, Patra and Tkachenko [85] showed that a desired diamond polymorph can be self-assembled. With a small modification of the coloring scheme, an alternative polymorph, hexagonal diamond can also be obtained (Figure 2.5b). More interestingly, the self-assembly of the DNA-functionalized patchy colloids can also be reprogrammable due to their thermal reversibility, which is a desired property for many applications such as in the intelligent materials field. For instance, Oh et al. [86] fabricated DNA-coated bifunctional Janus colloids consisting of a small

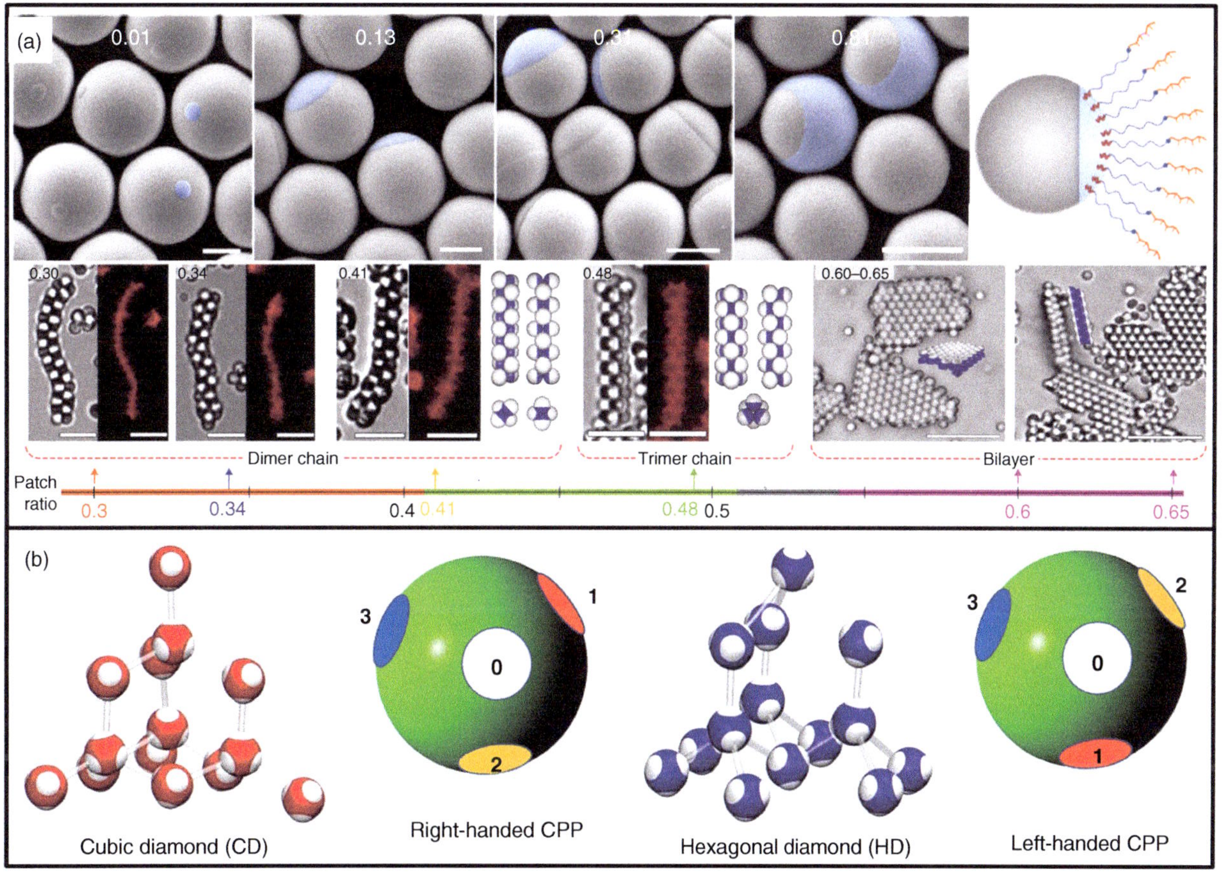

Figure 2.5 (a) Top: SEM images of PS-TPM Janus particles with various patch ratios, following an illustration image of the Janus particle with only one patch functionalized with DNA. Bottom: the assembled structures with patch ratios increase. Source: Oh et al. [84] / with permission of Springer Nature. (b) Computer-simulated structures assembled by right-handed and left-handed chromatic patchy particles. Source: Patra and Tkachenko [85]. Reproduced with permission of American Physical Society.

TPM patch with azide groups (TPMA) and a large PS patch. Each face of the Janus particles is coated with different self-complementary DNA strands and thus has different binding energies. Then, by using toehold strand displacement, which can activate or deactivate the DNA binding between small faces (i.e. TPMA patches) through temperature control, a reconfiguration of assembling structures between colloidal chains and bilayers have been successfully realized when the temperature switches between high and low.

2.3.2 Assembly in External Fields

2.3.2.1 Electric Field Assisted

Electric fields, which can generate orientational-dependent interparticle interactions, have been widely used to direct the assembly of colloids. Particularly, combining the field-directed assembly with anisotropic colloids that have unique orientational-dependent properties opens up a new realm to make a broad range of highly ordered superstructures. Ma et al. [87] studied the assembly of colloidal dumbbells in both external direct current (DC) and alternate current (AC) electric fields. In a DC electric field, they found that unlike the crystals formed by spherical particles, dumbbell particles lay on the substrate and formed amorphous aggregates with locally ordered structures, which is mainly caused by the electro-osmotic flow. By contrast, in an AC electric field, the dumbbells are subject to a complicated interplay between dipolar interactions and induced-charge electroosmotic flows that are both frequency dependent and thus showed a rich frequency-dependent assembled structures including 2D close-packed crystals of perpendicularly aligned dimers (Figure 2.6a) [87]. Due to the unique polarization properties of anisometric colloids resulted from their geometry, the orientation of these particles is not necessary to be aligned with the electric field and thus can assemble into intriguing structures. For instance, Cheng et al. [88] showed that when bowl-shaped colloids were put in an external electric field, their orientations (defined by the normal axis of their open end) were not aligned with the electric field due to the anisotropic distribution of mass, and the induced dipolar interactions organized them into stacking-bowl structures in which the open end of each bowl is facing the direction perpendicular to the electric field (Figure 2.6b).

Besides the pure geometrically anisotropic colloids, Janus or patchy anisotropic colloids have another degree of freedom via patchiness control to tune particle responses in electric fields and thus show even broader capabilities in assembling novel structures. By depositing metals on different parts of anisotropic-shaped particles fabricated by photolithography, Shields et al. [89] made a variety of patchy particles with different shapes. In an AC electric field, by tuning the strength and frequency of the field, they found that for top-, angle-, and side-coated cubes or cylinders, when they were attracted toward each other due to induced dipolar interactions, they reoriented and aligned with their edges parallel and orthogonal to the field direction to form chain-like structures, typically with their metallic patches aligned with the field direction. But the details of chain structures are closely associated with how the metal patches are distributed on particle surfaces.

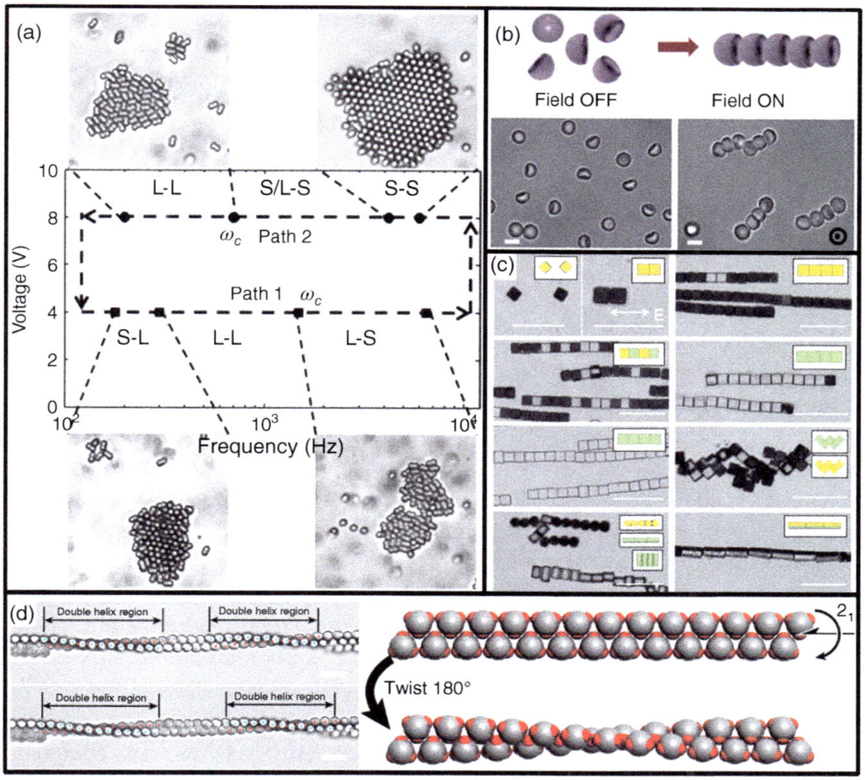

Figure 2.6 (a) The phase diagram of the assembly of dumbbell particles under an AC field. Source: Ma et al. [87]/with permission of John Wiley & Sons, Inc. (b) Schematic illustration and optical microscopic images showing the assembly of bowl-like colloids in an electric field. Source: Cheng et al. [88]/with permission of Royal Society of Chemistry. (c) Optical micrographs showing the assembly of microcubes and microcylinders with part of their surfaces coated with metals in an AC electric field. Source: Shields et al. [89]/with permission of Royal Society of Chemistry. (d) Left: optical microscopic images showing paired chains of three-patch particles containing two double-helix regions at different time points. Right: a model of a chain pair with head-to-head packing across the screw axis (upper) and the twist structure of three-patch particles when forming a helix region with a pitch of 13 particles. Source: Song et al. [90]/with permission of American Chemical Society.

For instance, for top-coated cubes, their chains were nearly perfectly aligned with the field direction, and the metallic face of each cube in the chains was orientated closest to the electric field maximum, whereas angle (or two-side)-coated cubes produced "kinked" chains with the particles oriented 45° in-plane from the field direction (Figure 2.6c) [89]. On the other hand, Song et al. [90] studied the electric field-directed assembly of patchy particles with two or more charged patches. In an AC electric field, these patchy particles with charged patches are subject to both dipolar interactions and dielectrophoretic forces and could assemble into different structures ranging from 1D chains, 2D layers, to 3D crystal packings [90].

Interestingly, some chain pairs assembled by three patchy particles could form unanticipated double-helix structures (Figure 2.6d), resulted from mutual twisting of the chains about each other, implying great potential of electric field-directed assembly of anisotropic colloids in making unconventional delicate ordered structures.

2.3.2.2 Magnetic Field Assisted

Compared with electric fields, magnetic fields have several advantages including that magnetic interactions are unscreened even in electrolyte solutions, and there is no heating damages to samples that may exist in the case of electric fields. So magnetic fields have also been widely used to direct the assembly of colloids in suspensions. When simple anisotropic paramagnetic or ferromagnetic colloids, for example, peanut-shaped colloids made of hematite cores and silica shells [91], are in a magnetic field, they tend to form chain-like structures along the field direction due to the dipolar interactions. However, by increasing the complexity of anisotropic colloids, various ordered structures can be obtained, which resulted from the competition between magnetic dipolar interactions and other types of interactions from the anisotropy of particles such as steric hinderance. For instance, Zerrouki et al. [92] showed that for dumbbell colloids composed by spherical particles binding with magnetic colloidal caps, they could be assembled into chain structures in a magnetic field, in which particles are orientated alternatively up and down (Figure 2.7a). By contrast, for symmetric dumbbell colloids consisting of two bound identical spheres with a solid magnetic ring located around the contact point between the two spheres, periodically twisted achiral chains were obtained. More interestingly, using asymmetric dumbbell colloids, which were fabricated in the same way as symmetric dumbbells except that two different sized spheres were used, helical structures with a single helicity were produced [92].

Similar to the case of electric fields, Janus or patchy anisotropic colloids with magnetic patches have also been used to make novel structures. By putting colloidal rods whose head is functionalized with magnetic particles in a magnetic field, Gao et al. [94] observed bottle-brush-like assembles, in which the magnetic head of each rod lines up in columns, whereas the nonmagnetic tail of each rod points out randomly in a plane perpendicular to the columns. By contrast, by coating half of the cylindrical surface of silica rods with a Ni magnetic layer, Yan et al. [95] found that these dipolar magnetic Janus rods assembled into ribbonlike structures in an external magnetic field. Moreover, by switching the direction of the magnetic field, linear ribbons were observed to be transformed into rings, which can be stable after the field was turned off.

Beyond the static structures that magnetic anisotropic colloids can be assembled, by employing rotating magnetic fields, a variety of dynamically stable configurations involving rich collective behavior of particles can also be realized. One representative example is the synchronization-selected microtubes (Figure 2.7b) formed by magnetic Janus colloids in a precessing magnetic field, a work reported by Yan et al. [93]. To form such dynamically stable microtubes, the synchronization between the rotating microtube and the motion of its constituent particles is the key, which are

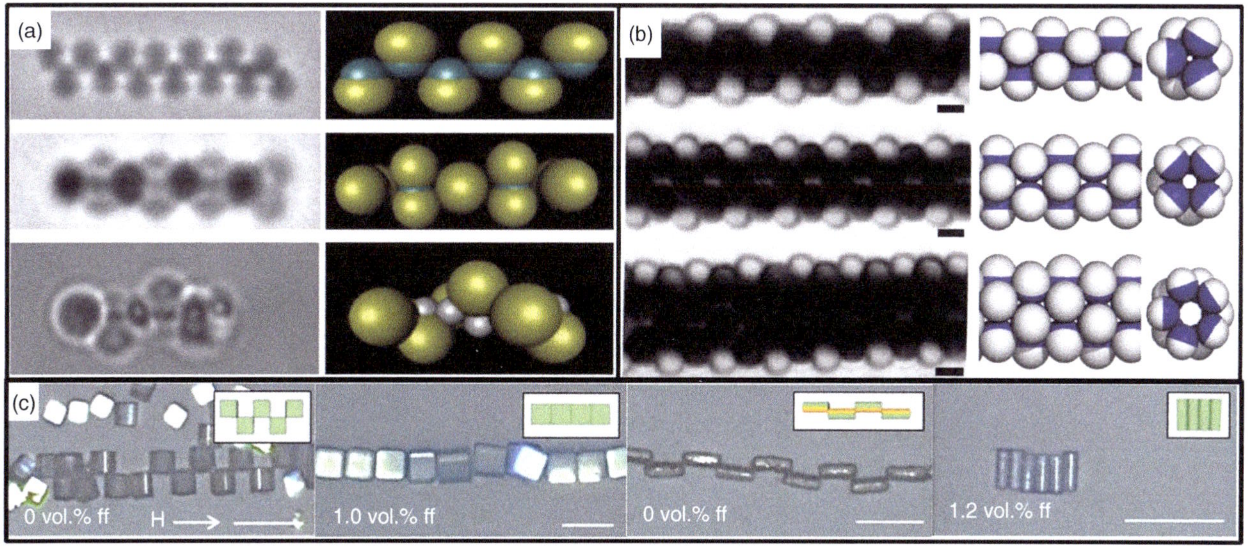

Figure 2.7 (a) Self-assembly of magnetic dumbbells into different chains under a magnetic field Source: Zerrouki et al. [92]/with permission of Springer Nature. (b) Experimental images (left) and corresponding models of microtubes formed in a precessing magnetic field. Source: Yan et al. [93]/with permission of Springer Nature. (c) Self-assembly of top-coated cubes and side-coated cylinders in water (left one, 0 vol. ff) and in a ferrofluid (right one, 1.0 vol. ff). Source: Shields et al. [89]/with permission of Royal Society of Chemistry.

all precessing magnetic field dependent. Later, the same group further studied in details the assembly behavior of such magnetic Janus colloids in external rotating magnetic fields and found rich new superstructures including footstep chains and other dicolloid-based lattice structures as well as intriguing cluster dynamics such as dislocation formation and rotation, periodic dynamic patterns of domain boundaries, etc. [96, 97]. Such time-dependent field-based methods offer new ways in controlling structures using dynamic criteria rather than static energy minimization.

Besides the aforementioned methods focusing on building blocks themselves, the properties of continuous phase in suspensions of building blocks (dispersed phase) can also be used to control their self-assembly in magnetic fields. Using anisotropic SU-8 colloids with magnetic patches, Shields et al. [89] showed that in a magnetic field, when top-coated cubes and side-coated cylinders were suspended in water, they assembled into zigzagged chains, whereas when they were suspended in a ferrofluid, edgewise chains were formed (Figure 2.7c) due to the efficient formation of magnetic quadrupoles resulting from the presence of the diamagnetic SU-8 component with respect to the ferrofluid magnetization.

2.3.2.3 Self-propelled Colloidal Motors

Self-propelled colloidal motors are a kind of colloids that can convert energies of surrounding environment into their motion in fluids. Different from passive matters where equilibrium structures are typically assembled as a consequence of entropy and potential interactions, in the system of active motors, the assembled structures are usually dynamic and out of equilibrium.

One widely used self-propelled colloidal system is catalytic micromotor system powered by chemical fuels such as hydrogen peroxide [98]. For instance, by putting Janus silica spheres with one hemisphere coated with platinum (Pt) into aqueous hydrogen peroxide solutions, they autonomously move, driven by Pt-catalyzed decomposition of hydrogen peroxide. However, for motors with such autonomous motion to self-assemble into dynamically stable organized structures, additional control is needed. For example, by modifying silica surfaces with octadecyltrichlorosilane, Gao et al. [99] fabricated hydrophobic Janus colloids that are capped with a catalytic Pt hemisphere patch. Due to the hydrophobic attraction force together with the self-propelled interactions between colloids, doublet, triplet, or quadruplet dynamic clusters were observed, which displayed controlled coordinated self-propulsion (Figure 2.8a) [99]. A more delicate control on the dynamic assembly of motors can be realized by using different materials whose catalytic activities can be triggered by external cues. Toward this goal, Palacci et al. [102] synthesized spherical polymer colloids with protruding hematite cube that are photoactivable. Then, when blue light was shining on an aqueous suspension of these colloids containing hydrogen peroxide, they found that particles assembled into 2D "living crystals," which can form, break, explode, and reform elsewhere [102]. The mechanism behind this dynamic assembly behavior is not clear yet, although it has been thought to be a result of the competition between self-propulsion of particles and an attractive interaction induced, respectively, by osmotic and phoretic effects, which are associated with some gradient generated by

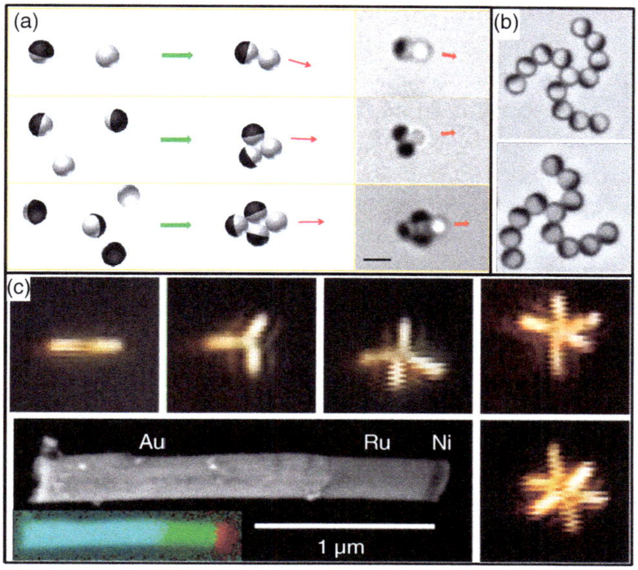

Figure 2.8 (a) Assemblies of Janus motor/nonmotor microparticles. Source: Gao et al. [99]/with permission of American Chemical Society. (b) Example of symmetric pinwheels (top) and asymmetric pinwheels (bottom) with clockwise motion in AC electric fields. Source: Zhang and Granick [100]/with permission of Royal Society of Chemistry. (c) FE-SEM image of an Au–Ru–Ni nanorod and optical images of dimer, trimer, and higher multimers assembled by these nanorod motors. Source: Ahmed et al. [101]/with permission of American Chemical Society.

the hematite cube during the decomposition of hydrogen peroxide and activated by light. Similarly, when peanut-shaped hematite colloids suspended in an aqueous hydrogen peroxide solution were used under blue light, 1D, slithering ribbon structures were observed [103].

Beyond the chemical-powered motors, one alternative way to induce the autonomous motion of colloids is by applying external fields. When half-metal-coated silica Janus spheres suspended in water are subject to an AC electric field, they can show autonomous motion at certain frequencies due to induced-charge electrophoresis [104]. For each of these silica Janus particles, the electric field also induces two dipoles with one in each hemisphere and shifted from the center of the particle. By tuning the AC electric field, the interaction between the two dipoles can be attractive. Thus, by mixing these active Janus particles with homogeneous silica spheres, various dynamically stable assemblies, ranging from multiarmed chiral pinwheels (Figure 2.8b) to rotating chiral clusters like tetrahedral rotors and square pyramidal rotors, can be obtained under appropriate AC electric fields [100, 105].

Despite micromotors driven by chemical fuels or electric field-induced mechanisms display all these fascinating assembly behaviors, they are limited for applications in biological environments. Hydrogen peroxide are toxic for biological samples, and biological fluids typically have high ionic strength that is incompatible with electrophoresis-based propulsion mechanisms. To circumvent these limitations, Wang et al. used ultrasonic standing waves in MHz to power autonomous micromotors [101, 106]. They found that, driven by a self-acoustophoresis mechanism, axially segmented metallic rods (Au–Ru or Au–Pt, with the two ends being different metals) suspended in water self-propelled in the levitation plane created by ultrasonic standing waves and formed long spinning chains with a head-to-tail alternating structure [106]. Such autonomous motion and assembled structures were also observed when particles were in solutions of high ionic strength. By adding thin Ni segments at the Ru end of Au–Ru rods (i.e. segmented Au–Ru–Ni rods), dimers, trimers, and higher multimers of Au–Ru–Ni nanorods were assembled (Figure 2.8c) due to the introduced magnetic interactions, and these multimer assemblies displayed several distinct modes of autonomous motion in the 4-MHz ultrasonic standing waves [101]. These studies open up new ways to manipulate micromotors in biologically relevant media using ultrasound.

2.4 Applications of Self-assembly of Anisotropic Colloids

With a rich and still expanding library of anisotropic colloids available, together with a variety of smart self-assembly strategies, new functional materials with custom-designed properties are continuing to be fabricated. These materials of anisotropic colloids can possess new types of functionalities that would not be given by their spherical counterparts and have been applied in fields as diverse as LCs, photonic crystals, sensors, electrode materials, etc. In this section, we will briefly overview some typical applications of the self-assembly of anisotropic colloids that highlight their anisotropic structures.

2.4.1 Liquid Crystals

LCs are a very important class of materials in both fundamental studies and technological applications [107]. Compared with spherical counterparts, anisometric particles have broken rotational symmetries, and this lends them a unique capability to form LC materials. Even for particles with a very simple rodlike shape, they can display a relatively rich LC behavior and can be assembled into nematic and smectic LCs depending on both the aspect ratio and concentration of rodlike particles [108, 109]. Besides nematic and smectic, other types of LC structures can also be assembled using appropriate mesogens, such as columnar phase formed by discotic mesogens [110] and chiral phases formed by bent-core mesogens [111]. In addition, discotic LCs of synthetic graphitic hydrocarbons are widely reported [12, 112, 113]. For instance, using graphene oxide LCs, a new display device without polarizing optics can be fabricated [12].

LCs of anisotropic colloids typically have slower responses to external cues than those consisting of molecules due to their relatively large size. Thus, colloidal LCs have long been used as model systems in fundamental studies. On the other hand, the properties and functions of colloids can be easily tailored.

Practically, by combining the desired properties of liquid crystalline structures with the rationally designed properties of anisotropic colloids, new functional materials have been fabricated, which has greatly expanded the applicable range of liquid crystalline materials. For example, choosing TiO_2 rutile nanorods that have photocatalytic properties as building blocks, a nematic liquid crystalline film has been self-assembled [114]. Interestingly, the resultant film shows anisotropic photocatalytic activities with a more efficient activity when exposed to UV light whose polarization is along the quadratic axis of the rutile nanorods (Figure 2.9a) [114]. Similarly, liquid crystalline materials of CdSe nanorods were also fabricated [118], which can have potential applications in electrooptical devices due to the unique properties of CdSe materials like linearly polarized photoluminescence and anisotropic nonlinear optical properties [119].

When anisotropic magnetic colloids can be used, their assembled structures are then expected to be magnetic field-responsive. Along the same lines, Wang et al. [115] fabricated LCs of ferrimagnetic inorganic nanorods, whose optical properties can be instantly and reversibly controlled by external magnetic fields. By dispersing the nanorods in a UV curable resin, they further showed that patterns with controllable light transmittance in designed areas can be created by combining magnetic alignment and lithography processes (Figure 2.9b) [115]. This work provides a new way to fabricate light-control devices. Beyond ferrimagnetic materials, paramagnetic materials in principle can also be used for similar applications, although a much higher magnetic field will be needed. For instance, using external magnetic fields to align zirconium phosphate (ZrP) nanoplates that have a positive anisotropy of diamagnetic susceptibility, a large monodomain of nematic superstructure can be obtained (Figure 2.9c) [116].

Beyond the previously mentioned methods that take advantage of the intrinsic properties of anisotropic colloids, one alternative way to make new functional materials is to directly functionalize the anisotropic colloids. For instance, by grafting poly(*N*-isopropylacrylamide) (PNIPAM) onto ZrP platelets, these functionalized particles can self-assemble into discotic liquid crystalline structures that display thermo-sensitive properties (Figure 2.9d) [117].

2.4.2 Photonic Crystals

Many studies in this field use spherical particles to fabricate colloidal photonic crystals due to their monodisperse property and simple preparation methods. However, because the self-assembly of isotropic spherical particles typically form FCC crystals or variants, the optical properties of the resultant photonic crystals are very limited. For instance, the diamond lattice, which can have a full 3D photonic bandgap [120], cannot be made through the self-assembly of isotropic spherical colloids. By contrast, in a very recent work done by He et al. [121], using partially compressed tetrahedral clusters with retracted sticky patches, cubic diamond structures have been successfully self-assembled through patch–patch adhesion interactions together with a steric interlock mechanism that favors the desired staggered bond orientation [121]. In general, the wide range of accessible

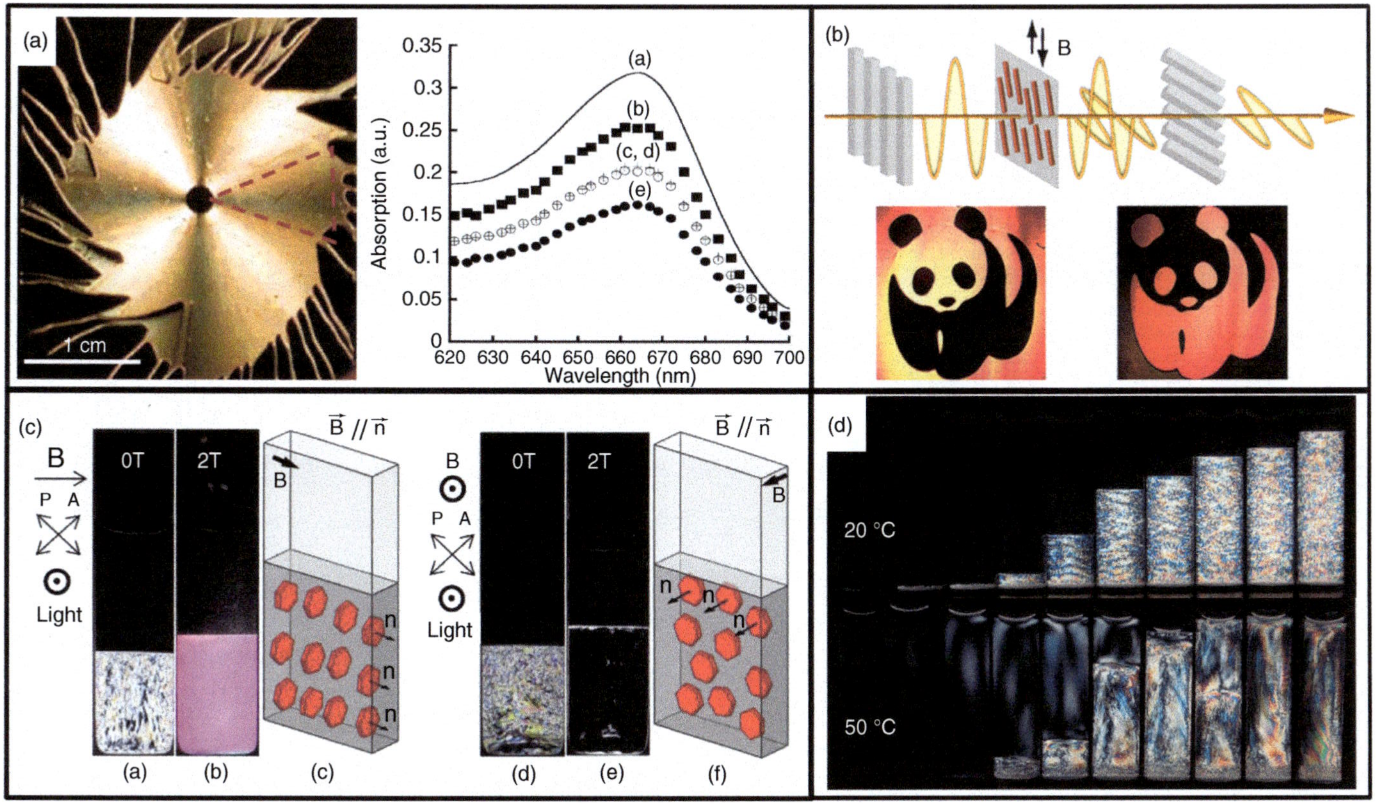

Figure 2.9 Liquid crystal materials formed through self-assembly of anisotropic colloids. (a) Photograph of the rutile film observed between crossed polarizers (left), which shows anisotropic photocatalytic activities as illustrated in the UV absorption curves (right). Source: Dessombz et al. [114]/with permission of American Chemical Society. (b) Polarized optical microscopy images of one example pattern shown under cross polarizers before and after shifting the direction of the transmission axis of the polarizers. Source: Wang et al. [115]/with permission of American Chemical Society. (c) A large monodomain of nematic superstructure formed by ZrP nanoplates that have a positive anisotropy of diamagnetic susceptibility. Source: Chen et al. [116]/with permission of IOP Publishing. (d) Cross-polarizing photographs of aqueous ZrP-PNIPAM suspensions at 20 and 50 °C. Source: Wang et al. [117]/with permission of Royal Society of Chemistry.

self-assembled structures of anisotropic colloids leads to the photonic crystals of anisotropic colloids to have a much better tunability in their optical properties and thus have drawn a lot of attention in recent years [5, 11, 121–123].

Photonic crystals of anisotropic colloids can be made from crystals of spherical colloids by deforming spheres into ellipsoids through either thermal stretching [123] or ion irradiation [122]. Beyond these post-crystallization methods, another way to fabricate them is through direct self-assembly. However, this requires a precise control on both positional and orientational order of anisotropic colloids. Toward this goal, Ding et al. [124, 125] first put the suspensions of ellipsoidal colloids with a magnetic core under an external magnetic field, which would align the colloids (i.e. all colloids have same orientations), and then assembled them into a 3D crystal by a convective assembly method (Figure 2.10a). Such crystals of ellipsoids can break the degeneracy at the W- and U-point of the photonic band structure in FCC crystals of spherical colloids [128]. Using Fe@SiO_2 nano-ellipsoids, Wang et al. [126] further developed a new class of magnetically responsive photonic structures whose optical properties can be dynamically tuned by controlling the direction of magnetic fields applied to the samples. Thus specific photonic patterns like a rainbow pattern can be created when magnetic fields with nonuniform directions are used (Figure 2.10b) [126]. Anisotropic colloids such as rods can be aligned in an electric field due to their anisotropic polarizability; thus similar to the aforementioned cases in magnetic fields, photonic crystals of anisotropic colloids have also been fabricated with the assistance of electric fields. For example, using AC electric fields, Forster et al. [127] fabricated photonic crystals of dumbbells, which not only show structural color due to the ordering of particles but also display birefringence and field-addressability of LCs presumably due to anisotropic shape of dumbbells (Figure 2.10c). Beyond ellipsoids and dumbbells, photonic crystals of other shaped colloids like mushroom cap-shaped colloids [129] have also been fabricated.

2.4.3 Sensors

With the excellent tunability in the properties of anisotropic colloids as well as their assembled structures, there have been much attention devoted to explore their applications as sensors for a variety of stimuli such as metal ions, biomolecules, gas, and environmental cues like temperature.

2.4.3.1 Metal Ions

Metal ions are a great concern in the quality control of environment. Many methods have been developed to detect heavy metal ions in samples such as atomic absorption spectrometry [130], inductively coupled plasma mass spectroscopy [131], and potentiometric ion selective electrode method [132]. However, these methods typically involve specific instruments and are time-consuming.

In recent years, LC-based sensors for metal ions have been developed. Such LC-based sensors typically contain two components: one is for specific recognition and binding of metal ions, and the other is LC mesogens, whose alignments are changed as a response to the binding events. As a result, the optical pattern of

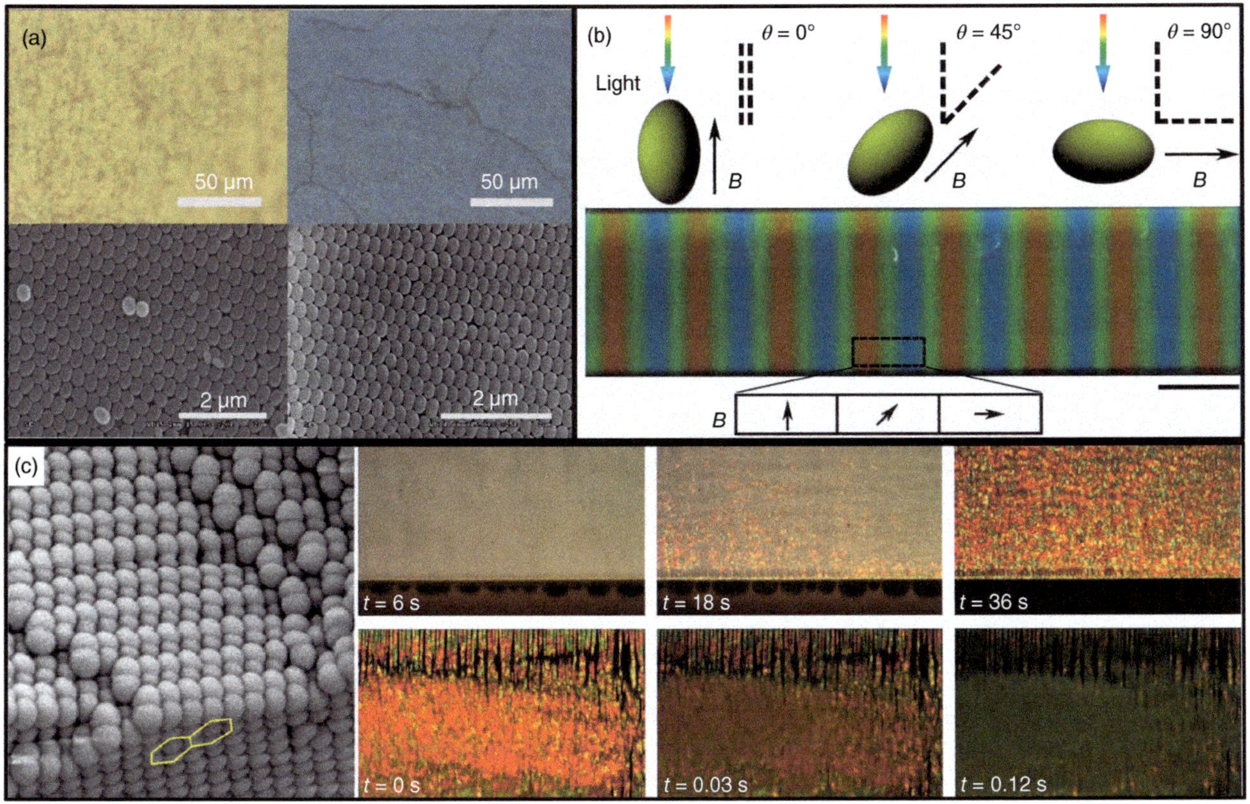

Figure 2.10 Examples of photonic crystals obtained by self-assembly of anisotropic colloids. (a) Optical microscope images of the assembled crystals of ellipsoids with aspect ratio of 1.3 and 1.5 (top row) and corresponding SEM images of the ellipsoids (bottom row). Source: Ding et al. [124]/with permission of John Wiley & Sons, Inc. (b) Schematic representations of the spontaneous alignment of nano-ellipsoids under magnetic fields (top row) and the bottom digital photo showing a rainbow photonic pattern under a nonideal linear Halbach array. Source: Wang et al. [126]/with permission of John Wiley & Sons, Inc. (c) Aqueous suspensions of dumbbells display reversible crystallization in AC electric fields, where both structural color and birefringence can be seen. The details of shape and packed crystal structures of dumbbells can be seen in the SEM image (left). Source: Forster et al. [127]/with permission of American Chemical Society.

LC is altered when metal ions are detected. For instance, by doping amphiphilic potassium *N*-methyl-*N*-dodecyldithiocarbamate (MeDTC) into LC, which align LC molecules at aqueous interfaces, Singh et al. [133] fabricated an LC-based sensor system for detecting Hg^{2+} ions. They showed that for solutions containing Hg^{2+} ions, the binding between dithiocarbamate chelating group of MeDTC with Hg^{2+} at aqueous interfaces led to a change in the orientation of LC molecules, which then resulted in a dark to bright change of the LC image [133]. However, the detection limit of their system is 0.5 μM for aqueous Hg^{2+} ions. In another work done by Yang et al. [134], a high-sensitivity LC-based sensor for Hg^{2+} was developed by enhancing the disruption for the orientation of LC through the target-induced DNA conformational change. Specifically, with the interaction between two thymine (T) bases and Hg^{2+}, T-Hg^{2+}-T, the specific oligodeoxy acid probe changed from hairpin structure to stable DNA duplexes, which then greatly distorted the orientation of LC and resulted in a visible change in the optical signal of sensor. This method can detect Hg^{2+} at a concentration of as low as 0.1 nM [134]. Similarly, LC-based sensors for other metal ions such as Ca^{2+} and Cu^{2+} have also been developed [135, 136].

Compared with those detection methods involving specific instruments, the pattern change of LC in response to metal ions is fast and can be easily observed by naked eyes. So LC-based sensors are suitable for a variety of environmental and industrial applications where simple, fast, and continuous measurements of metal ions are needed.

2.4.3.2 Biomolecules

Detection of specific biomolecules or biomarkers is critical in applications such as disease screening, diagnosis, and food safety monitoring. There have been methods developed for the detection of biomolecules using anisotropic particles as sensing platforms. Two types of anisotropic materials are often involved in these sensing platforms, which are LC-based and graphene-based.

Similar to those for metal ions, LC-based sensors for biomolecules also require appropriate component that can recognize and bind target biomolecules [137]. By incorporating cationic surfactants, myristoylcholine chloride (Myr), into LC at the aqueous/LC interface through self-assembly, an LC-based sensor for acetylcholinesterase (AChE) and its inhibitor has been developed [138]. In this sensing platform, when AChE are present in the solutions, they will hydrolyze Myr into myristic acid and choline [138], which will destabilize the assembled Myr monolayer at the LC/aqueous interface and cause the orientation change of LC molecules. Thus, the optical appearance of the sensor changes from dark to bright. AChE inhibitors can also be detected by using known AChE solutions incubated with test inhibitor samples [139]. Beyond the LC-based sensors constructed at the aqueous/LC interfaces, sensors based on the change of bulk LC materials have also been fabricated, which typically show a high sensitivity for target biomolecules. For example, through step-by-step fabrication, Zhao et al. [140] made an LC-based sensor for thrombin, which has a sandwich structure consisting of aptamer/thrombin/aptamer-functionalized gold nanoparticles in nickel nanoparticles (NiNSs)-doped LC. Because a thrombin molecule has two binding sites for aptamer, thrombin molecules in the samples will bridge the gold nanoparticles–aptamer conjugations together, which can distort the orientation of LC much more than thrombin molecules do, resulting in a clear change in the optical appearance of LC visible to naked eyes (Figure 2.11a). This method can detect thrombin with concentrations of 0.1–100 nM [140].

Another commonly used anisotropic material for sensors is graphene, due to their unique physical and chemical properties such as excellent electrical and thermal conductivity, high mechanical and chemical stability, etc. By immobilizing organophosphorus hydrolase (OPH) enzymes on the surfaces of reduced graphene oxide/Nafion (RGON) hybrid films, Choi et al. [13] developed an electrochemical RGON sensor for organophosphate detection. The RGON films were obtained through a directional convective-assembly of RGON particles that were assembled by RGON (Figure 2.11b). Owing to the excellent properties of reduced graphene oxide (rGO) together with their functionalization by Nafion, the as-prepared RGON film sensors exhibit superior qualities like high conductivity, fast electron transfer reaction, and low interfacial resistance, which all make them suitable for biosensor applications [13]. Similar graphene/rGO sensors for cancer or dopamine have also been fabricated through self-assembly methods [142, 143].

Beyond biomolecules, these biosensors can also be used to detect bacteria if the detection/binding component of biosensors can interact with bacteria cells. For instance, by incorporating lipopolysaccharides (LPS) into LCs at the aqueous/LC interface, which can cause the orientational change of LC when LPS interact with bacterial cells, Zafiu et al. [144] fabricated LC-based bacterial sensors. Due to the specificity of LPS on different bacterial species, such biosensors have a good sensitivity and efficiency for detecting LPS-specific bacterial cells.

2.4.3.3 Gases

For gas detections, graphene materials have been widely used due to their large surface areas for molecular adsorption as well as their outstanding electrical properties suitable for electrochemical sensing. Using layer-by-layer (LbL) self-assembly, Zhang et al. [145] fabricated a CO gas sensor based on copper oxide-decorated graphene hybrid nanocomposite. Due to the hierarchical porous structures as well as the heterojunctions formed between copper oxide nanoflowers and rGO nanosheets, the fabricated gas sensor exhibited fast response and recovery times, excellent repeatability, and, more importantly, a wide detection range of CO gas from 0.25 to 1000 ppm [145]. Similarly, humidity sensors can be fabricated when the target molecule is water molecule. By chemically reducing the graphene oxide/poly(diallyldimethylammonium chloride) (PDDA) composites that have a hierarchical nanostructure with alternating layers of graphene oxide and PDDA formed by LbL assembly, Zhang et al. [141] made a resistive-type film sensor of rGO/PDDA on flexible polyimide (PI) substrate with interdigital microelectrodes structures (Figure 2.11c), which showed good performance for humidity measurements. The sensitivity of gas sensors can be greatly enhanced when graphene (not rGO) is used as demonstrated in Wang et al. [146], where sensor devices were fabricated by self-assembling graphene sheets onto patterned gold structures via

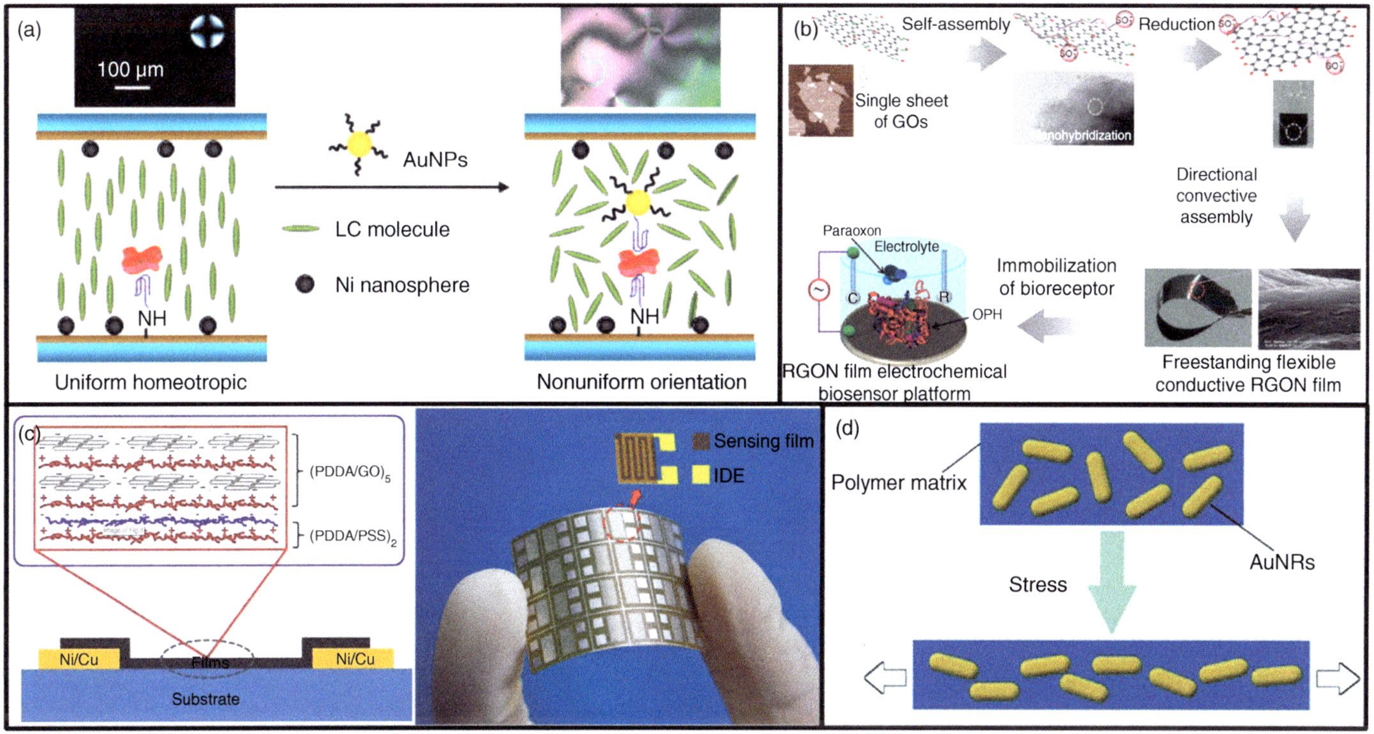

Figure 2.11 Applications of the self-assembly of anisotropic colloids in sensors. (a) Schematic illustration of the detection method for the NiNS-based thrombin LC sensor. Source: Zhao et al. [140]/with permission of American Chemical Society (b) An example procedure to design RGON hybrids and to apply RGON platform in electrochemical biosensors. Source: Choi et al. [13]/with permission of American Chemical Society. (c) Schematic diagram of the LbL-assembled sensor for humidity and an optical image showing a 4 × 6 sensor array on a flexible PI substrate. Source: Zhang et al. [141]/with permission of Elsevier. (d) Schematic illustration of a stress-responsive film based on direction-related change of localized surface plasmon resonance of Au nanorods. Source: Fu et al. [14]/with permission of American Chemical Society.

electrostatic interactions, and the obtained devices can detect ammonia at the ppb level.

2.4.3.4 Other Environmental Cues

Besides the aforementioned ones for metal ions, biomolecules, and gases, sensors based on anisotropic colloids for other environmental cues have also been demonstrated. For example, a novel pressure/strain sensor consisting of gold nanorods embedded in a polymer matrix have been reported [14]. Gold nanorods display two modes of localized surface plasmon resonance (LSPR) in extinction spectra [147], and the relative intensity ratio between the two modes is rod-orientational dependent. Thus, when an external pressure is applied on the composite of gold nanorods and polymer matrix, the alignment of gold nanorods due to polymer flow induces a change in the intensity ratio of two modes of LSPR, which can be quantitatively correlated to the applied pressure (Figure 2.11d) [14]. But this method requires complicated detection instruments and the materials used are also costly. To develop low-cost, flexible, and sensitive strain sensors that are highly in demand for wearable devices [148, 149], Li et al. [150] developed a simple technique for making graphene-based sensors. They first fabricated large-area ultrathin graphene films by single-step Marangoni self-assembly and then transferred the films to arbitrary substrates to construct strain sensors. The obtained sensors exhibited remarkably high sensitivity to small strains. The simple fabrication of these strain sensors together with their high performance in strain sensing make them an interesting candidate for applications in wearable devices such as electronic skin and health monitoring devices.

2.4.4 Electrode Materials

Flexible, high-energy/power-density devices, such as batteries and supercapacitors are in urgent demand due to the increasing need for the portable electronic devices and for the clean energy resources. Anisotropic colloids, due to their abilities to form complicated hierarchical structures, have drawn great attentions in this field. Particularly, assembled graphene-based materials have been widely studied in lithium-ion batteries as anode and/or cathode materials due to their unique properties like large surface area, superior electric conductivity, excellent mechanical flexibility, and chemical stability [16, 151]. One major problem in lithium-ion battery is the rapid capacity fade and poor cycling stability that resulted from the volume variation of electrode materials during lithium insertion and extraction. So many methods have been proposed to solve this problem.

One way is to introduce hollow structures into electrode materials to buffer the volume expansion of active materials during repeated charge–discharge cycles. For example, Wang et al. [152] fabricated a 3D $ZnSnO_3$ hollow cubes@reduced graphene oxide aerogels (ZGAs) by electrostatic self-assembly and subsequent hydrothermal and freeze-drying treatments. The porous structure of $ZnSnO_3$ hollow cubes and the flexibility of rGO together effectively buffer the volume change of active materials and also help the transport of charge carriers, thus the as-obtained ZGAs displayed an enhanced cycling stability and rate capability compared with bare $ZnSnO_3$ hollow cubes and rGO electrodes [152]. However, one apparent drawback of introducing hollow structures into electrode materials is that their volumetric energy density is reduced. To improve the packing density of hollow-structured materials, Liang et al. [153] used bowl-like SnO_2@carbon hollow particles, which can have a higher packing density than hollow spheres with the same diameter. Moreover, the interconnection between bowl-like particles is also better compared with spheres. These factors together improved the electrochemical performance of the fabricated device (Figure 2.12a). Similarly, Pei et al. [157] synthesized *N*-doped hollow porous carbon bowls (N-HPCBs), and the sulfur *N*-doped hollow porous carbon bowls (S/N-HPCBs) cathode in a lithium–sulfur (Li–S) battery showed excellent cycling stability and rate capability.

Beyond introducing hollow structures, another way to enhance the cycling stability is to form alternating structures of the target electrode material with graphene-based materials. For instance, by self-assembly, Zhao et al. [154] synthesized van der Waals heterostructures consisting of alternating nitrogen-doped graphene (NDG) and MoS_2 layers (Figure 2.12b). The fabricated materials as an anode for lithium-ion batteries showed excellent rate capabilities and cycling stability due to their low charge-transfer resistance, high sulfur reservation, and structural robustness [154]. Using surfactant-assisted self-assembly, Wang et al. [158] fabricated ordered metal oxide-graphene nanocomposites, which include two types of structures: one is an ordered alternating-layered structure of nanocrystalline metal oxides with graphene or graphene stacks, and the other is a high surface area conductive network formed by incorporating graphene or graphene stacks into LC-templated nanoporous structures. The obtained SnO_2–graphene nanocomposite films can achieve near-theoretical specific energy density without significant charge/discharge degradation [158]. Despite the superior performance that ordered structures can provide, to form such ordered structures, it typically requires a guided assembly process either by surfactants or other mechanisms in a precise manner, which limit its practical applications.

Another common way to improve the performance of electrode materials but not based on ordered structures is to simply wrap around (or anchor) electroactive particles using graphene networks. By this way, the agglomeration of electroactive materials can be reduced. Moreover, the conductive carbon matrices can absorb the volume changes due to their mechanical flexibility to keep the integrity of the electrode structure. For instance, Zhu et al. [159] fabricated graphene nanosheets-wrapped silicon nanowires (GNS@Si NWs) through electrostatic self-assembly. The as-prepared GNS@Si NWs as anode showed significantly improved rate capability and cycling performance. Composite films formed by silicon particles and rGO also show similar improvement in the electrochemical performance [160]. To gain a better performance, more complex hierarchical structures can be constructed. Using a facile two-step reduction approach, Sun et al. [161] synthesized hierarchically nanostructured $Mn_2Mo_3O_8$–graphene nanocomposites, which contain $Mn_2Mo_3O_8$ microspheres that were formed by $Mn_2Mo_3O_8$ nanosheets. Interestingly, through a template strategy, Liu et al. [155]

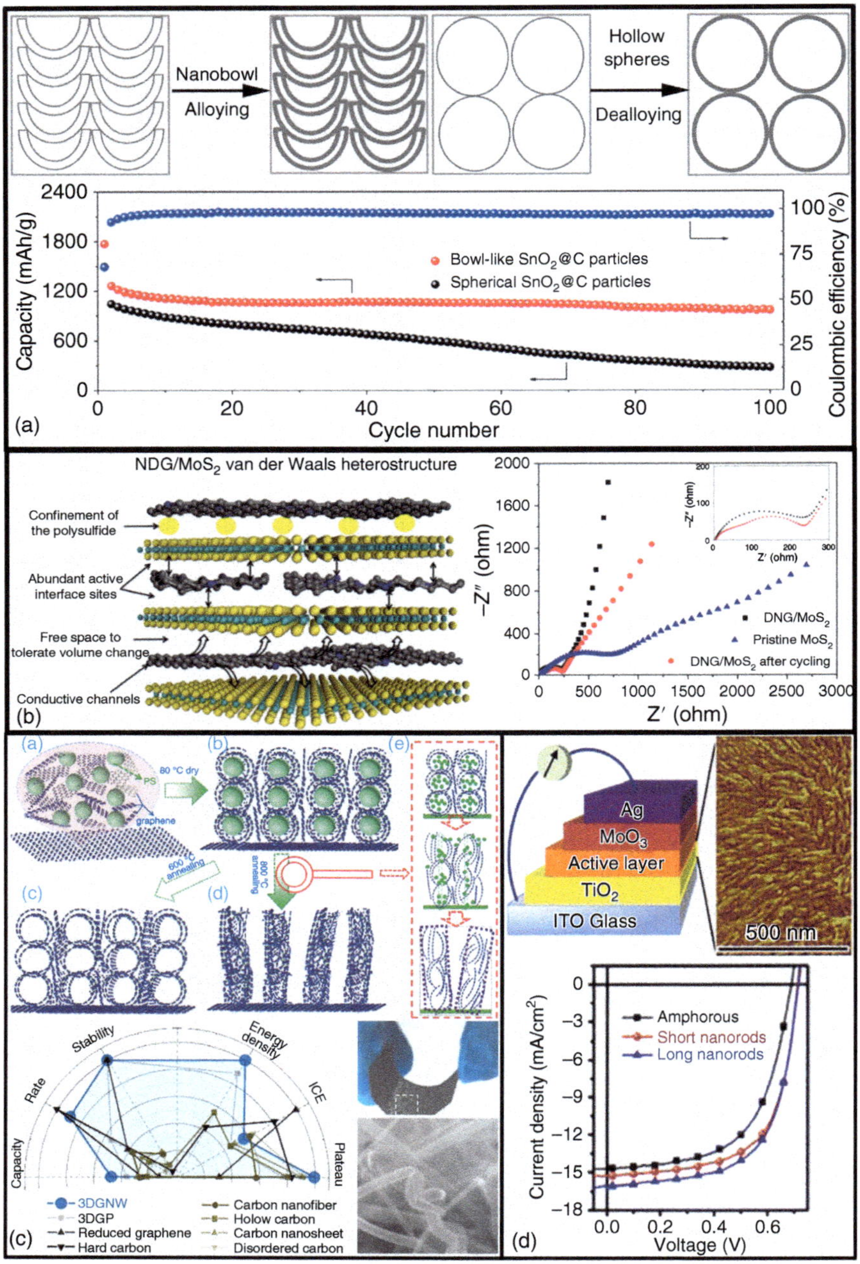
Nanobowl
Alloying
Hollow spheres
Dealloying
Capacity (mAh/g)
Coulombic efficiency (%)
Bowl-like SnO_2@C particles
Spherical SnO_2@C particles
Cycle number
(a)
NDG/MoS_2 van der Waals heterostructure
Confinement of the polysulfide
Abundant active interface sites
Free space to tolerate volume change
Conductive channels
(b)
−Z″ (ohm)
Z′ (ohm)
DNG/MoS_2
Pristine MoS_2
DNG/MoS_2 after cycling
80 °C dry
PS
graphene
600 °C annealing
800 °C annealing
Stability
Energy density
Rate
ICE
Capacity
Plateau
3DGNW
3DGP
Reduced graphene
Hard carbon
Carbon nanofiber
Holow carbon
Carbon nanosheet
Disordered carbon
(c)
Ag
MoO_3
Active layer
TiO_2
ITO Glass
500 nm
Current density (mA/cm^2)
Amphorous
Short nanorods
Long nanorods
Voltage (V)
(d)

Figure 2.12 Examples of electrode materials assembled by anisotropic colloids. (a) Comparison of packing and the cycling performance of anodes formed by bowl-like particles and spherical hollow particles, respectively. Source: Liang et al. [153]. Reproduced with permission of John Wiley and Sons). (b) Schematic illustration of the NDG/MoS_2 heterostructure and Nyquist plots of the pristine MoS_2 and NDG/MoS_2 anode before and after cycling, indicating the excellent cycling stability of the NDG/MoS_2 anode. Source: Zhao et al. [154]. Reproduced with permission of American Chemical Society. (c) Schematics for the fabrication of graphene nanowires. The comparison of 3DGNW electrode with reported pure carbonaceous anode show its superior performance in general. Source: Liu et al. [155]/with permission of Elsevier. (d) Self-assembled TiO_2 nanorods as electron extraction layer for inverted polymer solar cells, showing excellent performance. Source: Lv et al. [156]/with permission of American Chemical Society.

fabricated a new structured graphene material consisting of reduced graphene nanowires on three-dimensional graphene foam (3DGNW) (Figure 2.12c). Such materials when tested as anodes all showed improved performance with excellent reversible capacity, rate capability, and durable tolerance. Using similar methods, other composite materials constructed from MoS_2, $LiFePO_4$, NiO, and TiO_2 with graphene-based materials as potential electrode candidates for lithium-ion batteries have also been reported [155, 162–165].

Besides lithium-ion batteries, assembled structures of anisotropic colloids have also been used in other energy sources such as solar cells. Lv et al. [156] fabricated a film of highly crystalline TiO_2 nanorods (TiO_2-NRs), which showed high performance as an electron extraction layer in inverted polymer solar cells (Figure 2.12d). Moreover, their results also showed that long rods behaved better than short rods, presumably due to the better alignment of long rods in the film, implying the importance of the shape and orientation of anisotropic colloids in their functions.

2.4.5 Other Applications in Making Functional Materials

With the increasingly need for new functional materials, applications of anisotropic colloids have also been extended to other emerging fields. Just a few examples of these include superhydrophobic surfaces [166–168], stabilizer materials for Pickering emulsions [169–172], materials for tissue engineering [15], and catalyst materials [173]. As we cannot numerate all of them, here, we rather choose their applications in superhydrophobic surfaces and stabilizer materials for Pickering emulsions as representative examples, to highlight the recent progress in these two fields.

The wetting property of a surface is largely determined by its chemical composition and roughness [174], for which anisotropic particles provide easier tunability compared with their spherical homogeneous counterparts. Thus, to make superhydrophobic surfaces, which have water contact angles larger than 150°, methods based on surface roughness control and/or chemical property control of components have been developed. For example, using hydrophobic hybrid silica nanowires as building blocks, superhydrophobic functional materials with 3D microporous networks have been fabricated (Figure 2.13a) [168]. On the other hand, Li et al. [167] fabricated poly((glycidyl methacrylate)-*co*-(ethylene glycol dimethacrylate)) raspberry-like colloids with micro-/nanoscale surface roughness through one-pot

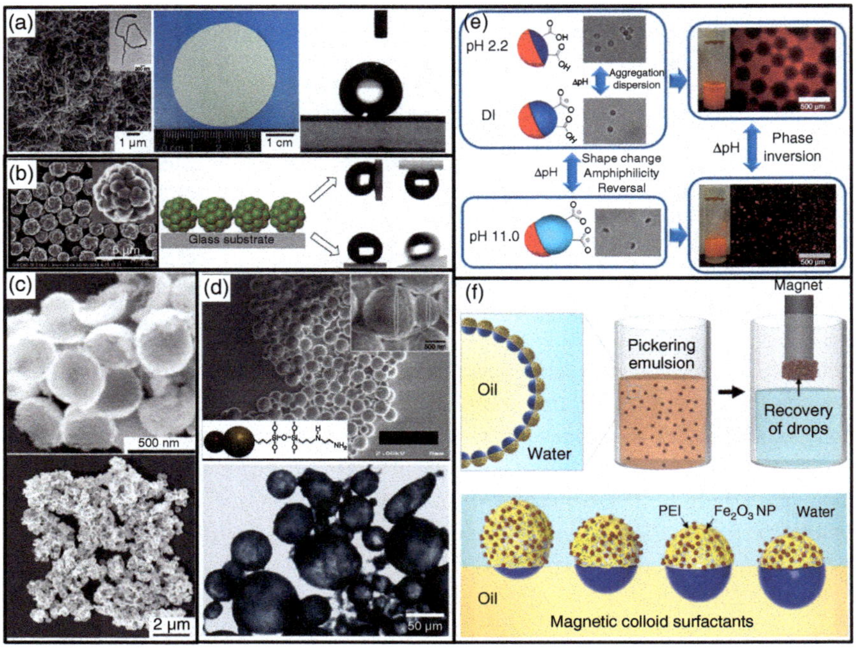

Figure 2.13 Examples showing applications of anisotropic colloids in superhydrophobic materials and stabilizers for Pickering emulsions. (a) Superhydrophobic membrane made of silica nanowires. Source: Yi et al. [168]/with permission of John Wiley & Sons, Inc. (b) Surfaces made of fluorinated raspberry-like particles, showing tunable superhydrophobic properties ranging from lotus-leaf-like hydrophobicity to rose-petal-like hydrophobicity. Source: Li et al. [167]/with permission of Royal Society of Chemistry. (c) SEM images of gold half-shells and its aggregation that is hydrophobic. Source: Love et al. [166]/with permission of American Chemical Society. (d) Stabilization of hexadecane drops (including nonspherical ones) with amphiphilic dimer particles. Source: Kim et al. [169]/with permission of John Wiley & Sons, Inc. (e) Amphiphilic Janus colloids whose morphology and particle–particle interactions are pH-responsive, which can lead to phase inversion at different pH. Source: Tu and Lee [175]/with permission of American Chemical Society. (f) Magnetic-patchy Janus colloids for both interfacial emulsification and magnetic responsiveness. Source: Kim et al. [171]. Reproduced with permission of American Chemical Society.

dispersion polymerization. The films made of these raspberry-like colloids showed excellent superhydrophobicity. More interestingly, by further chemically modifying the surfaces of these raspberry-like particles through fluorination, films with tunable superhydrophobic properties ranging from lotus-leaf-like hydrophobicity to rose-petal-like hydrophobicity can be obtained (Figure 2.13b) [167], demonstrating the great capability of combining the roughness and chemical control together to obtain desired superhydrophobicity. In addition to modifying the building blocks, superhydrophobic surfaces can also be fabricated by assembling anisotropic colloids into structures that have an enhanced surface roughness even anisotropic colloids themselves are not hydrophobic. For example, using gold half-shell colloids with nanometer-scale dimensions, which are difficult to form regular, densely packed arrays, superhydrophobic surfaces with a water contact angle of ~151° can be obtained by the aggregates of these colloids mainly due to their high degree of surface roughness (Figure 2.13c) [166]. Surfaces with even larger contact angle of ~163° could be fabricated by using modified gold half-shells whose surfaces were coated with a self-assembled monolayer of hexadecanethiolate [166].

Pickering emulsions are emulsions stabilized by solid colloids rather than by molecular emulsifiers and have broad applications in fields as diverse as food, cosmetics, and pharmaceutics [170, 176, 177]. In Pickering emulsions, solid colloids are adsorbed at the interface between two immiscible liquids to form a mechanically robust monolayer, which in turn stabilizes the emulsion droplets against coalescence. So these colloids act as colloidal surfactants, analogous to molecular surfactants. But due to their large size, they have a much stronger attachment to the interface than molecular surfactants do [178].

As an important member of the anisotropic particle's family, Janus particles that possess two sides with distinct chemistry and wettability (e.g. with amphiphilicity) are ideal candidates for Pickering emulsions. A variety of Janus particles have been fabricated and their great performance in stabilizing Pickering emulsions have been demonstrated [169, 172, 179]. Using seeded polymerization-based techniques, Kim et al. [169] synthesized amphiphilic dimer Janus particles, containing one hydrophilic bulb and one hydrophobic bulb (Figure 2.13d). Stable Pickering emulsions were obtained using these amphiphilic dimers as stabilizers. Different from molecular surfactants, these dimers can jam at the interface and thus are able to stabilize nonspherical emulsion droplets [169]. Amphiphilic nanosheets of α-ZrP-octadecyl isocyanate have also been fabricated and successfully stabilized toluene-in-water emulsions [179]. Compared with other Janus particles, these amphiphilic nanosheets are very thin (at the atomic scale) and have great flexibility to wrap around the droplets.

Benefiting from the easier controllability in their properties, Janus particles can have multiple functions by incorporating appropriate functionalized materials while keeping their amphiphilicity. For example, Tu et al. [175] fabricated amphiphilic Janus colloids whose morphology and particle–particle interactions are pH-responsive. The Janus colloids consist of styrene-rich hydrophobic part and acrylic acid-rich pH-sensitive hydrophilic part. When the pH of emulsion solution varies, induced changes in the particle shape lead to the reversal of the amphiphilicity of particles, which can result in phase inversion of emulsions (Figure 2.13e) [175]. In this example, the key is to use dual functional material as part of the Janus particles, which requires specific types of materials. Another more straightforward way is to simply incorporate a different type of material into Janus particles mainly for desired functions. For instance, taking advantage of the magnetic properties of Fe_2O_3 nanoparticles, Kim et al. [171] fabricated dual functional materials for both interfacial emulsification and magnetic responsiveness by selectively decorating the hydrophilic side of Janus colloids with Fe_2O_3 nanoparticles (Figure 2.13f), which show great potential for applications in oil recovery industries. Janus particle surfactants with catalytic functions can also be produced similarly [180].

By incorporating different functional materials into Janus particles, the applications of Pickering emulsions can be expected to be greatly expanded.

2.5 Summary and Outlook

We have overviewed the fabrication methods for anisotropic colloids, the main mechanisms for their self-assembly, and the applications of their assembled structures. Due to the extraordinary characteristics either from their unique shapes or well-designed particle interactions, they show a much broader range of assembled structures than isotropic colloids. With such a variety of accessible structures that could possess special properties, self-assembly of anisotropic colloids show great potential for applications in various fields. Recently, a growing trend in the field of self-assembly is toward preparing the well-defined stereotype structures, which is closely related to the development in fabrication techniques. Currently, three popular research directions in this field are controlling the distribution map of chemically functionalized patches on particles, fabricating particles with user-designed shapes and interaction potentials, and developing new assembly techniques. It is worth to note that some novel structures obtained in simulations are not observed in experiments mainly due to the polydispersity of experimentally fabricated anisotropic colloids. Thus, to test such simulation results, improving fabrication methods to get anisotropic particles with low enough polydispersity is essential. In addition, the environmental-responsive (including temperature, pH, and ion concentration) self-assembly is also a hot topic due to its huge potential applications in many areas. With all these advances, we anticipate that more and more novel assemblies from anisotropic colloids will be discovered and applied in our life.

Acknowledgment

This work is supported by the National Natural Science Foundation of China (Grants Nos. 11874277 and 11704276).

References

1 Whitesides, G.M. and Grzybowski, B. (2002). Self-assembly at all scales. *Science* 295 (5564): 2418–2421.
2 Glotzer, S.C. (2004). Some assembly required. *Science* 306 (5695): 419.
3 Pusey, P.N. and Megen, W.v. (1986). Phase behaviour of concentrated suspensions of nearly hard colloidal spheres. *Nature* 320 (6060): 340–342.
4 Frenkel, D. and Ladd, A.J.C. (1984). New Monte Carlo method to compute the free energy of arbitrary solids. Application to the FCC and HCP phases of hard spheres. *J. Chem. Phys.* 81 (7): 3188–3193.

5 Li, F., Josephson, D.P., and Stein, A. (2011). Colloidal assembly: the road from particles to colloidal molecules and crystals. *Angew. Chem. Int. Ed.* 50 (2): 360–388.

6 Glotzer, S.C. and Solomon, M.J. (2007). Anisotropy of building blocks and their assembly into complex structures. *Nat. Mater.* 6: 557–562.

7 Onsager, L. (1949). The effects of shape on the interaction of colloidal particles. *Ann. N. Y. Acad. Sci.* 51 (4): 627–659.

8 Noya, E.G., Kolovos, I., Doppelbauer, G. et al. (2014). Phase diagram of inverse patchy colloids assembling into an equilibrium laminar phase. *Soft Matter* 10 (42): 8464–8474.

9 Chen, Q., Whitmer, J.K., Jiang, S. et al. (2011). Supracolloidal reaction kinetics of Janus spheres. *Science* 331 (6014): 199–202.

10 Chen, Q., Bae, S.C., and Granick, S. (2011). Directed self-assembly of a colloidal kagome lattice. *Nature* 469 (7330): 381–384.

11 Lu, Y., Yin, Y., and Xia, Y. (2001). Three-dimensional photonic crystals with non-spherical colloids as building blocks. *Adv. Mater.* 13 (6): 415–420.

12 He, L., Ye, J., Shuai, M. et al. (2015). Graphene oxide liquid crystals for reflective displays without polarizing optics. *Nanoscale* 7 (5): 1616–1622.

13 Choi, B.G., Park, H., Park, T.J. et al. (2010). Solution chemistry of self-assembled graphene nanohybrids for high performance flexible biosensors. *ACS Nano* 4 (5): 2910–2918.

14 Fu, L., Liu, Y., Wang, W. et al. (2015). A pressure sensor based on the orientational dependence of plasmonic properties of gold nanorods. *Nanoscale* 7 (34): 14483–14488.

15 Du, Y., Lo, E., Ali, S., and Khademhosseini, A. (2008). Directed assembly of cell-laden microgels for fabrication of 3D tissue constructs. *Proc. Natl. Acad. Sci. U. S. A.* 105 (28): 9522–9527.

16 Forouzandeh, P., Kumaravel, V., and Pillai, S.C. (2020). Electrode materials for supercapacitors: a review of recent advances. *Catalysts* 10 (9): 969.

17 Rossi, L., Sacanna, S., Irvine, W.T.M. et al. (2011). Cubic crystals from cubic colloids. *Soft Matter* 7 (9): 4139–4142.

18 Sacanna, S., Irvine, W.T.M., Chaikin, P.M., and Pine, D.J. (2010). Lock and key colloids. *Nature* 464 (7288): 575–578.

19 Park, J.-G., Forster, J.D., and Dufresne, E.R. (2010). High-yield synthesis of monodisperse dumbbell-shaped polymer nanoparticles. *J. Am. Chem. Soc.* 132 (17): 5960–5961.

20 Wagner, C.S., Fischer, B., May, M., and Wittemann, A. (2010). Templated assembly of polymer particles into mesoscopic clusters with well-defined configurations. *Colloid Polym. Sci.* 288 (5): 487–498.

21 Xia, Y., Yin, Y., Lu, Y., and McLellan, J. (2003). Template-assisted self-assembly of spherical colloids into complex and controllable structures. *Adv. Funct. Mater.* 13 (12): 907–918.

22 Rothemund, P.W.K. (2006). Folding DNA to create nanoscale shapes and patterns. *Nature* 440 (7082): 297–302.

23 Wang, Y., Wang, Y., Breed, D.R. et al. (2012). Colloids with valence and specific directional bonding. *Nature* 491 (7422): 51–55.

24 Hernandez, C.J. and Mason, T.G. (2007). Colloidal alphabet soup: monodisperse dispersions of shape-designed lithoparticles. *J. Phys. Chem. C* 111 (12): 4477–4480.

25 Dendukuri, D., Pregibon, D.C., Collins, J. et al. (2006). Continuous-flow lithography for high-throughput microparticle synthesis. *Nat. Mater.* 5 (5): 365–369.

26 Champion, J.A., Katare, Y.K., and Mitragotri, S. (2007). Making polymeric micro- and nanoparticles of complex shapes. *Proc. Natl. Acad. Sci. U. S. A.* 104 (29): 11901–11904.

27 Liu, Y., Zhao, Z., Chen, Y. et al. (2008). Anisotropic deformation of polystyrene particles by MeV Au ion irradiation. *Nucl. Instrum. Methods Phys. Res., Sect. B* 266 (6): 894–898.

28 McMullen, A., de Haan, H.W., Tang, J.X., and Stein, D. (2014). Stiff filamentous virus translocations through solid-state nanopores. *Nat. Commun.* 5 (1): 4171.

29 Tombácz, E. and Szekeres, M. (2006). Surface charge heterogeneity of kaolinite in aqueous suspension in comparison with montmorillonite. *Appl. Clay Sci.* 34 (1-4): 105–124.

30 Murray, H.H. (2000). Traditional and new applications for kaolin, smectite, and palygorskite: a general overview. *Appl. Clay Sci.* 17 (5): 207–221.

31 Stöber, W., Fink, A., and Bohn, E. (1968). Controlled growth of monodisperse silica spheres in the micron size range. *J. Colloid Interface Sci.* 26 (1): 62–69.

32 Sacanna, S., Rossi, L., Kuipers, B.W.M., and Philipse, A.P. (2006). Fluorescent monodisperse silica ellipsoids for optical rotational diffusion studies. *Langmuir* 22 (4): 1822–1827.

33 Lee, S.H., Gerbode, S.J., John, B.S. et al. (2008). Synthesis and assembly of nonspherical hollow silica colloids under confinement. *J. Mater. Chem.* 18 (41): 4912–4916.

34 Niu, Z., Liu, J., Lee, L.A. et al. (2007). Biological templated synthesis of water-soluble conductive polymeric nanowires. *Nano Lett.* 7 (12): 3729–3733.

35 Kim, J.-W., Larsen, R.J., and Weitz, D.A. (2006). Synthesis of nonspherical colloidal particles with anisotropic properties. *J. Am. Chem. Soc.* 128 (44): 14374–14377.

36 Kraft, D.J., Vlug, W.S., van Kats, C.M. et al. (2009). Self-assembly of colloids with liquid protrusions. *J. Am. Chem. Soc.* 131 (3): 1182–1186.

37 Liu, M., Zheng, X., Grebe, V. et al. (2020). Tunable assembly of hybrid colloids induced by regioselective depletion. *Nat. Mater.* 19 (12): 1354–1361.

38 Perro, A., Duguet, E., Lambert, O. et al. (2009). A chemical synthetic route towards "colloidal molecules". *Angew. Chem.* 121 (2): 367–371.

39 Manoharan, V.N., Elsesser, M.T., and Pine, D.J. (2003). Dense packing and symmetry in small clusters of microspheres. *Science* 301 (5632): 483–487.

40 Zerrouki, D., Rotenberg, B., Abramson, S. et al. (2006). Preparation of doublet, triangular, and tetrahedral colloidal clusters by controlled emulsification. *Langmuir* 22 (1): 57–62.

41 Jones, M.R., Seeman, N.C., and Mirkin, C.A. (2015). Programmable materials and the nature of the DNA bond. *Science* 347 (6224): 1260901.
42 Seeman, N.C. (2010). Nanomaterials based on DNA. *Annu. Rev. Biochem.* 79 (1): 65–87.
43 Jones, M.R., Osberg, K.D., Macfarlane, R.J. et al. (2011). Templated techniques for the synthesis and assembly of plasmonic nanostructures. *Chem. Rev.* 111 (6): 3736–3827.
44 Xu, X., Rosi, N.L., Wang, Y. et al. (2006). Asymmetric functionalization of gold nanoparticles with oligonucleotides. *J. Am. Chem. Soc.* 128 (29): 9286–9287.
45 Brown, A.B.D., Smith, C.G., and Rennie, A.R. (2000). Fabricating colloidal particles with photolithography and their interactions at an air-water interface. *Phys. Rev. E* 62: 951–960.
46 Zhao, K. and Mason, T.G. (2009). Frustrated rotator crystals and glasses of Brownian pentagons. *Phys. Rev. Lett.* 103 (20): 208302.
47 Zhao, K., Bruinsma, R., and Mason, T.G. (2011). Entropic crystal–crystal transitions of Brownian squares. *Proc. Natl. Acad. Sci. U. S. A.* 108 (7): 2684–2687.
48 Zhao, K., Bruinsma, R., and Mason, T.G. (2012). Local chiral symmetry breaking in triatic liquid crystals. *Nat. Commun.* 3: 1–8.
49 Ho, C.C., Keller, A., Odell, J.A., and Ottewill, R.H. (1993). Preparation of monodisperse ellipsoidal polystyrene particles. *Colloid Polym. Sci.* 271 (5): 469–479.
50 van Dillen, T., Polman, A., van Kats, C.M., and van Blaaderen, A. (2003). Ion beam-induced anisotropic plastic deformation at 300 keV. *Appl. Phys. Lett.* 83 (21): 4315–4317.
51 Bawden, F.C., Pirie, N.W., Bernal, J.D., and Fankuchen, L. (1936). Liquid crystalline substances from virus-infected plants. *Nature* 138: 1051–1052.
52 Adams, M., Dogic, Z., Keller, S.L., and Fraden, S. (1998). Entropically driven microphase transitions in mixtures of colloidal rods and spheres. *Nature* 393 (6683): 349–352.
53 Lagaly, G. and Ziesmer, S. (2003). Colloid chemistry of clay minerals: the coagulation of montmorillonite dispersions. *Adv. Colloid Interf. Sci.* 100: 105–128.
54 Hong, L., Cacciuto, A., Luijten, E., and Granick, S. (2006). Clusters of charged Janus spheres. *Nano Lett.* 6 (11): 2510–2514.
55 Dempster, J.M. and de la Cruz, M.O. (2016). Aggregation of heterogeneously charged colloids. *ACS Nano* 10 (6): 5909–5915.
56 Cerbelaud, M., Lebdioua, K., Tran, C.T. et al. (2019). Brownian dynamics simulations of one-patch inverse patchy particles. *Phys. Chem. Chem. Phys.* 21 (42): 23447–23458.
57 Noya, E.G. and Bianchi, E. (2015). Phase behaviour of inverse patchy colloids: effect of the model parameters. *J. Phys. Condens. Matter* 27 (23): 234103.
58 Ferrari, S., Bianchi, E., and Kahl, G. (2017). Spontaneous assembly of a hybrid crystal-liquid phase in inverse patchy colloid systems. *Nanoscale* 9 (5): 1956–1963.

59 de Araujo, J.L.B., Munarin, F.F., Farias, G.A. et al. (2017). Structure and reentrant percolation in an inverse patchy colloidal system. *Phys. Rev. E* 95 (6): 062606.

60 Bianchi, E., Likos, C.N., and Kahl, G. (2014). Tunable assembly of heterogeneously charged colloids. *Nano Lett.* 14 (7): 3412–3418.

61 Sabapathy, M., Ann Mathews, K.R., and Mani, E. (2017). Self-assembly of inverse patchy colloids with tunable patch coverage. *Phys. Chem. Chem. Phys.* 19 (20): 13122–13132.

62 van Oostrum, P.D.J., Hejazifar, M., Niedermayer, C., and Reimhult, E. (2015). Simple method for the synthesis of inverse patchy colloids. *J. Phys. Condens. Matter* 27 (23): 234105.

63 Hong, L., Cacciuto, A., Luijten, E., and Granick, S. (2008). Clusters of amphiphilic colloidal spheres. *Langmuir* 24 (3): 621–625.

64 Chen, Q., Diesel, E., Whitmer, J.K. et al. (2011). Triblock colloids for directed self-assembly. *J. Am. Chem. Soc.* 133 (20): 7725–7727.

65 Repula, A., Menegon, M.O., Wu, C. et al. (2019). Directing liquid crystalline self-organization of rod-like particles through tunable attractive single tips. *Phys. Rev. Lett.* 122: 12.

66 Wang, Y., Wang, Y.F., Zheng, X.L. et al. (2014). Three-dimensional lock and key colloids. *J. Am. Chem. Soc.* 136 (19): 6866–6869.

67 Sacanna, S., Korpics, M., Rodriguez, K. et al. (2013). Shaping colloids for self-assembly. *Nat. Commun.* 4: 1688.

68 Tigges, T. and Walther, A. (2016). Hierarchical self-assembly of 3D-printed lock-and-key colloids through shape recognition. *Angew. Chem. Int. Ed.* 55 (37): 11261–11265.

69 Mihut, A.M., Stenqvist, B., Lund, M. et al. (2017). Assembling oppositely charged lock and key responsive colloids: a mesoscale analog of adaptive chemistry. *Sci. Adv.* 3 (9): e1700321.

70 Zhao, K. and Mason, T.G. (2007). Directing colloidal self-assembly through roughness-controlled depletion attractions. *Phys. Rev. Lett.* 99 (26): 268301.

71 Zhao, K. and Mason, T.G. (2008). Suppressing and enhancing depletion attractions between surfaces roughened by asperities. *Phys. Rev. Lett.* 101 (14): 148301.

72 Zhao, K. and Mason, T.G. (2012). Twinning of rhombic colloidal crystals. *J. Am. Chem. Soc.* 134 (43): 18125–18131.

73 Zhao, K. and Mason, T.G. (2015). Shape-designed frustration by local polymorphism in a near-equilibrium colloidal glass. *Proc. Natl. Acad. Sci. U. S. A.* 112 (39): 12063–12068.

74 Kraft, D.J., Ni, R., Smallenburg, F. et al. (2012). Surface roughness directed self-assembly of patchy particles into colloidal micelles. *Proc. Natl. Acad. Sci. U. S. A.* 109: 10787–10792.

75 Hou, Z., Zhao, K., Zong, Y., and Mason, T.G. (2019). Phase behavior of two-dimensional Brownian systems of corner-rounded hexagons. *Phy. Rev. Mater.* 3 (1): 015601.

76 Hou, Z., Zong, Y., Sun, Z. et al. (2020). Emergent tetratic order in crowded systems of rotationally asymmetric hard kite particles. *Nat. Commun.* 11 (1): 2064.

77 Wolters, J.R., Avvisati, G., Hagemans, F. et al. (2015). Self-assembly of "Mickey Mouse" shaped colloids into tube-like structures: experiments and simulations. *Soft Matter* 11 (6): 1067–1077.

78 Macfarlane, R.J., Lee, B., Jones, M.R. et al. (2011). Nanoparticle superlattice engineering with DNA. *Science* 334 (6053): 204–208.

79 Senesi, A.J., Eichelsdoerfer, D.J., Macfarlane, R.J. et al. (2013). Stepwise evolution of DNA-programmable nanoparticle superlattices. *Angew. Chem. Int. Ed.* 52 (26): 6624–6628.

80 Nykypanchuk, D., Maye, M.M., van der Lelie, D., and Gang, O. (2008). DNA-guided crystallization of colloidal nanoparticles. *Nature* 451 (7178): 549–552.

81 Park, S.Y., Lytton-Jean, A.K.R., Lee, B. et al. (2008). DNA-programmable nanoparticle crystallization. *Nature* 451 (7178): 553–556.

82 Di Michele, L., Varrato, F., Kotar, J. et al. (2013). Multistep kinetic self-assembly of DNA-coated colloids. *Nat. Commun.* 4: 2007.

83 Ben Zion, M.Y., He, X.J., Maass, C.C. et al. (2017). Self-assembled three-dimensional chiral colloidal architecture. *Science* 358 (6363): 633–636.

84 Oh, J.S., Lee, S., Glotzer, S.C. et al. (2019). Colloidal fibers and rings by cooperative assembly. *Nat. Commun.* 10: 3936.

85 Patra, N. and Tkachenko, A.V. (2018). Programmable self-assembly of diamond polymorphs from chromatic patchy particles. *Phys. Rev. E* 98 (3): 032611.

86 Oh, J.S., Yi, G.R., and Pine, D.J. (2020). Reconfigurable transitions between one- and two-dimensional structures with bifunctional dna-coated Janus colloids. *ACS Nano* 14 (11): 15786–15792.

87 Ma, F.D., Wang, S.J., Smith, L., and Wu, N. (2012). Two-dimensional assembly of symmetric colloidal dimers under electric fields. *Adv. Funct. Mater.* 22 (20): 4334–4343.

88 Cheng, Z.F., Luo, F.H., Zhang, Z.X., and Ma, Y.Q. (2013). Syntheses and applications of concave and convex colloids with precisely controlled shapes. *Soft Matter* 9 (47): 11392–11397.

89 Shields, C.W., Zhu, S., Yang, Y. et al. (2013). Field-directed assembly of patchy anisotropic microparticles with defined shape. *Soft Matter* 9 (38): 9219–9229.

90 Song, P.C., Wang, Y.F., Wang, Y. et al. (2015). Patchy particle packing under electric fields. *J. Am. Chem. Soc.* 137 (8): 3069–3075.

91 Pal, A., Zinn, T., Kamal, M.A. et al. (2018). Anomalous dynamics of magnetic anisotropic colloids studied by XPCS. *Small* 14 (46): e1802233.

92 Zerrouki, D., Baudry, J., Pine, D. et al. (2008). Chiral colloidal clusters. *Nature* 455 (7211): 380–382.

93 Yan, J., Bloom, M., Bae, S.C. et al. (2012). Linking synchronization to self-assembly using magnetic Janus colloids. *Nature* 491 (7425): 578–581.

94 Gao, Y., Romano, F., Dullens, R.P.A. et al. (2018). Directed self-assembly into low-density colloidal liquid crystal phases. *Phy. Rev. Mater.* 2 (1): 015601.

95 Yan, J., Chaudhary, K., Bae, S.C. et al. (2013). Colloidal ribbons and rings from Janus magnetic rods. *Nat. Commun.* 4: 1516.

96 Yan, J., Bae, S.C., and Granick, S. (2015). Rotating crystals of magnetic Janus colloids. *Soft Matter* 11 (1): 147–153.

97 Yan, J., Bae, S.C., and Granick, S. (2015). Colloidal superstructures programmed into magnetic Janus particles. *Adv. Mater.* 27 (5): 874–879.

98 Ebbens, S.J. and Gregory, D.A. (2018). Catalytic Janus colloids: controlling trajectories of chemical microswimmers. *Acc. Chem. Res.* 51 (9): 1931–1939.

99 Gao, W., Pei, A., Feng, X.M. et al. (2013). Organized self-assembly of Janus micromotors with hydrophobic hemispheres. *J. Am. Chem. Soc.* 135 (3): 998–1001.

100 Zhang, J. and Granick, S. (2016). Natural selection in the colloid world: active chiral spirals. *Faraday Discuss.* 191: 35–46.

101 Ahmed, S., Gentekos, D.T., Fink, C.A., and Mallouk, T.E. (2014). Self-assembly of nanorod motors into geometrically regular multimers and their propulsion by ultrasound. *ACS Nano* 8 (11): 11053–11060.

102 Palacci, J., Sacanna, S., Steinberg, A.P. et al. (2013). Living crystals of light-activated colloidal surfers. *Science* 339 (6122): 936–940.

103 Lin, Z.H., Si, T.Y., Wu, Z.G. et al. (2017). Light-activated active colloid ribbons. *Angew. Chem. Int. Ed.* 56 (43): 13517–13520.

104 Yan, J., Han, M., Zhang, J. et al. (2016). Reconfiguring active particles by electrostatic imbalance. *Nat. Mater.* 15 (10): 1095–1099.

105 Zhang, J., Yan, J., and Granick, S. (2016). Directed self-assembly pathways of active colloidal clusters. *Angew. Chem. Int. Ed. Engl.* 55 (17): 5166–5169.

106 Wang, W., Castro, L.A., Hoyos, M., and Mallouk, T.E. (2012). Autonomous motion of metallic microrods propelled by ultrasound. *ACS Nano* 6 (7): 6122–6132.

107 Lagerwall, J.P.F. and Scalia, G. (2012). A new era for liquid crystal research: applications of liquid crystals in soft matter nano-, bio- and microtechnology. *Curr. Appl. Phys.* 12 (6): 1387–1412.

108 Frenkel, D., Lekkerkerker, H.N.W., and Stroobants, A. (1988). Thermodynamic stability of a smectic phase in a system of hard rods. *Nature* 332: 822–823.

109 Maeda, H. and Maeda, Y. (2003). Liquid crystal formation in suspensions of hard rodlike colloidal particles: direct observation of particle arrangement and self-ordering behavior. *Phys. Rev. Lett.* 90 (1): 018303.

110 Wohrle, T., Wurzbach, I., Kirres, J. et al. (2016). Discotic liquid crystals. *Chem. Rev.* 116 (3): 1139–1241.

111 Takezoe, H. (2012). Spontaneous achiral symmetry breaking in liquid crystalline phases. In: *Liquid Crystals: Materials Design and Self-assembly* (ed. C. Tschierske), 303–330. Berlin, Heidelberg: Springer Berlin Heidelberg.

112 Dan, B., Behabtu, N., Martinez, A. et al. (2011). Liquid crystals of aqueous, giant graphene oxide flakes. *Soft Matter* 7 (23): 11154–11159.

113 Kim, J.E., Han, T.H., Lee, S.H. et al. (2011). Graphene oxide liquid crystals. *Angew. Chem. Int. Ed.* 50 (13): 3043–3047.

114 Dessombz, A., Chiche, D., Davidson, P. et al. (2007). Design of liquid-crystalline aqueous suspensions of rutile nanorods: evidence of anisotropic photocatalytic properties. *J. Am. Chem. Soc.* 129: 5904–5909.

115 Wang, M., He, L., Zorba, S., and Yin, Y. (2014). Magnetically actuated liquid crystals. *Nano Lett.* 14 (7): 3966–3971.

116 Chen, M., Shinde, A., Wang, L. et al. (2019). Rainbows in a vial: controlled assembly of 2D colloids in two perpendicular external fields. *2D Mater.* 6 (2): 025031.

117 Wang, X., Zhao, D., Diaz, A. et al. (2014). Thermo-sensitive discotic colloidal liquid crystals. *Soft Matter* 10 (39): 7692–7695.

118 Li, L.-s., Walda, J., Manna, L., and Alivisatos, A.P. (2002). Semiconductor nanorod liquid crystals. *Nano Lett.* 2: 557–560.

119 Brus, L. (1991). Quantum crystallites and nonlinear optics. *Appl. Phys. A Mater. Sci. Process.* 53: 465–474.

120 Ho, K.M., Chan, C.T., and Soukoulis, C.M. (1990). Existence of a photonic gap in periodic dielectric structures. *Phys. Rev. Lett.* 65 (25): 3152–3155.

121 He, M., Gales, J.P., Ducrot, É. et al. (2020). Colloidal diamond. *Nature* 585 (7826): 524–529.

122 Velikov, K.P., van Dillen, T., Polman, A., and van Blaaderen, A. (2002). Photonic crystals of shape-anisotropic colloidal particles. *Appl. Phys. Lett.* 81 (5): 838–840.

123 Lu, Y., Yin, Y., Li, Z.-Y., and Xia, Y. (2002). Colloidal crystals made of polystyrene spheroids: fabrication and structural/optical characterization. *Langmuir* 18: 7722–7727.

124 Ding, T., Song, K., Clays, K., and Tung, C.-H. (2009). Fabrication of 3D photonic crystals of ellipsoids: convective self-assembly in magnetic field. *Adv. Mater.* 21 (19): 1936–1940.

125 Ding, T., Song, K., Clays, K., and Tung, C.-H. (2010). Fabrication and multiangular optical characterization of ellipsoidal photonic crystal. *J. Nanosci. Nanotechnol.* 10 (11): 7571–7573.

126 Wang, M., He, L., Xu, W. et al. (2015). Magnetic assembly and field-tuning of ellipsoidal-nanoparticle-based colloidal photonic crystals. *Angew. Chem. Int. Ed.* 54 (24): 7077–7081.

127 Forster, J.D., Park, O.J.-G., Mittal, O.M. et al. (2011). Assembly of optical-scale dumbbells into dense photonic crystals. *ACS Nano* 5 (8): 6695–6700.

128 Haus, J.W., Sözüer, H.S., and Inguva, R. (1992). Photonic bands. *J. Mod. Opt.* 39 (10): 1991–2005.

129 Hosein, I.D. and Liddell, C.M. (2007). Convectively assembled nonspherical mushroom cap-based colloidal crystals. *Langmuir* 23 (17): 8810–8814.

130 Kunkel, R. and Manahan, S.E. (1973). Atomic absorption analysis of strong heavy metal chelating agents in water and waste water. *Anal. Chem.* 45 (8): 1465–1468.

131 Bings, N.H., Bogaerts, A., and Broekaert, J.A.C. (2010). Atomic spectroscopy: a review. *Anal. Chem.* 82 (12): 4653–4681.

132 Wilson, D., del Valle, M., Alegret, S. et al. (2012). Potentiometric electronic tongue-flow injection analysis system for the monitoring of heavy metal biosorption processes. *Talanta* 93: 285–292.

133 Singh, S.K., Nandi, R., Mishra, K. et al. (2016). Liquid crystal based sensor system for the real time detection of mercuric ions in water using amphiphilic dithiocarbamate. *Sensors Actuators B Chem.* 226: 381–387.

134 Yang, S., Wu, C., Tan, H. et al. (2013). Label-free liquid crystal biosensor based on specific oligonucleotide probes for heavy metal ions. *Anal. Chem.* 85 (1): 14–18.

135 Yeo, D.-H. and Park, S.-Y. (2019). Liquid-crystal-based biosensor for detecting Ca^{2+} in human saliva. *J. Ind. Eng. Chem.* 74: 193–198.

136 Hu, Q.Z. and Jang, C.H. (2011). Liquid crystal-based sensors for the detection of heavy metals using surface-immobilized urease. *Colloids Surf. B Biointerfaces* 88 (2): 622–626.

137 Luan, C., Luan, H., and Luo, D. (2020). Application and technique of liquid crystal-based biosensors. *Micromachines (Basel)* 11 (2): 176.

138 Li, Y., Bai, H., Li, C., and Shi, G. (2011). Colorimetric assays for acetylcholinesterase activity and inhibitor screening based on the disassembly-assembly of a water-soluble polythiophene derivative. *ACS Appl. Mater. Interfaces* 3 (4): 1306–1310.

139 Wang, Y., Hu, Q., Guo, Y., and Yu, L. (2015). A cationic surfactant-decorated liquid crystal sensing platform for simple and sensitive detection of acetylcholinesterase and its inhibitor. *Biosens. Bioelectron.* 72: 25–30.

140 Zhao, D., Peng, Y., Xu, L. et al. (2015). Liquid-crystal biosensor based on nickel-nanosphere-induced homeotropic alignment for the amplified detection of thrombin. *ACS Appl. Mater. Interfaces* 7 (42): 23418–23422.

141 Zhang, D., Tong, J., and Xia, B. (2014). Humidity-sensing properties of chemically reduced graphene oxide/polymer nanocomposite film sensor based on layer-by-layer nano self-assembly. *Sensors Actuators B Chem.* 197: 66–72.

142 Zhang, B. and Cui, T. (2011). An ultrasensitive and low-cost graphene sensor based on layer-by-layer nano self-assembly. *Appl. Phys. Lett.* 98 (7): 073116.

143 Yu, B., Kuang, D., Liu, S. et al. (2014). Template-assisted self-assembly method to prepare three-dimensional reduced graphene oxide for dopamine sensing. *Sensors Actuators B Chem.* 205: 120–126.

144 Zafiu, C., Hussain, Z., Kupcu, S. et al. (2016). Liquid crystals as optical amplifiers for bacterial detection. *Biosens. Bioelectron.* 80: 161–170.

145 Zhang, D., Jiang, C., Liu, J., and Cao, Y. (2017). Carbon monoxide gas sensing at room temperature using copper oxide-decorated graphene hybrid nanocomposite prepared by layer-by-layer self-assembly. *Sensors Actuators B Chem.* 247: 875–882.

146 Wang, H., Wang, X., Li, X., and Dai, H. (2009). Chemical self-assembly of graphene sheets. *Nano Res.* 2: 336–342.

147 Nikoobakht, B. and El-Sayed, M.A. (2003). Preparation and growth mechanism of gold nanorods (NRs) using seed-mediated growth method. *Chem. Mater.* 15: 1957–1962.

148 Akinwande, D., Petrone, N., and Hone, J. (2014). Two-dimensional flexible nanoelectronics. *Nat. Commun.* 5: 5678.

149 Zang, Y., Zhang, F., Di, C.-a., and Zhu, D. (2015). Advances of flexible pressure sensors toward artificial intelligence and health care applications. *Mater. Horiz.* 2 (2): 140–156.

150 Li, X., Yang, T., Yang, Y. et al. (2016). Large-area ultrathin graphene films by single-step marangoni self-assembly for highly sensitive strain sensing application. *Adv. Funct. Mater.* 26 (9): 1322–1329.

151 Nitta, N., Wu, F., Lee, J.T., and Yushin, G. (2015). Li-ion battery materials: present and future. *Mater. Today* 18 (5): 252–264.

152 Wang, Y., Li, D., Liu, Y., and Zhang, J. (2016). Self-assembled 3D $ZnSnO_3$ hollow cubes@reduced graphene oxide aerogels as high capacity anode materials for lithium-ion batteries. *Electrochim. Acta* 203: 84–90.

153 Liang, J., Yu, X.-Y., Zhou, H. et al. (2014). Bowl-like SnO_2@carbon hollow particles as an advanced anode material for lithium-ion batteries. *Angew. Chem. Int. Ed.* 53 (47): 12803–12807.

154 Zhao, C., Wang, X., Kong, J. et al. (2016). Self-assembly-induced alternately stacked single-layer MoS_2 and N-doped graphene: a novel van der Waals heterostructure for lithium-ion batteries. *ACS Appl. Mater. Interfaces* 8 (3): 2372–2379.

155 Liu, X., Chao, D., Su, D. et al. (2017). Graphene nanowires anchored to 3D graphene foam via self-assembly for high performance Li and Na ion storage. *Nano Energy* 37: 108–117.

156 Lv, L., Lu, Q., Ning, Y. et al. (2015). Self-assembled TiO_2 nanorods as electron extraction layer for high-performance inverted polymer solar cells. *Chem. Mater.* 27 (1): 44–52.

157 Pei, F., An, T., Zang, J. et al. (2016). From hollow carbon spheres to N-doped hollow porous carbon bowls: rational design of hollow carbon host for Li-S batteries. *Adv. Energy Mater.* 6 (8): 1502539.

158 Wang, D., Kou, R., Choi, D. et al. (2010). Ternary self-assembly of ordered metal oxide-graphene nanocomposites for electrochemical energy storage. *ACS Nano* 4 (3): 1587–1595.

159 Zhu, Y., Liu, W., Zhang, X. et al. (2013). Directing silicon–graphene self-assembly as a core/shell anode for high-performance lithium-ion batteries. *Langmuir* 29 (2): 744–749.

160 Tang, H., Zhang, Y.J., Xiong, Q.Q. et al. (2015). Self-assembly silicon/porous reduced graphene oxide composite film as a binder-free and flexible anode for lithium-ion batteries. *Electrochim. Acta* 156: 86–93.

161 Sun, Y., Hu, X., Luo, W., and Huang, Y. (2011). Hierarchical self-assembly of $Mn_2Mo_3O_8$–graphene nanostructures and their enhanced lithium-storage properties. *J. Mater. Chem.* 21 (43): 17229–17235.

162 Chao, Y., Jalili, R., Ge, Y. et al. (2017). Self-assembly of flexible free-standing 3D porous MoS_2-reduced graphene oxide structure for high-performance lithium-ion batteries. *Adv. Funct. Mater.* 27 (22): 1700234.

163 Kim, W., Ryu, W., Han, D. et al. (2014). Fabrication of graphene embedded $LiFePO_4$ using a catalyst assisted self assembly method as a cathode material for high power lithium-ion batteries. *ACS Appl. Mater. Interfaces* 6 (7): 4731–4736.

164 Huang, Y., Huang, X.-l., Lian, J.-s. et al. (2012). Self-assembly of ultrathin porous NiO nanosheets/graphene hierarchical structure for high-capacity and high-rate lithium storage. *J. Mater. Chem.* 22 (7): 2844–2847.

165 Zhang, Z., Xiao, F., Guo, Y. et al. (2013). One-pot self-assembled three-dimensional TiO_2-graphene hydrogel with improved adsorption capacities and photocatalytic and electrochemical activities. *ACS Appl. Mater. Interfaces* 5 (6): 2227–2233.

166 Love, J.C., Gates, B.D., Wolfe, D.B. et al. (2002). Fabrication and wetting properties of metallic half-shells with submicron diameters. *Nano Lett.* 2: 891–894.

167 Li, F., Tu, Y., Hu, J. et al. (2015). Fabrication of fluorinated raspberry particles and their use as building blocks for the construction of superhydrophobic films to mimic the wettabilities from lotus leaves to rose petals. *Polym. Chem.* 6 (37): 6746–6760.

168 Yi, D., Xu, C., Tang, R. et al. (2016). Synthesis of discrete alkyl-silica hybrid nanowires and their assembly into nanostructured superhydrophobic membranes. *Angew. Chem. Int. Ed.* 55 (29): 8375–8380.

169 Kim, J.-W., Lee, D., Shum, H.C., and Weitz, D.A. (2008). Colloid surfactants for emulsion stabilization. *Adv. Mater.* 20 (17): 3239–3243.

170 Schmitt, V., Destribats, M., and Backov, R. (2014). Colloidal particles as liquid dispersion stabilizer: pickering emulsions and materials thereof. *C. R. Phys.* 15 (8-9): 761–774.

171 Kim, H., Cho, J., Cho, J. et al. (2017). Magnetic-patchy Janus colloid surfactants for reversible recovery of pickering emulsions. *ACS Appl. Mater. Interfaces* 10 (1): 1408–1414.

172 Wei, D., Ge, L., Lu, S. et al. (2017). Janus particles templated by Janus emulsions and application as a pickering emulsifier. *Langmuir* 33 (23): 5819–5828.

173 Ong, W.-J., Tan, L.-L., Chai, S.-P. et al. (2015). Surface charge modification via protonation of graphitic carbon nitride (g-C_3N_4) for electrostatic self-assembly construction of 2D/2D reduced graphene oxide (rGO)/g-C_3N_4 nanostructures toward enhanced photocatalytic reduction of carbon dioxide to methane. *Nano Energy* 13: 757–770.

174 Adamsom, A. and Gast, A. (1997). *Physical Chemistry of Surfaces*. New York: Wiley.

175 Tu, F. and Lee, D. (2014). Shape-changing and amphiphilicity-reversing Janus particles with pH-responsive surfactant properties. *J. Am. Chem. Soc.* 136 (28): 9999–10006.

176 Yang, Y., Fang, Z., Chen, X. et al. (2017). An overview of pickering emulsions: solid-particle materials, classification, morphology, and applications. *Front. Pharmacol.* 8: 287.

177 Jiang, H., Sheng, Y., and Ngai, T. (2020). Pickering emulsions: versatility of colloidal particles and recent applications. *Curr. Opin. Colloid Interface Sci.* 49: 49.

178 Nonomura, Y., Komura, S., and Tsujii, K. (2004). Adsorption of disk-shaped janus beads at liquid–liquid interfaces. *Langmuir* 20 (26): 11821–11823.

179 Mejia, A.F., Diaz, A., Pullela, S. et al. (2012). Pickering emulsions stabilized by amphiphilic nano-sheets. *Soft Matter* 8 (40): 10245–10253.

180 Yang, J., Wang, J., Liu, Y. et al. (2020). Stimuli-responsive Janus mesoporous nanosheets towards robust interfacial emulsification and catalysis. *Mater. Horiz.* https://doi.org/10.1039/d0mh01260b.

3

Self-assembly Enabling Materials Nanoarchitectonics

Katsuhiko Ariga[1,2]

[1] National Institute for Materials Science (NIMS), World Premier International (WPI) Research Center for Materials Nanoarchitectonics (MANA), 1-1 Namiki, Tsukuba 305-0044, Japan
[2] The University of Tokyo, Graduate School of Frontier Sciences, 5-1-5 Kashiwanoha, Kashiwa, Chiba 277-8561, Japan

3.1 Introduction

Scientists and engineers today are challenged to solve many problems, including how to save resources, how to generate energy, how to reduce emissions, how to protect the environment, and how to process information efficiently. All of these problems can be solved through a common concept – the control of NANO and its associated functional structures. The development of materials and functional systems that are rationally manufactured with structural precision on the nanometer scale saves resources, generates energy, reduces emissions, protects the environment, and processes information efficiently. In the coming days, human society will be sufficiently supported by science and technology capable of regulating structures at the nanoscale. The relevant methodology was initiated under the name of nanotechnology and has now evolved under the name of nanoarchitectonics.

NANO is definitely an important keyword in today's science and technology. Especially in fields of application, nanotechnology becomes an extremely influential magic concept, with which we can achieve everything. It is said that this term, nanotechnology, was originally proposed in Feynman's prediction [1–3]. It anticipated enormous scientific and technological possibilities in nanoscale spaces. Later, the technological possibilities were re-evaluated, as evidenced by a 2001 statement by former US President Bill Clinton for the promotion of nanotechnology, called the National Nanotechnology Initiative. This historical flow is also supported by the innovation of various nanoscale object observation and evaluation tools, such as the invention of scanning tunneling microscopy (STM) and nanofabrication techniques.

Another important proposal was made historically by Eric Drexler [4]. In contrast to precise fabrication approaches, he proposed in his book entitled *Nanosystems: Molecular Machinery, Manufacturing, and Computation*. Indeed, similar bottom-up concepts have been performed in a field of chemistry, the supramolecular chemistry,

Functional Materials from Colloidal Self-assembly, First Edition. Edited by Qingfeng Yan and George Zhao.

which manipulates organized structures resulting from molecular assemblies and creates new properties beyond single components. Not limited to chemistry, the importance of molecular assembly is also recognized in biology. Many sophisticated functions in biological systems, including photosynthesis and signal transduction, are based on well-designed arrangements of biomolecules and related small molecules. Furthermore, the artificial assembly and organization of biomolecules has been extensively studied as seen in the layer-by-layer (LbL) assemblies of biomolecules and the programmed organization of DNA origami.

In addition to historical efforts in organic synthesis and materials fabrication, much research has been conducted on functional optimization of materials involving structure adjustment at the atomic/molecular and nanometer scales, and this research has been accompanied by corresponding developments in observational techniques and instrumentation at these length scales. Although it is widely believed that nanotechnology plays a major role in the development of materials involving nanoscale structures, nanotechnology also offers significant advantages in the development of nanoscale observation and fabrication techniques, such as a better understanding of new nanoscale phenomena and the physical principles underlying them. The construction of functional materials from nanoscale units requires significant contributions from fields unrelated to nanotechnology, such as supramolecular chemistry with self-organization, materials fabrication, and biotechnology. Therefore, obtaining materials from nanometric units in the nanoscale regime should follow a new concept, which includes the agglomeration of nanotechnology concepts with the aforementioned diverse research disciplines. The resulting concept of "nanoarchitectonics" [5–7].

The emergence of the concept of nanoarchitectonics can be seen as an inevitable event in the development of science and technology, and its historical background can be traced in scientific discourse. The importance of the concept of architectonics in nanoscale science was first proposed in 1999 in a paper by Heath and colleagues (University of California Los Angeles, UCLA) entitled "Architectonic Quantum Dot Solids" by Heath and coworkers [8]. Subsequently, a dedicated research center, Functional Engineered Nano Architectonics (FENA), was established at UCLA in 2003. In the same year, Stefan Hecht of the Freie Universität Berlin, Germany, published a paper entitled "Welding, Organizing, and Planting Organic Molecules on Substrate Surfaces: Promising Approaches Towards Nanoarchitectonics from the Bottom Up" [9], in which the term "nanoarchitectonics" is directly mentioned in the title of an international scientific journal for the first time.

Three years before these events, in 2000, Masakazu Aono organized the 1st International Symposium on Nanoarchitectonics Using Suprainteractions in Tsukuba, Japan. This was probably the first time that the term "nanoarchitecture" was used in the scientific community [10–12]. Masakazu Aono, the founder of the nanoarchitecture concept, initiated the establishment of the World Premier International Research Center for Materials Nanoarchitectonics (WPI-MANA) at the National Institute for Materials Science (NIMS) in Tsukuba in 2007. In the same city, Toshimi Shimizu of the National Institute of Advanced Industrial Science and Technology (AIST) also initiated a research center, the Interfacial Nanoarchitectonics. Thus, in

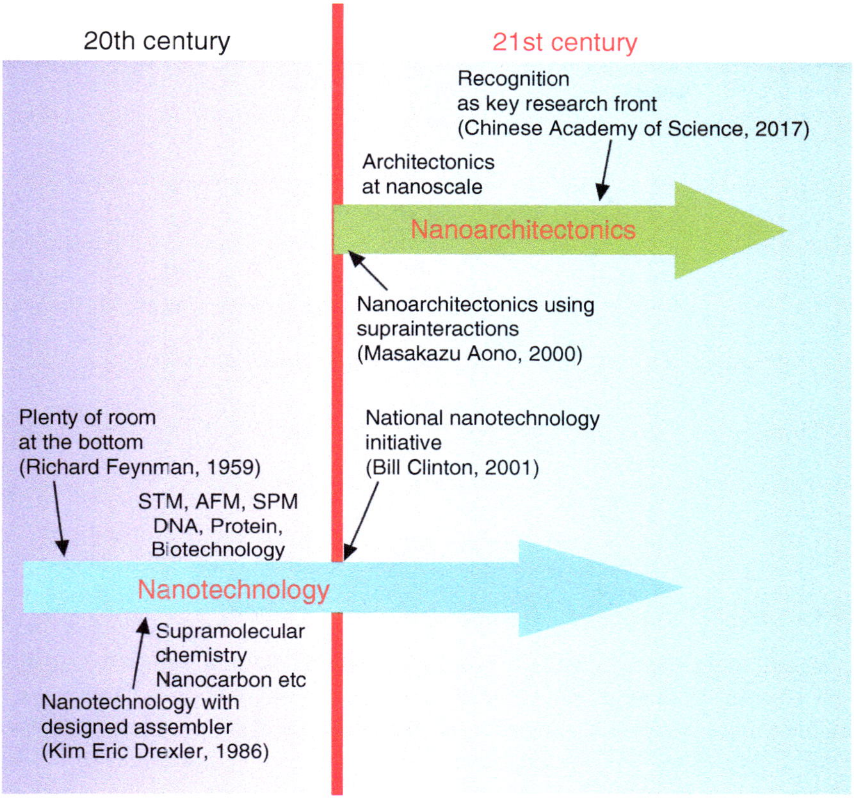

Figure 3.1 Historical backgrounds of nanotechnology and nanoarchitectonics.

the years around 2000, the concept of nanostructuring science was activated in countries around the world. These historical backgrounds are summarized in Figure 3.1.

According to Masakazu Aono, the guidelines of nanoeconomics strategies can be summarized as follows: (i) the organization of unitary nanoscale structures leads to the creation of reliable materials and systems, and some unavoidable uncertainties in nanoscale phenomena must be included in a balanced harmonization; (ii) interactions between nanometer components are often more important than the identities of individual nanocomponents for the creation of novel functionalities; and (iii) unexpected functions may arise from the assembly or organization of a large number of nanoscale components. Nanoarchitectonics constructions can be realized by known strategies, such as atomic/molecular manipulation, chemical synthesis, chemical nanomanipulation, field-induced material control and self-assembly, and self-organization (Figure 3.2) [13, 14]. However, in many cases, uncontrollable or unexpected perturbations arising from complex molecular interactions, thermodynamic perturbations, statistical uncertainties, and quantum effects cannot be avoided. The harmonization of these mostly obstructive effects is one of the important key [15].

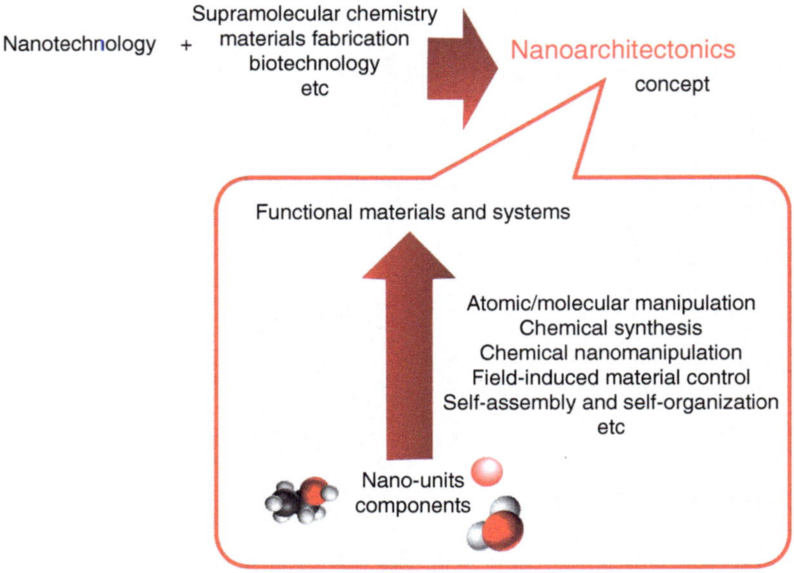

Figure 3.2 Outline of nanoarchitectonics processes.

In general, the nanoarchitecture concept represents a common standard strategy for the creation of materials and systems in many research areas. Therefore, it has been used in a wide range of research areas and applications, such as materials production [16–19], materials fabrication and organization [20–23], supramolecular assemblies [24], physical devices and systems [25–29], sensing [30–33], energy-related applications [34–37], environmental strategies [38–41], and biological and biomedical applications [42–49].

As an analogous process, the chemical self-assembly has been widely studied, particularly with regard to supramolecular concepts. The concept of self-assembly is not limited to conventional supramolecular chemistry and can also be applied to the formation of structures involving assembly processes of biomolecules and inorganic substances. In many cases, self-assembly processes are based on simple equilibria and displacements without energy. The formation of asymmetric or hierarchical structures by simple self-assembly processes is generally not favored. On the other hand, biological systems often use energy from self-organization processes far from equilibrium, which is advantageous for the construction of asymmetric, heterogeneous, and/or hierarchical structures. Therefore, the inclusion of non-balance and irreversibility aspects to conventional self-assembly concepts would be beneficial for the manufacture of very advanced functional systems as artificial materials.

According to this concept, functional materials and systems are constructed through a combination of atom and molecule manipulation, self-assembly, field-controlled organization, and nano-/microfabrication. It is similar to the preceding strategy of introducing nonequilibrium processes into self-assembly for the fabrication of advanced functional structures. Therefore, the nanoarchitectonics concept represents the next generation of methodology for preparing functional structures beyond traditional self-assembly [50].

The aforementioned nanoarchitectonics concept is an effective means for the construction of advanced material structures with asymmetric and hierarchical structural motifs as a method beyond self-assembly. Although the nanoarchitectonics concept has only recently been introduced, its essence has already emerged in many research papers and can be recognized as a modification of the self-assembly concepts. The reanalysis and re-evaluation of this research through the lens of the nanoarchitectonic concept is therefore important as a way to bridge the evolutionary process from "self-assembly" to "nanoarchitectonics."

In this chapter, developments in self-assembly research approaches based on nanoarchitectonics concept are explained. The typical two categories of nanoarchitectonics, (i) fullerene assemblies at liquid interface and (ii) assemblies based on LbL adsorption, are exemplified.

3.2 Fullerene Nanoarchitectonics

The basic strategies for fullerene nanoarchitectonics with assembly and shape changes are briefly explained in this section. One of the key concepts in these strategies is interface. Nanoarchitectonics processes that involve self-assembly and dynamic functions at interfaces have several specific features such as limited motional freedom, anisotropic structures, highly amplified molecular interaction, and coupling of macroscopic dynamic motion and molecular functions. Recent examples also demonstrate the importance of interfacial environments for various properties and functions, as evidenced by the control of surface domains by self-assembly of semi-fluorinated alkanes and related molecules [51], interface-regulated photoinduced motion of molecular arrays [52], photocatalytic conversion of organic compounds on surface complexation [53], heterogeneous low-temperature catalytic reactions on surface protons [54], and interfacial water regulation for function design of polymeric biomaterials [55].

For fullerene nanoarchitectonics, the liquid–liquid interfacial precipitation process is used in particular to fabricate self-assembled structures with specific shapes [56]. Although several methods have been described for fabricating fullerene-based composites, including slow evaporation of fullerene solutions, template synthesis, and vapor deposition, the liquid–liquid interfacial precipitation method is versatile in fabricating fullerene assemblies with controlled dimensions from nano- to microscale. As pioneers in fullerene assembly by the liquid–liquid interfacial precipitation method, Miyazawa and colleagues have demonstrated mostly one-dimensional whiskers, rods, and tubes with well-controlled structural dimensions [57, 58]. Not limited to typical one-dimensional structures, the liquid–liquid interfacial precipitation method can be extended to the fabrication of two-dimensional nanosheets, three-dimensional microcubes, ellipsoidal structures, and their modified structures.

The liquid–liquid interfacial precipitation method is based on the difference in solubility of fullerene molecules when they come in contact with two solvents. For example, fullerene molecules are first dissolved in a good solvent (with higher solubility), and then a poor solvent (with lower solubility) is quietly added to form a clear interface of these two liquids. This process is usually carried out under static, vibration-free conditions to produce fullerene assemblies with uniform shapes. The shapes of the resulting fullerene assemblies are selected by a combination of good and bad solvents. While the static liquid–liquid interfacial precipitation processes mentioned earlier are quite time-consuming (up to several days), the dynamic liquid–liquid interfacial precipitation method described in the following text uses short duration processes to precipitate fullerene assemblies by stirring processes such as agitation, gentle sonication, and vortexing. The formation of fullerene assemblies occurs quite rapidly, even within a few minutes.

In the interfacial regions of the liquid–liquid interfacial precipitation processes, the poor solvent diffuses into the good solvent phase, reducing the solubility of fullerene molecules in the interfacial region. This process causes the formation of fullerene clusters (nuclei), and subsequent solvent mixing promotes the growth of the fullerene clusters. The liquid–liquid interfacial precipitation processes involve only simple action parameters, such as solvent combination, fullerene concentration, volume ratio between good and bad solvent, temperature, etc. Adjusting these parameters creates many opportunities to produce assemblies of different shapes and sizes according to our design and sometimes with unexpected surprises.

As discussed in the following sections, exposure of complete fullerene assemblies formed to selected solvents may induce a change in the shape of the assemblies, dissolving the surface and reforming another assembly structure. In addition, selective etching of fullerene assemblies is possible to make holes and channels using certain types of reagents, such as amine derivatives. These processes would make it possible to modify the shape of preformed fullerene assemblies.

A variety of morphologies can be produced from the same molecule depending on the combination of good and poor solvents and various external conditions (Figure 3.3). For example, two-dimensional rhombic disks are obtained from the interface of *tert*-butyl alcohol and toluene; two-dimensional hexagonal sheets are obtained from the interface of isopropyl alcohol and carbon tetrachloride; and two-dimensional structures are obtained from the interface of *tert*-butyl alcohol and toluene, but the shapes are a mixture of various polygons [59]. According to the X-ray diffraction patterns, the C_{60} powder itself has a face-centered cubic lattice crystal system, while the two-dimensional hexagonal sheet has a hexagonal close-packed crystal lattice. From the latter example, we can see that the basic crystal lattice seems to determine the structure of the nanosheet. On the other hand, rhombic and polygonal nanosheets have a mixed structure of face-centered cubic lattice and hexagonal densest lattice.

When these nanosheets of various shapes are exposed to water, selective shape transformation occurs. A hexagonal nanosheet consisting of a single hexagonal closest lattice is a stable structure that does not change its morphology when exposed to water. On the other hand, rhombic and polygonal nanosheets with mixed regular

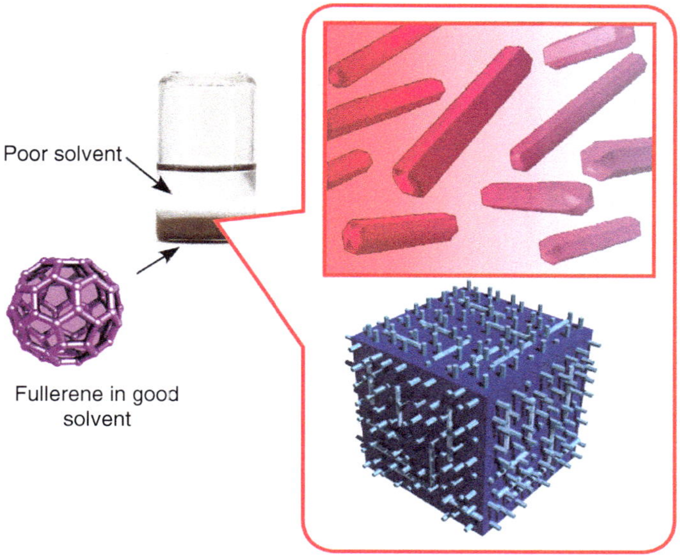

Figure 3.3 Production of various fullerene assembling objects by the liquid–liquid interfacial precipitation processes.

structures roll up and transform into one-dimensional nanorods when exposed to water. In this case, the regular structure is converted from a mixed lattice to a face-centered cubic lattice system. This example shows that the mixed crystalline system is in a less stable state, that the structure can be changed by external stimuli such as exposure to water, and that the change in the crystalline system at the molecular level determines the micrometer-sized structure. In this case, the control of the regular structure at the molecular level gives rise to polymorphic phenomena at the larger structural level, such as crystals and supramolecular assemblies.

It is also possible to fabricate cubic molecular assemblies using ellipsoidal fullerene C_{70} as a starting material (Figure 3.4) [60]. In this case, each face of the cube has a micropocket. This supramolecular assembly is obtained by dissolving ellipsoidal fullerene C_{70} in a good solvent, mesitylene, followed by the rapid addition of a poor solvent, *tert*-butyl alcohol, and allowing the mixture to stand (12 h) at room temperature. The pockets on each side of the cube can be opened and closed. When an excess amount of C_{70} molecules is added to the cubic structure with pockets, a crystalline thin film is formed on top of the pockets, which can be used as a lid to cover the pores of the pockets. When irradiated with a strong electron beam, only the fullerene sheet, which is the lid, is removed, and the pores of the pocket can be regenerated.

This pocket structure has the ability to selectively take in micrometer-sized particles. For example, when comparing the uptake into the pocket of microparticles derived from resin, which are almost the same size, and microparticles of carbon material, which are similar to graphite, it is clear that the latter microparticles of carbon material are better taken up. This can be attributed to the aromatic nature

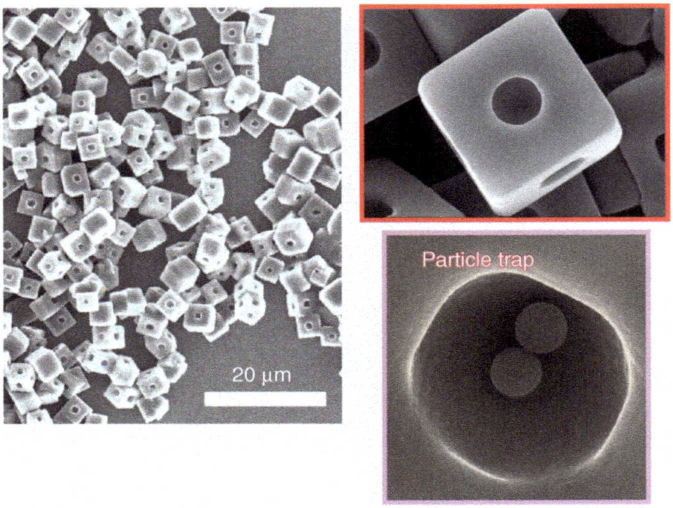

Figure 3.4 C_{70} cube with micropockets for particle trap.

of both the pocketed cubes and the carbon particles and the high affinity between them, resulting in such selective particle recognition. Such carbon particles are highly toxic, and the ability of this cube to selectively capture carbon particles is expected to be useful for PM2.5 detection and removal. The recognition of micrometer-sized substances could also be applied to biological applications such as the removal of bacteria.

Shrestha and coworkers have succeeded in developing a microhorn with a micro-sized hollow structure [61]. The microtubule structure is formed by the spontaneous assembly of a mixture of C_{60} and C_{70} molecules called fullerenes. When the mixture is placed on a solid substrate and waits for the solvent to evaporate, only the stable parts of the structure remain, and a carbon microhorn, a hollow horn structure with a micrometer-sized entrance diameter, is naturally formed (Figure 3.5).

Three model particles, carbon microparticles (model of PM2.5, hydrophobic), silica microparticles (model of hydrophilic bio-particles like virus particles), and polymer microparticles (model of microplastic particles, hydrophobic), were used. The results showed that silica microparticles, a model for viral particles, were taken up several to 10 times more selectively than carbon microparticles or polymer microparticles. It is thought that the weak charge on the surface of the carbon microhorn interacts electrostatically with the hydrophilic and charged silica microparticles and causes their uptake. The size of the microparticles and the size of the pores of the microhorn are exactly the same, and the microparticles are taken up very efficiently, just as bean grains are taken up into the pod of a pea.

The morphology of supramolecular assemblies of simple small molecules and their crystal structures can vary depending on various conditions (polymorphism phenomenon). This does not mean that such assemblies will always arrive at a particular structural form, but rather that they are likely to reach other morphological

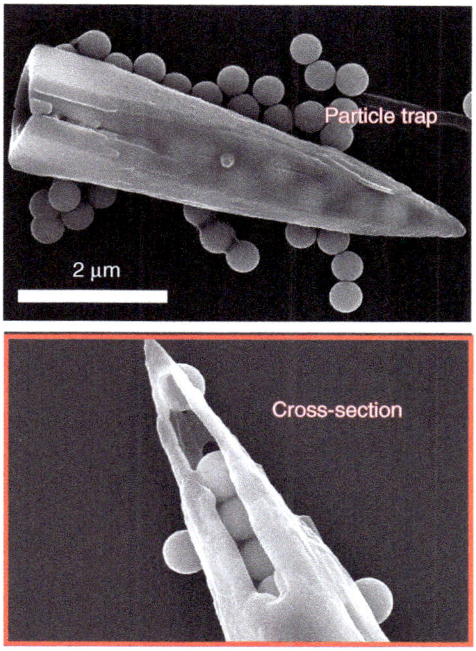

Figure 3.5 Fullerene microhorns with a micro-sized hollow structure for particle trap.

states due to kinetic traps. Furthermore, the various forms of structure are reasonably stable and can maintain their orientation. This situation is different from that of biological materials such as proteins, which ultimately take on a single structure through the sequencing of peptide chains. Supramolecular assemblies of simple small molecules have a degree of freedom and the ability to take on a variety of forms.

On the other hand, what makes biological systems far superior to artificial systems in terms of morphological change is the phenomenon of differentiation, in which a single cell changes into various forms of tissue over time, or the phenomenon of metamorphosis, in which the form of an individual changes. This is due to an elaborate mechanism based on the information programmed in DNA. It is difficult to reproduce this differentiation phenomenon in an artificial supramolecular system that does not contain genetic information such as DNA. In other words, it is easy to show polymorphic phenomena that produce different forms under various conditions, but it is a big challenge to realize differentiation and metamorphosis phenomena that change their forms with time in supramolecular systems.

When supramolecular assemblies are prepared by the liquid–liquid interfacial precipitation method using C_{60} and Ag(I) ions, micrometer-sized cubes are formed (Figure 3.6) [62]. This is a reflection of the basic shape of the organometallic complex ($C_{60}(AgNO_3)_5$). When exposed to lower alcohols, the mere cube mutates into a cube with nanorods. Further contact with alcohol transforms it into an object on a hedgehog. This is because the elution of Ag(I) ions changes the basic structure of the supramolecule from the cubic lattice of the organometallic complex to the one-dimensional nanorod structure of C_{60}. As a result, the micrometer-sized object

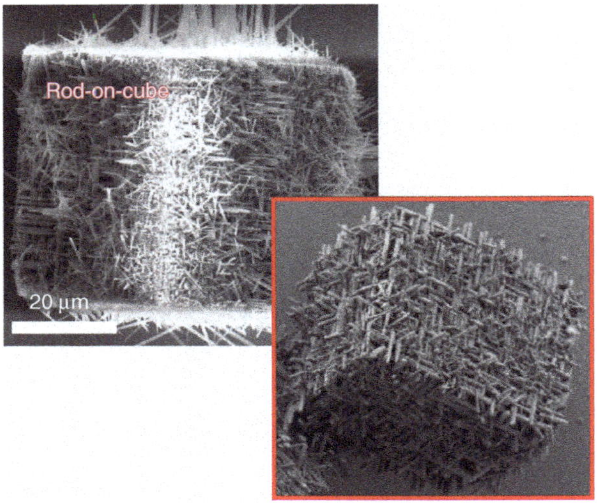

Figure 3.6 Micrometer-sized fullerene cubes with nanorods.

changes from a flat cube to a cube with intricate rods and then to a form with many protruding needles, as if the cell is differentiating into complex forms.

A similar phenomenon has been observed with C_{70} molecular assemblies; the liquid–liquid interfacial precipitation treatment of C_{70} molecules under appropriate conditions yields cubes with smooth surfaces [63]. When treated with alcohol, the surface layer partially dissolves and the aggregate morphology changes into nanorods. As a result, the nanorods elongate like antennae from the surface of the cube. This is similar to the metamorphosis of insects and the differentiation of cells and tissues, in which a simple cube is transformed into a structure with extended antennae.

In order to show the differentiation and metamorphosis phenomena similar to those of living organisms in artificial systems, it is necessary to develop supramolecular assemblies that change their morphology with time as if they were programmed to do so without external stimuli. In the following example, supramolecular assemblies of two fullerene derivatives initially produce egg-like molecular assemblies, from which a tubelike structure as a tail develops spontaneously over time (Figure 3.7) [64]. This morphological change from an egg to a tadpole is realized in an inanimate, nongenetic system at the micrometer level. Supramolecular differentiation, in which a supramolecular aggregate composed of identical components becomes either an egg or a tadpole, and supramolecular transformation, in which an egg transforms into a tadpole, are observed. This phenomenon can be considered to be achieved by the formation of domains by phase separation of the two components in the spherical aggregate at the interface and the elongation of the secondary structure from the domains.

Fullerenes and their derivatives, which are very simple unit molecules, can undergo biological-like morphological changes. Supramolecular assemblies that do not follow a strict structure formation mechanism, such as proteins whose final

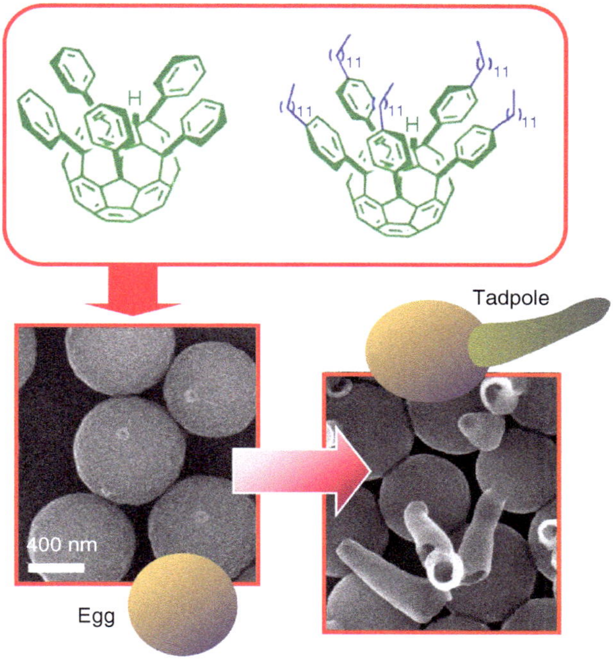

Figure 3.7 Supramolecular assemblies of two fullerene derivatives with shape transitions from egg-like molecular assemblies into tadpole-like structures. Source: Bairi et al. [64] / American chemical society.

three-dimensional structure is determined by the peptide sequence, may be able to take various forms by kinetic trap. By using them successfully, we can recreate the differentiation process of organisms that change their shape over time without genetic programming. These examples demonstrate the high morphogenetic potential of molecular assemblies and suggest that lipid molecules and their ancestors, which are simpler than RNA and DNA, may have played a very important role in the evolution of life.

3.3 Layer-by-Layer Nanoarchitectonics

The LbL assembly method is one of the most popular methods to fabricate layered structures [65–68]. The alternating adsorption method is characterized by a wider range of applicable materials than other thin film preparation methods such as Langmuir–Blodgett (LB) method [69–72]. A variety of polymers (especially polyelectrolytes), biomaterials such as proteins, DNA or viral particles, inorganic colloids and nanomaterials, and molecular assemblies can be stacked. The driving force for adsorption is mainly electrostatic interactions, but hydrogen bonds, coordination bonds, covalent bonds, bio-specific recognition, and supramolecular interactions can also be used. The adsorption process can be carried out easily with tweezers and a beaker in solution, or by spin coating, spray, or other methods. In other words, it is a method that can be used to make thin films of all kinds of materials in a variety of ways through various interactions.

As a strategy for constructing hierarchical structures, we consider the method of fabricating slightly complex nanostructures separately and then stacking them. Since the target of stacking is not necessarily a simple structure, the alternating adsorption method, which can be used to stack a wide range of materials as described earlier, is the most suitable manipulation method for this purpose. This chapter presents an attempt to create a hierarchical structure by stacking pre-synthesized nanostructures using the LbL alternating adsorption method, including the fabrication method and its application.

The use of natural phospholipids and synthetic amphiphilic compounds to create liposomes and vesicles structures with a bimolecular membrane structure has been widely studied as an artificial cell membrane model. The results of this research include the reproduction of bio-like functions in bilayer environments and the development of drug delivery functions such as controlled transport and release of drugs using these vesicles. These can be used as models for single cells, but there have been few attempts at multicellular models in which the cells are assembled or organized. For example, when lipid assemblies such as vesicles are further stacked and organized, the vesicle's endoplasmic reticulum structure may break down and become a mere lipid multilayer.

For this purpose, Katagiri et al. developed an organic–inorganic hybrid vesicle called cerasome, which has an inorganic silica structure on the vesicle surface [73–75]. The term cerasome was coined by combining the word ceramics (e.g. silica) and the word soma (vesicular structure). The cerasome consists of an amphiphilic substance with an alkoxysilanol group on its head dispersed in water to form a bilayer vesicular structure, and at the same time, a silica structure is formed on its surface by hydrolysis and condensation (Figure 3.8). Due to the presence of this silica film structure, the cerasome structure is stabilized and the fusion between the lipid bilayers is suppressed even in the aggregated/aggregated structure, and the individual endoplasmic reticulum structure is maintained.

As the next step, we attempted to laminate the cerasomes. Since the surface of the cerasome has a well-developed silica layer, the surface becomes negatively charged under appropriate pH conditions. Under these conditions, multilayered films of cerasomes can be fabricated on a substrate by alternately LbL adsorbing poly(diallyldimethylammonium chloride) (PDDA), a polycation [75]. The results of analysis of the amount of adsorbed material using a crystal oscillator and other techniques indicate that the cerasome is multilayered while maintaining the morphology of the endoplasmic reticulum. It has also been observed that when anionic vesicles without a silica backbone are adsorbed alternately with PDDA, the vesicles break down and become lipid multilayers. The latter finding indicates that the surface of lipid vesicles should be designed to be reinforced with an inorganic skeleton in order to form hierarchical structures that are stacked while maintaining the vesicular structure.

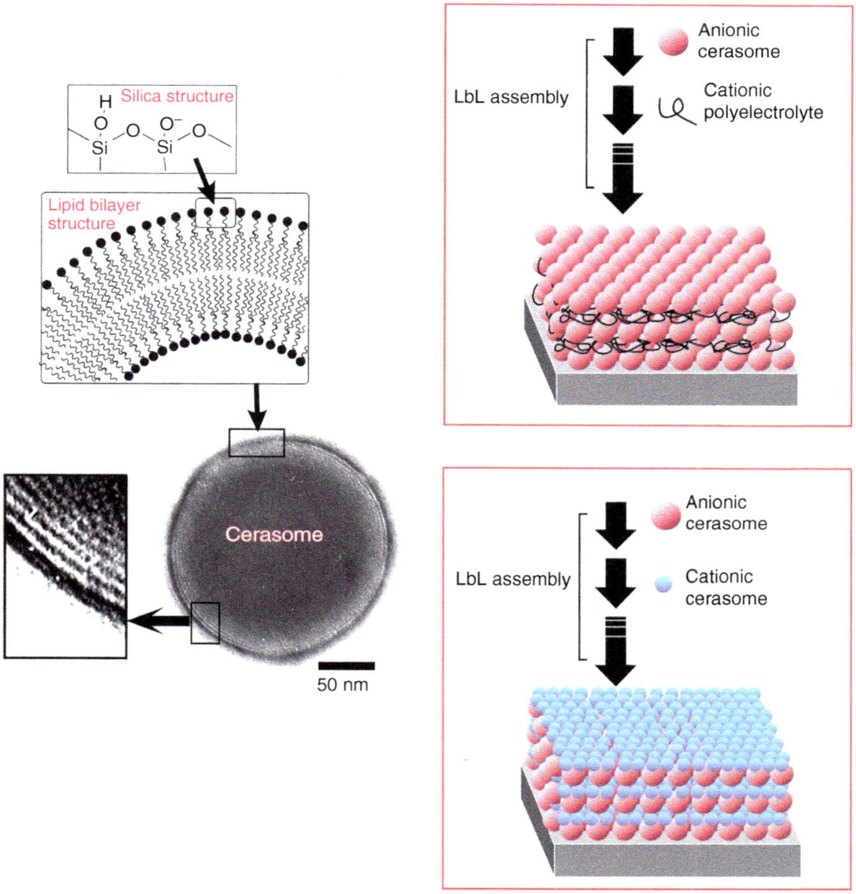

Figure 3.8 Organic–inorganic hybrid vesicles called cerasomes and their LbL assemblies.

It is also possible to fabricate cerasome multilayer films without polymers such as PDDA [76]. Cerasomes are prepared based on the self-assembled structure of amphiphilic compounds with alkoxysilane groups, and depending on the structure of the amphiphilic compounds, various surface states of the cerasomes can be prepared. The amphiphilic compounds that use the negative charge from the dissociation of silanol groups of alkoxysilane groups as hydrophilicity can produce anionic cerasomes. Cationic cerasomes can also be produced by using amphiphilic compounds that contain quaternary ammonium groups. Since these cerasomes have opposite charges on their surface, they can be adsorbed LbL alternately to form multilayer films. The size of the cationic cerasomes (20–100 nm in diameter) is slightly larger than that of the anionic cerasomes (70–300 nm in diameter), and the multilayer formation was observed by atomic force microscopy (AFM).

The preceding examples are of organic–inorganic hybrid vesicles, which are lipid vesicles with improved structural stability that are hierarchically structured by an alternating adsorption method. These make liposomes and vesicles, which were only models for single cells, into models for multicellular tissues. This is expected to be applied to artificial skin and tissue. The hierarchical structure obtained here is not the result of random assembly, but can be intentionally organized layer by layer. Therefore, it is expected to be possible to create functional hierarchical structures that cannot be obtained by other methods, such as multilayer structures containing drugs with different efficacy in each layer and drug delivery in other stages.

In the previous example, we described the preparation of alternatively adsorbed films of cerasome, a structurally stabilized vesicle with a silica surface layer. The key to successful lamination was the structural stabilization of lipid vesicles. If this is the case, we believe that various hierarchical structures can be obtained by alternating adsorption of more stable nanostructures. From this point of view, the next target was to create alternatively adsorbed films of nanostructures of silica itself.

Yu et al. reported a method to synthesize microcapsules of mesoporous silica and mesoporous carbon using zeolite as a template [77]. First, a thin film of mesoporous silica is synthesized around the zeolite using octadecyltrimethoxysilane. In the mesopores, phenol and paraformaldehyde are polymerized and carbonized, and then the silica portion and the zeolite core are dissolved and removed with HF to synthesize mesoporous carbon capsules. Next, silica is made in the pores of the mesoporous carbon, and only the carbon part is selectively removed this time by high-temperature treatment in air to obtain mesoporous silica capsules. The resulting capsules have a content area of $1000 \times 700 \times 300$ nm, and the silica walls have mesopores with an average pore size of 2.2 nm.

The mesoporous silica capsules thus obtained can be stacked by alternating adsorption, since they, like silica nanoparticles, are negatively charged under appropriate pH conditions. In particular, when mixed with silica nanoparticles having the same negative charge and then alternately adsorbed together with polycationic PDDA, a thin film containing an internal capsule structure was obtained [78, 79]. In this structure, the mesoporous silica structure with nanometer-level pores forms a hierarchical structure in which the capsule structure with micrometer-level inner space further forms a thin film structure (Figure 3.9).

This hierarchical thin film has an internal capsule space in which the drug can be incorporated and its release can be controlled by the mesopore structure. As a model experiment to investigate the material release behavior of the fabricated thin film, the hierarchical thin film was fabricated on a quartz crystal microbalance (QCM), which is known as a microbalance with nanogram level sensitivity, and water was absorbed into it. The evaporation behavior of water in air was measured over time. In the case of the alternating adsorption membranes of silica nanoparticles and PDDA without mesoporous silica capsules, the water uptake into the membranes was negligible, while in the case of the alternating adsorption membranes with capsules, the amount of water uptake was as expected from the internal volume of the capsules. When the latter membrane was exposed to air and the evaporation behavior of water was observed, surprising results were obtained. The water evaporation behavior was not smooth, and the water evaporation was observed in several steps. In other words, the evaporation of water was repeated periodically. The same ON/OFF

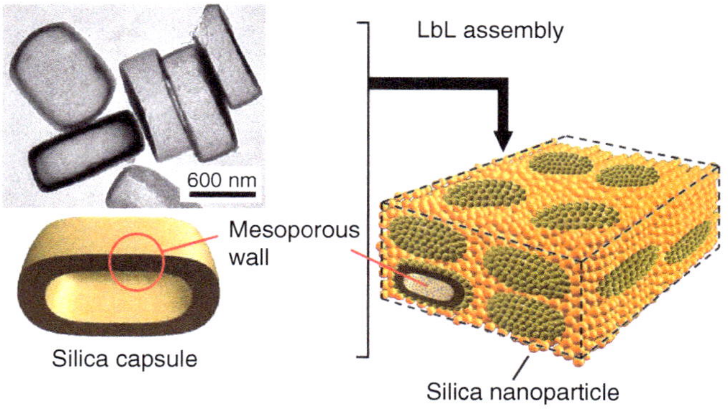

Figure 3.9 LbL assembly of mesoporous silica capsules with the aids of silica nanoparticles and polyelectrolytes.

behavior was observed when water was reintroduced to the evaporated membrane and the evaporation experiment was conducted again. This behavior was different from other material release systems in that it was not caused by external stimuli, but was automatically and spontaneously repeated.

Water exists in two places: the inner capsule space and the mesopore space on the outer wall. Water first evaporates from the outer mesopores. However, since water is not replenished from the inner capsule space to the mesopore space, the evaporation rate of water gradually decreases. At this point, the evaporation of water seems to stop once. In order for water to be replenished from the inside to the mesopores, it is necessary to create an air passage for air to replace it. Therefore, replenishment of water from the inside of the capsule to the mesopores is done by capillary osmosis only when the water in the mesopores has almost evaporated and air passages have been created. After this, the evaporation of water from the mesopores will occur again. The previously mentioned process is repeated to automatically turn on and off the evaporation of water. This behavior has been observed not only with water but also with slightly viscous liquid drugs. For example, it was confirmed that the release of liquids such as Anisole, Ionone, and Limonene into the air was automatically turned on and off without any external stimuli. This kind of drug delivery behavior has not been reported for other substances. This system can be used for the purpose of periodically and repeatedly administering drugs without external stimuli. For example, once the system is set up, it may be possible to develop a system for administering drugs morning, noon, and night without forgetting.

Mesoporous carbon capsule can be obtained during the synthesis of mesoporous silica capsules. A new type of sensor can be fabricated by using this as a hierarchical structure film by the alternating LbL adsorption method [80]. Since mesoporous carbon capsules do not have sufficient surface charge for alternating adsorption, a surfactant with an electrical charge is first adsorbed on the capsule surface. This complex can be alternatively adsorbed with a polyelectrolyte having the opposite surface charge. This alternately adsorbed film was fabricated on a quartz oscillator,

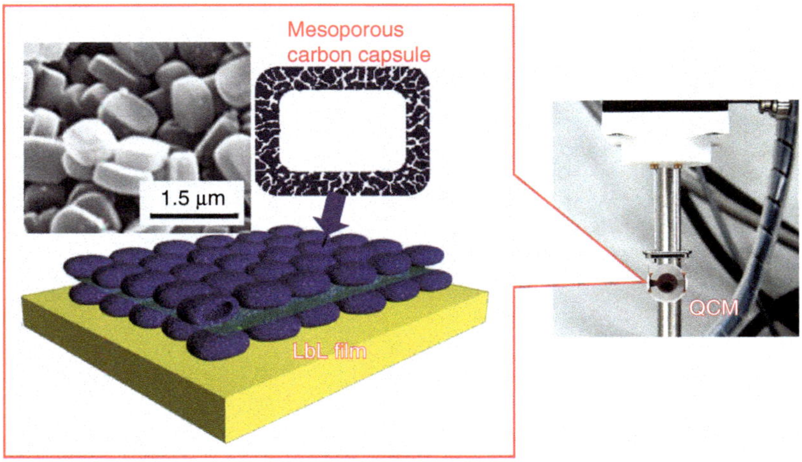

Figure 3.10 An LbL film of mesoporous carbon capsules prepared on a QCM plate.

which was then exposed to various solvent vapors, and its sensor function was investigated from the adsorption response (Figure 3.10). The sensor fabricated in this way was generally more sensitive to nonpolar solvents than to polar solvents, and the sensitivity to aromatic solvents was particularly pronounced. For example, there was no significant difference between cyclohexane and benzene in terms of their molecular diameter, molecular weight, and vapor pressure, but the latter showed about five times higher sensitivity (five times larger adsorption). This is thought to be due to the sp2 nature of the carbon in the mesoporous carbon capsule and the strong interaction with the aromatic guests.

The unique feature of this sensor membrane structure is that a capsule structure with an inner space exists in the thin film. By inserting a recognition element that specifically interacts with the guest in this inner space, the selectivity of the sensor can be changed. In the case of a normal mesoporous carbon capsule, which does not contain anything in the inner capsule space, the sensitivity to aniline and pyridine is high. When lauric acid is encapsulated in the inner capsule space, the sensitivity to aliphatic amines such as ammonia and butylamine is high. Also, when dodecylamine was encapsulated, the sensitivity to acetic acid was better. These results indicate that the selectivity of the sensor can be changed by acid–base interactions. We have developed a sensor that allows us to change the selectivity relatively freely by simply placing the recognition component inside the capsule.

In addition to this, it is possible to fabricate an alternating layer film-type sensor using various carbon materials. Figure 3.11 shows a sensor using mesoporous carbon, CMK-3, as an alternating adsorption film [81]. First, the surface of the carbon is oxidized to the extent that it does not destroy the structure of CMK-3, and carboxyl groups are introduced. Using the negative charge on the carbon surface, an alternatively adsorbed film of mesoporous carbon CMK-3 can be fabricated on the crystal oscillator using polycation. In this case, also, a hierarchical structure is formed with more layers of carbon having mesopore structure inside. The sensor film was placed

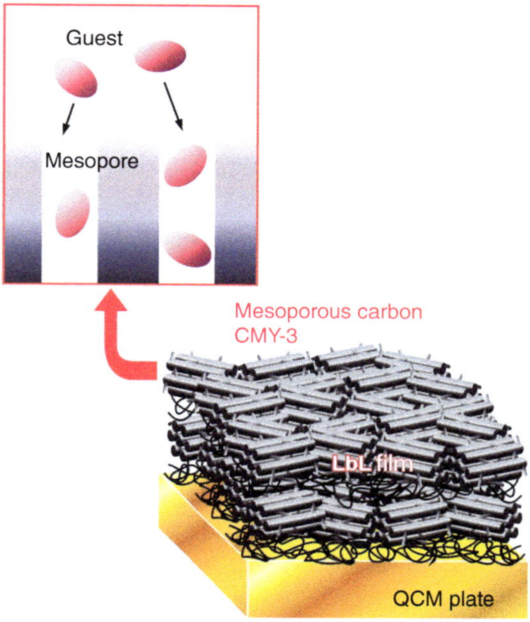

Figure 3.11 An LbL film of mesoporous carbon (CMK-3) with polyelectrolyte on a QCM plate for guest recognition.

in water and various tea components (caffeine, catechins, tannic acid, etc.) were injected into the film to detect them. Specifically, the detection sensitivity of tannic acid was 13.6 times and 3.9 times higher than that of caffeine and catechin, respectively. This is due to the fact that the molecular diameter of tannic acid matches the pore diameter of the mesopores of CMK-3 well, and the adsorption of tannic acid in the mesopores is considered to be a dense packing of tannic acid. In support of this is the fact that the adsorption of tannic acid is highly cooperative and that, above a certain concentration, the sensor sensitivity becomes very high. This is because the adsorption is stabilized by the stacking of tannic acid in the mesopores.

Graphene, a two-dimensional nanocarbon, can also be used as an alternatively adsorbed membrane with ionic liquids (Figure 3.12) [82]. Ionic liquids, such as imidazolium salts, interact with nanocarbons such as carbon nanotubes. First, graphite is oxidized and dispersed in aqueous solution as graphene oxide nanosheets. When reduced in the presence of imidazolium salts, which form ionic liquids, the graphene oxide is converted into graphene nanosheets, and at the same time, complexes with ionic liquid molecules adsorbed on the surface are obtained. This complex can be made into an alternatively adsorbed membrane with a suitable polyelectrolyte, since the charge derived from the ionic liquid molecules will be present on the surface. The ionic liquid molecules remain between the layers even after drying, resulting in an alternating layered film with ionic liquid molecules sandwiched between the graphene nanosheets. The layered film was fabricated on a quartz crystal oscillator, and the sensing of various gases was performed. This may reflect the fact that aromatic molecules can stably penetrate between the electron spaces of graphene due to their interactions. The film was also shown to exhibit sensitive adsorption

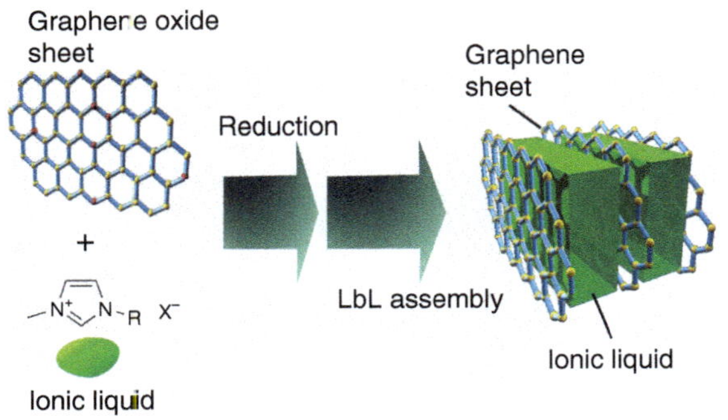

Figure 3.12 An LbL assembly between graphene nanosheet and ionic liquid. Source: Ji et al. [82] / John Wiley & Sons.

behavior for carbon dioxide, which may lead to the development of sensitive sensors for carbon dioxide.

As can be seen in the various functions of living organisms, high functionality requires the coordination of several actions, which is why many structures in living organisms have a hierarchical structure. Although it is not easy to artificially reproduce such structures and functions, it provides a guideline for the development of material systems with higher functions. It is difficult to construct all the hierarchical structures spontaneously as is done in living organisms, but we have shown here that the creation of hierarchical structures can lead to the development of drug delivery and highly sensitive and selective sensors that show ON/OFF behavior independent of external stimuli. Although it is difficult to construct all the hierarchical structures spontaneously as is done in living organisms, it is a feasible approach to create hierarchical structures by first creating nanostructures and then layering them, as shown in some of the examples in this paper. In such an approach, the use of alternating LbL adsorption method is very useful because it can produce stacked films with a large number of materials.

3.4 Conclusions

In this chapter, the two types of nanoarchitectonics processes, (i) fullerene assemblies at liquid interface and (ii) assemblies based on LbL adsorption, are exemplified. In both the cases, coupling of self-assembly processes and the other processes (such as posttreatments and LbL adsorption) is required to produce advanced hierarchical structures.

One of the ultimate goals of nanoarchitectonics is to create highly functional structures, such as those found in living organisms, from basic units such as functional molecules. In many cases of highly functional systems in biochemical systems, the functional structures have hierarchical and asymmetric structural motifs, where these hierarchical structures, which allow the combination of vector-like functional relays and processes with high efficiency and specificity, usually cannot be obtained via conventional self-assembly under equilibrium conditions. It is not possible. The introduction of nonequilibrium actions into nanoarchitectonic processes, such as self-organization processes involving nonequilibrium energy-consuming processes in biological systems, is important for fabricating such hierarchical structures. For example, hierarchical and asymmetric structures can be created by introducing artificial structure formation methods such as LB and alternating adsorption methods step by step into the equilibrium self-assembly process. Nanoarchitectonics, which harmoniously incorporates these processes, can be a universal methodology for constructing highly functional systems such as those found in living organisms.

Acknowledgments

This study was partially supported by JSPS KAKENHI Grant Number JP16H06518 (Coordination Asymmetry), JP20H00392, JP20H00316, and JP20K05590.

References

1 Feynman, R.P. (1960). *Eng. Sci.* 23: 22.
2 Roukes, M. (2001). *Sci. Am.* 285: 48.
3 Ariga, K. (2015). *J. Inorg. Organomet. Polym.* 25: 177.
4 Eric Drexler, K. (1992). *Nanosystems: Molecular Machinery, Manufacturing, and Computation*. New York: Wiley.
5 Ariga, K., Ji, Q., Nakanishi, W. et al. (2015). *Mater. Horiz.* 2: 406.
6 Ariga, K., Minami, K., Ebara, M., and Nakanishi, J. (2016). *Polym. J.* 48: 371.
7 Ariga, K. (2021). *Small Sci.* 1: 2000032.
8 Markovich, G., Collier, C.P., Henrichs, S.E. et al. (1999). *Acc. Chem. Res.* 32: 415.
9 Hecht, S. (2003). *Angew. Chem. Int. Ed.* 42: 24.
10 Aono, M. (2011). *Sci. Technol. Adv. Mater.* 12: 040301.
11 Aono, M., Bando, Y., and Ariga, K. (2012). *Adv. Mater.* 24: 150.
12 Ariga, K., Ji, Q., Hill, J.P. et al. (2012). *NPG Asia Mater.* 4: e17.
13 Ariga, K., Li, M., Richards, G.J., and Hill, J.P. (2011). *J. Nanosci. Nanotechnol.* 11: 1.
14 Ariga, K., Li, J., Fei, J. et al. (2016). *Adv. Mater.* 28: 1251.
15 Aono, M. and Ariga, K. (2016). *Adv. Mater.* 28: 989.
16 Ramanathan, M., Shrestha, L.K., Mori, T. et al. (2013). *Phys. Chem. Chem. Phys.* 15: 10580.
17 Cordier, S., Grasset, F., Molard, Y. et al. (2015). *J. Inorg. Organomet. Polym.* 25: 189.
18 Komiyama, M., Mori, T., and Ariga, K. (2018). *Bull. Chem. Soc. Jpn.* 91: 1075.
19 Ariga, K., Jia, X., and Shrestha, L.K. (2019). *Mol. Syst. Des. Eng.* 4: 49. Lee, Y.J. and Park, Y. (2020). *J. Nanosci. Nanotechnol.* 20: 2781.
20 Ariga, K., Lee, M.V., Mori, T. et al. (2010). *Adv. Colloid Interface Sci.* 154: 20.
21 Zhang, L., Wang, T., Shen, Z., and Liu, M. (2016). *Adv. Mater.* 28: 1044.
22 Ariga, K., Matsumoto, M., Mori, T., and Shrestha, L.K. (2019). *Beilstein J. Nanotechnol.* 10: 1559.
23 Ariga, K., Mori, T., Kitao, T., and Uemura, T. (2020). *Adv. Mater.* 32: 1905657.

24 Ariga, K., Nishikawa, M., Mori, T. et al. (2019). *Sci. Technol. Adv. Mater.* 20: 51.
25 Deka, R.C., Deka, A., Deka, P. et al. (2020). *J. Nanosci. Nanotechnol.* 20: 5153.
26 Wang, K.L., Galatsis, K., Ostroumov, R. et al. (2008). *Proc. IEEE* 96: 212.
27 Ariga, K., Watanabe, S., Mori, T., and Takeya, J. (2018). *NPG Asia Mater.* 10: 90.
28 Nayak, A., Unayama, S., Tai, S. et al. (2018). *Adv. Mater.* 30: 1703261.
29 Ariga, K., Ito, M., Mori, T. et al. (2019). *Nano Today* 28: 100762.
30 Ishihara, S., Labuta, J., Van Rossom, W. et al. (2014). *Phys. Chem. Chem. Phys.* 16: 9713.
31 Jackman, J.A., Cho, N.-J., Nishikawa, M. et al. (2018). *Chem. Asian J.* 13: 3366.
32 Ariga, K., Makita, T., Ito, M. et al. (2019). *Beilstein J. Nanotechnol.* 10: 2014.
33 Liu, J., Zhou, H., Yang, W., and Ariga, K. (2020). *Acc. Chem. Res.* 53: 644.
34 Chen, R., Zhao, T., Zhang, X. et al. (2016). *Nanoscale Horiz.* 1: 423.
35 Kim, J., Kim, J.H., and Ariga, K. (2017). *Joule* 1: 739.
36 Giussi, J.M., Cortez, M.L., Marmisollé, W.A., and Azzaroni, O. (2019). *Chem. Soc. Rev.* 48: 814.
37 Huang, H.J., Yan, M.M., Yang, C.Z. et al. (2019). *Adv. Mater.* 31: 1903415.
38 Ariga, K., Ishihara, S., Abe, H. et al. (2012). *J. Mater. Chem.* 22: 2369.
39 Pandeeswar, M., Senanayak, S.P., and Govindaraju, T. (2016). *ACS Appl. Mater. Interfaces* 8: 30362.
40 Paul, L., Banerjee, B., Bhaumik, A., and Ali, M. (2020). *J. Nanosci. Nanotechnol.* 20: 2858.
41 Pham, T.-A., Qamar, A., Dinh, T. et al. (2020). *Adv. Sci.*: 2001294.
42 Ariga, K., Naito, M., Ji, Q., and Payra, D. (2016). *CrystEngComm* 18: 4890.
43 Ariga, K., Leong, D.T., and Mori, T. (2018). *Adv. Funct. Mater.* 28: 1702905.
44 Dutta, S., Kim, J., Hsieh, P.-H. et al. (2019). *Small Methods* 3: 1900213.
45 Banerjee, S. and Pillai, J. (2019). *Expert Opin. Drug Metab. Toxicol.* 15: 499.
46 Nakanishi, W., Minami, K., Shrestha, L.K. et al. (2014). *Nano Today* 9: 378.
47 Zou, Q., Liu, K., Abbas, M., and Yan, X. (2016). *Adv. Mater.* 28: 1031.
48 Ariga, K. (2016). *ChemNanoMat* 2: 333.
49 Stulz, E. (2017). *Acc. Chem. Res.* 50: 823.
50 Ariga, K., Jia, X., Song, J. et al. (2020). *Angew. Chem. Int. Ed.* 59: 15269.
51 Liu, X., Riess, J.G., and Krafft, M.P. (2018). *Bull. Chem. Soc. Jpn.* 91: 846.
52 Seki, T. (2018). *Bull. Chem. Soc. Jpn.* 91: 1026.
53 Leow, W.R. and Chen, X. (2019). *Bull. Chem. Soc. Jpn.* 92: 505.
54 Torimoto, M., Murakami, K., and Sekine, Y. (2019). *Bull. Chem. Soc. Jpn.* 92: 1785.
55 Tanaka, M., Kobayashi, S., Murakami, D. et al. (2019). *Bull. Chem. Soc. Jpn.* 92: 2043.
56 Shrestha, L.K., Ji, Q., Mori, T. et al. (2013). *Chem. Asian J.* 8: 1662.
57 Miyazawa, K. (2009). *J. Nanosci. Nanotechnol.* 9: 41.
58 Miyazawa, K. (2015). *Sci. Technol. Adv. Mater.* 16: 013502.
59 Sathish, M., Miyazawa, K., Hill, J.P., and Ariga, K. (2009). *J. Am. Chem. Soc.* 131: 6372.
60 Bairi, P., Minami, K., Hill, J.P. et al. (2017). *ACS Nano* 11: 7790.
61 Tang, Q., Maji, S., Jiang, B. et al. (2019). *ACS Nano* 13: 14005.

62 Shrestha, L.K., Sathish, M., Hill, J.P. et al. (2013). *J. Mater. Chem. C* 1: 1174.
63 Bairi, P., Minami, K., Nakanishi, W. et al. (2016). *ACS Nano* 10: 6631.
64 Bairi, P., Minami, K., Hill, J.P. et al. (2016). *ACS Nano* 10: 8796.
65 Ariga, K., Yamauchi, Y., Rydzek, G. et al. (2014). *Chem. Lett.* 43: 36.
66 Rydzek, G., Ji, Q., Li, M. et al. (2015). *Nano Today* 10: 138.
67 Ariga, K., Ahn, E., Park, M., and Kim, B.-S. (2019). *Chem. Asian J.* 14: 2553.
68 Wang, B., Xu, Y.-J., Li, P. et al. (2020). *Appl. Surf. Sci.* 509: 145323.
69 Ariga, K., Yamauchi, Y., Mori, T., and Hill, J.P. (2013). *Adv. Mater.* 25: 6477.
70 Ariga, K. (2019). *Langmuir* 35: 3585.
71 Ariga, K. (2020). *Langmuir* 36: 7158.
72 Ito, M., Yamashita, Y., Tsuneda, Y. et al. (2020). *ACS Appl. Mater. Interfaces* 12: 56522.
73 Katagiri, K., Ariga, K., and Kikuchi, J. (1999). *Chem. Lett.*: 661.
74 Katagiri, K., Hashizume, M., Ariga, K. et al. (2007). *Chem. Eur. J.* 13: 5272.
75 Katagiri, K., Hamasaki, R., Ariga, K., and Kikuchi, J. (2000). *Langmuir* 18: 6709.
76 Katagiri, K., Hamasaki, R., Ariga, K., and Kikuchi, J. (2002). *J. Am. Chem. Soc.* 124: 7892.
77 Yu, J.-S., Yoon, S.B., Lee, Y.J., and Yoon, K.B. (2005). *J. Phys. Chem. B* 109: 7040.
78 Ji, Q., Miyahara, M., Hill, J.P. et al. (2008). *J. Am. Chem. Soc.* 130: 2376.
79 Ji, Q., Acharya, S., Hill, J.P. et al. (2009). *Adv. Funct. Mater.* 19: 1792.
80 Ji, Q., Yoon, S.B., Hill, J.P. et al. (2009). *J. Am. Chem. Soc.* 131: 4220.
81 Ariga, K., Vinu, A., Ji, Q. et al. (2008). *Angew. Chem. Int. Ed.* 47: 7254.
82 Ji, Q., Honma, I., Paek, S.-M. et al. (2010). *Angew. Chem. Int. Ed.* 49: 9737.

4

Self-assembly of Colloidal Crystals: Strategies

Junchao Liu[1], *Jingxia Wang*[1,2], *and Lei Jiang*[1,2]

[1] *Technical Institute of Physics and Chemistry, Chinese Academy of Sciences, Key Laboratory of Bio-inspired Materials and Interface Sciences, Beijing 100190, China*
[2] *University of Chinese Academy of Sciences, School of Future Technologies, Beijing 101407, China*

4.1 Introduction

Photonic crystal (PC) is composed of the periodic arrangement of the dielectric materials with different refractive index, which brings about a photonic stopband characteristic similar to electronic stopband [1–3]. PC structure prohibits the propagation of light with certain wavelengths or frequencies located in the stopband of PC, contributing to the ability to regulate light. PC has aroused wide research attention and important applications in sensing [4, 5], detecting [6, 7], and advanced optical devices [8–10].

A prominent example of PC in nature is opal (Figure 4.1a,a′), which contains a periodic structure of silica microspheres responsible for its iridescence [11–14]. Analogously, spherical opal structure has been found in the mosquito eye (Figure 4.1b,b′). Besides, rod-shaped particles can also effectively construct PC structure, such as drake feather (Figure 4.1c,c′) or pearl with flaky structure (Figure 4.1d,d′). PC structure not only endows nature creature a bright structure color but also provides special functions to better adapt to the environment. For example, the superhydrophobicity of mosquito compound eyes favors it with good vision in a humid environment because fog droplets easily roll off the superhydrophobic eye surface. The function of antifogging is useful for not only nature creature but also for the design of antifogging materials. It is worth noting that peacock feather and drake feather are representative examples of superhydrophobic biological PC. The superhydrophobicity can enable the rain droplets to be rolled off swiftly at the slightest tremble and shield them from the rainfall. Moreover, the surface can easily shake off the dust, achieving its purpose of self-cleaning.

Inspired by nature, colloidal particle assembly has become a primitive fabrication approach to PC materials. Ordered arrays of highly monodisperse silica or polymer microspheres form the most typical PC structure, which afford spatial, periodic variations in dielectric properties on the submicrometer scale. The self-assembly of colloidal PC can be easily achieved by direct dropping, vertical deposition, spin/spray

Functional Materials from Colloidal Self-assembly, First Edition. Edited by Qingfeng Yan and George Zhao.

coating, direct writing, Langmuir–Blodgett (LB), doctor blade, etc. Recently, some ingenious self-assembly approaches of cylinder assembly from capillary tube, electrospun, or microfluidics have been developed to construct PC structure. Herein, we summarize the research progress in the self-assembly of latex particles, including assembly mechanism, self-assembly techniques, and current research challenges. Understanding the self-assembly characteristics of latex particles is important for the construction of various functional PC films.

4.2 Assembly Mechanism of Latex Particles

4.2.1 General Assembly Process

The self-assembly of latex particles is a complex interaction process involving gravity, electrostatic repulsion, and Brownian motion [15]. The assembly of latex particles is mainly driven by capillary force and convective interaction, which is firstly proved by Nagayama and his coworkers in 1992 [16]. The main function of the capillary force between latex particles is to organize particles into two-dimensional (2D) hexagonal arrays [17]. The formation process of latex arrangement is demonstrated in Figure 4.1a. First, a colloidal nucleus is formed when the thickness of the liquid layer approaches the particle diameter. Then, more particles are pushed toward this nucleus by convective transport, eventually forming a 2D hexagonal latex array based on the capillary attraction. In this process, temperature and humidity play an important role in both the convection force and the assembly rate, and the electrostatic repulsion between the latex particles also affects the assembly rate. Besides, the assembly of the larger particles is

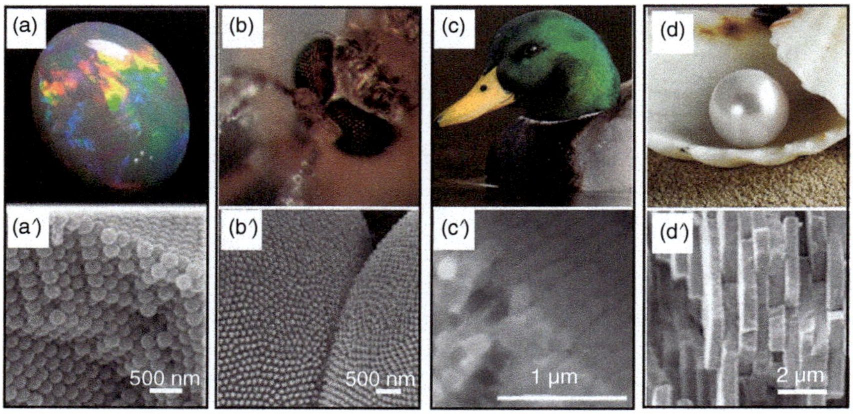

Figure 4.1 Typical colloidal PC nanostructures in nature creature and corresponding SEM image: (a, a′) Nature opal. Source: Zhao et al. [11]/with permission of American Chemical Society. (b, b′) Mosquito compound eyes. Source: Gao et al. [12]/with permission of John Wiley & Sons, Inc. (c, c′) Drake feather. Source: Khudiyev et al. [13]/with permission of American Chemical Society. (d, d′) Pearl. Source: Wang et al. [14]/with permission of American Chemical Society.

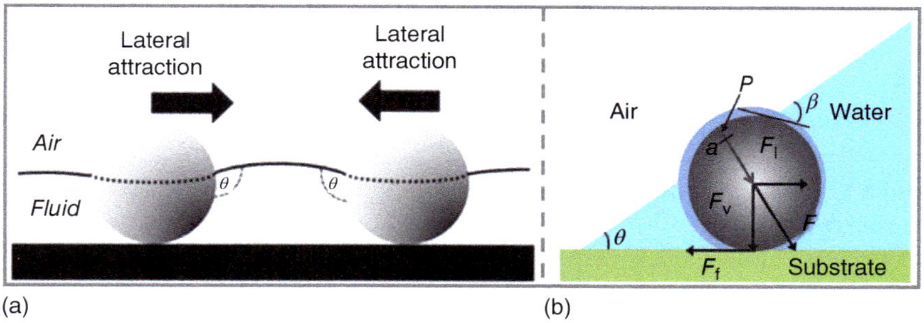

Figure 4.2 Mechanism of colloidal PC assembly: (a) colloidal particle assembly forced by the liquid flow as the evaporation of the solvent and (b) capillary forces (*F*) acting on a colloid trapped at an air–water interface. Source: Reproduced with permission from Morales et al. [17], copyright 2013 American Chemical Society.

restricted by gravity force. Figure 4.2b schemed the assembly situation where a capillary force (F) acts on a spherical particle located at the air–water–substrate interface. F is decomposed into lateral (F_l) and vertical (F_v) components. The friction force (F_f) between the sliding colloid and the substrate surface is used to offset F_l and pin the colloid to the substrate. For $F_l = F_f$, the colloid undergoes a stick-slip transition, separating the two different dynamic states. The case of $F_f \geq F_l$ corresponds to the stick state of the colloid trapped at air–water–substrate. Conversely, when $F_f < F_l$, the friction between the particles and the substrate is not enough to overcome the lateral capillary forces acting on the colloid, the colloid will slide as the contact line recedes. Understanding the force of colloidal particles in the assembly process is of great significance for developing novel assembly methods.

4.2.2 Influence of Substrate Wettability on the Assembly Process

Wettability is a basic property of solid surface reflecting the spreading behavior of liquid on solid surface, which is determined by the surface chemical composition and roughness of surface [18, 19]. The substrate wettability plays an important role in the colloidal assembly [20–22]. For example, the hydrophilic substrate with contact angle (CA) < 90° results in an easy spreading of droplet on solid surface. Hydrophobic substrate (CA > 90°) results in a less spreading of liquid on solid substrate. Superhydrophobic substrate (CA > 150°) restricts the droplet to spread on the surface. There shows distinct assembly behavior for colloidal particles on substrates with different wettability. For example, an excellent spreading behavior of the colloidal suspension on a superhydrophilic substrate contributes to a uniform assembly structure; thus superhydrophilicity becomes a necessary property for the assembled substrate of colloidal particles based on the assembly methods, such as vertical deposition and spin/spray coating. Comparatively, hydrophilic substrate results in a unique "coffee-ring" assembly structure. This is the attributed colloidal particles transferring from the interior toward the exterior, region owing to the faster evaporation in the exterior region of the droplet. On the contrary, assembly of colloidal particles on hydrophobic substrate has become an effective method for removing "coffee-ring" effect, where the receding three-phase contact line (TCL) accelerates the particle transfer from the interior toward the external region during droplet evaporating process. Superhydrophobic substrate is a special substrate with low adhesive surface, and rapid receding TCL will provide additional assembly force for the assembly process. Crack-free colloidal PC film can be obtained on the superhydrophobic substrate (Figure 4.3b–b_4) [23]. When assembled on the hydrophilic substrate, the stress aroused from the adhesive force between the particles, and the substrate restricts the free shrinkage of latex particles during drying process (Figure 4.3a,a_1) leading to the crack formation. In contrast, the low adhesive property of superhydrophobic substrate and continuous receding TCL released the stress contributes to the formation of crack-free colloidal PC assembly with narrow stopband (Figure 4.3b–b_4). As a result, crack-free single-crystalline colloidal PC (Figure 4.3b_1–b_3) demonstrated high reflectance intensity and narrow

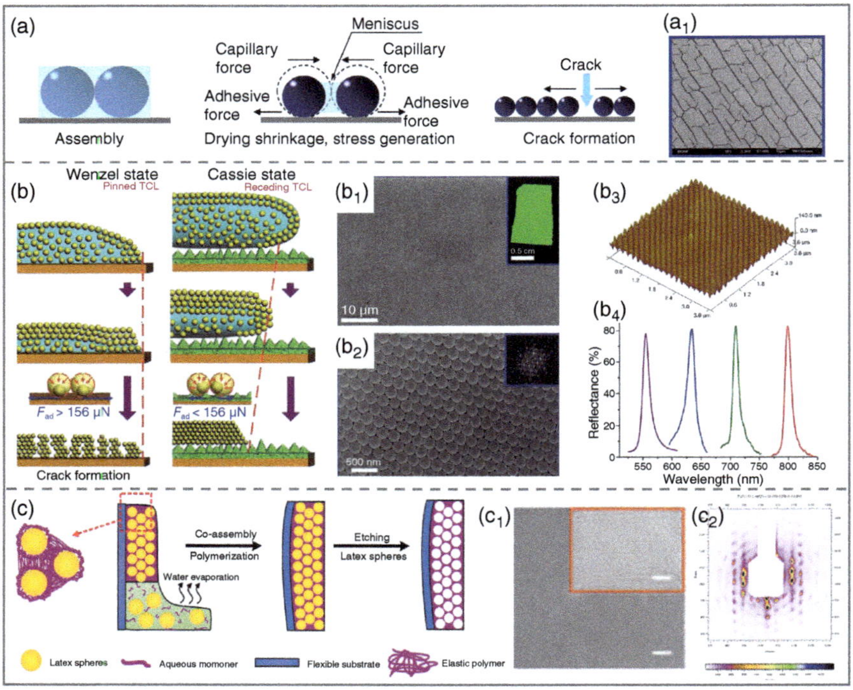

Figure 4.3 (a) Illustration of the reason for the formation of cracks in the colloidal PC. (a_1) Cracks in the assembly colloidal PC. (b) Schematic illustration of colloidal PCs assembled on high adhesive force substrate (left) and low adhesive force superhydrophobic substrate (right). (b_1, b_2) SEM images of sample without cracks and (b_3) corresponding AFM image. (b_4) Reflectance spectra of sample. Source: Huang et al. [23]/with permission of American Chemical Society. (c) Illustration of fabrication process for crack-free colloidal PC by polymerization-assisted assembly on flexible substrate. (c_1) SEM image of composite opal PCs. (c_2) Small-angle X-ray scattering diffraction pattern of the as-prepared sample without crack. Source: Zhou et al. [24]/with permission of Springer Nature.

full-width-at-half-maxima stopband of 12 nm (Figure 4.3b_4). Similarly, large-area crack-free single-crystalline colloidal PC (Figure 4.3c_1,c_2) could be also assembled on flexible hydrophilic substrate by the synergistic effects of substrate deformation and monomer infiltration/polymerization (Figure 4.3c–c_2) [24]. The tensile stress generated from colloid shrinkage was reduced, owing to the formation of an elastic polymer network from co-assembly polymerization process. Besides, residual stress was further released based on the timely transformation of the flexible substrate (Figure 4.3c).

4.2.3 Influence of Magnetic/Electric Field on the Assembly Process

Magnetic/electric field plays an important role in the self-assembly of colloidal particles. Similar to temperature and pressure, magnetic field is a physical parameter that can be used to transport the magnetic materials. Superparamagnetic colloidal particles can be successfully assembled into periodically arranged structure under magnetic field. In particular, the photonic properties of as-prepared colloidal PC can be reversibly tuned by manipulating the external magnetic field. One-dimensional colloidal chains containing a string of particles are obtained by a strong magnetic dipole–dipole interparticle attraction within the dispersion of superparamagnetic particles. Yin and coworkers fabricated colloidal PC with regulable bandgap, covering the entire visible spectrum based on superparamagnetic polyacrylate-capped Fe_3O_4 colloidal particles (Figure 4.4a–a_3) [25]. The as-prepared superparamagnetic colloidal particles had good monodispersity (Figure 4.4a). As shown in Figure 4.4a_1, polyacrylate bound to the particle surface based on the strong coordination of carboxyl groups and iron ions, while the non-coordinating carboxyl groups on the polymer chain extended into the aqueous solution to charge the particle surface. The superparamagnetic colloidal particles in deionized water were assembled to PC structure with brilliant structure color and, and the regulable bandgap could be achieved by changing the distance between the magnet and the sample (Figure 4.4a_2,a_3). Magnetic control can also be utilized to produce assembly structures with controllable shape. Chen and coworkers developed an available magnetic-directed assembly strategy to construct a series of colloidal PC (Figure 4.4b–b_2) [26]. Monodispersed core–shell colloids could be obtained by choosing monodispersed poly(styrene-*co*-2-hydroxyethyl acrylate) colloids as cores and poly(VI-*co*-HEA) hydrogels as shells based on seeded emulsion copolymerization. Then CdS quantum dots were introduced by the chelation between poly(*N*-vinylimidazole) and Cd^{2+} (Figure 4.4b). Several sophisticated Janus units were obtained by a triphase microfluidic device under different magnetic field regulation (Figure 4.4b_1,b_2). Kwon and coworkers invented a material called "M-Ink" with tunable color by changing the periodicity of the nanostructure under magnetic field and fixable in a polymer network by photopolymerization (the experimental apparatus in Figure 4.4c) [27]. Instantaneous exposure to patterned ultra-violet (UV) light allowed the fast production of structural color in pattern. Tunable color could be easily achieved by changing the intensity of magnetic field (Figure 4.4c_1). Finally, high-resolution patterns of multiple structure colors were accordingly

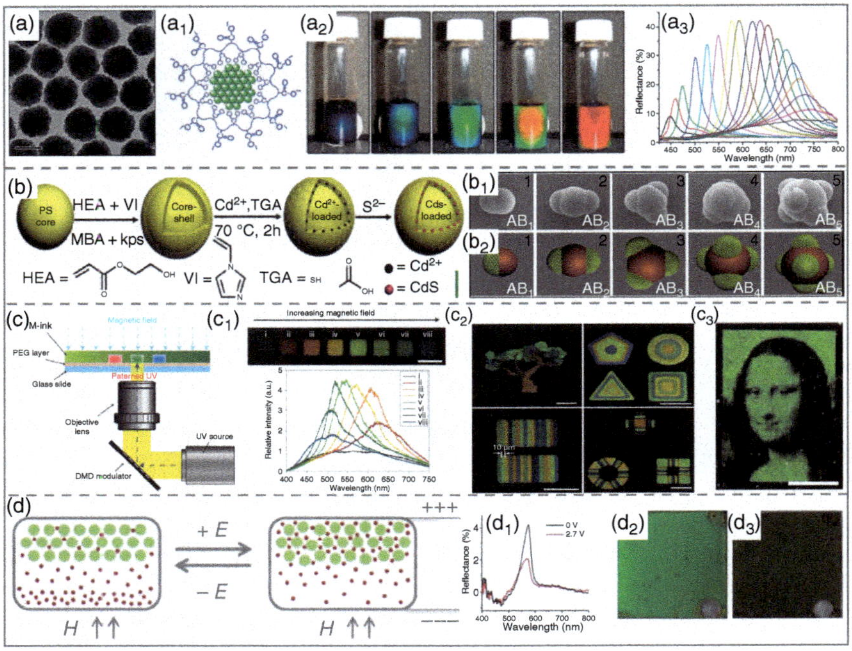

Figure 4.4 (a) TEM image and corresponding (a_1) schematic illustration of polyacrylate-capped Fe_3O_4 colloidal particles (scale bar: 100 nm). (a_2) Photos of colloidal PC formed under an external magnetic field with gradually decreasing magnet–sample distance from right to left. (a_3) Reflectance spectra of colloidal PC assembled under an external magnetic field at different magnet–sample distances. Source: Ge et al. [25]/with permission of John Wiley & Sons, Inc. (b) The chemical synthesis route toward quantum dots-loaded hybrid latex particles. (b_1) SEM image and corresponding (b_2) model of diverse structures of magnetic-controlled colloidal PC. Source: Yin et al. [26]/with permission of American Chemical Society. (c) Maskless lithography setup for the formation of structure color under magnetic field. (c_1) Photo and corresponding reflectance spectra of different structure colors generated by gradually increasing magnetic fields. (c_2, c_3) Different patterns fabricated under magnetic field and photopolymerization. Source: Kim et al. [27]/with permission of Springer Nature. (d) Migration and distribution of magnetic nanoparticles in the absence or presence of an electrical field. (d_1) The reversible change of reflectance for the as-prepared sample under alternating electrical field and (d_2) corresponding optical images. Source: Liu et al. [28]/with permission of Royal Society of Chemistry.

obtained (Figure 4.4c_2,c_3). Except magnetic field, electric field-assisted assembly has also been proved to be a highly efficient method. Ge and coworkers fabricated an electrically tunable colloidal PC with decreasing/recovering reflectance intensity when being on/off an electric field (Figure 4.4d–d_3) [28]. An increasing localized ionic strength in the upper zone was achieved owing to the movement of Fe_3O_4 nanoparticles (red dots) under application of an electric field (Figure 4.4d. As a result, the repulsion and attraction between polystyrene (PS) colloids was accordingly shielded, contributing to the decreasing of the order degree of colloidal PC and corresponding reflectance intensity (Figure 4.4d_1–d_3).

4.3 Assembly Strategies of Colloidal Crystal

Various assembly approaches are developed based on different application conditions and practical requirements. Self-assembly technique is originated from the most primitive gravity-induced natural deposition. Subsequently, a simple vertical deposition approach is developed and widely spread owing to its easy repeat. LB approach allows for the layer-by-layer fabrication of colloidal crystal that may not be easily assembled using any other approach. Furthermore, rapid and time-saving methods of spin/spray coating and doctor blade are accordingly developed. The approach of inkjet printing, magnetic/electric field and physical confinement further improve the controllability of colloidal assembly toward patterned PC. Recently, several newly developed assembly approaches are introduced, such as microfluidics, cylinder assembly from capillary tube, and wettability-assisted assembly. In the following sections, we will introduce in detail these assembly approaches in combination with literatures based on large-area, patterned, and specific structure colloidal PC (Figure 4.5).

4.3.1 Large-Area Colloidal PC

The fabrication of large-area and high-quality colloidal PC film is of significance for its extended application. Some approaches have been developed for the previously mentioned requirements, such as vertical deposition, LB, doctor blade, spin/spray coating, etc.

4.3.1.1 Vertical Deposition

Vertical deposition is a widely spread assembly approach owing to its simple steps and good repeatability. The approach was developed by Jiang et al. who simplified the convection assembly process by adding a vertical glass substrate to the latex suspension (Figure 4.6a) [29]. During the evaporation process, the drying latex particles were transferred to the glass substrate by capillary force, followed by a transfer of convective flow from the bulk of colloidal suspension to the drying particle layer. The balance of capillary force and convective particle flux during solvent

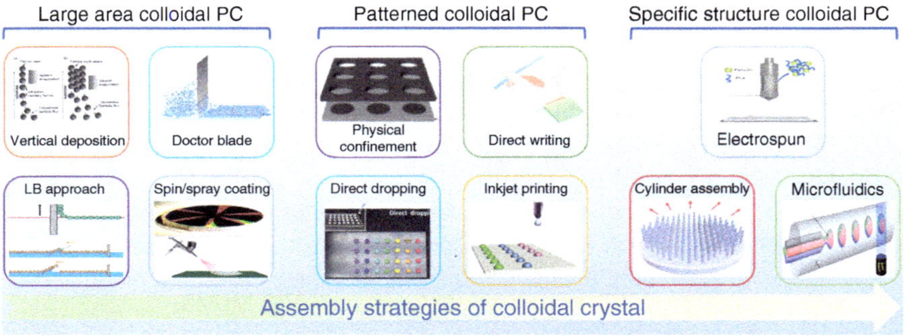

Figure 4.5 The summary chart of assembly strategies of colloidal crystal.

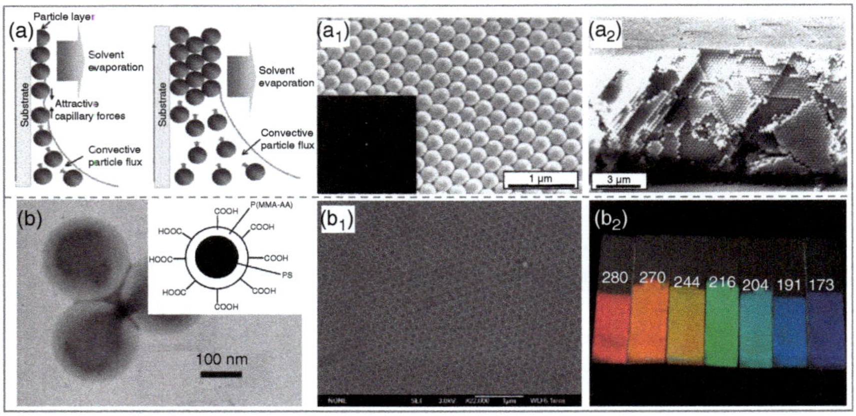

Figure 4.6 (a) The schematic of self-assembly latex particles by vertical deposition during solvent evaporation process. Typical top (a_1) and side (a_2) view SEM images of SiO_2 colloidal PC. Source: Jiang et al. [29]/with permission of American Chemical Society. (b) Scheme of the structure of the as-prepared latex sphere and corresponding TEM image. (b_1) SEM image of colloidal PC assembled of spheres with a diameter of 173 nm. (b_2) Photo of the as-prepared film deposited on a glass substrate (the width of the glass was *ca.* 1.2 cm). Source: Wang et al. [30]/with permission of John Wiley & Sons, Inc.

evaporation played an important role in the formation of 2D or three-dimensional (3D) colloidal PC. A well-assembled SiO_2 colloidal PC was obtained after the evaporation of ethanol within 3–4 days (Figure 4.6a_1). Usually 10–50-layer film could be prepared in one vertical deposition process (Figure 4.6a_2). To improve the assembly interaction among particles, some flexible particles were synthetized to enhance viscoelastic force among particles. For example, Wang et al. fabricated a large-area and high-quality colloidal PC film by synthetizing modified core–shell PS spheres (Figure 4.6b–b_2) [30], where PS domains are located on the core of the latex sphere and a thin shell rich in polyacrylic acid (PAA) and polymethyl methacrylate (PMMA) over the core polymer (Figure 4.6b). The enriched carboxyl groups upon the surface of latex spheres favored the formation of hydrogen bonds among latex spheres. A perfect package of hexagonally ordered structure (Figure 4.6b_1) was obtained by vertical deposition of the core–shell latex spheres. The structure color of colloidal PC film covered the whole visible range (Figure 4.6b_2), which could be modified by changing the size of latex spheres from 173 to 280 nm. The vertical deposition method provides a simple and effective approach for the fabrication of colloidal PC in the laboratory, which lays a foundation for its further research.

4.3.1.2 LB Approach

LB approach is a powerful method for the assembly and orientational control of molecular monolayers, which allows the production of a series of novel, layered PC that are not easily assembled by any other method. Heterostructured colloidal PC comprising of layers of spheres with different diameters can be easily fabricated by LB approach. LB process (Figure 4.7a) involves generally controlled compression of the monolayer film formed by the amphiphilic compound diffusion at the air–water

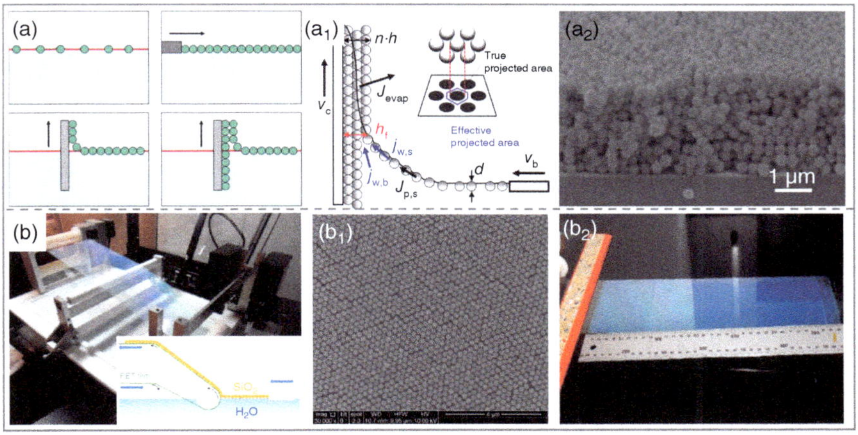

Figure 4.7 (a) Multilayers colloidal PC film formation during LB deposition process. (a_1) Schematic illustration of the LB assembly of microspheres and the definition of the assembly parameters. (a_2) Side-view SEM images of colloidal PC prepared by LB approach. Source: Atiganyanun et al. [31]/with permission of American Chemical Society. (b) Photo of trough, the roll-to-roll unit, and the poly(ethylene terephthalate) substrate with a LB monolayer of silica spheres deposited. (b_1) Top-view SEM images of the monolayer assembled by SiO_2 spheres. (b_2) Colloidal PC film with five LB layers covering 24 cm × 10 cm. Source: Parchine et al. [32]/with permission of American Chemical Society.

interface, thereby inducing the change of ordering and phase (orientation) [31]. A hydrophilic substrate is initially immersed in water and then drawn slowly upward; a layer of amphiphilic particles adheres to the substrate leaving a hydrophobic surface. An organized multilayer film is fabricated in which alternating layers are located in opposite directions. Repeating this process can prepare colloidal PC film with desired thickness. LB assembly mechanism is demonstrated in Figure 4.7a_1. The top of the microspheres near the water surface is hydrophilic at the appropriate level. The submerged microspheres with diameter d on the water surface are pushed laterally toward the vertical substrate by the two barriers on both sides of the substrate. Each barrier moves horizontally at the speed of v_b, and the substrate is pulled out of the water vertically at the speed of v_c. When the substrate is pulled out, the microspheres on the surface of the liquid are transferred to the substrate. From side-view scanning electron microscope (SEM) image in Figure 4.7a_2, the colloidal PC film assembled by LB approach is considerably disordered compared with the controlled evaporation film. On this basis, Pemble and coworkers presented an advanced roll-to-roll LB technique to fabricate large-area colloidal PC on flexible poly(ethylene terephthalate) substrate (Figure 4.7b–b_2) [32]. A trough (Figure 4.7b) was utilized to fabricate colloidal PC from SiO_2 spheres. The standard alternate dipper mechanism in the trough was replaced by a special roll-to-roll unit. SiO_2 spheres could be assembled by spreading monolayer SiO_2 spheres on the air–water interface then rotating poly(ethylene terephthalate) substrate, forming a high-quality 2D close-packed SiO_2 spheres (Figure 4.7b_1). Finally, Colloidal PC film with five LB layers covering 24 cm × 10 cm (Figure 4.7b_2) was successfully fabricated by repeating the deposition process several times.

4.3.1.3 Doctor Blade

Similar to LB approach, doctor blade technique is also a facile self-assembly method for high-volume and large-scale production. Doctor blade technique is widely used in the printing, photographic film and textile to create highly uniform film with large area. Also, doctor blade technique is utilized to assemble colloidal PC with variable thickness. Jiang and coworker fabricated 3D highly ordered colloidal PC-polymer nanocomposite by roll-to-roll doctor blade technique (Figure 4.8a–a_2) [33]. Firstly, monodisperse SiO_2 particles were dispersed in ethoxylated trimethylolpropane triacrylate (ETPTA) to form transparent colloidal suspension. Then the suspensions were dispensed along a sidewall of an immobilized, vertically beveled razor blade that gently touches a substrate. During the movement of substrate dragged by a syringe pump, a uniform shear force was applied to align the suspended SiO_2 particles to form periodic arrangement structure (Figure 4.8a_2). Finally, the ETPTA monomer was photopolymerized under UV irradiation to obtain highly ordered colloidal PC-polymer nanocomposite (Figure 4.8a_1). Yang and coworkers prepared colloidal PC-polymer nanocomposite (Figure 4.8b_1) and corresponding inverse opal structure (Figure 4.8b_2) by doctor blade technique [34].

4.3.1.4 Spin Coating

Spin coating is a mature micro-processing technology that can form a highly uniform thin film with adjustable thickness over a large area in a spontaneous

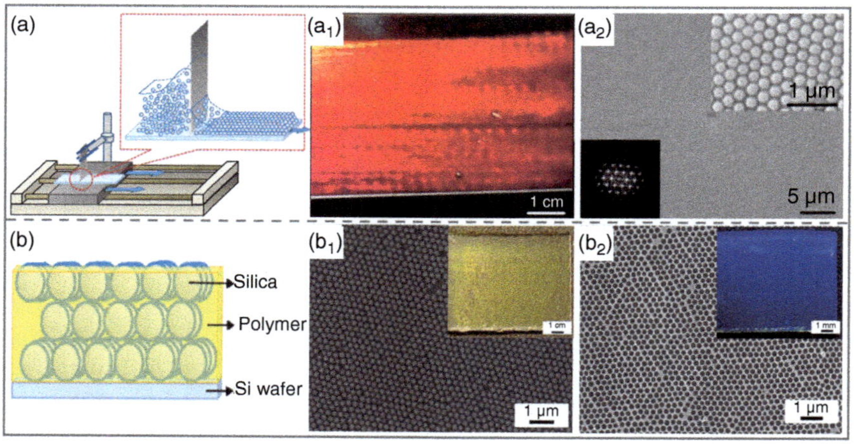

Figure 4.8 (a) Schematic illustration of assembly process of colloidal PC by doctor blade technique. (a_1) Photo of colloidal PC-polymer nanocomposite fabricated by doctor blade technique and corresponding SEM image (a_2). Source: Yang et al. [33]/with permission of American Chemical Society. (b) Schematic illustration of side-view structure of film prepared by doctor blade technique. (b_1) SEM/photo image of colloidal PC-polymer nanocomposite. (b_2) SEM/photo image of inverse opal structure film. Source: Lin et al. [34]/with permission of Elsevier.

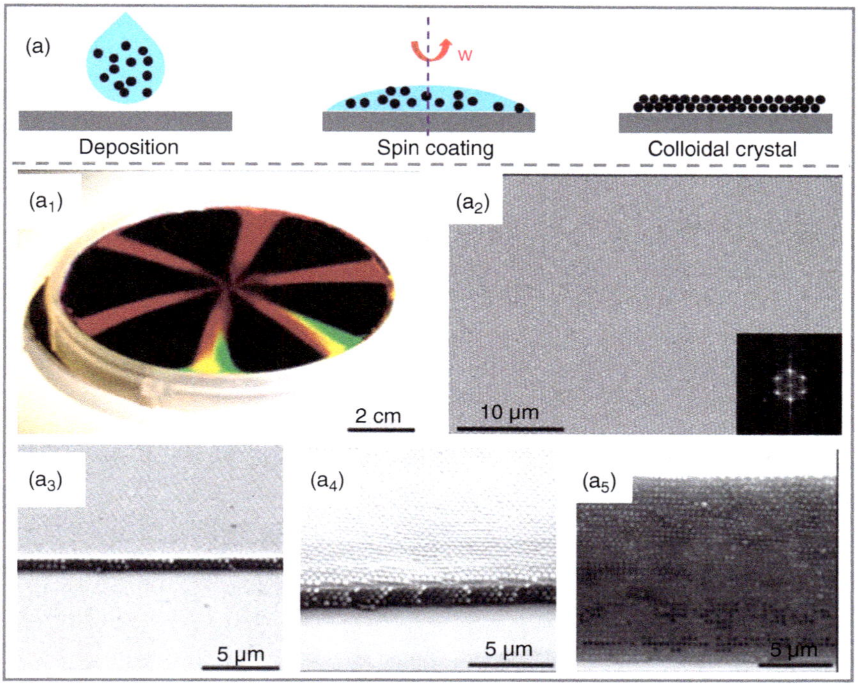

Figure 4.9 (a) Schematic of spin-coating process. (a_1) Photo of a four-inch sample assembled by 325 nm SiO_2 spheres under white light irradiation and corresponding SEM image (a_2). (a_3–a_5) Different thicknesses of sample controlled by spin coating: (a_3) two-layer, (a_4) five-layer, and (a_5) 41-layer. Source: Jiang et al. [35]/with permission of American Chemical Society.

crystallization process. Compatible with standard micromachining technology, wafer-scale colloidal PC can be fabricated from a simple, fast, and inexpensive technique of spin coating. Figure 4.9a showed a schematic drawing of the spin-coating process [35]. A rapid acceleration process was initiated to produce centrifugal forces after a colloidal suspension was dispersed onto a substrate in a spin-coater. This process causes mass to be ejected and the topographical "thinning" of the liquid film. Finally, ultrathin film with several tens of nanometers could be fabricated. However, when volatile solvent (such as alcohol) was used to disperse the colloidal particles in the traditional spin-coating approach, the rapid evaporation of solvent induced poor quality polycrystalline sample. Therefore, Jiang et al. further developed a novel spin-coating platform where colloidal particles could also be dispersed to form a stable suspension without solvent evaporation (Figure 4.9a_1–a_5) [35]. In this case, SiO_2 particles with diameters from 30 nm to 2 μm were dispersed in ETPTA at volume fraction of *ca.* 20% to form transparent and stable suspension. Bright monochromatic six-arm diffraction was observed in self-assembled SiO_2 colloidal PC-polymer nanocomposite film, indicating the formation of hexagonally packed spheres parallel to the wafer surface (Figure 4.9a_1), as confirmed by top-view SEM image (Figure 4.9a_2). The thickness of film was inversely proportional to the spin speed and the square root of the spin time and could be successfully changed from two-layer (Figure 4.9a_3), five-layer (Figure 4.9a_4), to 41-layer (Figure 4.9a_5), under different spin-coating conditions. The development of spin coating provides a rapid and simple fabrication approach for large-area colloidal PC film.

4.3.1.5 Spray Coating

General assembly approach of colloidal crystal is required of flat and hydrophilic substrate, which greatly restricts the fabrication of colloidal PC on irregular objects or highly curved surfaces and hinders the corresponding practical applications. Spray coating can successfully realize the assembly of latex particles on 3D irregular surface in a simple and rapid way. Besides, spray coating can contribute to a noniridescent structure color by breaking down their long-range ordered arrangement after adding other substances. Song and coworkers fabricated a short-range ordered amorphous colloidal PC with noniridescent structure color by introducing graphene nanosheets and graphene quantum dots into spray coating suspension [36]. Figure 4.10a showed a typical schematic illustration of spray

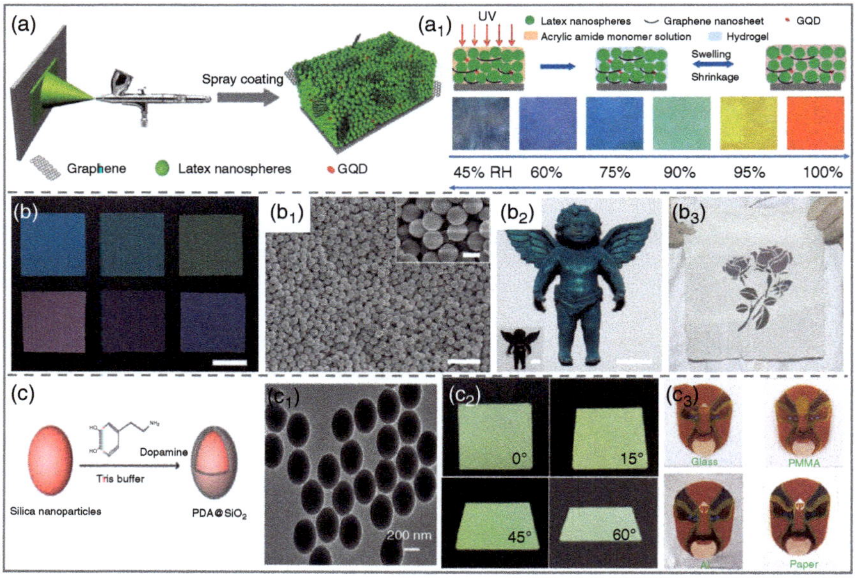

Figure 4.10 (a) Schematic illustration of the fabrication process by spray coating. (a_1) Photo of the colloidal PC sensor at reversible relative humidity variations. Source: Zhang et al. [36]/with permission of John Wiley & Sons, Inc. Photo (b) and corresponding SEM image (b_1) of the as-prepared noniridescent colloidal PC assembled by SiO_2 spheres by spray coating. (b_2) Cyan-color dyed angel toy using 212 nm SiO_2 spheres by spray coating. (b_3) Colored rose pattern coated on silk fabric using five kinds of fabricated noniridescent colors. Source: Li et al. [37]/with permission of American Chemical Society. (c) Schematic of the synthesis process of poly-dopamine@SiO_2. (c_1) TEM image of as-prepared composite particles. (c_2) Array images of colloidal PC assembled by 232 nm composite particles by spray coating at different observing angles. (c_3) The multicolor PC pattern on different substrates, such as glass, PMMA, Al, and paper. Source: Liu et al. [38]/with permission of Royal Society of Chemistry.

coating process for amorphous colloidal PC structure. An aqueous dispersion composed of two different sizes monodisperse PS particles and a small amount of graphene nanosheets and graphene quantum dots was thoroughly mixed by ultrasound. Subsequently, the mixture was spray-coated onto substrate by a commercial airbrush. Finally, brilliant structural color films could be uniformly deposited on the substrate after the water evaporation. Notably, the addition of graphene nanosheets and graphene quantum dots was favorable for the construction of amorphous colloidal PC structure and enhanced color saturation. As a result, a humidity sensor was obtained by introducing polyacrylic amide hydrogel into the interstice of as-prepared sample (Figure 4.10a_1). The structure color of sensor changed with the variation of humidity based on the swelling or shrinking of hydrogel. Analogously, Zhang and coworkers presented an atomization deposition of SiO_2 particles with the additive of poly(vinyl alcohol) (PVA) to fabricate large-area homogeneous amorphous colloidal PC structure with noniridescent structure color on ceramic or silk substrate based on spray coating [37]. The as-prepared sample demonstrated noniridescent structure colors (Figure 4.10b) and short-range ordered property (Figure 4.10b_1). A resin toy with uneven surface (Figure 4.10b_2) could be uniformly dyed by spray coating. Also, large-area silk fabrics (Figure 4.10b_3) were quickly dyed by atomization deposition of amorphous colloidal PC composed of SiO_2 particles with some carbon black and PVA. Unlike mixing latex particles with other substances, Gu and coworkers synthesized a poly-dopamine-coated SiO_2 nanoparticles (PDA@SiO_2) and assembled these particles on different substrates by spray coating [38]. The monodispersed PDA@SiO_2 were synthesized by polymerizing dopamine onto spherical SiO_2 particles (Figure 4.10c,c_1). The as-prepared sample demonstrated invariable color when observed at different angles (Figure 4.10c_2). A multicolor pattern (Figure 4.10c_3) could be easily spray-coated on different substrates, such as glass, PMMA, Al, and paper.

4.3.1.6 Other

Convective assembly is considered to be one of the most frequently used method to assemble colloidal particles because of its simplicity and robustness. This technique usually contains the formation of a TCL at the meniscus of the suspension film during an evaporating process. Then a laterally capillary force is induced between adjacent particles on the condition that the solvent thickness becomes comparable to the colloidal particle size. Finally, more and more free particles move toward the settled colloids to form close-packed triangular lattice driven by the synergistic effect of evaporation and convective flow. To fabricate uniform colloidal PC, convective assembly confined between two planar substrates has been demonstrated (Figure 4.11a–a_2) [39]. In this study, the colloidal PC film was fabricated by dragging a small volume of colloidal suspension at a constant velocity confined in a meniscus between two substrates (Figure 4.11a). As-prepared colloidal film demonstrated different colors (Figure 4.11a_1,a_2) under distinct light conditions. Furthermore, Wang and coworkers investigated the influence of assembly cell's shape on the self-assembly behavior of colloidal particles (Figure 4.11b–b_2) [40]. Obviously, distinct self-assembly behaviors could be obtained by changing the curvature of

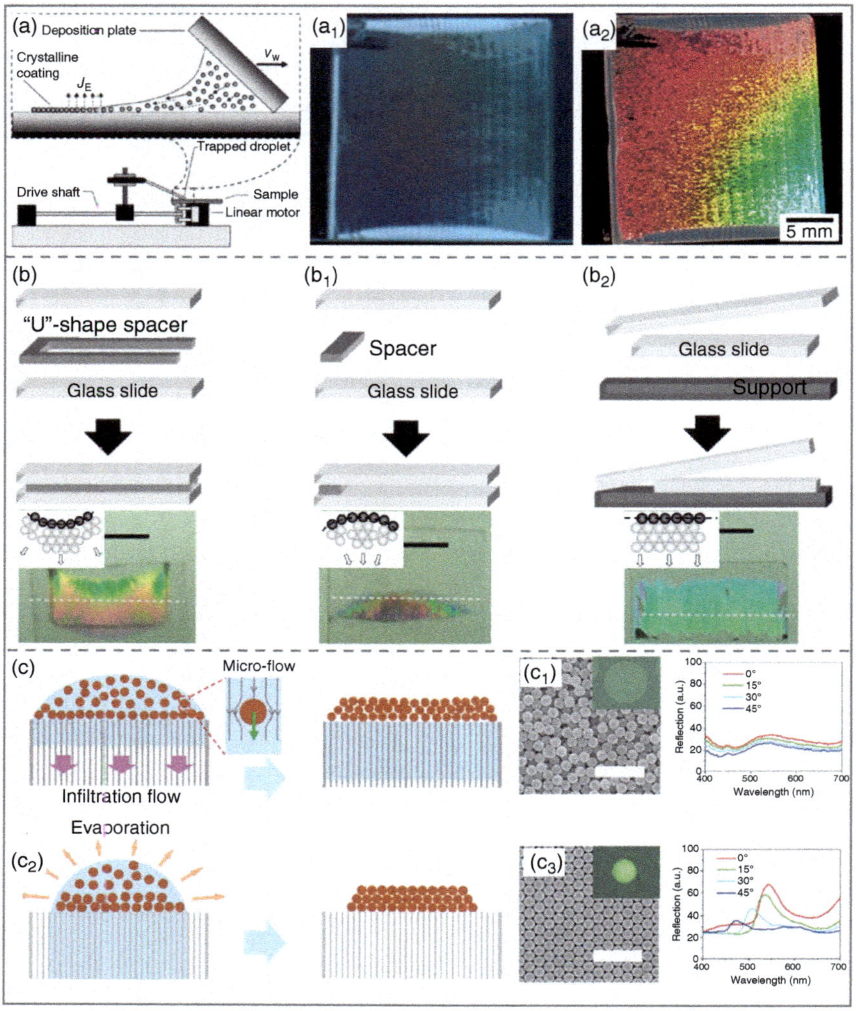

Figure 4.11 (a) Schematic illustration of convective assembly confined in two substrates. The as-prepared colloidal PC film observed in ambient light (a_1) or illuminated with a white light from behind (a_2). Source: Prevo and Velev [39]/with permission of American Chemical Society. Schematic of cells used to assemble colloidal PC: (b) a rectangular cell with one opening, (b_1) a rectangular cell with three open sides, and (b_2) a wedge-shaped cell with three open sides. Source: Sun et al. [40]/with permission of American Chemical Society. (c) Illustration of the formation of amorphous colloidal PC based on a strong downward infiltration flow and corresponding SEM image/reflectance spectra (c_1). (c_2) Illustration of the changes of order range in the colloidal arrays driven by evaporation and corresponding SEM image/reflectance spectra (c_3). Source: Bai et al. [41]/with permission of John Wiley & Sons, Inc.

drying front based on the shape of assembly cell. A single-domain colloidal PC with uniform structure color could be grown under a straight drying front formed in a wedge-shaped cell (Figure 4.11b_2). This technique would be conducive to expand colloidal PC applications in angle-resolved nanosphere lithography and high-quality plasmonic crystal.

The complete evaporation of colloidal suspension usually takes several minutes in the direct dropping approach. To shorten the assembly time of colloidal particles, Duan and coworkers demonstrated an infiltration-driven nonequilibrium assembly of colloidal particles on liquid-permeable and particle-excluding anodic aluminum oxide substrate (Figure 4.11c–c_3) [41]. When the suspension was hydrophilic on substrate, the whole assembly process could be completed at 900 μs owing to the rapid infiltration flow (Figure 4.11c), contributing to an amorphous colloidal array (Figure 4.11c_1). In contrary, close-packed PC structure (Figure 4.11c_3) was obtained on longer assembly time (810 ms) based on hydrophobic state of suspension (Figure 4.11c_2).

4.3.2 Patterned Colloidal PC

Patterned colloidal PC [42, 43] contributes to a novel approach to construct high-performance PC devices with unique structures and specific functions, which expand PCs' excellent applications in the field of sensors, displays, and anti-fake technology [44–46]. Accordingly, several self-assembly approaches have been developed for the patterned PC, such as physical confinement, direct dropping, direct writing, inkjet printing, and wettability assisted.

4.3.2.1 Physical Confinement: Geometric Defect

Conventional assembly substrate requires flatness and defect-free, so as-prepared colloidal PC film exhibits a uniform scale. Patterned PC can be obtained by introducing specific template (such as microchannels or capillary tubes) into the substrate, which is called physical confinement. This is an effective way to construct complex patterned PC structure at microscale based on its advantages, such as flexible choice of substrate with different shapes and corresponding different shapes of assembly colloidal PC structure. In details, planar capillary cells, which are composed of two close planar substrates, have been used recently as physical confinement to fabricate colloidal PC. Xia and coworkers demonstrated a strategy to assemble monodisperse spherical colloidal particles into uniform aggregations with well-controlled structure (Figure 4.12a–a_3) based on physical confinement [47]. When colloidal particles in suspension dewetted from an appropriate relief patterned solid surface, the recessed regions could trap colloidal particles and make them to assemble into desired structure based on the geometric confinement provided by the template (Figure 4.12a–a_1). A typical example of polygonal aggregations by physical confinement was demonstrated in Figure 4.12a_2, which showed good periodicity and precise controllability. Furthermore, a 2D array of six-membered ring (Figure 4.12a_3) was assembled by 2 μm PS particles on a glass substrate. The releasing of these fixed aggregations from the substrate could be easily

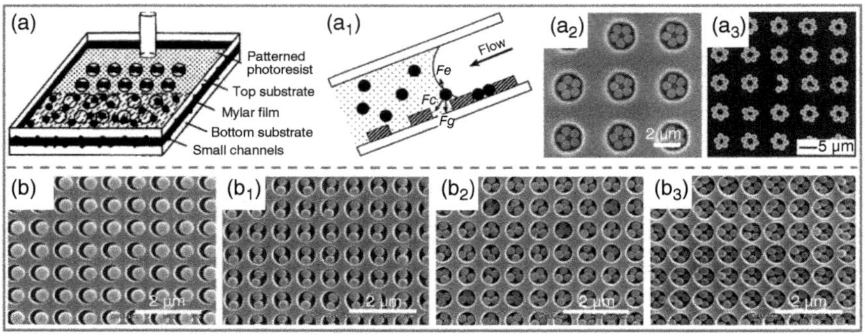

Figure 4.12 (a) Illustration for the process of assembling colloidal particles into well-defined aggregations. (a_1) Side view of the fluidic cell, illustrating the possible applied force on the colloidal particles located at the rear edge of the liquid slug. (a_2) SEM image of a 2D array of pentagons assembled by 0.7 μm PS particles. (a_3) SEM image of ring-shaped aggregations array formed from 2 μm PS particles. Source: Yin et al. [47]/with permission of American Chemical Society. Nonclosely packed square net arrays of (b) one, (b_1) two, (b_2) three, and (b_3) four SiO_2 particles. Source: Khanh et al. [48]/with permission of American Chemical Society.

achieved by sonication. The assembly of colloidal particles in liquid phase driven by capillary force is a general approach, but Yoon and coworker developed a dry manual assembly method where colloidal particles were placed into hole patterns by mechanical rubbing (Figure 4.12b–b_3) [48]. With the decreasing of particle's size, multiple numbers of particles could enter into one hole (Figure 4.12b_1–b_3).

4.3.2.2 Wettability-Induced Template Assembly

In addition to introducing geometric defect into substrate, physical confinement can be also realized by the modulation of chemical properties (such as wettability) to construct patterned colloidal PC. A typical example of wettability-assisted assembly is the sandwich assembly approach (Figure 4.13a) [49]. Firstly, structured Si template (1 cm × 1 cm) is treated with heptadecafluorodecyltrimethoxysilane (FAS) for six hours to ensure superhydrophobicity, while clean glass is treated with plasma for 60 s to make a superhydrophilic substrate. Secondly, colloidal suspension was sandwiched between superhydrophobic-structured Si template and hydrophilic clean glass to produce a fixed-gap sandwich assembly system. Then, the system is maintained at constant temperature and pressure for the evaporation of solvent. The groove structure on Si template could be served as wetting defects to control the rupture of colloidal suspension, yielding a micrometer-scale liquid film between the Si template and the top surface of the groove structure, which provides space for the aggregation of colloidal particles. Finally, precisely assembled colloidal PC structure can be generated after removing Si template by physical peeling. Wang et al. fabricated patterned assembly colloidal PC based on superhydrophobic groove-structured Si template (Figure 4.13b–b_3) [50]. As-prepared linear colloidal PC demonstrated a bright structure color (Figure 4.13b) and well-ordered latex arrangement (Figure 4.13b_1). As-prepared sample showed strong fluorescence (Figure 4.13b_2) and optical waveguiding behavior (Figure 4.13b_3) by introducing

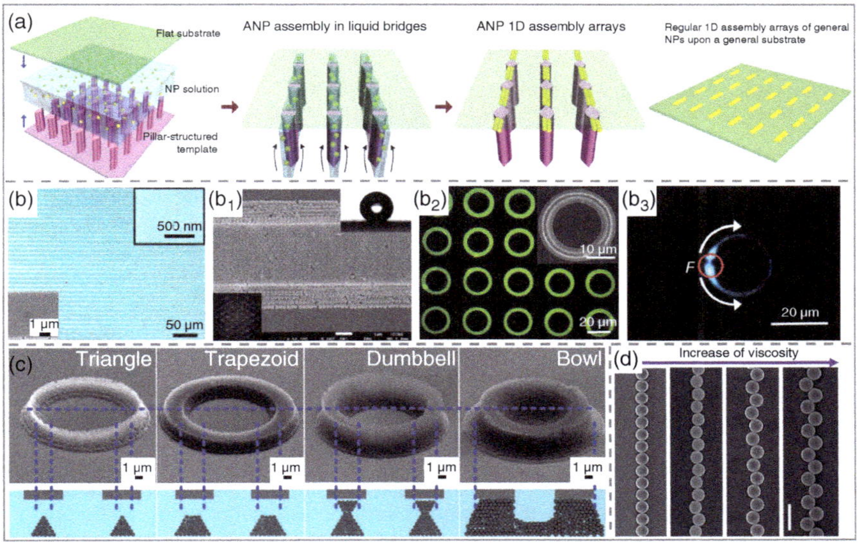

Figure 4.13 (a) Schematic illustration for the general strategy to assemble colloidal particles by superhydrophobic groove-structured Si template. Source: Reproduced with permission from Su et al. [49], copyright 2014 John Wiley and Sons. (b) Optical image and corresponding SEM image (b_1) of patterned colloidal latex assembly. (b_2) Fluorescence image of colloidal microrings. (b_3) Waveguide behavior of microring pattern under 400 nm laser irradiation. Source: Wang et al. [50]/with permission of American Chemical Society. (c) Side view of an annular 3D architecture with various cross sections. Source: Zhang et al. [51]/with permission of John Wiley & Sons, Inc. (d) Different assembly structures controlled by the viscosity of system . Source: Guo et al. [52]/with permission of John Wiley & Sons, Inc.

fluorescent dye into colloidal particle. Furthermore, Wu and coworkers developed a simple method for controllable cross sections and shapes of assembly colloidal PC by changing the physical parameters of the system and pinning point number (Figure 4.13c) [51]. Song and coworkers precisely controlled assembly colloidal PC by the viscosity of the assembly system. High viscosity system was beneficial for colloidal zigzag assembly, while low viscosity of the system contributed to the formation of linear structure (Figure 4.13d) [52].

4.3.2.3 Direct Dropping

Traditional techniques for preparing patterned PC often require the prefabrication of templates or masks and lithographic etching. Recently, scientists have used a direct dropping method for the patterning of colloidal PC film. Direct dropping is the simplest patterning method of colloidal PC without any auxiliary equipment, just dropping latex emulsion on the substrate and waiting for the solvent to evaporate naturally. Chen and coworkers proposed a general scheme for colloid film deposition with inhibition of the coffee-ring effect (Figure 4.14a–a_2) [53]. A layer of regular colloid matter was formed on the top surface of the dried droplets to change the evaporation flux and broke the outward capillary flow (Figure 4.14a), contributing to the formation of tightly packed polymer colloids. In addition, this strategy

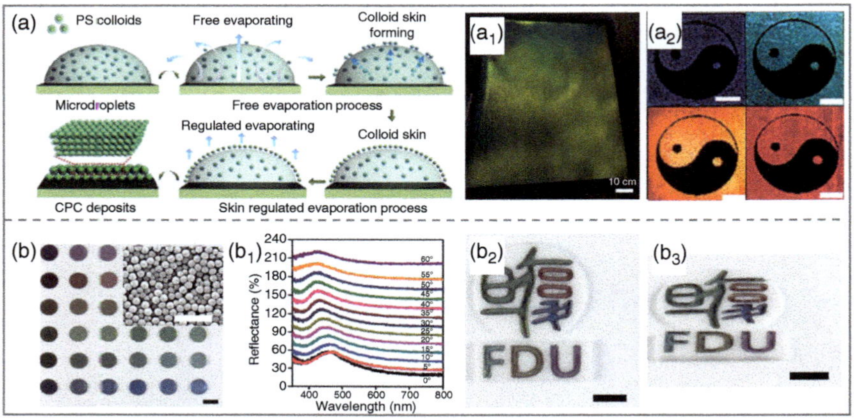

Figure 4.14 (a) Schematic diagram of self-assembly process by "colloid skin" regulation. (a_1) Photo of the as-prepared large-area PC film. (a_2) Tai Chi patterns with different structure colors. Source: Zhang et al. [53]/with permission of Royal Society of Chemistry. (b) Color palette offered by the fabricated amorphous photonic structures (scale bar: 5 mm). The insert is corresponding SEM images (latex particle: 200 nm). (b_1) Angle-resolved reflectance spectra. (b_2, b_3) University logo colored under normal and oblique observations, respectively (scale bar: 1 cm). Source: Zhang et al. [54]/with permission of John Wiley & Sons, Inc.

could combine with spray coating technology with high compatibility, making it possible to easily build large-scale colloidal PC film (*ca.* 90 × 70 cm) with uniform topography and rich color (Figure 4.14a_1). Furthermore, printed Tai Chi patterns (Figure 4.14a_2) was successfully fabricated on substrate with distinct wettability using latex particles with sizes of 195, 205, 215, and 272 nm, respectively. Zi and coworkers presented a new drop-coating strategy based on an additive of cuttlefish ink to fabricate amorphous colloidal PC (Figure 4.14b–b_3) [54]. Amorphous colloidal PC with different noniridescent colors could be obtained by changing the size of PS sphere and the proportion of cuttlefish ink (Figure 4.14b). Figure 4.14b(insert) demonstrated a typical SEM image of a self-assembled amorphous colloidal PC consisting of 200 nm PS spheres and cuttlefish ink particles, showing a very good structure quality with a homogeneous distribution of both particles. Cuttlefish ink particles played an important role in destroying the long-range ordered arrangement of PS particles, which eventually lead to the formation of amorphous structure based on size difference between PS particles and ink particles. The bandgap shift for sample when being observed at different angles was rather small, indicating a strong angle dependence property of as-prepared sample (Figure 4.14b_1). As a demonstration of the angle independence, these artificial structure colors were used to manifest the logo of the university, shown in Figure 4.14b_2–b_3. It was obvious that no color differences could be observed under normal and oblique observations.

4.3.2.4 Direct Writing

Although direct dropping can fabricate patterned colloidal PC film, it still lacks in accuracy and controllability. Direct writing of colloidal suspension is accordingly

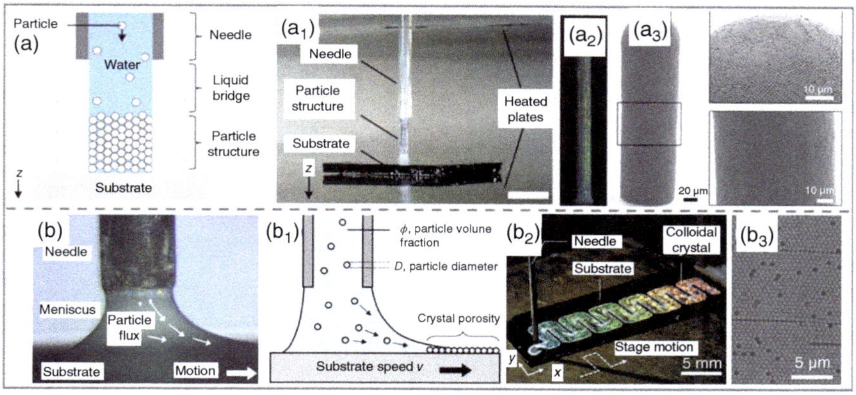

Figure 4.15 (a) Schematic illustration of direct writing device and corresponding photo (a_1). (a_2) Optical image of linear colloidal PC assembled by direct writing device and corresponding SEM image (a_3). Source: Tan et al. [55]/with permission of John Wiley & Sons, Inc. (b) Photo of in-plane direct writing self-assembly process and corresponding schematic illustration (b_1). (b_2) A serpentine colloidal PC drawn by direct writing device and corresponding SEM image (b_3). Source: Tan et al. [56]/with permission of John Wiley & Sons, Inc.

developed as a technique for fabrication of colloidal PC in arbitrary pattern with high precision. Hart and coworkers achieved freeform colloidal pattern based on direct writing technique (Figure 4.15a–a_3) [55]. Direct writing process was performed by precision dispense of a colloidal suspension from a needle, coupled with lateral substrate motion (Figure 4.15a_1,b,b_1). In this case, a fine needle was utilized to precisely dispense colloidal solution, and then the desired colloidal PC structure was obtained by the control of multi-axis substrate motion. A linear colloidal PC (Figure 4.15a_2) was fabricated and demonstrated a typical periodic photonic structure (Figure 4.15a_3). On this basis, Hart and coworkers further fabricated arbitrary colloidal PC pattern by direct writing technique (Figure 4.15b–b_3) [56]. A serpentine-shaped colloidal PC with iridescent structure colors was obtained by coordinated in-plane motion of the substrate (Figure 4.15b_2), reflecting periodic arrangement structure could be observed (Figure 4.15b_3).

4.3.2.5 Inkjet Printing

Inkjet printing has become a widely used method for assembly of latex particles toward large-scale complex colloidal PC patterns because of high throughput and low cost. The fabrication of patterned PC from inkjet printing can be summarized as follows: the pattern can be designed through the computer software program according to the requirements. The capillary and interparticle force are the main assembly-driven forces of printing process. Baumann and coworkers demonstrated a self-assembly method of spherical colloidal PC based on inkjet printing of colloidal suspension (Figure 4.16a,a_1) [57]. A self-assembly force inside tiny droplets of a stochastic mist contributed to the formation of colloidal particles into stable spherical colloidal PC (Figure 4.16a). Furthermore, 3D lineup of spherical

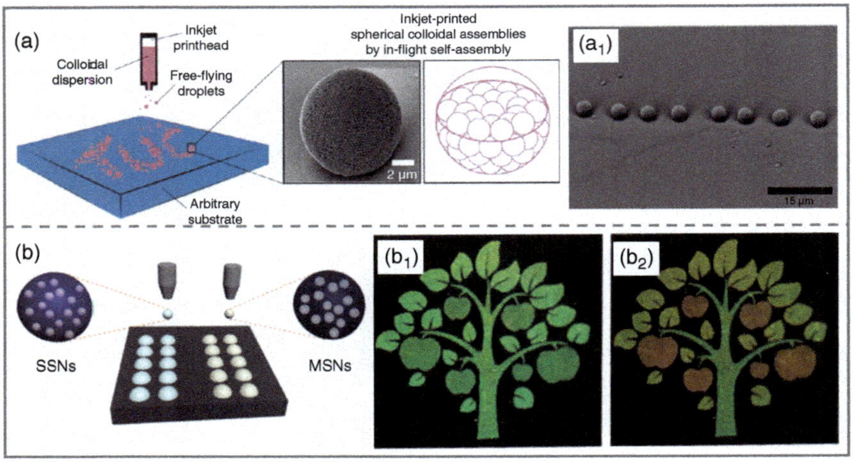

Figure 4.16 (a) Illustration for the assembling process of spherical PC by inkjet printing and corresponding SEM image. (a_1) Tilted-view SEM image of the lineup of spherical PC with similar size. Source: Sowade et al. [57]/with permission of John Wiley & Sons, Inc. (b) Illustration for inkjet printing process by two different colloidal particles. (b_1, b_2) Color changes of the as-prepared colloidal PC pattern in N_2 and saturated ethanol vapor atmospheres . Source: Bai et al. [58]/with permission of American Chemical Society.

colloidal PC with similar size (Figure 4.16a_1) could be obtained by fine-tuning of the jetting parameter. Inkjet printing is a good technology to fabricate PC pattern based on its high precision control. Gu and coworkers fabricated colloidal PC pattern with distinct vapor-responsive behaviors by inkjet printing of two different colloidal particles (Figure 4.16b–b_2) [58]. Mesoporous SiO_2 nanoparticles, solid SiO_2 nanoparticles, or their mixture of different ratios were selected as inkjet printing ink, respectively (Figure 4.16b). The original structure color and vapor-responsive color shift extent could be easily adjusted by the size and mesopores' proportion of nanoparticles. As a result, distinct color changes in different areas of as-prepared colloidal PC pattern (Figure 4.16b_1,b_2) were achieved when it was placed in ethanol vapor atmosphere.

4.3.3 Specific Structure Colloidal PC

Several assembly techniques are developed to fabricate colloidal PC with unique shape, such as cylinder assembly from capillary tube, electrospun for cylindrical colloidal PC, and microfluidics for spherical colloidal PC.

4.3.3.1 Cylinder Assembly from Capillary Tube

Different from the general assembly of colloidal PC on flat substrates, cylinder assembly from capillary tubes have unique significance for the fabrication of functional colloidal PC. Recently, PC fibers have attracted widespread attention. This kind of optical fibers are generally 2D structure, with different air holes arranged on the cross section, and light wave can be limited to the core area. The optical fiber

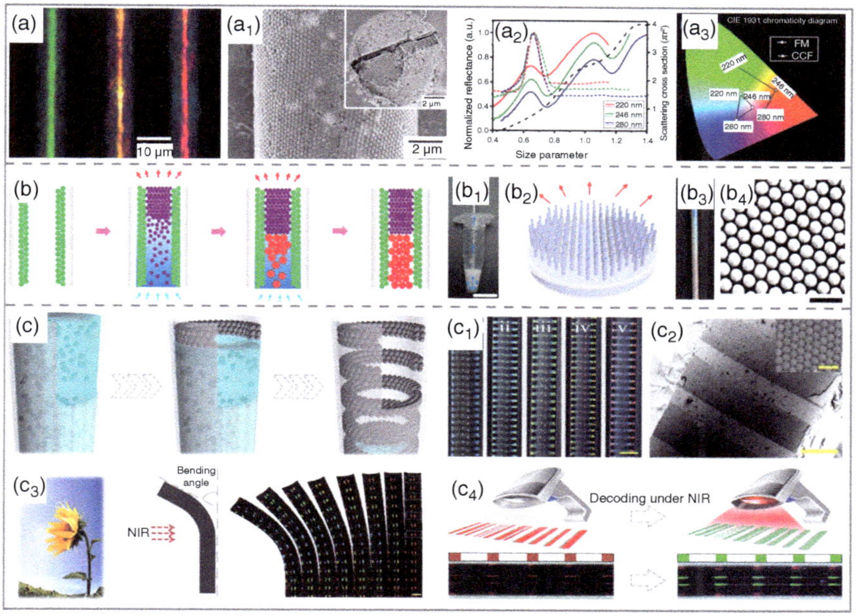

Figure 4.17 (a) Dark-field optical image of colloidal PC assembled in capillary and corresponding SEM image (a_1). (a_2) Reflectance spectra (dashed line) of the as-prepared sample and corresponding CIE tristimulus values and chromaticity values (a_3). Source: Yuan et al. [59]/with permission of American Chemical Society. (b) Schematic illustration of fabrication process for colloidal PC assembled in capillary with multiple heterostructures. (b_1) Photo of device for fabrication of opal capillary (scale bar: 1 cm). (b_2) Schematic illustration of mass fabrication for multiple opal capillaries. (b_3) Photo of opal capillary assembled with three kinds of SiO_2 and corresponding SEM image (b_4). Source: Gao et al. [60]/with permission of American Chemical Society. (c) Schematic illustration of colloidal PC with striped pattern assembled in capillary. (c_1) Optical image of five stripes with different structure colors and corresponding SEM image (c_2). (c_3) Biomimic actuator under NIR light. (c_4) Anti-counterfeiting application of hydrogel stripe as dynamic barcode label. Source: Zhao et al. [61]/with permission of John Wiley & Sons, Inc.

with colloidal PC structure can be fabricated by introducing colloidal particles into capillary. As a result, capillary, as a typical cylinder assembly from capillary tube, is widely used to research the assembly behavior of colloidal particles on its surface. Zhang and coworkers fabricated colloidal PC fiber by assembling colloidal particles in capillary (Figure 4.17a–a_3) [59]. As-prepared sample demonstrated bright structure color (Figure 4.17a) and high reflectance intensity (Figure 4.17a_2). The typical structure of face-centered cubic could be evidently observed (Figure 4.17a_1). The reflectance peaks were converted to Commission Internationale de L'Eclairage (CIE) tristimulus value (Figure 4.17a_3), which indicated better color saturation and gamut. Furthermore, colloidal PC fiber with multiple heterostructures could be easily achieved by assembling colloidal particles with different sizes step by step (Figure 4.17b) [60]. The experiment could be easily achieved based on a simple device (Figure 4.17b_1), which immersed the end of the capillary into colloidal suspension. Obviously, mass fabrication was readily realized based on a parallel apparatus (Figure 4.17b_2). Different colors of blue, green, and red were clearly observed at distinct portions of the capillary (Figure 4.17b_3). As-prepared sample demonstrated typical periodic photonic structure in the capillary (Figure 4.17b_4). Gu and coworkers fabricated striped colloidal PC based on fast self-assembly of colloidal particles in capillary (Figure 4.17c–c_2) [61]. Figure 4.17c presented a new strategy for the fabrication of striped colloidal PC in capillary. The formation reason of striped pattern was mainly attributed to the nonsynchronous process of the solid–liquid–air interface falling and colloidal assembly. Colloidal PC with stripe patterns (Figure 4.17c_1) was observed based on this synergistic effect. Obviously, colloidal PC structure with a striped pattern was formed on the inner surface of the glass capillary (Figure 4.17c_2). Based on the cylindrical assembly colloidal PC, some novel applications could be accordingly achieved. For example, hydrogel actuator could bend in any direction under near-infrared (NIR) light, thanks to the cylindrical structure (Figure 4.17c_3). Besides, stripe patterns showed dynamical color shift under NIR light scanning, which could be unitized for the anti-counterfeiting label (Figure 4.17c_4).

4.3.3.2 Electrospun

Although the cylinder assembly from capillary tube method can be used to fabricate cylindrical colloidal PC, the fabrication process is difficult with low yield, and the as-prepared sample must be attached to capillary, which limits its development and application. Electrospun is an emerging technology to fabricate large-area, free-standing colloidal PC fiber. Electrospun is a special form of electrostatic atomization of polymer fluids. In this case, the material split by atomization is not tiny droplets, but tiny polymer jets, which can run for a long distance and eventually solidify into fibers. Zhang and coworkers reported the use of electrospun strategy for the fabrication of colorful colloidal PC fiber (Figure 4.18a–a_3) [59]. The electrospun precursor solution was prepared by mixing PS colloidal suspension and PVA solution together. Then, the mixture was stirred and ultrasonically treated to obtain a homogeneous milky solution. The syringe was fixed vertically and the solution was fed by using a microinjection pump. The white electrospun film was collected on the surface of a grounded copper plate covered by aluminum foil. Finally, as-prepared film was immersed into water and demonstrated structure color after evaporation of water. The colloidal fiber was uniform in size distribution (Figure 4.18a_1), and colloidal particles were cylindrically packed with a local hexagonal order (Figure 4.18a_2). The as-prepared fiber demonstrated two characteristic peaks and bright structure color (Figure 4.18a_3), owing to reflection of the bandgap of PC structure and Mie scattering of the colloidal particles. Similarly, Wang and coworkers fabricated single colloidal PC fiber with different structure colors by electrospun technique (Figure 4.18b) [62]. The as-prepared sample presented typical PC structure (Figure 4.18b_1) and two reflectance peaks (Figure 4.18b_2). Furthermore, different patterns could be easily obtained by inkjet printing water onto the colloidal fiber (Figure 4.18b_3–b_5).

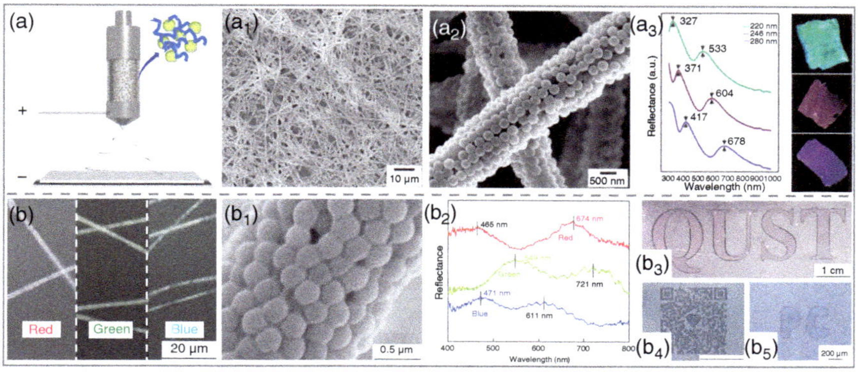

Figure 4.18 (a) Schematic illustration of electrospun set-up for fabricating colloidal PC fiber. (a_1, a_2) SEM images of the as-prepared fiber. (a_3) Reflectance spectra and corresponding photo of the as-prepared fiber with different structure colors. Source: Yuan et al. [59]/with permission of American Chemical Society. (b) Optical image of single colloidal PC fiber fabricated by electrospun. (b_1) SEM image of as-prepared fiber and corresponding reflectance spectra (b_2). (b_3–b_5) Different patterns on colloidal PC fiber fabricated by inkjet printing. Source: Yuan et al. [62]/with permission of American Chemical Society.

4.3.3.3 Microfluidics

Different from the traditional colloidal PC assembled on the plane substrate, macroscopic spherical colloidal PC has several unique advantages, such as large specific surface area and easy encapsulation of cells, proteins, drugs, etc. Two major techniques for fabricating spherical colloidal PC from monodisperse latex suspensions are proposed. One is assembling colloidal particles on superhydrophobic substrate. Velev and coworkers reported a method to assemble colloidal particles within droplets on superhydrophobic substrate with controllable final shape (Figure 4.19a–a_2) [63]. Superhydrophobic substrate kept the droplet spherical during the entire evaporation process. Evaporation-induced colloidal assembly could be regarded as a gradual concentration process. In this process, a spherically symmetrical compressive force was applied from capillary force, which eventually led to the formation of spherical colloidal PC (Figure 4.19a). The as-prepared sample demonstrated bright structure color (Figure 4.19a_1) and close-packed periodic structure (Figure 4.19a_2). However, this assembly method is time-consuming, obtains low yield, and is single component. Another method for fabricating spherical colloidal PC is microfluidics. Microfluidics is achieved by multiphasic fluid mixtures of emulsion droplet or gas bubble to assemble colloidal particles. Microfluidics can accurately control the composition and its distribution; as a result, Janus spherical colloidal PCs are obtained with high yield and good monodispersity. Gu and coworkers fabricated a new kind of Janus spherical colloidal PC by microfluidics with four channels (Figure 4.19b–b_2) [64]. Different oil phases were injected through the channels in an aqueous continuous phase environment with same flowing direction. Monodisperse oil-in-water emulsion droplets were formed at the tip of the four-barrel capillary (Figure 4.19b). Janus spherical colloidal PC with four

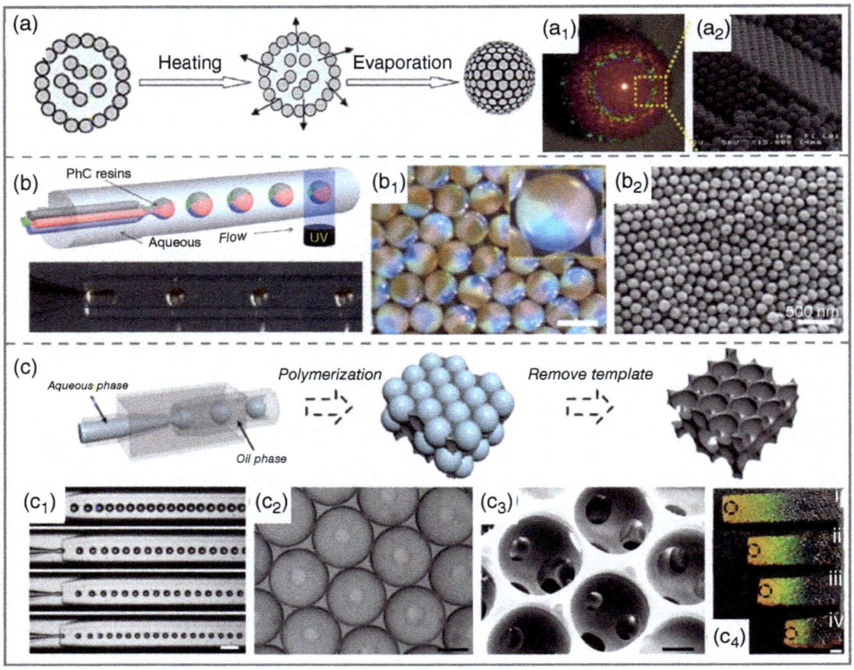

Figure 4.19 (a) Schematic illustration of the close-packed spherical colloidal PC generation and corresponding photo (a_1) and SEM image (a_2). Source: Rastogi et al. [63]/with permission of John Wiley & Sons, Inc. (b) Schematic illustration of generation of Janus spherical colloidal PC in a capillary microfluidic device and corresponding photo. (b_1) Photo of Janus spherical colloidal PC with four compartments and corresponding SEM image (b_2). Source: Zhao et al. [64]/with permission of American Chemical Society. (c) Schematic illustration of the fabrication of a microfluidic droplet array template. (c_1) Real-time photo of the generation of single emulsion droplets in different outer phase flow rates. (c_2) Order self-assembly of the droplet templates (scale bar: 100 μm). (c_3) Negatively replicating droplet templates. (c_4) Photo of the as-prepared PC film when being observed at different angles. Source: Chi et al. [65]/with permission of Royal Society of Chemistry.

compartments could be obtained after UV polymerization of these emulsion droplet (Figure 4.19b_1). The gray color corresponded to a magnetic component while red, blue, and green were the structure color from assembly of colloidal particles (Figure 4.19b_2). Monodisperse microfluidic droplet can also be utilized to assemble colloidal PC. Zhao and coworkers presented a colloidal PC with angle-independent structure color, which building blocks were monodisperse spheres prepared from microfluidics (Figure 4.19c–c_4) [65]. Firstly, monodisperse spheres were prepared by microfluidics. Then they were assembled into colloidal PC and its interstice was infiltrated by substance. Finally, negatively replicating droplet template was obtained after removing droplet template (Figure 4.19c). Monodisperse spheres with different sizes could be easily achieved by controlling the flow rate of outer phase (Figure 4.19c_1), contributing to the fabrication of colloidal PC with different bandgaps. The SEM image of droplet template (Figure 4.19c_2) and corresponding negatively replicating droplet template (Figure 4.19c_3) demonstrated periodic arrangement. The as-prepared sample presented angle-independent property (Figure 4.19c_4), owing to the spherical symmetry element of building block.

4.4 Conclusions

In this chapter, we summarize the various methods of colloidal assembly, which have distinct characteristics. The corresponding development process is also the epitome for understanding the assembly mechanism of colloidal PC. To solve the time-consuming drawback of the primitive direct dropping/vertical deposition method, the rapid spin/spray coating has been developed accordingly. LB approach allows for the fabrication of the layer-by-layer control of colloidal crystal with heterostructured structure. Inkjet printing provides the possibility for the fabrication of accurate and controllable patterned colloidal PC. Magnetic/electric field method makes the assembly process of colloidal PC reversible. Recently, newly developed assembly approach of wettability assisted are introduced, which provides patterned colloidal PC in micro-region. Various assembly methods will bring about new insight for the creation of novel-type PC optical devices.

A major challenge of the self-assembly method is how to fabricate colloidal PC without crack from micrometer scale to device dimensions. To obtain large-scale ideal colloidal PC through self-assembly, each parameter must be considered at the same time to achieve the best conditions, such as selection of colloidal particle, assembly condition/method, etc. Another challenge is how to introduce functional materials into colloidal PC and how to truly realize applications in the optical and other field.

Finally, developing new assembly approaches for novel colloidal PC is an attractive research topic, and it is expected that further development in assembly method will greatly boost the fabrication and application of functional colloidal PC.

References

1 Ito, M.M., Gibbons, A.H., Qin, D.T. et al. (2019). Structural colour using organized microfibrillation in glassy polymer films. *Nature* 570: 363.

2 Sunku, S.S., Ni, G.X., Jiang, B.Y. et al. (2018). Photonic crystals for nano-light in moiré graphene superlattices. *Science* 362: 1153.

3 Yang, B., Guo, Q.H., Tremain, B. et al. (2018). Deal weyl points and helicoid surface states in artificial photonic crystal structures. *Science* 359: 1013.

4 Zhu, B.T., Fu, Q.Q., Chen, K., and Ge, J.P. (2018). Liquid photonic crystals for mesopore detection. *Angew. Chem. Int. Ed.* 57: 252.

5 Heuser, T., Merindol, R., Loescher, S. et al. (2017). Photonic devices out of equilibrium: transient memory, signal propagation, and sensing. *Adv. Mater.* 29: 1606842.

6 Zhan, Y., Wang, Y., Cheng, Q.F. et al. (2019). A butterfly-inspired hierarchical light-trapping structure towards a high-performance polarization-sensitive perovskite photodetector. *Angew. Chem. Int. Ed.* 58: 16456.

7 Liu, C.C., Zhang, W.L., Zhao, Y. et al. (2019). Urea-functionalized poly(ionic liquid) photonic spheres for visual identification of explosives with a smartphone. *ACS Appl. Mater. Interfaces* 11: 21078.

8 Liu, J.C., Ren, J.K., Xie, Z. et al. (2018). Multi-functional organosilane-polymerized carbon dot inverse opals. *Nanoscale* 10: 4642.

9 Wu, X.J., Hong, R., Meng, J.K. et al. (2019). Hydrophobic poly(*tert*-butyl acrylate) photonic crystals towards robust energy-saving performance. *Angew. Chem. Int. Ed.* 58: 13556.

10 Fu, F., Shang, L., Chen, Z. et al. (2018). Bioinspired living structural color hydrogels. *Sci. Robot.* 3: eaar8580.

11 Zhao, Y.J., Shang, L.R., Cheng, Y., and Gu, Z.Z. (2014). Spherical colloidal photonic crystals. *Acc. Chem. Res.* 47: 3632.

12 Gao, X.F., Yan, X., Yao, X. et al. (2007). The dry-style antifogging properties of mosquito compound eyes and artificial analogues prepared by soft lithography. *Adv. Mater.* 19: 2213.

13 Khudiyev, T., Dogan, T., and Bayindir, M. (2014). Biomimicry of multifunctional nanostructures in the neck feathers of mallard (*Anas platyrhynchos* L.) drakes. *Sci. Rep.* 4: 4718.

14 Wang, J.X., Zhang, Y.Z., Wang, S.T. et al. (2011). Bioinspired colloidal photonic crystals with controllable wettability. *Acc. Chem. Res.* 44: 405.

15 Hunter, R.J. (1989). *Foundations of Colloid Science*. Oxford: *Oxford University Press*.

16 Nagayama, K. (1992). Mechanism of formation of 2-dimensional crystals from latex-particles on substrates. *Langmuir* 8: 3183.

17 Morales, V.L., Parlange, J.Y., Wu, M.M. et al. (2013). Surfactant-mediated control of colloid pattern assembly and attachment strength in evaporating droplets. *Langmuir* 29: 1831.

18 Li, C., Guo, R.W., Jiang, X. et al. (2009). Reversible switching of water-droplet mobility on a superhydrophobic surface based on a phase transition of a side-chain liquid-crystal polymer. *Adv. Mater.* 21: 4254.

19 Xu, L., Wang, J.X., Song, Y.L., and Jiang, L. (2008). Electrically tunable polypyrrole inverse opals with switchable stopband, conductivity, and wettability. *Chem. Mater.* 20: 3554.

20 Gu, Z.Z., Yu, Y.H., Zhang, H. et al. (2005). Self-assembly of monodisperse spheres on substrates with different wettability. *Appl. Phys. A Mater.* 81: 47.

21 Sun, Z.J., Bao, B., Jiang, J.K. et al. (2016). Facile fabrication of a superhydrophilic-superhydrophobic patterned surface by inkjet printing a sacrificial layer on a superhydrophilic surface. *RSC Adv.* 6: 31470.

22 Zhou, J.M., Yang, J., Gu, Z.D. et al. (2015). Controllable fabrication of noniridescent microshaped photonic crystal assemblies by dynamic three phase contact line behaviors on superhydrophobic substrates. *ACS Appl. Mater. Interfaces* 7: 22644.

23 Huang, Y., Zhou, J.M., Su, B. et al. (2012). Colloidal photonic crystals with narrow stopbands assembled from low-adhesive superhydrophobic substrates. *J. Am. Chem. Soc.* 134: 17053.

24 Zhou, J.M., Wang, J.X., Huang, Y. et al. (2012). Large-area crack-free single-crystal photonic crystals via combined effects of polymerization-assisted assembly and flexible substrate. *NPG Asia Mater.* 4: e21.

25 Ge, J.P., Hu, Y.X., and Yin, Y.D. (2007). Highly tunable superparamagnetic colloidal photonic crystals. *Angew. Chem. Int. Ed.* 46: 7428.

26 Yin, S.N., Yang, S.Y., Wang, C.F., and Chen, S. (2016). Magnetic-directed assembly from janus building blocks to multiplex molecular-analogue photonic crystal structures. *J. Am. Chem. Soc.* 138: 566.

27 Kim, H., Ge, J.P., Kim, J.H. et al. (2009). Structural colour printing using a magnetically tunable and lithographically fixable photonic crystal. *Nat. Photonics* 3: 534.

28 Liu, J., Mao, Y.W., and Ge, J.P. (2013). Electric field tuning of magnetically assembled photonic crystals. *J. Mater. Chem. C* 1: 6129.

29 Jiang, P., Hwang, K.S., Mittleman, D.M. et al. (1999). Template-directed preparation of macroporous polymers with oriented and crystalline arrays of voids. *J. Am. Chem. Soc.* 121: 11630.

30 Wang, J.X., Wen, Y.Q., Ge, H.L. et al. (2006). Simple fabrication of full color colloidal crystal films with tough mechanical strength. *Macromol. Chem. Phys.* 207: 596.

31 Atiganyanun, S., Zhou, M., Abudayyeh, O.K. et al. (2017). Control of randomness in microsphere-based photonic crystals assembled by Langmuir-Blodgett process. *Langmuir* 33: 13783.

32 Parchine, M., McGrath, J., Bardosova, M., and Pemble, M.E. (2016). Large area 2D and 3D colloidal photonic crystals fabricated by a roll-to-roll Langmuir-Blodgett method. *Langmuir* 32: 5862.

33 Yang, H.T. and Jiang, P. (2010). Large-scale colloidal self-assembly by doctor blade coating. *Langmuir* 26: 13173.

34 Lin, Y.H., Suen, S.Y., and Yang, H.T. (2017). Visual and reversible carbon dioxide sensing enabled by doctor blade coated macroporous photonic crystals. *J. Colloid Interface Sci.* 506: 319.

35 Jiang, P. and McFarland, M.J. (2004). Large-scale fabrication of wafer-size colloidal crystals, macroporous polymers and nanocomposites by spin-coating. *J. Am. Chem. Soc.* 126: 13778.

36 Zhang, Y.X., Han, P., Zhou, H.Y. et al. (2018). Highly brilliant noniridescent structural colors enabled by graphene nanosheets containing graphene quantum dots. *Adv. Funct. Mater.* 28: 1802585.

37 Li, Q.S., Zhang, Y.F., Shi, L. et al. (2018). Additive mixing and conformal coating of noniridescent structural colors with robust mechanical properties fabricated by atomization deposition. *ACS Nano* 12: 3095.

38 Liu, P.M., Chen, J.L., Zhang, Z.X. et al. (2018). Bio-inspired robust non-iridescent structural color with self-adhesive amorphous colloidal particle arrays. *Nanoscale* 10: 3673.

39 Prevo, B.G. and Velev, O.D. (2004). Controlled, rapid deposition of structured coatings from micro- and nanoparticle suspensions. *Langmuir* 20: 2099.

40 Sun, J., Tang, C.J., Zhan, P. et al. (2010). Fabrication of centimeter-sized single-domain two-dimensional colloidal crystals in a wedge-shaped cell under capillary forces. *Langmuir* 26: 7859.

41 Bai, L., Mai, V.C., Lim, Y. et al. (2018). Large-scale noniridescent structural color printing enabled by infiltration-driven nonequilibrium colloidal assembly. *Adv. Mater.* 30: 1705667.

42 Hou, J., Li, M.Z., and Song, Y.L. (2018). Patterned colloidal photonic crystals. *Angew. Chem. Int. Ed.* 57: 2544.

43 Wu, P.P., Wang, J.X., and Jiang, L. (2020). Bio-inspired photonic crystal patterns. *Mater. Horiz.* 7: 338.

44 Cho, S., Shim, T.S., Kim, J.H. et al. (2017). Selective coloration of melanin nanospheres through resonant Mie scattering. *Adv. Mater.* 29: 1700256.

45 Du, X., Wang, M., Welle, A. et al. (2018). Reparable superhydrophobic surface with hidden reactivity, its photofunctionalization and photopatterning. *Adv. Funct. Mater.* 28: 1803765.

46 Qi, Y., Chu, L., Niu, W.B. et al. (2019). New encryption strategy of photonic crystals with bilayer inverse heterostructure guided from transparency response. *Adv. Funct. Mater.* 29: 1903743.

47 Yin, Y.D., Lu, Y., Gates, B., and Xia, Y.N. (2001). Template-assisted self-assembly: a practical route to complex aggregates of monodispersed colloids with well-defined sizes, shapes, and structures. *J. Am. Chem. Soc.* 123: 8718.

48 Khanh, N.N. and Yoon, K.B. (2009). Facile organization of colloidal particles into large, perfect one and two-dimensional arrays by dry manual assembly on patterned substrates. *J. Am. Chem. Soc.* 131: 14228.

49 Su, B., Zhang, C., Chen, S.R. et al. (2014). A general strategy for assembling nanoparticles in one dimension. *Adv. Mater.* 26: 2501.

50 Wang, Y.Z., Wei, C., Cong, H.L. et al. (2016). Hybrid top-down/bottom-up strategy using superwettability for the fabrication of patterned colloidal assembly. *ACS Appl. Mater. Interfaces* 8: 4985.

51 Zhang, B., Meng, F.S., Feng, J.G. et al. (2018). Manipulation of colloidal particles in three dimensions via microfluid engineering. *Adv. Mater.* 30: 1707291.

52 Guo, D., Li, C., Wang, Y. et al. (2017). Precise assembly of particles for zigzag or linear patterns. *Angew. Chem. Int. Ed.* 56: 15348.

53 Zhang, J., Zhu, Z.J., Yu, Z.Y. et al. (2019). Large-scale colloidal films with robust structural colors. *Mater. Horiz.* 6: 90.

54 Zhang, Y.F., Dong, B.Q., Chen, A. et al. (2015). Using cuttlefish ink as an additive to produce non-iridescent structural colors of high color visibility. *Adv. Mater.* 27: 4719.

55 Tan, A.T.L., Beroz, J., Kolle, M., and Hart, A.J. (2018). Direct-write freeform colloidal assembly. *Adv. Mater.* 30: 1803620.

56 Tan, A.L.T., Nagelberg, S., Chang-Davidson, E. et al. (2019). In-plane direct-write assembly of iridescent colloidal crystals. *Small* 16: 1905519.

57 Sowade, E., Blaudeck, T., and Baumann, R.R. (2016). Self-assembly of spherical colloidal photonic crystals inside inkjet printed droplets. *Cryst. Growth Des.* 16: 1017.

58 Bai, L., Xie, Z.Y., Wang, W. et al. (2014). Bio-inspired vapor-responsive colloidal photonic crystal patterns by inkjet printing. *ACS Nano* 8: 11094.

59 Yuan, W., Zhou, N., Shi, L., and Zhang, K.Q. (2015). Structural coloration of colloidal fiber by photonic band gap and resonant Mie scattering. *ACS Appl. Mater. Interfaces* 7: 14064.

60 Gao, B.B., Tang, L., Zhang, D.G. et al. (2017). Transpiration-inspired fabrication of opal capillary with multiple heterostructures for multiplex aptamer-based fluorescent assays. *ACS Appl. Mater. Interfaces* 9: 32577.

61 Zhao, Z., Wang, H., Shang, L.R. et al. (2017). Bioinspired heterogeneous structural color stripes from capillaries. *Adv. Mater.* 29: 1704569.

62 Yuan, S.J., Meng, W.H., Du, A.H. et al. (2019). Direct-writing structure color patterns on the electrospun colloidal fibers toward wearable materials. *Chin. J. Polym. Sci.* 37: 729.

63 Rastogi, V., Melle, S., Calderón, O.G. et al. (2008). Synthesis of light-diffracting assemblies from microspheres and nanoparticles in droplets on a superhydrophobic surface. *Adv. Mater.* 20: 4263.

64 Zhao, Y.J., Gu, H.C., Xie, Z.Y. et al. (2013). Bioinspired multifunctional janus particles for droplet manipulation. *J. Am. Chem. Soc.* 135: 54.

65 Chi, J.J., Shao, C.M., Zhang, Y.L. et al. (2019). Magnetically responsive colloidal crystals with angle-independent gradient structural colors in microfluidic droplet arrays. *Nanoscale* 11: 12898.

5

2D and (2+1)D Colloidal Photonic Crystal

Lijing Zhang, Jiaqi Han, and Bofan Liu

Dalian University of Technology, School of Chemical Engineering, No.2 Linggong Road, Dalian, Liaoning 116024, China

5.1 Colloidal Photonic Crystals

Colloidal photonic crystals (CPCs) are an ordered array or packing of monodisperse colloidal particles, which afford periodic, spatial variations in dielectric properties. Bragg diffraction of light within colloidal crystals gives rise to a stopband, in which the propagation of light within a narrow range of wavelengths in specific directions is prohibited and presents a brilliant structural colors. Gem opals are natural CPCs (Figure 5.1), which are made from the ordered deposition of spherical silica particles after years of siliceous sedimentation and compression under hydrostatic and gravitational forces. Different from the real crystals that comprised from ions, atoms, and molecules, the crystalline elements at the lattice site for a colloidal crystal are micrometer and submicrometer scale colloids. The most commonly used monodisperse colloidal particles are inorganic silica spheres prepared via Stober method [2] and polymeric latex beads prepared through emulsion polymerization [3], such as polystyrene (PS), polymeric methyl methacrylate (PMMA), poly(styrene-acrylic acid) (P(St-AA)), and other corresponding copolymers.

According to the space structure, CPCs are usually classified into 2D and 3D colloidal crystals (Figure 5.2a,b). Recently, (2+1)D CPCs, as a unique structure between 2D and 3D, were proposed by Romanov et al. [4] as a new type of CPCs to be studied (Figure 5.2c). For the 3D CPCs, numerous reviews on their optical properties, preparation methods, and applications have been reported, while less attention was paid to 2D and (2+1)D CPCs. In this chapter, we give a systematic, balanced, and comprehensive summary of the main aspects of 2D and (2+1)D CPCs related to their preparation, optical property, and application, and we propose perspectives for their future development.

Functional Materials from Colloidal Self-assembly, First Edition. Edited by Qingfeng Yan and George Zhao.

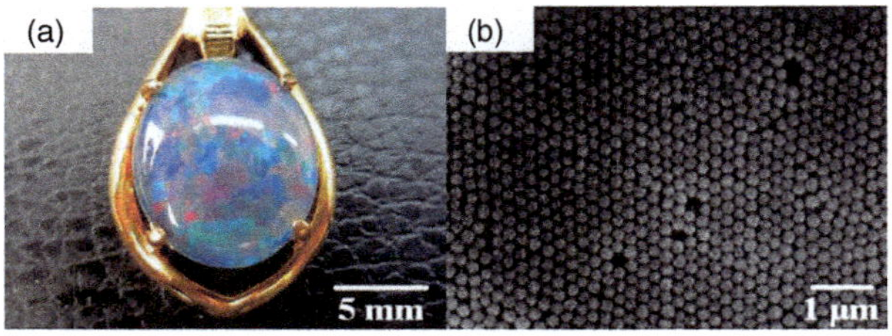

Figure 5.1 Optical and SEM images of natural colloidal crystals of gem opals to show structural colors (a) and packing structures (b). Source: Cong et al. [1]/with permission of Royal Society of Chemistry.

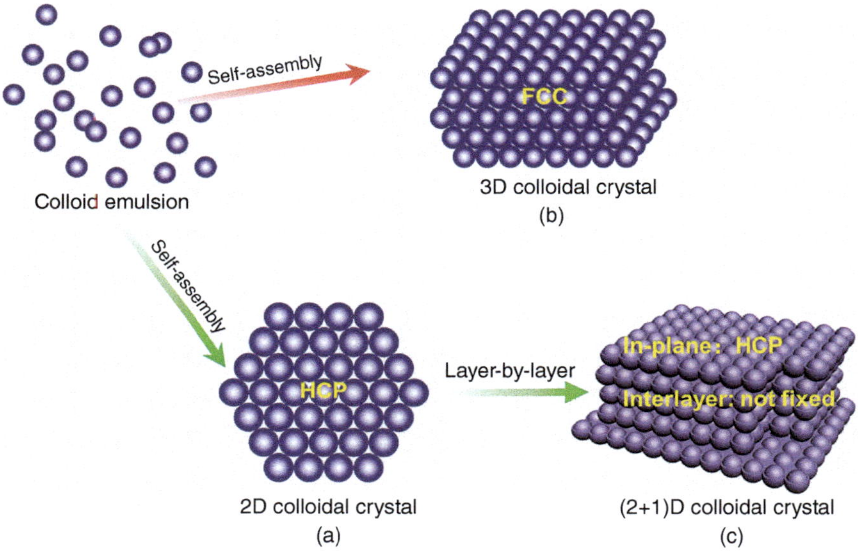

Figure 5.2 Types and structures of common colloidal photonic crystals formed from self-assembly.

5.2 2D Colloidal Photonic Crystal

2D CPCs (Figure 5.3) are composed of close-packed hexagonal colloidal particles, a single-layer of opal photonic crystal, which is the basic building unit of 3D and (2+1)D CPCs and is also a universal template for preparing a variety of 2D nanostructure arrays.

5.2.1 Preparation Methods

A variety of methods have been developed to the fabrication of 2D CPCs, including solvent evaporation-induced self-assembly (such as drop-coating and dip-coating), spin coating, electrophoretic deposition, and air/liquid interface self-assembly, as illustrated in Figure 5.4. The colloidal microspheres dispersed in the solvent will

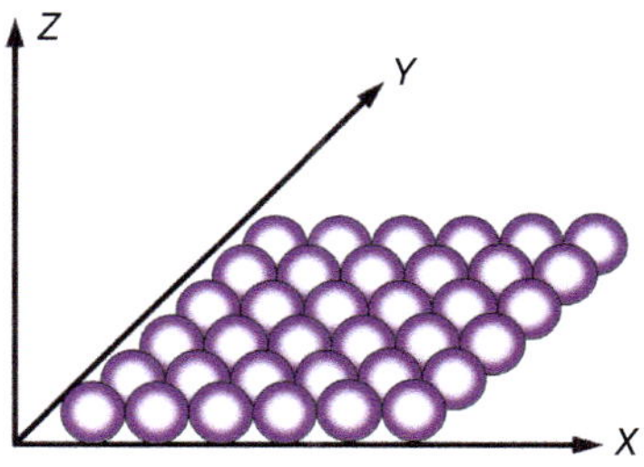

Figure 5.3 Schematic diagram of 2D colloidal photonic crystal structure.

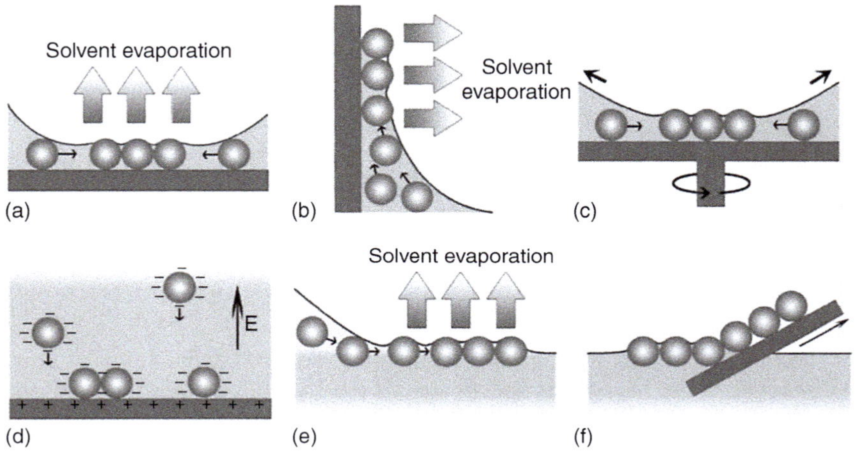

Figure 5.4 Diverse self-assembly strategies toward hcp 2D CPCs: (a) drop-coating, (b) dip-coating, (c) spin coating, (d) electrophoretic deposition, (e) self-assembly at the gas/liquid interface, and (f) transfer from the gas/liquid to the gas/solid interface. Source: Reproduced from Ye and Qi [5]. Copyright 2011, Elsevier.

be subjected to various forces, including van der Waals force, steric hindrance, and Coulomb repulsion. The self-assembly process of colloidal microspheres is the result of the balance between these forces. Hexagonal-close-packed (hcp) structure is the most commonly obtained and the most easily self-assembled 2D CPCs, since the hcp structure is a thermodynamically stable 2D arrangement. The conditions and parameters of self-assembly, such as the concentration of the colloidal emulsion, the composition of the solvent, the temperature and humidity of the environment, etc., are essential for controlling the morphology and quality of the colloidal crystals.

5.2.1.1 Solvent Evaporation-Induced Self-assembly

Nagayama and coworkers first observed the evaporation-induced convective assembly of randomly dispersed colloidal particles into hcp 2D CPCs by spreading a drop of dilute colloidal particle suspension on a horizontal flat substrate [6]. During such a drop-coating procedure (Figure 5.4a), the attractive capillary force and the convective transport of the particles arising from the continuous solvent evaporation are the main factors dominating the self-assembly process, the ordering and quality of the obtained arrays are largely determined by the evaporation rate. Later on, Dimitrov and Nagayama developed a dip-coating procedure for continuous formation of hcp 2D CPCs on a vertical flat substrate by accurately controlling the rates of water evaporation and substrate withdraw (Figure 5.4b), resulting in centimeter-sized, polycrystalline 2D CPCs [7]. Basically, the substrate can be inserted in the suspension at any tilted angles, and a slow and homogeneous evaporation is favorable for the formation of 2D colloidal crystal monolayer instead of multilayer colloidal crystals.

5.2.1.2 Spin Coating

During spin coating (Figure 5.4c), the solvent flows across a wettable substrate at high shear rates, and the colloidal spheres are densely packed into hcp 2D CPCs on the substrate rapidly. The quality and thickness of the resulting colloidal crystals are mainly affected by the spin speed, the concentration of the colloidal suspension, the rheology of the suspension, the wettability of the substrate, and the charges of the substrate and nanospheres [8]. Because the spin-coating process is rapid and compatible with wafer-scale processes, the spin-coating methods have an advantage for both scaling-up and mass production.

5.2.1.3 Electrophoretic Deposition

In electrophoretic deposition (Figure 5.5), a thin layer of colloidal suspension is confined between two electrodes, and an applied electrical field drives the charged nanospheres to move toward the electrodes, leading to their self-assembly into 2D CPCs on the electrode interfaces. Electrophoretic deposition in direct current fields [10] or alternating current fields [9] has been applied for the rapid self-assembly and facile manipulation of 2D CPCs because electrophoretic movement not only accelerates the sedimentation speed of nanospheres but also guides the growth of 2D CPCs over a large area in a controlled manner.

5.2.1.4 Air/Liquid Interface Self-assembly

The air/liquid interface self-assembly is considered as the main preparation method due to its easy operation, high efficiency, and high success rate. The alcohol–water mixed dispersion of colloidal particles spreads rapidly on the high surface tension liquid surface (usually the water surface or mercury surface) and spontaneously forms a hexagonal close-packed unit on the liquid surface, that is, 2D CPC structure. This is because the surface tension of the alcohol–water mixture is low, when the solvent comes into contact with a liquid surface with high surface tension (such as the water surface), under the action of the Marangoni effect caused by the surface tension gradient, the alcohol–water solution spreads out quickly on the water surface. Usually, at the end of self-assembly, external force or a small amount of surfactant will be added to help the colloidal array to be packed more tightly and orderly [11]. Based on the air–liquid interface self-assembly methods, researchers have developed a series of improved strategies, including LB method, glass slide-assisted air/liquid self-assembly, needle tip flow method, and gas flow-induced self-assembly:

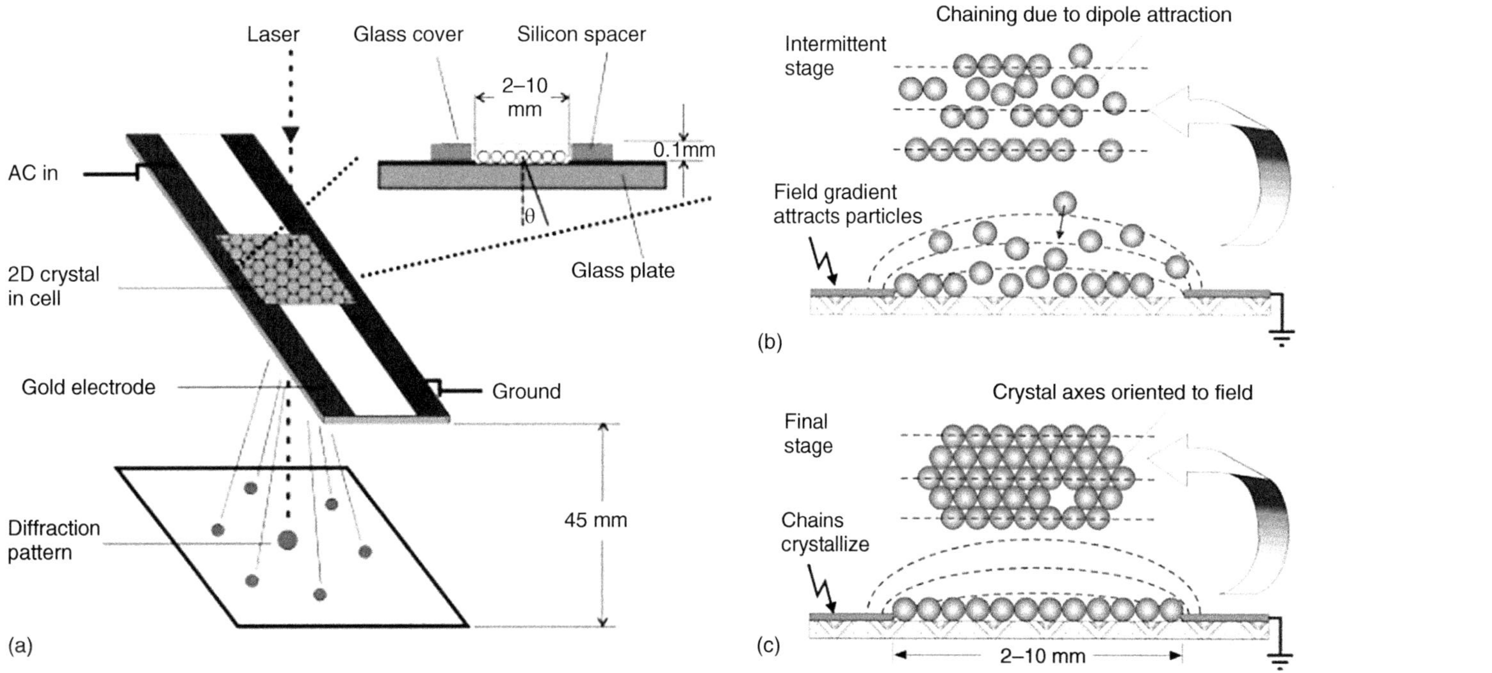

Figure 5.5 (a) Schematics of the experimental arrangement of electrophoretic deposition for colloidal assemble. (b) Particle chain formed immediately after the field is applied. (c) The particle chains confined on the surface crystallize to form 2D hexagonal crystals aligned by the field. Source: Reproduced from Lumsdon et al. [9]. Copyright 2004, American Chemical Society.

The Langmuir–Blodgett (LB) Method Langmuir–Blodgett (LB) method is introduced as a powerful method for the assembly and orientation control of colloidal particles to fabricated 2D CPCs (Figure 5.6). van Duffel et al. first published the results of depositing SiO_2 monolayers on solid substrates by the LB technique [13]. On one hand, the particles were rendered hydrophobic and were suspended in sodium dodecylsulfate (SDS) solution in chloroform/ethanol. On the other hand, the substrate (glass or Si) was treated so as to have a hydrophilic surface and is initially immersed in water, and deposition on the upstroke was obtained at a pressure of 5 mN/m. van Duffel et al. [13] also found that they were unable to deposit monolayers without surface hydrophobicity and SDS, because the spheres did not float at the air/water interface. Muramatsu et al. studied the formation of LB monolayers of SiO_2 and TiO_2 particles, which they purchased and then treated with octadecylamine (ODA). They varied the pH of the subphase and obtained good 2D films of SiO_2 at neutral pH [14]. Ábrahám et al. [15] prepare Langmuir films at the air/water interface and LB films on solid (glass, quartz, Si wafer) substrates in the Kibron MicroTrough. They found that the critical surface pressures for hydrophilic and hydrophobic microspheres are about 61 and 46 mN/m, respectively [16]. The decrease of the surface hydrophilicity of microspheres facilitates their self-assembly on the water surface, which further leads to higher coverage and less defects of the 2D CPCs. A coverage of as high as 97% was obtained when using hydrophobic microspheres.

Needle Tip Flow Method Smith et al. [17] fabricate 2D CPCs by spreading a dispersion of charged colloidal particles (diameters = 409, 570, and 915 nm) onto the surface of electrolyte solutions using a needle tip flow method. The alcohol–water suspension of colloidal particles is guided to the surface of the water through a narrow pipe such as a needle or capillary, and the interface self-assembly of the colloidal particles is realized under the action of the Marangoni effect (Figure 5.7). The advantage of this method is that the amount of effective colloidal particle suspension injected on the water surface can be arbitrarily controlled, thereby realizing the possibility of large-area preparation.

Gas Flow-Induced Self-assembly Meng et al. [18] fabricated centimeter-sized 2D colloidal single crystals of PS particles at the air/water interface by capillary-modulated self-assembly (Figure 5.8). Different from previous reports, in this work, emulsifier was used to facilitate the stress release during 2D colloidal crystal formation by adjusting the interparticle lateral interactions. With the assistance of compressed

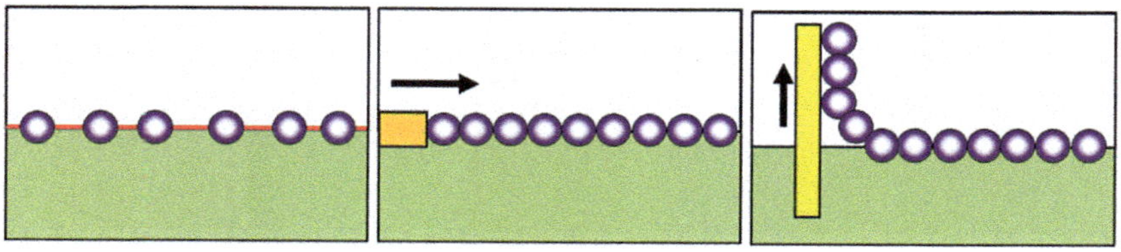

Figure 5.6 LB deposition process. Source: Reproduced with permission from Bardosova et al. [12]. Copyright 2010, WILEY-VCH.

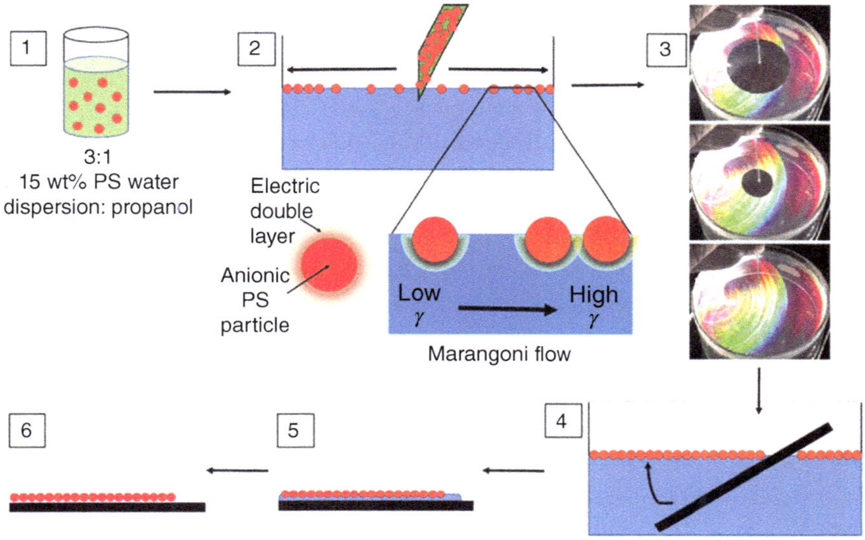

Figure 5.7 Schematic of needle tip flow 2D CPC fabrication at the air/water interface. Source: Smith et al. [17]/with permission of Royal Society of Chemistry.

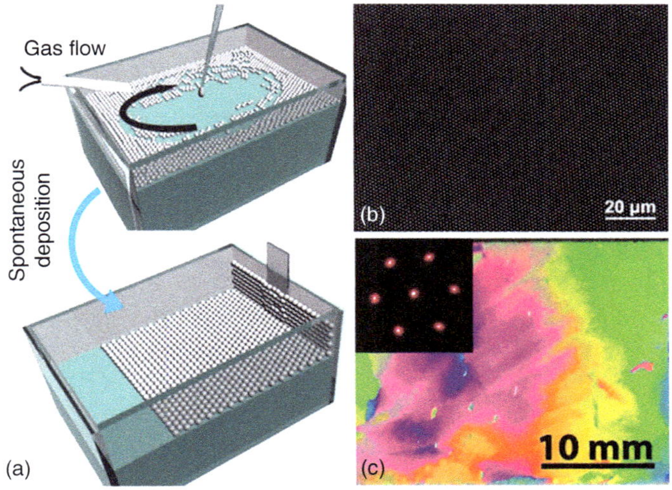

Figure 5.8 (a) Illustration of the formation of 2D colloidal monolayer on the water/air interface with the assistance of gas flow. (b) Optical image (c, image size $140 \times 105\ \mu m^2$) of the prepared mirrorlike colloidal monolayer. (c) Optical images and the corresponding diffraction patterns of the 2D colloidal crystals. Source: Meng and Qiu. [18]/with permission of American Chemical Society.

nitrogen flow, 2D hexagonal colloidal single crystals of centimeter size were obtained under appropriate emulsifier concentrations. A new method was also developed to transfer the 2D colloidal crystals from the air/water interface to the desired substrate without obvious disturbance. This new transferring method was proven not to be sensitive to surface wettability nor curvature; thus 2D colloidal single crystals with large areas could be obtained on different kinds of substrate.

The description of the air/liquid interface self-assembly process highlights three very important points. First, this method is relative simplicity of preparing 2D CPC films of colloidal particles. Second, the particles on the interface can only be assembled into a single layer, and no special measures are needed to control the number of layers of colloidal crystals. For other methods, such as solvent evaporation induction method, the single-layer control is difficult to achieve. Third, 2D CPC films growing on the interface can be transferred to any substrate without considering the surface properties of the substrate, such as hydrophilicity, roughness, polarity, or curvature [19]. However, one problem worth noting lies that the transferred 2D CPCs always show irregular cracks after the liquid is eventually evaporated, which attributes to the lateral shrinkage of the colloidal spheres with respect to the solid substrate [20–22]. To solve this problem, Yan et al. developed the *in situ* thermal annealing and solvent vapor annealing method to the treatment of a floating colloidal monolayer at the air/liquid interface, which resulted in high-quality colloidal crystal monolayer with few defects, especially when they need to be transferred to rigid and bent substrate (Figure 5.9) [23–25]. Such an annealed colloidal crystal monolayer presents less cracks and enhanced mechanical strength. The density of dislocations and cracks of the annealed monolayer transferred to a solid substrate is about 1/3 of that of the unannealed monolayer [24, 25].

5.2.1.5 Other Strategies

Attractive Force Gradient Method Sun et al. present a novel attractive force gradient method to fabricate high-quality, single-domain colloidal crystals for the first time (Figure 5.10). [26] Colloidal particles in a water–lutidine (WL) binary liquid mixture experience temperature-dependent attraction close to the mixture's demixing temperature. This temperature-tunable interaction can be potentially harnessed to assemble colloids and grow colloidal crystals. This is an efficient and robust method to prepare colloidal crystals with little or no defects, being suitable for applications such as colloidal lithography and the fabrication of perfect 3D colloidal crystals.

Direct Writing Method Leveraging the principles of convective assembly, the direct writing of colloidal suspensions is presented as a technique for fabrication of iridescent colloidal crystals in arbitrary 2D patterns. Tan et al. use this method to prepare 2D patterns (Figure 5.11) [27]. The process can be optimized for high writing speeds (≈600 μm/s) at mild process temperature (30 °C) while maintaining long-range (cm-scale) order in the colloidal crystals. However, the quality of 2D CPC is not very satisfactory with many defects occurring.

Roll-To-Roll (R2R) Technique Based on the LB method, Parchine and colleagues further developed a R2R LB technique to fabricate 2D CPCs on flexible PET film in an area as large as 340 cm^2 to meet the demand of high volume manufacturing [28].

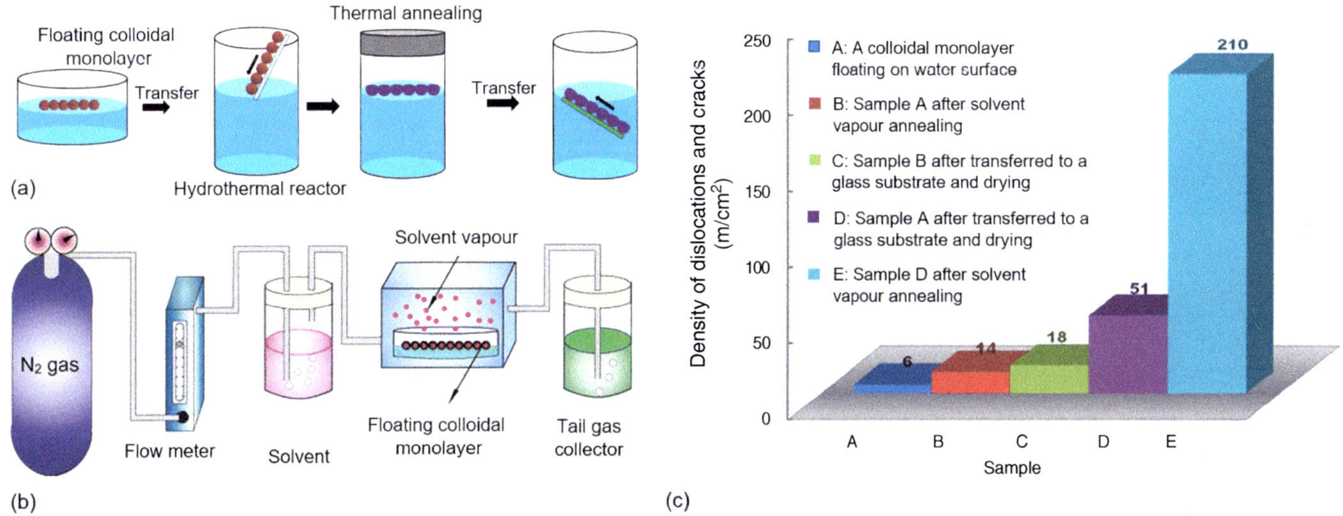

Figure 5.9 Schematics for the preparation of high-quality 2D CPCs by *in situ* annealing treatment: (a) thermal annealing treatment, (b) solvent vapor annealing treatment, (c) density of dislocations and cracks for 2D CPCs fabricated with and without annealing treatment. Source: (a) Reproduced with permission from Geng et al. [23]. Copyright 2012, from the RSC. (b, c) Reproduced with permission from Yu et al. [24]. Copyright 2012, from the American Chemical Society.

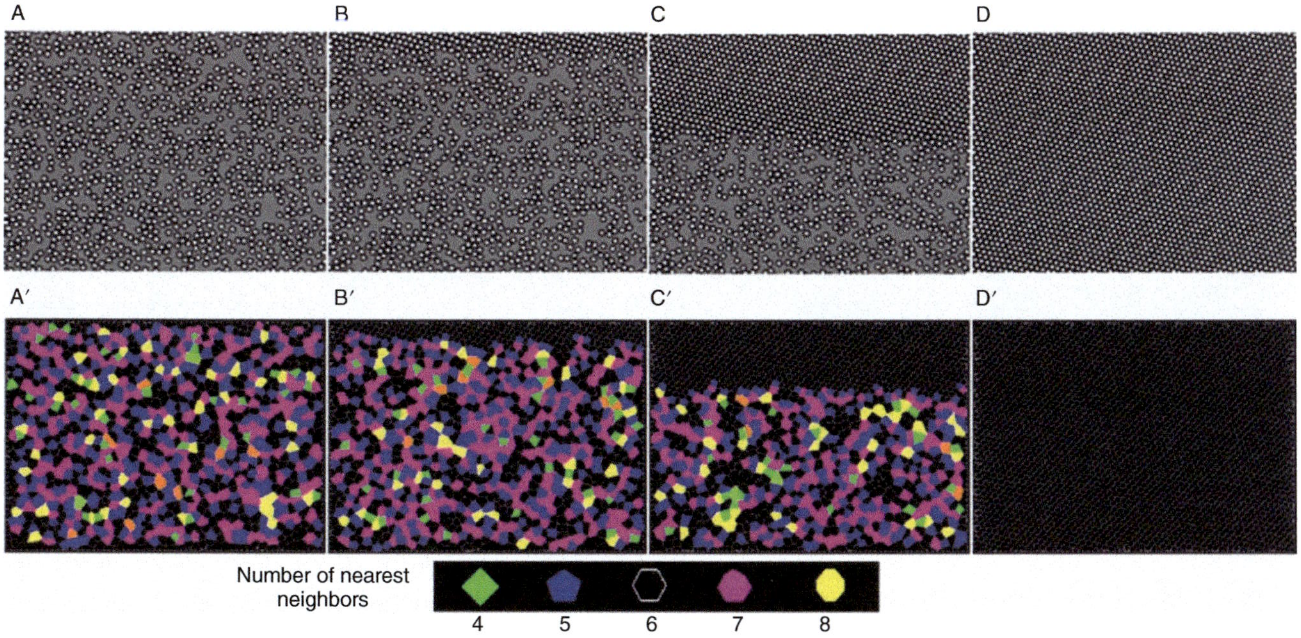

Figure 5.10 Crystallization process. Source: Sun et al. [26]/with permission of American Chemical Society.

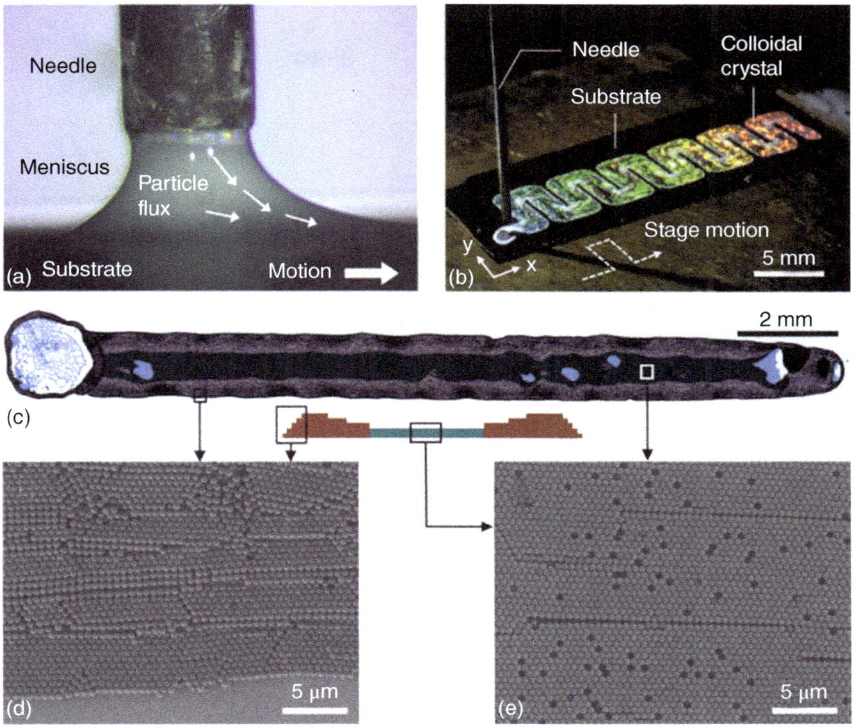

Figure 5.11 Fabrication of colloidal crystals by in-plane direct-write self-assembly. (a) In-plane direct-write self-assembly is performed by precision dispense of a colloidal suspension from a needle, coupled with lateral substrate motion. (b) A serpentine colloidal crystal is drawn by movement of the stage. (c) Optical image (top) and cross-section schematic (bottom) of an exemplary colloidal crystal trace. The edge of the crystal (brown coloration) is thicker than the middle. The middle consists of mostly particle bilayers (green coloration) and monolayers (blue coloration). (d) SEM image showing multilayer terraces at the edge of the crystal. (e) SEM image of the middle region of the crystal showing defects such as dislocations and vacancies. Source: Tan et al. [27]/with permission of WILEY-VCH.

Figure 5.12a shows the schematic illustration of the R2R LB method; the improvement lies in that a special R2R unit replaces the standard alternate dipper mechanism. Figure 5.12b,c shows large-area monolayer of SiO_2 with iridescent color deposited on the PET film. The SEM image in Figure 5.12d displays the high quality of the 2D hexagonal close-packed structure of the SiO_2 monolayer fabricated by R2R LB method. Cagnani et al. [29] investigated SiO_2 suspensions in ethanol with different concentrations and particle sizes (diameters between 150 nm and 520 nm) that were directly deposited onto PET, ITO/PET, and stainless steel substrates via wire-bar R2R. An empirical equation was proposed that one can predict the optimal concentration of the colloidal suspension as a function of the processing parameters necessary to produce a colloidal monolayer. The development of the R2R method promises to overcome the technological obstacles for the large-area fabrication of high-quality 2D CPCs.

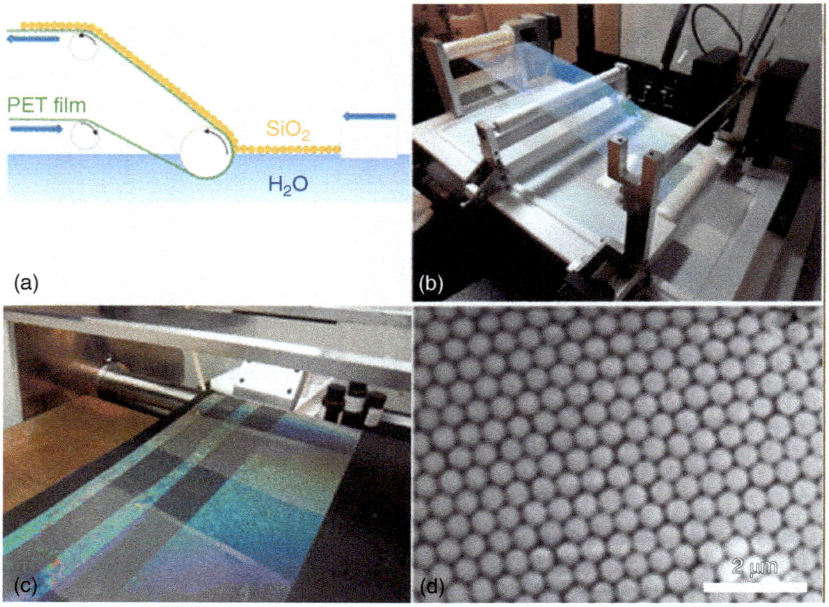

Figure 5.12 (a) Schematic illustration of the roll-to-roll LB method. (b) Photograph of the SiO_2 monolayer deposited on PET substrate by the roll-to-roll unit. (c) Photograph of the roll-to-roll slot-die coating with strips of SiO_2 spheres plus optical adhesive NOA164 on the PET. (d) SEM images of the monolayer of 550 nm SiO_2 spheres assembled on the PET film using the roll-to-roll LB method. Source: Parchine et al. [28]/with permission of American Chemical Society.

5.2.2 Optical Properties

5.2.2.1 Diffraction

In terms of optical properties, the 2D CPCs will also produce Bragg diffraction on the incident light. For incident light normally incident to the array, the diffraction follows Bragg's law:

$$m\lambda = \sqrt{3}d \sin\theta \tag{5.1}$$

where, m is the diffraction order, λ refers to the diffracted wavelength in air, d is the 2D particle spacing, and the θ is the angle between the incident light and the normal direction of the 2D colloidal crystal array. It can be seen from the formula that both θ and d can be adjusted to obtain diffraction in the visible spectral region [30]. Figure 5.13a shows the diffraction spectra of 2D arrays of 523 nm, 626 nm, 673 nm, and 748 nm PMMA particles, which are measured at 30° angle. Their diffraction peak redshifts from 470 to 621 nm as the particle diameter increases. This strong diffracted light can be observed by the naked eye by placing a mirror under the photonic crystal, which is the structural color of the 2D photonic crystal. Figure 5.13b shows the corresponding back-diffracted color from blue redshifts to green, yellow, and red. This feature is the basis for the application of 2D photonic crystals in visual detection and sensing.

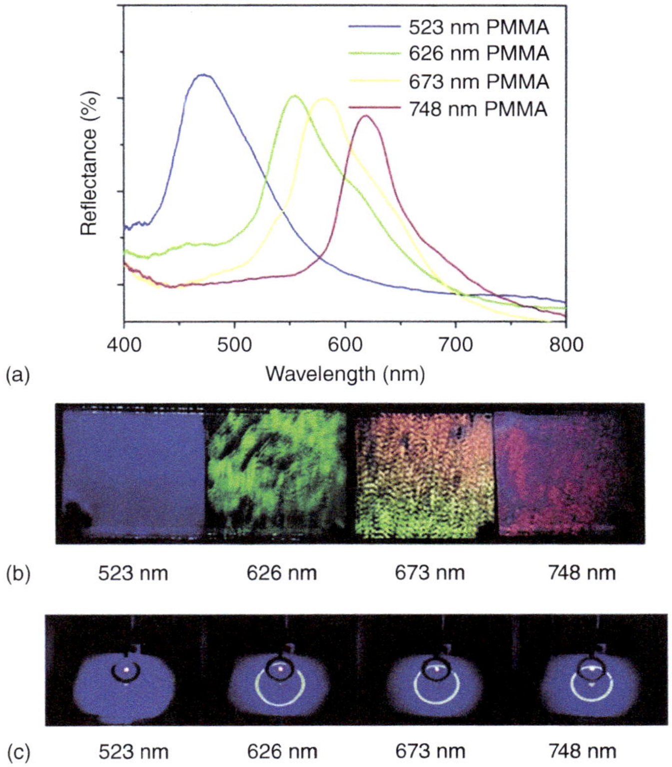

Figure 5.13 (a) Diffraction spectra measured in Littrow configuration of 2D monolayer arrays of PMMA particles with diameters of 523, 626, 673, and 748 nm on a mirror. (b) Photograph of diffracted color of forward diffracted white light at 30 degrees from the normal, and (c) Debye diffraction ring. Source: .Xue et al [30]/with permission of Royal Society of Chemistry.

5.2.2.2 Debye Diffraction Ring

Another important feature of 2D CPCs is the Debye diffraction ring. If the incident light is monochromatic and incident perpendicularly, Debye diffraction will occur on the back of the 2D CPCs. When the monochromatic light is irradiated perpendicularly on the surface of a strictly defect-free 2D CPC, the same side of the transmitted light will produce six bright spots (Figure 5.14a). But if the area of the monochromatic light beam is larger than the single crystal area of the 2D CPC, the diffracted spots will be connected together to form a bright ring, also known as the Debye diffraction ring (Figure 5.14b). These special optical properties of 2D CPC make them have potential applications in the fields of optical devices, optical waveguides, and biochemical sensing materials.

The Debye diffraction of the 2D colloidal array is as follows:

$$\sin \alpha = 2\lambda/\sqrt{3}d \tag{5.2}$$

where, α is the forward diffraction angle of the Debye diffraction, λ is the incident wavelength in air, and d is the particle spacing.

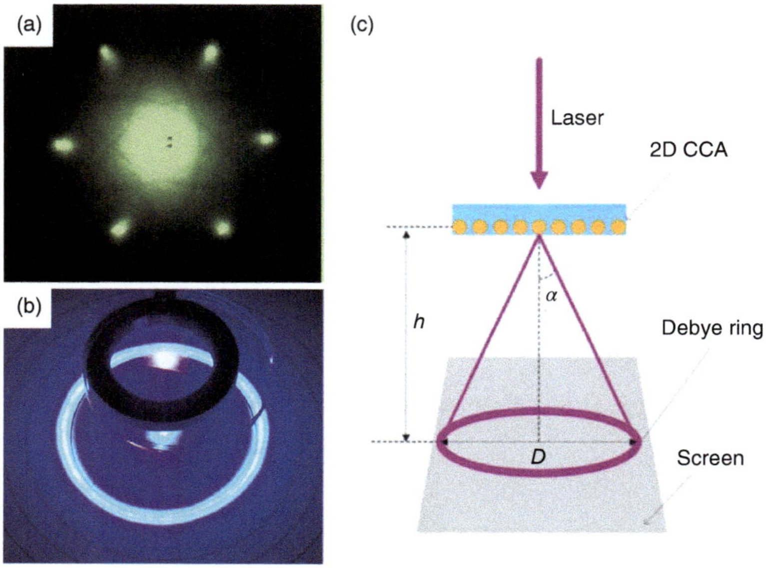

Figure 5.14 (a) Debye diffraction spot and (b) diffraction ring produced by a 2D photonic crystal to vertically incident monochromatic light. Source: Asher et al. [31]/with permission of American Physical Society. (c) Schematic illustration of the Debye diffraction ring pattern. Source: Reproduced with permission from Xue et al. [32]. Copyright 2014, RSC.

Figure 5.14c shows the illustration of Debye diffraction ring measurement. h is the distance between the 2D array plane and the screen. D is the average diameter of the diffraction ring on screen. The diffraction angle, α, can be calculated from

$$\tan \alpha = D/2h \tag{5.3}$$

Then,

$$d = 2\lambda\sqrt{(D^2 + 4h^2)}/\sqrt{3}D \tag{5.4}$$

Monitoring the 2D particle spacing is easy by measuring the Debye ring diameter; the diameter of the Debye ring is inversely proportional to the 2D particle spacing (Figure 5.13c). This feature also makes 2D CPC have potential applications in the fields of biochemical sensing.

5.2.3 Application

5.2.3.1 Templates

Due to the advantages of low preparation cost, high throughput, good repeatability, and easy control of chemical composition, 2D CPCs act as an efficient and universal template in fields of 2D nanostructure pattern design, colloidal lithography, nano-processing, and other surface patterning technologies (Figure 5.15) [5, 34, 35]. Especially, the combination of 2D colloidal crystal templates with

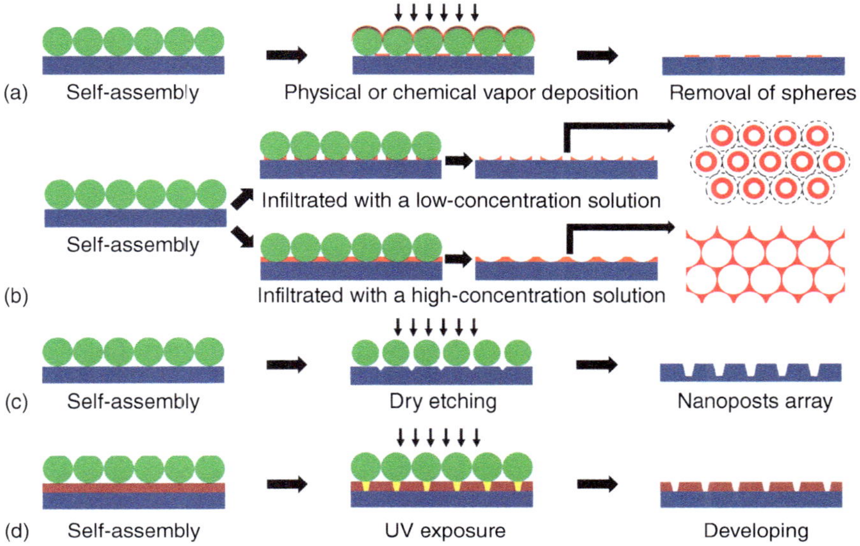

Figure 5.15 Schematic illustrations of the nanostructures derived from a self-assembled 2D CPC. 2D CPC is used as a physical mask for selectively depositing materials by using (a) physical vapor deposition or (b) wet chemical infiltration. 2D CPC is employed as an etching mask to prepare an ordered nanoposts array (c) and serves as an optical microlens array to focus ultraviolet light for generating sub-wavelength holes on a photoresist layer (d). Source: Reproduced with permission from Geng et al. [33]. Copyright 2014 Wiley-VCH.

different fabrication techniques, such as physical vapor deposition, wet chemical infiltration, dry etching, reactive ion etching, electrochemical deposition, UV lithography, etc. provides abundant means and strategies for the preparation of 2D patterned nanostructures. A variety of 2D patterned nanostructures with a high reproducibility have been fabricated, such as ordered nanowire (SiO_2, TiO_2, ZnO), nanorod array (SiO_2, TiO_2, ZnO), Si nanocones array, nanoparticle arrays (Au, Ag, Cu, CdS), nanorings (Au, CdSe, $CsPbBr_3$ [36], $C_4N_2H_{14}Pb_xMn_{1-x}Br_4$ [37]), GaSb "micro-candle stands," graphene nanomesh, AgS nononets, gold nanopatterns, 2D ordered nanobowl array (Ag film, metal–organic framework (MOF) [38]), PbS nanostars, etc. (Figure 5.16). These obtained 2D patterned nanostructures have been widely applied in various fields, such as photonics, plasmonics, SERS, antireflection, surface wetting, biological and chemical sensing, solar cells, photocatalysis, field emission, biomimetic fabrication, and other biological and electronic applications.

For example, Du et al. [52] prepared a series of 2D ordered arrays of silver half-shell by depositing a thin silver layer onto 2D CPCs composed of PS microspheres and studied their plasmonic properties (Figure 5.17). Using Rh6G as the probe molecules, they show that the SERS signal can be efficiently improved by controlling the silver layer thickness or the microsphere size. Under the excitation of a laser with 514 nm in wavelength, the SERS signal intensity is significantly higher than that of flat silver film; the largest spatially average enhancement factor

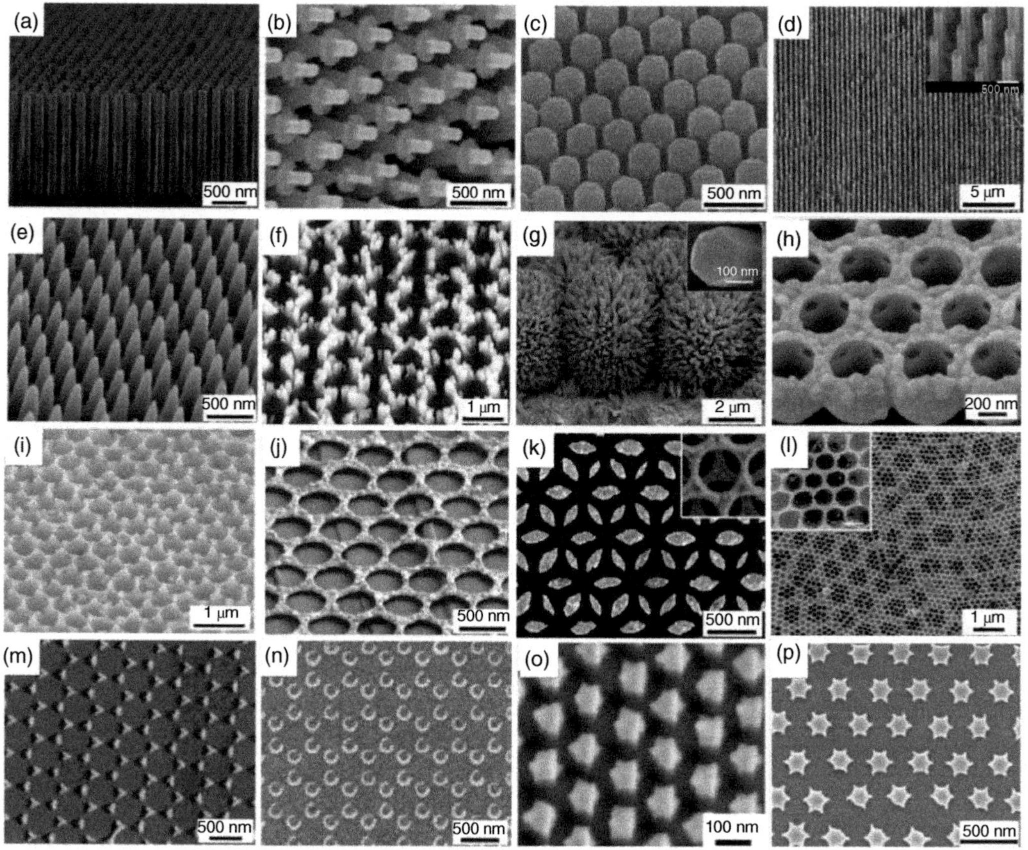
(a)
500 nm
(b)
500 nm
(c)
500 nm
(d)
500 nm
5 μm
(e)
500 nm
(f)
1 μm
(g)
100 nm
2 μm
(h)
200 nm
(i)
1 μm
(j)
500 nm
(k)
500 nm
(l)
1 μm
(m)
500 nm
(n)
500 nm
(o)
100 nm
(p)
500 nm

Figure 5.16 SEM images of patterned nanoarrays obtained by 2D photonic crystal template. (a) Ordered silicon nanowire radial p–n junction arrays made by deep reactive ion etching. Source: Garnett and Yang [39]/with permission of American Chemical Society. (b) GaSb "micro-candle stands" made by stepwise selective etching. Source: Sun et al. [40]/with permission of Royal Society of Chemistry. (c) TiO_2 nanorod array by PLD method. Source: Li et al. [41]/with permission of WILEY-VCH. (d) Periodic ZnO nanorod arrays by vapor growth. Source: Zhang et al. [42]/ with permission of Royal Society of Chemistry. (e) a-Si:H nanocone arrays. Source: Zhu et al. [43]/with permission of American Chemical Society. (f) Patterned ZnO nanorod arrays grown by electrochemical deposition. Source: Zhang et al. [42]/with permission of WILEY-VCH. (g) Ordered hollow urchin-like structure of ZnO nanowires fabricated by electrochemical deposition. Source: .Elias et al. [44]/with permission of WILEY-VCH. (h) 2DOM ZIF-8 thin films obtained by asymmetric growth based on 2D PCs. Source: Li et al. [38]/with permission of American Chemical Society. (i) Ag nanobowl array fabricated via the nanosphere lithography at the gas/liquid interface. Source: Hong et al. [45]/with permission of WILEY-VCH. (j) Ag_2S nanonets array on a silicon plate with terraces and (k) gold nanopatterns created by evaporation deposition with bilayer Ag_2S nanonets (inset) as the deposition masks. Source: Li et al. [46]/with permission of American Chemical Society. (l) In_2O_3 hierarchical porous structure obtained by solution dipping deposition. Source: Jia et al. [47]/with permission of American Chemical Society. (m) AFM image of periodic Ag particle arrays by thermal evaporation. Source: Haynes et al. [48]/with permission of American Chemical Society. (n) Periodic Au split ring resonator arrangement by angle-resolved deposition combining rotating the templates. Source: Gwinner et al. [49]/with permission of WILEY-VCH. (o) PbS nanostars array assembled by vertical deposition. Source: Huang et al. [50]/with permission of American Chemical Society. (p) PS nanostar arrays made by electron irradiation and subsequent heating of PS colloidal monolayers. Source: Li et al. [51]/with permission of American Chemical Society.

of SERS signal is on the order of $\sim 3 \times 10^7$ when the silver nanostructure is patterned by sputtering$\sim$100-nm-thick silver layer onto 2D colloidal microspheres of 360 nm in diameter. Numerical simulations revealed that this huge enhancement of Raman signal is attributed to the excitation of a dipolar plasmonic resonance of the silver nanostructure based on the 2D CPC template.

The emerging new materials combined with 2D CPC template technology can often enrich the types of new materials and improve their properties and functions. Li et al. [38] presented a facile approach toward producing multifunctional MOF superstructures by asymmetric growth of continuous MOFs thin films on 2D CPC arrays anchored at the air–solution interfaces, in which the control over spatial configuration, structural hierarchy, and overall dimensionality of MOF superstructures can be realized (Figure 5.18). Taking advantage of the resultant periodic and hierarchical porous structures, the as-grown MOF superstructures lend themselves to superior performance in fields of efficient vapor sensing, size-screening of nanoparticles, and removal of dye molecules from aqueous solutions. Xu et al. [36] developed a simple method using 2D colloidal crystal templates to achieve perovskite ring arrays of $CsPbBr_3$ and showed their application in the field of lasers. They further reported a $C_4N_2H_{14}Pb_xMn_{1-x}Br_4$ perovskite ring array [37], which has better stability in humid environments compared with perovskite thin films without a template. The LED chip based on a $C_4N_2H_{14}Pb_xMn_{1-x}Br_4$ ring array film showed high color quality.

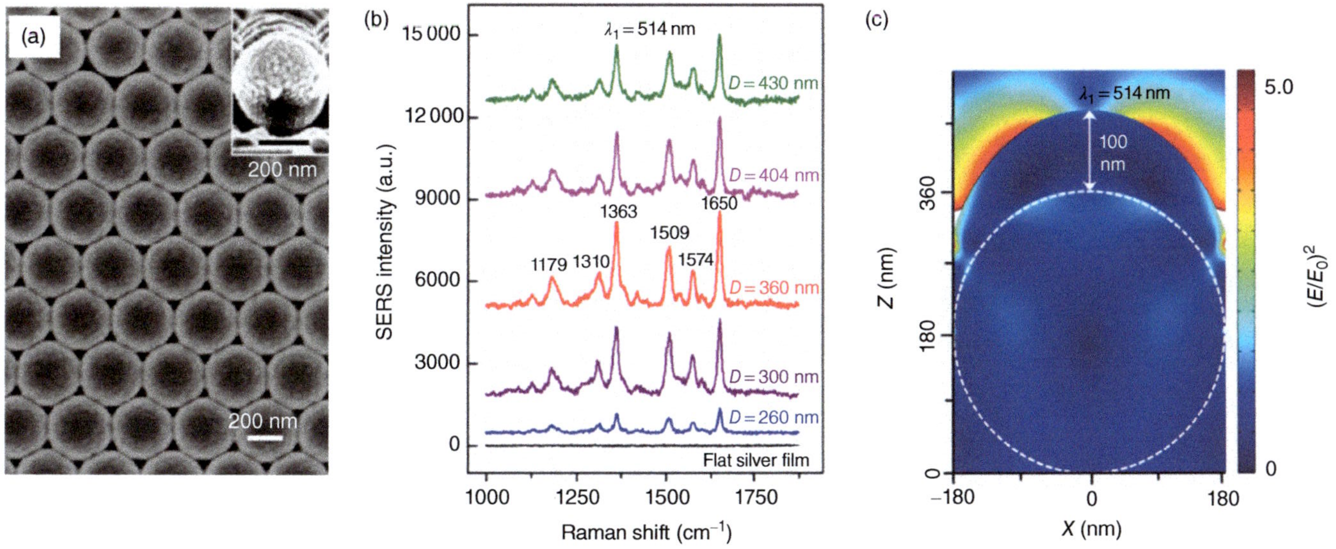

Figure 5.17 (a) Top-view and cross-sectional (inset) SEM images of a silver half-shell array patterned on a monolayer of colloidal crystal composed by PS microspheres of 360 nm in diameter. (b) Typical SERS spectra of R6G molecules on a series of silver half-shell array substrates with PS microspheres ranging from 260 to 430 nm but with a fixed silver thickness of ~100 nm. As a reference, black line is the Raman spectrum from the 100-nm-flat silver film deposited on a quartz substrate. (c) Calculated total electric field distribution at kl = 514 nm for silver half-shell array template by 360-nm-diameter PS microspheres. Source: Du et al. [52]/with permission of Springer Nature.

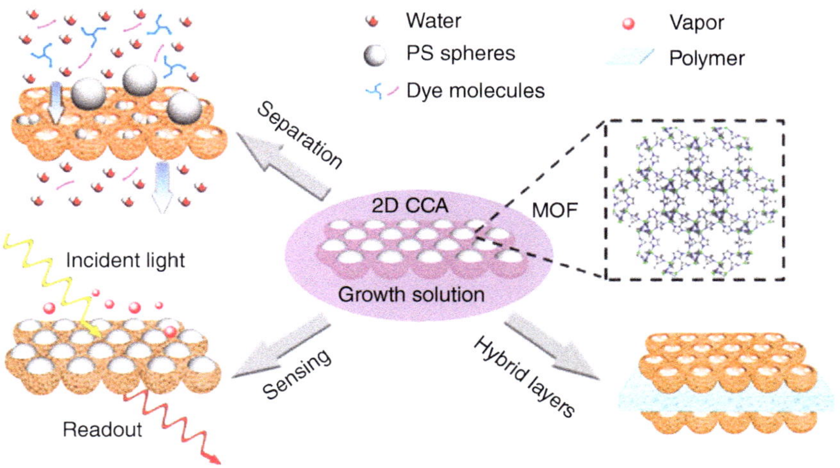

Figure 5.18 MOF superlattice based on 2D CPC template and their applications. Source: Reproduced with permission from Li et al. [38]. Copyright 2016, American Chemical Society.

5.2.3.2 Detection and Sensing

In recent years, the combination of 2D CPC structure and stimulus-responsive materials has gradually attracted the attention of researchers for the development of biochemical sensing materials. The theoretical basis for 2D CPCs to achieve analyte detection lies in the 2D array diffraction and Debye ring diffraction. In the 2D array diffraction phenomenon, the wavelength of the diffraction light is proportional to the particle spacing, while in Debye ring diffraction, the diameter of the Debye ring is inversely proportional to the particle spacing, i.e. *Formulas 5.1 and 5.4.* When used for analyte detection, responsive hydrogels with specific ligand were attached to the 2D CPC array, and the cross-links between analyte and ligand will cause the hydrogel volume to shrink or swell, which results in the change of diffraction wavelength and the diameter of Debye ring, so as to realize the quantitative detection of the analyte.

Professor Sanford Asher of the University of Pittsburgh has done a lot of pioneering work in the development of 2D CPC biochemical sensors. He embedded a 2D CPC array on the surface of the smart gel material to obtain a 2D CPC gel film. The response of the gel to the external environment causes the change of the lattice parameters of the 2D CPC array, thereby changing the wavelength of its diffracted light and realizing the output of optical signals. Based on the guidance of this idea, Asher has successively realized the detection of temperature [53], ionic surfactants [54], and avidin [55] (Figure 5.19), etc.

The advantages of Debye ring diffraction measurement lie in many aspects: First, it is a low-cost detection method that does not require expensive detection equipment, especially when the diffraction shifts to the near-IR region and the diffraction cannot be measured by a visible wavelength spectrometer. Second, the Debye diffraction ring method does not require careful control of the incident and diffracted light angles; this is especially important for delicate samples that are hard to handle. Third, the detection range is much wider. In addition, compared with the

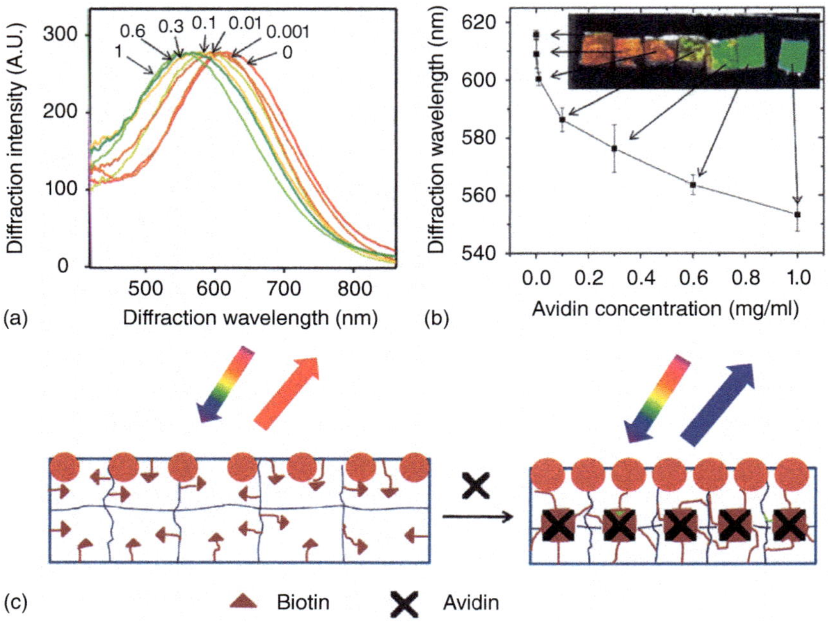

Figure 5.19 (a) Dependence of normalized diffraction spectra of 2D biotin hydrogel sensors upon avidin concentration in 0.1 M NaCl aqueous solution. Diffraction is measured in a Littrow configuration with an angle between the probe and the 2D array normal of ~19°. (b) Dependence of diffraction wavelength on avidin concentration. Inset shows photographs taken close to the Littrow configuration at an angle of 19° between the light source/camera and the 2D array normal. (c) Schematic showing avidin molecule binding to multiple hydrogel biotins. This increases the cross-linking, blue shifting the diffraction. Source: Zhang et al. [55]/with permission of Royal Society of Chemistry.

semiquantitative colorimetric detection, the Debye diffraction detection method can achieve accurate quantification. By measuring the diameter of Debye diffraction ring, Li et al. [56] developed a label-free urease coupled colloidal crystal hydrogel (UCCH) biosensor for the visual monitoring of urea and urease inhibitor phenyl phosphorodiamidate (PPD), and the detection limits for urea and PPD are 1 mM and 5.8 nM, respectively (Figure 5.20). Wang et al. [57] realized the high sensitivity and high selectivity detection of lysozyme, and the detection limit was confirmed to be 1.38×10^{-3} mg/ml based on the 2D functional hydrogel colloidal crystal array.

2D CPCs are an important part of the photonic crystal family. Although there have been encouraging developments in the preparation and application of 2D CPC, it still faces some challenges. For example, how to develop a simple and effective method for preparing large-area defect-free 2D CPC is still the direction that people are pursuing. Another problem is how to controllably self-assemble to prepare more complex and fine materials based on 2D photonic crystal structures. Therefore, expanding or optimizing its optical performance is also a common challenge faced by researchers.

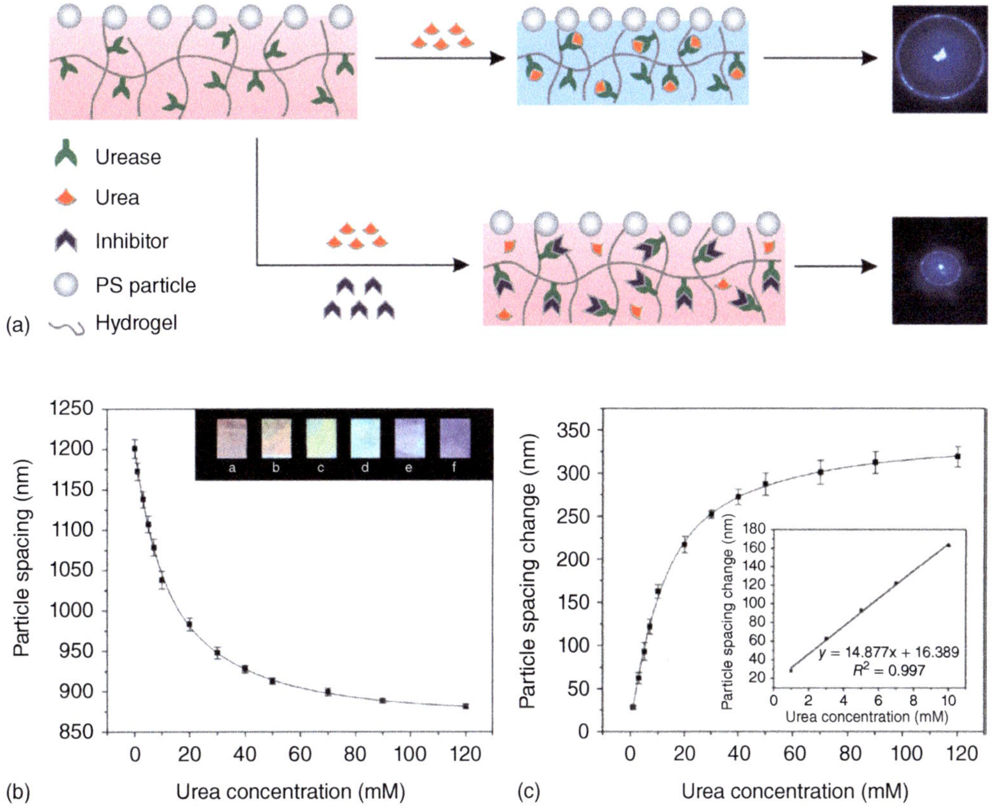

Figure 5.20 (a) Schematic representation of the detection of urea and urease inhibitor based on the UCCH biosensor. (b) Dependence of the particle spacing of the UCCH upon urea concentration. (c) The relationship between the particle spacing change of the UCCH and the urea concentration. Source: Li et al. [56]/with permission of Elsevier.

5.3 (2+1)D Colloidal Photonic Crystal

3D photonic crystal formed by self-assembly method generally results in the crystallization of opals with face-centered cubic (fcc) lattices of spheres, which ensures a minimum potential energy. However, the colloidal crystal obtained by the LB method is obviously different from the fcc structure, and it can be regarded as a stack of 2D colloidal crystal monolayers (CCM) along the 1D direction. Due to the uncertainty of the position between the layers, this photonic crystal structure not only maintains the ordered characteristics of the 2D CPCs but also breaks the restrictions of the specific structure inside the 3D CPCs. In 2006, Romanov et al. [4] classified the LB multilayers as (2+1)D CPCs, i.e. 2D colloidal crystals stacked along the 1D direction.

The key differences between the 3D CPCs and (2+1)D CPCs were investigated from the perspective of packing. Figure 5.21 shows SEM images of the two types of photonic crystals side by side [12]. The (2+1)D CPC was fabricated by LB method, while the 3D CPC was obtained by controlled evaporation (CE) self-assembly method. It can be seen apparently that the LB film is considerably more disordered compared with the CE film; there are many pores between the layers, and spheres collapse slightly, resulting in obvious structural distortions. Whereas the latter CE film shows clearly fcc packing. The thickness of the films made by the LB method can be estimated from a precise knowledge of the number of monolayers deposited. However, for CE films, it is not possible to rely on this method in terms of generating structures with a precise number of layers. Bertone et al. [58] described a calculation method of the number of layers by analyzing the Fabry–Pérot (FP) interference fringes. In the simple case that we are considering where the light impinges on the substance in a direction normal to the surface, it leads to the following expression for this amplitude: [59, 60]

$$T = (m_2 - m_1)\lambda_2\lambda_1/2n(\lambda_2 - \lambda_1) \tag{5.5}$$

where, T is the thickness of the material, m_2 and m_1 are the orders of the peaks under consideration, n is the refractive index, and λ_2 and λ_1 are the wavelengths

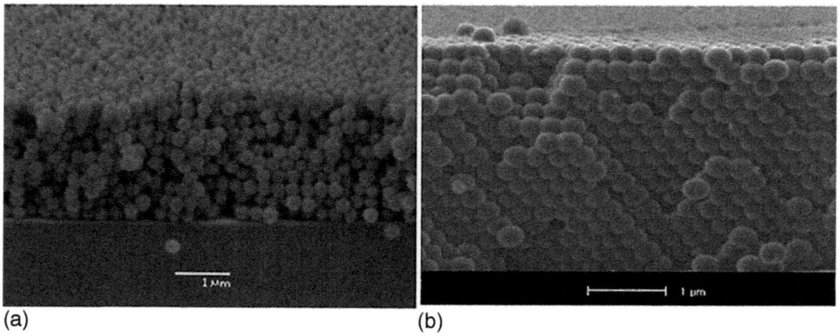

(a) (b)

Figure 5.21 SEM images of LB (a) and CE (b) samples; the difference in packing is clearly visible. Source: Bardosova et al. [12]/with permission of WILEY-VCH.

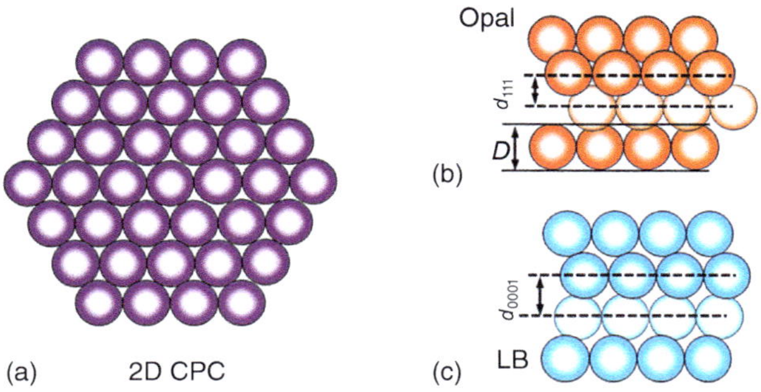

Figure 5.22 (a) Schematic drawing of the structure of a growth monolayer formed by spheres in an opal (the (111) plane) and an LB crystal (the (0001) plane). Arrangement of growth layers in the fcc lattices of (b) the opal and (c) the LB crystal. Designations: D is the sphere diameter. Source: Reproduced from Romanov et al. [61], Copyright 2010, Pleiades Publishing.

corresponding to these peaks. Thus plotting $(m_2 - m_1)$ versus $2n(\lambda_2 - \lambda_1)/\lambda_2\lambda_1$ should yield a straight line with a slope of T.

Both in LB and CE films, the monolayer of spheres close packed in hexagonal symmetry is the main structural unit (Figure 5.22a). The monolayers are the (111) growth planes of the fcc lattice in the CE film, and these planes alternate in triads, ...ABC..., grouped according to the mutual displacement of the (111) planes in the fcc lattice (Figure 5.22b). The interplanar distance along the [111] direction can be calculated by $d_{hkl} = a/(h^2 + k^2 + l^2)^{0.5}$, where $a = \sqrt{2}D$ is the parameter of the fcc lattice of the opal and D is the diameter of the spheres. Then $d_{111} = 0.816\ D$. While in LB case, the number of monolayers is determined by the number of transfer cycles. It turned out that it is difficult to introduce the notion of the lattice parameter for the LB film as a whole; however, the lattice constants can be separately introduced for the monolayer of spheres and the stack of the monolayers (Figure 5.22c) [4, 62]. The growth planes in the LB film can be considered as (0001) planes, due to the uncertainty of the colloidal ball stacking position between the layers; the interplanar distance is not a fixed value, and it is speculated that it should be between 0.816 and 1.0. Romanov et al. used s-polarized light transmission spectroscopy analysis to confirm that there is only weak interaction between the monolayers of LB colloidal crystals in the 3D direction and reported an interlayer spacing of 0.89 D in the vertical direction [4, 61, 62]. Pemble et al. used the simple linear relationship in Bragg law, analogy to FCC, and got the result of $d = 0.941$ D [63]. Zhang et al. reported that d = 0.91 D using a similar method [64].

5.3.1 Preparation Method

(2+1)D CPC is a stack of 2D CCMs along the 1D direction. The fabrication of the 2D CCM is of paramount importance to the optical properties of the resultant (2+1)D CPCs. Nowadays, several methods have been developed to fabricate (2+1)D CPCs on the basis of 2D CCM.

5.3.1.1 The Langmuir–Blodgett Method

The LB multiple deposition technique was rendered as the most important method for preparing (2+1)D CPCs (Figure 5.23) [59, 60, 65]. This method uses monodisperse SiO_2 colloidal balls to form CCM and then forms LB multilayer colloidal crystals by multiple depositions. In 2003, Reculusa and Ravaine first demonstrate that silica particles of various diameters can be organized in colloidal crystals through the LB technique. The main advantage of this approach is that a perfect control of the thickness (at the layer level) of the crystal is obtained in a very reasonable scale of time [59]. More importantly, this allows us to design (2+1)D CPCs layer by layer to obtain defective photonic crystal structures, including heterojunctions, superlattices, or other complex structures through the size selection of each layer of colloidal spheres and the design of deposition sequence.

By using the flexibility of the LB technology, the Ravaine group successfully introduced plane defects at any position of the (2+1)D CPCs by changing the sphere size of a certain CCM during multiple deposition processes. For example, Massé et al. [66] introduce a 200-layer behenic acid film in the middle of 10 layers of 390-nm colloidal crystals (Figure 5.24a) or introduce another colloidal monolayer with bigger sphere size of 980 nm in the middle (Figure 5.24b). The location of the introduced plane defects can also be flexibly designed according to actual needs [72]. In addition, the plane defects can be introduced periodically to form a (2+1)D photonic crystal superlattice (Figure 5.24c) [67]. In addition to constructing plane defects, the flexibility of LB multiple deposition technology can also be used to construct photonic crystal heterostructures. In the process of multiple deposition, the heterojunction structure

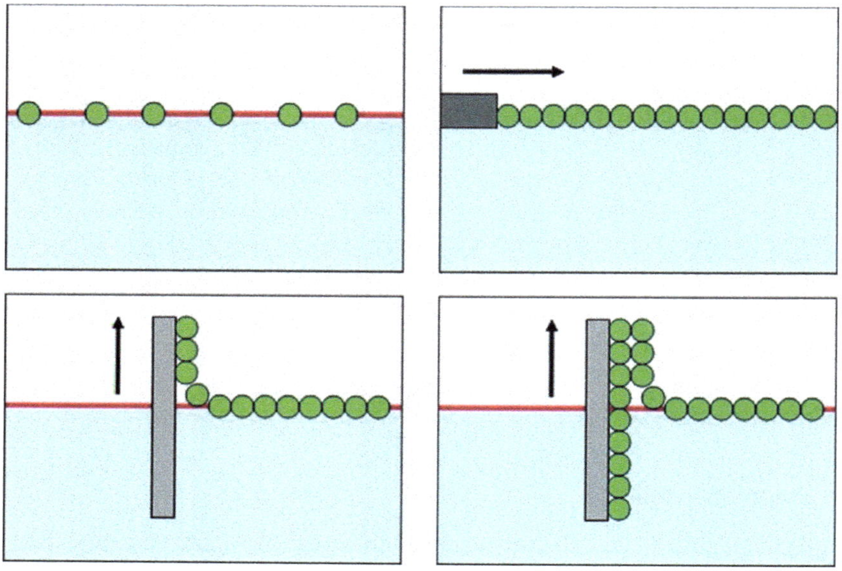

Figure 5.23 Multilayers formation during the LB deposition process. Source: Reproduced with permission from Bardosova et al. [12]. Copyright 2010, WILEY-VCH.

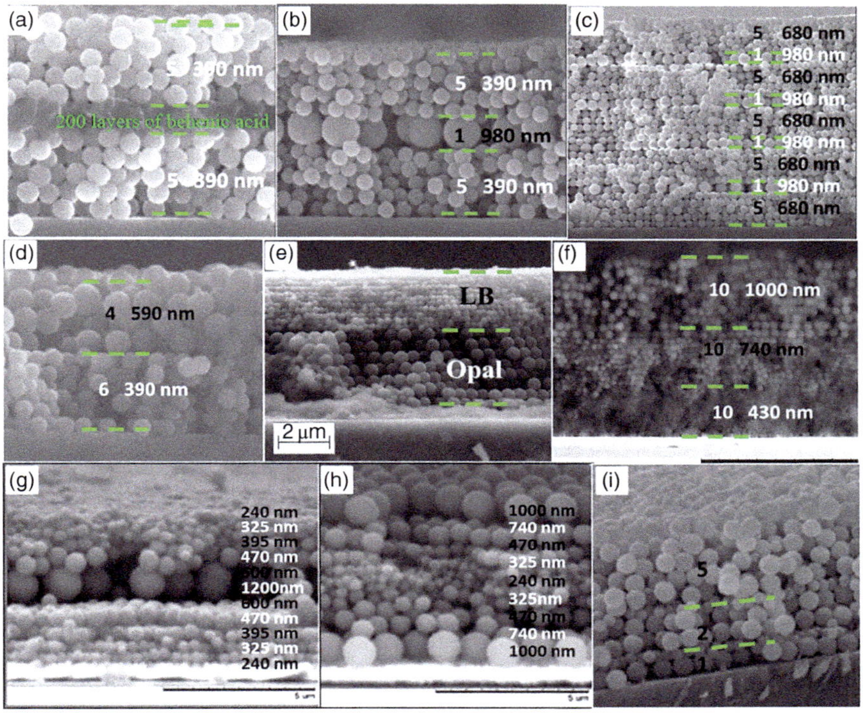

Figure 5.24 SEM images of complex (2+1)D photonic crystal structure prepared by LB multiple deposition technology: (a) $(390)_5$/(behenic acid)$_{200}$/$(390)_5$; (b) $(390)_5/(980)_1/(390)_5$; (c) $(680)_5[(980)_1/(680)_5]_4$; (d) $(390)_6/(590)_4$; (e) Opal(530)/LB$(316)_{10}$; (f) $(430)_{10}/(740)_{10}/(1000)_{10}$; (g) ascending–descending architecture with larger particles in the middle; (h) descending–ascending architecture with smaller particles in the middle; and (i) $(680)_1/(220)_1/(680)_2/(220)_1/(680)_5$. Source: (a–b) Massé et al. [66]/with permission of Elsevier. (c) Ding et al. [67]/with permission of WILEY-VCH. (d) Heim et al. [73]/with permission of American Chemical Society. (e) Heim et al. [68]/with permission of WILEY-VCH. (f–h) Bardosova et al. [69]/with permission of WILEY-VCH. (i) Li et al. [74]/with permission of Elsevier.

can be obtained by stacking several layers of colloidal monolayers of a certain size and then depositing several colloidal monolayers of another size (Figure 5.24d) [65, 70, 73]. LB technology can also be used to prepare (2+1)D photonic crystals on the basis of 3D photonic crystals to obtain a heterojunction structure (Figure 5.24e) [68]. Three or more colloidal monolayers of different sizes are deposited multiple times to obtain a double heterojunction structure (Figure 5.24f). More refined structures can also be prepared by the LB technique, such as the (2+1)D photonic crystal superstructure with a double gradient size of the colloidal sphere (the colloidal monolayer with the largest colloidal sphere size in the middle [Figure 5.24g] or the smallest size in the middle [Figure 5.24h]) [69], or the crystal structure of arbitrary monolayer sphere sizes and deposition times (Figure 5.24i) [74].

Although the LB technique is widely used and very flexible, it still has two shortcomings. On one hand, the dispersibility of SiO_2 colloidal spheres is not good, and the quality of 2D CCM is not good, which resulted in the (2+1)D CPCs with irregular structure, and poor crystal quality. One the other hand, deliberate surface modification of particles is necessitated to achieve a proper hydrophobicity for the particles to spread well at the air/water interface, [59] and the area of the (2+1)D CPCs finally obtained is limited.

5.3.1.2 PDMS Sheet-Assisted Layer-by-Layer Transfer Technique

Li et al. [71] demonstrate a new and versatile way by using PDMS sheets to layer-by-layer (LbL) transfer hcp colloidal monolayers and then transfer them onto solid substrates with poly(vinyl alcohol) (PVA) as glue to create (2+1)D CPCs. The SiO_2 colloidal monolayer used in this method is obtained by using a PDMS soft template to peel off from a 3D photonic crystal with fcc structure prepared in advance. After LbL peeling and stacking, a (2+1)D CPC structure with defined layer number and packing structure is obtained. The schematic diagram of the preparation method is shown in Figure 5.25a. Figure 5.25b shows the SEM image of the obtained crystal structure. Although this method can obtain (2+1)D CPCs more conveniently, it will inevitably cause damage to the order of the colloidal monolayer during the process of multiple peeling off and hot pressing transfer.

5.3.1.3 Layer-by-Layer Scooping Transfer Technique

Since the dispersibility of SiO_2 colloidal spheres is not good, the quality of CCM is not so perfect, resulting in the disorder of final crystal structure. In 2011, Oh et al. developed a LbL scooping technique to fabricated crack-free (2+1)D CPCs by using PS spheres as building blocks (Figure 5.26) [75]. The 2D PS colloidal monolayers were prepared by self-assembly at the air/water interface. The close-packed monolayer is achieved by the interaction between chemically modified PS nanospheres and surfactants, while the close-packed monolayer prepared by LB technique was physically packed by more deliberate instrumentation. The concept for fabricating a (2+1)D CPC with a controllable number of crack-free layers is very similar to the LB technique. The LBL scooping transfer technique allows much easier control of both the thickness and fabrication scale of the as-deposited PS colloidal crystal, as well as the ability to achieve crystallization with large scale by using inexpensive equipment. On the other hand, the LBL scooping transfer technique enables the PS monolayer be transferred to arbitrary substrate, including rigid, flexible, flat, or curved solid substrates.

5.3.1.4 *In Situ* Annealing Combined Layer-by-Layer Transfer Technique

In the preparation process of (2+1)D CPCs, it is inevitable to cause the accumulation of defects, especially in the stacking of small-sized colloidal monolayers onto the surface of large-sized colloidal monolayers. In this case, small spheres tend to sag into the hollows formed by the large spheres, hence distorting the structure at the interface between constituent monolayer colloidal crystals and resulting in a (2+1)D CPC with poor crystalline quality. Because the colloidal spheres are only densely packed, no strong interaction exist between them.

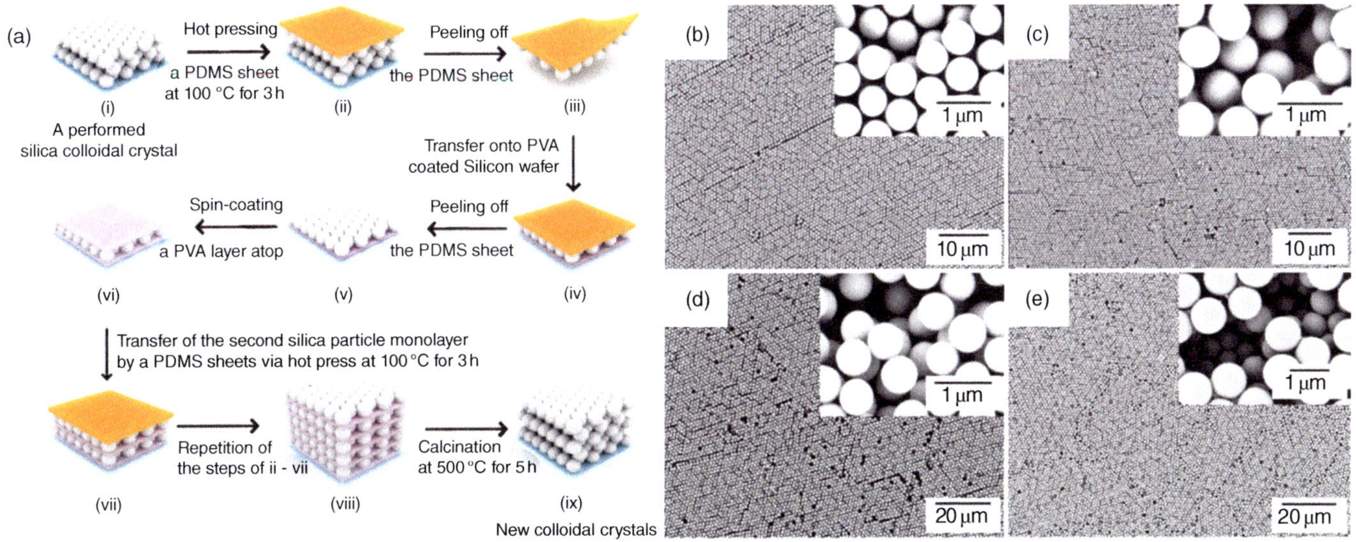

Figure 5.25 (a) Schematic illustration of LbL growth of (2+1)D colloidal crystals using PDMS sheets to transfer ordered closely packed particles into new colloidal crystals from a preformed crystal with the aid of PVA as glue. (b–e) SEM images of SiO_2 colloidal crystals obtained via LbL transfer of the monolayers using a PDMS sheet with the aid of PVA. Colloidal crystals composed of (b) $(632)_2$, (c) $(632)_3$, (d) $(632)_6$, and (e) $(632)_3/(841)_3/(1040)_3$. The insets show high magnification SEM images of the defect void in the crystals. Source: Oh et al. [71]/with permission of American Chemical Society.

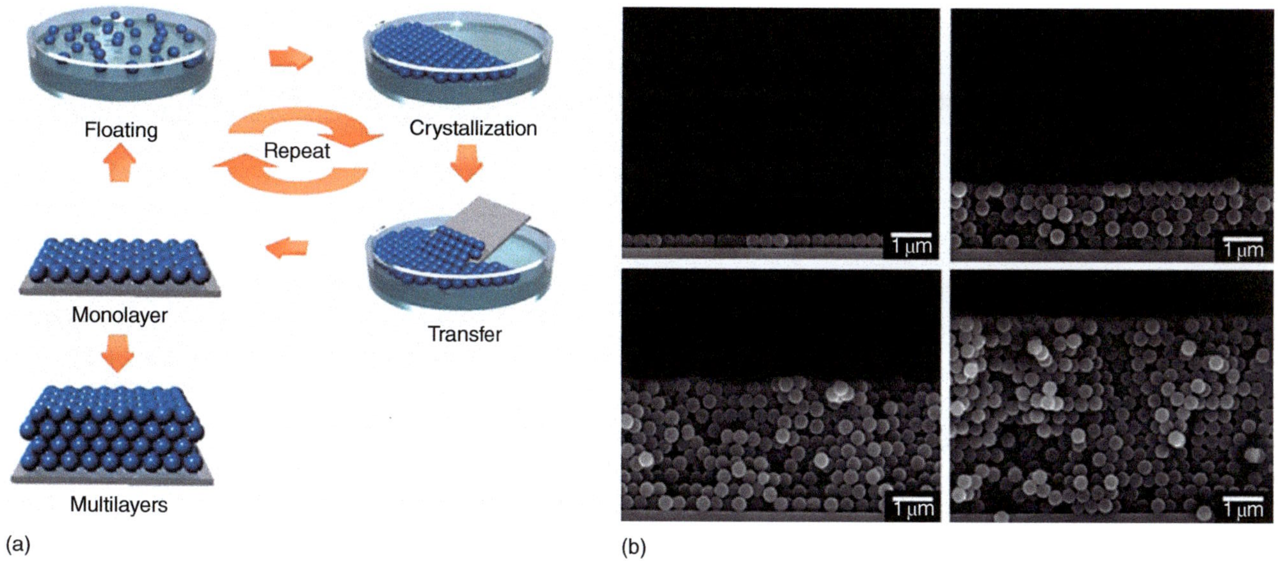

Figure 5.26 (a) Schematic diagram of fabrication process: dispersed PS nanospheres floating on the water surface/formation of a monolayer of polystyrene (PS) beads at the water–air interface by self-assembly/scooping transfer to the glass surface/repetition of the scooping transfer of the PS monolayer from the water–air interface to PS nanosphere-coated glass. (b) FE-SEM images showing side views of colloidal crystals ($d = 350$ nm) with different numbers of layers (1, 5, 10, and 15 layers). Source: Liu et al. [75]/with permission of Royal Society of Chemistry.

Yan's group have developed an air/water interface self-assembly combined with solvent vapor annealing method to the fabrication of large-area and high-quality PS colloidal crystal monolayer [24, 25, 76]. They further conducted the quantitative characterization of the mechanical properties of a PS CCM after *in situ* solvent vapor annealing treatment [25]. After annealing for a period of time, both the compressive and bending elastic modulus of PS CCM was improved, which lays a solid foundation for the fabrication of high-quality (2+1)D CPCs. In addition, the interstice size of the colloidal monolayer can be precisely controlled by manipulating the solvent vapor annealing time (Figure 5.27), which enables (2+1)D photonic crystals more abundant structural information and optical properties.

Based on the high-quality annealed colloidal monolayer, Yan's group further stack these annealed colloidal monolayers by using the LbL transfer technique to form (2+1)D CPCs with enhanced crystalline integrity (Figure 5.28a) [25, 64, 78]. Figure 5.28b,c show that the PS monolayer above the water surface experienced a uniform lattice stretching and the interstices between PS spheres shrank to achieve a close contact with each other, resulting in the high mechanical strength. Figure 5.28d demonstrates a (2+1)D CPC superlattice structure ABABAB by alternately stacking PS colloidal monolayers 420 nm (A) and 337 nm (B) perpendicular to the substrate. It can be seen that there are no obvious interface imperfections in this superlattice; even when stacking 337 nm spheres onto the monolayer of 420 nm spheres, the small spheres did not sag into the hollows formed by the large spheres, but maintain the individual integrity of each layer, resulting in a PC superlattice with high crystalline quality which is much better than superlattices reported in previous works [67, 69].

By manipulating the diameter of PS spheres and the repetition period of the colloidal monolayers, flexible control of the (2+1)D CPC structure has been realized, such as superlattice, heterostructure, or other complicated structures (Figure 5.29), which may afford new opportunities for engineering photonic bandgap materials. Moreover, the high-quality and abundant (2+1)D CPC structures also provide experimental model for systematic study of the relationship between the structure and optical properties, which lays a solid experimental foundation for studying its photonic banggap (PBG) and optical transmission characteristics.

5.3.2 Optical Properties

5.3.2.1 Comparison Between (2+1)D and 3D

The difference in optical response between LB film and CE film was systematically studied by Romanov and Pemble et al. [4, 62, 63] As seen in Figure 5.30a,b, the dispersion of the Bragg peak with angle for the LB film is simpler than that for CE film. Figure 5.30c shows the close examination of these data. For the CE sample, the Bragg peaks splits at angles of incidence close to 45° as had been predicted by Romanov et al. [77]. This splitting is absent from the spectra for the LB film, which shows the fact that LB film cannot be described as a true fcc structure. Figure 5.30c also reveals that, for the LB film and CE film composed of spheres of the same size, the Bragg peak of the LB film display an obvious redshift, broader stopband, and

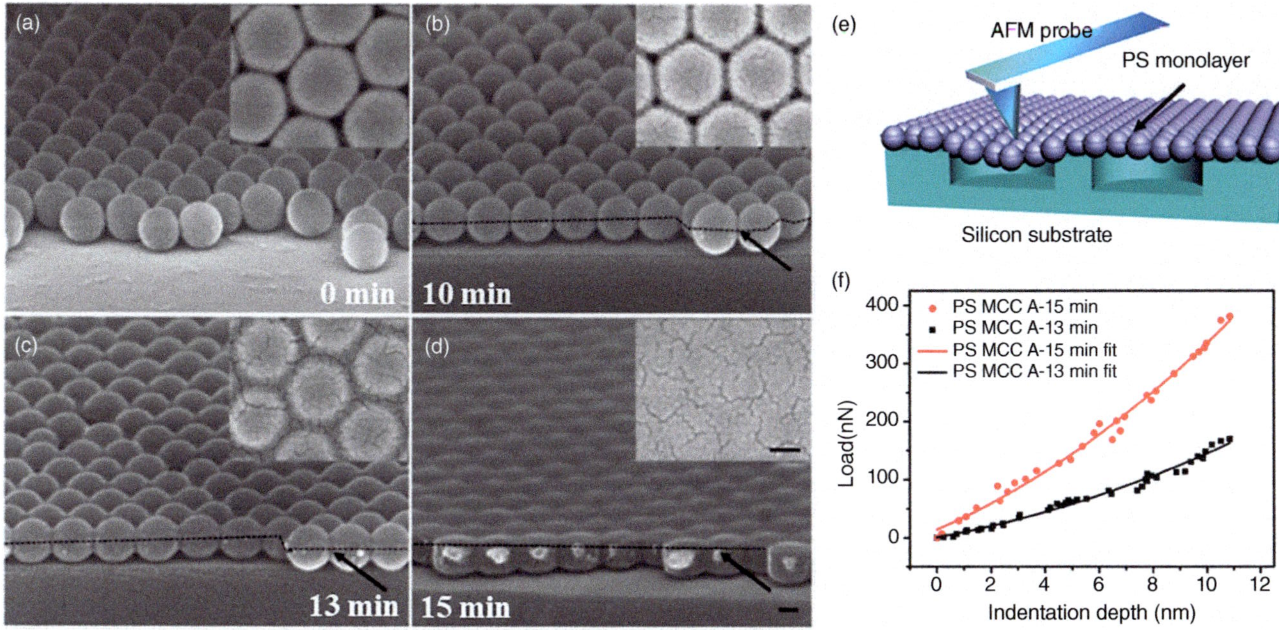

Figure 5.27 (a–d) PS colloidal monolayers after toluene vapor annealing with different annealing time. (e–f) Schematic depiction of the bending test and results for the suspended PS CCMs annealed for 13 and 15 min. The original PS sphere diameter is 235 nm. The scale bar is 200 nm. Source: Romanov et al. [25]/with permission of American Chemical Society.

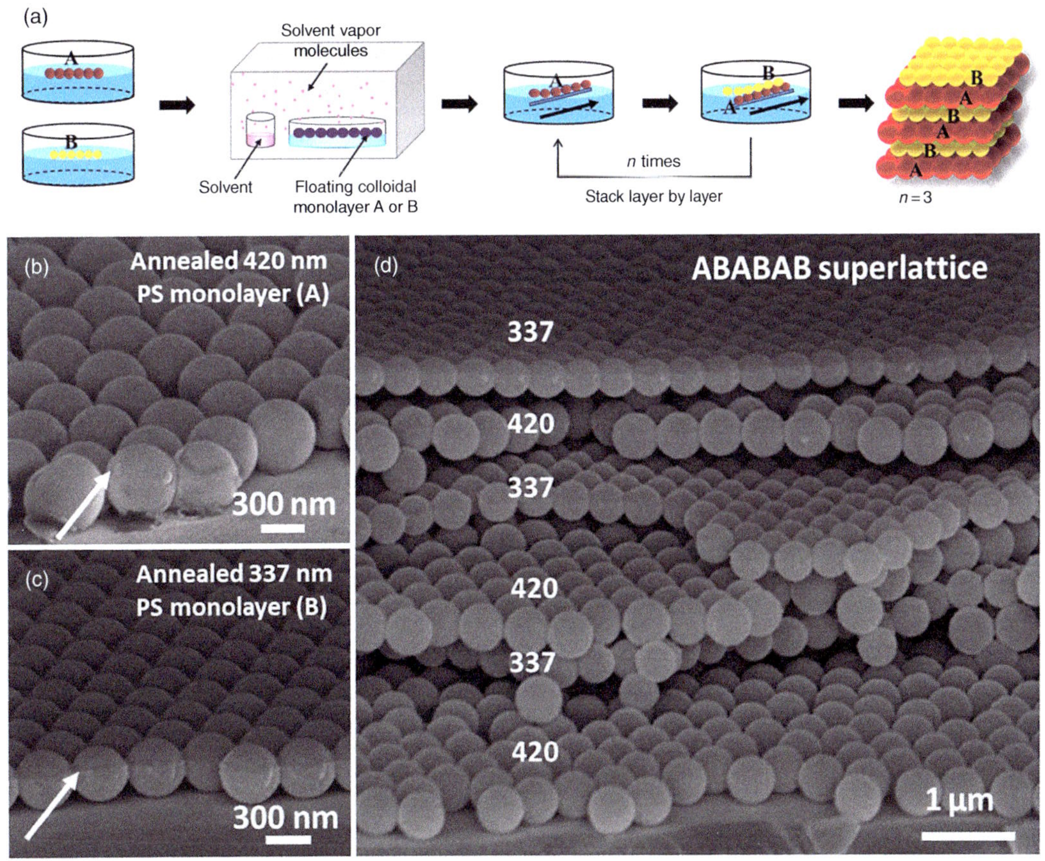

Figure 5.28 (a) Schematic illustration of the fabrication process for (2+1) D CPCs with complex structure. The colloidal monolayer with PS spheres diameter of 337 nm annealed for 20 min (b) and 420 nm annealed for 15 min (c) under toluene vapor at room temperature. (d) The (2+1) D colloidal photonic crystal superlattice formed by alternately stacking colloidal monolayers of 420 and 337 nm in ABABAB arrangement. Source: Zhang et al. [64]/with permission of WILEY-VCH.

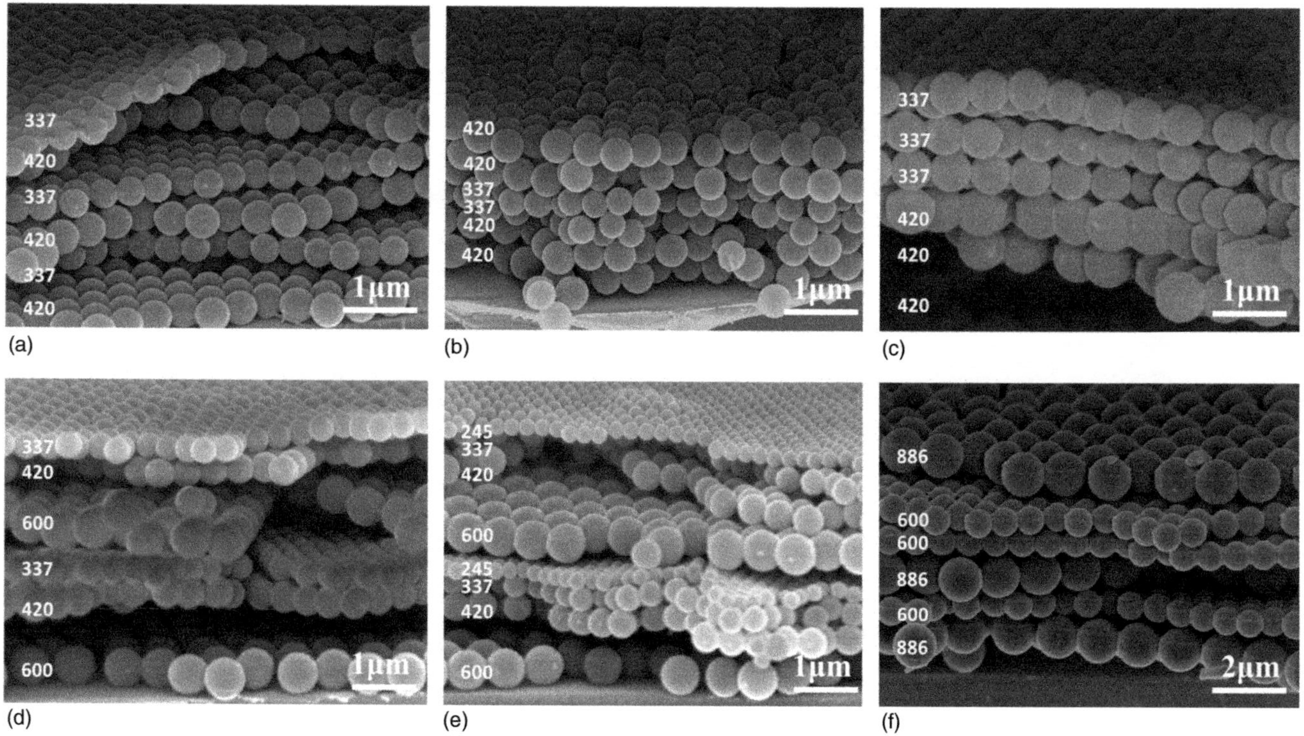

Figure 5.29 Typical cross-sectional SEM images of (2+1)D CPCs with flexible size chosen and controllable stacking ways: (a) superlatice ABABAB (A: 420 nm; B: 337 nm); (b) superlatice AABBAA (A: 420 nm; B: 337 nm); (c) heterojunction AAABBB (A:420 nm; B: 337 nm); (d) superlatice ABCABC (A: 600 nm; B: 420 nm; C: 337 nm); (e) superlatice ABCDABCD (A: 600 nm; B: 420 nm; C: 337 nm; D: 245 nm); and (f) ABABBA (A: 886 nm; B: 600 nm). Source: Zhang et al. [64]/with permission of WILEY-VCH.

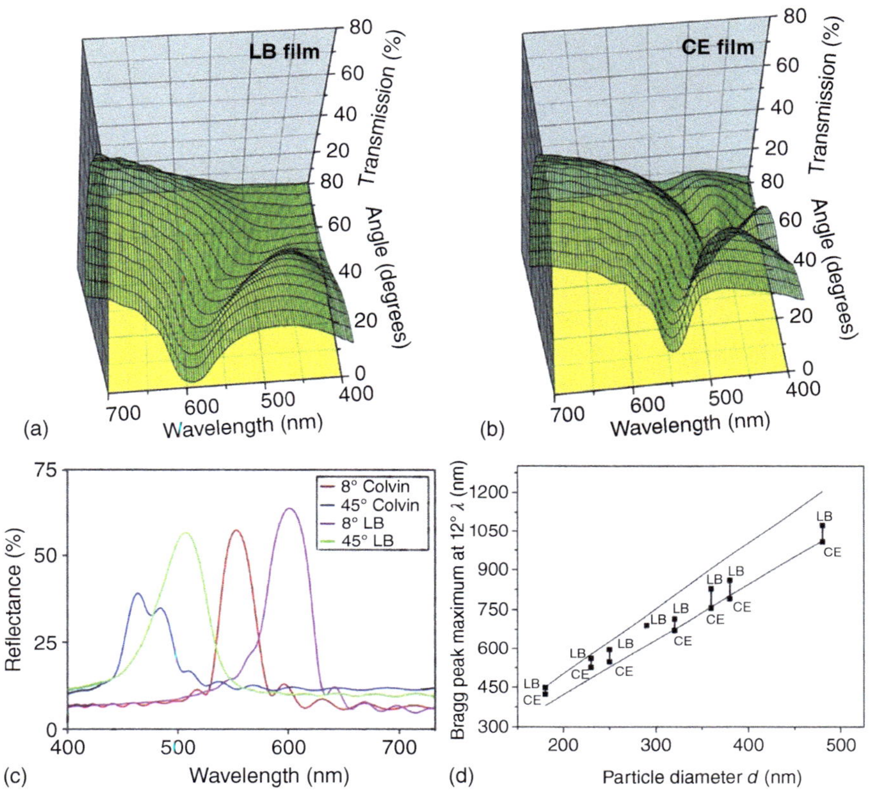

Figure 5.30 Angle dependence of the transmission for CPCs made from silica particles of diameter 250 nm: (a) the LB film and (b) the CE film. (c) Selected reflection spectra obtained from both CE- and LB-grown CPC thin films, for angles of incidence of 8° and 45°. Source: Reproduced from Pemble et al. [63]. Copyright 2007, Elsevier. (d) Bragg peak maxima for samples made of particles with different diameters. Source: Reproduced from Bardosova et al. [12]. Copyright 2010, WILEY-VCH.

higher peak intensity compared with the CE film (incidence angle at 8° and 45°). Figure 5.30d shows the maxima Bragg peak increases with increasing particle size for samples of CE and LB. It is obvious that both CE and LB samples meet the Bragg equation, and the relationship between λ and D is linear. It is worth noting that when considering the same diameter D, the Bragg peak of LB is always larger than CE, and the difference between them increases with the increasing particle size; e.g. for particles having diameters $D = 180$ nm, the difference between the Bragg peak maxima is 25 nm, which increases to 66 nm for particles having diameters $D = 480$ nm. In a word, the differences between the two types of samples are a comprehensible result when considering how the LB process was occurring.

5.3.2.2 Bandgap Engineering

Except the simple (2+1)D CPCs, Zhang et al. systematically studied the reflection spectra of (2+1)D CPC superlattice and heterojunction (Figure 5.31). They thought

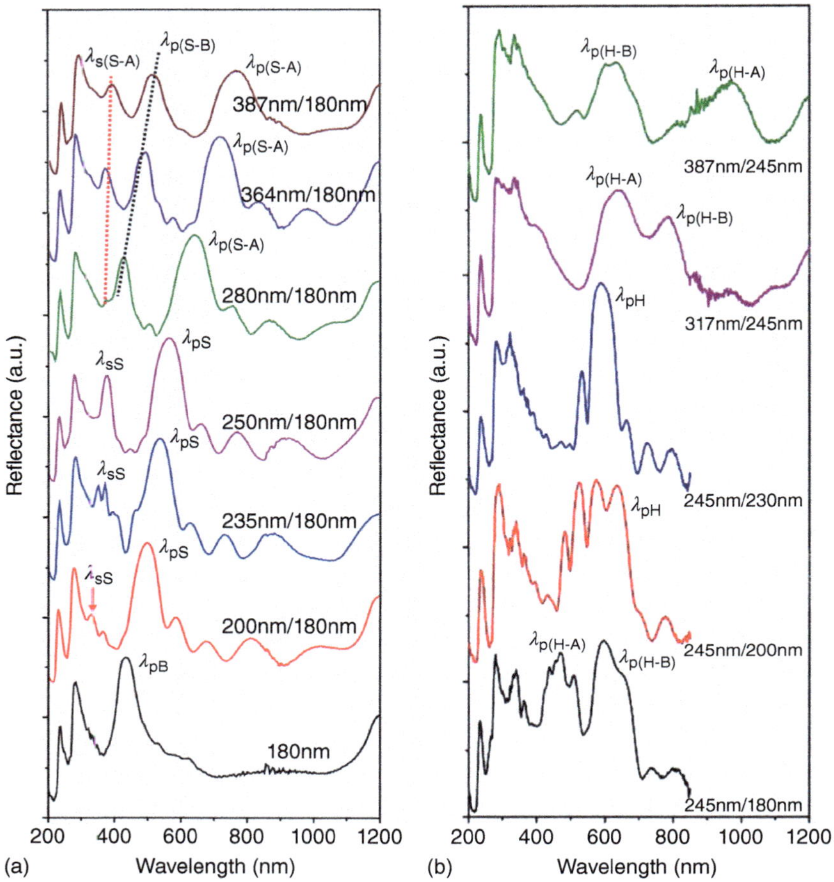

Figure 5.31 (a) The reflection spectra of (2+1)D PC superlattices ABABAB with fixed B of 180 nm and varying A. Source: Reproduced with permission from Zhang et al. [64]. Copyright 2015, WILEY-VCH. (b) The reflection spectra for the (2+1)D PC heterostructures AAABBB with fixed B of 245 nm and varying A. Source: Reproduced from Zhang et al. [78]. Copyright 2018, Elsevier.

that the PBG structure of the (2+1)D CPC superstructures and heterostructures has a close correlation with the bandgap overlapping degree of compositional two homo-crystals, and the conception of resolution (R_{AB}) is introduced to represent the stopband overlapping degree of homo-PCs of A and B, which is widely used in chromatographic analysis. The R_{AB} can be defined as $R_{AB} = 2(\lambda_A - \lambda_B)/(W_A + W_B)$, where the W_A and W_B stand for the full width of λ_A and λ_B, respectively.

For the (2+1)D CPC superlattice, when the value of $R_{AB} \leq 1$, the primary diffraction peaks of the original constituent homo-PCs overlapped with each other, leading to a coupling bandgap with enhanced intensity, such as superlattices of 200 nm/180 nm, 235 nm/180 nm, and 250 nm/180 nm. The periodicity of the superlattice will also cause significant modifications to the stopbands and exhibit pronounced modulation, leading to the formation of the FP fringes [59]. In the case

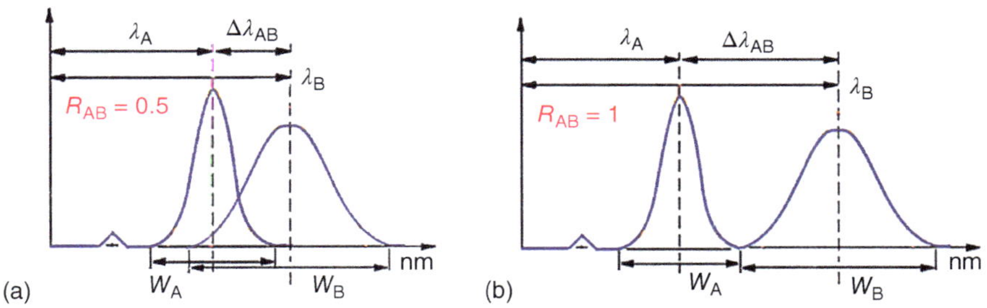

Figure 5.32 (a) Illustration for bandgap overlapping situation for (a) $R_{AB} = 0.5$ and (b) $R_{AB} = 1$. Source: Reproduced from Zhang et al. [78]. Copyright 2018, Elsevier.

of $R_{AB} > 1$, the interaction between monolayers A and B was not strong enough, and the two primary diffraction peaks of the constituent colloidal PCs did not merge into one enhanced coupling bandgap but just presented as two isolated peaks, such as superlattices of 280 nm/180 nm, 364 nm/180 nm, and 387 nm/180 nm.

For the (2+1)D CPC heterostructure, there are two special situations that were noted as shown in Figure 5.32; i.e. the two peaks coincide with half overlap ($R_{AB} = 0.5$) and just do not overlap ($R_{AB} = 1.0$). By comprehensive analysis, the optical properties of the heterostructure can be approximated from the spectra of the individual homo-crystals according to following simple rules. (i) When $R_{AB} > 1$, the two peaks of λ_A and λ_B do not overlap at all. The Bragg diffraction peak of heterostructure exhibits two separate peaks, and the peak's position are basically the same with original λ_A and λ_B. It is equivalent to the simple addition of two stopbands of A and B, such as heterostructures of 245/180 nm and 245/387 nm. (ii) When $0.5 < R_{AB} < 1$, the overlapping degree of λ_A and λ_B is between half and just overlapping. The stopband of heterostructure presents a single and broad peak by stopband superposition. Several "passbands" also appear within the stopband due to the mutual coupling of the original bandgaps, such as 245/200 nm and 245/317 nm. (iii) When R_{AB}<0.5, the overlapping degree of λ_A and λ_B is more than half peak width, and the interaction between A and B photonic crystals is very strong, resulting in the heterostructures with a coupled resonance strong peak. For example, the heterostructure of 245/230 nm. Based on this PBG variation rules, the controllable design and fine adjustment of bandgap structure of (2+1)D CPC heterostructures can be achieved, such as dual- or multi- stopband and width-extended stopband by stacking colloidal crystals with different stopband overlapping degree.

5.3.3 Application

Compared with the fcc structure 3D CPC, the most important feature of the (2+1)D CPC lies in the controllable flexibility in structural construction and bandgap engineering. Its unique structure and bandgap characteristics may play important roles in complex materials fabrication, special kind of wideband optical filters, high-performance reflectors, and highly efficient fluorescence (FL) signal enhancement.

5.3.3.1 Engineering of Complex Macroporous Materials

Heim et al. [69] fabricated macroporous materials with a sophisticated architecture by electrochemical deposition of gold or polypyrrole in (2+1)D CPC superstructures templates, which were fabricated by the LB technique. Figure 5.33 shows the SEM side views of the original superstructures and the inverse well-defined macroporous structure. Figure 5.33a shows a template with a five-layer stack of 430-nm particles separating two five-layer stacks of 1000-nm particles. Figure 5.33b is an "ascending–descending" architecture composed of six different diameters ranging from 240 to 1200 nm. In this case, the particle diameter was gradually increased for the first six layers and decreased for the following five ones. Figure 5.33c is a corresponding "descending–ascending" architecture. Figure 5.33d–f is the corresponding gold or polypyrrole inverse replicas of these superstructures that show highly regular macroporous materials with a uniform thickness and interconnected pores. (2+1)D photonic crystals provide the perfect template for the construction of complex porous structures.

5.3.3.2 High-Performance Back Reflector

Attaching a back reflector on the bottom of device is the most direct and easiest method to increase the light extraction efficiency (LEE) of GaN-based LEDs. Traditional 3D CPC-based back reflector has been studied and considered not to be the most ideal back reflectors due to their strong angle dependence. Based on the high intensity and wide PBG of the (2+1)D CPCs, Zhang et al. demonstrate their application in improving the LEE of GaN-based LED as a back reflector. They designed and fabricated a (2+1)D CPC superlattice that is composed of 24 layers of PS monolayers in $(ABCD)_6$ (A: 200 nm; B: 182 nm; C: 176 nm; D: 147 nm) stacking manner. The stopband is centered at 457 nm, with an intensity of 88% and an FWHM around 60 nm, which well matches with the emitted wavelength of GaN-based LED and can effectively reflect light with incident of $-30°\sim30°$. Finally, the LED containing this (2+1)D CPC superlattice exhibits significant improvement (97.3%) in the light output power as compared with a plain GaN-based LED and 74.5% improvement compared with GaN LED with an fcc 3D CPC reflector (Figure 5.34). The performance enhancement was attributed to the PBG enhancement as well as its good angle independence of the (2+1)D CPC superlattice.

5.3.3.3 FL Enhancement Substrate

PCs can achieve fluorescence enhancement in two ways: (i) the bandgap position matching with the fluorophore excitation wavelength (E) for enhancing excitation and (ii) the bandgap position matching with the emission wavelength (F) for enhanced light extraction. PC heterostructures with dual bandgaps can simultaneously match with the E and F of fluorescent medium, which may realize remarkable enhancement of fluorescence [79–81]. Zhang et al. [78] fabricated dual-bandgap (2+1)D CPC heterostructures 245/180 nm, whose stopbands just overlap the E and F of fluorescent dye $[Ru(dpp)_3]Cl_2$, respectively (Figure 5.35a,b). Compared with single E or F matching PC, the (2+1)D CPC heterostructure with

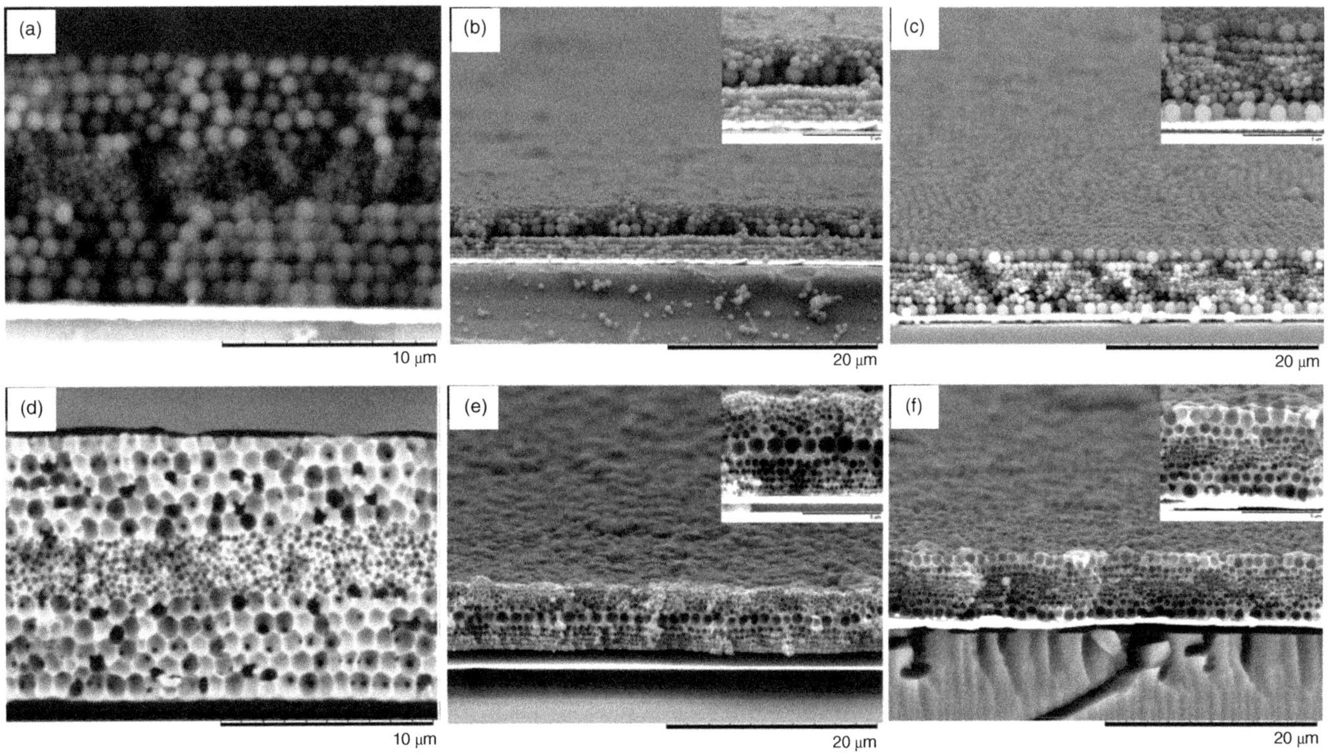

Figure 5.33 (a–c) Illustration for the original complex (2+1)D CPC structures and (d–f) the corresponding inversed macroporous gold structure after remove the templates. Source: Bardosova et al. [69]/with permission of WILEY-VCH.

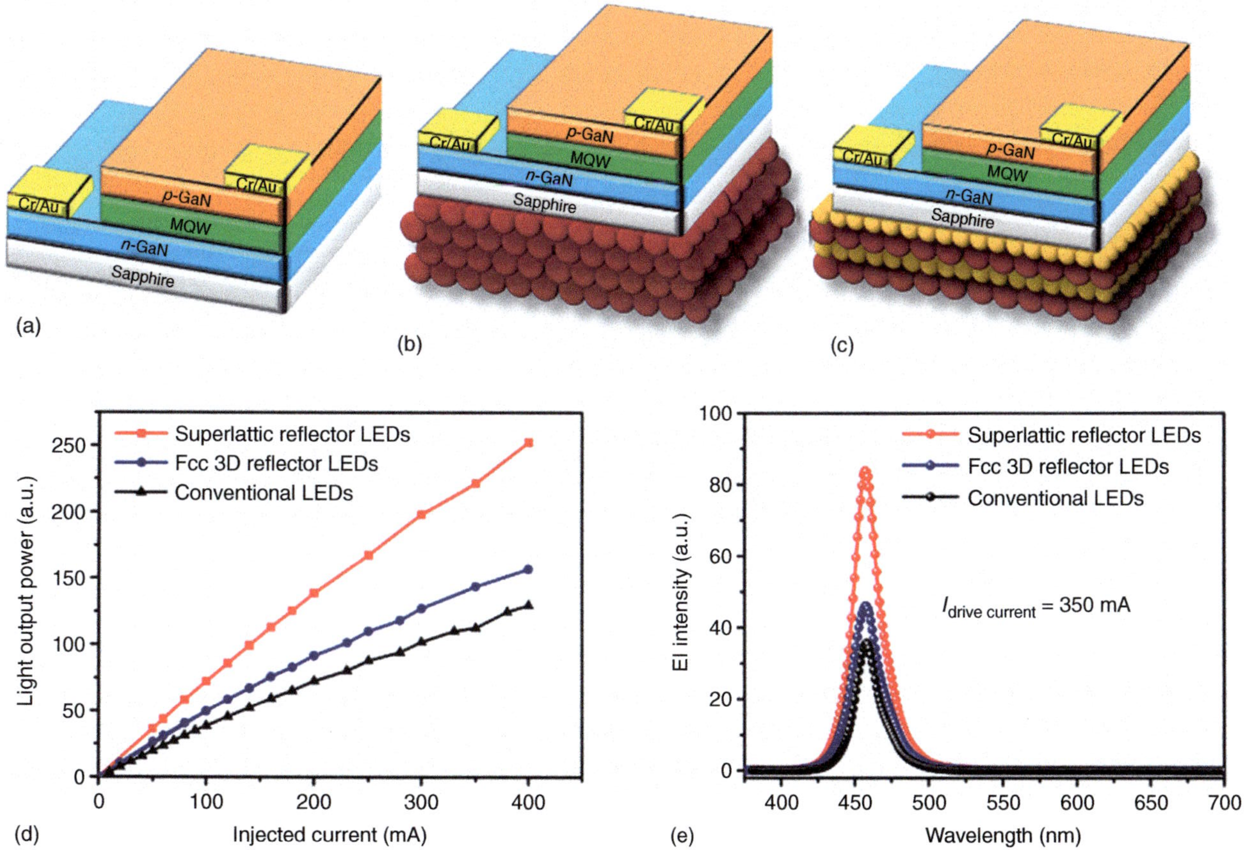

Figure 5.34 (a–c) Schematic diagram of traditional LED structure and LED with different bottom reflector structure. (d) Relative light output power versus injection current (L–I) characteristic for these three kinds of LEDs. (e) Electroluminescence spectra measured at 350 mA of these three kinds of LEDs. Source: Reproduced from Zhang et al. [64]. Copyright 2015, WILEY-VCH.

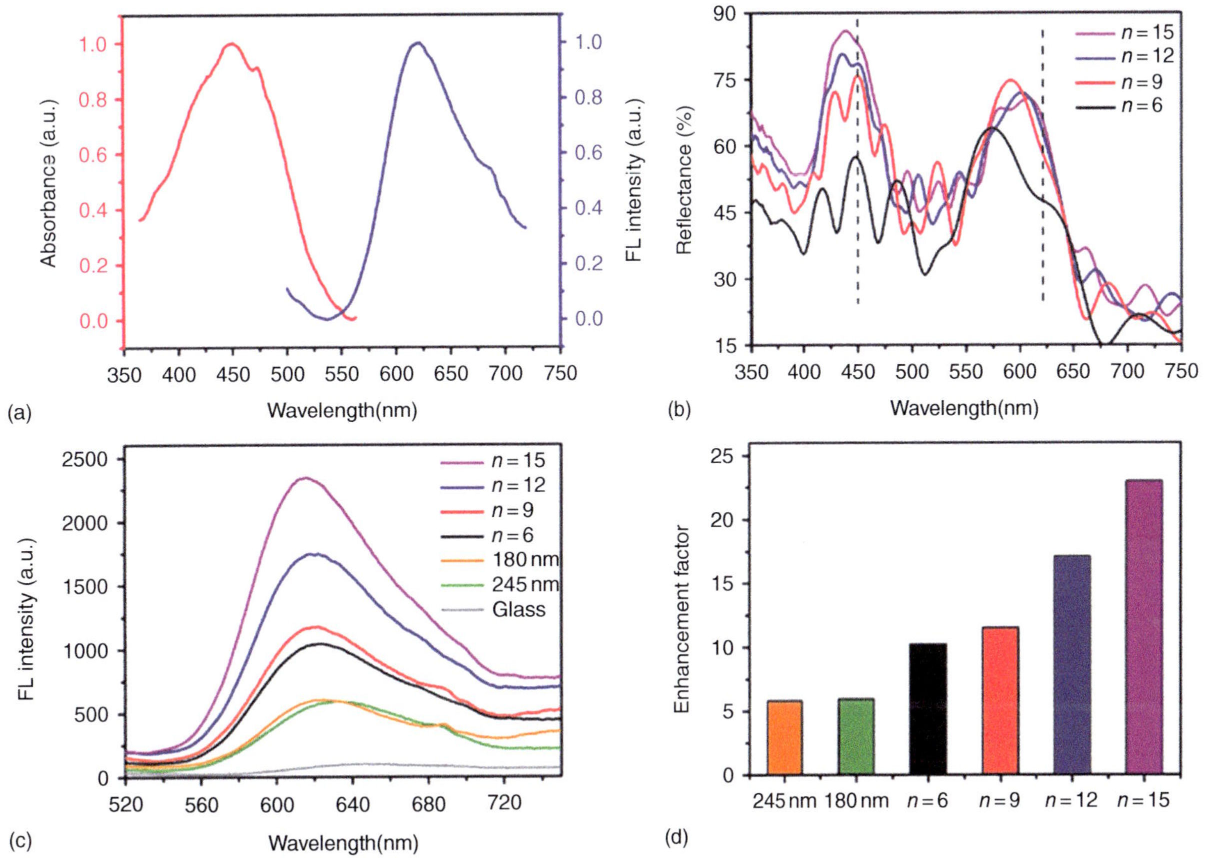

Figure 5.35 (a) UV–visible absorption (red) and fluorescence (blue) spectra of the $[Ru(dpp)_3]Cl_2$. (b) The normal-incidence reflection spectra of (2+1)D PC heterostructure 245 nm/180 nm of different layers with two separate bandgaps, one centered at 450 nm and the other red edge covering 620 nm. The fluorescence spectra (c) and enhancement factor (d) of $[Ru(dpp)_3]Cl_2$ on heterostructure of 245 nm/180 nm with different layers and on the homogeneous photonic crystal films compared to that on glass. Source: Reprinted from Zhang et al. [78], Copyright 2018, Elsevier.

E–F-matching dual bandgap leads to an extra FL enhancement of nearly two-fold, and the enhancement factor was further improved with the layer number increase.

Besides the dual-bandgap (2+1)D CPC heterostructures, Zhang et al. [82] further constructed a multiple heterostructure photonic crystal (MHPC) based on the stopbands overlapping and broadening strategy. This MHPC is fabricated by LbL deposition of four colloidal crystals with 250, 215, 195, and 180 nm PS spheres arranged from the bottom-up and presents a broadened stopband (from 280 to 650 nm) that can overlap with the E and F of most commonly used fluorescent media (Figure 5.36a,b) [82, 83]. In traditional PC structures used for FL enhancement, for each fluorescent medium, the central position of the stopband must be specifically prepared and regulated, which can be difficult and inconvenient

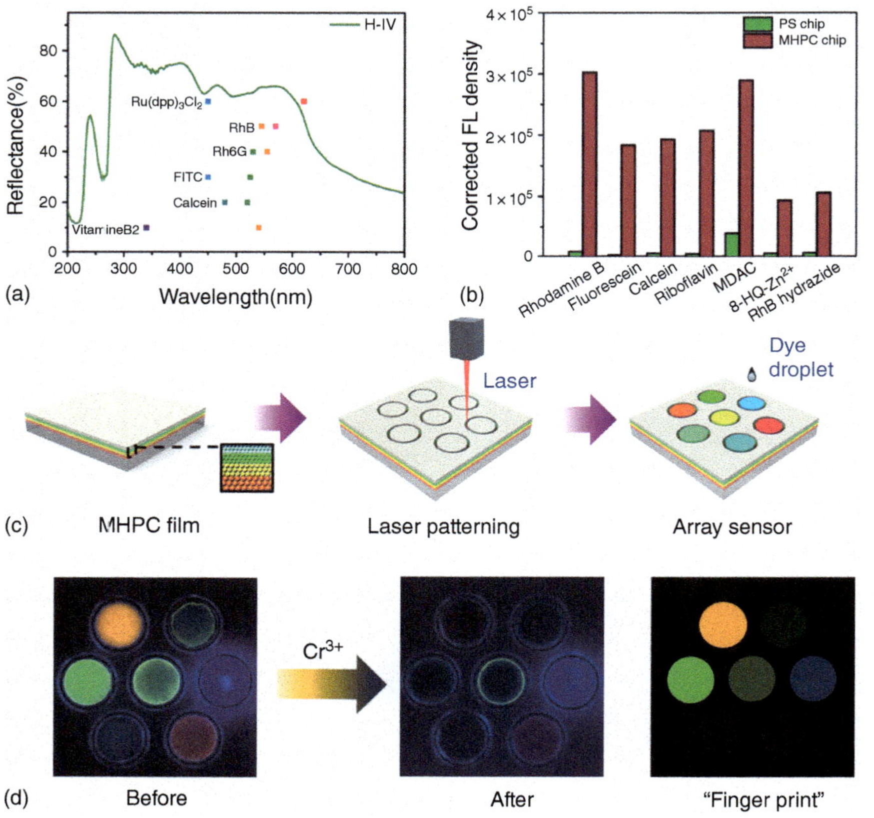

Figure 5.36 (a) The normal-incidence reflectance spectra for MHPC overlaid with E and F distributions of several commonly used fluorescent media. Source: Reproduced from Zhang et al. [82]. Copyright 2018, WILEY-VCH. (b) The FL density of seven fluorescent dyes on PS film and MHPC, respectively. (c) Schematic illustration of the preparation of patterned MHPC sensor array by laser direct writing. (d) A simple process illustrates the identification of Cr^{3+} by the MHPC sensor array. Source: Zhang et al. [83]/with permission of American Chemical Society.

in certain practical applications. This super-wide stopband MHPC can serve as a universal FL enhancement substrate and may find significant applications for augmenting FL intensity in chemical and biochemical sensing, disease diagnosis, and environmental monitoring. Recently, Zhang et al. [83] reported a straightforward method to realize the patterning of MHPC by direct laser writing technology (Figure 5.36c). The patterning mechanism and water confinement function were presented, which enables the patterned MHPC array to be used as a platform for the detection of aqueous solution. When integrated with seven cross-reactive fluorescent dyes, a differential fluorescent sensor array was obtained, and nine different metal ions were successfully recognized (Figure 5.36c,d).

5.4 Outlook

In the last three decades, 3D CPCs have been widely studied and received diversified applications, such as super-wettability surface [84], bioinspired photonic pigments [85], eco-friendly paintings [86–88], multiple anti-counterfeiting technologies [89, 90], novel display devices [91–93], high-performance sensors [94–96], and high-performance photovoltaic devices and lasers [36, 97, 98]. However, photonic devices based on 2D and (2+1)D CPCs are still proofs of concept. There is still a long way to go for the real applications of these photonic devices.

As the versatile template and basic building block of (2+1)D CPCs, the high quality and large area are the long-term goal to pursue in the area of 2D CPCs. Among these preparation methods, the air/liquid interface self-assembly method seems to be a versatile and effective method for fabricating crack-free 2D CPCs in a large area. More importantly, this method can combine with automated equipment to achieve the large-area continuous preparation of 2D CPCs on PET/ITO, for example, the R2R method. Of course, it is necessary to further refine the method to avoid the defects after the floating 2D CPCs transferred onto a flat substrate. Perhaps the introduction of *in situ* solvent vapor annealing is expected to improve this problem. The biggest characteristic of (2+1)D CPCs lies in the controllable design of structures along 1D direction and bandgap engineering. However, its development in functional devices is greatly hindered by the complexity and tediousness of the present LbL preparation method and the limitation of the small area. The combination of program control and automatic equipment is also a new idea to solve this problem.

In addition, for the expansion of new functional materials and devices based on 2D and (2+1)D CPCs, breakthroughs are expected to receive when focusing on the following two aspects: (i) introduce new materials into the CPC skeleton, such as the MOFs, perovskite, ionic liquids, shape memory polymer, and natural materials. The integration of new materials and CPC structure can often give full play to the advantages of both, break through the bottleneck of existing technology and get better solutions [99, 100]. (ii) Combine with advanced manufacturing technologies to achieve precise, fast, and large-scale preparation for practical applications, such as inkjet printing [101], 3D printing [102], laser engraving [83, 103] and microfluidics

[104–106]. These will bring CPC devices enhanced stability, repeatability, and low cost. CPCs will be continuing to enrich and beautify the world with their fascinating properties.

References

1 Cong, H., Yu, B., Tang, J. et al. (2013). *Chem. Soc. Rev.* 42: 7774.
2 Stöber, W., Fink, A., and Bohn, E.J. (1968). *J. Colloid Interface Sci.* 26: 62.
3 Kim, J.W., Joe, Y.G., and Suh, K.D. (1999). *Colloid Polym. Sci.* 277: 252.
4 Romanov, S.G., Bardosova, M., Pemble, M. et al. (2006). *Appl. Phys. Lett.* 89: 043105.
5 Ye, X. and Qi, L. (2011). *Nano Today* 6: 608.
6 N D Denkov, O., Velev, P., Kralchevsky, I. et al. (1992). *Langmuir* 8: 3183.
7 Dimitrov, A.S. and Nagayama, K. (1996). *Langmuir* 12: 1303.
8 Jiang, P. and McFarland, M.J. (2004). *J. Am. Chem. Soc.* 126: 13778.
9 Lumsdon, S.O., Kaler, E.W., and Velev, O.D. (2004). *Langmuir* 20: 2108.
10 Trau, M., Saville, D., and Aksay, I.A. (1996). *Science* 272: 706.
11 Zhang, J.-T., Wang, L., Lamont, D.N. et al. (2012). *Angew. Chem. Int. Ed.* 51: 6117.
12 Bardosova, M., Pemble, M.E., Povey, I.M. et al. (2010). *Adv. Mater.* 22: 3104.
13 van Duffel, B., Ras, R.H.A., De Schryver, F.C. et al. (2001). *J. Mater. Chem.* 11: 3333.
14 Muramatsu, K., Takahashi, M., Tajima, K. et al. (2001). *J. Colloid Interface Sci.* 242: 127.
15 Ábrahám, N., Sebők, D., Papp, S. et al. (2011). *Colloids Surf. A Physicochem. Eng. Asp.* 384: 80.
16 Wu, H., Niu, G., Ren, W. et al. (2020). *Langmuir* 36: 10061.
17 Smith, N.L., Coukouma, A., Dubnik, S. et al. (2017). *Phys. Chem. Chem. Phys.* 19: 31813.
18 Meng, X. and Qiu, D. (2014). *Langmuir* 30: 3019.
19 Retsch, M., Zhou, Z., Rivera, S. et al. (2009). *Macromol. Chem. Phys.* 210: 230.
20 Teh, L.K., Tan, N.K., Wong, C.C. et al. (2005). *Appl. Phys. A* 81: 1399.
21 Singh, K.B. and Tirumkudulu, M.S. (2007). *Phys. Rev. Lett.* 98: 218302.
22 Xu, P., Mujumdar, A.S., and Yu, B. (2009). *Dry. Technol.* 27: 636.
23 Geng, C., Zheng, L., Yu, J. et al. (2012). *J. Mater. Chem.* 22: 22678.
24 Yu, J., Geng, C., Zheng, L. et al. (2012). *Langmuir* 28: 12681.
25 Zhang, L., Wang, W., Zheng, L. et al. (2016). *Langmuir* 32: 451.
26 Sun, X., Li, Y., Zhang, T. et al. (2013). *Langmuir* 29: 7216.
27 Tan, A.T.L., Nagelberg, S., Chang-Davidson, E. et al. (2020). *Small* 16: e1905519.
28 Parchine, M., McGrath, J., Bardosova, M. et al. (2016). *Langmuir* 32: 5862.
29 Cagnani, G.R., Spada, E.R., Cagnani, L.D. et al. (2020). *Colloids Surf. A Physicochem. Eng. Asp.* 588: 124389.
30 Xue, F., Asher, S.A., Meng, Z. et al. (2015). *RSC Adv.* 5: 18939.
31 Asher, S.A., Weissman, J.M., Tikhonov, A. et al. (2004). *Phys. Rev. E* 69: 066619.

32 Xue, F., Meng, Z., Wang, F. et al. (2014). *J. Mater. Chem. A* 2: 9559.
33 Geng, C., Wei, T., Wang, X. et al. (2014). *Small* 10: 1668.
34 Ye, X. and Qi, L. (2013). *Sci. China Chem.* 57: 58.
35 Marlow, F., Muldarisnur, Sharifi, P. et al. (2009). *Angew. Chem. Int. Ed.* 48: 6212.
36 Xu, J., Xu, S., Qi, Z. et al. (2018). *Nanoscale* 10: 10383.
37 Xu, J., Qu, J., Shao, H. et al. (2020). *New J. Chem.* 44: 13619.
38 Li, L., Jiao, X., Chen, D. et al. (2016). *Cryst. Growth Des.* 16: 2700.
39 Garnett, E. and Yang, P. (2010). *Nano Lett.* 10: 1082.
40 Sun, C.-H., Min, W.-L., Jiang, P. et al. (2008). *Chem. Commun.* 27: 3163.
41 Li, Y., Fang, X., Koshizaki, N. et al. (2009). *Adv. Funct. Mater.* 19: 2467.
42 Zhang, X., Liu, D., Zhang, L. et al. (2009). *J. Mater. Chem.* 19: 962.
43 Zhu, J., Yu, Z., Burkhard, G.F. et al. (2009). *Nano Lett.* 9: 279.
44 Elias, J., Levy-Clement, R., Bechelany, M. et al. (2010). *Adv. Mater.* 22: 1607.
45 Hong, G., Li, C., and Limin, Q. (2010). *Adv. Funct. Mater.* 20: 3774.
46 Li, C., Hong, G., and Qi, L. (2010). *Chem. Mater.* 22: 476.
47 Jia, L., Cai, W., Wang, H. et al. (2009). *ACS Nano* 3: 2697.
48 Haynes, C.L. and Van Duyne, R.P. (2001). *J. Phys. Chem. B* 105: 5599.
49 Gwinner, M.C., Koroknay, E., Fu, L. et al. (2009). *Small* 5: 400.
50 Huang, T., Zhao, Q., Xiao, J. et al. (2010). *ACS Nano* 4: 4707.
51 Li, Y., Lee, E.J., Cai, W. et al. (2008). *ACS Nano* 2: 1108.
52 Du, W., Yan, Z., Gu, P. et al. (2015). *Appl. Phys. A* 120: 11.
53 Zhang, J.-T., Wang, L., Chao, X. et al. (2011). *Langmuir* 27: 15230.
54 Zhang, J.-T., Smith, N., and Asher, S.A. (2012). *Anal. Chem.* 84: 6416.
55 Zhang, J.-T., Chao, X., Liu, X. et al. (2013). *Chem. Commun.* 49: 6337.
56 Li, G., Xiao, F., Liao, S. et al. (2018). *Sens. Actuat. B Chem.* 277: 591.
57 Wang, Z., Meng, Z., Xue, M. et al. (2020). *Microchem. J.* 157: 105073.
58 Bertone, J.F., Jiang, P., Hwang, K.S. et al. (1999). *Phys. Rev. Lett.* 83: 300.
59 Reculusa, S. and Ravaine, S. (2003). *Chem. Mater.* 15: 598.
60 Reculusa, S., Perrier-Cornet, R., Agricole, B. et al. (2007). *Phys. Chem. Chem. Phys.* 9: 6385.
61 Romanov, S. and Bardosova, M. (2010). *Phys. Solid State* 52: 533.
62 Romanov, S.G., Bardosova, M., Whitehead, D.E. et al. (2007). *Appl. Phys. Lett.* 90: 133101.
63 Pemble, M.E., Bardosova, M., Povey, I.M. et al. (2007). *Phys. B Condens. Matter.* 394: 233.
64 Zhang, L., Xiong, Z., Shan, L. et al. (2015). *Small* 11: 4910.
65 Massé, P., Reculusa, S., and Ravaine, S. (2006). *Colloids Surf. A Physicochem. Eng. Asp.* 229: 284–285.
66 Massé, P., Reculusa, S., Clays, K. et al. (2006). *Chem. Phys. Lett.* 422: 251.
67 Massé, P., Pouclet, G., and Ravaine, S. (2008). *Adv. Mater.* 20: 584.
68 Ding, B., Bardosova, M., Povey, I. et al. (2010). *Adv. Funct. Mater.* 20: 853.
69 Heim, M., Reculusa, S., Ravaine, S. et al. (2012). *Adv. Funct. Mater.* 22: 538.
70 Bardosova, M., Pemble, M.E., Povey, I.M. et al. (2006). *Appl. Phys. Lett.* 89: 093116.

71 Li, W., Yang, B., and Wang, D. (2008). *Langmuir* 24: 13772.
72 Dechézelles, J.-F., Massé, P., Cloutet, E. et al. (2009). *Colloids Surf. A Physicochem. Eng. Asp.* 343: 8.
73 Massé, P. and Ravaine, S. (2005). *Chem. Mater.* 17: 4244.
74 Reculusa, S., Massé, P., and Ravaine, S. (2004). *J. Colloid Interface Sci.* 279: 471.
75 Oh, J.R., Moon, J.H., Yoon, S. et al. (2011). *J. Mater. Chem.* 21: 14167.
76 Liu, X., Zhao, D., Geng, C. et al. (2014). *J. Colloid Interface Sci.* 434: 98.
77 Romanov, S.G., Maka, T., Sotomayor Torres, C.M. et al. (2001). *Phys. Rev. E* 63: 056603.
78 Zhang, L., Wang, W., Zheng, L. et al. (2018). *J. Colloid Interface Sci.* 509: 318.
79 Ganesh, N., Zhang, W., Mathias, P.C. et al. (2007). *Nat. Nanotechnol.* 2: 515.
80 Li, H., Wang, J., Lin, H. et al. (2010). *Adv. Mater.* 22: 1237.
81 Zhou, P., Zhou, D., Tao, L. et al. (2014). *Light Sci. Appl.* 3: e209.
82 Zhang, L., Wang, J., Tao, S. et al. (2018). *Adv. Opt. Mater.* 6: 1701344.
83 Zhang, L., Liu, B., Yang, W. et al. (2021). *ACS Appl. Mater. Interfaces* 13: 4330.
84 Kuang, M., Wang, J., and Jiang, L. (2016). *Chem. Soc. Rev.* 45: 6833.
85 Goerlitzer, E.S.A., Klupp Taylor, R.N., and Vogel, N. (2018). *Adv. Mater.* 30: 1706654.
86 Cui, L., Li, Y., Wang, J. et al. (2009). *J. Mater. Chem.* 19: 5499.
87 Tian, D., Song, Y., and Jiang, L. (2013). *Chem. Soc. Rev.* 42: 5184.
88 Zhou, J., Han, P., Liu, M. et al. (2017). *Angew. Chem. Int. Ed.* 56: 10462.
89 Zhang, J., Yang, S., Tian, Y. et al. (2015). *Chem. Commun.* 51: 10528.
90 Heo, Y., Kang, H., Lee, J.-S. et al. (2016). *Small* 12: 3819.
91 Lee, S.Y., Kim, S.-H., Hwang, H. et al. (2014). *Adv. Mater.* 26: 2391.
92 Zhang, Y., Dong, B., Chen, A. et al. (2015). *Adv. Mater.* 27: 4719.
93 Yang, J., Choi, M.K., Kim, D.-H. et al. (2016). *Adv. Mater.* 28: 1176.
94 Ge, J. and Yin, Y. (2011). *Angew. Chem. Int. Ed.* 50: 1492.
95 Zhao, Y., Xie, Z., Gu, H. et al. (2012). *Chem. Soc. Rev.* 41: 3297.
96 Phillips, K.R., England, G.T., Sunny, S. et al. (2016). *Chem. Soc. Rev.* 45: 281.
97 Chang, Y., Liou, J., and Liu, W. (2013). *IEEE Electron. Dev. Lett.* 34: 777.
98 Chen, S., Roh, K., Lee, J. et al. (2016). *ACS Nano* 10: 3959.
99 Xu, H., Lu, Y.F., Xiang, J.X. et al. (2018). *Nanoscale* 10: 2090.
100 Shang, L., Fu, F., Cheng, Y. et al. (2017). *Small* 13: 1600286.
101 Wang, L., Wang, J., Huang, Y. et al. (2012). *J. Mater. Chem.* 22: 21405.
102 Liu, X., Gu, H., Wang, M. et al. (2018). *Adv. Mater.* 30: 1800103.
103 Yue, Y. and Kurokawa, T. (2019). *ACS Appl. Mater. Interfaces* 11: 10841.
104 Shen, W., Li, M., Ye, C. et al. (2012). *Lab Chip* 12: 3089.
105 Gao, B., Tang, L., Zhang, D. et al. (2017). *ACS Appl. Mater. Interfaces* 9: 32577.
106 Gao, B., Liu, H., and Gu, Z. (2016). *Anal. Chem.* 88: 5424.

6

Structural Color due to Self-assembly

Yong Qi and Shufen Zhang

Dalian University of Technology, School of chemical engineering, State Key Laboratory of Fine Chemicals, West Campus, 2# Linggong Rd, Dalian 116024, China

6.1 Structural Color in Nature

Colors have been important throughout human history and biological evolution. They provide critical functions in both recognition and communication, linking interactions between the animate and inanimate worlds. Several mechanisms for generating color in materials are known, including preferential absorption, emission, birefringence, photochromism, and other mechanisms [1]. From the principle of color rendering, the color can be divided into chemical color and physical color. Chemical colors refer to all kinds of colors caused by the absorption of coordination field-effect transition (such as turquoise, ruby, emerald, and other gems with fluorescence or phosphorescence), absorption of transitions between molecular orbitals (such as dyes and biological colorants), and absorption of band transitions (such as metal, pure semiconductor, and doped semiconductor). Therefore, chemical colors are determined by the constituting materials and fade out as the chemicals are transformed or decomposed. Physical colors, also known as structural colors, are all kinds of colors produced by the microscopic structures corresponding to the light wavelength through interference, diffraction, scattering, and other ways to reflect the incident light at a specific wavelength and in a specific direction [2]. Unlike organic pigments, structural colors vary according to the structural dimension and refractive index so that various colors can be obtained using a single set of materials. They never fade as long as the structures persist. They developed by inorganic materials that can survive at high temperatures where conventional pigments die out. As the colors can be purely developed by structures without the use of chemical pigments, low toxicity can be achieved by selecting a biocompatible set of dielectric materials [3–5]. These features render the structural coloration more promising than chemical-based coloration for various applications.

For a long time, the formation mechanism of the structural color has been gradually revealed from the beginning of distinguishing the order of structure [6, 7]. One is an ordered structure, in which the spatial arrangement of the scatterer has

Functional Materials from Colloidal Self-assembly, First Edition. Edited by Qingfeng Yan and George Zhao.

translational symmetry, such as grating, monolayer, photonic crystal, etc. The other is disordered structure, in which the spatial arrangement of scatterers loses translational symmetry, such as white clouds in the droplet, coating in the titanium white particles. The scatterers with ordered structures have spatial coherence between them. The scattering of light can be analyzed qualitatively by photon energy bands. The scatterers with disordered structures lose spatial coherence. The scattering effect is equal to the arithmetic sum of the scattering of all the monomers. It is generally believed that the physical origin of color produced by ordered structure is coherent scattering, while the physical origin of color produced by disordered structure is monomer scattering. With the further study of structural color, however, the order of structure is no longer the essential feature of the formation mechanism of the structural color [8]. The coherence of the interaction between light and structure is the essential characteristic of the formation mechanism of the structural color.

Structural color derives from the interactions of light with microstructures with a featured size comparable to visible wavelengths via optical phenomena. The study of biological structure color can not only reveal many interesting color regulation methods in nature but also give us a lot of enlightenment in the design and preparation of photon structures. For plantae, the petals of *Hibiscus trionum* flower form hierarchical structures when nanoscale cuticular ridges overlay bulged epidermal cells (see Figure 6.1a) [9]. These hierarchical structures can broaden the observable angles of iridescence, enhancing the foraging efficiency of pollinators. The pointillist structural color in *Pollia* fruit is caused by the Bragg reflection of helicoidally stacked cellulose microfibrils that form multilayered structures in the cell walls (see Figure 6.1b) [10]. Because the multilayers form with both helicoidicities, the reflected light from every epidermal cell is polarized circularly either to the left or to the right.

For insects, the transparent wings of small Hymenoptera and Diptera produce wing interference patterns due to thin film interference (see Figure 6.1c) [11]. The visibility of these patterns is affected by the way the insects display their wings against various backgrounds with different light properties. Peacock spiders (*Maratus robinsoni* and *Mitostoma chrysomelas*) court females by using their tiny structured scales that reflect the full visual spectrum (see Figure 6.1d) [12]. Those scales have two-dimensional (2D) nanogratings on microscale three-dimensional (3D) convex surfaces with at least twice the resolving power of a conventional 2D diffraction grating of the same period. Throughout their wildly successful proliferation across the Earth, organisms have evolved enviable discoloring adaptations to their diverse habitats.

Across animals, many species of birds have feathers that are brilliantly colored without the use of pigments. An example is the crown of *Lepidothrix coronata* (see Figure 6.1e) [13], in which the light of specific wavelengths is selectively scattered from nanostructures with variations in index of refraction on length scales of the order of visible light. Coloration serves as a dynamic form of information, [15] such as the broad-tailed hummingbird, which feathers are structurally colored and change dramatically in appearance from black to magenta depending on the viewing angle and/or the angle of illumination. Colorful body parts are moved

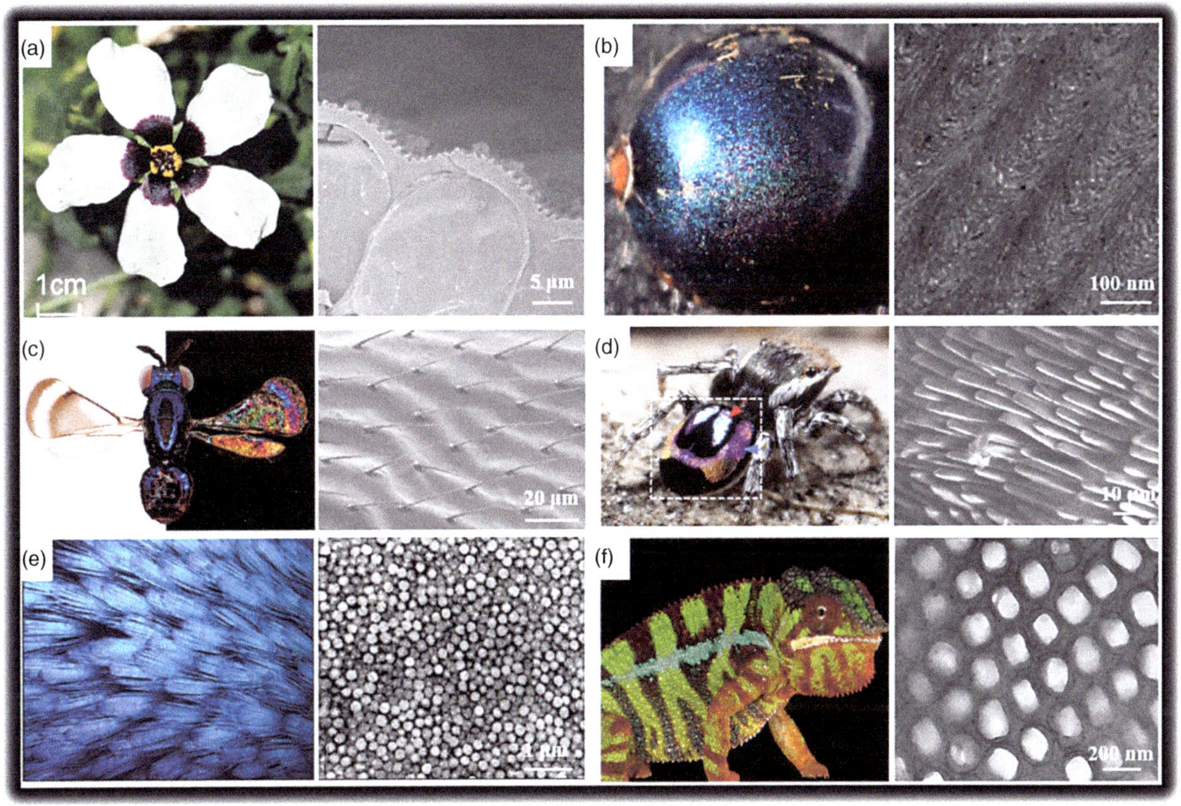

Figure 6.1 Structural colors and their corresponding microstructures in nature. (a) A blooming *Hibiscus trionum* flower and its cryo-SEM imaging of a cross section. Source: Chen et al. [9]/with permission of John Wiley & Sons. (b) *Pollia* fruit and its layered microstructure. Source: Vignolini et al. [10]/with permission of National Academy of Sciences. (c) Female Eulophidae Closterocerus coffeellae and its corrugation ridges with rows of setae. Source: Shevtsova et al. [11]/with permission of National Academy of Sciences. (d) Miniature peacock spider and its abdominal scales. Source: Hsiung et al. [12]/with permission of Springer Nature. (e) Photograph of crown feathers from *Lepidothrix coronata* and its TEM image of air spheres in beta-keratin. Source: Forster et al. [13]/with permission of John Wiley & Sons. (f) Chameleon and its guanine nanocrystal microstructures in the skin layer. Source: Teyssier et al. [14]/Springer Nature/CC BY 4.0.

in behavior, and both pigments and structural colors change at various temporal resolutions [16].

Colors produced by wavelength-dependent light diffraction, scattering, and interference are key components of visual communication in nature and acts particularly strongly in visual signaling by structurally colored animals during courtship, camouflage, and evolution. Particular interesting is chromotropic effects, which refer to reversible color transformations in a material due to external chemical or physical influences. Chameleons change color by actively adjusting the lattice of guanine nanocrystals in a thick skin (see Figure 6.1f) [14]. A deeper population of iridophores with larger crystals reflects a substantial proportion of sunlight, especially in the near-infrared range. A variety of external stimuli have been demonstrated to produce such structural color changes. Although many nanostructures with excellent optical properties have been shown in nature, the way to obtain artificial microstructures and a largely untapped pool of unique self-assembly strategies remains a challenge. With the intent of providing materials scientists and engineers with a muse for the next generation of structural color materials (SCMs), here, a selection of self-assembly strategies is summarized and categorized.

6.2 The Type of the Structural Color and Its Formation Mechanism

6.2.1 Structural Color due to Interference

The biological structure colors in nature mostly come from thin film interference, which can be divided into monolayer and multilayer. The theory of interference mainly refers to the interference of thin films. When a light beam strikes the upper surface of the film, one part is reflected and the other part is transmitted into the interior of the film. When a light beam strikes the top surface of the film, one part is reflected and the other is transmitted into the inner of the film. The transmitted part is reflected on the bottom surface of the thin film. The light reflected from the top and bottom surfaces of the film are mutually parallel and coherent and therefore interferes. For example, the color of soap bubbles in sunlight is produced by thin film interference. Many kinds of the iridescent phenomenon in the biological world are mainly caused by interference [17]. The iridescent phenomenon refers to the fact that the color of the film changes according to the angle.

Compared with monolayer, the structure color produced by multilayer is more bright, with higher saturation and more diverse forms. A multilayer reflector consists of parallel-sided layers of a transparent material separated by layers of material of a different refractive index (see Figure 6.2) [18]. Such a structure can act as a reflector when the optical thicknesses (the product of the actual thickness and the refractive index) of all layers are comparable with the wavelength of the incident light. The reflected beams from all layers interfere, and the color is observed when this interference is constructive for a given wavelength of light. As the angle of incidence changes, different wavelengths constructively interfere due to variations in path lengths through the layers. White light could be produced when

Figure 6.2 Optical/histologic properties, TEM cross-section composite image, and the corresponding reflection simulation of the white stripe of the *Pyjama Squid*. Source: Bell et al. [18]/with permission of John Wiley & Sons.

the multilayer reflectors exist nanostructures that reflect all kinds of wavelengths. Multilayer reflectors from *Pyjama Squid* are composed of several stacks. Each stack reflects a different wavelength, but collectively they reflect all wavelengths, showing white color. Large variation in plate orientation reduces specular effects, making this optical system a weakly polarized broadband diffuser over a wide range of viewing angles.

6.2.2 Structural Color due to Scattering

Scattering refers to the phenomenon that some light waves deviate from the original propagation direction and scatter in different directions due to the existence of random inhomogeneity in the medium. The light that spreads out in all directions is scattered light. The nonuniformity of the medium may be the internal structure of the medium loose undulation, which may also be the presence of impurities in the medium particles. The scattering of light generally falls into two broad categories. One is that the wavelength frequency of the scattered light changes, such as Raman scattering. The other is that the wavelength frequency of the scattered light does not change, such as Rayleigh scattering and Mie scattering. The color of the scattered light depends on the size of the particles and the refractive index difference between the particles and the surrounding medium. When the particle size is smaller than the wavelength of light, the ratio of the scattered light intensity to the incident light intensity is inversely proportional to the fourth square of the wavelength, and the scattering is Rayleigh scattering. In this case, the short-wavelength blue light will be preferentially scattered, such as the blue sky. Blue Rayleigh scattering was observed for particles between 300 nm and 1 µm in size. When the size of the scattered particles is close to or larger than the wavelength of the light wave, Mie scattering can be observed. The scattering color is no longer blue, but mainly red and green. When the particle size is close to 1 nm, most of the visible light is scattered and the scattered light appears white.

The scattering can be divided into incoherent scattering and coherent scattering according to the order of the dielectric system [19]. Incoherent scattering refers to the scattering of a disordered system, in which each scatterer interacts with the incident light separately and has no influence on each other, such as Rayleigh scattering and Mie scattering. Coherent scattering means that the system has a certain order and periodicity, and there will be an interaction between each scatterer. One difference between coherent scattering and incoherent scattering is that associated scattering has a certain directivity. Structural color change as a function of humidity has been reported for the flower of *Diphylleia grayi* [20]. On sunny days, 3D nano/micropore arrays of the loose cell structure in the white petals play a role in generating Mie scattering of visible light. On rainy days, water fills the nano/micropore cavity array of the loose cell structure of the petals, changing it to a homogeneous clear phase. Inspired by the Mie scattering effect in *D. grayi*, white-generating glazing films are simply fabricated by patterning ordered air hole inverse opal arrays on the surface of a PDMS film (see Figure 6.3a) [21].

Likewise, Hyun and coworkers present an alternative pathway to achieve color with strong scattering near-micron-sized titanium dioxide (TiO_2) particles through the use of polydispersity [22]. While a single TiO_2 particle of near-micron size is expected to scatter light indiscriminately over the visible range, resulting in a white appearance, a collection of such particles with different sizes and population results in the display of visible colors, as schematically shown in Figure 6.3b. The generation of polydisperse particles with a normal size distribution is naturally achieved with a hydrothermal procedure, presenting a cost-effective, scalable, and practical route toward displaying color without the use of clean-room methods. The visible colors are shown to result from smooth variations in wavelength in the extinction spectrum due to the ensemble averaging of Mie scattering from individual particles of different sizes. Contrary to structures of subwavelength sizes where the use of single resonant modes demands identically sized nanostructures for color control, strong scattering, large-sized particles that exhibit multiple resonant modes demand a polydisperse size distribution to give rise to a unique combination of colors.

6.2.3 Structural Color due to Diffraction

Diffraction refers to the phenomenon that light wave deviates from the straight propagation direction when it passes through the edge of an obstacle or pore in the process of propagation. Just like the interference phenomenon, diffraction is essentially based on the linear superposition principle of the wave field. Diffractions caused by structures in nature can be divided into two categories. The first is the surface with a regular structure. Some structures appear as a series of parallel or approximately parallel grooves or projections with regular intervals on the epidermis, showing a bright rainbow color [23]. Bioinspired artificial microstructures have greatly expanded the research progress of functional SCMs. Pisignano and coworkers demonstrated a bioinspired dual photonic structure that contains the combination of microscopic periodic 1D surface grating and nanoscopic helical organization, giving rise to programmable color mixing and polarization rotation (see Figure 6.4) [24].

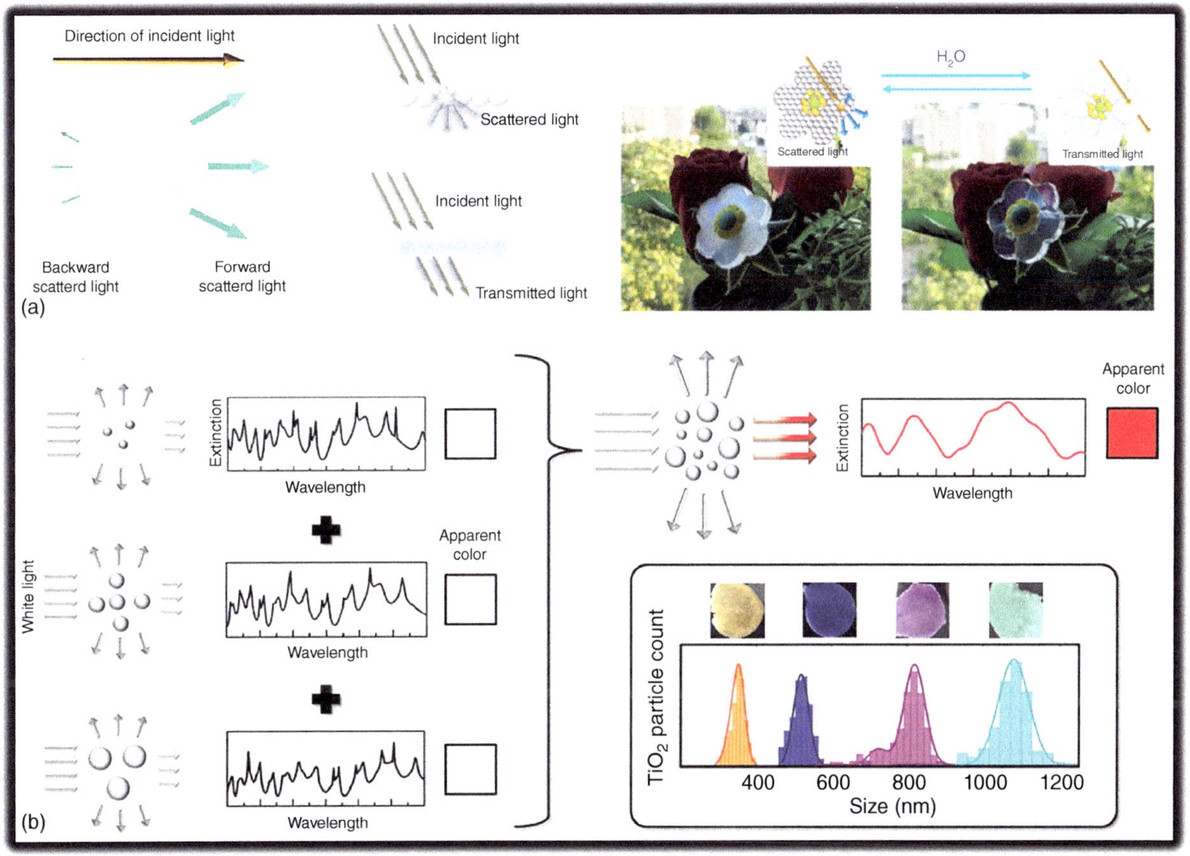

Figure 6.3 (a) Schematic diagram of Mie scattering on single nano/microcavity with incident light and the Mie scattering effect on nano/microcavities in scattered and bleached states as incident light induces. Source: Yoo et al. [21]/with permission of American Chemical Society. (b) Schematic illustration of the color generating mechanism for a polydisperse collection of particles. Inset: Optical images of particles in ethanol dried onto a glass slide, viewed before a white LED light source. Source: Reproduced with permission from Alam et al. [22], copyright 2017 American Chemical Society.

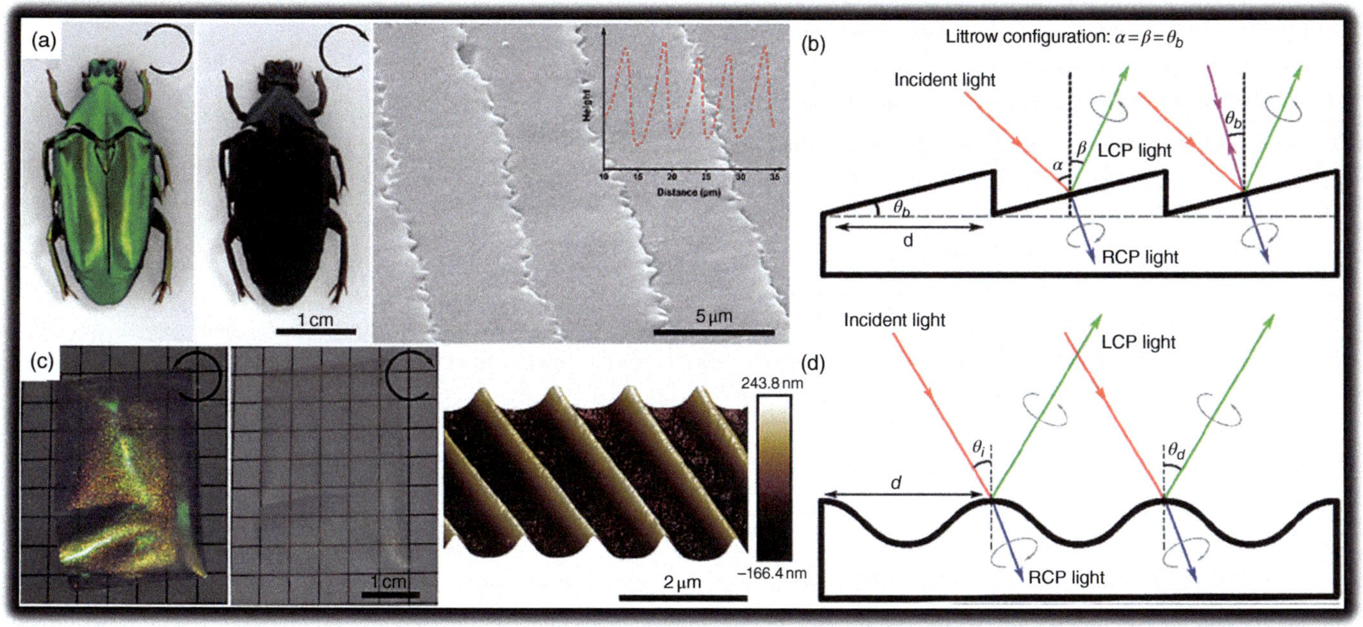

Figure 6.4 (a) Photographs of the beetle *Chalcothea smaragdina* captured in a left-handed circularly polarized (LCP) and a right-handed circularly polarized (RCP) filter, respectively, and the periodic surface structure of the beetle; (b) the diffraction schematic of the polarization-selective surface blazed grating for the beetle exocuticle; (c) photographs of a bionic chiral nematic patterned film captured using a LCP (left) and a RCP (right) filter, respectively, and the periodic surface structure of the bionic film; (d) the diffraction schematic of the chiral nematic ordered holographic surface grating. Source: Chu et al. [24]/with permission of Royal Society of Chemistry.

The second is a structure that produces Bragg diffraction effects at optical wavelengths, sometimes referred to as a diffraction grating or a photonic crystal. Materials with such a structure are often referred to as photonic bandgap materials. When the bandgap range falls within the visible light range, visible light of a particular wavelength will not penetrate the material. This non-propagating light will be reflected by the photonic crystal, causing coherent diffraction on the surface of the periodic crystal, producing a structural color that the eye can perceive. The coloring mechanisms of photonic crystals have been outlined in other chapters and will not be repeated here.

6.3 The Assembly Methods of the SCMs

Self-assembly of nanoparticles is an important method for the biomimetic construction of SCMs. The complex interaction process involves gravity, electrostatic repulsion, and Brownian motion. The relevant force analysis and action mechanism have been described in detail in Chapter 4, Self-assembly of Colloidal Crystals: Strategies, and will not be repeated here. Several self-assembly methods will be introduced for constructing SCMs, including evaporation self-assembly, membrane separation-assisted assembly (MSAA), air–liquid interface self-assembly, oil–oil interface self-assembly, oil–water interface self-assembly, controlled micellization self-assembly, layered hydrogels self-assembly, spray coating self-assembly, unidirectional rubbing self-assembly, edge-induced rotational shearing self-assembly, screen printing self-assembly, magnetic-induced self-assembly, photoinduced self-assembly, atomic layer deposition (ALD) self-assembly, physical vapor deposition self-assembly, and surface wrinkling. The corresponding assembly process is introduced in detail. Different assembly methods are suitable for different types of SCMs. These methods can enlighten the construction of large-area structural color films.

6.3.1 Evaporation Self-assembly

Self-assembly methods have attracted considerable interest as potential alternatives to conventional up–down preparation processes for the scalable, low-cost syntheses of nanoporous solid structures. Opals and inverse opals are the most common SCMs. The construction of inverse opal is usually prepared based on the opal template. Here, we will introduce the construction method of large-area inverse opals based on evaporation self-assembly. Three sequential steps were included during the process of conventional self-assembly (see Figure 6.5a) of inverse opals: (i) assembly of the colloidal crystal template, (ii) infiltration and deposition of a matrix phase, or a precursor to a solid matrix phase, and (iii) selective removal of the template [25]. However, the use of sedimentation [26], shear flow [27], spin-coating [28], and evaporative [29] or flow-controlled deposition [30] self-assembly processes to form large-area structural color films has very typically resulted in the formation of cracks, domain boundaries, colloid vacancies, and other defects. The conventional

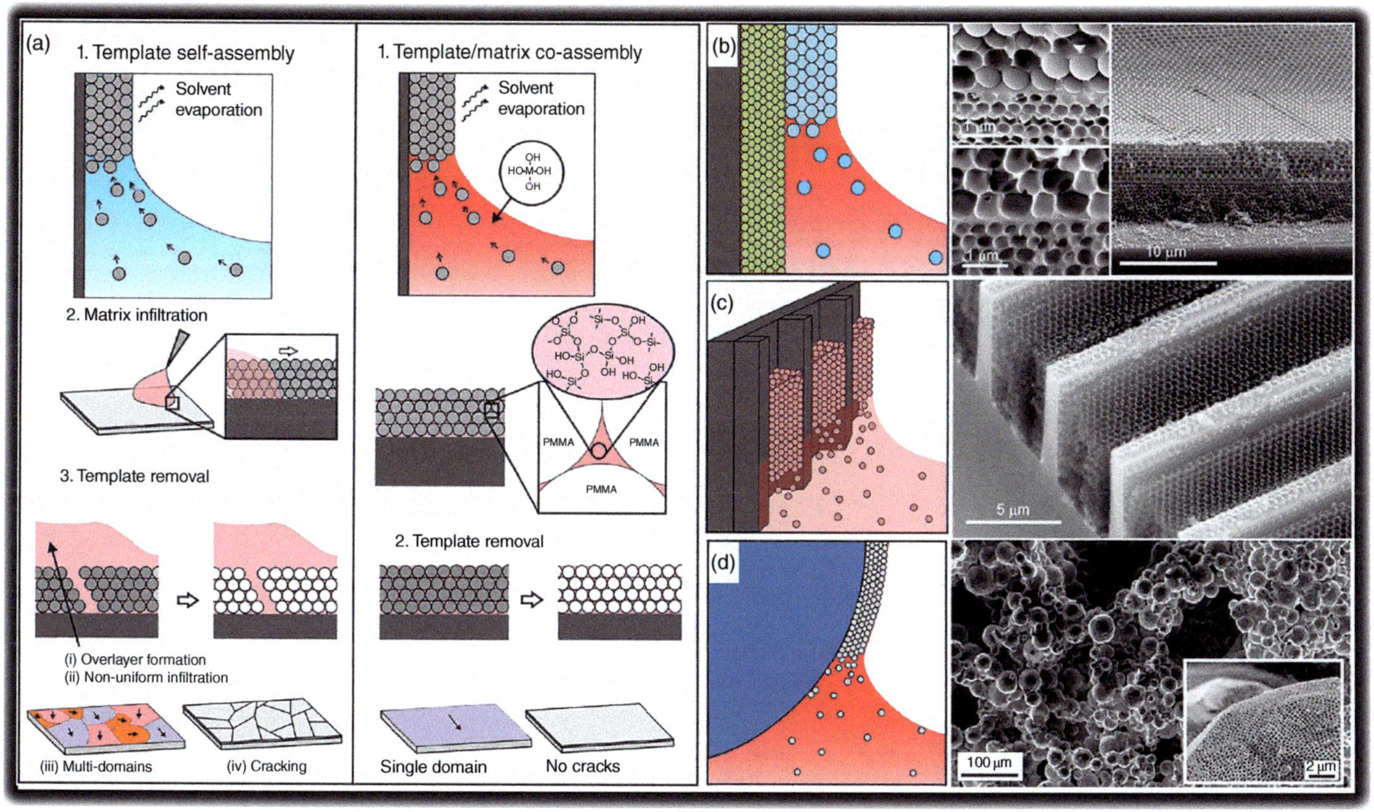

Figure 6.5 (a) Schematics of template self-assembly and template/matrix co-assembly for the syntheses of inverse opal films. Processes schematics and the representative scanning electron microscope (SEM) images of the novel co-assembly silicon dioxide (SiO_2) inverse opals; (b) multilayered hierarchical films; (c) co-assembly on topologically patterned substrates; (d) co-assembly onto curved surfaces. Source: Hatton et al. [25]/with permission of National Academy of Sciences.

self-assembly has yielded ordered inverse opals over modest (≤10 μm) length scales. The uncontrolled formation of defects over larger length scales limits their real-world applications.

Aizenberg and coworkers demonstrate the evaporation co-assembly of a sacrificial colloidal template with a matrix material in a single step to yield a colloidal composite (see Figure 6.5) [25]. Multilayered composite colloidal crystal films have been generated via evaporative deposition of polymeric colloidal spheres suspended within a hydrolyzed silicate sol–gel precursor solution. It avoids the infiltration process and minimizes the associated cracking and inhomogeneities of the resulting inverse opal films.

To determine a well-ordered colloidal deposition, varying amounts of tetraethyl orthosilicate (TEOS) solution were added to suspensions of colloidal polymethyl methacrylate (PMMA) in water. The ratio of TEOS was found to have a significant influence on the film structure and defect formation. Insufficient silicate additions did not allow for the formation of a continuous silica network throughout the colloidal template, leading to considerable tearing/cracking of the film, similar to that observed for a PMMA colloidal crystal. Highly ordered crack-free films formed at the condition of approximately 0.15 ml TEOS solution per 20 ml PMMA suspension. Further additions of TEOS to the deposition mixture (>0.20 ml) caused the initial formation of an overlayer, which increased in thickness with further TEOS additions.

The co-assembly mechanisms involve the preferred crystallographic orientation, the absence of defects, and the remarkable atypical orientations of the crack patterns. The <112> growth orientation is expected to dominate the <110> orientation under conditions of slow conventional growth for colloidal crystals. During this co-assembly process, however, the strong preference for the <110> orientation was made particularly evident by a unique defect self-healing phenomenon. Domains that occasionally nucleate and grow with a different orientation gradually become corrected to the default <110> orientation, and the defect eventually disappears. This co-assembly approach allows the fabrication of hierarchical structures not achievable by conventional methods, such as multilayered films and deposition onto patterned or curved surfaces (see Figure 6.5b–d). These robust SiO_2 inverse opals obtained by removing the PMMA can be transformed into various materials that retain the morphology and order of the original films. Colloidal co-assembly based on evaporation is available for a range of organometallic sol–gel and polymer matrix precursors and represents a simple, low-cost, and scalable method for generating high-quality, chemically tailorable inverse opal films for a variety of applications.

This kind of assembly method is usually suitable for preparing orderly nanoarray from colloidal particle dispersion. Several advantages of this approach include (i) few template defects and crack density, (ii) controlled growth of large, highly ordered domains via a scalable process, (iii) prevention of overlayer formation and nonuniform infiltration, (iv) simplified preparation process, and (v) the ability to form multilayered, hierarchical, patterned, and curved structures that are not easily possible by any other method. Although it takes a long time, the area and thickness of the nanoarray can be accurately controlled. SCMs can be prepared on many kinds of substrates.

6.3.2 Membrane Separation-assisted Assembly

The preparation of large-area structured color films is usually accompanied by cracks and defects. The conventional evaporative self-assembly method is not suitable for the construction of large-area SCMs due to the structural defects and long assembly time. The self-assembly of nanoparticles is constantly optimized.

Chen and Weitz and coworkers reported a MSAA strategy for the construction of angle independent structural colors (see Figure 6.6) [31]. By taking advantage of the water transport channels in sulfonated reduced graphene oxide (rGO), water and submicron particles could be separated easily, along with colloidal particle assembles into amorphous structures. The color contrast can be greatly enhanced by the reduced graphene oxide-sulfopropyl methacrylate potassium salt (rGO-SPM) background. The dark color rGO-SPM results in a decrease in coherent light scattering and reduction of light reflection from the substrate. Importantly, the

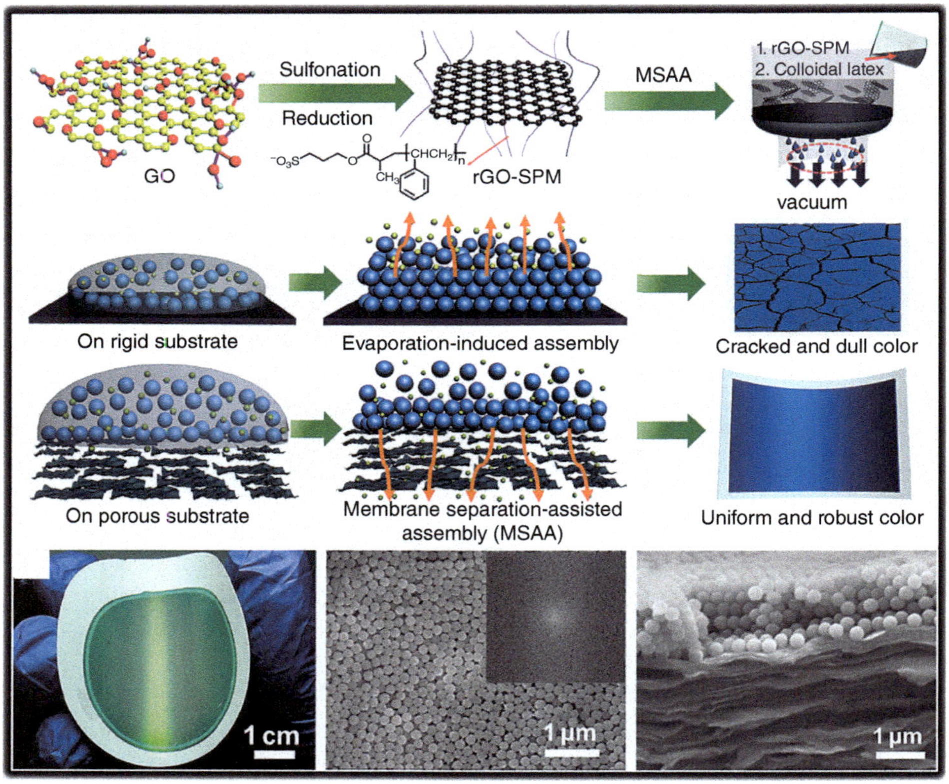

Figure 6.6 (a) Schematic illustration of the fabrication of rGO-SPM and assembly progress of the rGO-SPM/poly(styrene-methylmethacrylate-acrylic acid) (poly(St-MMA-AA)) film via MSAA; (b) schematic illustration of evaporation-induce assembly; (c) schematic illustration of MSAA method; (d) photograph of a prepared structural color film and the corresponding SEM images. Source: Zhu et al. [31]/with permission of American Chemical Society.

obtained structural color films are crack-free and could be bent in any directions (see Figure 6.6b–c).

Briefly, 2–4 ml of the as-synthesized rGO-SPM aqueous solution was filtered through a mixed cellulose ester filter membrane (47 mm in diameter, 0.2 μm pore size) (see Figure 6.6a). Then, 4–6 ml of the diluted colloidal suspension with a concentration of 0.2 wt% was dropped on the rGO-SPM film carefully and filtered until dry (see Figure 6.6c). Moreover, a stepwise filtration method can be used to prepare structural color patterns. After filtration and assembly of poly(styrene-*co*-methyl methacrylate-*co*-acrylic acid [St-MMA-AA]) microspheres, the obtained films are flexible and have uniform morphologies with improved structural colors (see Figure 6.6d). The rGO-SPM substrate greatly decreases the transverse tensile stress across the colloidal film. Nanoscale wrinkles and wavelike ripples on rGO-SPM caused by in-plane defects endow colloidal particles with amorphous arrangements.

Different from conventional evaporation self-assembly of structural colors, MSAA is suitable for rapid self-assembly (less than 2 min) of uniform nanoarrays, which usually appear to be short-range ordered and crack-free. The outward capillary flow was broken by the soft and wrinkle-like rGO-SPM structures, forming a consecutive and symmetrical colloidal film. It is suitable for preparing the large-area structural color film on flexible substrates.

6.3.3 Air–Liquid Interface Self-assembly

The preparation of large-area structural color films can effectively satisfy the requirements of large-size display applications. Some self-assembly methods have also been developed, such as inkjet printing [32], spray coating [33], and bar coating [34]. During the preparation process, the so-called "coffee-ring" effect (produced by the contact angle of the assembled liquid droplet in a hard substrate–liquid–air interface) [35] has always been a difficult problem in the preparation of large color films. As an alternative to a hard substrate, an air–liquid interface can be used as a soft substrate to construct structural color films [36]. Two different methods of air–liquid interface self-assembly are shown in the following text.

Zhang and coworkers presented a new water-rewritable photonic crystal film as large as the A4 size ($210 \times 300\,mm^2$) with a high-quality structure color (see Figure 6.7a–c) [37]. The water-rewritable photonic crystal paper was obtained by thermally assisted self-assembly at the air–liquid interface. During the assembly process, partially deformed and coalesced nanoparticles with a low glass transition temperature (T_g) were the key to the formation of transparent and structural colorless films (see Figure 6.7a). The low glass transition temperature makes the ordered microspheres partially fused into a whole. Although the original 3D ordered structure is maintained, it does not have optical diffraction properties. Importantly, utilizing the hydrophilic of the assembled block, the transparent and structural colorless film can be switched to a structural color state by touching the water (see Figure 6.7c). Once the deformation–coalescence and continuous spheres met water, the nanostructures undergo geometrical morphological transformations to form

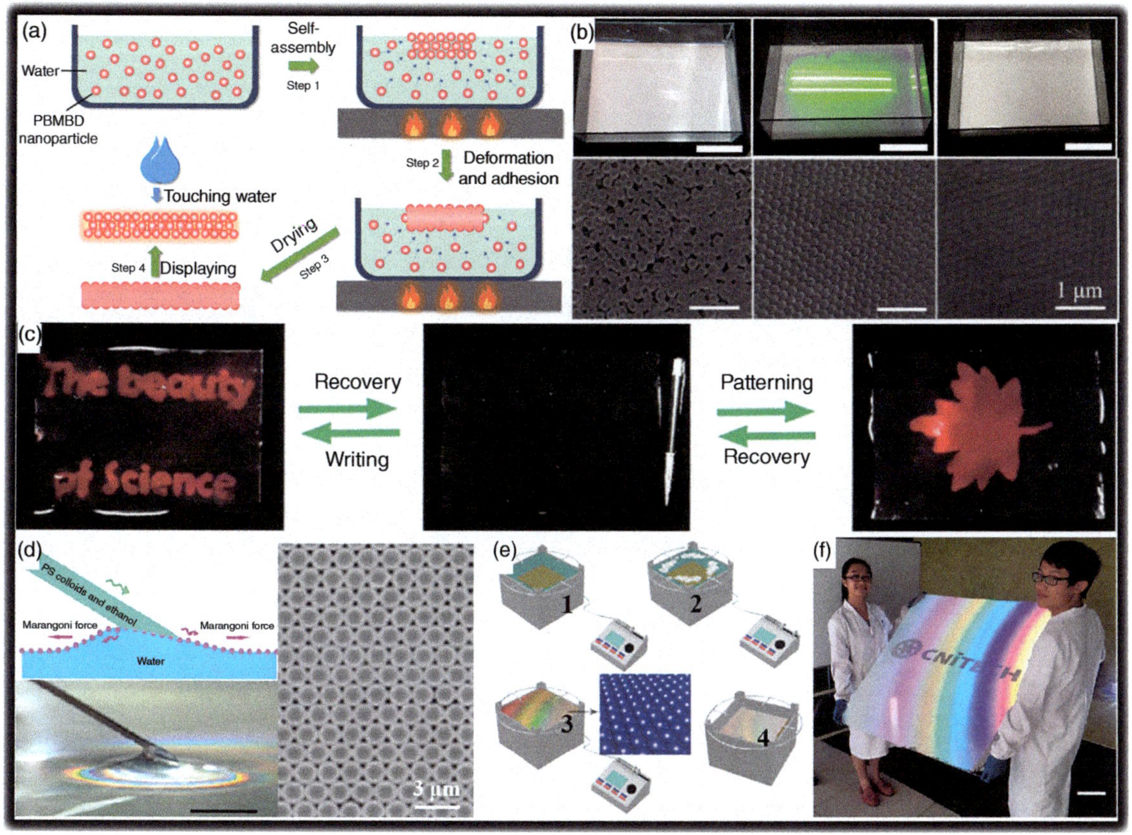

Figure 6.7 (a) Schematic illustration of the fabrication of the large-area water-rewritable structural color film; (b) digital photos and the corresponding SEM images of the three stages in the fabrication process; (c) the process of the water writing and patterning. Source: Yu et al. [37]/with permission of American Chemical Society. (d) Procedure of preparation of close-packed PS monolayer on a substrate, including injector tip injection, PS driving by Marangoni forces, monolayer formation, and the corresponding SEM images; (e) schematics of the micro-propulsive injection systems and the formation processes for large-area PS nanosphere arrays; (f) photographs of as-deposited PS monolayer on 36 wafers and 1 m [2] glass substrate. Source: Gao et al. [38]/with permission of American Chemical Society.

the separate spheres, showing corresponding brilliant structural color. This kind of large-area photonic crystal film using water writing or patterning is a potential candidate for replacing traditional one-time use papers.

The following self-assembly stages should be included: (i) heating a 15 wt% poly(butyl methacrylate-*co*-methylmethacrylate-*co*-butyl acrylate-*co*-diacetone acrylamide) (PBMBD) water dispersion placed in an open glass vessel (a size of $220 \times 310\,mm^2$) at an evaporating temperature slightly above T_g (see Figure 6.7b). A close-packed layer of PBMBD particles was assembled at the air–liquid interface via thermal-assisted evaporating driving force. (ii) High-reflectance and brilliant iridescent film floated on the air–liquid interface. As the heating went on, the interstices between the PBMBD nanoparticles would shrink due to deformation against one another between soft PBMBD particles above their T_g. (iii) The brilliant iridescent color disappeared, and a translucent film was obtained with a lacking refractive index contrast.

The air–liquid interface self-assembly was controlled by the evaporating temperature, which directly determined the quality of the ordering photonic arrays. Air–liquid interface self-assembly is usually suitable for systems with a large difference in density. The method requires that the microsphere is incompatible with the solvent. For the self-assembly of soft microspheres (nanoparticles with low T_g, <60 °C), the assembly temperature is generally required to be greater than the T_g of the microspheres and lower than 90°. Higher evaporation temperatures will lead to excessive driving forces, resulting in a disordered arrangement. Evaporation temperatures (higher than T_g) can promote the integration of soft microspheres. The assembly area is determined by the area of the air–liquid interface, and the large area of the functional structure color film can be prepared by preparing microspheres with different properties. The thickness of the prepared film is related to the concentration of dispersion, temperature, and assembly time.

Nowadays, a monolayer of the colloidal template produced by self-assembly has been verified to be an effective and versatile route toward functional 2D patterned nanostructures. Ye and Cui and coworkers presented a high-throughput surface texturing process for optical and optoelectronic devices based on large-area self-assembly of nanospheres via a low-cost micropropulsive injection (MPI) method (see Figure 6.7d–f) [38]. The formation process of the polystyrene (PS) monolayer is schematically shown in Figure 6.7e, including (i) presetting the substrates underneath the water and connecting the injectors with the MPI system; (ii) tuning the contact states between the injectors and the water surface and then initiating the MPI for the injection of PS colloids; (iii) turning off the MPI until a colorful interference pattern is formed on the entire water surface when illuminated by a white light source; and (iv) slowly declining the water level or raising the substrate to transfer the PS monolayer onto the preset substrates. The mechanism of forming PS monolayer on the water surface is based on the well-known Langmuir–Blodgett (LB) effect.

Briefly, to obtain well-spread PS nanospheres on the water surface, PS colloidal solution with low surface tension should be prepared by adding ethanol. The addition of ethanol is determined by the size and concentration of the PS nanospheres.

After sonicating for several minutes, the mixture is continuously injected onto the water surface by using the MPI system (see Figure 6.7d). The PS colloidal solution together with ethanol is spread on the water surface through the nozzle and then spread out under the driving of the Marangoni forces (a mass transfer along with the interface between two fluids as a result of surface tension gradient) [39]. Due to the surface tension of the water, the formed meniscus damps the vertical flow forces to prevent the break of the water surface, enabling the layered PS nanospheres immediately spreading out with the help of the dispersing agent of ethanol. The PS monolayer can be fabricated as large as possible by enlarging the water surface areas and as fast as possible by using more injectors. The formed PS monolayer is transferred to submerged substrates by draining the water bath or raising the substrates slowly. During the self-evaporation or low-temperature baking process, the capillary forces draw the nanospheres together and made the nanospheres pack in a hexagonally close-packed pattern and tightly attached on the substrates. This self-assembly method is suitable for the preparation of large-area structural color films and lays a foundation for industrial production.

The air–liquid interface self-assembly of monolayer arrays is widely used in the preparation of 2D SCMs. Interestingly, the assembled 2D templates can be further improved, such as Janus colloidal crystal film [40] and 2D array-hydrogel composite film [41]. Asymmetric Janus materials consist of objects with two sides of different chemistry or morphology. Fujii et al. described a facile method to make soft Janus-type 2D colloidal crystal films with an area over $4 \times 10^2\ cm^2$ by using air–water interface self-assembly (see Figure 6.8a,b) [40]. The key steps include (i) the synthesis of monodispersity particles, (ii) the formation of the regular 2D arrays at the air–water interface, and (iii) chemical oxidative polymerization of pyrrole, which was selected as a model reaction to coat 2D colloidal crystals facing the aqueous phase and connect the particles in a continuous film. Furthermore, 2D femtoliter cup array can be prepared by removing the particles from the Janus film. The films are hydrophilic and hydrophobic on opposite sides and spontaneously fold in both air and aqueous phases. The Janus film and femtoliter cup-array film were soft and deformable, enabling the introduction of patterned structures. In addition, the liquid used for the self-assembly is not limited to the water, such as Hg (see Figure 6.8c,d) [41]. The interface of Hg also provides specular reflection to further enhance diffraction while assembling the 2D array. After combining with hydrogel, the 2D array-hydrogel composite film can be obtained. It can transform the environmental response of hydrogel into structural color information.

This air–liquid interface self-assembly approach was an effective and versatile route toward functional 2D patterned nanostructures, owing to its high reproducibility, low cost, and good controllability on both of the chemical composition and structural parameters. It is suitable for the construction of Janus nanoarrays that depend on density difference to regulate the structure difference between the gas-phase side and the liquid-phase side. The microspheres were required to be incompatible with the solvent and less dense than the solvent.

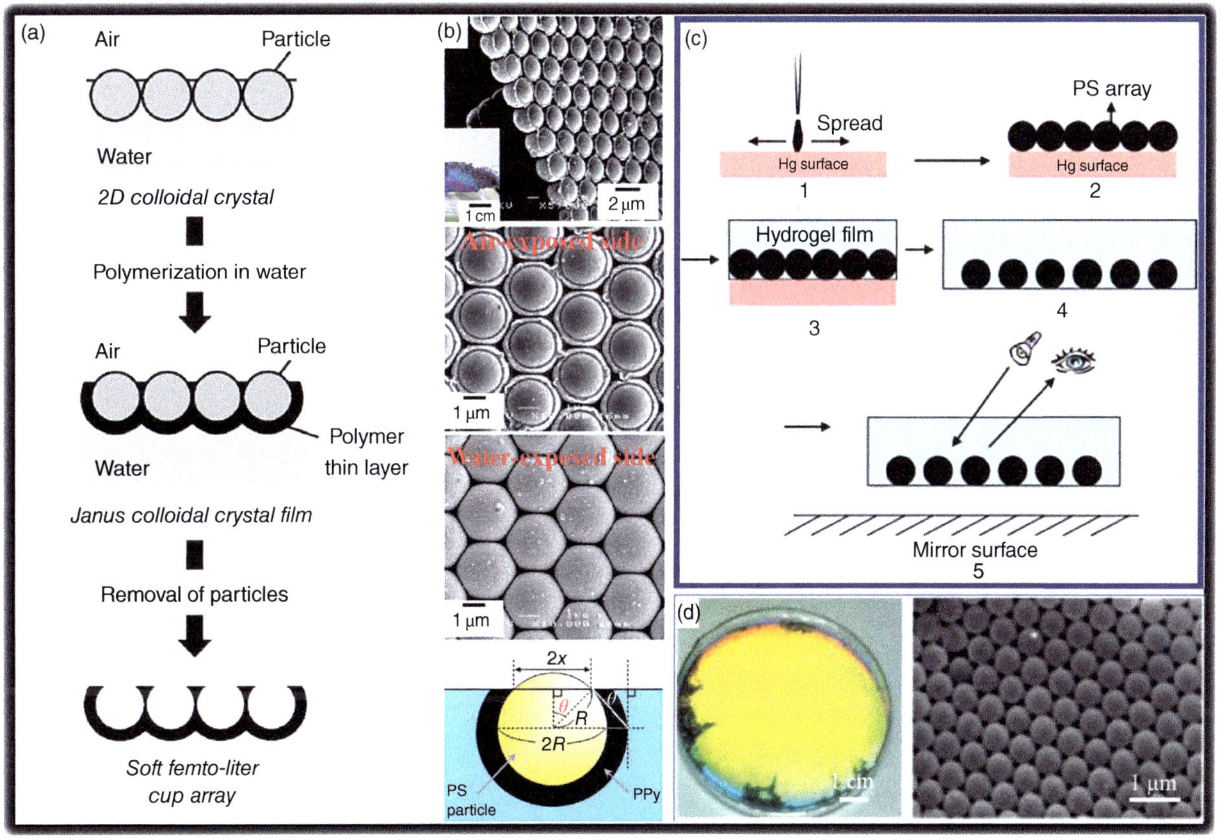

Figure 6.8 (a) Schematic outline for preparing soft Janus colloidal crystal film; (b) SEM images of a Janus colloidal crystal film and determination of the contact angle (through water) of the PS latex particle at the air–water interface; the inset shows the Janus colloidal crystal films floating on the water surface under sunlight. Source: Fujii et al. [40]/with permission of John Wiley & Sons. (c) Fabrication of a 2D photonic crystal for sensing applications; (d) photograph showing bright diffraction of yellow light from a PS particle array on a Hg surface and the corresponding SEM image. Source: Zhang et al. [41]/with permission of American Chemical Society.

6.3.4 Oil–Oil Interface Self-assembly

As we know, scattering, diffraction, and interference mechanisms are inherent to spherical droplets, which contribute to atmospheric phenomena such as glories, coronas, and rainbows [42]. Coloration by total internal reflection (TIR) and interference at microscale concave interfaces is a special coloring mechanism [43]. It can be used for the production of large-area structural color films through the air–solid interface or oil–oil interface self-assembly. Light traveling along different trajectories of TIR at a concave optical interface can interfere to generate brilliant patterns of color (see Figure 6.9a,b). Only the partially open concave interfaces that transition from high to low refractive index can lead to this coloration. It is readily observed in systems as simple as water drops condensed on a transparent substrate. As shown in Figure 6.9d, water droplets condensed onto the underside of a PS Petri dish lid, producing structural color at the three-phase contact line. The color emanates from near the contact line and varies with droplet diameter. The water drops were created by placing warm water in a dish under the room temperature Petri dish substrate and allowing water to condense. The macroscale patterned reflectance image of the elephant was created using selected area ultraviolet–ozone treatment to increase the hydrophilicity of the PS surface. By controlling the hydrophilicity of the air–solid interface, large-area patterned structure color can be assembled.

Furthermore, the iridescent biphasic oil droplets can be fabricated by manipulating the curvature of the oil–oil interface. Briefly, biphasic droplet emulsions were created by using a four-channel microfluidics. Varying ratios of surfactants were used to tune the droplet shape. And then 1 wt% sodium dodecyl sulfate in water was added to the droplet solution until the droplets exhibited a Janus shape that reflected light. Biphasic oil droplets were polymerized into solid particles. Two key requirements to generate structure color are (i) an interfacial refractive index contrast and (ii) a microscale geometry that supports multiple trajectories for TIR. By introducing a light-responsive azobenzene surfactant, light-responsive structure color can be created (see Figure 6.9c). Those Janus-shape droplets prepared by oil–oil interface self-assembly can be assembled on the water surface to form a large-area structural color film. Tunable droplet colors could be of interest for sensors or displays.

Oil–oil interface self-assembly is usually applied to incompatible oil phases with different refractive indices, and the coloring mechanism of SCMs is realized by TIR. The Janus and double emulsion morphologies can be switched by tuning the balance of interfacial tensions of the iridescent biphasic oil droplets. Then, the curvature of the oil–oil interface can be manipulated, regulating the TIR and the reflected color. This assembly method is not suitable for the preparation of polymer films with structural color, which are usually kept in a solution. The triple-phase line of the oil–oil interface is easily affected by external environmental conditions. However, such responsive, tunable droplet colors could be of interest for sensors or displays.

6.3.5 Oil–Water Interface Self-assembly

In the previous section, we described TIR from self-assembled structures. Interference from TIR is due to light interacting with microstructures that have dimensions

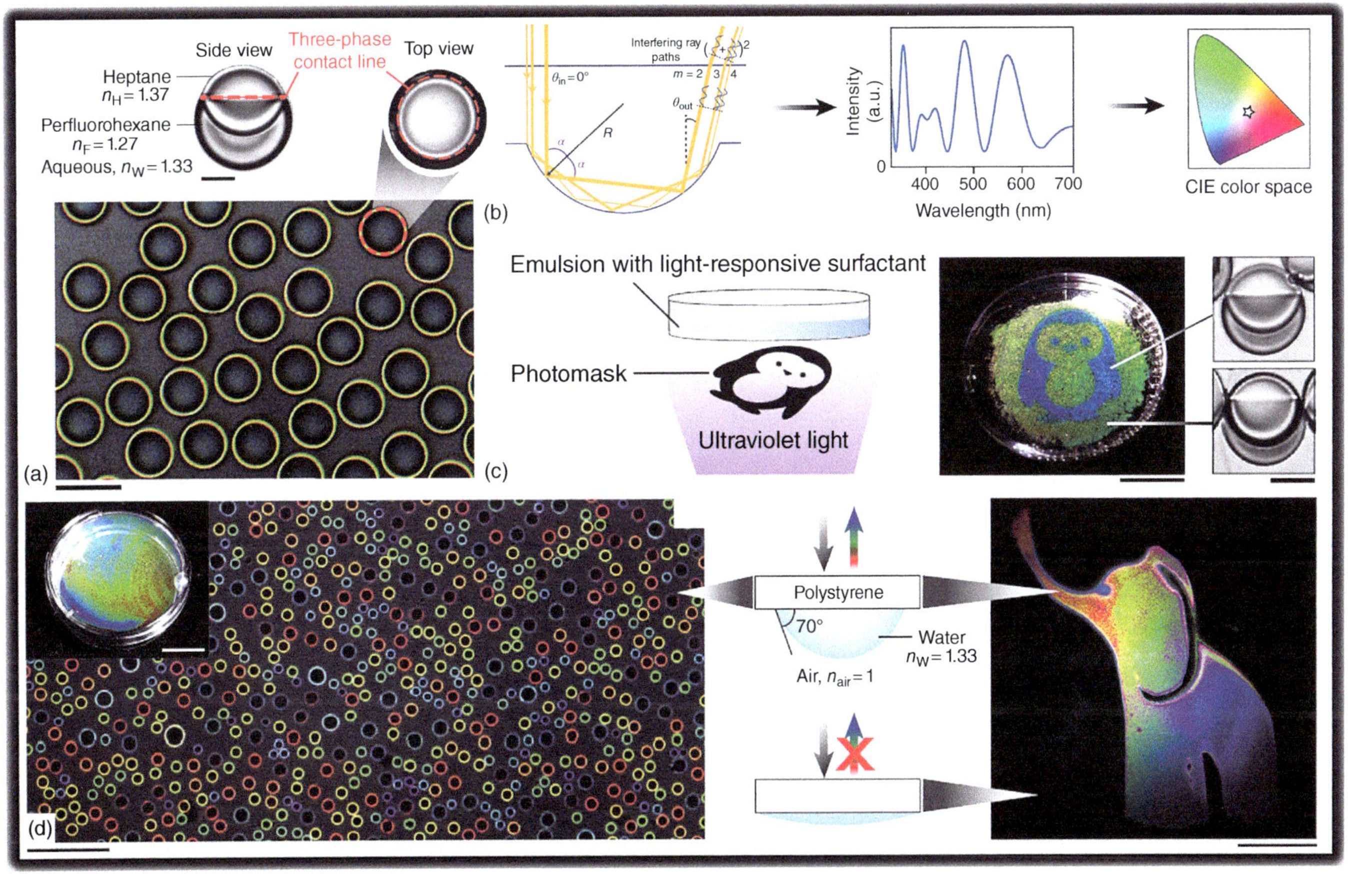

Figure 6.9 (a) Schematic and optical micrograph showing the biphasic droplet geometry and composition; (b) diagram of three rays taking different trajectories along the concave interface that interfere, causing the observed coloration; (c) iridescence of biphasic oil droplets depends on the curvature of the oil–oil interface; (d) reflectance optical micrograph of water droplets condensed onto a polystyrene Petri dish, and a colorful elephant created by condensing water onto a polystyrene Petri dish with surface hydrophilicity patterned. Source: Goodling et al. [43]/with permission of Springer Nature.

orders of magnitude larger than the wavelength of visible light. Only the three-phase contact line that transition from high to low refractive index can lead to TIR. Self-assembly techniques can be used to produce colloidal crystals onto surfaces, which can act as replicative masks or templates in mold-replication methods [44]. Utilizing oil–water interface self-assembly, the preparation process of structural color films with TIR can be greatly simplified. Zarzar and coworkers reported a simple route to the fabrication of polymeric microwell and microdome-array surfaces that display structural color via TIR interference (see Figure 6.10) [45]. The fabrication approach provides control over the radius of curvature and contact angle of the resultant microstructures by utilizing the assembly of glass particles at oil–water interfaces. The microstructure can be replicated via soft lithography into many kinds of polymeric materials of varying refractive indices and mechanical properties as a means to manipulate structural color and responsivity.

Briefly, a thin layer of uncured Norland Optical Adhesive 71 (NOA 71) was poured into a Petri dish followed by an aqueous suspension of soda-lime glass particles in hexadecyltrimethylammonium bromide aqueous surfactant (see Figure 6.10a). The dense particles sank to the water–monomer interface, forming a particle monolayer. Then, the particles were fixed by curing the NOA 71 via ultraviolet (UV). An array of glass spheres partially embedded in the polymer was formed when the aqueous phase was washed away (see Figure 6.10b). Varying concentrations of isopropyl alcohol could be added to tune the depth of the particles settled at the aqueous–monomer interface. Particles that are more deeply embedded into the monomer would thus result in domes with a lower effective contact angle. Both the contact angle, radius of curvature, refractive index contrast, and structure symmetry are important parameters that affect the structural color. By dynamically changing such parameters, for example, by deforming the surfaces with mechanical force or using temperature to change the refractive index, it could create stimuli-responsive color-changing surfaces and structurally colored patterned images (see Figure 6.10d,e). Moreover, an inverse replica of this domed surface was fabricated in cross-linked polydimethylsiloxane elastomer (PDMS). Importantly, the PDMS wells could be used directly to create the structural color by filling the wells with a high refractive index oil or polymer, or the wells could be used as a mold to fabricate domes of various polymers (see Figure 6.10c,d).

Oil–water interface self-assembly is suitable for the construction of embedded nanoarray, and the density of nanoparticles is generally required to be less than that of the oil phase. The water phase is not limited to water and should be incompatible with the oil phase. The oil phase can be selected as PDMS or other monomer mixtures. The nanoarrays assembled by this method are well organized. It is suitable for the preparation of large-area 2D composite structures and bowl structures.

6.3.6 Controlled Micellization Self-assembly

Brush block copolymers (BBCPs) enable the rapid fabrication of self-assembled 1D photonic crystals with photonic bandgaps that are tunable in the UV–vis–IR [46]. The structural color can be readily tuned by either altering the molecular

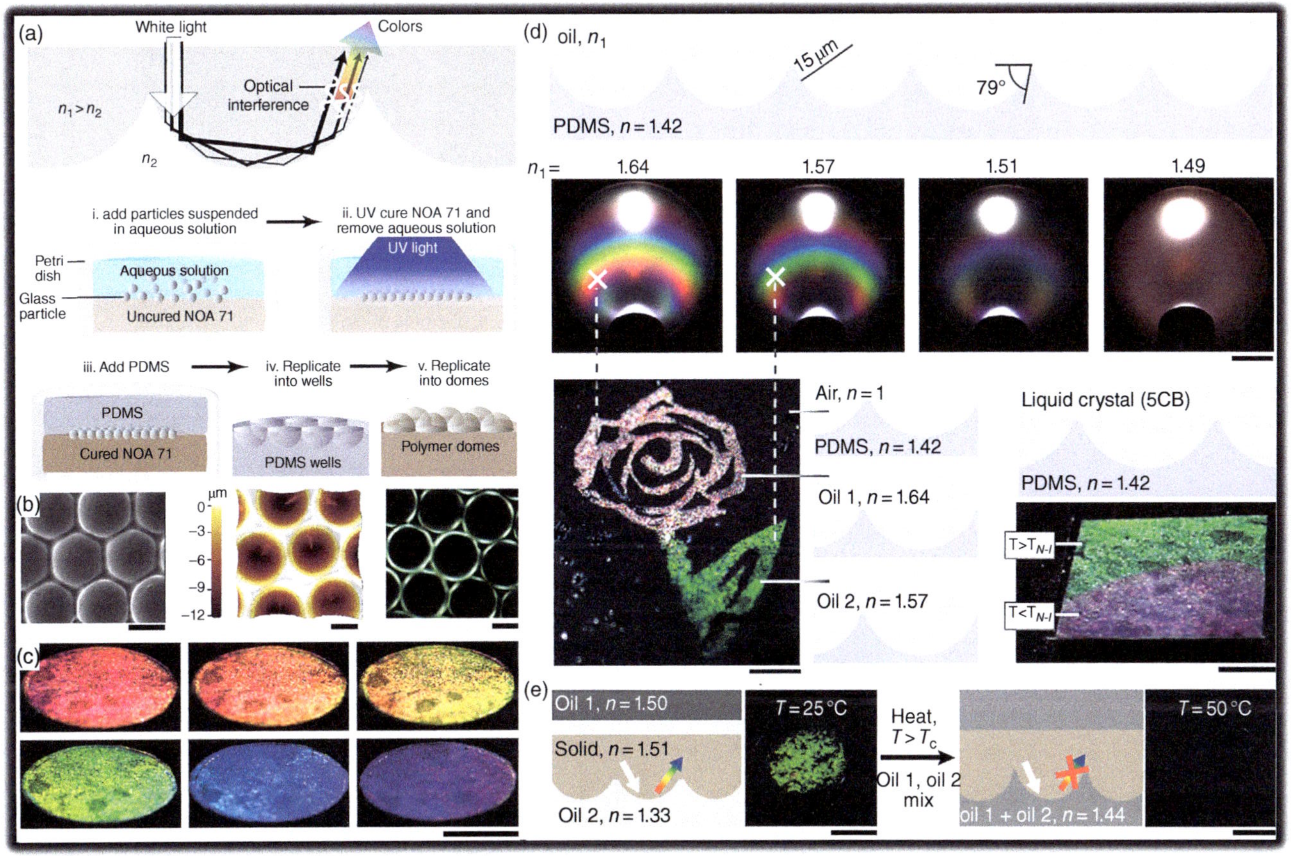

Figure 6.10 (a) Schematic of a concave geometry generating interference from TIR, and the method of fabrication for monodisperse microwells and domes; (b) SEM, optical profilometry images, and the reflection optical micrograph; (c) photographs of the film at different viewing angle; (d) patterned and responsive structural color patterns achieved by varying refractive index contrast; (e) iridescent color switched on and off in response to temperature by varying the index contrast at the optical interface through oil mixing. Source: Goodling et al. [45]/with permission of American Chemical Society.

weight or by blending the bottlebrush block copolymers [47]. The fabrication of the stimuli-sensitive structural colors by block copolymers has potential applications in displays, media boards, and sensors [48].

Song, Vignolini, and Parker and coworkers report a controlled micellization self-assembly mechanism that exploits the ability of amphiphilic BBCPs to behave as giant surfactants, to produce isotropic photonic materials (see Figure 6.11) [49]. By inducing a controlled swelling of reverse BBCP micelles via soft confinement within a toluene–water microdroplet, the dimensions of internal aqueous droplets were regulated comparable to the wavelengths of visible light. The tightly packed

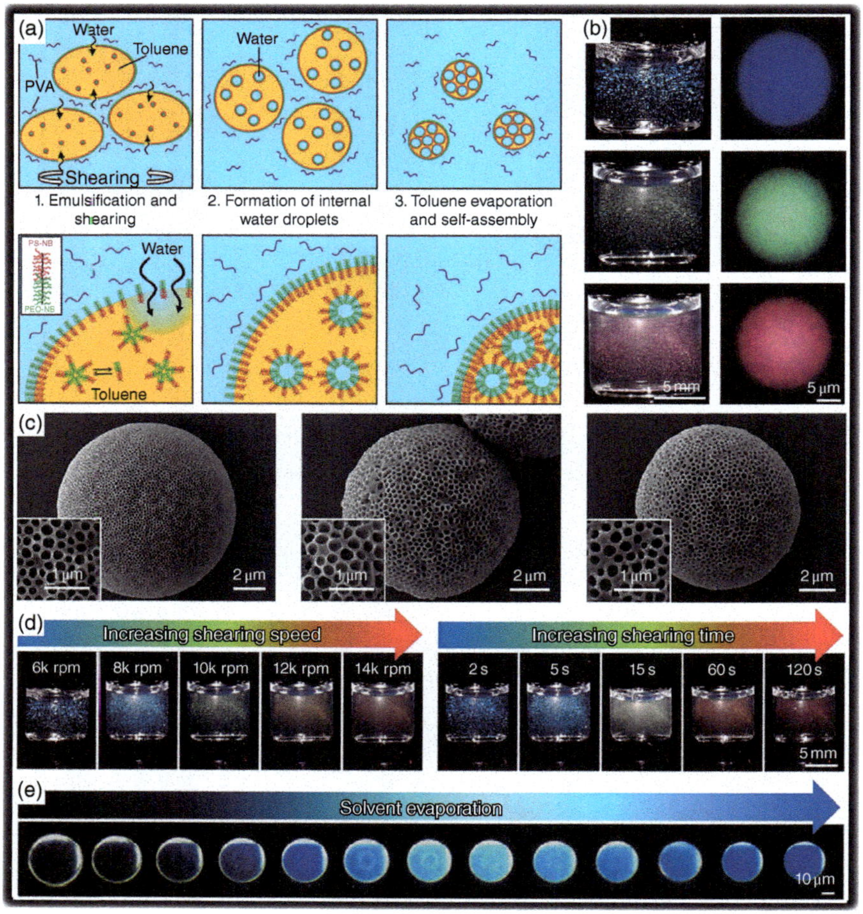

Figure 6.11 (a) Formation schematic of the microspheres from an evaporating toluene-in-water emulsion containing the amphiphilic BBCP; (b) photographs and the corresponding microscopy images of the same 0.03 wt% dispersions under direct illumination; (c) SEM images of the porous surface of the BBCP microspheres; (d) the reflected color from the porous BBCP microspheres can be tuned by either the homogenization speed or homogenization time; (e) time-lapse microscopy image series recorded in darkfield. Source: Zhao et al. [49]/with permission of John Wiley & Sons.

nanodroplets self-assemble to form highly porous microparticles after the toluene evaporation. The short-range distribution of pores within the BBCP scaffold leads to the creation of structural coloration by coherent scattering effects. Moreover, the amount of water available to the micelles can be controlled by temporarily disrupting the microdroplet interface with high shear, allowing the pore size and therefore the color of the resulting structure to be tuned (see Figure 6.11b,d). The final structural color was determined by the intensity of the homogenization process.

Briefly, a transparent and colorless stock solution was afforded by mixing (polynorborneneg-polystyrene)-*b*-(polynorbornene-*g*-polyethylene oxide) BBCP (20 mg) and toluene (1.0 ml) (see Figure 6.11a). The aqueous phase was obtained by dissolving poly(vinyl alcohol) (PVA; 200 mg, as a stabilizer) in Milli-Q water (10 ml). Polydisperse microdroplets were generated using an IKA T25 Digital Ultra-Turrax homogenizer. The BBCP solution (10 μl) was injected into the aqueous PVA solution (10 ml) under the rotation of 6000–14 000 rpm, an emulsion of toluene-in-water microdroplets was immediately formed. The formed microemulsion floated to the air–water interface after a controlled period of time (2–120 s). Colored microspheres formed after the evaporation of toluene over approximately 30 min. The final loss of toluene from the microspheres caused them to sediment and for the color to further evolve (see Figure 6.11e). The produced photonic microspheres were washed with Milli-Q water and dried under nitrogen flow to yield a dispersible powder (see Figure 6.11c). SCMs dependent on BBCPs can also be assembled by inkjet printing [50] and direct writing [51].

The controlled micellization self-assembly is suitable for the SCMs constructed by block copolymers, which requires high environmental factors. The self-assembly in this method was an isotropic process within the volume of the droplet, leading to a rapid fabrication time, high tolerance to structural defects, and color-independence from the microparticle dimensions. Direct writing with block copolymers was another important self-assembly method to construct SCMs [48]. The difficulty of this kind of material is the design and synthesis of block copolymers.

6.3.7 Layered Hydrogels Self-assembly

Soft photonic materials prepared by layered hydrogels self-assembly have broad color tunability and fast color switching. The most typical is poly(dodecyl glyceryl itaconate) (PDGI) amphiphilic bilayers with thick of ~5 nm. Inspired from the structure of colorful tropical fishes, Gong and coworkers presented ultrafast-response photonic hydrogels based on a photonic hydrogel consisting of uniaxially aligned bilayers and chemically cross-linked hydrogel matrix named as PDGI/PAAm hydrogel (see Figure 6.12a) [52]. PAAm is polyacrylamide hydrogel layers of a hundred-nanometer thickness. The response of the gel can be improved by introducing an interpenetrating network of the PAAm hydrogel layers. Ultrafast response and broad wavelength-tunable photonic hydrogel can be obtained by splitting the rigid, continuous bilayers into small domains via converting some amide groups in PAAm gel layers into carboxyl groups.

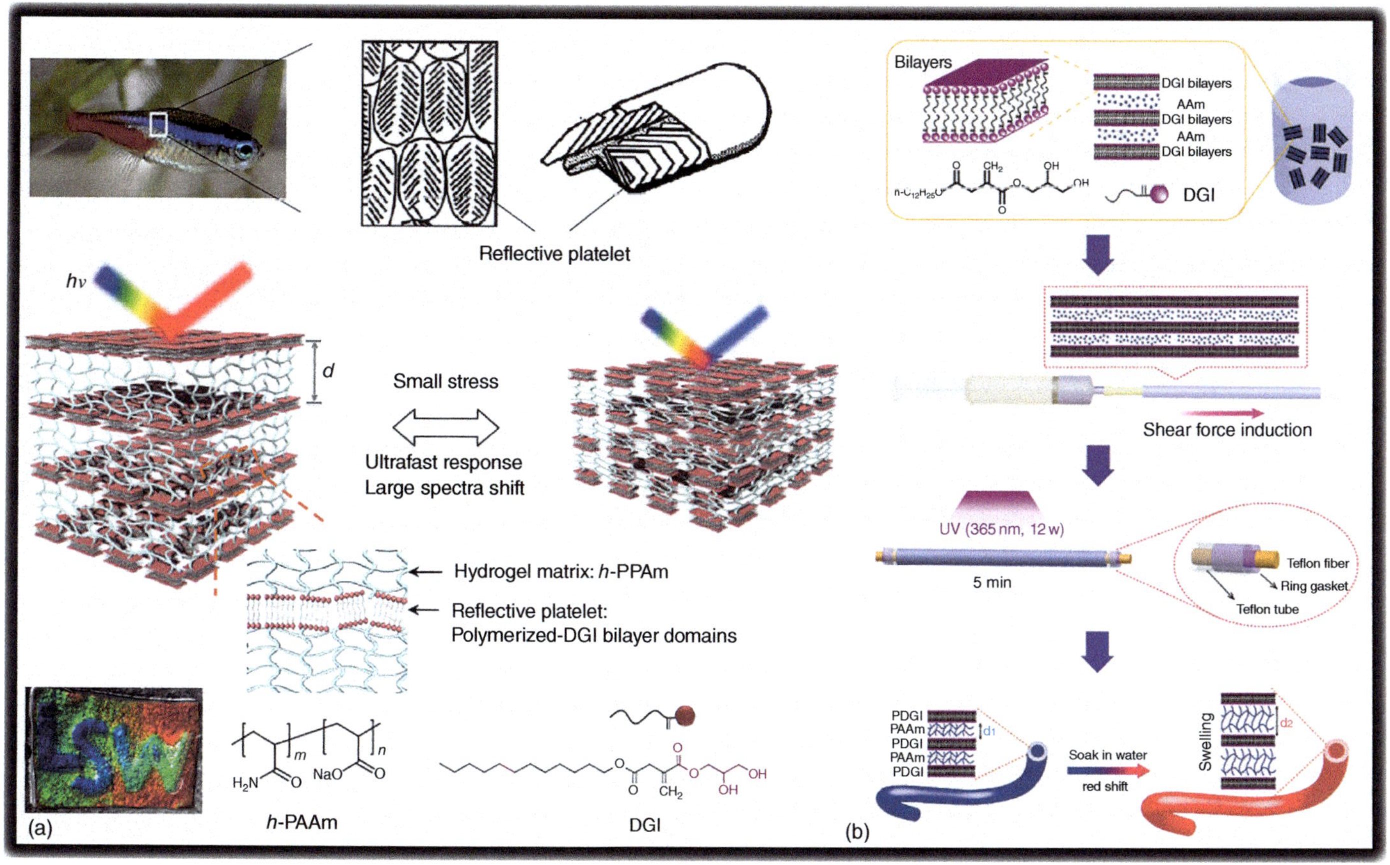

Figure 6.12 (a) Tropical fish *neon tetra*-inspired ultrafast color tuning of the photonic hydrogel. Photonic hydrogel contains alternative stacking of rigid PDGI bilayer domains. Source: Yue et al. [52]/with permission of Springer Nature. (b) Schematic illustration of the fabrication process of PDGI photonic crystal hydrogel microtubule. Source: Reproduced with permission from Wang et al. [53], copyright 2020 John Wiley and Sons.

Briefly, the shear flow was applied to the precursor during the injection of the solution into a reaction cell of 0.5-mm spacing. The lamellar bilayers of self-assembled dodecyl glyceryl itaconate (DGI) were aligned in the direction parallel to the cell wall. The self-assembled DGI bilayers were polymerized by irradiating ultraviolet light for 8 hours at 50 °C under an argon gas atmosphere. Then the gel was immersed in a large amount of water for one week to reach an equilibrium swelling state. The PAAm network of the gel was hydrolyzed by applying 1 M sodium hydroxide aqueous solution for various times at 50 °C. The obtained hydrogel, which contains multi-bilayer structures (PDGI) stacked periodically in partially hydrolyzed PAAm, was then extensively washed with lots of deionized water and immersed in water for 41 weeks to reach an equilibrium swelling state.

Niu and coworkers further designed and manufactured a stretchable photonic hydrogel hollow microtubule as an optical sensing element (see Figure 6.12b) [53]. Firstly, a stretchable photonic crystal hydrogel hollow microtubule composed of several thousands of rigid PDGI lamellar bilayers and PAAm hydrogel networks was manufactured. Secondly, a bi-sheath piezoresistive fiber was fabricated by coating a silicone core with a multiwall carbon nanotubes layer and a silicone insulating layer. Subsequently, the piezoresistive fiber was coupled with the photonic microtubule to form an interactive electronic fiber sensor. The layered hydrogels self-assembly opens a new way to prepare soft photonic materials.

The layered hydrogels self-assembly is the preferred method to construct flexible SCMs, but it is time-consuming and difficult to lock solvents. This type of SCM is usually designed for environmental sensing. The assembly of the DGI bimolecular layer is time-consuming and unstable. The introduction of hydrogels can fix the bimolecular layer, and the function of the SCM varies with the type of hydrogels.

6.3.8 Spray Coating Self-assembly

Large-area SCMs can also be obtained by spray coating self-assembly. It offers the benefits of rapid patterning over a large area on both planar and curved surfaces. Importantly, the modification of the SCM can be completed before spray coating. Several one-step spray coating self-assembly methods are described in the following text, including superhydrophobic SCMs [54], self-healing SCMs [55], spray-hot-press dyeing [56], and special coloring of high refractive index nanoparticles [57]. Yang and coworkers created superhydrophobic angle-independent colors by one-step spray coating of monodispersed, fluorosilane-functionalized silica (F-SiO_2) nanoparticles suspended in isopropanol (see Figure 6.13a) [54]. Briefly, F-SiO_2 nanoparticles were hydrophobilized by fluoroalkyl silane. The F-SiO_2 nanoparticles were then mixed with isopropanol as the dispersion solvent and a low concentration of glass resin for spray coating. Poly(methylsiloxane) precursor consisting of methyl and hydroxyl groups as the glass resin was introduced here to enhance the wetting and adhesion of the hydrophobic F-SiO_2 nanoparticles on the hydrophilic glass slide after curing. The F-SiO_2 nanoparticles self-assembled into a quasi-amorphous photonic structure (APS), which showed angle-independent color, while the dual-scale roughness generated by the nanoparticle assembly made the film superhydrophobic.

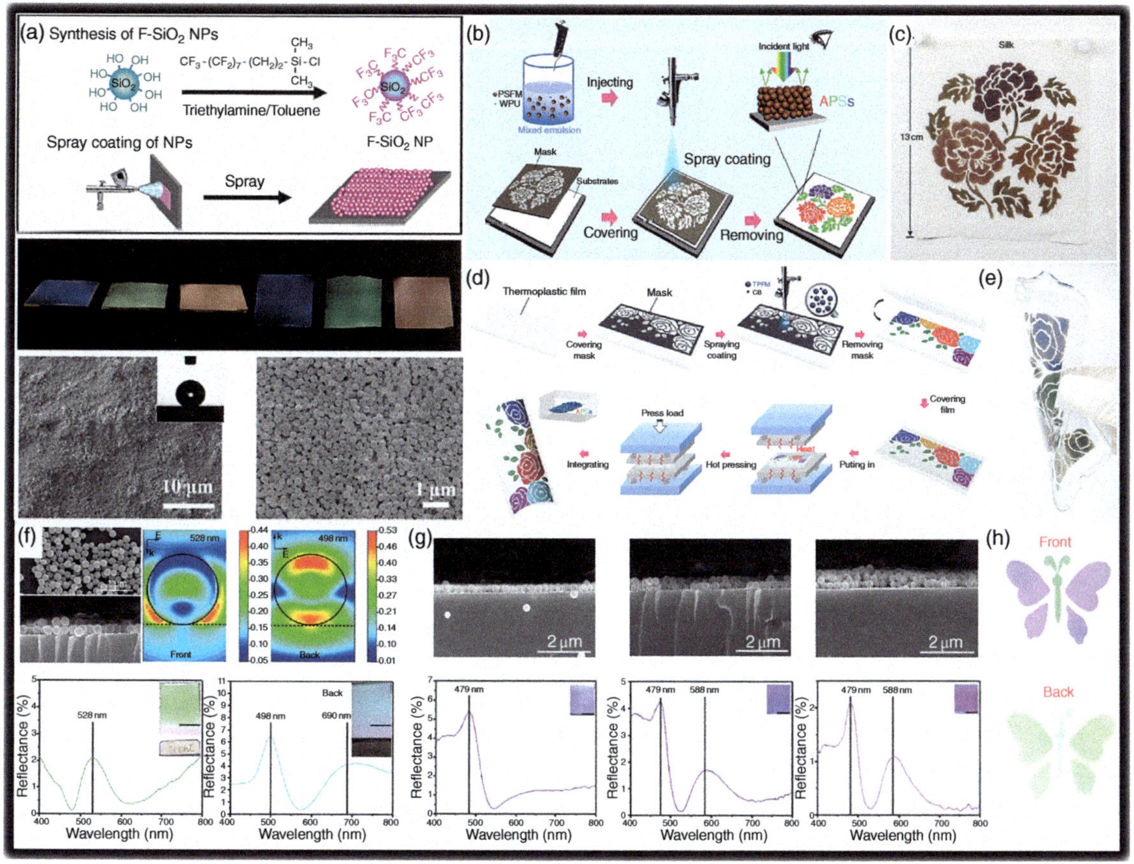

Figure 6.13 (a) Schematic illustration of preparing superhydrophobic angle-independent colored film by using spray coating self-assembly, and the structural color film and the corresponding SEM images. Source: Ge et al. [54]/with permission of Royal Society of Chemistry. (b) Schematic of the formation of amorphous photonic structures on various substrates by spray coating; (c) complex structural color pattern on silk. Source: Meng et al. [55]/with permission of American Chemical Society. (d) Schematic illustration to fabricate the thermoplastic composite films with noniridescent structural color patterns. (e) The colored thermoplastic polyurethanes composite phone shell. Source: Meng et al. [56]/with permission of Royal Society of Chemistry. (f) The distributions of electric field intensity of a single crystal Cu_2O sphere on a transparent glass slide calculated by a finite-difference time-domain (FDTD) method from front incidence at 528 nm and back incidence at 498 nm; (g) cross-sectional SEM images and reflection spectra of the films with different coverage density; (h) photographs of the front and back of butterfly pattern. Source: Bi et al. [57]/with permission of Royal Society of Chemistry.

Tang and coworkers introduced waterborne polyurea (WPU) into the amorphous polysulfide microspheres (PSFMs) to fabricate noniridescent structural color coating with high color visibility, good structural stability, and self-healing properties (see Figure 6.13b,c) [55]. Significant improvement in the color visibility of APS coatings can be achieved by a one-step assembly without adding black light-absorbing materials. The high-efficient adhesive of WPU can be filled in the gaps among the microspheres to achieve structural locking after the assembly. Furthermore, the resprayed WPU imparts self-healing properties to the coating. Briefly, the concentration of PSFM and WPU emulsion was adjusted to 10 wt% in deionized water. The resprayed WPU emulsion was adjusted to 3 wt%. The dispersed mixture can be used in spray coating self-assembly. The pressure, distance, and temperature of the substrate surface, emulsion concentration, and solvent characteristics all affect the optical properties of the assembled APSs.

Subsequently, they present an innovative concept of thermal-guided interfacial confinement for fabricating flexible composites with noniridescent structural color patterns (see Figure 6.13d) [56]. The mixed emulsion of thiodiphenol-formaldehyde microspheres and carbon black was spray-coated on a thermoplastic polymer film to form a homogeneous APS pattern. A sandwich structure was shaped by covering it with another thermoplastic film, and the intermediate APS layer was locked by interfacial confinement via hot-pressing at a certain temperature and pressure. The obtained sandwich composites possessed evident structural colors, excellent mechanical strength, and flexibility (see Figure 6.13e).

Spray coating self-assembly can usually obtain angle-independent structural colors. Interestingly, asymmetric structural colors can be obtained by spray coating self-assembly of cuprous oxide (Cu_2O) nanoparticles with a high refractive index (see Figure 6.13h) [57]. Its special coloring makes it necessary to introduce it here. It was generated by the asymmetric scattering of Cu_2O single-crystal spheres. Briefly, the ethanol suspension of 2 wt% single-crystal Cu_2O spheres was formed via ultrasonic treatment. Then, the suspension was sprayed onto the transparent substrates by a 100-kPa air pump. Nanoparticles were self-assembled on the substrate, forming structural color. The reflection peaked at 528 nm (front side) is ascribed to the spectrally overlapped electric and magnetic dipole resonances (see Figure 6.13f). The reflection peak position on the backside of the film was left-shifted to 498 nm. The FDTD simulations confirmed that the asymmetric colors were ascribed to the inhomogeneous distribution of the electric field intensity. Besides, the coverage density affects the color on the front side of the Cu_2O film but does not affect the backside. There is only one reflection peak (479 nm, front side), which is originated from electric dipole resonance, while the other reflection peak disappeared due to the lower overlapping electric and magnetic dipole resonances (see Figure 6.13g). With the coverage density increasing, the front side shows the vivid blue-purple color originating from the electric dipole resonance and the overlapped electric and magnetic dipole resonances.

Spray coating self-assembly is suitable for the rapid construction of SCMs, which usually results in disordered or short-range ordered nanoarrays with shallow color, and generally requires adding light-absorbing materials. But for high refractive index nanoparticles, absorbent materials can be avoided. Complex structural color patterns can be constructed by using masks. It is suitable for fabricating SCMs by using most kinds of monodisperse microspheres or nanoparticles.

6.3.9 Unidirectional Rubbing Self-assembly

Assembly of a colloidal suspension on a solid substrate has been the most commonly used approach. Most of these techniques rely on the use of convective flow and require delicate evaporation conditions to obtain monolayers over large areas. Recently, by unidirectional rubbing dry powder of particles between two rubber plates, Myoung and Jeong and coworkers reported a self-assembly method for fabricating single-crystal colloidal monolayer on a flat or curved substrate (see Figure 6.14) [58].

Briefly, a dry powder of spherical particles, such as PS beads 1 μm in diameter, was placed on top of one poly(dimethylsiloxane) (PDMS) substrate (20 × 20 cm) and rubbed with another PDMS substrate along a randomly chosen direction (see Figure 6.14a). After being rubbed for 5 s and detaching the rubber substrates, a small number of residual particles were blown off by the compressor air gun. The particles had assembled into hexagonally close-packed monolayers on the surfaces of both PDMS substrates (see Figure 6.14e,f).

The particles collectively roll along the rubbing direction so that the initial multiple grains merge to form a large single domain (see Figure 6.14a). The nanoparticles start to roll in a steady state when new contact adhesion at the front edge and the pull-out at the rear edge are balanced (see Figure 6.14b). During the rubbing self-assembly, aggregated particles absorb the impact and reduce the impulse; meanwhile, the impacts are simply transmitted to the other particles, and the particles are scattered. Close-packing is initiated at the relatively slow particles through consecutive collisions of faster particles with the slow particles. The practical key variables for the monolayer assembly are the adhesion energy of the substrate, the rubbing speed, and the normal pressure (see Figure 6.14c,d). The following tenets apply: (i) the self-assembly applies to any rubber surface that is not too sticky or too hard; (ii) high pressure and slow rubbing for submicrometer particles, while low pressure and fast rubbing for microparticles; and (iii) repeated unidirectional rubbing.

Unidirectional rubbing self-assembly is a fast method to obtain 2D SCMs with simple operation. The order of nanoparticles depends on the conditions of friction. One of the advantages is that it is suitable for constructing SCMs on curved surfaces.

6.3.10 Edge-Induced Rotational Shearing Self-assembly

Self-assembly on macroscopic length scales presents new routes for assembling solid ordered photonic materials. Finlayson and Baumberg and coworkers produced flexible opals by using melting and shear ordering under compression of core/shell polymer nanoparticles [59]. The preparation process includes a combination of extrusion, rolling (linear shear), and edge-induced rotational shearing (see Figure 6.15a). By using an allyl-methacrylate as a grafting agent,

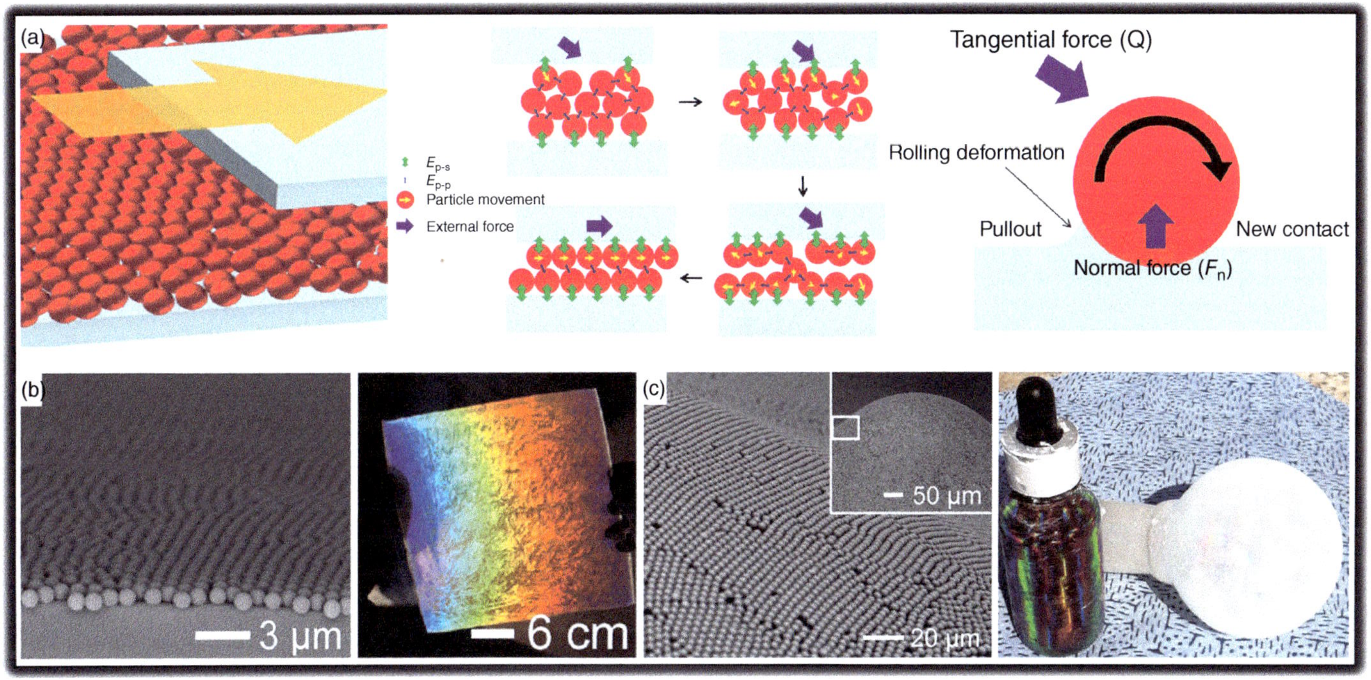

Figure 6.14 (a) Schematic illustration of the unidirectional rubbing self-assembly, the formation of the monolayer, and the rolling condition; structural color film and the corresponding SEM image prepared by unidirectional rubbing self-assembly. (b) Rubbery surfaces and (c) glass bottle and a 100 ml flask surface. Source: Park et al. [58]/with permission of John Wiley & Sons.

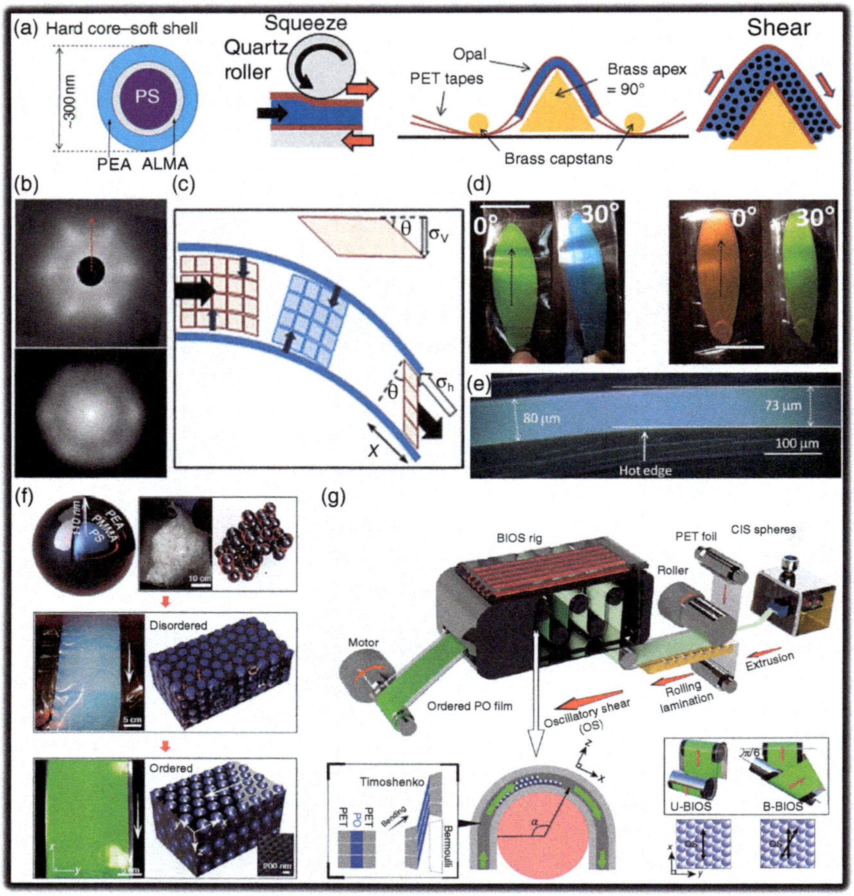

Figure 6.15 (a) Schematic of the edge-induced rotational shearing self-assembly, including roller process via squeeze/linear shear and hot-edge process via edge-shearing; (b) difference in transmission diffraction pattern with (top) and without (bottom) edge-shearing; (c) schematic of bend-induced shearing (blue arrows) of opal between rigid polyester films; (d) corresponding photographs of green and red opal sample viewed in reflected white-light at angles of 0° and 30° to the normal; (e) microscope image of the green-opal film in cross-section transverse to the edge-shear processing direction, during the edge-shear process. Source: Finlayson et al. [59]/with permission of John Wiley & Sons. (f) Fabrication of opal films and mechanism, core–shell sphere, the film after rolling-lamination, and the film after bending-induced oscillatory shear (BIOS) with improved sphere packing; (g) production line of a large-scale film with two types of BIOS. Source: Zhao et al. [60]/with permission of Springer Nature.

rigid cross-linked PS spheres are capped by a soft polyethylacrylate shell. The core–interlayer–shell material and a small amount of carbon-black powder were extruded into millimeter-sized ribbons using a twin co-rotating-screw mini-extruder at 150 °C. The translation stage was moved horizontally at 1 mm/s relative to the rotating roller, pressing the sample into uniform thin films in a squeeze/shear mechanism. The polyester tapes were secured at both ends, and the translation stage was used to draw the thin film tapes over the heated brass edge at typical speed of 5 mm/s. The obtained opaline samples show brilliant visual iridescence (see Figure 6.15d). The sharpening diffraction spots represent the improved ordering following the edge-induced rotational shearing process (see Figure 6.15b). The thickness of the opal film depends on the bending angle after the film relaxes (see Figure 6.15c,e).

As opposed to normal reflective iridescence based on Bragg diffraction, the color generation of the low-refractive-index contrast regime arises through spectrally resonant scattering inside a 3D fcc lattice photonic crystal [61]. An edge-induced rotational shearing process produces reproducible highly uniform samples and enhances both the intensity and chromaticity of the structural color. The novel use of soft nanomaterials in the design of photonic structures, with macroscale bulk ordering, presents opportunities for a step-change away from the monolithic architectures. Subsequently, they improved the self-assembly method. Large-area flexible films of stacked polymer nanoparticles can be directly assembled in a roll-to-roll process using a bending-induced oscillatory shear technique (see Figure 6.15f,g) [60]. The resulting structure of random hexagonal close-packed layers is improved by shearing bidirectionally, alternating between two in-plane directions.

The edge-induced rotational shearing self-assembly is suitable for soft microsphere assembly and can be easily mechanized. A controllable "roll-to-roll" process, together with the interchangeable nature of the base composite particles, opens potentially transformative possibilities for mass manufacture of nano-ordered materials, including advances in optical materials, photonics, and metamaterials/plasmonics [62].

6.3.11 Screen Printing Self-assembly

The application of structural color in the textile printing and dyeing industry remains a challenge. Zhang and coworkers fabricated large-scale APSs on white fabric with vivid noniridescent structural colors by a fast screen printing technique (see Figure 6.16) [63]. The multicolor pattern on fabrics can be obtained via a multistep printing process. Due to the absorption of the incoherently scattered light by carbon black, the obtained structural colors on various white substrates show high color visibility. Meanwhile, the strongly cohesive polyacrylate contributes to the structural stability of the structural colors on fabrics.

Briefly, carbon black (3 wt% to microspheres) and the polyacrylate waterborne adhesive (8 wt% to microspheres) were added into the PS emulsion (48 wt%). The well-mixed printing paste (the viscosity was about 0.5672 Pa s) was obtained by ultrasonication for 3 hours. The screen printing frame and the polyester fabric were separated by polyurethane spacers with a thickness of ~2 mm (see Figure 6.16a). Then the prepared printing paste was poured onto the screen mask with different patterned stencils (100 meshes). A polyurethane squeegee (~45° angle with the printing mask) was used to brush the printing paste at a speed of ~5 cm/s. Then, the coated fabrics were dried in an oven at 70 °C for about 1 min. A thin film

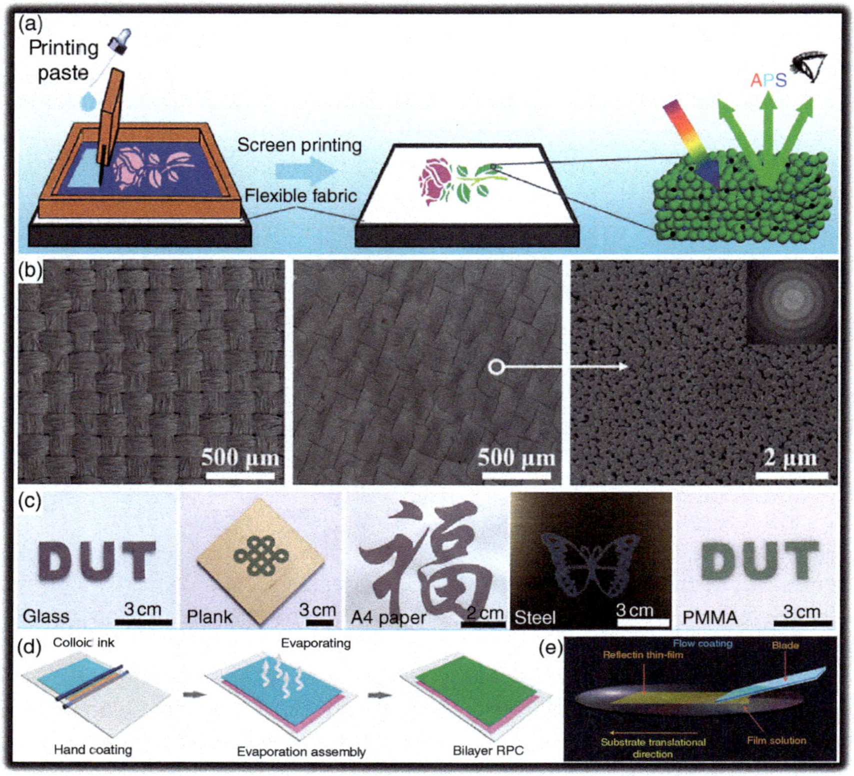

Figure 6.16 (a) Schematic of fabricating amorphous photonic structures on fabrics by screen printing; (b) SEM image of the original polyester fabric and the coated fabric after screen printing; (c) photographs of the brilliant noniridescent structural colors fabricated by screen printing on various substrates. Source: Zhou et al. [63]/with permission of Elsevier. (d) The hand coating process for creating optical code films on flexible substrates. Source: Reproduced with permission from Zhang et al. [64], copyright 2015 Royal Society of Chemistry. (e) Schematic diagram of the flow-coating technique for casting thin films of reflectin protein onto a silicon-wafer substrate. Source: Kramer et al. [65]/with permission of Springer Nature.

covered the surface of the fabric forms (see Figure 6.16b). Due to the action of polyacrylate adhesive, colloidal microspheres are stuck to each other and firmly attached to the fibers. The facile and fast dyeing technique on fabrics will facilitate the practical applications of large-area noniridescent structural colors in textile coloring and other color-related fields (see Figure 6.16c). Similar preparation methods of the large-area structure color films also include hand coating and fluid coating self-assembly (see Figure 6.16d,e) [64, 65].

Screen printing self-assembly is suitable for the construction of large-area patterned SMCs and generally requires adding light-absorbing materials. The obtained nanoarrays are usually short-range ordered and long-range disordered. The color shows angle-independent. It is worth noting that the concentration of the printing paste used in this method is much greater than that in the evaporation assembly, and the thickness of the prepared structure color film is related to the paste concentration and screen height. This low cost, short time, and practical dyeing method shows great potential in fabric coloring and other color-related fields.

6.3.12 Magnetic-Induced Self-assembly

Nowadays, plasmonic color is a kind of man-made structural color, which originates from collective oscillations of the conduction electrons caused by resonant interactions between visible light and manufactured metal nanostructures [66], and is emerging as an important solution for ink-free color printing [67]. The magnetically induced assembly of anisotropic microspheres can obtain ordered photonic structures and provide an assembly method for structural color materials with plasmonic anisotropy. Yin and coworkers developed programmable mechanochromic films with precise colorimetric by preparing magnetic–plasmonic hybrid nanorods (see Figure 6.17) [68]. The plasmonic excitation of the hybrid nanorods can be collectively regulated by using magnetic fields. It facilitates the incorporation between hybrid nanorods and polymer films with a well-controlled orientation and enables sensitive colorimetric changes in response to linear and angular motions. Fixed by UV and magnetic fields, the alignment of concave Au nanorods (cAuNRs) was evidenced by the uniform red and green color across the films under perpendicular and parallel polarization, respectively (see Figure 6.17b,c). The asymmetric alignment of anisotropic plasmonic nanostructures about the active axis of external mechanical stimuli induces the excitation of different resonance modes, producing readable color changes (see Figure 6.17d).

Briefly, cAuNRs were obtained by centrifuging 1 ml of cAuNRs colloidal dispersion at 9000 rpm for 3 min (see Figure 6.17a). Then, 200 μl of the precursor solution was added and sonicated for ~ 15 seconds to fully disperse the colloidal nanoparticles. The magnetic–plasmonic nanorods solution was sandwiched between glass slides and exposed to UV light (254 nm) for 1 min. The alignment of magnetic–plasmonic nanorods in specific locations can be programmed by the sequential magnetic alignment and UV exposure. PDMS films were used to prepare the mechanochromic devices.

Moreover, magnetic-induced self-assembly is also regulated by environmental conditions. Ma and Guan and coworkers demonstrated responsive hydrogel-based photonic nanochains with high-resolution and real-time response, by developing a general hydrogen bond-guided template polymerization method (see Figure 6.18) [69]. The hydrogen bonds formed between monomers, and the polyvinyl pyrrolidone anchored to the surfaces of uniform superparamagnetic ferroferric oxide (Fe_3O_4) nanoparticles concentrates the monomers on the adjoining area of 1D periodical structures of magnetically responsive photonic crystals, which are formed under a magnetic field, forming responsive hydrogel-based photonic nanochains after UV light-initiated copolymerization. The original red color (diffracted under the magnetic field) immediately turns into deep blue after the acid droplet falls (see Figure 6.18d).

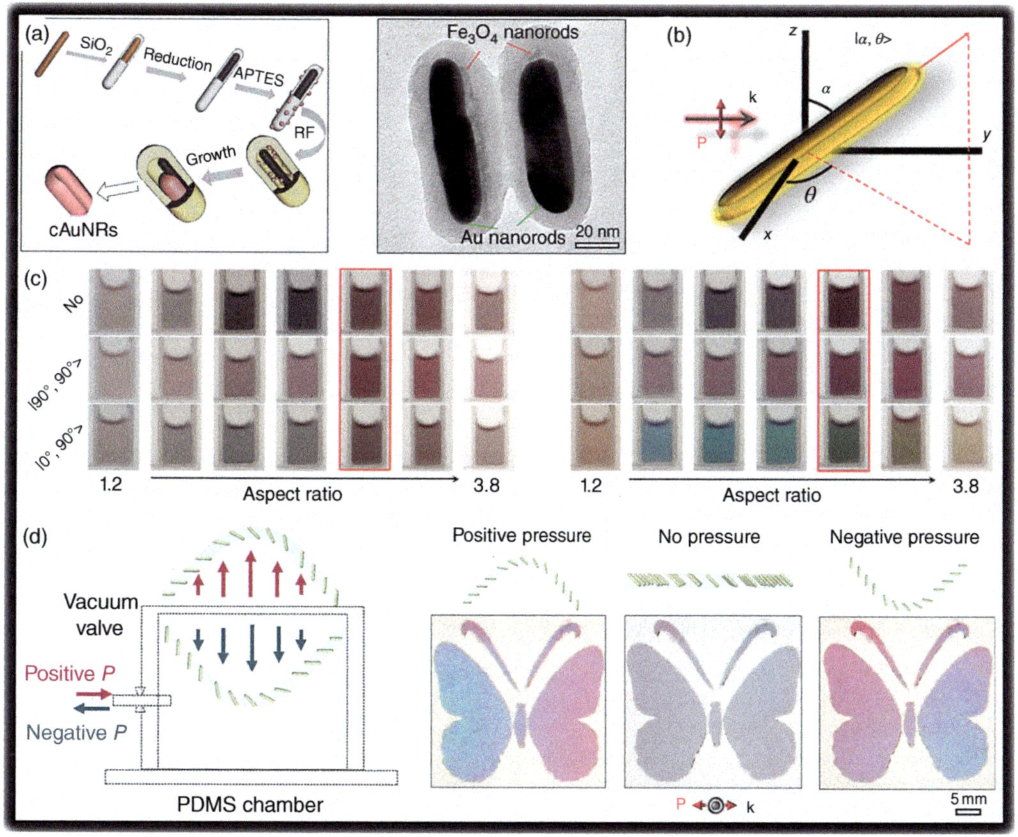

Figure 6.17 (a) Scheme of the confined growth toward magnetic–plasmonic hybrid nanorods and the corresponding TEM images; (b) schematic illustration of cAuNRs under the orientational state $|\alpha, \theta>$ with respect to the polarization of light; (c) digital photos of cAuNRs dispersions under normal (left) and polarized light (right); (d) scheme of the mechanochromic pressure sensor, and the cAuNRs are aligned 45° to the surface normal in the butterfly patterns (middle panel). Source: Li et al. [68]/with permission of Springer Nature.

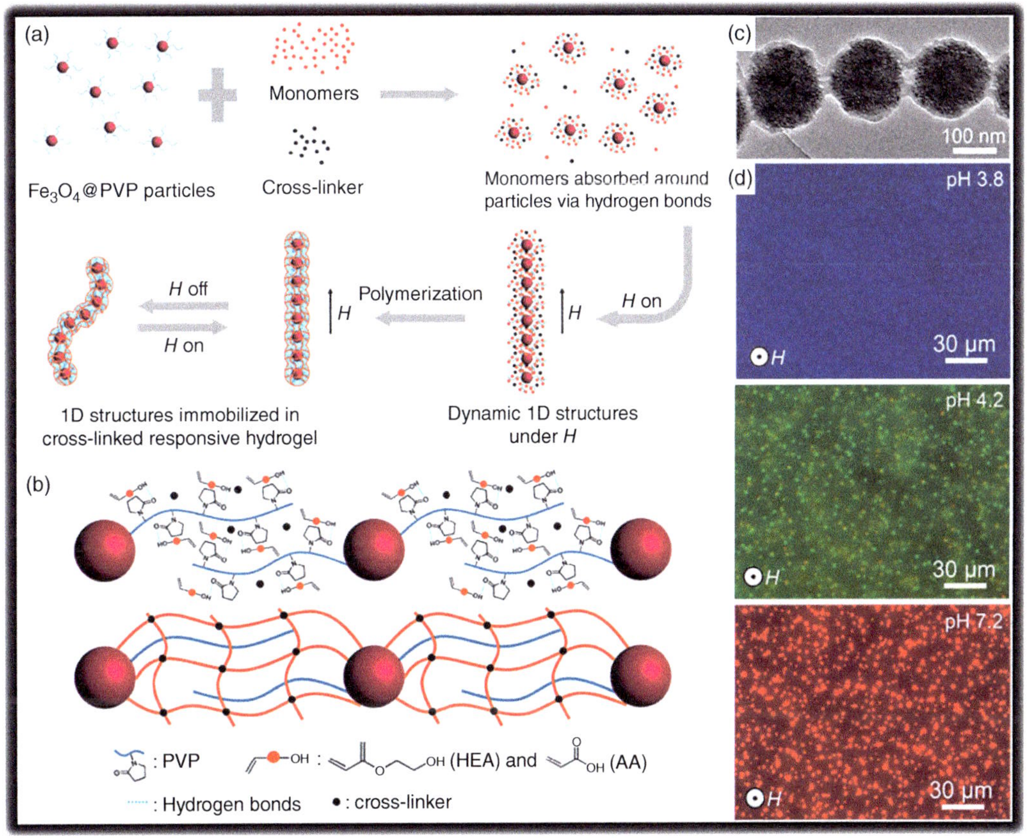

Figure 6.18 (a) The formation mechanism of the pH-responsive photonic chains; (b) schematic illustration of hydrogen bonds formed between monomers and PVP polymer chains and the schematic illustration of cross-linked poly(HEA-*co*-AA) hydrogel pod-coated chains; (c) TEM of the as-obtained Fe_3O_4@PVP@poly(HEA-*co*-AA) pH-responsive photonic nanochains; (d) optical properties of pH-responsive hydrogel-based photonic chains. Source: Luo et al. [69]/with permission of American Chemical Society.

Briefly, monomers 2-hydroxyethyl acrylate (HEA) and acrylic acid (AA), cross-linker ethylene glycol dimethacrylate, photoinitiator 2-hydroxy-2-methylpropiophenone, and monodispersed superparamagnetic Fe_3O_4@ PVP (polyvinylpyrrolidone) particles were all added to ethylene glycol (see Figure 6.18a). The mixture was vortexed for 15 min to form a homogeneous brown precursor solution. The nanoparticles are aligned linearly by the magnetic field into dynamic 1D periodical structures. Then, the polymerization was initiated by UV irradiation under the magnetic field and lasted for 5 min, forming separated pH-responsive 1D photonic nanochains (see Figure 6.18c). During the self-assembly, the PVP shells of the nanoparticles provide steric repulsion to counterbalance the magnetic attraction of the superparamagnetic nanoparticles. At the same time, they concentrate the monomers containing hydroxyl or carboxyl groups in the vicinity of the 1D periodic nanoparticles via hydrogen bonds in the precursor solution (see Figure 6.18b), resulting in a higher monomer concentration around those structures.

Magnetic-induced self-assembly is suitable for the rapid assembly of magnetic nanoparticles and the assembly of composite composed of ordered magnetic nanoparticles and gels. The photonic structures assembled by magnetic nanoparticles can be encapsulated in polymer gels, whose chemical and physical properties can be demonstrated in the form of structural color. Highly sensitive SCMs can be prepared by using this method.

6.3.13 Photoinduced Self-assembly

Photoinduced self-assembly is a special kind of environment-induced self-assembly method because it involves the transformation of molecular structure. It is well-known that certain bent-core liquid crystalline molecules spontaneously form helical nanofilaments (HNFs) or bundles of twisted crystal layers, usually with a diameter of ~40 nm and a helical pitch of 200–300 nm depending on the molecular length [70, 71]. The aromatic cores and aliphatic tails in bent-shaped molecules appear a typical bookshelf-like arrangement in the smectic layers. However, an in-layer mismatch occurs upon cooling from the smectic phase to the HNF phase. During this transition, local saddle-splay deformation of layers regulated the twisted structures of the HNFs. The HNFs can rotate geometrically under external stimulation and show reversible physical properties.

Yoon and coworkers reported structural color reflectors by directed self-assembly of an HNF liquid crystal phase via UV irradiation (see Figure 6.19) [72]. Aligned arrays of helical nanostructures can be obtained by using the photoisomerization-induced Weigert effect [73]. As a result, structural colors can be observed due to the extrinsic chiral reflection in the visible wavelength range, and the reflected color can be tuned by adjusting the molecular length of the azobenzene derivative. Moreover, the complex helical nanostructures can be easily written on with UV light and erased by heating. It can be used for fabricating large, reversible, and patternable color reflectors, providing a new platform for interference-based structural coloration.

Briefly, the sandwich-type cells were made of a silicon (Si) wafer and glass. The spacing between the substrates was maintained at 3 μm using silica particle spacers.

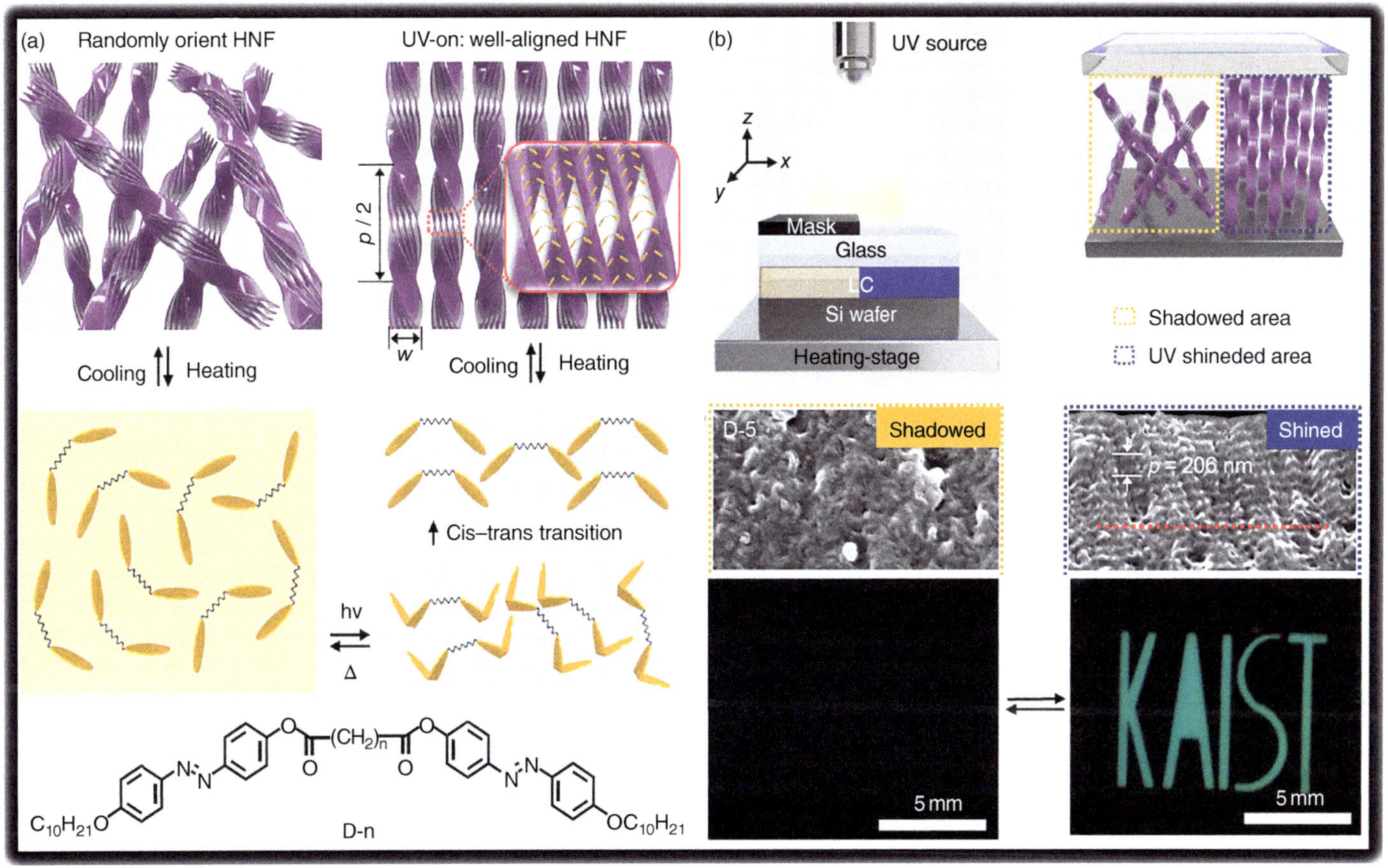

Figure 6.19 (a) Randomly oriented HNFs grown from randomly oriented molecules in an isotropic, nematic, or smectic phase, and oriented HNFs observed upon illuminating with unpolarized UV light. Alignment of the mesogenic cores along the light propagation direction by the Weigert effect, which causes the growth of the helical filaments vertically concerning the sample substrate; (b) schematic illustration of the UV irradiation experiment, the corresponding SEM images of HNFs, and the corresponding digital photos. The heating stage is placed beneath the sample to control the phase transition. Source: Park et al. [72]/with permission of Springer Nature.

The liquid crystal materials were injected into the cells in the isotropic liquid phase (~170 °C) via capillary forces. The temperature and cooling rates of the samples were precisely controlled by a heating stage. The cells were irradiated with UV light from a diode lamp at an effective power density. During irradiation, the sample was cooled to the temperature range of the B4 phase (HNFs, ~140 °C) at a rate of 1 °C/min.

Photoinduced self-assembly is suitable for liquid crystal SCMs and the self-assembly of photo-responsive nanoparticles. The design and synthesis of liquid crystal molecules is difficult, but this type of SCMs can provide effective methods for preparing the flexible display.

6.3.14 Atomic Layer Deposition Self-assembly

ALD is known to be a nonsolution nanotechnology for the conformal deposition of nanoscale thin films and surface layers down to atomic layers. It shows high uniformity and is well controllable in thickness and interface. ALD self-assembly of SCMs makes it an exciting and promising strategy in realizing the coloration of carbon fibers. Zhang and Tok and coworkers designed and fabricated new photonic crystal carbon fiber yarns and fabrics with tunable structural colors by utilizing surface chemical interaction in which a densified 1D photonic crystal was coaxially grown on the carbon fiber surface (see Figure 6.20) [74]. The periodic stack of the 1D photonic crystal was fabricated by using zinc oxide (ZnO) and aluminum oxide (Al_2O_3) layers with large refractive index contrasts, which were chemically and conformally deposited onto the plasma-activated carbon fiber surface via the ALD technique.

Briefly, carbon fiber yarns were treated with plasma (15 Pa, 25 °C, 10.2 W) for 10 min after ultrasonic washing in acetone and deionized water. Then, the Al_2O_3 layer was first deposited at a substrate temperature of 200 °C using deionized water and trimethyl aluminum (TMA) as precursors (timing sequence: 0.02, 8, 0.02, 8 seconds and TMA, N_2 purge, H_2O, N_2 purge). An ALD cycle of 400 was set to grow a layer of Al_2O_3 with a certain thickness (see Figure 6.20a). Subsequently, the ZnO layer was deposited at 170 °C using deionized water and diethyl zinc (DEZ) as precursors (timing sequence: 0.02, 5, 0.02, 5 seconds and DEZ, N_2 purge, H_2O, N_2 purge). ZnO layers with different thicknesses were controlled by the number of ALD cycles (from 200 to 600). The alternating layers of Al_2O_3/ZnO comprised one unit cell of 1D photonic crystal coating (see Figure 6.20b). The structure color of the carbon fiber is regulated by the deposition thickness of ZnO (see Figure 6.20c).

ALD self-assembly has expanded the construction methods of SCMs and endowed carbon fiber with new coloring methods. It is suitable for preparing SCMs on the inert substrate surface, but the cost is high. It is also suitable for constructing the structure color on many kinds of curved surfaces.

6.3.15 Physical Vapor Deposition Self-assembly

In Section 6.3.12, we introduce the plasma coloring mechanism. The key feature of plasmonics is the inherent property of geometrical shape-dependent enhanced light wavelength selective resonance coupling [66]. The plasma structural color can also

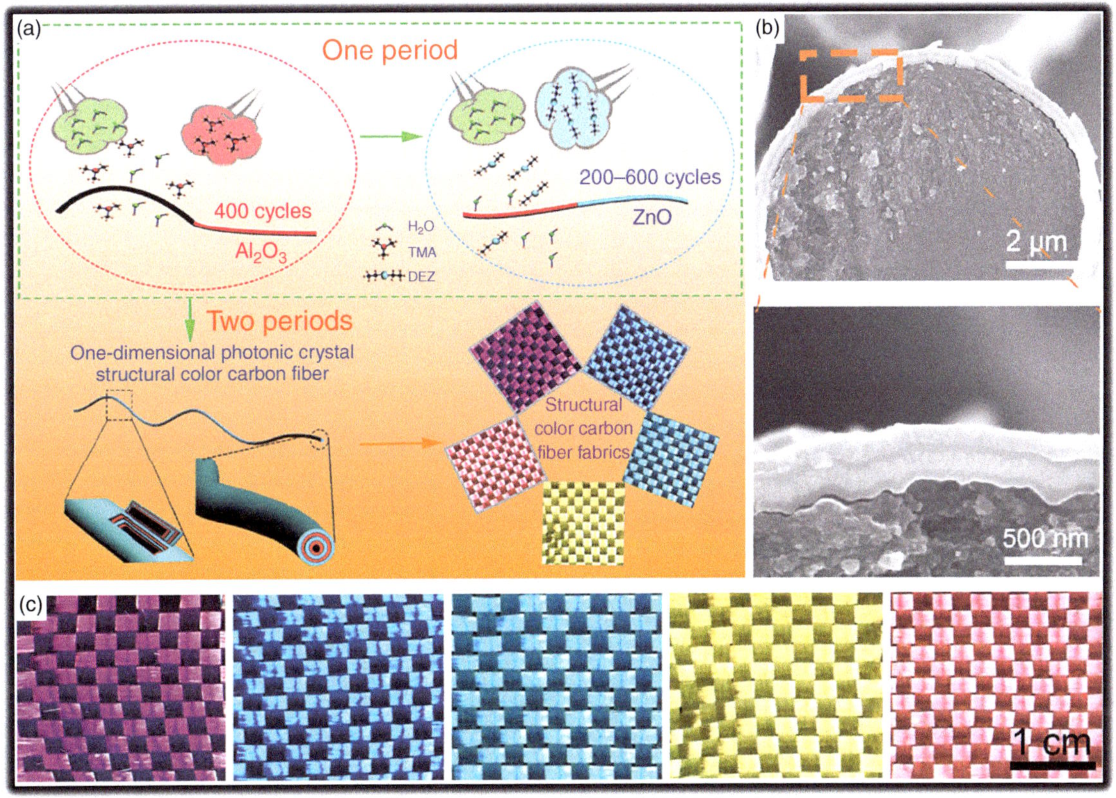

Figure 6.20 (a) Schematic describing the fabrication process of carbon fiber yarns and fabrics; (b) SEM images of 1D photonic crystal carbon fiber; (c) optical photos of the colored carbon fiber. Source: Niu et al. [74]/with permission of American Chemical Society.

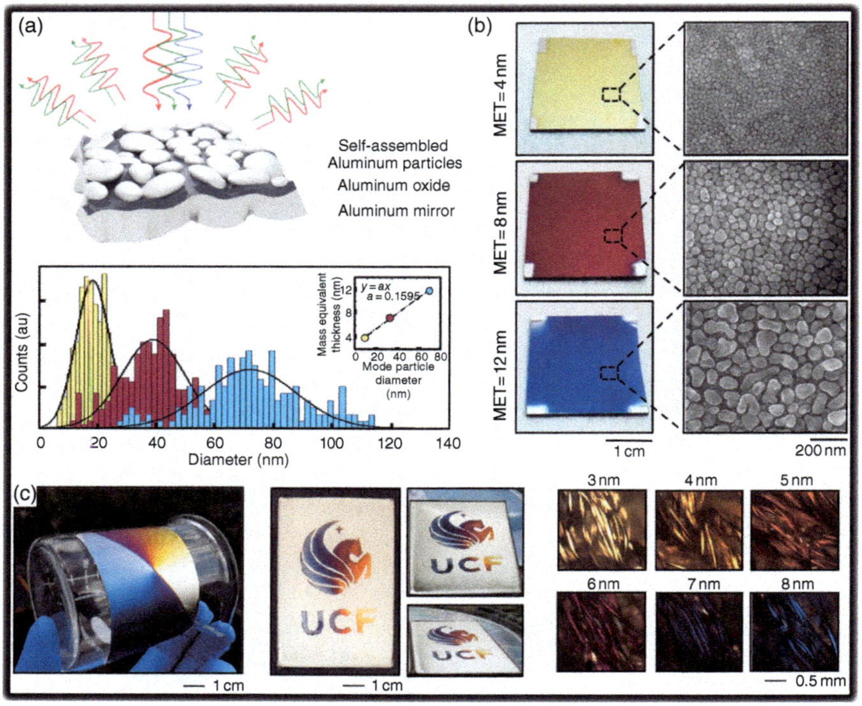

Figure 6.21 (a) Self-assembled plasmonic system and particle statistics; (b) structural color and the corresponding wavelength of plasmon resonance scales with particle size; (c) fabrication performed on a thin, flexible layer of PDMS and masked fabrication of the aluminum particles in the form of the UCF Pegasus logo. The microscope images of the surface fabricated on a 100% polyethylene terephthalate fabric. Source: Franklin et al. [75]/with permission of National Academy of Sciences.

be obtained by physical vapor deposition self-assembly. Ambient white light excites resonances within the structure, which are predominantly confined to the gaps between particles (see Figure 6.21a). These resonances demonstrate a high degree of angle independence. The spectral location is related to the size distribution of the particles (such as aluminum particles), the surrounding refractive index, and the optical distance from each other and the mirror. The non-absorbed light is reflected to result in a perceived color [75].

The particles form through a temperature- and pressure-dependent thin film growth mechanism in an ultrahigh vacuum electron beam evaporator. Island formation during physical vapor deposition is generally categorized into three modes: Frank–van der Merwe (layer by layer), Volmer–Weber (island), and Stranski–Krastanov (layer and island). Initiated by the adsorption of atoms to the substrate surface, aluminum exhibits the Volmer–Weber growth mode. The energetic molecules are free to diffuse along the surface until they settle on a site with a local energy minimum. Subsequently, aluminum atoms adsorb, diffuse, and condense about these low-energy sites during the nucleation process. Particles anchored at adjacent nucleation sites grow and form grain boundaries, or coalesce to form larger particles. The shape of the combined particles depends on the equilibrium condition between the aluminum's free energy and interfacial stress with the substrate.

Chanda and coworkers demonstrate an actively addressed plasmonic display with angle-independent diffuse color enabled by multidirectional resonance hybridization that originates from a dense array of aluminum nanoparticles in the near-field proximity of a mirror (see Figure 6.21b) [75]. Dense arrays of nanoparticles were created in near-field proximity to a mirror by exploiting the thin film growth mechanisms of aluminum during ultrahigh vacuum physical vapor deposition. The sub-10-nm gaps between adjacent nanoparticles and mirror lead to strong multidimensional coupling of localized plasmonic modes, resulting in a singular resonance with negligible angular dispersion and ~98% absorption of incident light at a desired wavelength. The large-area self-assembly process is compatible with arbitrarily structured substrates and can produce diffusive, angle-independent, and flexible plasmonic color surfaces (see Figure 6.21c).

Briefly, the 150-nm Al mirrors was deposited on glass by using a Thermionics electron beam evaporation system. The deposition was performed at $\sim 5 \times 10^{-6}$ T at a rate of ~0.1 nm/s and room temperature. The aluminum oxide layer was deposited by ALD at 100 °C using pulses of trimethylaluminum and water. Then, the aluminum particles were evaporated in an AJA electron beam evaporator at $\sim 5 \times 10^{-8}$ T, 100 °C, and ~0.05 nm/s.

Physical vapor deposition self-assembly is suitable for the construction of plasmonic SCMs. It is usually used in combination with ALD. So, the assembly conditions are harsh and the cost is high. However, the large-area self-assembly process is compatible with arbitrarily structured substrates. Non-iridescent structural colors with superior color contrast can be obtained by nanostructured surfaces with high-index dielectrics. Importantly, a halftoned black color could be obtained by combining a highly saturated cyan–magenta–yellow [66]. Moreover, the resonance-mediated structural color printing governed by photothermal reshaping can be achieved by using the selectivity of resonant laser printing. Laser-printable structural color metasurfaces are scalable to a large area and provide a new way for full-color printing and decoration with nonfading and vibrant colors.

6.3.16 Surface Wrinkling

Wrinkles as a special surface geometrical deformation are ubiquitous in many biological systems and artificial materials that bring epidermis and surfaces with various functions. The mechanism of surface wrinkling can be explained by the pre-strained soft substrate coated with a thin elastic film undergoing a buckling instability due to a mismatch in the equilibrium states of both layers when releasing the pre-strain [76]. To arrive at a new equilibrium state, various wrinkling patterns on different length scales were formed. To minimize the total elastic energy in the bilayer system, wrinkling relaxes the compressive strain in the soft substrate to reduce the energy and bends the thin elastic layer to increase the elastic strain energy [77].

Surface wrinkling can be produced by different elastic layers or by local rigid strains, and its assembly method is simpler than that of nanoparticles. The structural colors produced by surface wrinkles are generally associated with gratings, interference, and 1D or 2D arrays. In particular, surface wrinkling usually requires additional environmental conditions to produce strain differences in the assembled material. Therefore, it is easy to combine it with environmental sensing.

Yin and coworkers reported a remarkable thickness-dependent wrinkling behavior of PDMS films used to prepare smart optical devices with desirable mechanochromic responses, in which an energy barrier separates the wrinkling mechanics into two regimes (see Figure 6.22a) [78]. Influenced by interfacial strains, uniform and nonuniform wrinkles were formed on the plasma-treated thick (>1 mm) and thin (<1 mm) films, respectively. For the thick films, bright structural colors with angle-dependent properties can be observed by applying

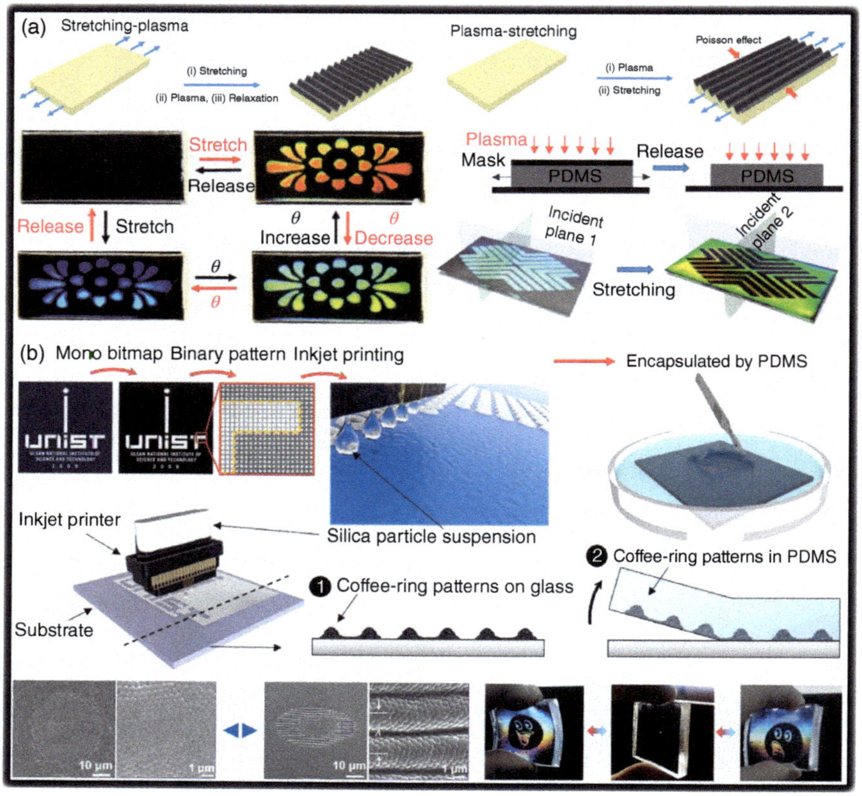

Figure 6.22 (a) Flexible fabrication processes for encryption and mechanochromic PDMS films. The insets are the corresponding digital photos. Source: Li et al. [78]/with permission of Springer Nature. (b) Schematic of the fabrication processes of the multimodal and covert–overt convertible structural coloration transformed by mechanical stress. Insets show the corresponding SEM images and digital photos. Source: Zhou et al. [79]/with permission of John Wiley & Sons.

strains. The complementary and reversible coloration process could be realized by sequentially treating PDMS films with strain and plasma. The first and second plasma treatments were run with and without strain and mask, respectively.

The PDMS substrate was first stretched along its length and plasma-treated under a commercial mask. Then, uniform wrinkles were formed with perpendicular alignment to the stretching direction after releasing the pre-strain, producing a positive pattern under incident plane 1. The second plasma treatment was the same as the first one. The remaining surface of the PDMS substrate was oxidized to form a thin and rigid film. The patterns that were formed in the first plasma treatment disappeared after suffering from pre-strain again. Meanwhile, uniform wrinkles were formed with the same alignment to the stretching direction at the areas that were treated by the second plasma, producing a negative pattern in incident plane 2. The reversible geometrical transformation of surface wrinkles and structural color patterns can be realized by stretching. These stimuli-responsive optical properties have promising applications in sensors, information encryption, anti-counterfeiting, smart windows, optical devices, and soft robotics.

Surface wrinkles produced by local rigid strain can be programmed by a device such as ink-jet printing. Song and Kim and coworkers developed a multimodal and covert–overt convertible structural coloration that is structurally switchable by applying external mechanical stimuli (see Figure 6.22b) [79]. Due to the difference in mechanical strength between PDMS and SiO_2, 1D or 2D arrays of wrinkles are formed on the surface of the laminate when subjected to different amounts of compressive stress based on buckling-type instability. Two independent structurally tunable colors can be incorporated into a single platform. When the platform is bent from one side, a colorful smiling face is revealed, while the other (sad) face remains hidden.

Briefly, the self-assembled nanoparticle structures were patterned on the piranha-treated substrates by using a piezoelectric drop-on-demand inkjet printer. The SiO_2 nanoparticle self-assembled monolayer structure was prepared on a substrate (contact angle, 10°) with a temperature of 28 °C. While the coffee-ring patterns were obtained by using inkjet printing on pristine glass (contact angle, 17°) with a substrate temperature of 38 °C, and the dome structures were printed on chlorotrimethylsilane-treated glass (contact angle, 90°) with a substrate temperature of 28 °C. All the samples were dried at 90 °C for 30 minutes after the complete self-assembly of SiO_2 nanoparticles. Then, the PDMS mortar was poured on the self-assembled pattern of the glass substrate and heated at 75 °C for 12 hours. The printed SiO_2 pattern on the glass was transferred and encapsulated in the PDMS matrix.

Surface wrinkling is suitable for the color construction on the soft surface. Bilayer structures with different elastic strains or flexible materials with local rigid structures are obtained by self-assembly, which forms wrinkles under external stimulation and then diffracts incident light to generate structural colors. Surface wrinkling caused by mechanical instability has been harnessed to create reversible patterns that are responsive to external stimuli and provide unique dynamic characteristics of surface morphology and optical properties. This method is suitable for the preparation of large-area SCMs. The precise preparation of arbitrary complex structure color patterns can be achieved by inkjet printing. In particular, the diffraction shows a strong angle dependence.

6.4 Typical Applications of Structural Color

Self-assembled structural color is a bridge between natural iridescence phenomenon and practical optical applications. The self-assembly of nanoparticles can obtain templates for the preparation of functional materials and then transform the microstructure into macro-devices for energy conservation [80], smart window [33], traffic warning [81], anti-counterfeiting [82], complex information coding [83], dynamic color display [84], environmental sensing [85], clinical medicine [86], fabric coloring [87], optical path control [88], and so on.

For example, Wu and Chen and coworkers constructed the magnetic-optics dual-functional Janus superballs via the microfluidic strategy, which demonstrates that the poly(*tert*-butyl acrylate) photonic crystals can vastly enhance photoluminescence intensity of quantum dots film (see Figure 6.23a) [80]. Yang et al. prepared a composite film consisting of a thin layer of quasi-amorphous array of silica nanoparticles embedded in bulk elastomeric PDMS, and presented a smart optical window that could be reversibly switched from a highly transparent state to opaqueness and display angle-independent reflective colors (see Figure 6.23b) [33].

Zhang and coworkers fabricated the wettability-controlled smart indicator for traffic warning based on bilayer photonic crystals with both selective retroreflection and response reflection optical properties (see Figure 6.23c) [81]. It can simultaneously display iridescent vivid structural colors to drivers and pedestrians at different viewing directions relative to the illumination direction and evaluate the risk of landslides on rainy nights, making it especially useful for smart traffic warning applications at night. Zhang and coworkers prepared a bottom patterned bilayer inverse heterostructure photonic crystals by using the self-assembled silica template, and the hidden structural colored QR code could be reversibly revealed by wiping with ethanol (see Figure 6.23d) [82]. Subsequently, they encoded complex information in a bilayer photonic crystal by introducing shape memory polymer to the system, revealing the hidden complex information through surface wettability (see Figure 6.23e) [83].

Duan and Liu and coworkers demonstrated a dynamic color printing scheme using magnesium-based pixelated Fabry–Pérot cavities by grayscale nanolithography (see Figure 6.23f) [84]. With controlled hydrogenation and dehydrogenation, the magnesium layer can be reversibly switched between a reflective metal state and a dielectric hydride state, enabling vivid structural color changes from individual Fabry–Pérot resonators. Niu and coworkers developed an intelligent chromotropic ionic skin by constructing a biomimetic ultrastructure with anisotropic electrostatic repulsion, realizing electrical response and optical visualization to mechanical strain and tactile sensation (see Figure 6.23g) [85]. Jin, Liu, and Zhao and coworkers designed a cardiomyocyte-driven structural color actuation with self-driven

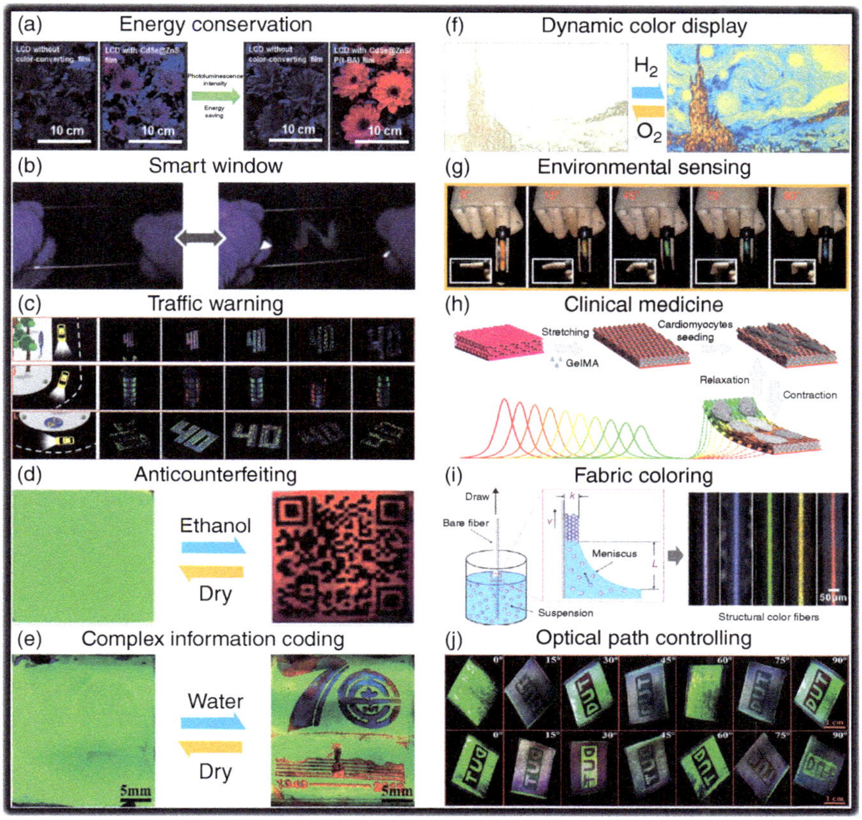

Figure 6.23 (a) Hydrophobic poly(*tert*-butyl acrylate) photonic crystals toward robust energy-saving performance. Source: Wu et al. [80]/with permission of John Wiley & Sons. (b) A robust smart window: reversibly switching from high transparency to angle-independent structural color display. Source: Qi et al. [81]/with permission of John Wiley & Sons. (c) Retroreflection and wettability controlled smart indicator based on responsive bilayer photonic crystals for traffic warning. Source: Qi et al. [81]/with permission of John Wiley & Sons. (d) New encryption strategy of photonic crystals with bilayer inverse heterostructure guided from transparency response. Source: Qi et al. [82]/with permission of John Wiley & Sons. (e) Encoding and decoding of invisible complex information in a dual-response bilayer photonic crystal with tunable wettability. Source: Qi et al. [83]/with permission of John Wiley & Sons. (f) Dynamic color displays using stepwise cavity resonators. Source: Chen et al. [84]/with permission of American Chemical Society. (g) Cephalopod-inspired chromotropic ionic skin with rapid visual sensing capabilities to multiple stimuli. Source: Wang et al. [85]/with permission of American Chemical Society. (h) Cardiomyocyte-driven structural color actuation in anisotropic inverse opals. Source: Reproduced with permission from Shang et al. [86], copyright 2019 American Chemical Society. (i) Structural color fibers directly drawn from colloidal suspensions with controllable optical properties. Source: Yuan et al. [87]/with permission of American Chemical Society. (j) Rotational periodicity display of the tunable wettability pattern in a photoswitch based on a response bilayer photonic crystal. Source: Qi et al. [88]/with permission of American Chemical Society.

actuation and self-reported feedback (see Figure 6.23h) [86]. Cardiomyocytes were planted on the surface to give the elastic substrate autonomic beating ability. Indirectly, the life characteristics of cells can be determined by the change of the structure color. It could be widely applied in integrated cell monitoring, drug testing, and biomedical engineering.

Yuan and Zhang and coworkers prepared structural color fibers by directly drawing bare fibers from colloid dispersions (see Figure 6.23i) [87]. Colloid coatings with different layers can be obtained by varying experimental parameters such as drawing speeds, suspension concentration, and diameters of fibers. Zhang and coworkers fabricated a rotational photoswitch based on a bilayer 3D photonic crystal with backward diffractions similar to those of 2D photonic crystals by vertically combining different thicknesses of nanoparticle templates (see Figure 6.23j) [88]. When rotating the bilayer photonic crystal, the opening or closing of the rotational photoswitch shows periodic reproducibility values of 30° for the thin side and 60° for the thick side. In conclusion, starting from the self-assembly of nanoparticles, a variety of functional SCMs can be developed. The development of self-assembly methods opens a new way to fabricate these types of materials.

6.5 Conclusions

In this chapter, we summarized the various self-assembly methods of structural color. Inspired by nature, various functional SMCs can be obtained through the self-assembly of nanoparticles. These methods have their advantages and disadvantages, and flexible use of them can effectively avoid structural defects in the process of construction. Among them, evaporation self-assembly is the most basic method to construct SMCs. The nanoarrays obtained are usually high- or short-range ordered, which are more suitable for assembling SMCs from colloidal dispersion. Although it takes a long time, the area and thickness of the nanoarray can be accurately controlled. SCMs can be prepared on many kinds of substrates. MSAA is an improvement over the conventional evaporation self-assembly, which can effectively obtain the crack-free nanoarray in a short time. It is suitable for rapid self-assembly (less than 2 min) of uniform nanoarrays, which usually appear to be short-range ordered and crack-free.

Air–liquid interface self-assembly is usually suitable for systems with a large difference in density. It mainly relies on the surface tension to drive the spread of nanoparticles and forms ordered nanoarrays assembled at the air–liquid interface. For the self-assembly of soft microspheres (nanoparticles with low T_g, <60 °C), the assembly temperature is generally required to be greater than the T_g of the microspheres and lower than 90°. It is also suitable for the construction of Janus nanoarrays, which depend on density difference to regulate the structure difference between the gas-phase side and the liquid-phase side. The oil–oil interface is usually applied to incompatible oil phases with different refractive indices, and the coloring mechanism of SCMs is realized by TIR. The Janus and double emulsion morphologies can be switched by tuning the balance of interfacial tensions of the iridescent biphasic oil droplets and then regulating the TIR and the reflected color. Oil–water interface self-assembly is suitable for the construction of embedded nanoarray, and the density of nanoparticles is generally required to be less than that of the oil phase. The nanoarrays assembled by this method are well organized. It is suitable for the preparation of large-area 2D composite structures and bowl structures. The controlled micellization self-assembly is suitable for the SCMs constructed by block copolymers, which requires high environmental factors. The self-assembly in this method was an isotropic process within the volume of the droplet, leading to a rapid fabrication time, high tolerance to structural defects, and color-independence from the microparticle dimensions. The layered hydrogels self-assembly is the preferred method to construct flexible SCMs, but it is time-consuming and difficult to lock solvents. So that PDMS is usually used for packaging and preservation.

Spray coating self-assembly is suitable for the rapid construction of SCMs, which usually results in disordered or short-range ordered nanoarrays with shallow color, and generally requires adding light-absorbing materials. But for high refractive index nanoparticles, absorbent materials can be avoided. Unidirectional rubbing self-assembly is a fast method to obtain 2D SCMs with simple operation. The order of nanoparticles depends on the conditions of friction. One of the advantages is that it is suitable for constructing SCMs on curved surfaces. The edge-induced rotational shearing self-assembly is suitable for soft microsphere assembly and can be easily mechanized. Screen printing self-assembly is suitable for the construction of patterned SMCs and generally requires adding light-absorbing materials. The obtained nanoarrays are usually short-range ordered and long-range disordered. The color shows angle independence.

Magnetic-induced self-assembly is suitable for the rapid assembly of magnetic nanoparticles and the assembly of composite composed of ordered magnetic nanoparticles and gels. The photonic structures assembled by magnetic nanoparticles can be encapsulated in polymer gels, whose chemical and physical properties can be demonstrated in the form of structural color. Photoinduced self-assembly is suitable for liquid crystal SCMs and the self-assembly of photo-responsive nanoparticles. The difficulty is the design and synthesis of liquid crystal molecules. ALD self-assembly is suitable for preparing SCMs on the inert substrate, but the cost is high. It is also suitable for constructing the structure color on many kinds of curved surfaces. Physical vapor deposition self-assembly is suitable for the construction of plasmonic SCMs. The process is compatible with arbitrarily structured substrates and can produce wafer-scale, diffusive, angle-independent, and flexible materials. Importantly, a halftoned black color could be obtained by combining a highly saturated cyan–magenta–yellow. Surface winkling is suitable for the preparation of large-area SCMs on the soft surface. The precise preparation of arbitrary complex structure color patterns can be achieved by inkjet printing.

A fascination for color and its use for artistic expression is deeply rooted in all human civilizations. The quest to develop inspiring color effects has led to exciting discoveries, which, in turn, have stimulated fundamental research toward the understanding of the underlying physics of the observed phenomena. The development of SCMs is always developing in a form of exploring, assembling, applying, and re-exploring. Exploiting new self-assembly approaches for nanoparticles makes a great contribution to developing functional SCMs. Mastering these self-assembly methods is conducive to combining advanced theories from different disciplines, recognizing microscopic dynamics, and understanding the nature of functional materials, decoding the laws of nature from the structure, and so on.

Acknowledgments

This work is supported by the State Key Program of the National Natural Science Foundation of China (21536002), the Program of the National Natural Science Foundation of China (21978041), the Science Fund for Creative Research Groups of the National Natural Science Foundation of China (21421005), the Liaoning Revitalization Talent Program (1801006), and the Fundamental Research Funds for the Central Universities (DUT19TD28).

References

1 Nassau, K. (2001). The physics and chemistry of color. In: *The Fifteen Causes of Color*, 2. New York: Wiley.
2 Fukata, Y. (2007). *Inorganic Chromotropism. Basic Concepts and Applications of Colored Materials*. Tokyo: *Kodansha Springer*.
3 Andreeva, Y.I., Drozdov, A.S., Fakhardo, A.F. et al. (2017). The controllable destabilization route for synthesis of low cytotoxic magnetic nanospheres with photonic response. *Sci. Rep.* 7: 11343.
4 Kamita, G., Frka-Petesic, B., Allard, A. et al. (2016). Biocompatible and sustainable optical strain sensors for large-area applications. *Adv. Opt. Mater.* 4: 1950–1954.
5 Kim, J.-W., Lee, J.-S., and Kim, S.-H. (2018). Biodegradable inverse opals with controlled discoloration. *Adv. Mater. Interfaces* 5: 1701658.
6 Kinoshita, S. and Yoshioka, S. (2005). Structural colors in nature: the role of regularity and irregularity in the structure. *Chem. Phys. Chem.* 6: 1442–1459.
7 Parker, A.R. (2000). 515 million years of structural colour. *J. Opt. A Pure Appl. Opt.* 2: R15–R28.
8 Yin, H., Shi, L., Sha, J. et al. (2006). Iridescence in the neck feathers of domestic pigeons. *Phys. Rev. E* 74: 051916.
9 Chen, C., Airoldi, C.A., Lugo, C.A. et al. (2021). Flower inspiration: broad-angle structural color through tunable hierarchical wrinkles in thin film multilayers. *Adv. Funct. Mater.* 31: 2006256.
10 Vignolini, S., Rudall, P.J., Rowland, A.V. et al. (2012). Pointillist structural color in Pollia fruit. *Proc. Natl. Acad. Sci. U. S. A.* 109: 15712–15715.
11 Shevtsova, E., Hansson, C., Janzen, D.H., and Kjærandsen, J. (2011). Stable structural color patterns displayed on transparent insect wings. *Proc. Natl. Acad. Sci. U. S. A.* 108: 668–673.

12 Hsiung, B.-K., Siddique, R.H., Stavenga, D.G. et al. (2017). Rainbow peacock spiders inspire miniature superiridescent optics. *Nat. Commun.* 8: 2278.

13 Forster, J.D., Noh, H., Liew, S.F. et al. (2010). Biomimetic isotropic nano-structures for structural coloration. *Adv. Mater.* 22: 2939–2944.

14 Teyssier, J., Saenko, S.V., van der Marel, D., and Milinkovitch, M.C. (2015). Photonic crystals cause active colour change in chameleons. *Nat. Commun.* 6: 6368.

15 Cuthill, I.C., Allen, W.L., Arbuckle, K. et al. (2017). The biology of color. *Science* 357: eaan0221.

16 Duarte, R.C., Flores, A.A.V., and Stevens, M. (2017). Camouflage through colour change: mechanisms, adaptive value and ecological significance. *Philos. Trans. R. Soc. B* 372: 20160342.

17 Parker, A.R. and Martini, N. (2006). Structural colour in animals-simple to complex optics. *Opt. Laser Technol.* 38: 315–322.

18 Bell, G.R.R., Mäthger, L.M., Gao, M. et al. (2014). Diffuse white structural coloration from multilayer reflectors in a squid. *Adv. Mater.* 26: 4352–4356.

19 Prum, R.O., Torres, R.H., Williamson, S., and Dyck, J. (1998). Coherent light scattering by blue feather barbs. *Nature* 396: 28–29.

20 Cai, G., Wang, J., Eh, A.L.S. et al. (2018). *Diphylleia grayi*-inspired stretchable hydrochromics with large optical modulation in the visible-near-infrared region. *ACS Appl. Mater. Interfaces* 10: 37685–37693.

21 Yoo, G.Y., Lee, S.J., Ko, M. et al. (2020). *Diphylleia grayi*-inspired intelligent hydrochromic adhesive film. *ACS Appl. Mater. Interfaces* 12: 49982–49991.

22 Alam, A.-M., Baek, K., Son, J. et al. (2017). Generating color from polydisperse, near micron-sized TiO_2 particles. *ACS Appl. Mater. Interfaces* 9: 23941–23948.

23 Whitney, H.M., Kolle, M., Andrew, P. et al. (2009). Floral iridescence, produced by diffractive optics, acts as a cue for animal pollinators. *Science* 323: 130–133.

24 Chu, G., Qu, D., Camposeo, A. et al. (2020). When nanocellulose meets diffraction grating: freestanding photonic paper with programmable optical coupling. *Mater. Horiz.* 7: 511–519.

25 Hatton, B., Mishchenko, L., Davis, S. et al. (2010). Assembly of large-area, highly ordered, crack-free inverse opal films. *Proc. Natl. Acad. Sci. U. S. A.* 107: 10354–10359.

26 Holland, B.T., Blanford, C.F., and Stein, A. (1998). Synthesis of macroporous minerals with highly ordered three-dimensional arrays of spheroidal voids. *Science* 281: 538–540.

27 Xia, Y., Gates, B., Yin, Y., and Lu, Y. (2000). Monodispersed colloidal spheres: old materials with new applications. *Adv. Mater.* 12: 693–713.

28 Jiang, P. and McFarland, M.J. (2004). Large-scale fabrication of wafer-size colloidal crystals, macroporous polymers, and nanocomposites by spin-coating. *J. Am. Chem. Soc.* 126: 13778–13786.

29 Wong, S., Kitaev, V., and Ozin, G.A. (2003). Colloidal crystal films: advances in universality and perfection. *J. Am. Chem. Soc.* 125: 15589–15598.

30 Gu, Z., Fujishima, A., and Sato, O. (2002). Fabrication of high-quality opal films with controllable thickness. *Chem. Mater.* 14: 760–765.

31 Zhu, Z., Zhang, J., Tong, Y. et al. (2019). Reduced graphene oxide membrane induced robust structural colors toward personal thermal management. *ACS Photonics* 6: 116–122.

32 Huang, Y., Li, F., Qin, M. et al. (2013). A multi-stopband photonic-crystal microchip for high-performance metal-ion recognition based on fluorescent detection. *Angew. Chem. Int. Ed.* 52: 7296–7299.

33 Ge, D., Lee, E., Yang, L. et al. (2015). A robust smart window: reversibly switching from high transparency to angle-independent structural color display. *Adv. Mater.* 27: 2489–2495.

34 Zhang, J., Zhu, Z., Yu, Z. et al. (2019). Large-scale colloidal films with robust structural colors. *Mater. Horiz.* 6: 90–96.

35 Deegan, R.D., Bakajin, O., Dupont, T.F. et al. (1997). Capillary flow as the cause of ring stains from dried liquid drops. *Nature* 389: 827–829.

36 Im, S.H., Lim, Y.T., Suh, D.J., and Park, O.O. (2002). Three-dimensional self-assembly of colloids at a water-air interface: a novel technique for the fabrication of photonic bandgap crystals. *Adv. Mater.* 14: 1367–1369.

37 Yu, S., Cao, X., Niu, W. et al. (2019). Large-area and water rewriteable photonic crystal films obtained by the thermal assisted air-liquid interface self assembly. *ACS Appl. Mater. Interfaces* 11: 22777–22785.

38 Gao, P., He, J., Zhou, S. et al. (2015). Large-area nanosphere self-assembly by a micro-propulsive injection method for high throughput periodic surface nanotexturing. *Nano Lett.* 15: 4591–4598.

39 Scriven, L.E. and Sternling, C.V. (1960). The marangoni effects. *Nature* 187: 186–188.

40 Fujii, S., Kappl, M., Butt, H.-J. et al. (2012). Soft janus colloidal crystal film. *Angew. Chem. Int. Ed.* 51: 9809–9813.

41 Zhang, J.-T., Wang, L., Luo, J. et al. (2011). 2-D array photonic crystal sensing motif. *J. Am. Chem. Soc.* 133: 9152–9155.

42 Nussenzveig, H.M. (1969). High-frequency scattering by a transparent sphere. II. Theory of the rainbow and the glory. *J. Math. Phys.* 10: 125–176.

43 Goodling, A.E., Nagelberg, S., Kaehr, B. et al. (2019). Colouration by total internal reflection and interference at microscale concave interfaces. *Nature* 566: 523–527.

44 van Dommelen, R., Fanzio, P., and Sasso, L. (2018). Surface self-assembly of colloidal crystals formicro- and nano-patterning. *Adv. Colloid Interf. Sci.* 251: 97–114.

45 Goodling, A.E., Nagelberg, S., Kolle, M., and Zarzar, L.D. (2020). Tunable and responsive structural color from polymeric microstructured surfaces enabled by interference of totally internally reflected light. *ACS Mater. Lett.* 2: 754–763.

46 Macfarlane, R.J., Kim, B., Lee, B. et al. (2014). Improving brush polymer infrared one-dimensional photonic crystals via linear polymer additives. *J. Am. Chem. Soc.* 136: 17374–17377.

47 Song, D.-P., Zhao, T.H., Guidetti, G. et al. (2019). Hierarchical photonic pigments via the confined self-assembly of bottlebrush block copolymers. *ACS Nano* 13: 1764–1771.

48 Eoh, H., Kang, H.S., Kim, M.J. et al. (2019). Nonvolatile, multicolored photothermal writing of block copolymer structural color. *Adv. Funct. Mater.* 29: 1904055.

49 Zhao, T.H., Jacucci, G., Chen, X. et al. (2020). Angular-independent photonic pigments via the controlled micellization of amphiphilic bottlebrush block copolymers. *Adv. Mater.* 32: 2002681.

50 Kang, H.S., Lee, J., Cho, S.M. et al. (2017). Printable and rewritable full block copolymer structural color. *Adv. Mater.* 29: 1700084.

51 Patel, B.B., Walsh, D.J., Kim, D.H. et al. (2020). Tunable structural color of bottlebrush block copolymers through direct-write 3D printing from solution. *Sci. Adv.* 6: eaaz7202.

52 Yue, Y., Kurokawa, T., Haque, M.A. et al. (2014). Mechano-actuated ultrafast full-colour switching in layered photonic hydrogels. *Nat. Commun.* 5: 4659.

53 Wang, Y., Niu, W., Lo, C.-Y. et al. (2020). Interactively full-color changeable electronic fiber sensor with high stretchability and rapid response. *Adv. Funct. Mater.* 30: 2000356.

54 Ge, D., Yang, L., Wu, G., and Yang, S. (2014). Spray coating of superhydrophobic and angle-independent coloured films. *Chem. Commun.* 50: 2469–2472.

55 Meng, F., Umair, M.M., Iqbal, K. et al. (2019). Rapid fabrication of noniridescent structural color coatings with high color visibility, good structural stability, and self-healing properties. *ACS Appl. Mater. Interfaces* 11: 13022–13028.

56 Meng, F., Umair, M.M., Zhang, S. et al. (2019). Thermal-guided interfacial confinement to fabricate flexible structural color composites for durable applications. *J. Mater. Chem. C* 7: 11258–11264.

57 Bi, J., Wu, Y., Li, L. et al. (2020). Asymmetric structural colors based on monodisperse single-crystal Cu_2O spheres. *Nanoscale* 12: 3220–3226.

58 Park, C., Lee, T., Xia, Y. et al. (2014). Quick, large-area assembly of a single-crystal monolayer of spherical particles by unidirectional rubbing. *Adv. Mater.* 26: 4633–4638.

59 Finlayson, C.E., Spahn, P., Snoswell, D.R.E. et al. (2011). 3D bulk ordering in macroscopic solid opaline films by edge-induced rotational shearing. *Adv. Mater.* 23: 1540–1544.

60 Zhao, Q., Finlayson, C.E., Snoswell, D.R.E. et al. (2016). Large-scale ordering of nanoparticles using viscoelastic shear processing. *Nat. Commun.* 7: 11661.

61 Baumberg, J.J., Pursiainen, O.L., and Spahn, P. (2009). Resonant optical scattering in nanoparticle-doped polymer photonic crystals. *Phys. Rev. B* 80: 201103.

62 Finlayson, C.E. and Baumberg, J.J. (2017). Generating bulk-scale ordered optical materials using shear-assembly in viscoelastic media. *Materials* 10: 688.

63 Zhou, C., Qi, Y., Zhang, S. et al. (2020). Rapid fabrication of vivid noniridescent structural colors on fabrics with robust structural stability by screen printing. *Dyes Pigments* 176: 108226.

64 Zhang, J., Yang, S., Tian, Y. et al. (2015). Dual photonic-bandgap optical films towards the generation of photonic crystal-derived 2-dimensional chemical codes. *Chem. Commun.* 51: 10528–10531.

65 Kramer, R.M., Crookes-Goodson, W.J., and Naik, R.R. (2007). The self-organizing properties of squid reflectin protein. *Nat. Mater.* 6: 533–538.
66 Zhu, X., Yan, W., Levy, U. et al. (2017). Resonant laser printing of structural colors on high-index dielectric metasurfaces. *Sci. Adv.* 3: e1602487.
67 Hail, C.U., Schnoering, G., Damak, M. et al. (2020). A plasmonic painter's method of color mixing for a continuous red-green-blue palette. *ACS Nano* 14: 1783–1791.
68 Li, Z., Jin, J., Yang, F. et al. (2020). Coupling magnetic and plasmonic anisotropy in hybrid nanorods for mechanochromic responses. *Nat. Commun.* 11: 2883.
69 Luo, W., Cui, Q., Fang, K. et al. (2020). Responsive hydrogel-based photonic nanochains for microenvironment sensing and imaging in real time and high resolution. *Nano Lett.* 20: 803–811.
70 Le, K.V., Takezoe, H., and Araoka, F. (2017). Chiral superstructure mesophases of achiral bent-shaped molecules-hierarchical chirality amplification and physical properties. *Adv. Mater.* 29: 1602737.
71 Kim, H., Ryu, S.H., Tuchband, M. et al. (2017). Structural transitions and guest/host complexing of liquid crystal helical nanofilaments induced by nanoconfinement. *Sci. Adv.* 3: e1602102.
72 Park, W., Ha, T., Kim, T.-T. et al. (2019). Directed self-assembly of a helical nanofilament liquid crystal phase for use as structural color reflectors. *NPG Asia Mater.* 11: 45.
73 Toshchevikov, V., Petrova, T., and Saphiannikova, M. (2017). Kinetics of light-induced ordering and deformation in LC azobenzene-containing materials. *Soft Matter* 13: 2823–2835.
74 Niu, W., Zhang, L., Wang, Y. et al. (2019). Multicolored photonic crystal carbon fiber yarns and fabrics with mechanical robustness for thermal management. *ACS Appl. Mater. Interfaces* 11: 32261–32268.
75 Franklin, D., He, Z., Ortega, P.M. et al. (2020). Self-assembled plasmonics for angle-independent structural color displays with actively addressed black states. *Proc. Natl. Acad. Sci. U. S. A.* 117: 13350–13358.
76 Yang, S., Khare, K., and Lin, P.-C. (2010). Harnessing surface wrinkle patterns in soft matter. *Adv. Funct. Mater.* 20: 2550–2564.
77 Huang, Z.Y., Hong, W., and Suo, Z. (2005). Nonlinear analyses of wrinkles in a film bonded to a compliant substrate. *J. Mech. Phys. Solids* 53: 2101–2118.
78 Li, Z., Liu, Y., Marin, M., and Yin, Y. (2020). Thickness-dependent wrinkling of PDMS films for programmable mechanochromic responses. *Nano Res.* 13: 1882–1888.
79 Zhou, Q., Park, J.G., Bae, J. et al. (2020). Multimodal and covert-overt convertible structural coloration transformed by mechanical stress. *Adv. Mater.* 32: 2001467.
80 Wu, X., Hong, R., Meng, J. et al. (2019). Hydrophobic poly(tert-butyl acrylate) photonic crystals towards robust energy-saving performance. *Angew. Chem. Int. Ed.* 58: 13556–13564.
81 Qi, Y., Zhou, C., Niu, W. et al. (2020). Retroreflection and wettability controlled smart indicator based on responsive bilayer photonic crystals for traffic warning. *Adv. Opt. Mater.* 8: 2001367.

82 Qi, Y., Chu, L., Niu, W. et al. (2019). New encryption strategy of photonic crystals with bilayer inverse heterostructure guided from transparency response. *Adv. Funct. Mater.* 29: 1903743.

83 Qi, Y., Niu, W., Zhang, S. et al. (2019). Encoding and decoding of invisible complex information in a dual-response bilayer photonic crystal with tunable wettability. *Adv. Funct. Mater.* 29: 1906799.

84 Chen, Y., Duan, X., Matuschek, M. et al. (2017). Dynamic color displays using stepwise cavity resonators. *Nano Lett.* 17: 5555–5560.

85 Wang, Y., Cao, X., Cheng, J. et al. (2021). Cephalopod-inspired chromotropic ionic skin with rapid visual sensing capabilities to multiple stimuli. *ACS Nano* 15: 3509–3521.

86 Shang, Y., Chen, Z., Fu, F. et al. (2019). Cardiomyocyte-driven structural color actuation in anisotropic inverse opals. *ACS Nano* 13: 796–802.

87 Yuan, W., Li, Q., Zhou, N. et al. (2019). Structural color fibers directly drawn from colloidal suspensions with controllable optical properties. *ACS Appl. Mater. Interfaces* 11: 19388–19396.

88 Qi, Y., Niu, W., Zhang, S. et al. (2020). Rotational periodicity display of the tunable wettability pattern in a photoswitch based on a response bilayer photonic crystal. *ACS Appl. Mater. Interfaces* 12: 9664–9672.

7

Colloidal Photonic Crystal Sensors

Zhongyu Cai[1], Xiaoying Xu[1], Zihui Meng[2], Bushra Rafique[2], and Ruixiang Liu[3]

[1]*Beihang University, Beijing Advanced Innovation Center for Biomedical Engineering, Research Institute for Frontier Science, School of Space and Environment, 37 Xueyuan Road, Beijing 100191, China*
[2]*Beijing Institute of Technology, School of Chemistry and Chemical Engineering, 5 Zhongguancunnan Street, Beijing 100081, China*
[3]*Tianjin University, School of Materials Science and Engineering, 135 Yaguan Road, Tianjin 300350, China*

7.1 Introduction

Photonic crystals (PhCs) are periodic optical structures that are designed to control the propagation of photons in a similar way that a semiconductor to control the flow of electrons. One distinct feature of PhCs is the existence of photonic bandgaps (PBGs) where light with frequencies falling within the PBG cannot be propagated [1]. The position of a PBG depends on the spatial periodicity of the PhCs. Though historically, as long ago as 1887, Lord Rayleigh investigated the strong reflectivity of a periodically arranged multilayer structure over a defined wavelength range [2] and derived a formula on the reflectivity in one-dimensional (1D) PhCs in 1917 [3], it was not until 1954 that Alfrey et al. first reported on optical properties of PhCs and electronic photographs of highly ordered structure consisted of polystyrene (PS) colloidal photonic crystals (CPCs) [4]. Later, O'Neill and coworkers investigated the Bragg diffraction from two-dimensional (2D) CPCs in 1968 [5]. In 1983, Asher filed the first patent on CPCs [6]. Shortly after this invention, Yablonovitch and John proposed the concept of "PhCs" independently to describe a class of material structure, which consists of dielectric materials with different refractive index (RI) arranged with three-dimensional (3D) periodicity in 1987 [7, 8]. Their seminal work further prompted huge interest and intense efforts in this exciting field to develop photonic devices for a wide variety of applications, such as optical filters [9], optical waveguides [10], displays [11], light emitting diodes (LEDs) [12], photocatalysis [13, 14], nanolithography [15, 16], solar cells [17], sensing [18, 19], and many others [20, 21].

PhCs can be fabricated via both top-down microfabrication approaches [22] and bottom-up self-assembly methods [23]. Top-down approaches start from bulk materials to fabricate PhCs with well-controlled features by using complex and expensive equipment. Comparing with top-down approaches, bottom-up methods such as sedimentation [24–26], interfacial assembly [27, 28], co-assembly

Functional Materials from Colloidal Self-assembly, First Edition. Edited by Qingfeng Yan and George Zhao.

[14, 29–32], vertical deposition [33–35], spin coating [36, 37], blade coating [38], and inkjet printing [39, 40] use self-assembly of colloidal particles to fabricate versatile CPCs. The bottom-up self-assembly methods offer a simple and low-cost strategy with tunable optical properties. Besides artificial PhCs, there are many examples of PhCs in nature showing bright colors, such as bristles of sea mouses, feathers of peacocks, scales of butterflies, gem opals, and wings of beetles [41, 42]. These iridescences are structural colors caused by the periodic optical microstructures within these biological organisms (Figure 7.1a–d). Similar to these natural examples, PhCs can be classified into 1D, 2D, and 3D according to their periodicity (Figure 7.1e). Nature utilizes these structural colors for intimidation (warning), or camouflage, or mimicry [41, 45–48]. Chameleons, for instance, can change their skin colors when they respond to danger surrounding them [49]. The change in the periodicity caused by (bio)chemical stimulus leads to the shift in the structural

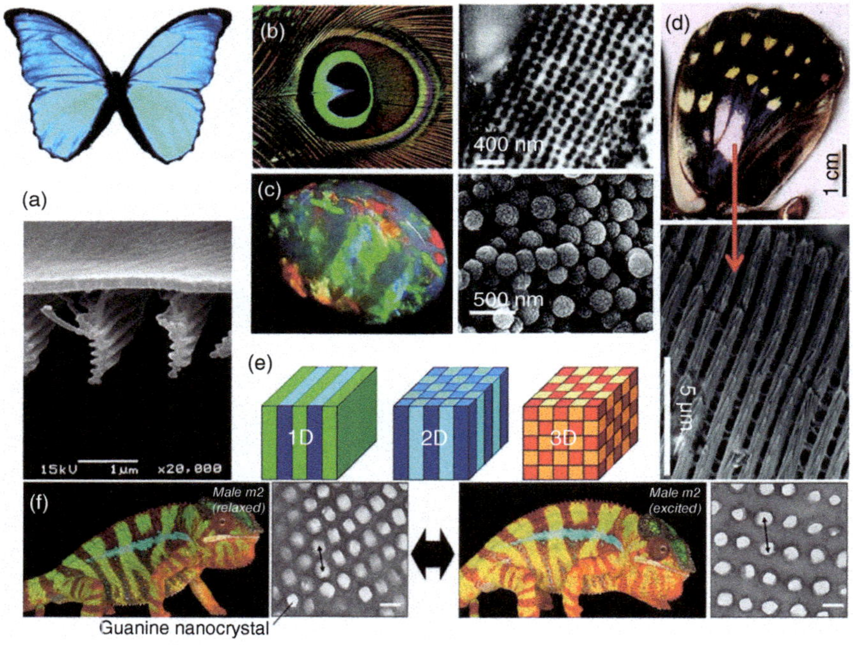

Figure 7.1 (a–d) Natural photonic crystals: (a) the blue iridescence and scanning electron microscope (SEM) image of the 1D structure of the morpho butterfly. (b) Multicolored peacock feather and transmission electron microscope (TEM) image of transverse cross section of the 2D structure of the blue area of a wing. (c) Natural opal gemstone and SEM image of the SiO_2 sphere structure within. (d) Wing of the male Sasakia Charonda butterfly and SEM image of the 3D structure of the white iridescent area. (e) Schematic representation of 1D, 2D, and 3D PhC structures, with different colors representative of different dielectric constants, indicating periodicity in one, two, or three directions. Source: Amstrong et al. [43]/with permission of Royal Society of Chemistry. (f) Reversible color change in the skin of a male panther chameleon from relaxed (left) to excited (right) state. TEM images show the change of periodicity in a regular array of guanine nanocrystals. Source: Teyssier et al. [44]/Springer Nature/CC BY 4.0.

colors (Figure 7.1f). This color change mechanism can be used as the principle for PhC sensing, where highly ordered PhCs are used as the signal transduction scheme to generate readouts [44].

Inspired by structural color changes of these natural examples of PhCs, scientists developed PhC sensors, where the Bragg diffraction changes were used as the readout of sensors. The Asher research group pioneered the development of PhC sensors using polymerized crystalline crystal arrays (PCCA) sensors for the detection of glucose and other analytes [50]. The 3D PCCA sensors utilized volume phase transition (VPT) caused by the interaction between intelligent materials and analytes. The VPT of the intelligent materials varies the lattice spacing of the incorporated PhCs, which shift the Bragg diffraction along with structural color change [18]. Later, Asher and coworkers further developed 2D PhC sensors by polymerizing intelligent materials on 2D CPCs, which tackled many challenges of 3D PCCA sensors [18, 51–55]. From then on, a huge number of CPC sensors have been explored for the detection of various analytes.

PhC sensors have attracted intense research interest due to their simplicity, low cost, high signal-to-noise ratio, and ease of operation [18, 19, 56–59]. CPCs are an excellent platform for colorimetric and fluorescent sensing. In the last two decades, PhC sensors have been intensively investigated, and significant progress has been achieved. In this book chapter, we will discuss the progress made in this field and shed our light on the future development of high-performance CPC sensors. This book chapter is structured as follows. We first briefly introduce the concept and fundamentals of CPC sensors. Then, we will review the types, structures, and materials of CPC sensors. Later, we will summarize CPC sensors outlining with an emphasis on their target analytes types and the features of emerging CPC sensors from physical sensors to chemical sensors and biosensors. In the end, we will shed our perspective on the future development of CPC sensors.

7.2 Fundamentals of Colloidal Photonic Crystal Sensors

For CPC sensors, the CPCs are utilized as a signal transducing platform, and the diffraction changes or visual color variations from the PhCs are used as readouts. In general, intelligent materials or recognition moieties are integrated onto the CPCs. PhCs diffract light at certain wavelengths, which can be explained with the model of X-ray diffraction as shown in Figure 7.2a. The diffraction from CPCs follows both the Bragg–Snell law of diffraction and Bragg diffraction. The diffraction equations of PhC sensors with different periodicities are discussed in the succeeding text.

The Bragg's law (Eq. (7.1)) details constructive interference of light within regularly arranged nano/microstructure, where d is the interplanar spacing, θ is the angle of incident light, m is the order of diffraction, λ is the wavelength of incident light (in vacuum), and n_{eff} is the average effective RI. The RI is defined in Eq. (7.2), where n_{p} and n_{m} represent the RI of particles and the medium and f_{p} represents the volume fraction of particles. The Snell's law (Eq. (7.3)) describes the relationship between the angles of incidence and refraction indices, where n_1 is the incident

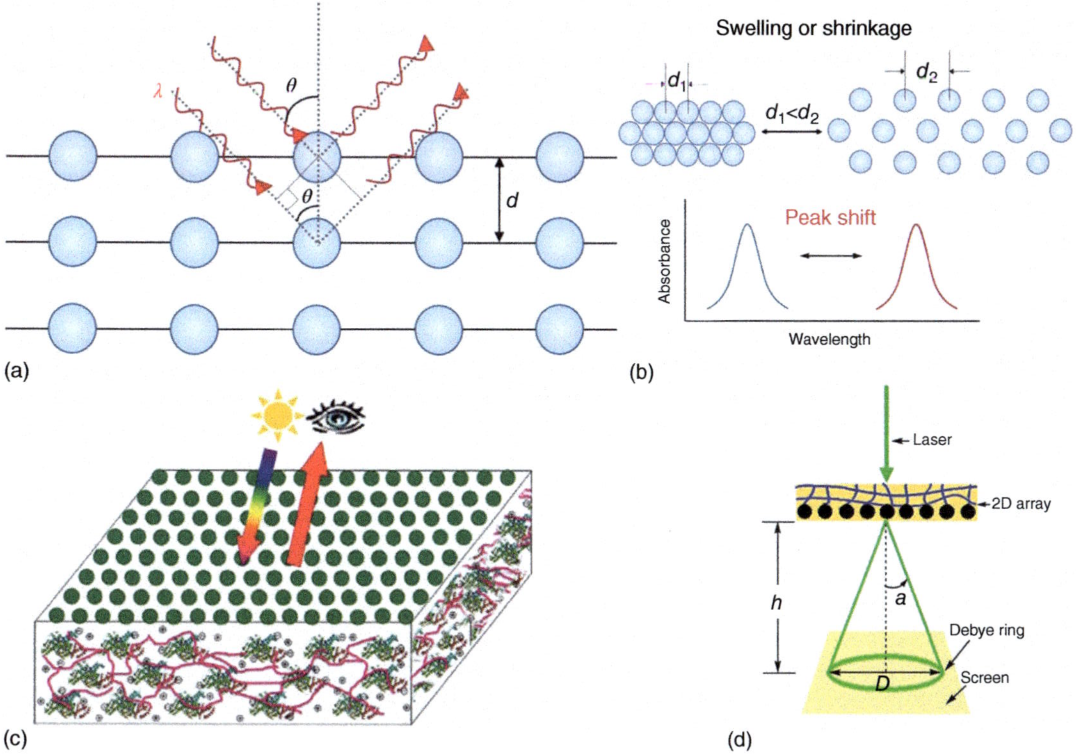

Figure 7.2 (a) Scheme of light reflection from photonic crystal by analogy to the explanation of X-ray diffraction. (b) Detection principle of a 3D CPC sensor using the swelling or shrinkage of polymer. Source: Reprinted with permission from Endo et al. [60]. Copyright 2007, Elsevier. (c) Scheme of a 2D PhC sensor. (d) Debye ring diffraction from 2D PhCs. Source: Reprinted with permission from Cai et al. [51]. Copyright 2014, American Chemical Society.

index, n_2 is the reflected index, θ_1 is the incident angle, and θ_2 is the reflected angle. The combination of Bragg's law and Snell's law gives rise to Bragg–Snell law, which is described as Eq. (7.4). For a typical 3D CPC sensor, the change in the effective RI or the lattice spacing, or local RI surrounding the cavities, causes Bragg diffraction shift of the CPCs, which can be used as a readout of the CPC sensor.

$$2n_{\mathrm{eff}} d \cos\theta = m\lambda \tag{7.1}$$

$$n_{\mathrm{eff}} = \sqrt{{n_{\mathrm{F}}}^2 f_{\mathrm{p}} + {n_{\mathrm{m}}}^2 (1 - f_{\mathrm{p}})} \tag{7.2}$$

$$n_1 \sin\theta_1 = n_2 \sin\theta_2 \tag{7.3}$$

$$2d\sqrt{{n_{\mathrm{eff}}}^2 - \sin^2\theta} = m\lambda \tag{7.4}$$

Therefore, external stimuli induce the changes of the average RI or the lattice particle spacing of PhC sensors, which cause Bragg diffraction shift of the PhCs (Figure 7.2b). When the PhC Bragg diffraction wavelength is within the visible spectrum range, structural color of PhCs can be observed with naked eyes. For self-assembled 3D CPC sensors, the colloidal particles arrange in face-centered cubic (*fcc*) format, and the Bragg diffraction can be simplified as Eq. (7.5), where D is the center-to-center distance of colloidal particles.

$$\sqrt{\frac{8}{3}} D({n_{\mathrm{eff}}}^2 - \sin^2\theta)^{1/2} = \lambda \tag{7.5}$$

The 2D PhCs diffraction is different from 3D PhCs diffraction. In 2D PhCs, multiple wavelengths are forward scattered and backscattered at different angles. For a 2D hexagonal lattice, the Bragg diffraction conditions meet Eq. (7.6), where d is the 2D colloidal particle spacing, α is the angle of incident light along the normal, and β is the angle of diffracted light [61, 62]. In the Bragg diffraction of 2D PhCs, there is no light path difference created within the crystal lattices.

$$\frac{\sqrt{3}}{2} d(\sin\alpha + \sin\beta) = m\lambda \tag{7.6}$$

In the Littrow configuration, the probe irradiates the sample with a polychromatic light and collects the back-diffracted wavelengths at the same angle as the incidence. This leads to the 2D Bragg diffraction equation with a single angle, as shown in Eq. (7.7).

$$\sqrt{3} d \sin\alpha = m\lambda \tag{7.7}$$

where α is the angle of incident light along the normal, m is the order of reflection, and λ is the wavelength of the reflected light (in vacuum). The 2D PhC Bragg diffraction is usually measured with spectroscopy instrument, which limits the application of 2D PhC sensors for remote sensing. Another way is to measure Debye ring diffraction with laser diodes or pointers (Figure 7.2c,d) [18, 51].

In this configuration, the Debye ring diameter, D, depends upon the interparticle spacing, d. In Eq. (7.8), λ is the laser wavelength, h is the distance between the screen and the 2D PhCs, and α is the angle subtended by the Debye diffraction from the 2D PhCs normal. Thus, one can easily determine the 2D PhCs particle spacing from Eq. (7.8).

$$d = \frac{4\lambda_{\text{laser}}\sqrt{(D/2)^2 + h^2}}{\sqrt{3}\text{D}} \tag{7.8}$$

1D PhCs, also known as Bragg stacks, the Bragg's condition of 1D PhCs, is shown in Eq. (7.9), where m is the order of Bragg diffraction, n_{L} and n_{H} are the respective RIs, and h_{L} and h_{H} are the thicknesses of the low-refractive-index and high-refractive-index materials [63, 64].

$$2(n_{\text{L}}h_{\text{L}} + n_{\text{H}}h_{\text{H}}) = m\lambda \tag{7.9}$$

Though 1D, 2D, and 3D PhC sensors show different diffraction properties, these PhC sensors possess their own merits and disadvantages. Therefore, PhC structures need to be optimized to achieve high-performance sensors.

7.3 Responsive Materials and Novel Photonic Structures

PhC sensors can be classified into 1D, 2D, and 3D PhC sensors according to their periodicity. Typically, according to the materials composed of the sensors, there are three kinds of PhC sensors, including (i) photonic resonators or waveguides, (ii) mesoporous PhC Bragg stacks, and (iii) colloidal PhCs. In this book chapter, we will focus on CPC sensors.

Besides conventional responsive materials such as polymers (hydrogels), a large number of novel materials have been explored for sensing applications with enhanced performance in the past decade. PhC sensors composed of metastable microcrystals [65, 66], mesoporous materials [67, 68], metal–organic frameworks (MOFs) [69, 70], or ionic liquids (ILs) [71–73] were developed for sensitive colorimetric detection of different analytes. These responsive materials endowed the PhC sensors with fast response speed, excellent selectivity, and unique sensitivity. To date, a number of colloidal photonic structures have been employed to fabricate CPC sensors. The most widely used CPC sensors are PhC hydrogel sensors. The Asher research group pioneered PhC hydrogel sensors, in which non-close-packed crystalline colloidal arrays (CCAs) are polymerized within responsive hydrogel materials [50]. The other types of frequently used CPC sensors are based on close-packed CPCs (opals or inverse opals), whose surfaces are chemically functionalized with recognition moieties. In addition, other types of novel CPCs, like spherical colloidal crystal beads (SCCBs) [74] and patterned PhCs [75], have been developed for sensing applications.

Velve et al. developed an evaporative method to fabricate SCCBs at the interface of air and liquid [76]. These SCCBs are intrinsically viewing angle-independent property owing to their super symmetry, which is very useful for sensing applications. Afterward, both the Yang and the Gu research groups developed other methods to produce SCCBs as platforms of barcodes for disease diagnosis, sensing, cell culture, drug screening, and many other applications [74, 77, 78]. The Song and the Gu research groups have independently developed multiple different patterned CPCs for sensing applications using an inkjet printing method [67, 79]. Song and coworkers explored patterned PhC sensors for water vapor sensing [79], while Gu and coworkers reported an apple treelike patterned PhC sensor for the detection of ethanol vapor, where the vivid color change can be observed upon exposure of the patterned PhCs to ethanol vapor [67].

7.4 Colloidal Photonic Crystal Sensors Responsive to Physical Stimuli

CPCs are very sensitive to physical stimuli in surroundings, such as temperature [80], pH [81], humidity [82], light [83], and mechanical strength [84]. There are many electrically and magnetically responsive CPCs [85–88]. However, few of them can be used to measure the electric or magnetic strengths. Therefore, in this section, we will focus on humidity and strain CPCs sensors.

7.4.1 Colloidal Photonic Crystals Responsive to Humidity

Humidity plays a significant role in chemical, agriculture, pharmaceutical industry, weather forecasting, and food storage. The development of efficient, real-time, and economical humidity monitoring methods has attracted intense attention.

Scientists have reported multiple photonic structures and responsive materials for humidity sensors. By embedding a poly(styrene-methyl methacrylate-acrylic acid) (P(St-MMA-AAc)) CPC template into poly(acrylamide) (PAAm) hydrogels, Song and coworkers developed a CPC humidity sensor, which changed its color from transparent to violet, blue, cyan, green, and red when varying relative humidity (RH) from 20% to 50%, 70%, 80%, 90%, and 100%, respectively, spanning the whole visible range (Figure 7.3a). This CPC sensor was stable and completely reversible (Figure 7.3b) [82]. For real-time humidity sensing, Gómez and coworkers fabricated 3D CPC sensors through self-assembly of bare SiO_2 spheres into opaline structure without functionalization (Figure 7.3c). The porosity and hydrophilicity of colloidal SiO_2 spheres enabled rapid water absorption/desorption, which imparted a fast response rate of 60 ms for the CPC sensor. The CPC sensors are highly sensitive to the entire RH range (0.1–88%), especially at low humidity range (<5% RH) (Figure 7.3d) [89]. Li et al. demonstrated a poly(ionic liquid) (PIL) PhC sensor for humidity sensing by using an imidazolium-based inverse opal structure [91]. The PIL exhibited strong affinity for water vapor because of the use of Br^- as counteranions. Therefore, the PIL PhC sensor can be used for humidity sensing. Apart from visible color change, which spanned the whole visible range, the PIL PhC sensor showed rapid response rate (8 seconds for swelling and 20 seconds for shrinking) due to the fast diffusion into the ordered porous structure [91]. Chen and coworkers prepared SCCBs as humidity sensors (Figure 7.3e) [90]. They prepared SCCBs via

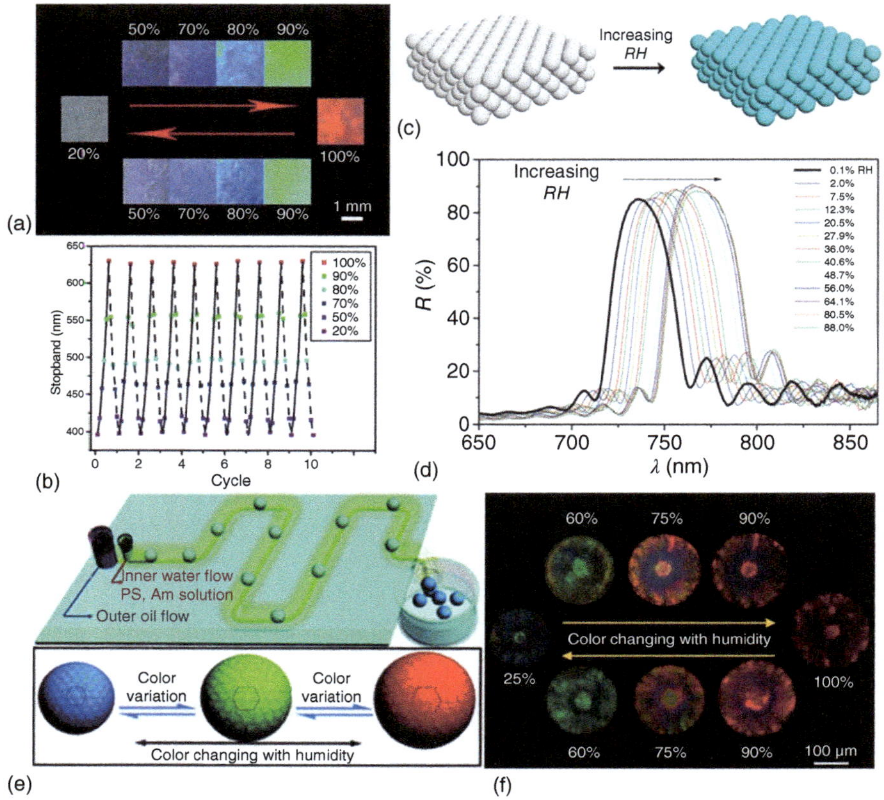

Figure 7.3 (a) Reversible changes of the color/stopband of the PAAm–P(St–MMA–AA) CPC hydrogel with relative humidity. Photographs of the as-prepared CPC hydrogel corresponding to relative humidities of 20%, 50%, 70%, 80%, 90%, and 100%, respectively. (b) Reversible conversion of the stopband position of the CPC hydrogel via alternate exposure to increasing and decreasing relative humidity conditions (between 20% and 100%). Source: Tian et al. [82]/with permission of Royal Society of Chemistry. (c) Scheme of the effect of increasing *RH* in a bare SiO_2 opal. The initially empty micropores of the hydrophilic SiO_2 spheres (light gray) become progressively filled (light blue) by the growing uptake of water from the humid air, leading to instantaneous increase of their refractive index. (d) Redshift of the Bragg peak of a SiO_2 opal upon increasing humidity from 0.1% to 88%. Source: Reprinted with permission from Gallego-Gómez et al. [89]. Copyright 2019, Wiley-VCH Verlag GmbH & Co. KGaA, Weinheim. (e) Protocol of the fabrication of PAAm hydrogel SCCBs via a microfluidic system. (f) Reversible color changes of the PAAm hydrogel SCCBs corresponding to relative humidities of 25%, 60%, 75%, 90%, and 100%. Source: Yin et al. [90]/with permission of Royal Society of Chemistry.

microfluidic-assisted self-assembly of PS colloids and immersed acrylamide (AAm) monomers into the voids of the SCCBs. Then, they polymerized AAm *in situ* to form PAAm hydrogel with the evaporation of the solvent. These SCCBs displayed good sensitivity to humidity as indicated by the reversible distinct color change at different humidities (Figure 7.3f). In addition, the special shape of SCCBs enabled angle-dependent colorimetric detection of humidity [90].

In addition to 3D CPC sensors, 1D CPC sensors also attracted intense attention for humidity sensing due to their simple structure and facile fabrication process. By using 2D $HSbP_2O_8$ nanosheet and TiO_2 nanoparticles, Lotsch et al. fabricated 1D PhCs for humidity sensing. The $HSbP_2O_8/TiO_2$ CPC sensor exhibited superior sensitivity for low humidity monitoring and excellent selectivity toward water vapor. This humidity sensor detected as low as 2000 ppm water in water/alcohol mixture within 23 seconds. The exceptional performance of this humidity sensor can be ascribed to the good diffusion of the analyte within the $HSbP_2O_8$ nanosheet and the dispersive interactions between analyte and $HSbP_2O_8$ nanosheet [92].

7.4.2 Mechanically Responsive Colloidal Photonic Crystals

In recent years, the emerging fields, including human–machine interfaces, human motion detection, and soft robotics, have called for increasing demands for strain sensors [93, 94]. Structural colors of PhCs derived from periodic arrangement of micro/nanostructures provide a straightforward way for stain sensing. Yoshino et al. pioneered the development of mechanically responsive CPCs by self-assembling poly(styrene-butadiene) (PSB) colloids into opals, which redshifted their reflection peaks by increasing the applied uniaxial compression strain [95]. The plastic deformation of the PSB CPC stain sensor, however, led to the hysteresis and reduced reversibility. To solve this issue, Fudouzi and Sawada explored CPC sheets by embedding CPCs into a silicone rubber poly(dimethylsiloxane) (PDMS), which reversibly tuned the diffracted color via elastic deformation (Figure 7.4a) [96]. The PDMS elastomer CPCs sheets were stretched along horizontal direction, causing the size decrease in the vertical direction along with the reduction in lattice constants. Consequently, the reflection peaks blueshifted, which was inversely proportional to the sheet elongation when the deformation was smaller than 20% (Figure 7.4b). Both the diffracted color and reflection peaks were reversely tuned to their original values upon the release of the mechanical strain [96]. Recently, Inci et al. designed a trilayer SiO_2 CPCs that encapsulated within poly(ethylene glycol) phenyl ether acrylate (PEGPEA) elastomers [97]. Upon applying a uniaxial tension to the PEGPEA-SiO_2 CPC film, the reflection peaks blueshifted from 568 to 496 nm with tunable structural colors. Therefore, elastomeric PhCs are very promising for strain sensing.

Current pressure sensors for the diagnosis of blast-induced traumatic brain injury (TBI) are expensive, sophisticated, time-consuming, and operated by power [98, 99]. An ideal sensor for TBI diagnosis should be fast, real-time, cost-effective, lightweight, and power-free for recording the MPa impact pressure. Yang and coworkers fabricated inverse opals PhC sensor by filling the SiO_2 CPCs with uncross-linked thermoplastic SU-8 and subsequently polymerizing SU-8 and removing the CPC templates (Figure 7.4c) [84]. The SU-8 inverse opals CPC sensor was deformed when measuring the impact loading without destroying the structural integrity. As the increase of applied force, the structural color of the CPC sensor blueshifted. The SU-8 inverse opals detected mechanical loading pressure ranging from 17.6 to 20.4 MPa due to the stiffness of SU-8, with the highest

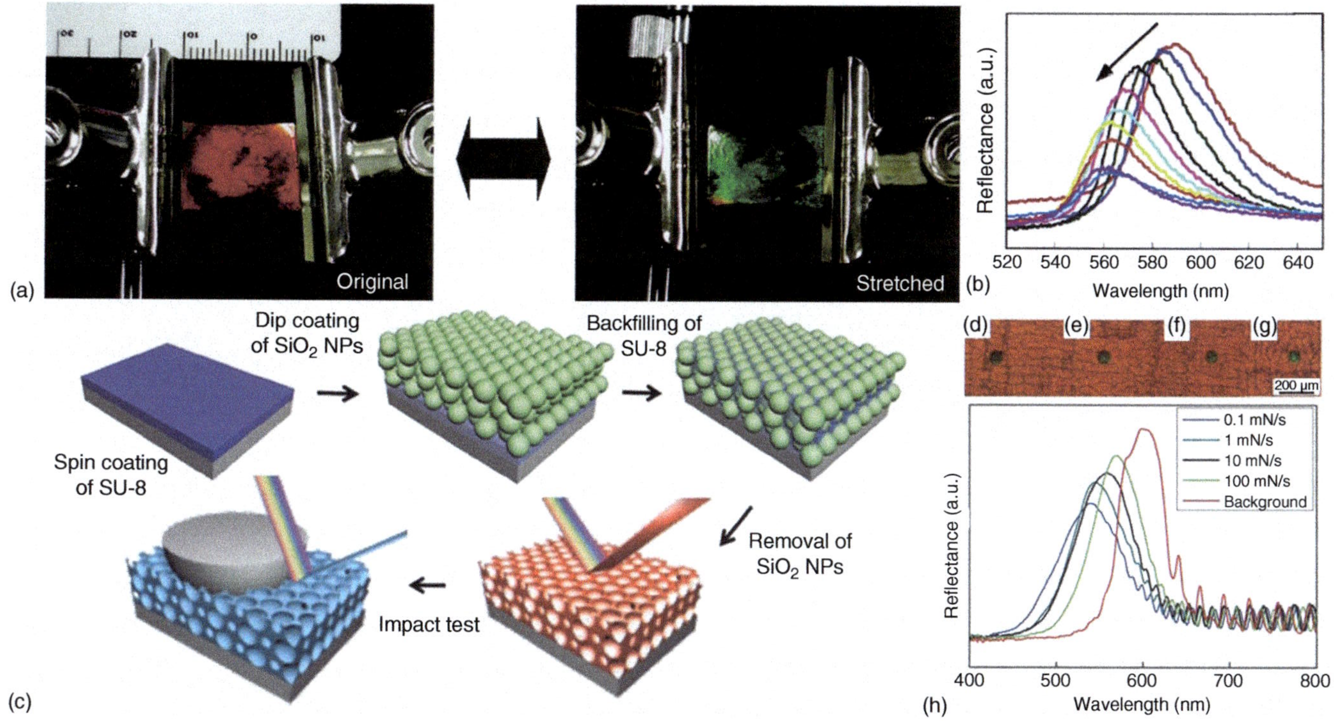

Figure 7.4 (a) Scheme of CPC strain sensor. (b) Relationship between the peak positions and elongation of the silicone rubber sheet owing to stretching. Source: Fudouzi and. Sawada [96]/with permission of American Chemical Society. (c) Scheme of the process for the preparation of an SU-8 inverse opal. (d–g) Optical microscopy images of inverse opal structures indented at a loading rate of (d) 0.1 mN/s, (e) 1 mN/s, (f) 10 mN/s, and (g) 100 mN/s. The load is fixed at 30 mN for all the samples. (h) Reflectance spectra of inverse opal structures after indentation. Source: Reprinted with permission from Cho et al. [84]. Copyright 2015, Wiley-VCH Verlag GmbH & Co. KGaA, Weinheim.

mechanochromic sensitivity ($\Delta\lambda/\Delta\varepsilon = 5.7\,nm/\%$) ever reported (Figure 7.4d–h). This SU-8 inverse opals are promising to be created into wearable patches or devices equipped with helmets and jackets for shockwave assessment [84].

7.5 Colloidal Photonic Crystal Sensors Responsive to Chemicals

7.5.1 Colloidal Photonic Crystals Responsive to Solvents

Most organic solvents are harmful to human health [100]. The monitoring of solvents is of great importance in chemical/biological industry [101], food industry [102], petroleum industry [103], etc. The study of using CPC sensors for the detection of these solvents is growing faster.

Chen et al. prepared physically cross-linked poly(vinyl alcohol) (PVA) hydrogel PhC sensors through embedding CPCs into PVA hydrogel with the aid of ethanol and subsequently modifying the hydrogel with carboxylates [104]. The diffraction maximum of the PVA hydrogel PhC sensor blueshifted from 673 to 466 nm in a few seconds, and the structural color changed from red to green when varying the ethanol volume fraction from 0 to 100 vol% ethanol in water [104]. Cai et al. fabricated SiO_2 inverse opals loading with Au nanocrystals for the detection of alcohols [105]. Both LSPR peaks and Bragg diffraction of the SiO_2 CPCs shifted upon exposure to alcohols with different RIs, which provided more confirmative information for sensing. PDMS can also be used to fabricate 3D PhCs for organic solvents detection. Chiappini et al. developed a simple and cost-effective method to fabricate a PDMS-based PhC sensor by embedding well-ordered CPCs in PDMS matrix [106]. The 3D PDMS-based PhC sensor sensitively detected methanol, terbutyl alcohol, ethylene glycol, and silicon fluid in less than one second.

Due to its responsivity to organic solvents, polymethylmethacrylate (PMMA) has been copolymerized with hydrophilic acrylic acid (AAc) to construct 1D PhCs with TiO_2 for organic vapor detection [107]. However, this 1D P(MMA-AAc)/TiO_2 CPC sensor was not stable in liquid organic solvents. To solve this issue, EGDMA was introduced to achieve P(MMA-AAc-EGDMA) microemulsions. Alternate deposition of P(MMA-AAc-EGDMA) microemulsions and TiO_2 nanoparticles via spin coating yielded 1D P(MMA-AAc-EGDMA)/TiO_2 CPC sensors (Figure 7.5a,b). The variation in the RI contrast between P(MMA-AAc-EGDMA) and TiO_2 upon exposure to organic solvents caused diffracted colors to change (Figure 7.5c). This CPC sensor was applied to detect both pure organic solvents, such as ethanol, ethyl acetate, chloroform, methylene chloride, and carbon tetrachloride, with a rapid response less than two seconds (Figure 7.5d–f). This CPC sensor also detected aqueous solution of organic solvents, like acetone, methanol, and ethanol [108].

MOFs-based 1D PhCs showed several advantages in pure solvents detection, such as excellent sensitivity, fast response time, and good recoverability. Lotsch

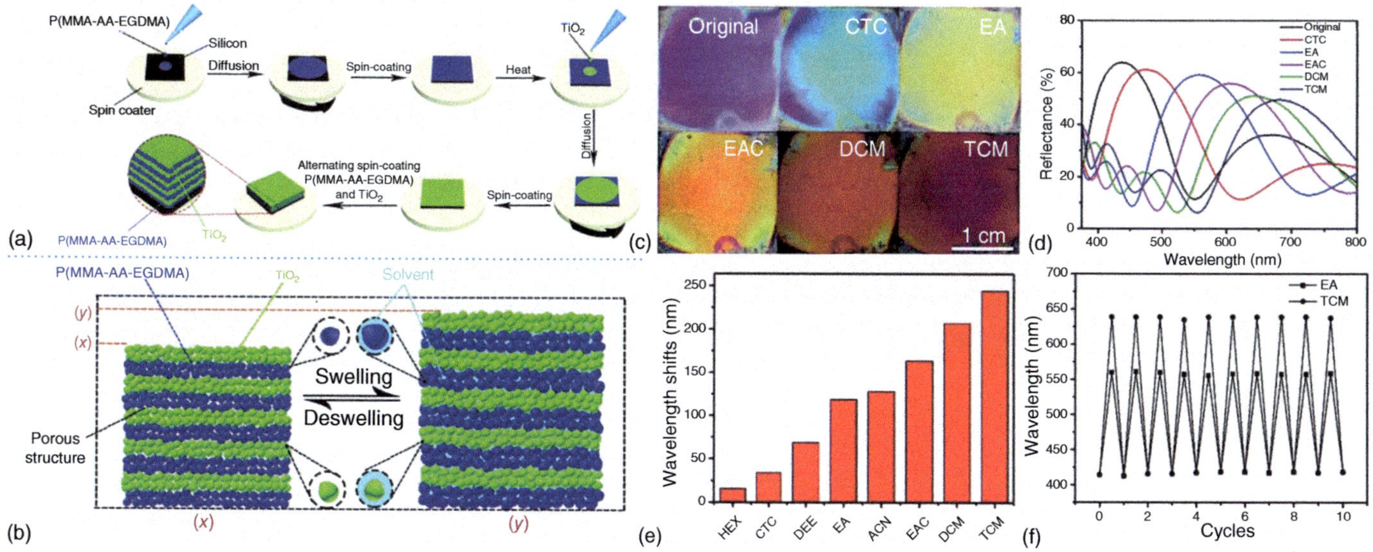

Figure 7.5 (a) Scheme of protocol to fabricate P(MMA–AAc–EGDMA)/TiO_2 1D **CPCs** s with structural colors via the layer-by-layer spin-coating process; (b) the illustration of 1D CPCs swelling and deswelling in an organic solvent: (*x*) original 1D CPCs; (*y*) the 1D CPCs soaked in organic solvent. (c) Photographs of the 1D CPC sensor in different solvents; (d) reflectance spectra of the sensor in various solvents; (e) shifts of the photonic stopband of the sensor in various solvents. (f) Recycling detection test of the 1D CPC sensor in ethanol and chloroform. Source: Kou et al. [108]/with permission of Royal Society of Chemistry.

and coworkers chose three prototypic MOFs, including ZIF-8, HKUST1, and CAU-1-NH_2, to fabricate multilayer Bragg stacks (1D CPCs) for selective detection of chemical solvents. They explored TiO_2/ZIF-8, TiO_2/HKUST1, TiO_2/CAU-1-NH_2, ZIF-8/HKUST1, and ZIF-8/CAU-1-NH_2 CPC sensors through alternate spin coating of respective colloidal nanoparticles. Selective host–guest interactions between target analytes and MOFs were simultaneously encoded in these 1D MOFs-based CPC sensors. These tandem sensors using MOFs with different polarities, e.g. ZIF-8/HKUST1 and ZIF-8/CAU-1-NH_2, along with principal component analysis enabled combinatorial sensing, which significantly enhanced their selectivity for the detection of pure and mixed solvents [70].

7.5.2 Colloidal Photonic Crystals Responsive to Vapors (VOCs)

Volatile organic compounds (VOCs) are hazardous chemicals, and the release of VOCs can cause severe environmental pollution and consequently harm to human health. For example, long-time exposure to BTEX threatens both the environment and public health, resulting in nerve system disease, cancer, leukemia, and liver and kidney damage [109].

Surface-modified 3D macroporous (3DOM) PhCs are frequently used for VOCs detection due to their relatively large surface area [110–112]. Cai and coworkers recently prepared SiO_2 inverse opals for the detection of VOCs [110]. Typically, SiO_2 inverse opals were firstly prepared and then functionalized with different functional groups. Cai et al. infiltrated the SiO_2 inverse opals with poly(4-vinylbenzyl chloride-*co*-methyl methacrylate) (P(VBC-*co*-MMA)), which has the biggest solution parameters with xylene. This P(VBC-*co*-MMA) infiltrated SiO_2 CPCs selectively detected xylene among various VOCs with a LoD of 0.51, 0.41, and 0.17 μg/ml for *o*-xylene, *m*-xylene, and *p*-xylene, respectively [110]. Using the identical strategy, Zhang et al. developed several CPC sensors using SiO_2 inverse opals to detect different types of VOCs, such as diethyl ether/petroleum ether, tetrahydrofuran/acetone, and alcohols [111–113]. For example, they infiltrated the SiO_2 inverse opals with poly(2-hydroxyethyl methacrylate-*co*-methyl acrylate) (P(HEMA-*co*-MA) for the selective detection of alcohols [112].

MOFs have attracted increasing research interests in chemical sensing due to their structural tunability, chemical functionality, high porosity, and large specific surface area, which endowed them with good adsorption capacity, size exclusion, and selectivity on the material [114]. Hupp and coworkers deposited HKUST-1 crystal on SiO_2 CPCs to fabricate MOF-SiO_2 colloidal crystal (MOF-SCC) films [115]. The MOF-SCC sensors redshifted their stopbands in response to vapors, such as water, ethanol, and carbon disulfide (CS_2). The LoDs of the MOF-SCC sensors for water, ethanol and CS_2 were 2.6, 0.3, and 0.5 ppm, respectively. Li and coworkers fabricated MOFs-based inverse opals CPCs using HKUST-1 and ZIF-8 [116]. When exposed to organic vapors, the capture of vapor molecules caused changes in the RI of both 3DOM MOFs CPCs. The HKUST-1 inverse opal CPC sensor quickly responded in 30 seconds and selectively detected ethanol as low as 10 ppm.

Apart from MOFs-based 3D CPCs, 1D MOF CPCs were also used for monitoring vapors. Lotsch and coworkers alternately spin coated microporous ZIF-8 and mesoporous TiO_2 to prepare 1D ZIF-8 PhC chemoselective signal transduce scheme for alcohol vapor monitoring. The combination of different materials into one system imparted size/chemical selectivity and functionality to the ZIF-8 CPC sensor and enabled the discrimination of different alcohols vapor [117]. In the ZIF-8 CPC sensor, the adsorption of ethanol vapor caused the wavelength of 1D CPC to redshift ~100 nm, which was much larger than that measured from a 3D HKUST-1 CPC sensor (9 nm) [116]. In another study, ZIF-8 MOFs were coated on the surface of monolayer CPCs to fabricate ultrathin 2D MOF CPC sensor for vapor sensing (Figure 7.6a–c) [118]. The 2D CPC sensor selectively detected between alcohols, water, and acetonitrile in a broad range of concentrations with excellent sensitivity and a fast response time less than five seconds (Figure 7.6d,e). The high performance of the 2D CPC sensor can be attributed to the improved signal transduction and rapid mass transportation derived from the special optical design of the sensor. The microporous 2D ZIF-8 CPC film had a thickness lower than 100 nm, which allowed fast diffusion of analytes and high accessible pore volumes [118]. Though these MOFs-based CPC sensors showed good sensitivity in discriminating different vapor analytes, their selectivity remains a major challenge.

7.5.3 Colloidal Photonic Crystals Responsive to Ions

Heavy metal ions and other toxic ions can cause harmful effects on both environment and human health [119]. Many CPC sensing materials have been developed for the detection of metal ions, such as Pb^{2+}, Cu^{2+}, and Hg^{2+} [120].

By polymerizing CCAs within responsive hydrogels, the Asher research group developed a variety of PCCA sensors for the detection of Pb^{2+} in aqueous solution [50, 121–123]. Among these PCCA sensors, different crown ethers were chosen as recognition agents. For example, 4-acryloylaminobenzo-18-crown-6 (AAB18C6) was incorporated into PAAm hydrogels and P(NIPAM-AAm) hydrogels, respectively, for the detection of Pb^{2+} in solution. The binding of Pb^{2+} formed metal ions–crown ether complexes, which created a Donnan potential within the hydrogels and increased the osmotic pressure. This led to the swelling of the hydrogels and redshift of Bragg diffraction wavelength, which was proportional to Pb^{2+} concentrations ranging from 0.1 μM to10 mM [50] and 0.1 μM to 20 mM [121], respectively. However, the crown ether-functionalized PCCA sensors were not suitable for the detection of Pb^{2+} in high ionic strength solutions or in body fluids owing to the swamp in the Donnan potential. To address this issue, Asher and coworkers explored a method by measuring the transient diffraction shift of PCCA sensors [122, 123]. Using the AAB18C6-functionalized PCCA sensors, Pb^{2+} in fetal bovine serum with a concentration less than 500 nmol/l was determined with the transient diffraction method [122]. With the same method, they integrated the PCCA sensors to explore an optrode device for real-time Pb^{2+}sensing in high ionic strength solutions (100 mM NaCl) [123].

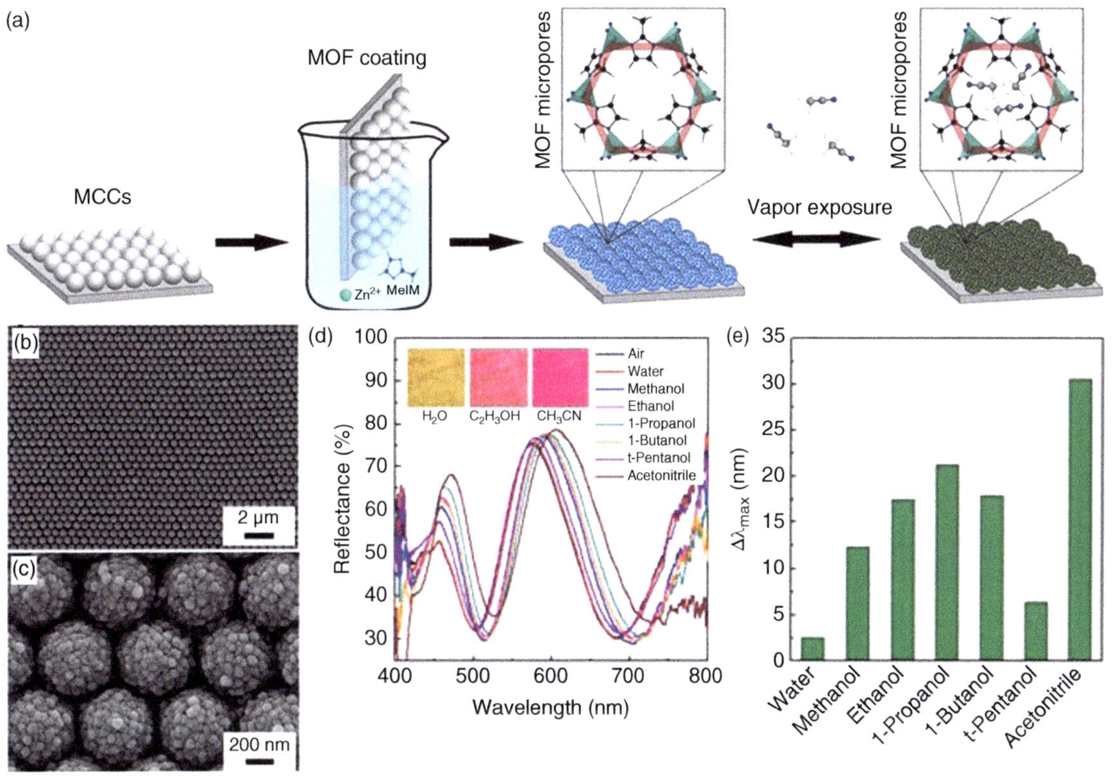

Figure 7.6 (a) Scheme of the fabrication of ultrathin MOF-coated monolayer colloidal crystals (MCCs) for optical sensing; (b) morphologies of ultrathin MOF-coated MCCs at low magnification; (c) morphologies of ultrathin MOF-coated MCCs at high magnification; (d) reflectance spectra of the sample prepared with four cycles of ZIF-8 growth on the silicon wafer; insets are optical micrographs upon exposure to water, ethanol, and acetonitrile vapor. (e) Corresponding shifts of the reflectance peak (originally at 575 nm). Source: Li et al. [118]/with permission of American Chemical Society.

Apart from crown ethers, biomolecules like aptamers have also been used as the recognition reagents for sensitive and selective detection of heavy metal ions owing to their excellent capability of selectively and sensitively binding to their specific target molecules or ions. In addition, they are versatile, reproducible, and easy for production, storage, and modification [124, 125]. Ye et al. developed an aptamer CPC sensor to selectively detect Hg^{2+} and Pb^{2+} ions [126]. PAAm-based colloidal photonic crystal hydrogels (CPCHs) were first prepared and were hydrolyzed with NaOH and TEMED and followed by activation with EDC/NHS. In the end, Hg^{2+} aptamer (5′-NH_2–$(CH_2)_6$–TTCTTTCTTCCCCTTGTTTGTT–$(CH_2)_6$–NH_2-3′) and the Pb^{2+} aptamer (5′-NH_2–$(CH_2)_6$–GGTTGGTGTGGTTGG–$(CH_2)_6$–NH_2-3′) were selected to react with the activated CPCHs to prepare aptamer-functionalized CPCHs. The binding of metal ions to aptamers formed metal ion-aptamer complex, which caused conformational change of aptamers. The conformation changes of aptamers shrank the hydrogels, which blueshifted the Bragg diffraction of the CPCs. The aptamer-functionalized CPCHs showed a wide detection range, which was 10 nM to 0.1 mM for Hg^{2+} and 1 nM to 1 mM for Pb^{2+}. In addition, dried aptamer-functionalized CPCHs can be regenerated via rehydration [126]. Ye and coworkers explored responsive SCCBs to selectively detect Hg^{2+} and Ag^+ [127]. They fabricated agarose SCCBs hydrogels, which were functionalized with ssDNA probes. The ssDNA-functionalized SCCB hydrogel sensors detected Hg^{2+} and Ag^+ with high selectivity and sensitivity. This sensor showed a LoD of 1 nM for Hg^{2+} [127].

Li and coworkers developed inverse opaline SCCBs using PILs for label-free discrimination of different anions (Figure 7.7a) [72]. The SCCBs were achieved via microfluidic-assisted self-assembly of SiO_2 colloids followed by the infiltration of ILs monomers. After UV polymerization of the ILs monomers followed by etching SiO_2 with hydrogen fluoride (HF), PILs-based inverted structure of SCCBs was obtained. The PILs-based SCCBs showed excellent sensitivity in detecting anions due to the counterion exchange reaction with PILs, which caused rapid shrinkage of the SCCBs along with blueshift of the stopband of the SCCBs. Moreover, the PILs-based SCCBs were also used to construct a cross-reactive sensor array for colorimetric detection of the mixture of different anions (ClO_4^-, NO_3^-, BF_4^-, PF_6^-, and Tf_2N^-) and organic solvents (Figure 7.7b,c). This label-free sensing method is mainly based on the anion exchange of PILs, providing a straightforward way to construct rapid and efficient multifunctional sensor arrays [72].

Poor mechanical properties of CPC hydrogel sensors hindered their practical applications for sensing. To enhance their mechanical properties, Liu et al. fabricated CPC sensors with the combination of UV-curable polyurethane (PPU) (Figure 7.7d). They prepared well-ordered CPCs via self-assembly of polystyrene-*co*-poly(*N*-isopropylacrylamide) (PS-*co*-PNIPAM) and then infiltrated PPU precursor into the voids of the CPCs followed by UV polymerization. The resulting CPC sensor film was robust with a tensile strength between 0.1 and 0.43 MPa due to its homogeneous structure. The Bragg diffraction peaks of PPU@PS-*co*-PNIPAM CPC sensor gradually blueshifted with the increasing SCN^-

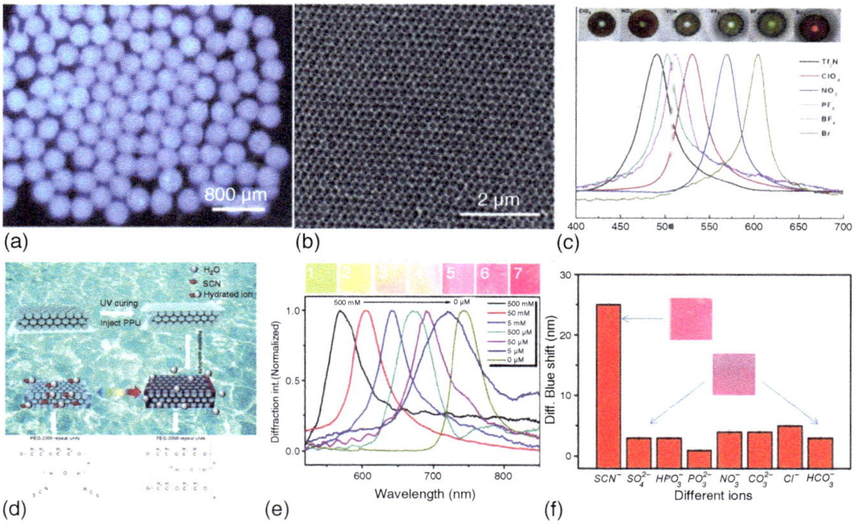

Figure 7.7 (a) Optical image of the prepared PILs-based SCCBs; (b) SEM image of the PILs-based SCCBs; (c) optical response of the prepared PILs-based SCCBs upon exposure to aqueous solution of six different counterions. Inset: optical images of the PIL-IOMS after the exposure to different counterions. Source: Cui et al. [72]/with permission of John Wiley & Sons. (d) A schematic illustration showing the preparation of PPU@PS-*co*-PNIPAM materials and the principle of swelling and deswelling in SCN^- solutions. (e) The diffraction wavelengths of PPU@PS-*co*-PNIPAM films in SCN^- solutions, with the concentrations as indicated from 0 μM to 500 mM. (1–7: the concentrations of SCN^- were 500 mM, 50 mM, 5 mM, 500 μM, 50 μM, 5 μM, and 0 M, respectively.) (f) The specificity of the PPU@PS-*co*-PNIPAM hydrogel response to SCN^- (the concentration of each ion was 8.7 μM). Source: Liu et al. [128]/with permission of Royal Society of Chemistry.

concentrations in solution, and the color changed from pink to green (Figure 7.7e). The CPC sensor selectively detected SCN^- over many anions in water through deswelling of the hydrogels with a LoD of 5 μM (Figure 7.7f) [128].

The development of chemical sensors for the multiple metal ions detection has also attracted intense research interest. Hong et al. fabricated polyelectrolyte inverse opal hydrogels for colorimetric monitoring of anions, cations, and zwitterions, respectively [129]. The polyelectrolyte PhC sensors were robust, inexpensive, and suitable for multiple ions detection with less interference when varying pH and ionic strength. The reduced interference from pH and ionic strength mainly can be attributed to stronger ion–ion interactions between polyelectrolyte and analytes than those noncovalent interactions [129].

7.5.4 Colloidal Photonic Crystals Responsive to Organophosphates

Organophosphorus (OP) compounds, such as parathion, paraoxon, dipterex, and dimethoate, have been widely used in agriculture as pesticides and insecticides worldwide [130]. OP compounds are very toxic and can inhibit the function of human nervous system, thus causing paralysis or death [131]. Asher and coworker explored a PhC hydrogel sensor to detect parathion by using acetylcholinesterase (AChE) as the recognition element [132]. The AChE-functionalized CPC hydrogel sensor selectively detected parathion since the parathion inhibited the AChE with concomitant formation of anionic species. The Donnan potential generated within the hydrogels caused the hydrogels to swell, which redshifted the Bragg reflection peaks. The AChE-functionalized hydrogel CPC sensor easily detected 4.26 fM parathion at low ionic strength. However, it was not reusable, and would cause false negatives because AChE was irreversibly inhibited by parathion and several other OP compounds [132].

To solve this issue, Asher et al. developed a reversible CPC sensor for the selective detection of methyl paraoxon by using organophosphorus hydrolase (OPH) as recognition agent (Figure 7.8a) [133]. They fabricated OPH-immobilized poly(2-hydroxyethylacrylate) (PHEMA) hydrogels. At basic pH, OPH catalyzed the hydrolysis of methyl paraoxon, giving rise to dimethylphosphate, *p*-nitrophenolate, and two protons, which decreased the local pH and formed a pH gradient. The free energy of mixing of the OPH-modified PHEMA hydrogel became less favorable due to the attachment of the protons to the phenolates. The hydrogel shrank and blueshifted the Bragg reflection (Figure 7.8b). The OPH-modified PHEMA hydrogel PhC sensor selectively detected methyl paraoxon with a LoD of 0.2 μM [133].

Meng and coworkers also developed an AChE-functionalized 2D CPC sensor for the detection of dipterex [134]. They first prepared a P(AAm-AAc) hydrogels on top of the 2D PhCs, which were then modified with AChE as recognition reagent (Figure 7.8c). The AChE-functionalized 2D PhC sensor selectively detected dipterex concentration ranging from 10^{-14} mol/l to 10^{-4} mol/l with the colors varying from yellow to blue (Figure 7.8d,e). The LoD of the sensor is estimated to be 7.7×10^{-15} mol/l. The use of CPC sensors is proven to be a promising approach for fast and onsite sensing of organophosphates.

Other OP compounds, such as sarin, soman, cyclosarin, tabun, and VX, can be used as chemical warfare agents due to their neurotoxicity. Meng and coworkers exploited a method by combining PhCs with molecular imprinted technique to discriminate different OP compounds [135]. They prepared P(HEMA-NIPA) inverse opal PhC hydrogels imprinted with methyl phosphonic acid (MPA), which is the hydrolysate of nerve agents. In the molecular imprinted PhC (MIPhC) sensors, when the water in matrix was replaced by MPA, pore volume changed, resulting in the change of diffraction intensity and the shift in wavelength. The MIPhC sensors monitored the concentrations of sarin, soman, VX, and R-VX with LoDs of 3.5×10^{-6}, 2.5×10^{-5}, 7.5×10^{-5}, and 7.5×10^{-5} M, respectively [135]. Meng and coworkers also utilized butyrylcholinesterase (BuChE) to functionalize PAAm hydrogels-based CPCs to monitor sarin [136]. The binding of sarin to BuChE resulted in the formation of HF, which decreased pH and increased ionic strength. This caused the blueshift of reflection peak wavelength. The BuChE-functionalized CPC sensor showed high sensitivity to sarin with a LoD of 10^{-15} mol/l.

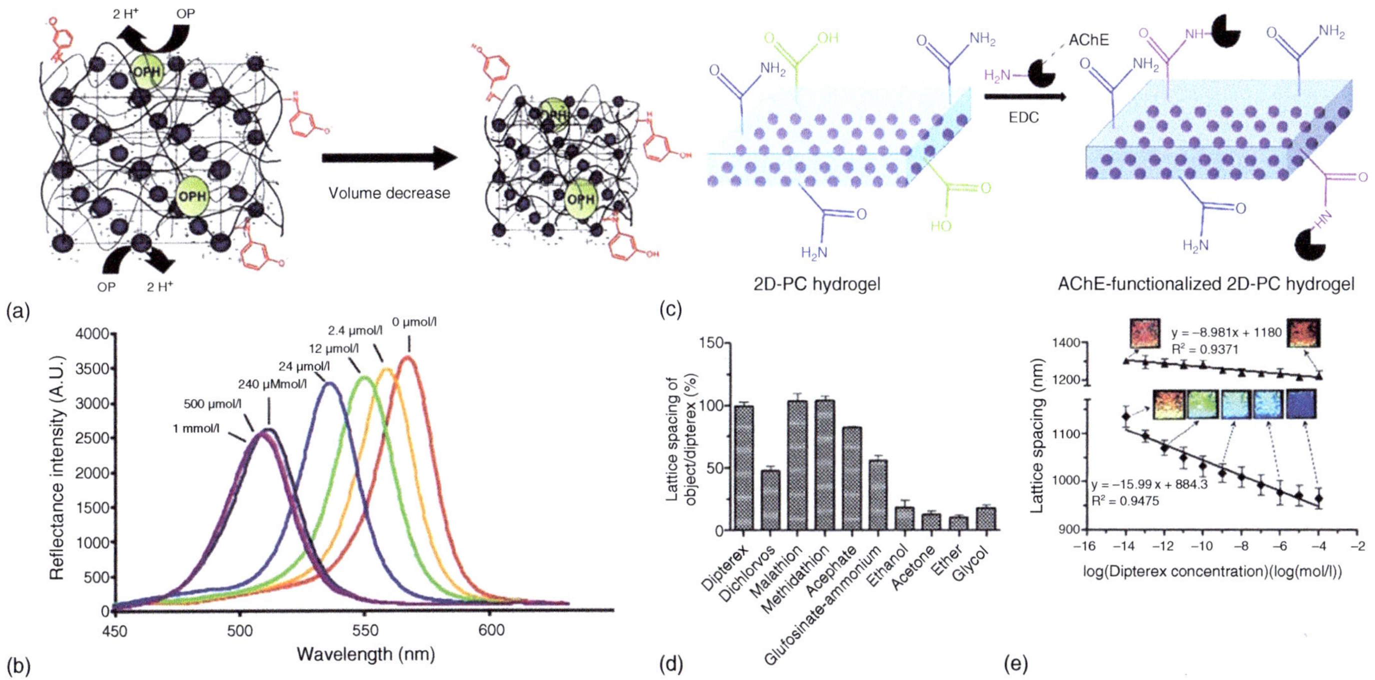

Figure 7.8 (a) Scheme of the working protocol of the organophosphorus hydrolase-modified PCCA sensor for reversible methyl paraoxon detection. (b) Representative diffraction spectra of the PCCA at several methyl paraoxon concentrations. Source: Reprinted with permission from Walker et al. [133]. Copyright 2007, Springer. (c) Scheme of the binding mechanism of AChE and the 2D CPC hydrogel. (d) The selectivity of AChE-functionalized 2D CPC to organophosphates and common organic compounds with the concentration of 10^{-4} mol/L. (e) AChE-functionalized 2D CPC (◆) and unfunctionalized 2D CPC (▲x025B2;) for dipterex detection. The photographs were taken with the camera along the normal while having the light source below at an angle of 45° to the normal of the 2D CPC sensor. Source: Qi et al. [134]/with permission of Royal Society of Chemistry.

7.5.5 Colloidal Photonic Crystals Responsive to Surfactants

Surfactants are widely used as emulsifiers, detergents, and dispersants. Surfactants containing wastewater discharged to the environment can cause damage and risks to the ecosystem [137, 138]. Zhang et al. reported a poly(*N*-isopropylacrylamide) (PNIPAM) hydrogel PhC sensor for the discrimination of surfactants [139]. They embedded 2D CPCs into a PNIPAM hydrogel, which responded to different types of surfactants, such as sodium dodecyl sulfate (SDS), sodium dodecylbenzene sulfonate (SDBS), and cetyl trimethylammonium bromide (CTAB). Charged groups on these surfactants were attached to the hydrogel due to the binding of charged surfactants to PNIPAM. The bound charges on the hydrogel and their counterions caused a Donnan potential, which increased the osmotic pressure within the hydrogel. This led to the swelling of the hydrogel and the increase of interparticle spacing. Thus, a redshift in the Bragg diffraction occurred, and vivid color changes were observed with increasing SDS concentrations.

Cai et al. developed a whole protein hydrogel PhC sensor using bovine serum albumin (BSA) to detect charged surfactants that bind to BSA monomers [51]. In this study, the lysin group of a protein molecule was moderately cross-linked with glutaraldehyde to form intermolecular network, which yielded a transparent BSA hydrogel. The charged surfactants such as SDS and dodecyltrimethylammonium bromide (DTAB) bound to BSA monomer, causing the number of protein charges to change. This electrostatic interaction caused the VPT of the BSA protein hydrogel. The microscopic change of a protein monomer resulted in the macroscopic change of a BSA hydrogel, enabling the detection of charged surfactants. Owing to different charges on the head of surfactants, the BSA protein hydrogel sensor showed different responses to SDS and DTAB. This study provided a new approach to fabricating pure protein hydrogel, which can enable a broad range of novel applications of protein hydrogel sensors.

Yang and coworkers fabricated curved macroporous CPCs using doctor blade coating method for the detection of surfactants [140]. The CPCs were modified to show different hydrophobicities at different angular positions. When exposed to surfactants, such as SDS and CTAB, the surfactant solutions infiltrated into the macropores of the CPCs, causing the changes in RI and diffraction color at different angular positions with different hydrophobicities. However, this CPC sensor is less sensitive and selective than the two hydrogel CPC sensors [139].

7.6 Colloidal Photonic Crystal Sensors Responsive to Biological Species

Proteins (enzymes and antibodies), carbohydrates, DNA, peptides, amino acids, and many other biomolecules are of paramount significance in many biological processes and human health [141–143].The detection of biomolecules for point-of-care (POC) testing is beneficial to health management [57, 142, 143]. In recent years, intense efforts have been devoted to the development of sensors to detect those biological analytes and PhCs as emerging biosensors have shown promise for many POC applications [57, 144].

7.6.1 Colloidal Photonic Crystals Responsive to Carbohydrates

Carbohydrates play significant roles in energy production, sparing protein, building macromolecules, energy storage, and assisting in lipid metabolism. Owing to the continuously increasing number of patients diagnosed with diabetes, there has been an ever-increasing demand for glucose monitoring [145].

The Asher research group has devoted intense efforts to the exploration of glucose sensors for many years. They developed several different CPCs-based glucose sensors to detect glucose in solution, blood, and tears [50, 121, 146–150]. Asher and coworkers pioneered the proof-of-concept PCCA glucose sensor, which is also the first sensor using CPCs as the transduction scheme [50]. The PCCA glucose sensor utilized glucose oxidase (GOx) as the recognition element, which was incorporated into the PAAm hydrogel [50]. GOx in the PCCA sensor converted glucose into gluconic acid and reduced flavin ions. The covalent binding of negatively charged flavin ions to the hydrogel generated a Donnan potential and an osmotic pressure within the hydrogel, which caused the swelling of the hydrogel accompanied with redshift of the Bragg diffraction peaks. However, it remains a challenge to use this Gox-functionalized hydrogel PCCA sensor for continuous and noninvasive monitoring of glucose due to its poor stability and the consumption of substrate by the catabolic enzymes [151].

Afterward, a class of PCCA sensor motifs via attaching phenylboronic acid (PBA) as the recognition moieties were developed for continuous detection of glucose [146–148]. Asher et al. developed a PCCA sensor to detect glucose at low ionic strength [148]. The sensing mechanism is illustrated in Figure 7.9a. 3-Aminophenylboronic acid was attached to the backbone of PAAm hydrogel as the recognition molecule [148]. The diffraction of the PCCA sensor redshifted with increasing glucose concentration in 2 mM Tris–HCl buffer at pH 8.5 (Figure 7.9b). The obtained PCCA hydrogel sensor selectively and reversibly detected glucose <50 μM concentrations in low ionic strength. To detect glucose in high ionic strength, they explored PCCA sensors through attaching pendent PBA groups to a polyacrylamide-poly(ethylene glycol) (PAAm-PEG) hydrogel or a polyacrylamide-15-crown-5 hydrogel [128]. This PCCA hydrogel sensor involved in the formation of a *bis*-bidentate glucose-boronic acid complex, which was stabilized by PEG or crown ether-capped sodium cations. The *bis*-bidentate complex formed cross-links within the hydrogel, which induced the shrinkage of the hydrogel and subsequent blueshift of the PhC Bragg diffraction (Figure 7.9c). This pendent PBA-functionalized PAAm-PEG (AFBA-AAm-PEG) hydrogel PCCA sensor selectively detected glucose over galactose, mannose, and fructose, which was also optimized to selectively detect glucose in tear fluids. The sensed glucose concentration ranged from 0 to 100 μmol/l, and the LoD of the CPC sensor was 1 μmol/l for synthetic tear fluid (Figure 7.9d) [152]. Moreover, the diffracted color spanned from red to blue over the physiologically relevant glucose concentrations

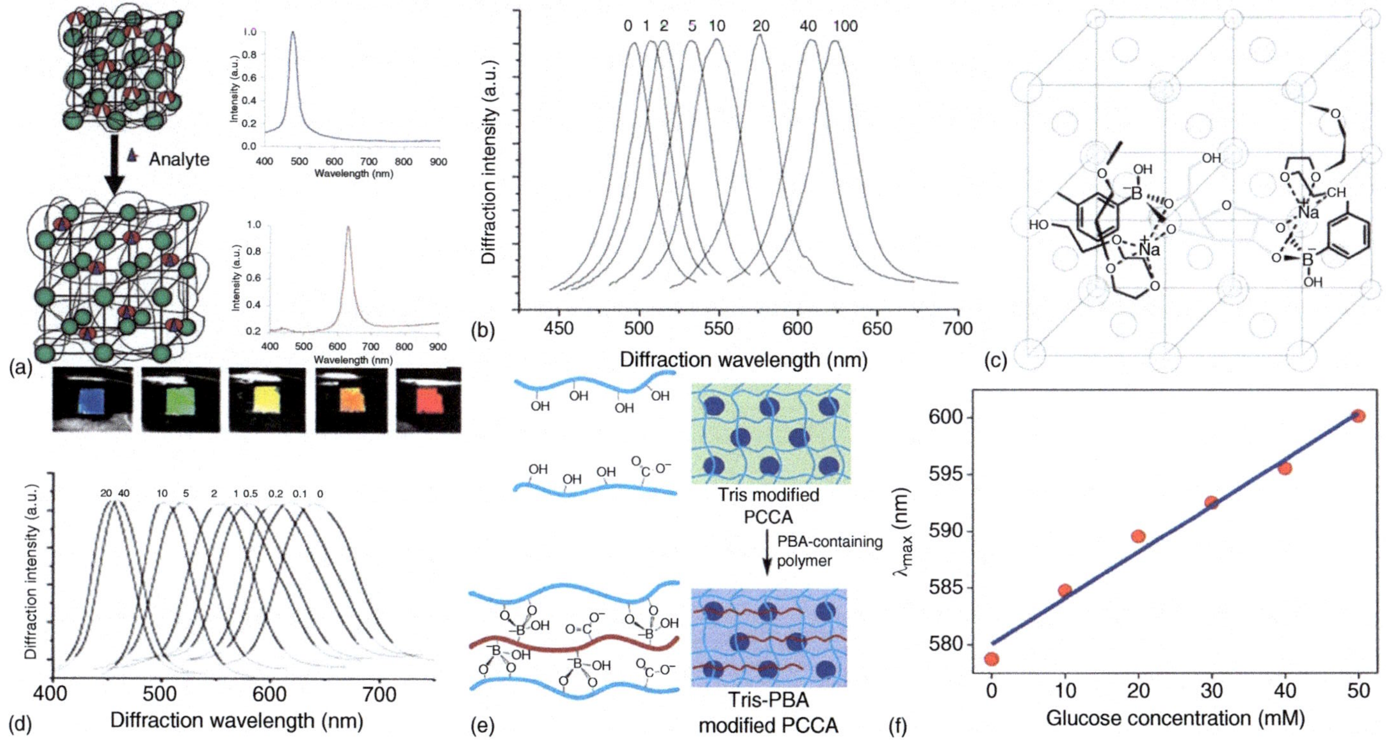
Analyte
Intensity (a.u.)
Wavelength (nm)
(a)
Diffraction intensity (a.u.)
0 1 2 5 10 20 40 100
Diffraction wavelength (nm)
(b)
(c)
Diffraction intensity (a.u.)
20 40 10 5 2 1 0.5 0.2 0.1 0
Diffraction wavelength (nm)
(d)
Tris modified PCCA
PBA-containing polymer
Tris-PBA modified PCCA
(e)
λmax (nm)
Glucose concentration (mM)
(f)

Figure 7.9 (a) Illustration of polymerized crystalline colloidal array sensing materials consisting of an embedded crystalline colloidal array surrounded by a polymer hydrogel network functionalized with a recognition element. As shown by spectra on the right, diffracted wavelength redshifts result from hydrogel volume increases induced by the interaction of the analyte with the recognition element. (b) Glucose concentration dependence of diffraction of the boronic acid PCCA sensor. The diffraction peaks are labeled with their glucose concentrations (mM). Source: Aher et al. [148]/with permission of Oxford University Press. (c) *Bis*-bidentate complex formation between glucose (in furanose form) and two boronates stabilized by PEG-Na^+ complex. (d) Dependence of diffraction of the AFBA-AAm-PEG PCCA sensor on glucose concentration. (Top) Diffraction color changes from red to blue with increasing glucose concentration. The spectra show dependence of diffraction wavelength on glucose concentration. The diffraction peaks are labeled with the corresponding glucose concentration (mmol/l). Source: Reprinted with permission from Alexeev et al. [152]. Copyright 2004, Oxford University Press. (e) Sensor material based on an inverse design. Tris-modified polyacrylamide PCCA is associated with a PBA-containing polymer volume resetting agent. (f) Diffracted wavelength as a function of glucose concentration of the sensor material depicted in (e). The line is a linear fit. λ_{max} is the wavelength corresponding to the maximum of the first-order diffraction peak. Source: Reprinted with permission from Zhang et al. [153]. Copyright 2014, Wiley-VCH Verlag GmbH & Co. KgaA, Weinheim.

in tear fluid. This sensor was developed into ocular inserts and diagnostic contract lenses for glucose sensing with diabetic patients [152].

The Braun research group also developed glucose sensors by polymerizing PhCs within PBA-functionalized hydrogels for the continuous detection of glucose [153, 154]. They investigated the interactions between the PCCA sensor hydrogel with different PBA structures and glucose under physiological conditions [154]. The volumetric response of the hydrogel was found strongly dependent on PBA chemistry. Hydrogels modified with PBAs with boronic acids groups positioned ortho to the AAm linker showed a slow and linear response. The hydrogel matrix showed hysteretic volumetric response (e.g. 2APBA and 2A5FPBA). In contrast, some PCCA sensors demonstrated a nonlinear response but fast response kinetics. Therefore, different types of PBAs-modified PAAm hydrogel sensors significantly impacted the magnitude of response to glucose.

Though the PBA-modified hydrogel PCCA sensors can be used for continuous glucose monitoring, they suffered from a significant nonlinear glucose response because the combination of 1 : 1 and 2 : 1 PBA–glucose complexes leads to a nonlinear correlation between glucose concentration and the hydrogel volume [155]. Braun and coworkers introduced poly(vinyl alcohol) (PVA), which acted as a volume resetting agent, to the PBA-modified hydrogel PCCA sensors to achieve a linear and fast response to glucose binding with minimal hysteresis (Figure 7.9e) [153]. PVA was attached to a preformed PBA-containing PAAm hydrogel, which caused cross-linking of the PBA moieties and shrinkage of the hydrogel. During the fabrication process, *tris*(hydroxymethyl)aminomethane (Tris) was also incorporated into the composite hydrogel to form PBA–Tris–PVA composite hydrogel PCCA sensor. Consequently, glucose binding with the PBA–Tris–PVA composite hydrogel mainly formed 1 : 1 PBA–glucose complexes, leading to linear and fast response of the sensor (Figure 7.9f). The swelling of the hydrogel redshifted the Bragg diffraction from the embedded CCAs in the hydrogels.

The fabrication of 2D CPC sensors is much simpler and faster. The analyte concentration can be conveniently determined with 2D CPC sensors by measuring the diameter of Debye diffraction ring. Cai et al. developed a 2D PhC glucose/galactose binding protein (GGBP) hydrogel sensor for the detection of glucose [53]. GGBP is a glucose-specific protein, in which the C- and N-domains of the protein rotate toward each other upon the binding of a glucose molecule into the binding pocket. The binding of glucose led to the conformational change of GGBP, which in turn shrank the GGBP hydrogel and blueshifted the diffraction from the 2D CPCs. The 2D CPC GGBP hydrogel sensor selectively detected glucose among various carbohydrates, such as galactose, lactose, glucose, and fructose, with a LoD of 1.5×10^{-7} M [53].

Intense efforts have been devoted to improve the performance of CPC glucose sensors, e.g. increasing its response rate by using microgel colloidal spheres [156] or ultrathin hydrogel films [157]. Zhang et al. prepared ultrathin glucose sensing films with thickness on sub-micrometer or micrometer scale via layer-by-layer assembly. The response time of a 30-bilayer film to 2-mM glucose was 0.47 minutes at 25 °C and 0.27 minutes at 37 °C [157]. Wu and coworkers also designed a PCCA glucose sensor by embedding poly(styrene-*co*-acrylamide-*co*-3-acrylamidophenylboronic acid) microgels colloidal spheres into poly(acrylamide-*co*-2-(dimethylamino)ethyl acrylate) hydrogel matrix, where the kinetics of the sensors depended on the VPT of the microgels. Glucose binding swelled the microgels, which reduced the intensity of reflected peaks. This design decreased the response time to 22.1 ± 0.2 seconds at a 7.5 mg/dl glucose concentration [158].

7.6.2 Colloidal Photonic Crystals Responsive to Proteins (Enzyme, Antibodies)

Proteins and carbohydrate are major building blocks of life and play significant roles in many biological processes, such as cell adhesion, proliferation, cell recognition, and in the etiology of a plenty of diseases [159–161]. CPC sensors are powerful in the detection of proteins. Avidin-biotin system was usually selected as model molecules in protein detection since this system possesses ultrahigh binding affinity (Ka = 10^{15} M^{-1}) and excellent specifity [162]. Qi and coworkers developed 2D ordered macroporous (2DOM) Ag [163] and ZnS [164] CPC films, for the highly sensitive detection of avidin, respectively. The 2DOM Ag CPC sensor could detect avidin with a concentration as low as 100 pM and a linear range from 100 pM to 200 nM [163], while the 2DOM ZnS CPC sensor showed a LoD less than 100 pM with a broad working range [164]. Zhang et al. developed a biotin containing hydrogel PhC sensor to detect avidin [165]. The binding of avidin to biotin within the hydrogel formed cross-links, which significantly shrank the hydrogel and blueshifted the PhCs diffraction. Subsequently the avidin concentration was determined.

Comparing to the detection of avidin, the recognition of protein with lower binding affinity remains a challenge. For example, the detection of lectin proteins is generally difficult due to low lectin–carbohydrate interactions. Cai et al. recently developed a novel strategy to discriminate lectin proteins using the "cluster glycoside effect," [166, 167], which can enhance the protein–carbohydrate interactions [52, 54]. They prepared several CPC hydrogel sensors through copolymerizing AAm, AAc, and vinyl-linked carbohydrate monomer to obtain protein-specific hydrogel sensors (Figure 7.10a). They chose mannose as the recognition molecule for selective detection of Concanavalin A (Con A) while lactose for ricin detection and galactose for jacalin detection. These recognition molecules were attached to the backbone of the copolymer to form responsive hydrogels. This CPC sensor design enabled multivalent protein–carbohydrate interaction, which led to the formation of cross-links within the hydrogel. This caused the hydrogel to shrink and blueshifted Bragg diffraction peaks of the 2D CPCs, which was used to determine the lectin protein concentration. These CPC hydrogel sensors showed specific "multivalent" binding to lectin proteins [52, 54]. The carbohydrate hydrogel sensors exhibited LoDs of 7.5×10^{-8} M for ricin, 3.8×10^{-8} M for Con A, and 2.3×10^{-7} M for jacalin, respectively (Figure 7.10b–d) [54]. The selective detection of Con A is mainly due to the specific binding of mannose toward Con A through van der Waals interactions and hydrogen bonding (Figure 7.10e) [168].

Enzymes are essential biocatalysts. Karr and coworkers demonstrated the development of a 3D PhC hydrogel biosensor for screening kinase activities, namely, enzymatic phosphorylation (Figure 7.11a) [169, 170]. They first fabricated a PAAm hydrogel containing CPCs and subsequently functionalized the hydrogel with a target peptide substrate (LRRASLG) for PKA recognition (Figure 7.11b). The phosphorylation changed the electrostatic environment of the hydrogel, which gave rise to a Donnan potential. The Donnan potential led to the swelling of the hydrogel and, in turn, the increase in the lattice spacing of the PhC inside and redshift of the Bragg diffraction peaks (Figure 7.11c). This work pioneered the use of CPC sensor to screen enzyme activity quantitatively [169]. Later, this type of sensor was extended to monitor kinase activity both in lysate of HEK293 cells and *in vitro* and examined the effect of hydrogel properties on the sensitivity of the PhC biosensor [170]. In this study, an azide–alkyne click chemistry was utilized to attach the target peptide LRRASLG onto the hydrogel for the screening of PKA. The sensitivity of the CPC sensor for PKA can be improved with a LoD of 0.1 U/μl comparing with 4 U/μl reported in previous study (Figure 7.11d) [169].

Direct functionalization is the most frequently used method to fabricate CPC hydrogel sensors for the detection of target analytes. However, this method usually requires large responses (change in volume or osmotic pressure or charge) actuated by the binding between analytes and the intelligent hydrogels, which significantly limits the types of detectable analytes [19, 50]. Qin et al. developed novel competition-based CPC biosensors using antibody–antigen interaction to recognize enzyme activities, such as protein kinase A [171]. The competition-based PhC sensors demonstrated a broader detection concentration range despite physical and chemical properties of analytes, higher sensitivity, and excellent selectivity comparing with their direct counterparts due to sole antibody–antigen interaction dependence of the displacement process [171].

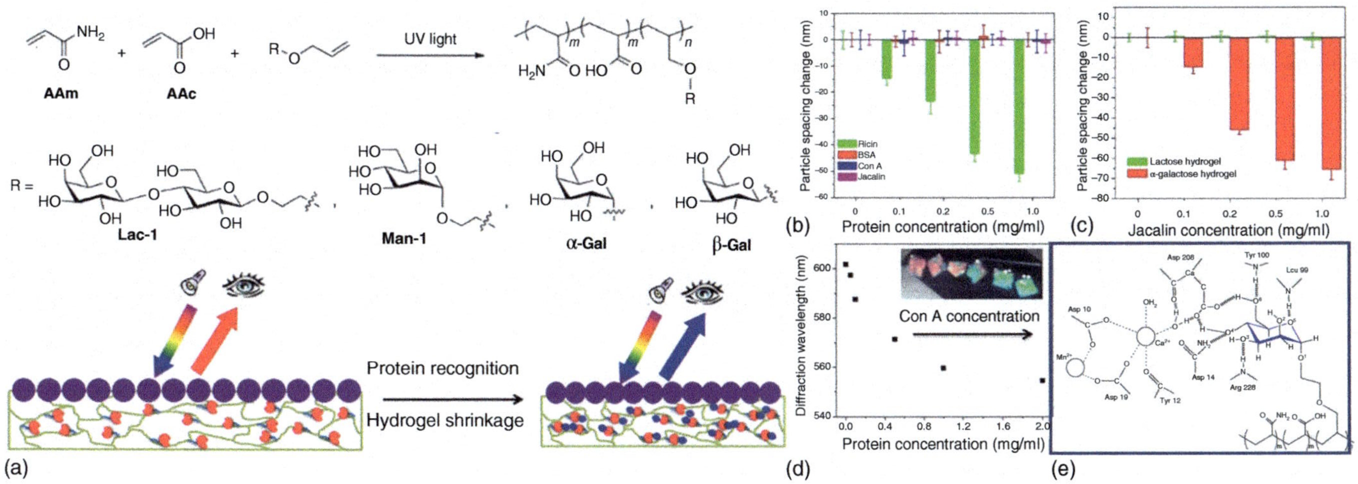

Figure 7.10 (a) Scheme of the fabrication of 2D CPC carbohydrate sensor for lectin proteins detection. (b) Dependence of 2D CPC PAAm-AAc-Lactose hydrogel particle spacing on BSA and the lectin proteins ricin, Con A, and jacalin. (c) 2D CPC PAAm-AAc-α-Galactose sensors for the selective detection of jacalin. Source: Cai et al. [54]/with permission of American Chemical Society. (d) Dependence of normalized diffraction spectra of the 2D mannose hydrogel sensors on Con A concentration. The diffraction is measured in a Littrow configuration with an angle of ~18° between the probe and the 2D CPC array normal. The inset shows photographs taken in the Littrow configuration where the camera lens axis is at an angle of ~18° from the normal. (e) Mechanism of Con A binding to mannose attached to hydrogel. Source: Reprinted with permission from Zhang et al. [52]. Copyright 2017, American Chemical Society.

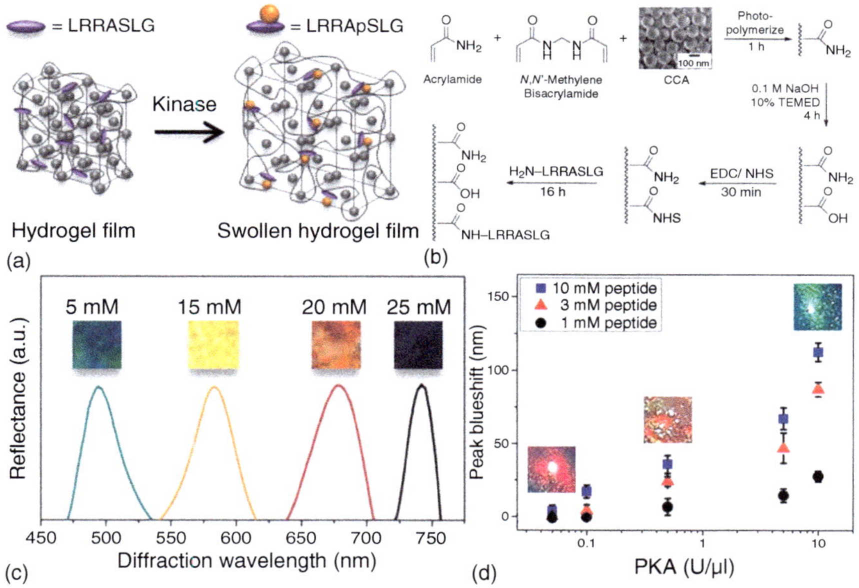

Figure 7.11 (a) Mechanism of the CPC kinase biosensor. (b) Fabrication of a kinase responsive PhC biosensor. (c) Redshift in peak reflectance of hydrogel-encapsulated CCAs with increasing concentration of immobilized negatively charged groups (at pH 5.5). The inset images (~5×5 mm^2) show visual color changes of the hydrogels. The total concentrations of negative charge from 5 to 25 mM were measured by colorimetric staining. Source: MacConaghy et al. [169]/with permission of American Chemical Society.
(d) Biosensor response as a function of peptide and PKA concentrations for hydrogels with a modulus of 0.8 kPa and χ of 0.587. The reported peptide concentrations correspond to the peptide concentration in the reaction mixture during hydrogel functionalization. Inset images (~3×3 mm^2) show the visual color difference between hydrogels reacted with 10 mM peptide and treated with 0.05, 0.5, and 10 U/μl PKA. Source: MacConaghy et al. [170]/with permission of American Chemical Society.

Immunoglobulin G (IgG) is a type of antibody that binds antigen specifically. Inspired by previous work [172], Song and coworkers utilized inkjet printing method to fabricate microfluidic chips for label-free colorimetric detection of IgG. Y-shape patterned channels with different functions and stopbands were integrated into microfluidic systems to fabricate surface-tension-confined microfluidic immune assays. Apart from successful detection of IgG without interference caused by nonspecific adsorption, this Y-shape PhC chip showed several advantages, such as elimination of contamination, high efficiency, and reliability [173]. PEG-DA hydrogel inverse opals immobilized with protein A were developed for label-free IgG detection. The immobilized protein A bound to IgG with high specificity, leading to visible color change from green to dark orange and peak wavelength shift from 563 nm (control) to 613 nm as the IgG concentration increased from 0 to 10 mg/ml [174].

7.6.3 Colloidal Photonic Crystals Responsive to Amino Acids

Amino acids are basic units of proteins and important biological substances for chemical, food, and pharmaceutical industries [175]. There is ever-increasing importance for the discrimination of amino acids. PhC sensors have been also employed to detect amino acids. Li and coworkers reported the development of an aggregation induced emission (AIE)-doped PIL PhC sensor to discriminate multiple types and concentrations of amino acids in H_2O [71]. In this method, they prepared inverse opal SCCBs using PIL via microfluidic synthesis, in which AIE luminogens were doped to monitor environmental changes. Owing to strong multiple intermolecular interactions between ILs and analytes, the AIE-doped PIL SCCBs differentially interacted with different amino acids, giving rise to diverse response patterns. This sensor element successfully discriminated four multi-analyte systems, including all 20 natural amino acids, 10 metal ions, 9 important phosphate derivatives, and 3 pairs of enantiomers. Moreover, the AIE-doped PIL PhC beads accomplished selective detection of 20 amino acids in human urine and discriminated 26 unprecedented complex mixtures [71].

The combination of MIPs with PhC sensing technique provides an effective method for the detection of amino acids. Wu and coworkers fabricated a molecular imprinted photonic hydrogel (MIPH) sensor to detect L-histidine (L-His) [176]. In this study, they prepared SiO_2 CPCs as a sacrificed template via vertical self-assembly and infilled the interstices of the CPCs with precursor solution containing L-His, MMA, EGDMA, H_2O, and methanol. The precursor solution was polymerized under UV light irradiation. Afterward, both the SiO_2 colloidal particles and L-His were selectively removed, leaving behind PMMA inverse opal PhCs imprinted with L-His. The resulting MIPH selectively and sensitively detected L-His with a LoD of 10 pM and a fast response rate of 60 seconds [176].

The Yu research group integrated MIP with PhC sensing technique to fabricate highly sensitive and selective PhC sensors for amino acids detection [177, 178]. Yu and coworkers developed an MIP-based CPC sensor to discriminate L-pyroglutamic acid (L-Pga). They polymerized the mixture of AAm, methacrylic acid (MAA), and L-Pga within the interstices of PS CPCs followed by removing the CPCs template and the template molecule L-Pga, resulting in an MIP PhC sensor. When exposed to L-Pga, the binding of L-Pga molecules to the imprinted cavities caused the swelling of the CPC hydrogel sensor, along with a redshift in the reflectance peaks of the MIP PhC sensor. This MIP CPC sensor selectively detected L-Pga over L-Thr, L-Phe, and L-Ala in the range of 0.01–0.50 mM in monosodium glutamate samples at pH 4 [177]. In another study, Yu and coworkers utilized a similar method to fabricate β-cyclodextrin (β-CD)-based MIP PhC sensor for selective detection of L-Phe [178]. AAc and maleic anhydride-modified β-CD were mixed with L-Phe and polymerized into the PS CPCs. L-Phe-imprinted CPC hydrogel sensors were achieved after removing both the PS CPCs template and L-Phe molecules. This L-Phe imprinted CPC hydrogel sensor selectively detected L-Phe in the range of 10^{-8}–10^{-4} M at pH 6 within 30 seconds. Apart from stronger response to L-Phe than D-Phe, multiple interactions between hydrophobic cavities of β-CDs and phenyl moiety [179–182] allowed for higher sensitivity [178].

7.6.4 Colloidal Photonic Crystals Responsive to Biomarkers

Biomarkers are indicators of biological and pathogenic processes, which play significant roles in therapy monitoring, medical diagnosis, and health management. The detection of biomarkers is extremely important in order to mitigate and diagnose cancers in the early stage [183]. However, during the detection process, impurities and other analytes can cause background noise, which dramatically limits the sensitivity and accuracy.

Tang and coworkers demonstrated a bioinspired PhC chip to detect cancer biomarker miRNA-21 in physiological conditions [184]. A target-recycling amplification process was constructed on PhCs with polydopamine spheres and DNase I, which improved the selectivity and reduced the amplification time. The combination of enhancement arising from the PhCs and enzymatic cycle amplification led to the highly sensitive detection of miRNA-21 with a LoD of 55 fM, which was four-order magnitudes lower comparing with the chip without amplification. The detection concentration ranged from 1 pM to 10 nM, making this method promising for early diagnosis of cancer [184].

To improve the detection sensitivity of bladder cancer biomarkers, Yuan and coworkers explored a luminescent biochip to detect bladder cancer-related miRNA(Figure 7.12a) [185]. They constructed an optical biochip based on PhCs to detect the hybridization of miRNA-21, in which luminescence signals were enhanced, while the background fluorescence interference was suppressed owing to the combination of PhCs (Figure 7.12b,c). They first functionalized the long lifetime luminescence nanoparticles with ssDNAs that are complementary to miRNA-21, and the ssDNAs was partially hybridized with black-hole-quencher-labeled DNAs (BHQDNAs), leading to the quench of the luminescence via fluorescence resonance energy transfer (FRET). Upon the addition of miRNAs to the nanoparticles, the strong hybridization of miRNA-21 with ssDNAs caused the BHQDNAs detachment from ssDNAs, followed with the restoration of the luminescence. The PhC fluorescent sensor sensitively detected the miRNA-21 concentration by measuring the fluorescent intensity and showed a LoD of 26.3 fM for miRNA-21(Figure 7.12d,e) [185]. In addition, the PhC fluorescent sensor demonstrated good selectivity for miRNA-21 detection (Figure 7.12f,g).

Recently, Zhao and coworkers developed a CPC sensor based on SCCBs and hybridization chain reaction (HCR) for high-throughput and high-sensitivity detection of bladder cancer associated miRNA [186]. They first prepared CPC barcodes labeling with three types of probes specific to three biomarkers of bladder cancer, including miR-143, miR-133a, and miR-200b. Then they combined CPC barcodes with enzyme-free HCR technique to amplify detection signals. This CPC sensor exhibited good sensitivity with a LoD of 10^{-3} μM using 20 μl miRNAs [186].

Fibrinopeptide B (FPB) is a critical biomarker of venous thromboembolism (VTE) [187]. Frasco et al. developed a novel sensor for FPB detection by combining PhCs and molecularly imprinted polymers (MIPs). The poly(methacrylic acid) (PMAA) hydrogels embedding with 3D SiO_2 CPCs were imprinted with FPB and subsequently removed to create specific recognition sites. The FPB imprinted PhC

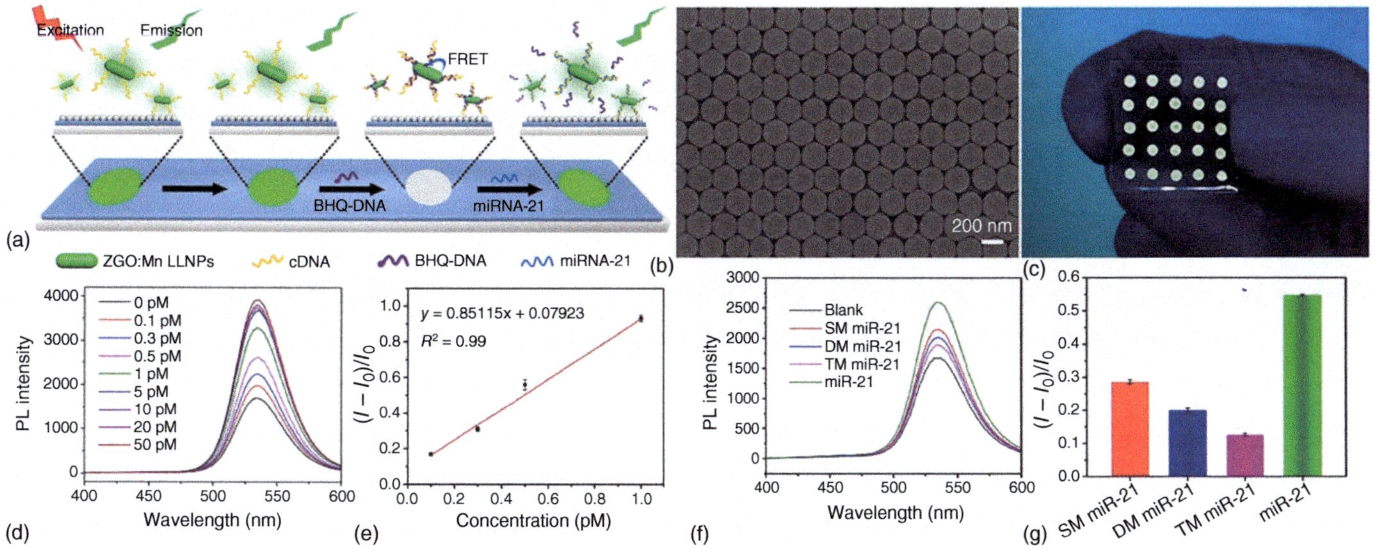

Figure 7.12 (a) Scheme of miR-21 detection by the biochip. (b) SEM image of CPCs substrate. (c) Photograph of the CPCs substrate after exposure to sunlight. (d) Long lifetime luminescence recovery of the sensing biochip in the presence of different concentrations of miR-21 within the range of 0.1–50.0 pM. (e) Linear relationship between the long lifetime luminescence recovery and the concentration of target miR-21 ranging from 0.1 to 1 pM. (f) Long lifetime luminescence spectrum after adding miR-21 and 100-fold concentration of single-mismatched miR-21 (denoted as SM miR-21), double-mismatched miR-21 (denoted as DM miR-21), and triple-mismatched miR-21 (denoted as TM miR-21), respectively. Buffer was used as the control. (g) Histogram of luminescence varieties corresponding to part d. Source: Wang et al. [185]/with permission of American Chemical Society.

sensor selectively and sensitively detected FPB in human urine with a LoD of 0.13 ng/ml. Moreover, this sensor showed a linear range from 0.2 to 22 ng/ml, which allowed the assessment of FPB at clinical relevant level. This method provides an alternative for real-time and noninvasive POC application [188].

7.6.5 Colloidal Photonic Crystal Sensors for Microorganisms Detection

Infectious disease is the leading cause of death worldwide, which is often caused by microorganisms, such as bacteria, fungi, and viruses. Therefore, it is extremely important to detect infectious diseases-related microorganisms. However, the detection of microbes is challenging. This is because conventional methods, including fluorescence method, filtration method, and DNA microarrays, are suffering from low sensitivity, while other highly sensitive approaches, like ELISA and PCR methods, are generally tedious and complex and require well-trained personnel [189–191].

The binding to surface carbohydrates is a pathway through which many pathogens infect host cells [192]. Cai et al. developed a PhC hydrogel sensor for the detection of microbes (Figure 7.13a) [55]. They prepared a Con A protein CPC sensor for the selective and sensitive detection of *Candida albicans*. This Con A hydrogel CPC sensor utilized the multivalent interaction between Con A and mannan on the cell surface of *C. albicans* to discriminate *C. albicans* [167, 194]. The binding of mannan to Con A formed cross-links, which shrank the protein hydrogel along with the blueshift of the Bragg diffraction of PhCs within the sensor. The Con A hydrogel PhC sensor selectively detected *C. albicans* over gram-negative bacterium *E. coli*. Moreover, this PhC sensor quickly detected (<2 hours) *C. albicans* in solutions with a concentration as low as 32 CFU/ml, which was much lower than physiological relevant concentration (Figure 7.13b).

Chen et al. demonstrated the utilization of an aptamer-based hydrogel PhC sensor for the detection of bacteria. They first self-assembled SiO_2 SCCBs as a template. Then spherical PhCs inverse opal poly(ethylene glycol) (PEG) hydrogel was prepared through replicating the SCCBs with a pre-gel solution consisting of PEG-DA and AAc. The PEG inverse opals SCCBs were functionalized with two different types of aptamers (amino-modified $Apt_{S.\ aureus}$ and amino-modified $Apt_{E.\ coli}$), respectively (Figure 7.13c). The PEG scaffold significantly decreased the nonspecific adsorption, while the aptamers specifically identified and captured different kinds of bacteria (Figure 7.13d). The resulting PhC sensors showed good performance in the selection of both *S. aureus* and *E. coli* in solution and could capture and identify bacteria at a low concentration of 100 CFU/ml with a response time of 2.5 hours (Figure 7.13e) [193].

Qu and coworkers demonstrated a 2D PhC hydrogel sensor based on glycated albumin for the detection of gram-negative bacteria [195]. In this study, they developed two different strategies to detect *E. coli*. One way is to directly prepare glycated albumin hydrogel through cross-linking with glutaraldehyde on top of self-assembled CPCs. This PhC hydrogel sensor showed a LoD of 153 CFU/ml for

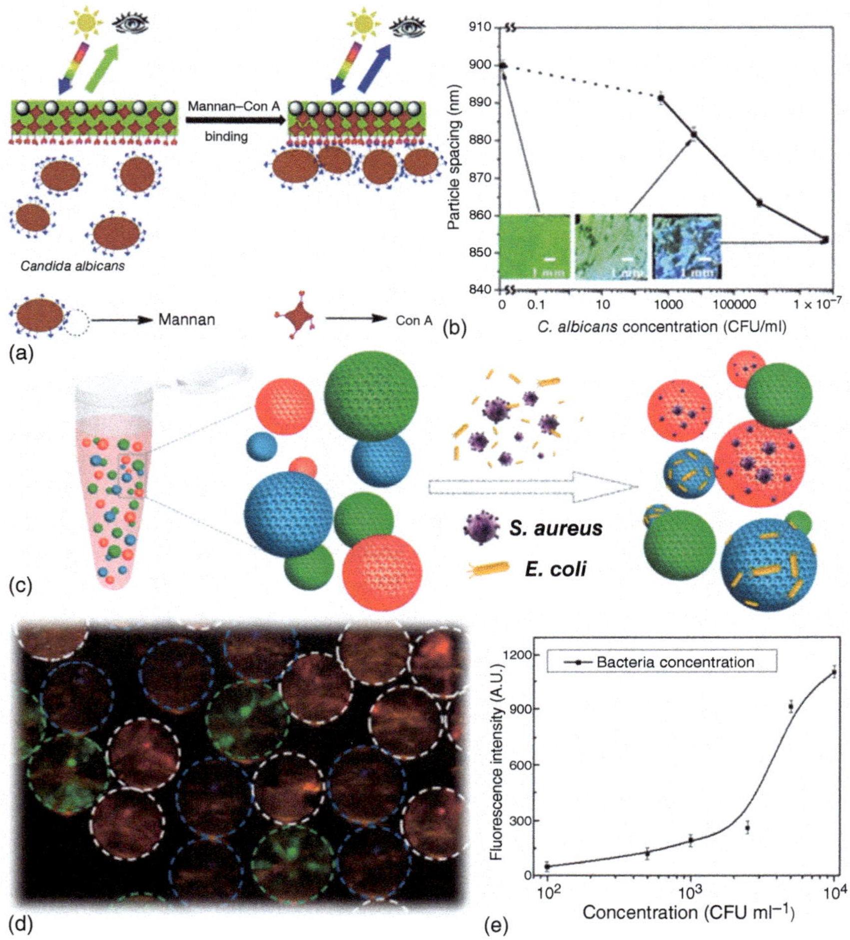

Figure 7.13 (a) Scheme of the binding mannan on the surface of *C. albicans* toward Con A protein hydrogel. (b) The particle spacing and photographs of Con A protein hydrogel at different *C. albicans* concentrations. Source: Cai et al. [55]. (c) Schematic illustration of the aptamer-decorated barcodes for capturing multiple types of bacteria. (d) Optical microscopy image of three kinds of barcodes after *S. aureus* capture. The red, green, and blue barcodes were decorated with $Apt_{S.\ aureus}$, $Apt_{E.\ coli}$, and no probes, respectively. Scale bar is 200 μm. (e) The bacteria fluorescence intensity as a function of the bacteria concentration. Source: Xu et al. [193]/with permission of Elsevier.

the selective detection of *E. coli* in solution. The other method is to use lipopolysaccharides to imprint glycated albumin hydrogel to detect *E. coli*, which displayed an improved sensitivity with a LoD of 58 CFU/ml. Recently, Yang and coworkers also prepared a 1D PhC sensor using alternate SiO_2 and Si layers for monitoring of *E. coli* colony [196].

7.7 Summary and Outlooks

In this book chapter, we summarized the advances made in CPC sensors with an emphasis on their target analytes such as chemical and biological species. CPC sensors have achieved significant progress during the past two decades, with aspect of photonic structures, responsive materials, and analysis strategies. 1D and 2D CPC sensors and their fundamentals were developed besides conventional 3D CPC sensors. Some patterned CPCs and microchips based on CPCs were also exploited for sensing applications. The development of these novel structures facilitated and opened more opportunities for novel PhC sensors. The most extensively investigated responsive sensor materials are based on polymers, such as hydrogels. Some inorganic materials were also used but generally suffered from lacking of selectivity. In recent years, several new responsive materials, such as MOFs and 2D materials (e.g. MoS_2, $HSbP_2O_8$, and graphene oxide), were also applied to CPC sensors, which introduced new functionalities to CPC sensors. The exploration of cross-reactive analysis enabled multi-substrate detection. The combination of novel analysis methods with unique structure of CPC sensors gave rise to PhC sensors with high sensitivity and excellent selectivity for multiple analytes detection.

Though significant advance has been achieved, the development of high-performance CPC sensors for the detection of analytes remains difficult. These challenges cover many aspects of sensors, such as the stability, selectivity, readout schemes, signal changes, and responsive materials. The future development of CPC sensors should focus on tackling these challenges. In addition, more responsive materials toward different analytes need to be explored for highly sensitive sensors. Novel intelligent materials, such as biomolecular organogels, 2D materials can expand the applications of PhC sensors or improve their performance. The novel photonic structures can endow the CPC sensors with unique properties, which in turn can improve the performance of the sensor. Several novel CPC structures such as diamond-structure CPCs [197] have been developed in recent years, which may further enhance the performance of PhC sensors. The development of CPC sensors can also be combined with the recent development of materials processing technology (additive manufacturing) and novel advances in materials, chemistry, or biological science. In the future, high-resolution 3D printing technology is expected to precisely fabricate PhC with demanded geometry for PhC sensors.

Acknowledgment

This work was supported by National Natural Science Foundation of China under award No. 22076008 and BUAA Faculty Research Grant under Grant Nos. ZG216S2094 and ZG216S2153 (Z.C.).

References

1 Joannopoulos, J.D., Johnson, S.G., Winn, J.N., and Meade, R.D. (2008). *Photonic Crystals: Molding the Flow of Light*. Princeton, NJ: Princeton University Press.
2 XVII Rayleigh, L. (1887). *London, Edinburgh Dublin Philos. Mag. J. Sci.* 24: 145–159.
3 Rayleigh, L. (1917). *Proc. R. Soc. London, Ser. A.* 93: 565–577.
4 Alfrey, T., Bradford, E.B., Vanderhoff, J.W., and Oster, G. (1954). *J. Opt. Soc. Am.* 44: 603–609.
5 Krieger, I.M. and O'Neill, F.M. (1968). *J. Am. Chem. Soc.* 90: 3114–3120.
6 Asher, S.A. (1986). Crystalline colloidal narrow band radiation filter. US Patent 46,325,17A, 9 December 1986.
7 Yablonovitch, E. (1987). *Phys. Rev. Lett.* 58: 2059–2062.
8 John, S. (1987). *Phys. Rev. Lett.* 58: 2486–2489.
9 Maigyte, L. and Staliunas, K. (2015). *Appl. Phys. Rev.* 2: 011102.
10 Lidorikis, E., Povinelli, M.L., Johnson, S.G., and Joannopoulos, J.D. (2003). *Phys. Rev. Lett.* 91: 023902.
11 Arsenault, A.C., Puzzo, D.P., Manners, I., and Ozin, G.A. (2007). *Nat. Photon.* 1: 468–472.
12 Tian, Y., Chen, M., Zhang, J. et al. (2018). *Small Methods* 2: 1800104.
13 Cai, Z., Teng, J., Xiong, Z. et al. (2011). *Langmuir* 27: 5157–5164.
14 Cai, Z., Xiong, Z., Lu, X., and Teng, J. (2014). *J. Mater. Chem. A* 2: 545–553.
15 Zhang, Z., Geng, C., Hao, Z. et al. (2016). *Adv. Colloid Interface Sci.* 228: 105–122.
16 Wang, Y., Zhang, M., Lai, Y., and Chi, L. (2018). *Nano Today* 22: 36–61.
17 Xie, K., Guo, M., and Huang, H. (2015). *J. Mater. Chem. C* 3: 10665–10686.
18 Cai, Z., Smith, N.L., Zhang, J.-T., and Asher, S.A. (2015). *Anal. Chem.* 87: 5013–5025.
19 Fenzl, C., Hirsch, T., and Wolfbeis, O.S. (2014). *Angew. Chem. Int. Ed.* 53: 3318–3335.
20 Heo, Y., Kang, H., Lee, J.-S. et al. (2016). *Small* 12: 3819–3826.
21 Mayer, M., Schnepf, M.J., Konig, T.A.F., and Fery, A. (2019). *Adv. Opt. Mater.* 7: 17.
22 López, C. (2003). *Adv. Mater.* 15: 1679–1704.
23 von Freymann, G., Kitaev, V., Lotsch, B.V., and Ozin, G.A. (2013). *Chem. Soc. Rev.* 42: 2528–2554.
24 Pusey, P.N. and van Megen, W. (1986). *Nature* 320: 340–342.
25 Holland, B.T., Blanford, C.F., and Stein, A. (1998). *Science* 281: 538–540.
26 Salvarezza, R.C., Vazquez, L., Miguez, H. et al. (1996). *Phys. Rev. Lett.* 77: 4572–4575.
27 Griesebock, B., Egen, M., and Zentel, R. (2002). *Chem. Mater.* 14: 4023–4025.
28 Zhang, J.-T., Wang, L., Lamont, D.N. et al. (2012). *Angew. Chem. Int. Ed.* 51: 6117–6120.
29 Wang, L. and Zhao, X.S. (2007). *J. Phys. Chem. C* 111: 8538–8542.
30 Cai, Z., Liu, Y.J., Leong, E.S.P. et al. (2012). *J. Mater. Chem.* 22: 24668–24675.

31 Cai, Z., Liu, Y.J., Teng, J., and Lu, X. (2012). *ACS Appl. Mater. Interfaces* 4: 5562–5569.
32 Cai, Z., Liu, Y.J., Lu, X., and Teng, J. (2014). *ACS Appl. Mater. Interfaces* 6: 10265–10273.
33 Jiang, P., Bertone, J.F., Hwang, K.S., and Colvin, V.L. (1999). *Chem. Mater.* 11: 2132–2140.
34 Vlasov, Y.A., Bo, X.Z., Sturm, J.C., and Norris, D.J. (2001). *Nature* 414: 289–293.
35 Cai, Z., Teng, J., Wan, Y., and Zhao, X.S. (2012). *J. Colloid Interface Sci.* 380: 42–50.
36 Jiang, P. and McFarland, M.J. (2004). *J. Am. Chem. Soc.* 126: 13778–13786.
37 Wang, D. and Möhwald, H. (2004). *Adv. Mater.* 16: 244–247.
38 Yang, H. and Jiang, P. (2010). *Langmuir* 26: 13173–13182.
39 Kuang, M., Wang, J., Bao, B. et al. (2014). *Adv. Opt. Mater.* 2: 34–38.
40 Wu, S., Liu, B., Su, X., and Zhang, S. (2017). *J. Phys. Chem. Lett.* 8: 2835–2841.
41 Sun, J., Bhushan, B., and Tong, J. (2013). *RSC Adv.* 3: 14862–14889.
42 Lu, T., Peng, W.H., Zhu, S.M., and Zhang, D. (2016). *Nanotechnology* 27: 14.
43 Armstrong, E. and O'Dwyer, C. (2015). *J. Mater. Chem. C* 3: 6109–6143.
44 Teyssier, J., Saenko, S.V., van der Marel, D., and Milinkovitch, M.C. (2015). *Nat. Commun.* 6: 6368.
45 Meadows, M.G., Butler, M.W., Morehouse, N.I. et al. (2009). *J. R. Soc. Interface* 6: S107–S113.
46 Liu, F., Dong, B.Q., Liu, X.H. et al. (2009). *Opt. Express* 17: 16183–16191.
47 Schultz, T.D. (1986). *Bull. Entomol. Soc. Am.* 32: 142–146.
48 Srinivasarao, M. (1999). *Chem. Rev.* 99: 1935–1962.
49 Stuart-Fox, D., Whiting, M.J., and Moussalli, A. (2006). *Biol. J. Linn. Soc.* 88: 437–446.
50 Holtz, J.H. and Asher, S.A. (1997). *Nature* 389: 829–832.
51 Cai, Z., Zhang, J.-T., Xue, F. et al. (2014). *Anal. Chem.* 86: 4840–4847.
52 Zhang, J.-T., Cai, Z., Kwak, D.H. et al. (2014). *Anal. Chem.* 86: 9036–9041.
53 Cai, Z., Luck, L.A., Punihaole, D. et al. (2016). *Chem. Sci.* 7: 4557–4562.
54 Cai, Z., Sasmal, A., Liu, X., and Asher, S.A. (2017). *ACS Sens.* 2: 1474–1481.
55 Cai, Z., Kwak, D.H., Punihaole, D. et al. (2015). *Angew. Chem. Int. Ed.* 54: 13036–13040.
56 Ge, J. and Yin, Y. (2011). *Angew. Chem. Int. Ed.* 50: 1492–1522.
57 Inan, H., Poyraz, M., Inci, F. et al. (2017). *Chem. Soc. Rev.* 46: 366–388.
58 Hou, J., Li, M., and Song, Y. (2018). *Nano Today* 22: 132–144.
59 Qi, F., Meng, Z., Xue, M., and Qiu, L. (2020). *Anal. Chim. Acta* 1123: 91–112.
60 Endo, T., Yanagida, Y., and Hatsuzawa, T. (2007). *Sens. Actuators B* 125: 589–595.
61 Goldenberg, L.M., Wagner, J., Stumpe, J. et al. (2002). *Langmuir* 18: 3319–3323.
62 Yamasaki, T. and Tsutsui, T. (1999). *Jpn. J. Appl. Phys.* 38: 5916–5921.
63 Alfrey, T. Jr., Gurnee, E.F., and Schrenk, W.J. (1969). *Polym. Eng. Sci.* 9: 400–404.
64 Bonifacio, L.D., Lotsch, B.V., Puzzo, D.P. et al. (2009). *Adv. Mater.* 21: 1641–1646.

65 Yang, D., Ye, S., and Ge, J. (2013). *J. Am. Chem. Soc.* 135: 18370–18376.
66 Yang, D., Ye, S., and Ge, J. (2014). *Adv. Funct. Mater.* 24: 3197–3205.
67 Bai, L., Xie, Z., Wang, W. et al. (2014). *ACS Nano* 8: 11094–11100.
68 Xiong, C., Zhao, J., Wang, L. et al. (2017). *Mater. Horiz.* 4: 862–868.
69 Cui, C., Liu, Y., Xu, H. et al. (2014). *Small* 10: 3672–3676.
70 Ranft, A., Niekiel, F., Pavlichenko, I. et al. (2015). *Chem. Mater.* 27: 1961–1970.
71 Zhang, W., Gao, N., Cui, J. et al. (2017). *Chem. Sci.* 8: 6281–6289.
72 Cui, J., Zhu, W., Gao, N. et al. (2014). *Angew. Chem. Int. Ed.* 53: 3844–3848.
73 Zhang, W., Li, Y., Liang, Y. et al. (2019). *Chem. Sci.* 10: 6617–6623.
74 Zhao, Y., Shang, L., Cheng, Y., and Gu, Z. (2014). *Acc. Chem. Res.* 47: 3632–3642.
75 Hou, J., Li, M., and Song, Y. (2018). *Angew. Chem. Int. Ed.* 57: 2544–2553.
76 Velev, O.D., Lenhoff, A.M., and Kaler, E.W. (2000). *Science* 287: 2240.
77 Yi, G.-R., Moon, J.H., and Yang, S.-M. (2001). *Chem. Mater.* 13: 2613–2618.
78 Kim, S.-H., Lee, S.Y., Yi, G.-R. et al. (2006). *J. Am. Chem. Soc.* 128: 10897–10904.
79 Wang, L.B., Wang, J.X., Huang, Y. et al. (2012). *J. Mater. Chem.* 22: 21405–21411.
80 Takeoka, Y. and Watanabe, M. (2003). *Langmuir* 19: 9104–9106.
81 Lee, Y.J. and Braun, P.V. (2003). *Adv. Mater.* 15: 563–566.
82 Tian, E., Wang, J., Zheng, Y. et al. (2008). *J. Mater. Chem.* 18: 1116–1122.
83 Liu, Y.J., Cai, Z., Leong, E.S.P. et al. (2012). *J. Mater. Chem.* 22: 7609–7613.
84 Cho, Y., Lee, S.Y., Ellerthorpe, L. et al. (2015). *Adv. Funct. Mater.* 25: 6041–6049.
85 Zhang, Y., Li, K., Su, F.Y. et al. (2019). *Opt. Express* 27: 15391–15398.
86 Han, M.G., Shin, C.G., Jeon, S.-J. et al. (2012). *Adv. Mater.* 24: 6438–6444.
87 Zhang, Y., Jiang, Y., Wu, X., and Ge, J. (2017). *J. Mater. Chem. C* 5: 9288–9295.
88 Cai, Z., Leong, E.S.P., Wang, Z. et al. (2015). *J. Mater. Chem. C* 3: 11645–11652.
89 Gallego-Gómez, F., Morales, M., Blanco, A., and López, C. (2019). *Adv. Mater. Technol.* 4: 1800493.
90 Yin, S.-N., Wang, C.-F., Liu, S.-S., and Chen, S. (2013). *J. Mater. Chem. C.* 1: 4685–4690.
91 Huang, J., Tao, C.-A., An, Q. et al. (2010). *Chem. Commun.* 46: 4103–4105.
92 Ganter, P., Szendrei, K., and Lotsch, B.V. (2016). *Adv. Mater.* 28: 7436–7442.
93 Rus, D. and Tolley, M.T. (2015). *Nature* 521: 467–475.
94 Mengüç, Y., Park, Y.-L., Pei, H. et al. (2014). *Int. J. Robot. Res.* 33: 1748–1764.
95 Yoshino, K., Kawagishi, Y., Ozaki, M., and Kose, A. (1999). *Jpn. J. Appl. Phys.* 38: L786–L788.
96 Fudouzi, H. and Sawada, T. (2006). *Langmuir* 22: 1365–1368.
97 Inci, E., Topcu, G., and Demir, M.M. (2020). *Sens. Actuators B.* 305: 127452.
98 Budil, K.S., Gold, D.M., Estabrook, K.G. et al. (1999). *Rev. Sci. Instrum.* 70: 806–809.
99 Chu, J.J., Beckwith, J.G., Leonard, D.S. et al. (2012). *Ann. Biomed. Eng.* 40: 203–212.

100 Li, M., Harten, P.F., and Cabezas, H. (2002). *Ind. Eng. Chem. Res.* 41: 5867–5877.
101 Potyrailo, R.A. (2006). *Angew. Chem. Int. Ed.* 45: 702–723.
102 Franke, M.E., Koplin, T.J., and Simon, U. (2006). *Small* 2: 36–50.
103 Li, H., Chang, L., Wang, J. et al. (2008). *J. Mater. Chem.* 18: 5098–5103.
104 Chen, C., Zhu, Y., Bao, H. et al. (2011). *Chem. Commun.* 47: 5530–5532.
105 Cai, Z., Liu, Y.J., Lu, X., and Teng, J. (2013). *J. Phys. Chem. C* 117: 9440–9445.
106 Chiappini, A., Armellini, C., Carpentiero, A. et al. (2013). *Opt. Mater.* 36: 130–134.
107 Wang, L., Zhang, S., Lutkenhaus, J.L. et al. (2017). *J. Mater. Chem. C* 5: 8266–8272.
108 Kou, D., Zhang, S., Lutkenhaus, J.L. et al. (2018). *J. Mater. Chem. C* 6: 2704–2711.
109 Mirzaei, A., Kim, J.-H., Kim, H.W., and Kim, S.S. (2018). *J. Mater. Chem. C* 6: 4342–4370.
110 Zhang, Y., Sun, Y., Liu, J. et al. (2019). *Sens. Actuators B* 291: 67–73.
111 Zhang, Y., Qiu, J., Gao, M. et al. (2014). *J. Mater. Chem. C* 2: 8865–8872.
112 Liu, J., Zhang, Y., Zhou, R., and Gao, L. (2017). *J. Mater. Chem. C* 5: 6071–6078.
113 Zhang, Y., Qiu, J., Hu, R. et al. (2015). *Phys. Chem. Chem. Phys.* 17: 9651–9658.
114 Kreno, L.E., Leong, K., Farha, O.K. et al. (2012). *Chem. Rev.* 112: 1105–1125.
115 Lu, G., Farha, O.K., Kreno, L.E. et al. (2011). *Adv. Mater.* 23: 4449–4452.
116 Wu, Y.-N., Li, F., Zhu, W. et al. (2011). *Angew. Chem. Int. Ed.* 50: 12518–12522.
117 Hinterholzinger, F.M., Ranft, A., Feckl, J.M. et al. (2012). *J. Mater. Chem.* 22: 10356–10362.
118 Li, L., Jiao, X., Chen, D. et al. (2015). *Chem. Mater.* 27: 7601–7609.
119 Vardhan, K.H., Kumar, P.S., and Panda, R.C. (2019). *J. Mol. Liq.* 290: 111197.
120 Wang, F., Meng, Z., Xue, F. et al. (2014). *Trends Environ. Anal. Chem.* 3–4: 1–6.
121 Holtz, J.H., Holtz, J.S.W., Munro, C.H., and Asher, S.A. (1998). *Anal. Chem.* 70: 780–791.
122 Asher, S., Peteu, S., Reese, C. et al. (2002). *Anal. Bioanal. Chem.* 373: 632–638.
123 Reese, C.E. and Asher, S.A. (2003). *Anal. Chem.* 75: 3915–3918.
124 Ellington, A.D. and Szostak, J.W. (1990). *Nature* 346: 818–822.
125 Tuerk, C. and Gold, L. (1990). *Science* 249: 505–510.
126 Ye, B.-F., Zhao, Y.-J., Cheng, Y. et al. (2012). *Nanoscale* 4: 5998–6003.
127 Yan, Z., Tian, C., Qu, X. et al. (2017). *Colloids Surf. B* 154: 142–149.
128 Liu, W., Li, L., Liu, S. et al. (2019). *J. Mater. Chem. C* 7: 8946–8953.
129 Hong, W., Hu, X., Zhao, B. et al. (2010). *Adv. Mater.* 22: 5043–5047.
130 Eddleston, M., Street, J.M., Self, I. et al. (2012). *Toxicology* 294: 94–103.
131 Radic, Z., Pickering, N.A., Vellom, D.C. et al. (1993). *Biochemistry* 32: 12074–12084.
132 Walker, J.P. and Asher, S.A. (2005). *Anal. Chem.* 77: 1596–1600.
133 Walker, J., Kimble, K., and Asher, S. (2007). *Anal. Bioanal. Chem.* 389: 2115–2124.
134 Qi, F., Lan, Y., Meng, Z. et al. (2018). *RSC Adv.* 8: 29385–29391.
135 Liu, F., Huang, S., Xue, F. et al. (2012). *Biosens. Bioelectron.* 32: 273–277.

136 Yan, C., Qi, F., Li, S. et al. (2016). *Talanta* 159: 412–417.

137 Liwarska-Bizukojc, E. (2009). *Fresenius Environ. Bull.* 18: 1666–1673.

138 Rout, P.R., Zhang, T.C., Bhunia, P., and Surampalli, R.Y. (2021). *Sci. Total Environ.* 753: 141990.

139 Zhang, J.-T., Smith, N., and Asher, S.A. (2012). *Anal. Chem.* 84: 6416–6420.

140 Chiu, C.-F., Tsai, H.-P., Chen, Y.-C. et al. (2017). *ACS Appl. Mater. Interfaces* 9: 26333–26340.

141 Hood, L., Heath, J.R., Phelps, M.E., and Lin, B. (2004). *Science* 306: 640–643.

142 Labib, M., Sargent, E.H., and Kelley, S.O. (2016). *Chem. Rev.* 116: 9001–9090.

143 Sinha, A., Lu, X., Wu, L. et al. (2018). *TrAC Trends Anal. Chem.* 98: 174–189.

144 Jung, S., Kaar, J.L., and Stoykovich, M.P. (2016). *Mol. Syst. Des. Eng.* 1: 225–241.

145 van Enter, B.J. and von Hauff, E. (2018). *Chem. Commun.* 54: 5032–5045.

146 Alexeev, V.L., Sharma, A.C., Goponenko, A.V. et al. (2003). *Anal. Chem.* 75: 2316–2323.

147 Muscatello, M.M.W., Stunja, L.E., and Asher, S.A. (2009). *Anal. Chem.* 81: 4978–4986.

148 Asher, S.A., Alexeev, V.L., Goponenko, A.V. et al. (2003). *J. Am. Chem. Soc.* 125: 3322–3329.

149 Ben-Moshe, M., Alexeev, V.L., and Asher, S.A. (2006). *Anal. Chem.* 78: 5149–5157.

150 Kamenjicki, M. and Asher, S.A. (2005). *Sens. Actuators B* 106: 373–377.

151 Lee, M.C., Kabilan, S., Hussain, A. et al. (2004). *Anal. Chem.* 76: 5748–5755.

152 Alexeev, V.L., Das, S., Finegold, D.N., and Asher, S.A. (2004). *Clin. Chem.* 50: 2353–2360.

153 Zhang, C., Cano, G.G., and Braun, P.V. (2014). *Adv. Mater.* 26: 5678–5683.

154 Zhang, C., Losego, M.D., and Braun, P.V. (2013). *Chem. Mater.* 25: 3239–3250.

155 Siegel, R.A., Gu, Y., Lei, M. et al. (2010). *J. Control. Release* 141: 303–313.

156 Liu, Y., Zhang, Y., and Guan, Y. (2009). *Chem. Commun.*: 1867–1869.

157 Zhang, X., Guan, Y., and Zhang, Y. (2012). *Biomacromolecules* 13: 92–97.

158 Hu, Y., Jiang, X., Zhang, L. et al. (2013). *Biosens. Bioelectron.* 48: 94–99.

159 Lis, H. and Sharon, N. (1998). *Chem. Rev.* 98: 637–674.

160 Dwek, R.A. (1996). *Chem. Rev.* 96: 683–720.

161 Wang, S.-K. and Cheng, C.-M. (2015). *Chem. Commun.* 51: 16750–16762.

162 Green, N.M. and Avidin (1975). *Advances in Protein Chemistry*, vol. 29 (eds. C.B. Anfinsen, J.T. Edsall and F.M. Richards), 85–133. Academic Press.

163 Hong, G., Li, C., and Qi, L. (2010). *Adv. Funct. Mater.* 20: 3774–3783.

164 Ye, X., Li, Y., Dong, J. et al. (2013). *J. Mater. Chem. C* 1: 6112–6119.

165 Zhang, J.-T., Chao, X., Liu, X., and Asher, S.A. (2013). *Chem. Commun.* 49: 6337–6339.

166 Lundquist, J.J. and Toone, E.J. (2002). *Chem. Rev.* 102: 555–578.

167 Mammen, M., Choi, S.-K., and Whitesides, G.M. (1998). *Angew. Chem. Int. Ed.* 37: 2754–2794.

168 Naismith, J.H., Emmerich, C., Habash, J. et al. (1994). *Acta Crystallogr. Sect. D: Struct. Biol.* 50: 847–858.

169 MacConaghy, K.I., Geary, C.I., Kaar, J.L., and Stoykovich, M.P. (2014). *J. Am. Chem. Soc.* 136: 6896–6899.

170 MacConaghy, K.I., Chadly, D.M., Stoykovich, M.P., and Kaar, J.L. (2015). *Anal. Chem.* 87: 3467–3475.

171 Qin, J., Li, X., Cao, L. et al. (2020). *J. Am. Chem. Soc.* 142 (1): 417–423.

172 Shiu, J.-Y., Kuo, C.-W., and Chen, P. (2004). *J. Am. Chem. Soc.* 126: 8096–8097.

173 Shen, W., Li, M., Ye, C. et al. (2012). *Lab Chip* 12: 3089–3095.

174 Choi, E., Choi, Y., Nejad, Y.H.P. et al. (2013). *Sens. Actuators B* 180: 107–113.

175 Liu, M. and Yu, L.-P. (2013). *Analyst* 138: 3376–3379.

176 Chen, Q., Shi, W., Cheng, M. et al. (2018). *Microchim. Acta* 185: 557.

177 Zhang, Y.-X., Zhao, P.-Y., and Yu, L.-P. (2013). *Sens. Actuators B* 181: 850–857.

178 Liu, X.-Y., Fang, H.-X., and Yu, L.-P. (2013). *Talanta* 116: 283–289.

179 Piletsky, S.A., Andersson, H.S., and Nicholls, I.A. (1999). *Macromolecules* 32: 633–636.

180 Qin, L., He, X.-W., Li, W.-Y., and Zhang, Y.-K. (2008). *J. Chromatogr. A* 1187: 94–102.

181 Yang, Y., Long, Y., Cao, Q. et al. (2008). *Anal. Chim. Acta* 606: 92–97.

182 Piletsky, S.A., Andersson, H.S., and Nicholls, I.A. (1998). *J. Mol. Recognit.* 11: 94–97.

183 Catto, J.W.F., Alcaraz, A., Bjartell, A.S. et al. (2011). *Eur. Urol.* 59: 671–681.

184 Li, Q., Zhou, S., Zhang, T. et al. (2020). *Biosens. Bioelectron.* 150: 111866.

185 Wang, Y., Li, Z., Lin, Q. et al. (2019). *ACS Sens.* 4: 2124–2130.

186 Wei, X., Bian, F., Cai, X. et al. (2020). *Anal. Chem.* 92: 6121–6127.

187 Beckman, M.G., Hooper, W.C., Critchley, S.E., and Ortel, T.L. (2010). *Am. J. Prev. Med.* 38: S495–S501.

188 Resende, S., Frasco, M.F., and Sales, M.G.F. (2020). *Sens. Actuators B* 312: 127947.

189 Scheld, W.M., Brown, R.S. Jr., Harding, S.A., and Sande, M.A. (1980). *J. Clin. Microbiol.* 12: 679–683.

190 Xiang, H., Xiong, L., Liu, X., and Tu, Z. (2007). *J. Microbiol. Methods* 69: 282–287.

191 Miyakawa, Y., Mabuchi, T., and Fukazawa, Y. (1993). *J. Clin. Microbiol.* 31: 3344–3347.

192 Nizet, V. and Esko, J.D. (2009). Bacterial and Viral Infections. In: *Essentials of Glycobiology* (eds. A. Varki, R.D. Cummings, J.D. Esko, et al.). La Jolla, California: Cold Spring Harbor Laboratory Press Copyright © 2009, The Consortium of Glycobiology Editors.

193 Xu, Y., Wang, H., Luan, C. et al. (2018). *Biosens. Bioelectron.* 100: 404–410.

194 Hall, R.A. and Gow, N.A.R. (2013). *Mol. Microbiol.* 90: 1147–1161.

195 Murtaza, G., Rizvi, A.S., Irfan, M. et al. (2020). *Anal. Chim. Acta* 1117: 1–8.

196 Hao, J.-J., Xie, X., Gu, K.-D. et al. (2019). *Plasmonics* 14: 1919–1928.

197 He, M., Gales, J.P., Ducrot, É. et al. (2020). *Nature* 585: 524–529.

8

Self-assembled Graphene Architectures for Electrochemical Energy Storage

Pei Li, Junwei Han, and Quan-Hong Yang

Tianjin University, School of Chemical Engineering and Technology, Nanoyang Group, State Key Laboratory of Chemical Engineering, No. 92 Weijin Road, Nankai District, Tianjin 300072, China

8.1 Introduction

Graphene is a carbon monolayer packed into a two-dimensional (2D) honeycomb lattice. Before Geim and Novoselov first isolated single-layer carbon atoms in 2004, graphene was considered to be only a component of other dimensionalities, such as graphite, fullerene, and carbon nanotubes (CNTs) [1]. From then on, interest in graphene has been growing constantly, leading to the "graphene gold rush." [2] Usually, the dimension of graphene sheets in the solvent is hundreds of nanometers, thus classified as colloids and show the attractive properties of colloids.

In the past few decades, many researchers have studied graphene-based materials for use in electrochemical energy storage (EES) devices, such as lithium-ion batteries (LIBs), lithium–sulfur (Li–S) batteries, and supercapacitors (SCs), which also promotes the rapid development of portable electronics and electric vehicles [3]. Graphene can play roles of the active material, electrically conductive additive, and electrode framework in EES. Graphene has a theoretical specific surface area (SSA) that can reach 2630 m^2/g, excellent electric and thermal conductivity, good flexibility, high mechanical strength, and good chemical stability [4]. In addition, there are several morphological advantages of its 2D nanosheet structure, for example, the highly accessible surface, abundant active sites, and rapid ion storage kinetics, potentially promoting energy density and power density. Its good flexibility also makes it hopeful to be used as the active material and current collector in flexible EES devices. Graphene nanosheets can be regarded as building units, and their assembly offers opportunities to form special structures and hybrid materials, which have better performance and can be used in various EES applications [5].

In the processes of electrode fabrication and electrochemical cycling, the spontaneous agglomeration of graphene layer seriously reduces the SSA for charge storage. Self-assembly is one of the most promising methods to transform a single graphene sheet into a hierarchical structure. It can effectively prevent the restacking between layers and construct new functional structures simultaneously. Due to the stable

Functional Materials from Colloidal Self-assembly, First Edition. Edited by Qingfeng Yan and George Zhao.

honeycomb structure of monolayer carbon, graphene lacks of active site to build desirable structures. Fortunately, chemically modified graphenes (CMGs), including graphene oxides (GOs), reduced graphene oxides (rGOs) and their derivatives, are ideal alternatives with excellent processability to assemble bulk graphene for practical macroscopic applications.

As shown in Figure 8.1, a number of hydroxyl, epoxy, carboxyl, and carbonyl groups exist at the defect positions of GO sheets. These oxygen-containing functional groups are hydrophilic; thus, the GO sheets are easy to be dispersed in water to form a colloidal suspension, which is important to provide a green solvent for the assembly process. Besides, by using chemical reducing agents, high-temperature thermal reduction, or electrochemical reduction methods, rGO sheets with different oxygen content can be prepared after controllably removing the oxygenated groups. There are several advantages for CMGs as building units of self-assembly. First, CMGs are easily accessible because they are usually the intermediate products of graphene preparation when we exploit strong oxidation of pristine graphite and stirring or ultrasonication to achieve the stripping of graphite, which is one of the mass production methods with high yield and low cost. Second, the rich functional groups facilitate the uniform dispersion in water and prevent the layer restacking by breaking conjugation of GO sheets. Such stable suspension is basic for various types of self-assembly. Third, CMGs are considered as amphiphilic surfactants that can disperse inorganic/organic active particles and then achieve uniform composite assembly, providing a flexible way to design multifunctional materials. Finally, post reduction treatment of assemblies is easy and effective to recover the conjugated structure of rGO sheets, retaining graphene's superior features like high conductivity, high SSA, and outstanding flexibility. Hence, CMGs are promising units to construct advanced graphene-based functional materials [6–8].

This chapter provides a broad overview of the self-assembly strategies of graphene and the methodology to optimize the nanostructure of assemblies. Furthermore, we discuss the growing graphene demand for EES devices, mainly focusing on the recent achievements in the development of graphene assemblies in SCs, LIBs, and Li–S batteries and attempt to discuss the opportunities and challenges for graphene assemblies in these devices comprehensively (Figure 8.2).

8.2 Self-assembly Strategies

G, GO, rGO, and other CMGs self-assembly can occur in the solution or at interfaces, driven by π–π interaction between their conjugated plane and electrostatic interaction (EI) between the functional groups on sheets. Due to intrinsic 2D structure, it is easy for CMGs to assemble into large-area, flat monolayer films by Langmuir–Blodgett (LB) technique and multilayer films by layer-by-layer (LbL) or flow-directed self-assembly method. As a colloid, CMGs exhibit many excellent interface chemistry properties, and 2D plane or three-dimensional (3D) network structure can form at solid/liquid/air interface by interface-induced self-assembly. Besides, template-directed self-assembly and hydrothermal processes are one-step

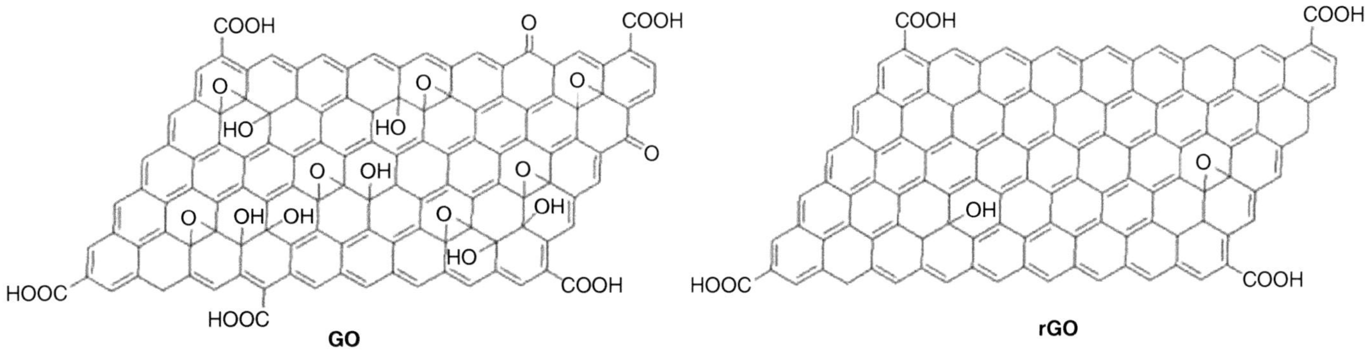

Figure 8.1 Schematic models of a GO or rGO sheet. Source: Xu and Shi [6]. Reproduced with permission of Royal Society of Chemistry.

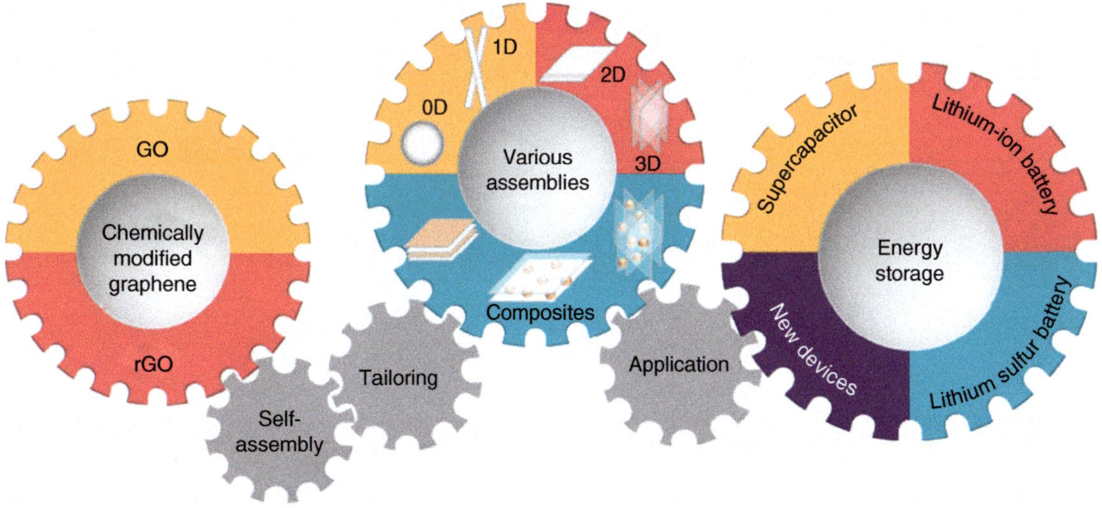

Figure 8.2 Self-assembled graphene structures for EES.

and effective methods for the construction of 3D CMG-based architectures. Interestingly, one-dimensional (1D) fibers derived from CMGs have been produced by spinning and space confinement self-assembly. These advanced strategies designed for CMG nanosheet assembly are discussed in detail in this section.

8.2.1 Langmuir–Blodgett (LB) Technique

LB assembly is the key technology to prepare the large-area, flat, and monolayer graphene films. LB technique was invented by I. Langmuir and K. Blodgett to make amphiphilic molecules with both hydrophilic head and hydrophobic tail into single-layer film with ordered microstructure, which is quite suitable for graphene and its derivative macromolecules [9]. For the first time, Cote et al. successfully applied the LB technique for a monolayer film preparation with amphiphilic GO sheets as a building block [10]. An ordered monolayer architecture was constructed on hydrophilic basements (such as silicon wafer) at a liquid–air interface without GO sheets overlapping because of strong electrostatic repulsion. As indicated in Figure 8.3a–d, the density of nanosheets at the interface could be precisely regulated in different regions of isotherm, from dilute, close-packed, to over-packed monolayer. Besides, Li et al. prepared hydrophobic rGO sheets monolayer films on a variety of transparent basements via LB assembly, and the high-quality graphene sheets show superior electrical conductivity both at room and cryogenic temperatures [11].

LB assembly approach for constructing a 3D graphene film with rich porosity was proposed by Vengadesh and coworkers, different from the conventional LB technique for 2D membrane [12]. The 3D porous graphene films were produced by compressing the monolayer beyond the collapse pressure state, and its dimension could be tuned from nanoscale to macroscale, achieving the applications of graphene in bioelectrodes with improved adhesion and biocompatibility.

8.2.2 Layer-by-Layer (LbL) Assembly Approach

Through successive adsorption of various macromolecular components, LbL-assembled thin films are synthesized. They can show attractive forces such as

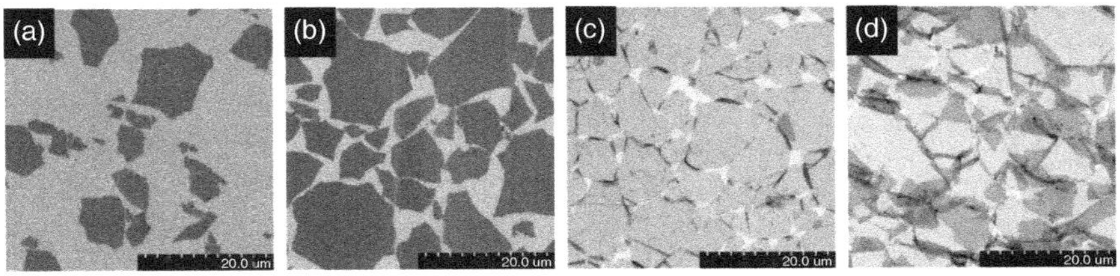

Figure 8.3 LB technique. SEM images showing the (a) isolated graphene monolayer, (b) close-packed GO monolayer, (c) edge-interconnected monolayer, and (d) partially overlapped monolayer on a silicon wafer. Source: Cote et al. [10] / with permission from American Chemical Society.

EI, hydrogen bonding, van der Waals forces, and charge-transfer complexes. This assembly method is attractive for fabricating highly tunable multilayer graphene structures because it has a good control of assembled thickness, composition, and physicochemical properties at the nanoscale through changing the parameters (such as concentration, ionic strength, and pH) of multilayer films [13].

So far, it has been proved that a range of various CMGs can be assembled into multilayer structures by the LbL assembly method. To obtain an effective LbL multilayer assembly, a key demand is that two constituents with opposite charges are well dispersed in the solution. Yoo and coworkers synthesized multilayer neat graphene films via LbL assembly by exploiting EI between well dispersed negatively (RGO-COO^-) and positively (RGO-NH^{3+}) charged graphene nanosheets. By changing the layer number, that is, film thickness, the transmittance and resistance of the film are easily modulated [14]. Furthermore, the LbL method is multifunctional, and it allows assembly on various basements, especially on polymer beads, metal particles, biomolecules, and other 3D colloidal particles. The removal of the colloid templates, following by LbL multilayer deposition, which can lead to hollow capsules, thus, opens up a new way to construct hollow graphene structures with multiple functions [15].

8.2.3 Flow-Directed Self-assembly

Before Dikin et al. firstly introduced the flow-directed self-assembly strategy to assemble GO sheets [16], it has been a mature technology employed in the preparation of commercial inorganic films composed of nano components. The GO was firstly exfoliated into individual sheets and well dispersed in water by ultrasonication. Afterward, vacuum filtration was used to drive the sheets in the dispersion to form a parallel oriented film. Directional flow of water during vacuum filtration drives this orientation of GO sheets. The resulting paper-like materials with good flexibility and layered structure also show outstanding mechanical characteristics. The tensile strength is about 80 MPa, and the Young's modulus is about 32 GPa. Such paper-like film combines flexibility and stiffness due to the unique arrangement of GO sheets.

Through vacuum filtration of rGO aqueous dispersion, Li et al. prepared a rGO paper, and electrostatic repulsion was adopted to keep its stability [17]. It showed higher electrical conductivity and thermal stability than GO paper. Alternatively, chemical and thermal reduction can be used to improve the conductivity of GO paper after vacuum filtration of GO sheets, such as low-temperature annealing and hydrazine reduction [18]. The mentioned method could maintain the flexibility and strength of the film after treatment to a large extension. Furthermore, by adjusting the concentration and the dispersion volume, the thickness of the film could be precisely tuned. As a result, a series of highly conductive rGO films with different thickness were used to fabricate transparent conducting electrodes and flexible transistors. In addition to proceeding in water, the flow-directed self-assembly of CMGs is extended to various organic solvents, such as filtration of rGO dispersed in propylene carbonate [19].

8.2.4 Interface-Induced Self-assembly

CMGs colloids show many interesting interfacial chemical properties. The self-assembly at liquid–air, liquid–liquid, and liquid–solid interfaces is also considered as a promising method to construct graphene assemblies [20]. Interface-induced self-assembly mainly comes from the GO self-aggregation at different interfaces, due to fantastic properties including amphiphilicity, negatively charged properties, abundant oxygen-containing groups, and flexibility. Such a strategy is a straightforward and controllable approach, paving the way to the scalable production of functional graphene macroscopic materials.

Inspired by milk skin formed by heating milk, Chen et al. found an interesting evaporation-induced GO assembly at a liquid–air interface [21]. When GO solution was heated in the air, GO sheets move upward as the water evaporated, gathered, and stayed steadily at the 2D interface, and finally a macroscopic GO film was formed (Figure 8.4a). In comparison with the conventional filtration approach, the process is relatively time-saving. The GO membrane with a LbL microstructure as shown in Figure 8.4b is freestanding and flexible after drying. The size of the membrane is tunable by selecting a container with a different diameter (Figure 8.4c), and the thickness alters with the heating time. In this work, a 0.5–20-μm thick freestanding film was obtained.

Artificial liquid–liquid interfaces were designed by Luo et al. Because of the water-miscible isopropanol introduction, the gelation of GO is achieved in the water under an ultralow concentration (0.1 mg/ml, the lowest ever reported) [22]. In essence, because the intercalation energy of water was much lower, the attraction of water to GO was much stronger than that of isopropanol, resulting in microphase separation in the miscible mixture of isopropanol and water. During the process of solvent heat, the partially reduced GO films interact with each other along with the water isopropanol interface and assemble into the interconnected skeleton. As shown in Figure 8.4d, due to restricted fraction of water, GO films are further concentrated around water molecules to keep a stable dispersion; thus the microphase separation between water and isopropanol molecules was introduced, and the micro liquid–liquid interfaces were formed around the GO sheets clusters. These interfaces allow rGO films gradually to precipitate and assemble into the connected frame. The separated water molecules are trapped in the network and act as spacers to avoid parallel re-accumulation to form a complete hydrogel. Therefore, at ultralow concentration the rGO hydrogels were successfully prepared (Figure 8.4e).

At a liquid–solid interface of a porous anodic aluminum oxide (AAO) film and a GO suspension, Shao et al. produced a GO hydrogel by the interface-induced self-assembly strategy [23]. As shown in Figure 8.4f, simply by inserting porous AAO film into the GO solution, a GO hydrogel can be investigated on the porous side after a time. The obtained hydrogel network was composed of randomly overlapping GO films with excellent mechanical characteristics. Further analysis shows that the chemical interaction between carboxyl groups on GO films and hydroxyl groups on the AAO surface contributes to the interlinking of the GO sheets. Besides, the Al ion was also proved to exist in the hydrogel, which is very

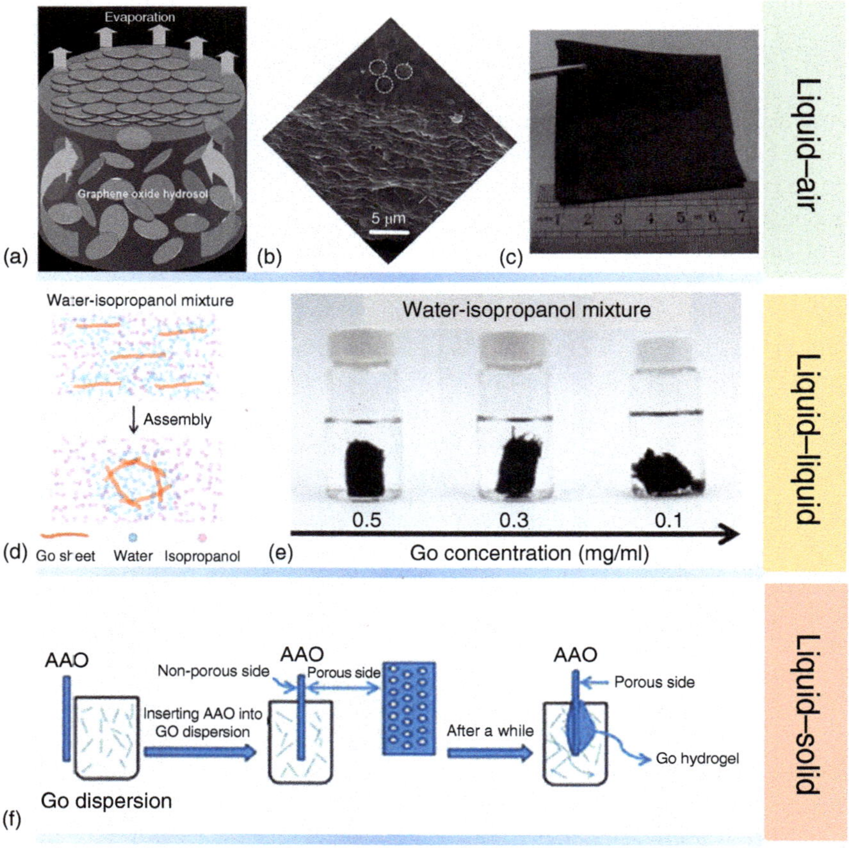

Figure 8.4 Self-assembly at liquid–air, liquid–liquid, and liquid–solid interfaces. (a) Schematic illustration of evaporation-induced assembly at liquid–air interface. (b) SEM image of the layer-by-layer structure of graphene assembly. (c) A large-area substrate-free membrane prepared at liquid–air interface. Source: Chen et al. [21] / with permission from John Wiley and Sons. (d) Schematic of GO assembly at an ultralow concentration at liquid–liquid interface. (e) Photography of rGO hydrogels synthesized in water–isopropanol mixture at different GO concentration. Source: Luo et al. [22] / with permission from John Wiley and Sons. (f) Schematic of the GO hydrogel formation on AAO liquid–solid surface. Source: Shao et al. [23]. Reproduced with permission of Royal Society of Chemistry.

important to enhance the strength of 3D macroform by forming a coordination bond with lone pair electrons of supermolecule structure during self-assembly of the hydrogel.

8.2.5 Template-Directed Self-assembly and Hydrothermal Process

3D CMG-based self-assemblies in the form of aerogels, foams, and hydrogels can also be obtained by the template-directed assembly and hydrothermal process [24, 25]. Ever since, it is very important to transform the excellent properties of

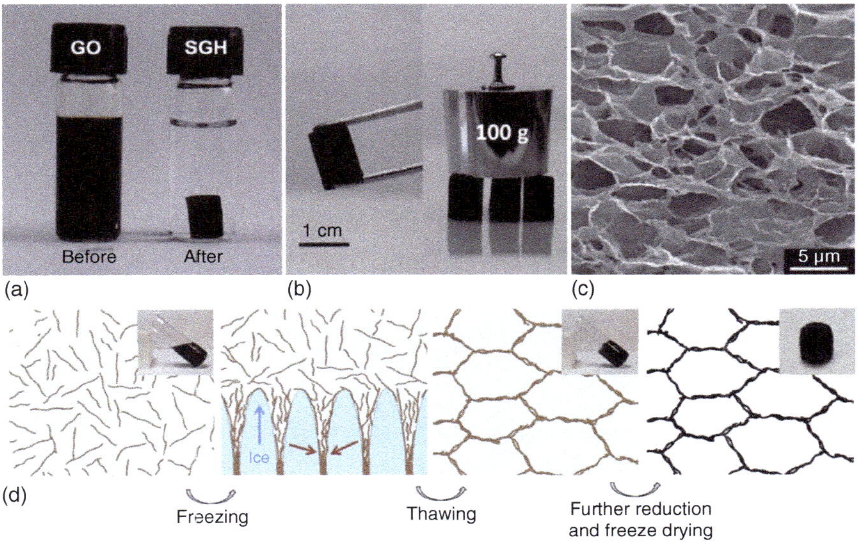

Figure 8.5 Hydrothermal process and ice-templated self-assembly. (a) Photographs of GO aqueous dispersion before and after hydrothermal reduction. (b) Intensity display of a strong self-assembly hydrogel. (c) SEM images of the interior hydrogel microstructures. Source: Xu et al. [25] / with permission from American Chemical Society. (d) Schematic of the cork-like monolith formation mechanism by freeze casting. Source: Qiu et al. [26] / with permission from Springer Nature.

single graphene into the macroscopic superstructure, because it can promote promising materials with unique properties for practical applications.

Since Shi and coworkers proposed an effective graphene hydrogel self-assembly through a one-step hydrothermal process in 2010, this method has been popular and widely used by researchers for constructing the 3D graphene assemblies. Typically, a 2 mg/ml GO aqueous dispersion was sealed in a Teflon-lined autoclave at 180 °C for 12 h; then a 3D porous and mechanically strong graphene hydrogel was generated (Figure 8.5a,b). A highly interconnected graphene network full of water (graphene: water ≈ 2 : 98 wt.%) was consisted of a stacking graphene wall and possessed a pore distribution from nanometers to several micrometers (Figure 8.5c) [25]. As illustrated by the authors, when GO is reduced by the hydrothermal method, the oxygen-containing functional groups on the graphene sheet decrease obviously, leading to a large recovery of π-conjugation. When the GO concentration was high enough, the hydrophilic rGO sheets with stronger π–π interaction would overlap with each other to construct an interconnected 3D network, trapping water inside using residual oxygen-containing groups. An ultralow density, compressible graphene aerogel was prepared by hydrothermal induction assembly and functional freeze-drying microwave treatment by Qiu and coworkers. The resulting aerogels exhibit an ultralow density of only 3 mg/cm^3, but after 90% compression, the original structure can be restored without obvious breakage [27].

Template-directed assembly is an effective method to prepare hollow CMG structures. For instance, Guo et al. exploited a soft template to assemble GO films into hollow GO spheres with a diameter of 2–10 μm, which is on the basis of the water–oil interface from a water/oil emulsion amphiphilic [28]. Li and coworkers found a graphene monolith with a cork-like hierarchical structure fabricated by an ice-templated approach [26]. By accurately designing the freeze-casting process of partially reduced GO, the obtained structure has significant mechanical characteristics due to its biomimetic architecture similar to natural cork (Figure 8.5d). This method can also be applied to other dispersible graphene-based materials.

8.2.6 Spinning and Space Confinement Self-assembly

Spinning and space confinement self-assembly are well-developed methods for the 1D assemblies of CMGs. Compared with the 2D or 3D self-assembly of CMGs mentioned previously, it was used to be a challenging work to assemble CMGs into 1D macroscopic fiber due to intrinsic 2D structure with the strong restacking tendency of CMG sheets. Taking advantage of the spontaneous formation of GO liquid crystals and intrinsic ordering of GO LCs, a simple yet effective wet spinning method was developed for continuously producing macroscopic graphene fibers. Gao and coworkers combined spinning technology of ordered GO LCs and ice-templating strategy with liquid nitrogen as the coagulation bath to prepare fiber graphene aerogel. The resulting fiber has a core–shell framework with a porous core and dense shell. The specific tensile strength is 188 kN m/kg, and the compressive strength is 3.3 MPa. Besides, reduction and annealing could further provide the aligned porous graphene fibers with improved SSA and excellent electrical conductivity. Based on spinning self-assembly, continuous production technology of graphene fiber was created [29].

Compared with wet spinning, space-confined hydrothermal processing free of the limit of equipment is a more convenient method. It is realized by confining the one-step hydrothermal process in a thin cylindrical reactor, which controls the architecture of hydrogel by the shape of the reactor. Qu and coworkers filled the concentrated aqueous GO suspension into the glass microtube as the space-constrained microreactor, which was then thermally treated at 230 °C for 120 min. During the process, GO films aligned along the fiber axis, attributed to the sheer force generated by the capillary and the tension generated by the surface tension. As a result, a graphene fiber with a tensile strength of 180 MPa was prepared, and after a further vacuum thermal treatment at 800 °C, a much higher strength up to 420 MPa was obtained [30].

8.3 Methods for Tailoring the Assemblies

Self-assembly is a useful approach to establish various pure graphene-based structures, and there are methods available for tailoring the hierarchical structure of the graphene assemblies during the assembly or drying process and tuning surface chemistry via heteroatom doping. Besides, inorganic/organic active components can be introduced during self-assembly due to the amphiphilicity of CMGs, which can role as a surfactant to make inorganic/organic components evenly dispersed and form functional hybrid architectures. Therefore, graphene is an ideal support material that can significantly improve the durability and stability of hybrid materials. Composite materials prepared by self-assembly can both inherit intrinsic properties of the individual component and generate new characteristics and functions derived from the cooperative effect. The following section describes the structural tuning and surface modification of graphene assemblies and the built-up of hybrid structures by exploiting the previously discussed self-assembly methods.

8.3.1 Structural Tuning and Surface Modification

8.3.1.1 Compact Assembly via Capillary and Mechanical Compression

Based on the intrinsic corrugated 2D structure and self-assembly behavior of chemically transformed graphene, Li and coworkers indicated that porous yet densely packed carbon materials can be formed by capillary compression of graphene gel films in the presence of a nonvolatile liquid electrolyte [31]. The film thickness can be regulated by changing the volume percent of the nonvolatile component in the liquid electrolyte. It was also observed that the graphene sheets stacked in a compact face-to-face way in the films, leading to a high packing density up to ~1.33 g/cm^3, nearly double that of the conventional activated porous carbon (0.5–0.7 g/cm^3). More importantly, the liquid electrolyte-mediated film possessed a continuous ion transport network because of the fluid nature of liquid electrolytes. As a result, the dense film achieved an incredibly high volumetric energy density.

Tao et al. reported a novel carbon whose density was 1.58 g/cm^3 (70% of the graphite density) and was composed of closely linked graphene nanosheets, which was generated by an evaporation-induced drying of a graphene hydrogel [32]. Usually, freeze drying is selected to remove the water confined in the hydrogel network after the hydrothermal treatment of GO suspension. This drying method completely retains the 3D network constructed in the wet hydrogel and resulted in a porous foam assembly whose apparent density is very low at 0.02 g/cm^3 (porous graphene in the form of a macromolecule, called PGM, the upper part of Figure 8.6a). In contrast, when evaporation-induced drying as shown in the lower part of Figure 8.6a was used, a stiff rodlike material was created with a large volume shrinkage. Although similar in density to graphite and the soft pencil lead, high-density porous graphene macroform (HPGM) showed different X-ray diffraction pattern without sharp (002) peak at 26.5° (Figure 8.6b), which meant that arrangement of the graphene nanosheets in the graphene-derived porous carbons were disordered instead of layered structure. Highly wrinkled nanosheets of HPGM could be interlinked with each other to form a porous but highly compact structure (Figure 8.6c). The novel graphene derived HPGM exhibited an extremely high density of 1.58 g/cm^3 and a high SSA of 367 m^2/g, which perfectly balanced the high density and rich porosity in the structure.

Mechanical compression is a straightforward method to change the pore structure of graphene assemblies. Macropores and large mesopores are firstly reduced, while micropores and small mesopores mostly remain to keep the considerable SSA,

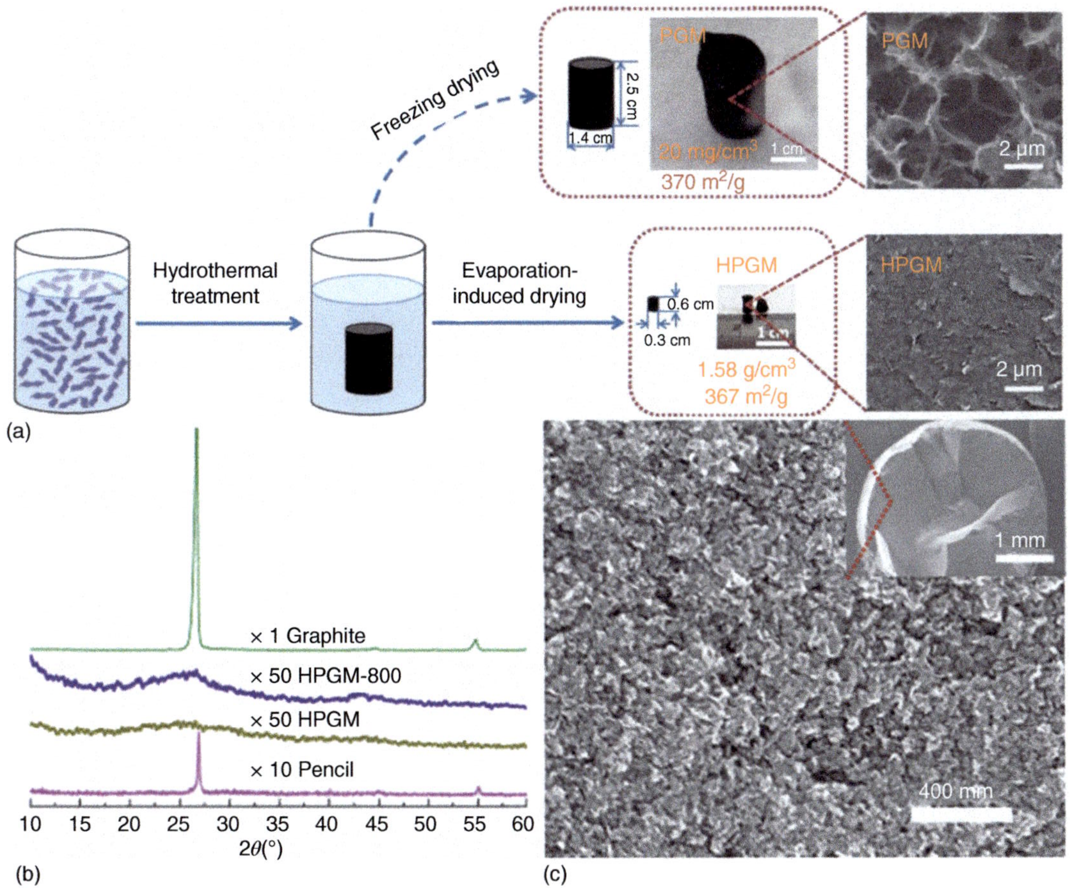

Figure 8.6 Dense graphene assemblies. (a) HPGM produced by a self-assembly and the subsequent evaporation-induced drying. (b) XRD patterns of graphite, soft pencil lead, HPGM, and HPGM-800. (c) Cross-sectional SEM image of HPGM. Inset: cross section of HPGM at low magnification showing the dense texture. Source: Tao et al. [32] / with permission from Springer Nature.

particularly for aerogels. Ruoff et al. proposed that the simple compression was used to a low density and porous microwave-exfoliated graphite oxide, resulting in higher density and smaller pore size after compression. When used in SCs, this compressed material could deliver higher volumetric capacitance. Yang and coworkers prepared porous graphene particles via KOH activation, which is also highly compressible. By adjusting the pressure within 0–40 MPa, electrodes with thickness ranged within 55–120 μm, and a minor decrease in gravimetric capacitance was obtained [8].

8.3.1.2 Tunable Porous Structure and Surface Chemistry

Chemical activation cannot only improve the SSA but also tune the pore structure of CMG assemblies, which is considered as an effective method to optimize the properties of assemblies. To prepare holey graphene framework (HGF) film with an effective ion transport channel, Xu et al. exploited a one-step method with simultaneous chemical corrosion and hydrothermal graphene self-assembly into 3D network frameworks (Figure 8.7a). H_2O_2 molecules, which could etch the carbon atoms at the defect area of GO sheets, were added into the suspension; then they created rich nanopores for the 3D interconnected structure during the self-assembly process. Compared with the highly tortuous ion transport path in the non-holy film, such holy 3D graphene film provided shortcuts for ions across the layers and led to fast ion transport (Figure 8.7b). The self-freestanding holy film held great promise to obtain high volumetric electrical capacity and excellent rate capability while it was highly compressed and used as a self-supported electrode in SCs [33].

Li et al. developed a new method to fabricate a monolithic ultra-thick and compact carbon electrode for symmetric SCs, beginning with graphene assembly and optimizing the porosity by activation [34]. A hydrogel was immersed in $ZnCl_2$ aqueous solution with concentrations within 0.1–2 M followed by heat treatment from 400 to 800 °C and dilute acid washing, obtaining a porosity adjustable graphene monolith (PaGM) with a dense but porous framework (Figure 8.7c). The porosity and density can be well adjusted by controlling the evaporation of $ZnCl_2$ at various temperatures and concentrations.

Besides the pore structure tuning, graphene assemblies surface chemistry is also an important and flexible factor to adjust the assembly properties by creating new active sites and altering the electron distribution of the films at the same time. Various heteroatoms such as nitrogen, oxygen, sulfur, phosphor, boron, etc. have been doped into graphene by substituting the carbon atom in the hexagonal lattice or covalently bonding with carbon atoms at the defect area. The electromagnetic, physicochemical, optical, and structural characteristics of CMG assemblies can be changed by heteroatom doping [35]. Chemical vapor deposition (CVD), ball milling, wet chemical methods, thermal annealing plasma, and arc discharge approaches are feasible to achieve the heteroatom doping during or after the self-assembly process.

Boron (B) and nitrogen (N) atoms are easy to be incorporated into graphene sheets due to similar sizes and valence electron numbers as carbon atoms. For instance, adopting boric acid and polystyrene as the B and C sources in a CVD preparation process, a 4.3 at.% (atomic percentage) B-doped graphene was grown on a copper foil at 1000 °C under the H_2/Ar atmosphere [36]. Zhao et al. employed a pyrrole-assisted

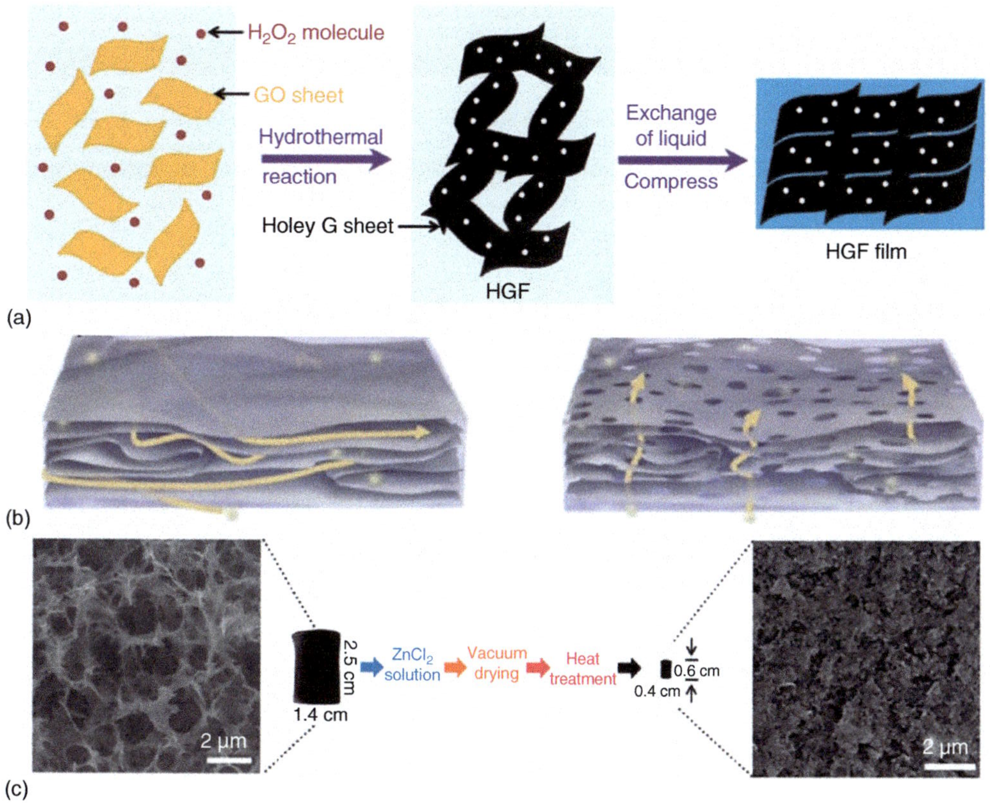

Figure 8.7 Activation of graphene assemblies. (a) Preparation process diagram of HGFs and HGF films. (b) Comparison of ion transport pathway in graphene film and HGF film. Source: Xu et al. [33] / with permission from Springer Nature. (c) Schematic of the graphene monolith adopting $ZnCl_2$ as a pore-forming agent. Source: Li et al. [34] / with permission from Royal Society of Chemistry.

hydrothermal method to produce a swollen hydrogel much bigger than common hydrogel, following via freeze drying and thermal annealing to develop ultralight N-doped graphene aerogel with ~2.1 mg/cm^3 density. The N-doped 3D graphene framework possesses ultralow density, which maximizes the exposure of active surfaces and high conductivity comparable to that of graphene fiber [37].

8.3.1.3 Assembly in the Electrostatic and Magnetic Field

Self-assembly with electric field guidance is a flexible and effective method for conductive CMGs. Electrodeposition, electrophoretic deposition, and electrospinning are common electrostatic-based techniques. For instance, a 3D porous molybdenum sulfide/graphene (MoSx/G) composite film was prepared via an electrodeposition method [38]. The self-supporting film exhibited a porous structure and superior electrical performance. Electrospinning is an electric force-assisted method used for fiber production. Niu et al. developed mesoporous nanotubes via a gradient electrospinning method. Compared with traditional nanowires, the designed fiber showed larger SSA as well as higher ionic and electronic conductivities [39].

The magnetic field is an external and adjustable force that are able to align ferromagnetic materials, so magnetic field-induced self-assembly is effective to construct ordered microstructure in CMG assemblies. A variety of materials with ordered structure can be produced simply by regulating the magnetic intensity and direction [40]. For example, an rGO/Fe_3O_4 composite membrane was prepared at the air–liquid interface under the assistance of a magnetic field, leading to a uniform dispersibility of Fe_3O_4 nanoparticles and preventing GO sheets from the agglomeration [41]. In addition, Kawamori et al. developed a NiO-covered nickel nanowire nonwoven cloth via electroless deposition in the magnetic field. After annealing, the highly conductive cloth could deliver high cycling stability when used as a self-supported electrode [42]. In addition to electrostatic and magnetic fields, other external forces, take microwaves as an example, have also been used to induce assembly processes to produce desired structures [43].

8.3.2 Composite Materials Prepared by Self-assembly

8.3.2.1 Graphene/Carbon Material Assembly

Carbonaceous nanomaterials besides graphene, such as CNTs and fullerenes, show lots of attractive properties in their assembled structure with CMGs. Due to the strong interactions between assembled components and their synergic effect, the physicochemical properties of these composite assemblies are usually superior to that of a pure CMG assembly [44].

With the use of 1D CNTs and 2D CMGs as building blocks, self-assembly is promising to prepare carbon-based materials with large SSA, excellent mechanical properties, and high electrical and thermal coefficients. Kim and coworkers exploited the wet spinning method to prepare rGO and single-walled carbon nanotube (SWCNT) hybrid fiber by mixing the two components into a uniform dispersion with assistance of the surfactant sodium dodecylbenzene sulfonate before being injected into a poly(vinyl alcohol) coagulation bath. As a result, the rGO/SWCNT hybrid fiber showed gravimetric toughness up to 1000 J/g [45]. In addition to CNTs, Zhang and coworkers introduced mesoporous carbon nanosphere into graphene dispersion and produced 2D hybrid films by the means of filtration [46]. A 3D graphene macroform (GM) with the interconnected conductive network is considered as an ideal framework for constructing composite materials. For instance, Zhang and coworkers reported a supercritical CO_2 dried graphene/multi-walled carbon nanotubes (MWCNTs) hybrid aerogel, produced by chemical reduction of well-dispersed GO sheets and MWCNTs [47]. More simply, without formation of the hydrogel, the GO-CNT aqueous mixture could be directly freeze dried to produce aerogel. Sun et al. developed ultralight hybrid aerogels with tunable density by controlling the mass ratio of CNTs in the mixture [48]. The combination of the graphene wall and the elasticity CNT ribs as reinforcements guaranteed a stable and ultralight 3D structure, resulting in outstanding mechanical and thermal properties.

8.3.2.2 Graphene/Polymer Assembly

The previously mentioned LBL assembly can construct ordered films with controllable components, architectures, and morphologies, so this technique is widely used to prepare different nanocomposites of CMGs and polymers. Negatively charged CMG sheets can composite with cationic polyelectrolyte via EI. The subsequent chemical or electrochemical reduction can obviously enhance the composite assembly's conductivity. For instance, because of EI, the negatively charged GO and positively charged poly(ethyleneimine) (PEI) were assembled into GO/PEI composite paper. Grunlan's group produced GO/PEI on a polyethylene terephthalate film through the LbL assembly approach, and such film showed great potential for gas barrier application. Interestingly, a tightly packed graphene/polymer assembly could still provide sufficient gas diffusion channels. Moreover, to enhance the mechanical strength of CMG/graphene film, two or more kinds of polymer components can be used simultaneously [49, 50].

Yang and coworkers adopted static adsorption and capillary drying to fill the porous structure of graphene assemblies with pseudocapacitive polyaniline and thus construct a dense graphene network. The polyaniline as a filler fully exploits the space in the porous graphene network and improves the density of the 3D graphene-based assemblies, contributing to an extremely high volumetric capacitance over 800 F/cm^3 in SCs [51]. Polymers can play the role of cross-linker to improve the mechanical characteristics of 3D graphene/polymer composites. Zhao et al. introduced conducting polymer polypyrrole (PPy) to produce compressible and conductive graphene aerogels [37]. Due to the abundant active sites on the surface and conjugated structure of PPy, the prepared aerogels were highly compressible and conductive (conductivity up to $\approx$30 S/cm). Even the deformation reaches 80% of the breaking strain, the PPy-linked graphene aerogel could recover without any structural network collapse or loss in electrical conductivity.

8.3.2.3 Graphene/Metal or Metal Oxide Assembly

Hybrid assembly of graphene and multiple inorganic components is widely investigated. For one thing, the 2D/3D graphene network can provide a conductive network and stable structure for the inorganic components. For another thing, the diverse inorganic components can bring the assemblies with new properties in a flexible way. Among the inorganic materials, metal and metal oxide such as MnO_2, RuO_2, TiO_2, Pd, Pt, Au, etc. have been introduced into graphene assembly due to special physical and chemical properties.

Zhou and coworkers adopted the LbL self-assembly method to prepare a film with sandwich-like graphene wrapping SnO/SnO_2. The well-designed double protection from graphene layers successfully constructed a hierarchical framework that could stabilize the metal oxide upon lithiation-induced volume change, leading to a good cycling and rate performance when applied in Li-ion batteries [52]. The LbL-assembly approach was also applied to prepare other graphene-based metal and metal compound nanostructures [53], such as Au [54] and MnO_2 [55]. Besides the LbL technique, the flow-induced self-assembly approach, another mature method for preparing membranes, was also used to assemble CMG and inorganic nanoparticles into a 2D hybrid thin film with ordered structure. Furthermore, based on template-directed self-assembly and hydrothermal strategy, 3D graphene-based assemblies with abundant porosity, high conductivity, and outstanding mechanical properties could provide a frame for metal or metal oxide particles. Li et al. adopted a template-directed self-assembly method to prepare a 3D GO/Fe_3O_4 composite [56]. Specifically, GO/Fe_3O_4 composite dispersion was deposited on a 3D Ni foam by using vacuum filtration, and then plasma treatment was used to achieve nitrogen doping and reduction, contributing to a high-conductivity and high-pseudocapacitance SC electrode. Yang and coworkers exploited disassembly–reassembly and capillary-drying approaches to introduce pseudocapacitive metal oxide (ruthenium oxide, RuO_2) into a dense graphene network, partially filling the porous structure while retaining the sufficient ion transport channel. Benefiting from the dense and porous graphene network combined with the high theoretical capacitance and density of RuO_2, the RuO_2/3D graphene electrode delivered a rather high volumetric performance in SCs even at an ultrahigh mass loading of as high as 12 mg/cm^2 [57].

8.4 Applications of Self-assembled Graphene Architecture

Graphene assemblies have indicated promises in EES. There are three typical EES devices: SCs, LIBs, and Li–S batteries, which are high-rate, long-life, and next-generation ultrahigh-energy devices. The preparation and typical energy storage applications of graphene assemblies cataloged by dimensions (1D fiber, 2D membrane, and 3D macroform) are studied in detail in this section.

8.4.1 SCs

There are two storage mechanisms for SC capacitance production: one is the electric double-layer capacitance (EDLC) obtained by interfacial separation of the charges between the electrode and electrolyte, and the other is the pseudocapacitance obtained by the redox reaction. There are two obvious advantages of graphene assemblies as the electrode materials of SCs: one is the high SSA, and the other is various controllable nanostructures, which can provide rapid ion transmission paths and the accommodating ability for conducting polymers or metal oxides and other active materials to achieve a high energy density. Different dimensions graphene assemblies can offer specific advantages when applied as the electrode in a SC. Flexible 1D graphene fiber is conducive to the preparation of all-solid-state SC, while 2D graphene membranes can be easily assembled into a freestanding electrode, although it is necessary to make the ion diffusion perpendicular to the plane possible to further enhance the capacitance. Due to the interconnected pore architecture and continuous conductive network of 3D graphene monolith, efficient charge transportation and storage can be achieved, and the utilization of pseudocapacitive materials would be enhanced as well.

For highly flexible 1D assembly, highly compressible and stretchable SCs can be prepared by graphene fibers. Meng et al. designed and prepared a unique all-graphene core–sheath fiber, in which the graphene fiber core was covered with a sheath of a 3D porous network-like graphene framework (GF@3D-G). The assembled flexible SC on the basis of GF@3D-G has a length-specific capacitance as high as 20 μF/cm [58]. To further enhance the capacitance of fiber-based SCs, many hybrid frameworks on the basis of graphene fibers have been designed. An all-solid symmetric fiber SC prepared from the hollow GO/conducting polymer composite fibers (HCFs) and a PVA (polyvinyl alcohol)/H_3PO_4 gel electrolyte shows an ultrahigh charge storage property and outstanding bending fatigue resistance, as the hollow fiber framework has a larger internal surface to improve electrolyte accessibility and accelerate ion transport (Figure 8.8a–c) [59, 60].

2D graphene assembly also called as graphene paper or graphene membrane has been widely studied for SCs [61]. The removal of trapped water in a wet GO membrane was used to introduce internal forces to adjust the membrane microstructure. Here, water evaporation and ice sublimation were applied as the two stages of the drying process, and a reduced GO membrane with an open and graded framework was achieved, which indicated that the surface utilization rate is high and consequently gave a high rate performance as electrodes for SCs [62]. Chen et al. proposed a 3D macroporous "bubble" graphene film (MGF) through template-directed vacuum filtration by a template of polymethyl methacrylate (PMMA) spheres that was later removed by pyrolysis. Therefore, this graphene film with uniform macropores provided a high specific capacitance of 58 F/g at 1 A/g [63]. 2D graphene assemblies can be directly adopted as the electrode and thus are promising in practical application, since the commercial SCs are made of stacked or rolled up electrode sheets [64, 65].

3D interconnected structure prepared from graphene is very significant to EES because most graphene films are exposed to the electrolyte in comparison with rolled graphene in 1D fibers and stacked graphene in 2D films, and they offer open mass transfer channels and sufficient space to accommodate active materials. Zheng and coworkers fabricated a GM by the hydrothermal reduction in GO dispersions, then further reduced by hydrazine or hydroiodic acid to enhance its conductivity, and the

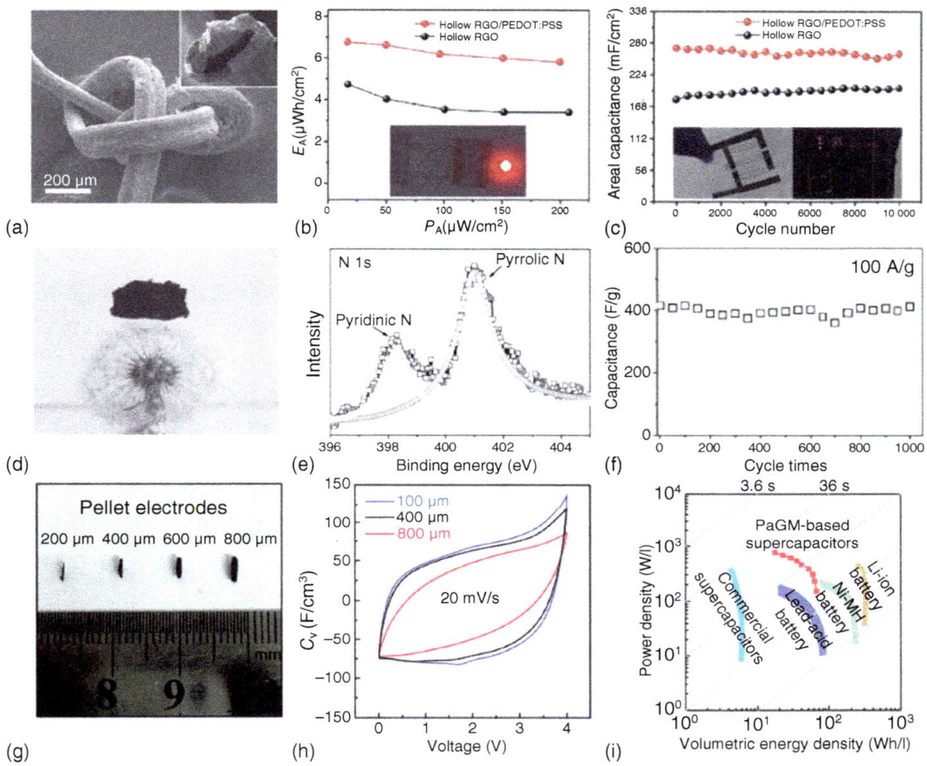

Figure 8.8 Applications of self-assembled graphene architecture in SCs. (a) Knotted fiber SEM image. The inset: cross-sectional SEM image of the knotted fiber. (b) Ragone plots based on the devices. The insert indicates that a fiber SC lights up an LED. (c) Cycling performance of the fiber SCs at 0.48 mA/cm^2. The insert indicates three fiber SCs in serial power three LEDs. Source: Qu et al. [59] / with permission from John Wiley and Sons. (d) Photography of the superlight GF on a dandelion. (e) XPS pattern and the high-resolution N1s peak of the superlight GF. (f) GF-based capacitor cyclic stability at 100 A/g. Source: Zhao et al. [37] / with permission from John Wiley and Sons. (g) Graphene pellet electrodes with various thicknesses. (h) Electrode CV curves with various thicknesses. (i) Ragone plot of the symmetric SC whose electrode thickness is 400 mm. Source: Li et al. [34] / with permission from Royal Society of Chemistry.

capacitance would be about ~200 F/g at 1 A/g [66]. The 3D graphene monolith can reduce the ion-diffusion resistance and enhance the transport dynamics, so it could be adopted as a promising template for the synthesis of nanomaterials with a 3D pore structure.

Because the surface chemical properties of graphene assembly can be adjusted by heteroatom doping, the conductivity and pseudocapacitance can be effectively improved. Zhao et al. [37] studied a multifunctional, N-doped, ultralight GM by a pyrrole-assisted hydrothermal approach, which produced 484 and 415 F/g capacitance at 1 and 100 A/g, respectively (Figure 8.8d–f). The design and optimization of pore structure play an important role in further enhancing energy density and power density. Due to the synergistic combination of mesoporous and macroporous materials, the specific capacitance of 3D graphene aerogel mesoporous carbons (GA-MC) is 226 F/g, which has an outstanding rate and cycle performance [66]. Chemical activation is another way to enhance the capacitive performance of graphene by introducing enough porosity and adjusting the pore framework [67]. When zinc chloride was adopted as activation to adjust the pore framework of a 3D compact graphene assembly, the ion transport path can be ensured even if the thickness is 400 μm. Therefore, this graphene pallet electrode provided a record-high bulk energy density of 64.7 Wh/l with a high operating voltage (4 V) taking $BMIMBF_4$ as an electrolyte (Figure 8.8g–i) [38].

Non-carbon materials, such as metal oxides, metal nanoparticles, and conductive polymers, can be used as active materials to enhance the capacitance of graphene-based films and can be used as spacers to create ion transport channels for high-performance SC. For instance, Peng et al. proposed an ultrathin (40 nm) MnO_2/graphene hybrid film generated by vacuum filtration, and planar SCs based on such a hybrid sheet showed capacitances of 267 F/g at 0.2 A/g and 208 F/g at 10 A/g. It also indicated outstanding rate capability and cycling stability with a capacitance retention up to 92% after 7000 charge/discharge cycles [65].

8.4.2 LIBs

With high energy density and environmental friendliness, LIBs are the most commonly used rechargeable battery for portable electronic products and electric vehicles in modern society. The frameworks and properties of the anode and cathode materials are the key factors to determine the performance of the battery. $LiFePO_4$, $LiCoO_2$, and $LiMn_2O_4$ are the most widely used cathode materials at present, but their conductivity is low, and the preparation of electrode needs a lot of conductive carbon, which limits the further enhancement of their rate performance. Graphene assemblies can provide an ideal conductive transport network for the cathode. Besides, non-carbon materials, such as Si, Sn, Ge, Al, and transition metal oxides, are promising high-capacity anode materials to replace the graphite in current LIBs. However, these materials have some problems such as low conductivity and large volume change during charging and discharging, which limit their practical application [68].

A carbon structure established from graphene nanosheets can provide the electrochemically active materials with a mechanically strong and flexible network, a continuous conductive scaffold, and an interpenetrating ion-diffusion path, which is very important to maintain the integrity of the whole electrode in the cycle process. A 2D assembly can be directly used in LIBs, but the performance of tightly packed graphene films is easily degraded due to the long ion transfer distance and low utilization of active sites on the surface of graphene [69]. Wang et al. [70] firstly proved a graphene film fabricated by vacuum filtration for LIBs that indicated a very low discharge capacity of 84 mAh/g at 50 mA/g. However, graphene cannot be directly used as the anode in LIBs because of its low Coulombic efficiency, which is the result of the irreversible solid electrolyte interphase (SEI) reactions in defects and nanoholes.

In graphene-based hybrid electrodes, the stacking frame can effectively stabilize the high-capacity electrochemically active materials, such as Si, Sn, and transition metal oxides, to accommodate the volume expansion/contraction of noncarbon particles by providing an elastic buffer space and limit them in the Li^+ insertion/extraction process. At the same time, non-carbon active particles act as spacers, which prevent the graphene sheet from face-to-face aggregation, thus forming a smooth ion diffusion channel. For instance, Lee et al. employed Si into the interlayers of graphene paper through vacuum filtration. The Si-graphene paper exhibits a capacity of 2200 mAh/g after 50 cycles and 1500 mAh/g after 200 cycles because of the anchoring regions formed for Si particles [71]. Yang and coworkers further put forward a "concrete"-like structure, in which the flexible graphene network offered a mechanically strong scaffold, and carbonized polyacrylonitrile (PAN) acts as an adhesive to combine silicon nanoparticles and graphene. This Si-porous carbon/graphene (Si–C/G) electrode showed a high specific capacity of 1711 mAh/g with good capacity retention [72].

3D graphene assemblies can enhance the graphene surface utilization in LIBs [73, 74]. Our team has confirmed a well-designed approach to introduce a precise amount of void space in 3D graphene networks for non-carbons, taking sulfur as a template [75]. In a typical synthesis, flowable soft sulfur and hard SnO_2 nanoparticles grew together in the shrinking graphene hydrogel and control the space around the SnO_2 particles accurately by adjusting the surrounding and mobile sulfur content (Figure 8.9a). As demonstrated in the XPS results and a high-resolution transmission electron microscopy (HRTEM) image in Figure 8.9b,c, the sulfur template tightly wrapped the SnO_2 nanoparticles and also caused a chemical environment change of SnO_2. After sulfur removal, a uniform distribution of void space around the ultrafine SnO_2 nanoparticles (sizes of 5–10 nm) was clearly seen (Figure 8.9d). An ultrahigh volumetric capacity of 2123 mAh/cm^3 and excellent cyclic stability was obtained by packing the graphene network to a high density but with adequate void space for SnO_2.

8.4.3 Li–S Batteries

Li–S battery is a next-generation rechargeable battery with a much higher energy density than LIBs. It has attracted extensive attention recently. However, due to the insulation properties of sulfur, the "shuttle effect" caused by soluble polysulfide (lithium polysulfides [LiPS]) intermediates and the volume expansion of sulfur cathode in the cycle process, the specific capacity and cycle life of these batteries cannot meet the requirements of practical application. GO was used to immobilize the soluble polysulfide intermediates, and active functional groups are used on them, and the strong interaction between them enabled the Li–S battery to achieve a stable cycling performance [76].

A 2D graphene film can confine sulfur particles in the interlayer space, which has the advantages of retaining highly conductive channels for active sulfur materials and preventing the shuttle of polysulfides. A flexible freestanding graphene sheets/sulfur (GS/S) sheet was prepared by vacuum infiltration. The layered

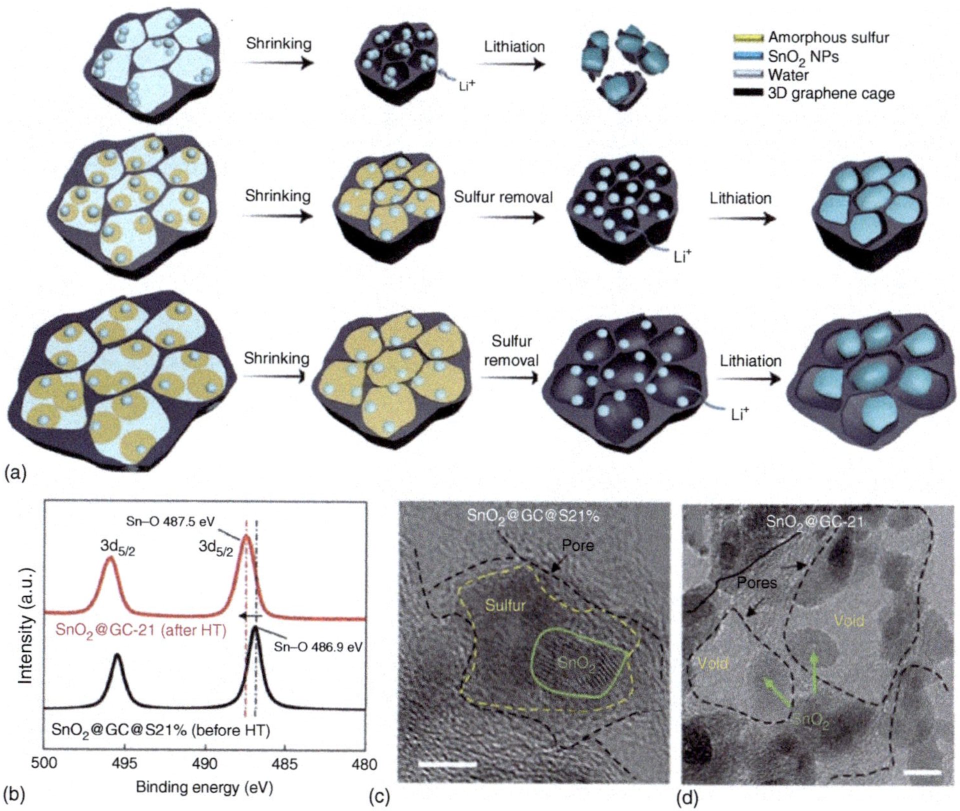

Figure 8.9 3D graphene buffering structure for Li-ion anode materials. (a) Schematic of the limited space produced after the shrinkage of a 3D graphene assembly. (b) Sn 3d XPS spectra before and after sulfur removal. (c) A HRTEM image of amorphous sulfur and the distribution of SnO_2 nanoparticles. (d) A HRTEM image of the uniformly distributed void space around SnO_2 nanoparticles after sulfur removal. Source: Han et al. [75]/ with permission from Springer Nature/CC BY 4.0.

framework effectively inhibits the dissolution of LiPSs in the electrolyte. The self-supporting electrode does not contain collector and adhesive, and shows high specific capacity (600 mAh/g) with good capacity retention up to 83% after 100 cycles [77]. But this compact structure is not conducive to the contact between electrolyte and sulfur, which limits the utilization of sulfur. Huang et al. took mesoporous graphene as the support scaffold for sulfur and prepared a mesoporous graphene/sulfur paper (MGP-S) electrode. In addition to the immobilization of soluble polysulfides, the mesoporous graphene also ensures the full penetration of electrolytes and the unimpeded diffusion of ions. Therefore, a high reversible capacity of 1393 mAh/g was obtained, and the cycle and rate performance are improved [78].

In addition to directly compounding with sulfur in the cathode, graphene can also form an intermediate layer between the sulfur cathode and the separator, which can obviously limit the shuttling of LiPSs. Graphene-based films were used as such an interlayer and the cycling performance of the Li–S battery was obviously enhanced. Liu et al. proposed a naphthalimide-functionalized poly(amidoamine) dendrimer and GO composite (Naph-Den-MGO) film, which combines chemical capture and physical blocking to effectively inhibit LiPS shuttling [79]. Our team also introduced a graphene film to load a TiO_2 and titanium nitride (TiN) heterostructure to form a thin and efficient polysulfide blocking layer (TiN : TiO_2-G), thus providing a longer cycle life (2000 cycles) at a high mass loading of 4.3 mg/cm^2 [80].

The use of 3D graphene assemblies of a Li–S battery could enhance the capacity performance since there are two significant advantages. One is the high conductivity of graphene nanosheets [81, 82]. Zhao et al. [83] prepared a 3D graphene/single-walled carbon nanotube (G/SWCNT) hybrid network to construct an electrode without other conductive additives, and the resulting graphene/CNT/sulfur hybrid showed outstanding performance with a high reversible capacity of 650 mAh/g at 5 C. Another advantage of a 3D GM is the porous nanostructure, which can effectively limit the polysulfides shuttling. By modifying the 3D graphene network with heteroatoms (N, O, S) and introducing chemical anchoring sites, the shuttle can be further inhibited.

8.4.4 Challenges for Practical EES Applications

Compared with the traditional carbons, such as graphite, activated carbons, and hard carbons, the graphene assemblies with well-designed nanostructures inherit the outstanding properties of graphene sheets and produce new unique characteristics from nanostructure design, which improves the energy density and power density of EES when used in electrodes [84–92]. Graphene assembly has broad application prospects in flexible EES devices and has great potential in future flexible mobile devices [93–101]. However, the graphene module still has a long way to go in the practical application of EES devices [102–105]. For instance, large-scale production of high-quality GM is still hard to achieve, and the cost also needs to be reduced. In addition, materials scientists and engineers shall pay attention to the further design and functionalization of graphene components for practical applications. The evaluation conditions of materials are usually different from those in practical use, which leads to many problems that are often ignored, such as low volumetric energy density caused by the low density of nanocarbons in electrodes, low Coulombic efficiency caused by the use of porous carbon, and high curvature and large interparticle resistance limit the mass loading of carbon nanomaterials, which are the main challenges in the application of carbon nanomaterials in EES.

8.5 Conclusions

As colloids, G and CMGs demonstrate attractive surface/interface properties, and thus many self-assembly strategies are developed. Based on the self-assembly techniques, multidimensional graphene assemblies are constructed with precisely controlled microstructures, and structural tuning and surface modification methods have been studied to tailor the hierarchical structure. Compared to bare graphene and CMG assemblies, composite materials show superior mechanical, thermal and chemical properties due to the synergistic effect of the graphene substrate and the active component. As a result, various graphene assemblies with tailored nanostructures are of great importance for achieving high-performance applications in EES systems.

Although graphene assembly has been widely studied and achieved great success, the following basic problems still need to be further studied to promote its practical application. First, the effects of electronic structure, surface structure, layer size, and sheet number and properties of graphene assemblies should be systematically studied. This is a basic problem for different applications. Second, it is necessary to systematically study the formation mechanism of assemblies from the perspective of surface/interface chemistry. Third, the relationship between nanostructures and properties is to be studied in detail from the aspects of electron mobility, mass transport, interfacial reaction kinetics and thermodynamics. With the solution to these problems, it is believed that graphene-derived carbons will have a large potential use toward the high-energy and high-power EES in the near future.

Acknowledgments

This work was supported by National Natural Science Foundation of China (Nos. 51525204, 51872195 and U1710256).

References

1 Geim, A.K. and Novoselov, K.S. (2007). *Nat. Mater.* 6: 183.
2 Raccichini, R., Varzi, A., Passerini, S., and Scrosati, B. (2015). *Nat. Mater.* 14: 271.
3 Lv, W., Li, Z., Deng, Y. et al. (2016). *Energy Storage Mater.* 2: 107.
4 Bonaccorso, F., Colombo, L., Yu, G. et al. (2015). *Science* 347: 1246501.

5 Han, S., Wu, D., Li, S. et al. (2014). *Adv. Mater.* 26: 849.
6 Xu, Y. and Shi, G. (2011). *J. Mater. Chem.* 21: 3311.
7 Wu, D., Zhang, F., Liang, H., and Feng, X. (2012). *Chem. Soc. Rev.* 41: 6160.
8 Li, P., Shang, T., Dong, X. et al. (2021). *Small*: 2007548.
9 Fendler, J.H. and Meldrum, F.C. (1995). *Adv. Mater.* 7: 607.
10 Cote, L.J., Kim, F., and Huang, J. (2009). *J. Am. Chem. Soc.* 131: 1043.
11 Li, X., Zhang, G., Bai, X. et al. (2008). *Nat. Nanotechnol.* 3: 538.
12 Jaafar, M.M., Ciniciato, G.P.M.K., Ibrahim, S.A. et al. (2015). *Langmuir* 31: 10426.
13 Lee, T., Min, S.H., Gu, M. et al. (2015). *Chem. Mater.* 27: 3785.
14 Park, J.S., Cho, S.M., Kim, W.-J. et al. (2011). *ACS Appl. Mater. Interfaces* 3: 360.
15 Hong, J., Char, K., and Kim, B.-S. (2010). *J. Phys. Chem. Lett.* 1: 3442.
16 Dikin, D.A., Stankovich, S., Zimney, E.J. et al. (2007). *Nature* 448: 457.
17 Li, D., Müller, M.B., Gilje, S. et al. (2008). *Nat. Nanotechnol.* 3: 101.
18 Eda, G., Fanchini, G., and Chhowalla, M. (2008). *Nat. Nanotechnol.* 3: 270.
19 Zhu, Y., Stoller, M.D., Cai, W. et al. (2010). *ACS Nano* 4: 1227.
20 Shao, J., Lv, W., and Yang, Q. (2014). *Adv. Mater.* 26: 5586.
21 Chen, C., Yang, Q., Yang, Y. et al. (2009). *Adv. Mater.* 21: 3007.
22 Luo, C., Lv, W., Qi, C. et al. (2019). *Adv. Mater.* 31: 1.
23 Shao, J.-J., Wu, S.-D., Zhang, S.-B. et al. (2011). *Chem. Commun.* 47: 5771.
24 Zhang, L. and Shi, G. (2011). *J. Phys. Chem. C* 115: 17206.
25 Xu, Y., Sheng, K., Li, C., and Shi, G. (2010). *ACS Nano* 4: 4324.
26 Qiu, L., Liu, J.Z., Chang, S.L.Y. et al. (2012). *Nat. Commun.* 3: 1.
27 Hu, H., Zhao, Z., Wan, W. et al. (2013). *Adv. Mater.* 25: 2219.
28 Guo, P., Song, H., and Chen, X. (2010). *J. Mater. Chem.* 20: 4867.
29 Xu, Z., Zhang, Y., Li, P., and Gao, C. (2012). *ACS Nano* 6: 7103.
30 Dong, Z., Jiang, C., Cheng, H. et al. (2012). *Adv. Mater.* 24: 1856.
31 Yang, X., Cheng, C., Wang, Y. et al. (2013). *Science* 341: 534.
32 Tao, Y., Xie, X., Lv, W. et al. (2013). *Sci. Rep.* 3: 2975.
33 Xu, Y., Lin, Z., Zhong, X. et al. (2014). *Nat. Commun.* 5: 1.
34 Li, H., Tao, Y., Zheng, X. et al. (2016). *Energy Environ. Sci.* 9: 3135.
35 Wang, X., Sun, G., Routh, P. et al. (2014). *Chem. Soc. Rev.* 43: 7067.
36 Wu, T., Shen, H., Sun, L. et al. (2012). *New J. Chem.* 36: 1385.
37 Zhao, Y., Hu, C., Hu, Y. et al. (2012). *Angew. Chem. Int. Ed.* 51: 11371.
38 Ge, Y., Wang, C., Zhao, Y. et al. (2018). *Small* 14: 1703096.
39 Qin, J., Pasko, V.P., McHarg, M.G., and Stenbaek-Nielsen, H.C. (2014). *Nat. Commun.* 5: 1.
40 Erb, R.M., Libanori, R., Rothfuchs, N., and Studart, A.R. (2012). *Science* 335: 199.
41 Zhang, K., Zhao, W., Lee, J.-T. et al. (2014). *J. Mater. Chem. A* 2: 9636.
42 Kawamori, M., Asai, T., Shirai, Y. et al. (2014). *Nano Lett.* 14: 1932.
43 Gohier, A., Laïk, B., Kim, K. et al. (2012). *Adv. Mater.* 24: 2592.
44 Shearer, C.J., Cherevan, A., and Eder, D. (2014). *Adv. Mater.* 26: 2295.
45 Shin, M.K., Lee, B., Kim, S.H. et al. (2012). *Nat. Commun.* 3: 1.
46 Lu, X., Dou, H., and Zhang, X. (2016). *Mater. Lett.* 178: 304.

47 Sui, Z., Meng, Q., Zhang, X. et al. (2012). *J. Mater. Chem.* 22: 8767.
48 Sun, H., Xu, Z., and Gao, C. (2013). *Adv. Mater.* 25: 2554.
49 Yang, Y., Bolling, L., Priolo, M.A., and Grunlan, J.C. (2013). *Adv. Mater.* 25: 503.
50 Xiang, F., Parviz, D., Givens, T.M. et al. (2016). *Adv. Funct. Mater.* 26: 2143.
51 Xu, Y., Tao, Y., Zheng, X. et al. (2015). *Adv. Mater.* 27: 8082.
52 Tang, J., Yang, J., Zhou, L. et al. (2014). *J. Mater. Chem. A* 2: 6292.
53 Xiao, F.-X., Miao, J., and Liu, B. (2014). *J. Am. Chem. Soc.* 136: 1559.
54 Xi, Q., Chen, X., Evans, D.G., and Yang, W. (2012). *Langmuir* 28: 9885.
55 Li, Z., Wang, J., Liu, X. et al. (2011). *J. Mater. Chem.* 21: 3397.
56 Li, D., Li, X., Wang, S. et al. (2014). *ACS Appl. Mater. Interfaces* 6: 648.
57 Ma, H., Kong, D., Xu, Y. et al. (2017). *Small* 13: 1701026.
58 Meng, Y., Zhao, Y., Hu, C. et al. (2013). *Adv. Mater.* 25: 2326.
59 Qu, G., Cheng, J., Li, X. et al. (2016). *Adv. Mater.* 28: 3646.
60 Cheng, H., Dong, Z., Hu, C. et al. (2013). *Nanoscale* 5: 3428.
61 Yoo, J.J., Balakrishnan, K., Huang, J. et al. (2011). *Nano Lett.* 11: 1423.
62 Lv, W., Li, Z., Zhou, G. et al. (2014). *Adv. Funct. Mater.* 24: 3456.
63 Su, D. (2012). *Chem. Commun.* 48: 7149.
64 Zhang, L.L., Zhao, X., Stoller, M.D. et al. (2012). *Nano Lett.* 12: 1806.
65 Peng, L., Peng, X., Liu, B. et al. (2013). *Nano Lett.* 13: 2151.
66 Wu, Z., Zheng, Y., Zheng, S. et al. (2017). *Adv. Mater.* 29: 1602960.
67 Li, H., Tao, Y., Zheng, X. et al. (2015). *Nanoscale* 7: 18459.
68 Freunberger, S.A. (2017). *Nat. Energy* 2: 1.
69 Chen, X.-C., Wei, W., Lv, W. et al. (2012). *Chem. Commun.* 48: 5904.
70 Wang, C., Li, D., Too, C.O., and Wallace, G.G. (2009). *Chem. Mater.* 21: 2604.
71 Lee, J.K., Smith, K.B., Hayner, C.M., and Kung, H.H. (2010). *Chem. Commun.* 46: 2025.
72 Yun, Q., Qin, X., Lv, W. et al. (2015). *Carbon N. Y.* 93: 59.
73 Li, N., Chen, Z., Ren, W. et al. (2012). *Proc. Natl. Acad. Sci.* 109: 17360.
74 Xu, Y., Lin, Z., Zhong, X. et al. (2015). *Angew. Chem. Int. Ed.* 54: 5345.
75 Han, J., Kong, D., Lv, W. et al. (2018). *Nat. Commun.* 9: 1.
76 Ji, L., Rao, M., Zheng, H. et al. (2011). *J. Am. Chem. Soc.* 133: 18522.
77 Jin, J., Wen, Z., Ma, G. et al. (2013). *RSC Adv.* 3: 2558.
78 Huang, X., Sun, B., Li, K. et al. (2013). *J. Mater. Chem. A* 1: 13484.
79 Liu, W., Jiang, J., Yang, K.R. et al. (2017). *Proc. Natl. Acad. Sci.* 114: 3578.
80 Zhou, T., Lv, W., Li, J. et al. (2017). *Energy Environ. Sci.* 10: 1694.
81 Hu, G., Xu, C., Sun, Z. et al. (2016). *Adv. Mater.* 28: 1603.
82 Zhou, G., Paek, E., Hwang, G.S., and Manthiram, A. (2015). *Nat. Commun.* 6: 1.
83 Zhao, M.-Q., Liu, X.-F., Zhang, Q. et al. (2012). *ACS Nano* 6: 10759.
84 Tung, T.T., Nine, M.J., Krebsz, M. et al. (2017). *Adv. Funct. Mater.* 27: 1702891.
85 Zhou, M., Li, X., Wang, B. et al. (2015). *Nano Lett.* 15: 6222.
86 Jin, H., Bu, Y., Li, J. et al. (2018). *Adv. Mater.* 30: 1707424.
87 Cong, H.-P., Chen, J.-F., and Yu, S.-H. (2014). *Chem. Soc. Rev.* 43: 7295.
88 Li, D. and Kaner, R.B. (2008). *Science* 320: 1170.
89 Liu, D., Jia, Z., Zhu, J., and Wang, D. (2015). *J. Mater. Chem. A* 3: 12653.
90 Qi, B., Ren, K., Lin, Y. et al. (2021). *Particuology* 60: 13.

91 Schütt, F., Rasch, F., Deka, N. et al. (2021). *Mater. Today* 48: 7.
92 Worsley, M.A., Pauzauskie, P.J., Olson, T.Y. et al. (2010). *J. Am. Chem. Soc.* 132: 14067.
93 Wang, G.-F., Qin, H., Gao, X. et al. (2018). *Chem* 4: 896.
94 Chao, Y., Jalili, R., Ge, Y. et al. (2017). *Adv. Funct. Mater.* 27: 1700234.
95 Li, W., Li, Y., Su, M. et al. (2017). *J. Mater. Chem. A* 5: 16281.
96 Chen, W., Liu, P., Liu, Y. et al. (2018). *Chem. Eng. J.* 353: 115.
97 Fan, Y.-M., Song, W.-L., Li, X., and Fan, L.-Z. (2017). *Carbon N. Y.* 111: 658.
98 Zhao, M., Trainor, N., Ren, C.E. et al. (2019). *Adv. Mater. Technol.* 4: 1800639.
99 Yan, J., Ren, C.E., Maleski, K. et al. (2017). *Adv. Funct. Mater.* 27: 1701264.
100 Zeng, W., Tao, X.-M., Lin, S. et al. (2018). *Nano Energy* 54: 163.
101 Chen, K., Wang, Q., Niu, Z., and Chen, J. (2018). *J. Energy Chem.* 27: 12.
102 Han, J., Wei, W., Zhang, C. et al. (2018). *Electrochem. Energy Rev.* 1: 139.
103 Fang, B., Chang, D., Xu, Z., and Gao, C. (2020). *Adv. Mater.* 32: 1902664.
104 Tao, Y., Sui, Z.-Y., and Han, B.-H. (2020). *J. Mater. Chem. A* 8: 6125.
105 Li, P., Han, D., Shang, T. et al. (2019). *Adv. Sci.* 6: 1802355.

9

Patterning Assembly of Colloidal Particles

Minxuan Kuang[1] and Yanlin Song[2]

[1] *Beijing Institute of Fashion Technology, School of Materials Design & Engineering, Beijing Key Laboratory of Clothing Materials R & D and Assessment, Beijing Engineering Research Center of Textile Nanofiber, No.2 East Yinghua Road, Chaoyang District, Beijing 100029, China*
[2] *Institute of Chemistry, Chinese Academy of Sciences, Beijing National Laboratory for Molecular Sciences (BNLMS), Key Laboratory of Green Printing, Key Laboratory of Organic Solids, Zhongguancun North First Street 2, Beijing 100190, China*

9.1 Introduction

Assembly is a method wherein a delicate structure is constructed with or without external forces, which is considered as a promising method in fabrication of nanostructures with a high degree of precision [1, 2]. Achieving highly accurate assembly, assembling efficiency and assembling quality are of great importance for many applications such as catalysis [3], chemical sensing [4], optoelectronics [5], biomedicine [6], etc. Over the past decades, extensive efforts have been devoted to explore the assembly mechanisms [7–9]. The main driving force for colloidal self-assembly without an external field is an attractive capillary force originating from formation of liquid bridges between particles during evaporation, also called meniscus effects [7]. This capillary force drives the colloidal particles to form a hexagonal assembly structure after liquid drying. Besides self-assembly by evaporation, assembly also can be realized based on accurate control of external fields, such as electronic field [10], magnetic field [11], or DNA origami template [12]. Patterning assembly of colloidal particles shows unique applications in the materials science and the manufacture of functional devices [13, 14]. In particular, photonic crystals (PCs) assembled by colloidal particles have been rapidly developed due to the advantage of facile and cost-effective fabrication for photonic devices [15–17]. For practical applications, the patterning assembly process should be simple, fast, and reproducible. Although many achievements have been reported, one of the main challenges in fabricating devices that assembled from particles is precise control of particle placement and integration on surfaces.

In this chapter, methods of patterning assembly are summarized, and applications of patterns consisting of colloidal PCs are introduced. We aim to provide various approaches that can be used to obtain assembly patterns of colloidal crystals and

Functional Materials from Colloidal Self-assembly, First Edition. Edited by Qingfeng Yan and George Zhao.

offer potential applications of the assembly structures. In the last section, a brief summary and some perspectives of this research field are proposed.

9.2 Strategies of Patterning Assembly

9.2.1 Inkjet Printing of Assembly Patterns

Inkjet printing technology, with the advantages of flexibility and low cost, has been widely applied for patterning functional materials [18, 19]. This patterning technology is based on the drop-on-demand and drop-jetting technique without any etching process. So it is a material-saving and environment-friendly technology. Patterns are firstly designed using typical layout software in computer, which then are transported to a printer. Droplets with fixed volumes are ejected from a nozzle via piezoelectric actions. An inkjet printer usually has tens to hundreds of nozzles and can eject 6000 droplets per second, providing high efficient printing of patterns. Various devices can be fabricated by the inkjet printing technology. PCs are periodically structured materials with different refractive indexes [20, 21]. Assembly from colloidal particles is an important method for constructing PCs. PC patterns can be obtained by printing PC inks and assembling the colloidal particles with or without external field. Printing inks for colloidal PC patterns are usually based on colloidal suspensions. Colloidal droplets are formed and ejected by exerting a piezoelectric pulse on the inks contained in a chamber. When the colloidal droplets are deposited and evaporate on the substrate, the colloidal particles assemble into an ordered structure by a convective capillary force.

The colloid assembly structures and the quality of PC patterns are influenced by the ink formulation and the surface wettability. A typical ink formulation for printing PC patterns contains colloidal particles, solvents, and additives, which determines the viscosity and surface tension of the inks, and the capillary flow within the droplets. These physical properties cooperatively influence the printing process and the particle assembly structures. The surface wettability including static or dynamic contact angles influences the morphology of printed droplets on the substrate as well as the evaporation mode of printed ink droplets. All these factors jointly affect the quality of printed PC patterns and their optical properties.

9.2.1.1 Ink Formulations

Challenges in inkjet printing of PC patterns include synthesis of monodispersed particles and development of ink formulations, which not only influence the ejecting process but also dictate the assembly process of the particles. Monodispersed particles with good dispersion stability are essential for inkjet printing of high-quality PC patterns. Any aggregations within the printing inks will clog the nozzles, reducing the fluency of the printing process. In order to avoid particle precipitations, the printing inks should be treated by ultrasonic for 30 min and filtrated through a mesh before printing. The mesh size depends on the diameter of the colloidal particles. Colloidal particles with narrow distribution are critical for assembly of long-range periodic structures and obtaining high-quality PC patterns. Cui et al. reported that the desired polydispersity of colloidal particles for the PC ink is 0.005 [22]. Monodispersed colloidal particles can be synthesized by emulsion polymerization [23, 24], dispersion polymerization [25], Stöber synthesis [26], hydrothermal synthesis [27], etc. The particle size and distribution can be tuned by changing the monomers and the synthesis condition.

In principle, all types of printing inks contain similar components. Printing inks for PC patterns are formulated using four main components: monodispersed colloidal particles, main solvents, cosolvents (also named secondary solvents), and additives. Colloidal particles are used to create long-range periodic structures. Main solvents keep the colloidal particles stably dispersed. They should be good dispersants for the colloidal particles, ensuring a long shelf life of the ink. The ink should be stable without precipitation for six months in general. Cosolvents regulate the viscosity, wetting behavior, evaporation mode, and drying time of the inks. Solvents with high boiling point and low surface tension, such as ethylene glycol, *N*,*N*-dimethylformamide, glycerol, diethylene glycol, or acetophenone, are always used as the cosolvents. Using multiple solvents in the formulation adjusts the viscosity and surface tension to a desired value and alters the capillary flow within the printed droplets. Lastly, additives can change the ink's chemical/physical properties to suit different situations, e.g. promoting the leveling property, sensing the environmental stimulation, enhancing the flexibility of the printed film, increasing scrub resistance, or promoting adhesion.

Surface tension of a PC ink affects properties such as the wettability and foaming of an ink. The wettability of the printing ink is important for coating on different substrates. Generally, the higher the surface tension of the ink, the more difficult it will wet the substrate. Similarly, the lower the surface tension of the substrate, the more difficult it will be coated. When the surface tension of the ink is lower than that of the substrate, it will coat the substrate. For water-based inks, there is a problem of wetting surface, as water has a very high surface tension of 72 mN/m, which is higher than that of most substrates. To solve this problem, typically a surfactant will be added to the formulation of a water-based ink to lower the surface tension. Generally, an ink formulation with viscosity of 8–14 cps and surface tension of 25–40 dyn/cm is desired for a fluent printing process.

The colloidal stability of an ink is a critical factor of printing fluently and fabricating high-quality PC patterns. Without stabilizing colloids of the ink, the particles would aggregate within a short time. There are two methods to stabilize inks: adding surfactants and adding polymers. In some cases, both surfactants and polymers are added. Because the added surfactants or/and polymers will adsorb at the particle/liquid interface and result in a repulsion of particles within the ink, the colloidal particles could be stabilized for a long time. However, the addition of surfactants or/and polymers may affect the assembly process of particles, as they could change the capillary flow within the printed droplets, which is a driving force for particle assembling during solvent evaporation. Therefore, the dosage of surfactants and polymers should be carefully considered when formulating a PC ink.

Ink formulations determine the particle depositions as well as the assembly structures. A ringlike deposition is usually formed when the printing ink contains only one kind of solvent. Particles may aggregate at the edge after solvent drying. This phenomenon is known as the "coffee-ring" effect, which is induced by the unsymmetrical evaporation. However, uniform films are desirable for the application of painting, printing technology, etc. A uniform film is necessary for printing patterns with high precision and resolution and obtaining high-performance devices [28]. Depositions with well-ordered assembly structures are also necessary to achieve good optical properties of a PC devices. Eliminating the "coffee-ring" effect becomes a main challenge in printing functional devices. Over the past decades, numerous researches have been devoted to understanding the mechanism of the "coffee-ring" effect and avoiding this inhomogeneous deposition. Deegan et al. firstly revealed the mechanism of forming the ring-shaped depositions. They found that the evaporation flux was highest at the edge of a drying droplet, which induced an outward radial capillary flow. Particles migrated to the droplet's edge and deposited there driven by this capillary flow (Figure 9.1) [29–31]. A ring-shaped deposition was formed after drying.

Adding cosolvents to printing formulations can effectively eliminate the ring-shaped deposition. Solvents with low surface tension and high boiling point are preferred [28, 32–34]. Ethylene glycol is a commonly used cosolvent in formulating water-based inks. A concentration gradient of the ethylene glycol will be formed between the center and edge of the droplet during evaporation because the evaporation rate of ethylene glycol is lower than that of water. The concentration gradient will generate a gradient of surface tension along the liquid surface, where the surface tension in the center is higher than that at the edge. This surface tension gradient will drive the fluids inward, which is opposite to the outward capillary flow. The inward capillary flow is named as Marangoni flow [35]. Particles within a drying droplet will be brought from the edge to the center by the Marangoni flow.

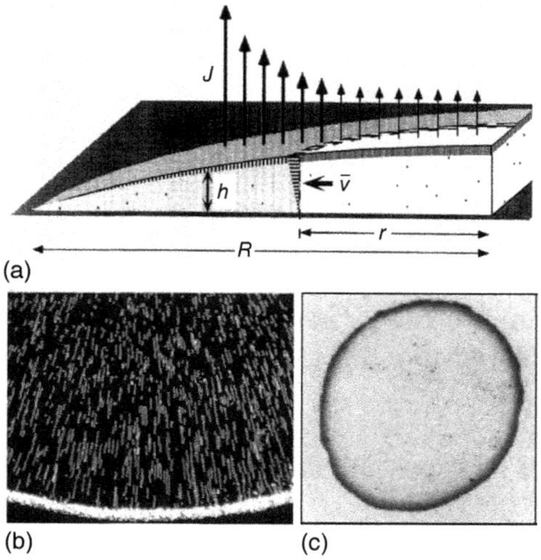

Figure 9.1 (a) Schematic illustration of the evaporative flux distribution of the droplet surface. (b) Particles migrating to the edge of the droplet during evaporating. Source: Deegan et al. [29]/with permission from Springer Nature. (c) A ring-shaped deposition on substrate. Source: Deegan [30]/with permission from AIP Publishing LLC.

It is reported that a uniform deposition of silver nanoparticles can be obtained by using an ink with mixed solvents containing 32 wt% ethylene glycol. Other researches also show that adding ethylene glycol into the inks can significantly improve the assembly of particles. Song and coworkers investigated the influence of ethylene glycol on the particle deposition and the optical properties of printed PCs [22]. As shown in Figure 9.2a, the particles formed a ring-shaped deposition when the concentration of ethylene glycol was 0%. The reflectivity of the printed PC patterns was ~6%, because the particles did not assemble into well-ordered structures. With the concentration of the ethylene glycol increasing, the reflectivity increased gradually. When the concentration of ethylene glycol increased to 40%, a uniform deposition was obtained, and the reflectivity of printed PC patterns reached ~46%. Scanning electron microscope (SEM) image showed that the particles were assembled in a close-packed face-centered cubic (fcc) structure. The printed PC patterns exhibited a bright color due to the Bragg scattering from the periodic lattice. However, the reflectivity decreased with the concentration of ethylene glycol further increasing. If the concentration of ethylene glycol is too high, the viscosity of the droplet will be too large, which impedes particle assembling. Besides ethylene glycol, formamide is another commonly used cosolvent in the water-based inks, which also results in a homogeneous particle assembly. Moon and coworker found that a self-assembled monolayer of particles with well-ordered structure was obtained by using inks with water–formamide mixed solvent, while ring-shaped deposition with scattered particles was obtained by using inks with a single solvent (Figure 9.2b) [32]. This transition is mainly attributed to an inward Marangoni flow within the drying droplet of the mixed solvent, which directs the particles to the center of the droplet. The inward Marangoni flow counters the outward convective flow and results in homogeneous colloidal crystals.

Adding surfactants into the ink formulations can induce Marangoni flows as well [36, 37]. The surfactants created localized swirling flows named as "Marangoni eddies" in a drying droplet. The direction of these eddies is opposite to the outward capillary flow, which prevents particles depositing at the drop's perimeter. A rather uniform deposition was obtained by adding surfactants, such as sodium dodecyl sulfate [36].

The concentration of colloidal particles is another important factor for the particle assembly and the reflective intensity of printed PC patterns. It is reported that an ink with very low concentration, such as 0.5 vol%, formed a ring-shaped deposition even when a mixed solvent was employed. Increasing the particle concentration, the width of the ring became thicker, and the particles eventually formed a uniform monolayer colloidal crystal at 4 vol%. The self-assembled monolayer colloidal crystal consisted of large grains of well-ordered hexagonal lattice of particles, with several point and line defects inside. Further increasing the particle concentration, assembly structure with double or multiple particle layers was formed [32].

Functional monomers are introduced into the formulation of the printing ink in some cases. Acrylic acid, *N*-isopropyl acrylamide, and acrylamide are the commonly used functional monomers [38, 39]. These monomers could regulate the surface tension and viscosity of the printing inks and further influence the printing fluency.

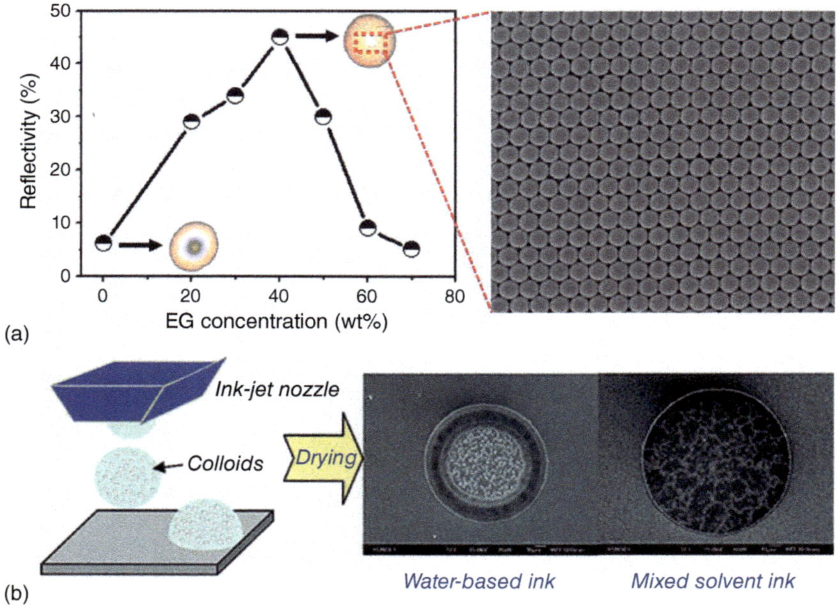

Figure 9.2 Influence of ink compositions on the particle deposition and assembly structure. (a) Reflective intensities of the printed PCs from inks with different ethyl glycol concentrations. Well-ordered assembly structure was obtained by using a mixed solvent containing 40 wt% ethylene glycol. Source: Cui et al. [22]/with permission from Royal Society of Chemistry. (b) Particle depositions printed from a water-based ink and a water/formamide-based ink. Source: Park and Moon [32]/with permission from American Chemical Society.

Adding monomers into the ink formulations also affects the wetting and spreading behavior of the ink droplets and the assembly behavior of the particles. The monomer's polymerization rate plays an important role in particle assembling. A perfect match between the polymerization rate and particle assembly rate is necessary. If the rate of particle assembly is faster than the rate of monomer polymerization, the particles will form a ring-shaped deposition. The fluid flow within the drying droplets is similar to that in the droplets with a single solvent and brings particles to the droplet's edge. If the rate of monomer polymerization is faster than the rate of particle assembly, the quickly increased viscosity will impede the particle assembly [38].

9.2.1.2 Surface Wettability

In a typical printing process, uniform ink drops are ejected to arbitrary substrates. For printing PC patterns, the colloidal particles in each drop assemble into a well-ordered structure on the substrate. The substrate's wettability plays a critical role for the assembly structure of the colloidal particles. It influences the assembly time and the interactive force between particles. The surface wettability includes the static wettability and the dynamic wettability. The static wettability of the

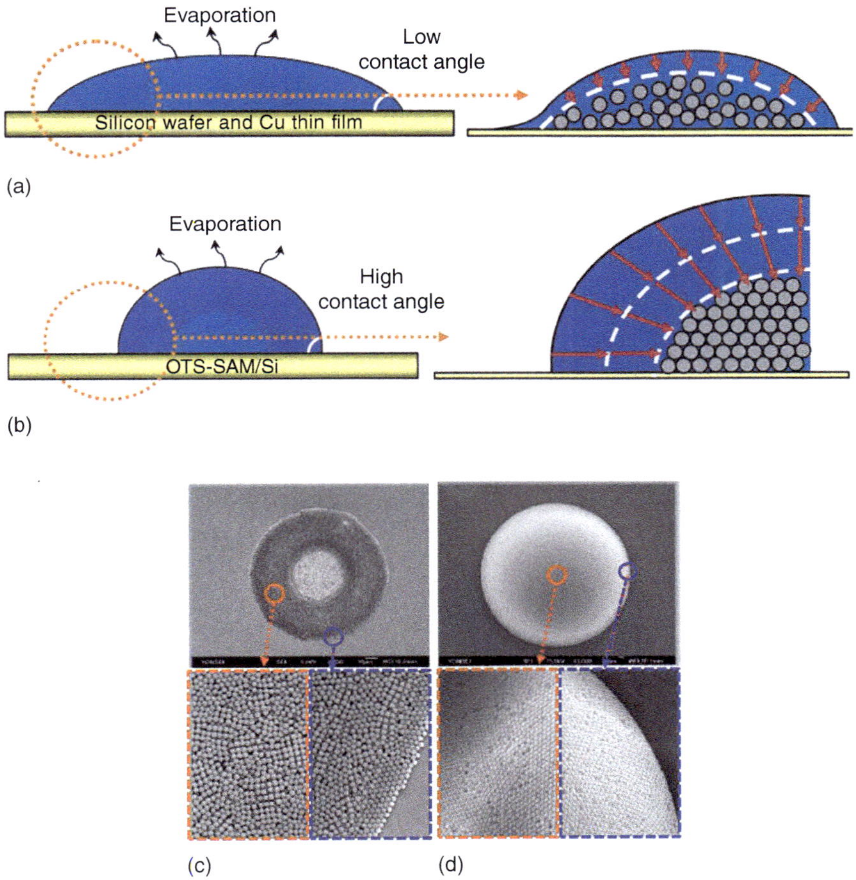

Figure 9.3 (a, b) Schematic illustration of colloidal particles assembly with a droplet printed on a (a) hydrophilic and (b) hydrophobic substrate. (c, d) SEM images of the printed PC dots on a (c) hydrophilic and (d) hydrophobic substrate. Source: Ko et al. [43]/with permission from American Chemical Society.

substrates usually determines the size of printed droplets, while the dynamic wettability governs the evaporating mode of droplets [40–42].

The static wettability of a surface is characterized by apparent contact angles. A droplet will form different shapes on the hydrophilic and hydrophobic substrate. Figure 9.3 shows the influences of the substrate wettability on the shape of droplets and the morphology of particle deposition [43]. The results show that the drying mode of droplets on a hydrophilic substrate and a hydrophobic substrate is different. On a hydrophilic substrate, printed droplets will spread on the surface and form disklike shapes (Figure 9.3a). The contact angle gradually decreases, and the base diameter remains consistent on the hydrophilic surface because of a pinning three phase contact line (TCL) during evaporation. The evaporation rate at the periphery of the droplet is much higher than that at the center. An outward capillary flow is induced, which drives the colloidal particles moving to the periphery. Furthermore, the droplets on the hydrophilic surface will dry at a short time because of the thin liquid layer. The assembly time for the colloidal particles is short. Thus, it is difficult to obtain well-ordered assembly structures on the hydrophilic surface. The particles usually randomly deposit after drying (Figure 9.3c). PCs printed on the hydrophilic surface may show a weak reflective intensity. In order to print PC patterns with good optical property, cosolvents are usually added to the ink formulations, which could change the capillary flow within the drying droplets and improve particle assembly [22]. In contrast, the drying mode on a hydrophobic substrate is different. The droplets printed on the hydrophobic substrate will form a hemispherical shape (Figure 9.3b). The liquid layer of the hemispherical-shaped droplets is thick, resulting in a lower evaporation rate and a longer assembly time. The colloidal particles are gradually concentrated during drying. When the particle concentration reaches a critical value, at which the particles' mobility is significantly reduced, the colloidal particles start to crystalline at the liquid surface. After drying, the colloidal particles assemble into a long-range ordered structure, displaying large domains of hexagonally packed structure (Figure 9.3d). PCs printed on such substrates always show strong reflective intensity. Song and coworkers printed hydrophilic monodispersed poly(styrene-methylmethacrylate-acrylic acid) (poly(St-MMA-AA)) nanoparticles on a hydrophobic polydimethylsiloxane (PDMS) substrate and fabricated patterned colloidal PC microchips. The hydrophilic nanoparticles assembled into a long-range ordered structure, which gave rise to strong reflective intensity of the printed PC dots [44].

The dynamic wettability of a surface is characterized by the advancing contact angle, the receding contact angle, the hysteresis contact angle, or the adhesion force of a liquid droplet on the substrate. The dynamic wettability also influences the evaporating mode of printed droplets. Extensive studies have been conducted on the evaporation process of droplets and the particle assembly structures on substrates with different dynamic wettabilities [45–48]. On a hydrophobic substrate with high hysteresis, the evaporation mode of printed droplets is similar to that on a hydrophilic substrate. The TCL pins on the surface, and the contact angle gradually decreases during drying. In contrast, the TCL of droplet continuously recedes on a hydrophobic substrate with low hysteresis, resulting in a consistent contact angle during the evaporation process. The different TCL behaviors exert great influences on the assembly structure of the colloidal particles. On the substrate with low hysteresis contact angle, the adhesion force between the droplet and the surface is low. Perfect well-ordered, crack-free PCs could be obtained on a low-adhesive superhydrophobic substrate (Figure 9.4a) [16]. There are three reasons why perfectly assembled structure could be obtained on this substrate. Firstly, the receding TCL of the latex droplet on such substrate provides enough time for the particle assembly. Secondly, the tensile stress between the particles can be released due to the receding TCL. Thirdly, the receding TCL squeezes the latex particles, which results in compact structure with low air fraction. PCs fabricated on such surfaces have a perfect well-ordered, crack-free assembly structure and show a very narrow

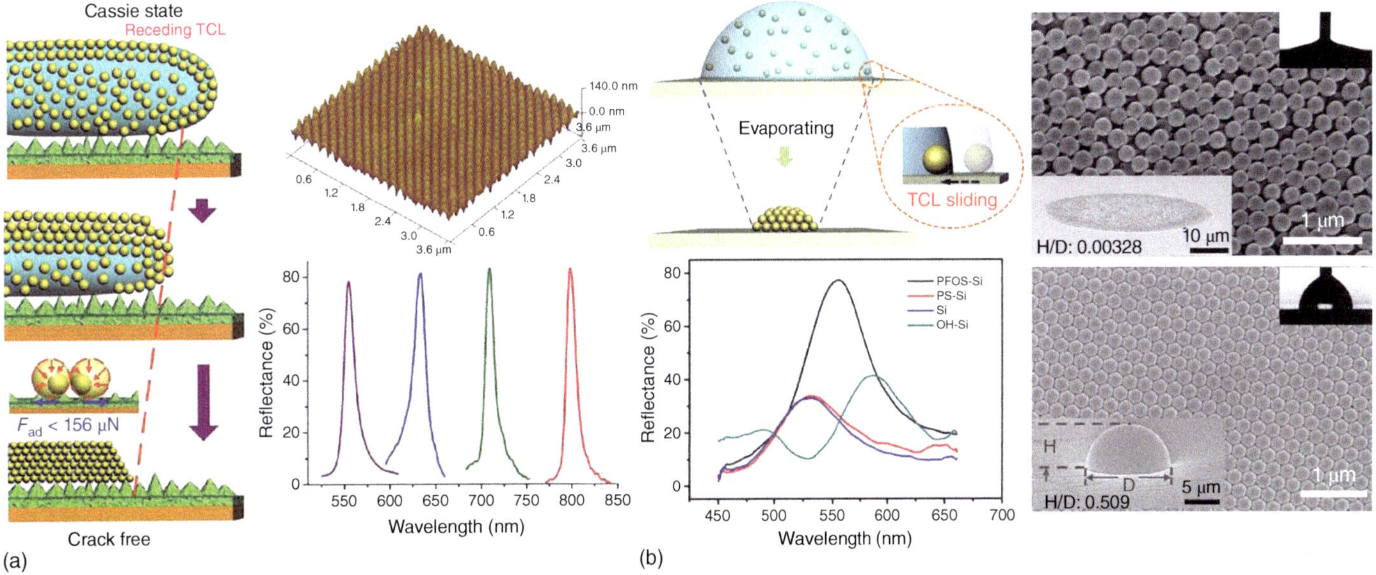

Figure 9.4 (a) Colloidal droplets on a superhydrophobic substrate with low adhesion generating PCs with narrow stopband [16]. Source: Reproduced with permission from the American Chemical Society, 2012. (b) Particle assembly structures and reflection intensities of printed PC dots on substrates with receding contact angle of 18.0° ± 5.7° and 93.3° ± 1.9°. Source: Kuang et al. [45]/with permission from John Wiley & Sons, Inc.

stopband (12 nm). Besides the hysteresis angle, the receding contact angle also exerts significant influences on the particle assembly structures and the optical property of printed PCs (Figure 9.4b–d) [45]. For example, on a substrate with a receding contact angle of $18.0 \pm 5.7°$, only the particles at the edge assemble into ordered structure after drying. The reflection intensity of printed PCs is weak. However, on a substrate with a receding contact angle of $93.3 \pm 1.9°$, the particles assemble into hexagonal ordered structure, resulting in strong reflective intensity. The particles in each ink droplet are accumulated at the center of the droplet and assemble into compact ordered structure by the receding TCL.

9.2.2 Patterned Substrate-Induced Assembly

Assembling particles on substrates with topographical or chemical patterning is a powerful strategy to obtain assembly patterns [2, 8, 49, 50]. The patterned substrates can induce heterogeneous TCL behaviors on the surfaces, which further direct the assembly of colloidal particles. On patterned surfaces, the pattern provides well-designed spatial distribution of forces to guide the deposition of particles. The assembly process usually involves three capillary forces, namely, the interactive force between the particles, the interactive force between the particle and the substrate, and the convective force induced by the evaporation. These capillary forces guide the assembly of particles in the micro-region, forming well-ordered dots, lines, and planes. By delicately designing the experimental parameters (such as the substrate temperature [8], the dewetting direction [8, 51], the receding contact angle [51], the particle concentration [8], the viscosity of colloidal suspension [52], etc.) and the surface topography, assembly patterns from single particle placement to two-dimensional (2D) arrangements can be realized [8, 51]. The experimental parameters and the surface topography provide full control of the assembly process and the final particle deposition. Song and coworkers precisely and facilely manipulated particles into different assembly patterns by utilizing a confined template-induced assembly technology [51–53]. The liquid confined assembly was realized in the gap between the pillar top of the template and a flat substrate. A 2D liquid membrane was broken into linear liquid bridges during the evaporation process. The particles arranged in the drying liquid bridges to form well-ordered assembly structures. By tuning the viscosity and the size confine ratio of the assembly system, the obtained patterns could change from linear assembly to zipper assembly, zigzag assembly, and twined zigzag assembly (Figure 9.5a). Substrates with microstructures, such as microchannels, can also direct nanoparticle assembly entirely in the geometrically confined spaces. Ozin and coworkers obtained regular rectangular-shaped PC arrays by using a microchannel template (Figure 9.5b) [50]. A strong capillary flow directs the nanoparticles to migrate into the spatial confines of the microchannels instead of on the top surface. Besides the physical structures, chemical patterns on the surfaces also be used to fabricate assembly patterns. On a substrate with chemical patterns, colloidal suspensions are always confined and assemble in the hydrophilic area because of its higher surface energy [49]. The baseline of these droplets copies the hydrophilic outline. The nanoparticles

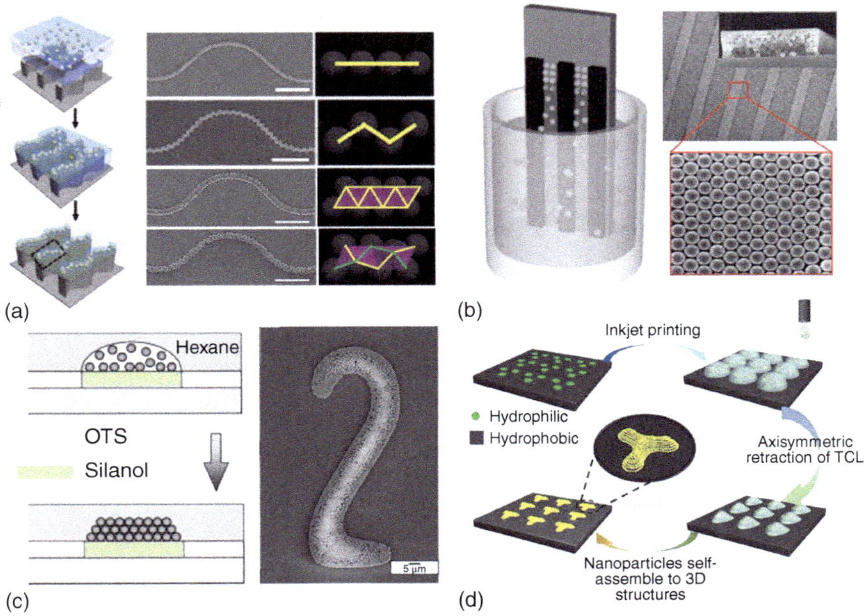

Figure 9.5 (a) Control of zigzag or linear assembly by utilizing a confined template-induced assembly technology. Scale bar: 20 μm. Source: Guo et al. [52]/with permission from John Wiley & Sons, Inc. (b) Colloidal PC pattern induced by the capillarity of the channel on the substrate. Source: Yang et al. [50]/with permission from John Wiley & Sons, Inc. (c) Fabricating patterned colloidal photonic crystals by using a two-solution self-assembly method. Source: Masuda and Koumoto [49]/with permission from American Chemical Society. (d) Three-dimensional PCs induced by asymmetrically dewetting on a hydrophilic–hydrophobic patterned substrate [54]. Source: Reproduced with permission from the Wiley-VCH Verlag GmbH & Co. KGaA, 2015.

confined in the hydrophilic area assemble into close-packed structures in the drying process driven by the meniscus force and eventually form patterns identical to the hydrophilic pattern on surface (Figure 9.5c) [49]. However, in some cases, the colloidal suspensions are not completely confined in the hydrophilic area. When a colloidal droplet has a diameter larger than the dimension of the hydrophilic area, it may crossover two or more hydrophilic zone and adapt deformed shapes. As shown in Figure 9.5d, Song and coworkers constructed three-dimensional (3D) PC arrays with various shapes on substrates with different chemical patterns [54]. The hydrophilic regions serve as pinning points, preventing TCL receding on the surface during the evaporation process. However, the TCL continuously recedes on the hydrophobic regions. Therefore, an axisymmetric TCL retraction occurs on the surface. The colloidal particles move with the TCL and assemble along the curved surface of the droplet, resulting in lines, quadrilaterals, stars, hexagons, octagons, and other shapes. Although assembly patterns with high resolutions can be obtained by utilizing the substrates with topographical or chemical patterning, the complicate and cost fabrication of the template limits its applications.

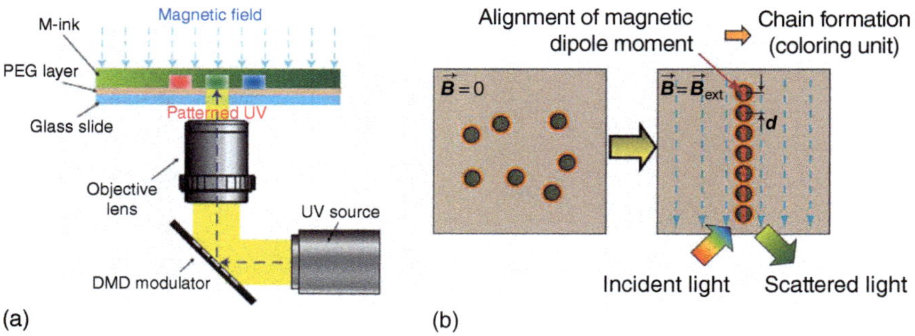

Figure 9.6 (a) Schematic illustration of fabricating PC patterns by using a magnetic field and UV curing; (b) the Fe_3O_4 nanoparticles assemble to form a chainlike photonic nanostructure upon the application of an external magnetic field. The color can be tuned by varying the interparticle distance by modulation of the external magnetic field intensity [55]. Source: Reproduced with permission from Springer Nature Limited, 2009.

9.2.3 External Stimuli-Induced Assembly

For a colloidal solution drying on a substrate, the colloidal particles assemble into ordered structure driven by a convective capillary force. Multicolor PC patterns can be obtained after drying. With the help of external stimulation, well-assembled structures can be constructed from a solution containing responsive nanoparticles. The exterior excitation guides the assembling of colloidal nanoparticles. Yin and coworkers fabricated multicolor PC patterns by assembling Fe_3O_4 nanoparticles under a magnetic field (Figure 9.6) [55]. They used a curable polymer to fix the arrangement of the nanoparticles under magnetic exposure. By regional photocuring the assembly structure individually, PC patterns with versatile colors were facilely obtained. The resolution of the PC patterns was determined by the resolution of the optical system, which achieved 17 μm. The selective UV-exposure technology also was employed to construct a series of patterned PC devices for anticounterfeiting measures [56], displays [57], and sensors [58].

The color of a PC film is determined by the interval space of the periodic structure and the dielectric coefficient of the materials. For the PC films assembled from monodispersed nanoparticles, different colors can be obtained by using particles with different diameters or incorporating responsive materials into the PCs. Responsive PC patterns show different colors under varied exterior excitations, such as magnetic field [59–61], electrical field [62], humidity [63, 64], etc. Yin and coworkers achieved multicolor PCs by changing the intensity of the magnetic exposure, which controlled the interval space between nanoparticles [55, 61].

9.3 Applications of PC Patterns

9.3.1 Displays

Bright reflective colors and low-energy consumption make PCs promising materials for the next-generation display devices. Numerous attempts to realize PC-based display devices have been demonstrated. Kim and coworkers developed a patterned PC RGB (red, green, and blue) device for reflection-mode displays by using the photolithography technology (Figure 9.7a) [57]. Yin and coworkers obtained high-resolution PC patterns by selectively immobilizing a colloidal Fe_3O_4 solution containing curable polymer (Figure 9.7b) [55]. Among these methods, the inkjet printing is a facile technology for patterning assembly of colloidal particles, as well as fabricating PC-based display devices. PC-based display devices are usually consisted of PC dots or lines. The brightness of the PC patterns is determined by the reflective intensity of the PC dots or lines and the effective coverage on the substrate. As discussed in Section 9.2.1.2, well-assembled structures can be obtained by printing inks on the substrates with a larger contact angle. However, the printed PC patterns on such surfaces may not be bright because the coverage area is relative low. In order to print PC patterns with good optical property, both the assembly structure and the coverage area should be considered. Song and coworkers obtained various PC patterns with bright colors by using a commercial printer (Figure 9.7c). They used colloidal suspensions with different diameters to formulate the inks. The core of the latex particles was synthesized from the hydrophobic polystyrene, and the shell was synthesized from the hydrophilic poly(methyl methacrylate) (PMMA)/polyacrylic acid (PAA). The hydrophilic PAA shell has abundant –COOH groups, which could promote the formation of hydrogen bonding between the latex particles. The interaction among the latex particles is strong and accelerates the assembly rate. In order to realize fluent ejecting of ink droplets and facilitate ink spreading, the viscosity of the ink was designed to ~1 mPa s, and the surface tension was about 57 mN/m. Moreover, the ink formulation contained a cosolvent of ethylene glycol to avoid the "coffee-ring" effect. The proportion of the ethylene glycol in the mixed solvents is ~40 wt%. The contact angle on the substrate was about 62°. Under the optimized conditions, PC patterns with homogenous deposition, well-ordered assembly structure, and enough surface coverage were fabricated. The printed patterns exhibited brilliant colors and good optical properties. The display colors of the printed PCs patterns were regulated by changing the diameter of the latex particles. PCs of red, green, and blue colors were printed from the particles with diameter of 280, 220, and 180 nm, respectively. The inks were held in the different cartridges and printed from different nozzles. To avoid particle precipitations, the inks were treated by ultrasonic for 30 min and filtered through a filter membrane before printing. The synthesized latex suspension may contain coagulations, which will clog the printing nozzles. Besides, the latex nanoparticles' diameter is better not to exceed 500 nm because large particles are easy to aggregate together and clog the nozzles.

PC films display different colors from the varied viewing angles according to the theory of Bragg diffraction [65]. This intrinsic angle-dependence character of PCs would restrain their applications in the wide viewing angle displays. Many strategies have been developed to achieve angle-independent PCs [66–69]. A spherical PC shows consistent color from different angles because the incident light is always perpendicular to the <111> plane of the fcc lattice. The spherical PCs are usually obtained by using a microfluidic technology. It is difficult to obtain it from the

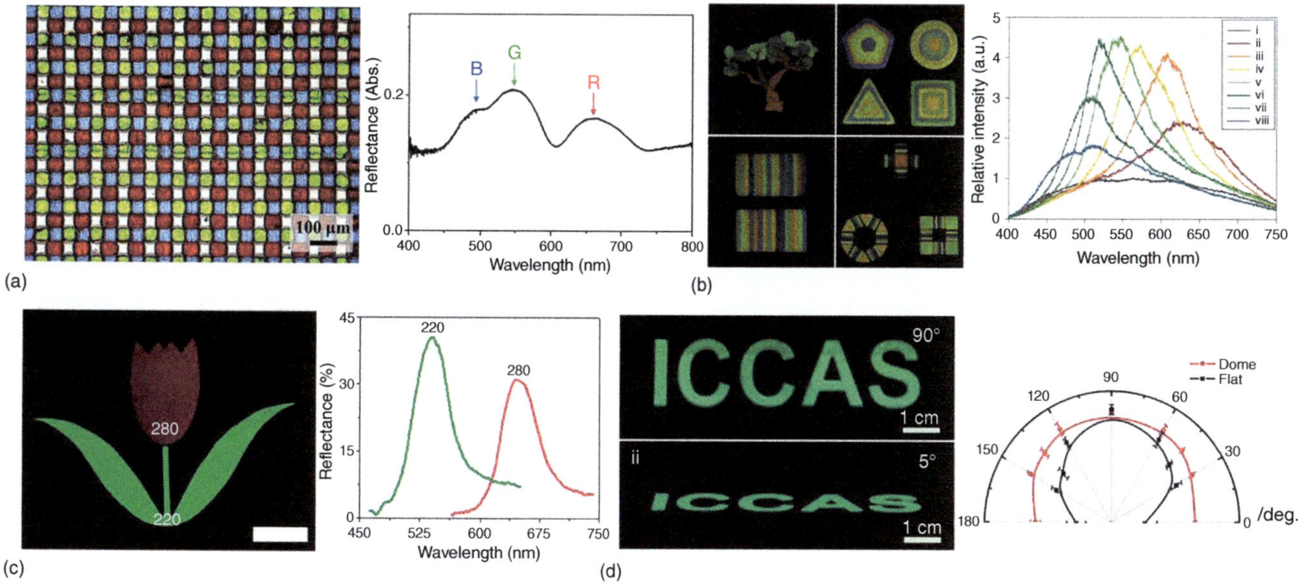

Figure 9.7 (a) Patterned PC RGB device for reflection-mode display. Source: Lee et al. [57]/with permission from John Wiley & Sons, Inc. (b) PC patterns fabricated by using a magnetic field and UV curing. Source: Kim et al. [55]/with permission from Springer Nature. (c) Inkjet-printed colloidal PC patterns. Source: Cui et al. [22]/with permission from Royal Society of Chemistry. (d) Inkjet-printed wide viewing angle PC patterns. Source: Kuang et al. [45]/with permission from John Wiley & Sons, Inc.

vertical deposition method. The inkjet printing technology has the advantages of controlling the morphology of PCs by adjusting the surface tension of the inks and the wettability of substrates. Song and coworkers printed angle-independent PC patterns by utilizing a sliding TCL on the surface (Figure 9.7d) [45]. They found that the height–diameter ratio (H/D) value increased from 0.00328 to 0.509 when the receding contact angle (θ_R) of the substrates increased from $18.0° \pm 5.7°$ to $93.3° \pm 1.9°$. Additionally, the colloidal nanoparticles arranged more compactly on the substrate with a higher θ_R value. They further demonstrated that TCL continuously retracted on the substrate with a high θ_R value and pushed the colloidal nanoparticles moving inward. The stopband of a flat PC film shifts obviously with the changing of the detecting angle. With the H/D ratio increasing, the stopband change of the printed PCs became less when changing the detecting angles. The stopband of the printed PCs reached a constant value when the H/D ratio increased to 0.375. The unchanged stopband of PC patterns was derived from a well-assembled hexagonal structure extending across the entire surface. This unique assembly structure resulted in isotropic reflective colors from 0° to 90° because the incident light was invariably perpendicular to their <111> plane of the fcc lattice. Beside, an isotropic fluorescent enhancement was achieved on this unique assembly structure. Due to the large surface area and slow photon effects, the printed spherical PCs exhibited enhanced fluorescent intensity more than 40 times when incorporating fluorescent molecules into the colloidal nanoparticles. This strategy provides an efficient way for fabricating advanced display and other optical devices. Besides incorporating fluorescent molecules into the colloidal nanoparticles, Song and coworkers also developed a reactive printing ink, which contained colloidal nanoparticles and cadmium chloride [70]. The patterns were treated with hydrogen sulfide gas after printing. Cadmium sulfide (CdS) quantum dots were synthesized *in situ* during the assembling process of the nanoparticles. The synthesized CdS quantum dots homogeneously distributed over the surface of the assembled colloidal nanoparticles. They printed 2D code patterns for anticounterfeiting, which were consisted of PC dot matrix.

9.3.2 Sensors

Colloidal PC-based sensors have been developed for decades because the colloidal PCs can be facilely fabricated by various assembly methods and can respond to the external stimulation if responsive materials are incorporated to the colloidal PCs. Colloidal PCs constructed from responsive materials change their reflective colors under corresponding stimulations, including temperature [71, 72], humidity [60, 73], pH [74], mechanical deformation [75, 76], electric field [62, 77], magnetic field [60, 61], and chemical analytes [78, 79]. Song and coworkers obtained a humidity sensitive PC hydrogel by infiltrating hydrophilic polyacrylamide (PAAm) into the opal template [73]. The PAAm PC hydrogel swelled under the humidity environment, resulting in color changing. The color reversibly varied from transparent to violet, blue, cyan, green, and red according to the humidity. The hydrophilic PCs were quickly swelled by the water vapor, contributing to the fast sensing of the humidity change.

Although a lot of colloidal PC-based sensors have been developed, achieving fast response rate and high detecting sensitivity and selectivity remain challenges. Patterned colloidal PCs show remarkable advantages and unique properties in sensors. Patterned colloidal PCs usually consist of well-assembled microdots, which exhibit fast response rate because of their large specific surface area. In order to realize a fast response rate, Song and coworkers constructed a PC sensor by printing microdots with diameters of 20–30 μm (Figure 9.8a) [38]. Each PC microdot was co-assembled from colloidal particles and humidity-responsive polymers (PAAm and poly(*N*-isopropyl acrylamide) (PNIPAm)). Introducing the PNIPAm segments into the PC microdots improved the response rate because the

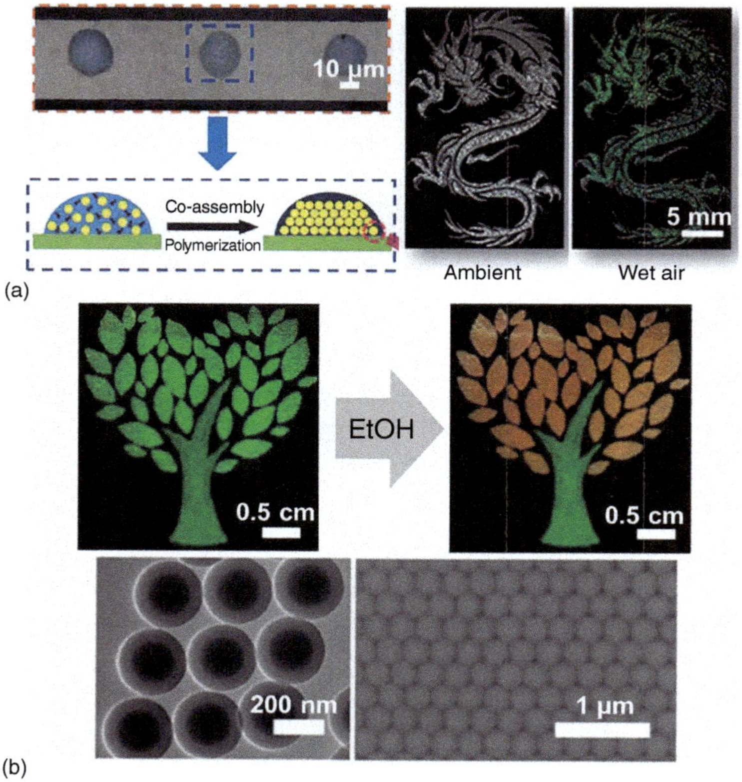

Figure 9.8 Vapor sensors based on patterned PCs. (a) Fast humidity sensor based on PC microdots co-assembled from colloidal nanoparticles and responsive polymers. Source: Wang et al. [38]/with permission from Royal Society of Chemistry. (b) Fast alcohol sensor based on patterned PC consisting of core–shell colloidal nanoparticles. Source: Bai et al. [80]/with permission from American Chemical Society.

wetting states of water on the polymer segments changed from the Wenzel state to the Cassie state when the temperature were above the lower critical solution temperature (LCST) of PNIPAm. Based on this strategy, fast response rates of 1.2 s in swelling process and 1.8 s in deswelling process were achieved. Gu and coworkers constructed a patterned responsive PC sensor for alcohol detecting by inkjet printing. The sensor consisted of PC microdots assembled from core–shell colloidal nanoparticles. The mesoporous shell absorbed alcohol vapor due to its large surface area, resulting in a volume expansion and redshift of the stopband (Figure 9.8b) [80]. This patterned PC sensor can response to an alcohol vapor in only 0.5 s.

Analyte detection is widely applied in early diagnosis [81], explosives detection [82], drugs testing [83], contamination detection [84], etc. Designing sensors with high sensitivity and efficiency has attracted a broad interest over the past years. A detector with high sensitivity and accuracy is one of the development trends of analyte detection. The sensitivity of ultratrace detection can be enhanced by well designing the patterned colloidal PC microchips. Song and coworkers constructed hydrophilic PC microfluidic channels on a superhydrophobic substrate (Figure 9.9a) [85]. The solution spontaneously flows along the channel, but cannot wet the superhydrophobic substrate. The hydrophilic PC on the superhydrophobic substrate serves as a surface-tension-confined microfluidic channel for guiding the fluidic transport. A prototype human IgG antigen–antibody colorimetric detection was realized on the patterned colloidal PC microchip. It is worth noting that this PC microfluidic channel will not be disassembled by the aqueous flows because the surface of the colloidal nanoparticles is plenty of carboxyl, and hydrogen bonding forms between the nanoparticles. On a hydrophilic–hydrophobic patterned surface, chemical analytes can be enriched from the solution to the hydrophilic area by a retracting TCL during evaporation. This strategy could significantly raise the concentration of the chemical analyte in the detection area. Combining the enriching effect with the fluorescence enhancement effect of the PCs, Song and coworkers realized a detection limitation down to 10^{-16} mol/L on a patterned PC (Figure 9.9b) [44]. They fabricated a hydrophilic–hydrophobic micropattern PC by printing hydrophilic PC microdots on a hydrophobic PDMS substrate. The fluorescent molecules were enriched in the hydrophilic PC microdots due to the directional TCL receding on the surface during evaporation. They also used the same strategy for colorimetric determination and ultratrace detection for tetracycline [86]. Both the volume of the droplet and the size of the hydrophilic PC microdot influence the enriching degree. The detection range can be easily regulated by changing the size of hydrophilic PC microdot.

Patterned PC microchips can realize not only high-sensitive detection but also high-throughput analyte identification [87]. For a composition of complex analyte mixture, analyte identification cannot be achieved by using a single sensor. More information from the sensors is needed. Thus the pattern recognition methods have been developed. In this sensor architecture, an array of different sensors is used. Every element in the sensor array is chosen to respond to a number of

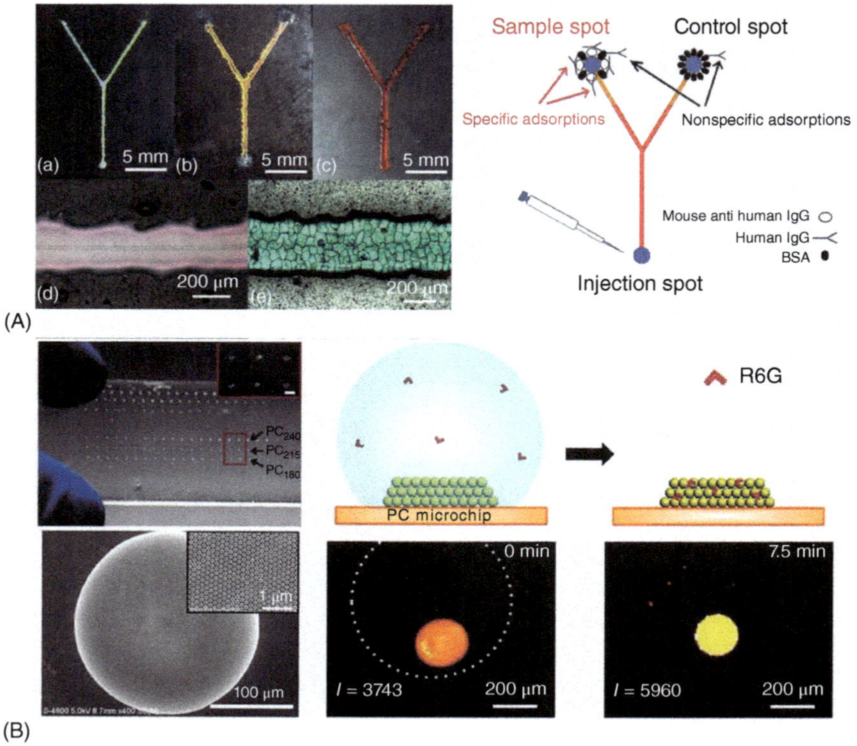

Figure 9.9 (A) Wettability confined microfluidic PC chips fabricated by inkjet printing. Human IgG antigen–antibody could be detected by the signal differences between the sample spot and control spot data. Source: Shen et al. [85]/with permission from Royal Society of Chemistry. (B) Hydrophilic–hydrophobic patterned PC microchip based on liquid enrichment. The target molecules were enriched into the PC microsensors due to the wettability difference. Source: Hou et al. [44]/with permission from John Wiley & Sons, Inc.

different chemicals. A distinct pattern of responses produced from the collection of sensors in the array provides a fingerprint for classification and identification of the analyte. By designing different sensor elements, the array of sensors exhibit specific encoding and determination. PCs with varied stopbands and responsibilities can be used as different sensors, so the array of different PCs is an ideal platform for cross-reactive analysis systems [88–90]. Gu and coworkers designed a PC chip for vapor identification in a cross-reactive system. This chip consists of arrays of mesoporous PC microbeads (Figure 9.10a) [91]. Each microbead was surface functionalized with an alkoxysilane to acquire different surface properties for vapor adsorption. After vapor exposure, the microbeads showed different shifts of the Bragg diffraction peak and exhibited a shift map. By this strategy, different vapors were successfully identified. By combining the advantages of the cross-reactive analysis and the fluorescence-enhancement effect of PCs, Song and coworkers recognized 12 metal ions on a patterned PC chip (Figure 9.10b) [92]. The patterned

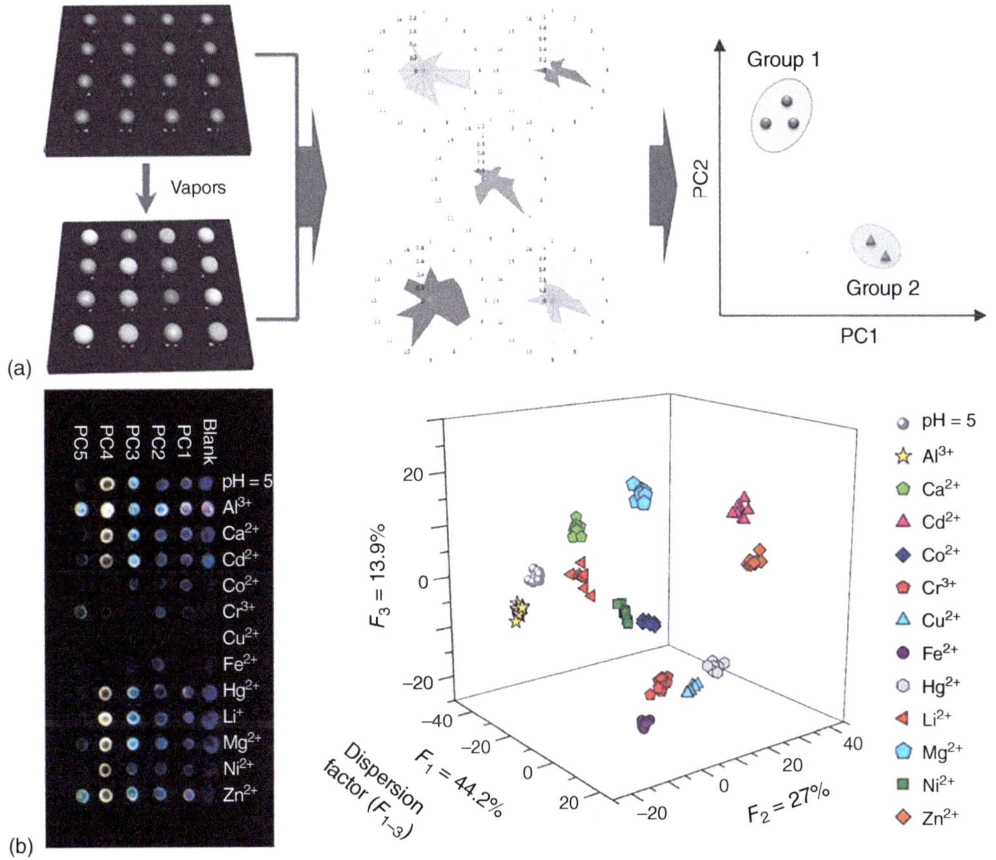

Figure 9.10 PC arrays for cross-reactive analysis. (a) Mesoporous colloidal PC beads for vapor sensing and recognition. Source: Xie et al. [91]/with permission from John Wiley & Sons, Inc. (b) Multi-stopband PC arrays for metal ion identification. Source: Huang et al. [92]/with permission from John Wiley & Sons, Inc.

PC chip selectively enhanced the sensing fluorescence on different PC elements and performed a high-efficient metal ion identification.

9.3.3 Anticounterfeiting

Anticounterfeiting patterns have attracted abroad attentions due to their wide applications in identifications [93–95]. Patterns from colloidal PCs show unique advantages for anticounterfeiting because the displaying color can be facilely regulated by changing the diameters of the colloidal nanoparticles or by incorporating responsive materials into the PCs, and the displaying color is unfading and difficult to copy. The displaying color of PCs comes from their periodic structures, so the brilliant color of a PC is different from the color of pigments or dyes and is more stable [96]. PC patterns for anticounterfeiting can be prepared on various substrates such as paper or plastic, from colloidal suspensions by photolithography or printing technologies [93, 95]. Patterned colloidal PCs provide a new way for anticounterfeiting. In recent years, two types of anticounterfeiting PC patterns have been developed. First, anticounterfeiting codes come from responsive PCs, which appear only under a stimulus. Ge and coworkers constructed a PC anticounterfeiting film, which consisted of a soft PC and a hard PC (Figure 9.11a) [97]. The two PCs showed green color without any stimulation; no pattern was seen. When a stretching stimulus or a squeezing stimulus was exerted on the PC film, the soft PC deformed and showed a blueshift or redshift; meanwhile, the hard PC remained their original color. Thus an anticounterfeiting pattern appeared. Besides the mechanical stimulus, vapor stimulus also were employed for preparing anticounterfeiting PC patterns [101, 102]. Chen and coworkers fabricated a PC pattern for anticounterfeiting, which consisted of responsive PCs and nonresponsive PCs (Figure 9.11b) [98]. The anticounterfeiting film was made of PDMS, which contained many vesicles with magnetic nanoparticles and glycol solution. UV light was exposed to cure PDMS and induce vesicle rupture and glycol leakage. The color of uncured region changed under a magnetic field, but the cured region lost the ability to change colors so that PC patterns appeared under the external magnetic field.

By incorporating responsive materials into PCs, the previously mentioned anticounterfeiting PC patterns are overt under the external magnetic fields. Besides the strategy of using responsive PCs, other strategies also have been developed for PC pattern-based anticounterfeiting. Nam et al. constructed anticounterfeiting PC patterns by using a monolayered, self-assembled PC pattern [99]. They inkjet-printed colloidal solutions on a white substrate and allowed the colloids forming a monolayered PC film after drying. This PC pattern was covert under daylight but became overt under strong illumination with smartphone flashlight and/or on a black background (Figure 9.11c). Multidimensional screening techniques provide an unparalleled approach to improve the information security and hold great potential in anticounterfeiting applications. Song and coworkers prepared anticounterfeiting patterns by utilizing PC arrays with various shapes (Figure 9.11d) [54, 100]. The PC shapes were regulated by the surface wettability, which included bumps, plates, coffee rings, square, star, hexagon, and octagon. Utilizing the diverse morphology

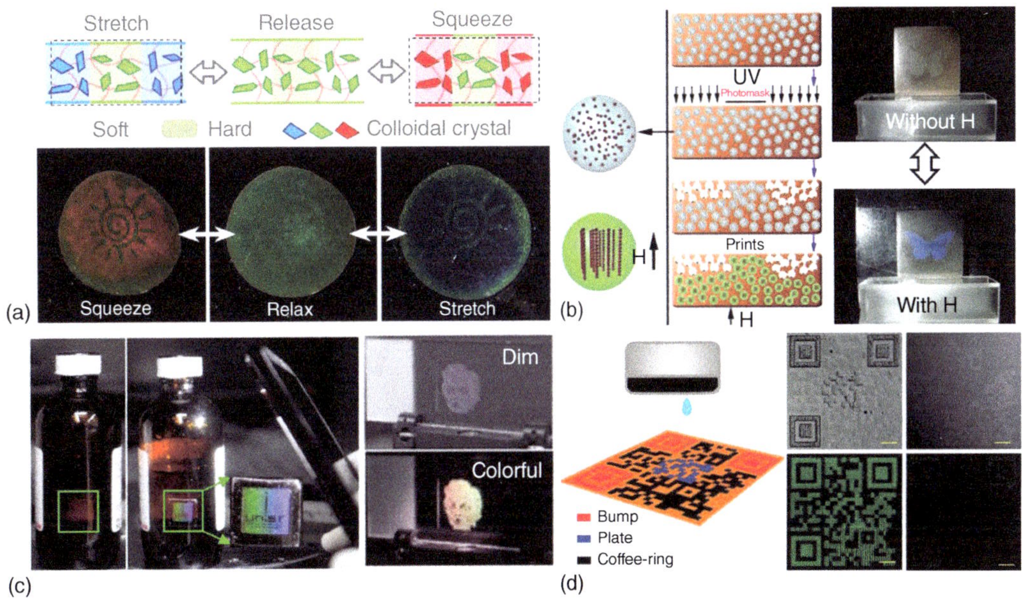

Figure 9.11 Anticounterfeiting PC patterns based on responsive or nonresponsive PCs. (a) Mechanical-responsive anticounterfeiting PC patterns. Source: Ye et al. [97]/with permission from John Wiley & Sons, Inc. (b) Magnetic-responsive anticounterfeiting PC pattern. Source: Hu et al. [98]/with permission from Springer Nature. (c) Inkjet-printed monolayer PC pattern, which was invisible in daylight but became overt under strong illumination. Source: Nam et al. [99]/Springer Nature/CC BY 4.0. (d) Four-dimensional screening anticounterfeiting pattern by inkjet-printed PCs, which can show four images under different lighting conditions. Source: Hou et al. [100]/with permission from John Wiley & Sons, Inc.

and the distribution of the PC microdots with different shapes, multi-information can be carried by the anticounterfeiting PC patterns.

9.4 Summary

Devices with assembly patterns of colloidal particles have been rapidly developed and show potential applications in displays, sensors, and anticounterfeiting areas. In this chapter, we summarized the fabrication strategies and potential applications of the colloidal assembly patterns. The fabrication methods include the inkjet printing technology, the template confining method, and the external stimuli method. In particular, we introduced the factors affecting the assembly quality by using the inkjet printing technology, including the ink composition and the surface wettability. Although great progress has been achieved on the fabrication of assembly patterns, fabricating assembly patterns of colloidal particles with controlled architectures, stable periodic structures, and high assembly efficiency and quality remains challenges. Furthermore, extending the applications of the assembly patterns also remains on the scientific and technical agenda for the researchers. The practical applications of colloidal assembly patterns will need the cooperation of chemists, physicist and materials scientists, and engineers. There is tremendous potential for future research in fabricating patterns assembled from colloidal particles. We believe that future research will bring more exciting insights for the development of patterning assembly, as well as their applications.

References

1 Sacanna, S., Pine, D.J., and Yi, G.R. (2013). *Soft Matter* 9: 8096–8106.
2 Dziomkina, N.V. and Vancso, G.J. (2005). *Soft Matter* 1: 265–279.
3 Kang, Y.J., Ye, X.C., Chen, J. et al. (2013). *J. Am. Chem. Soc.* 135: 42–45.
4 Cecchini, M.P., Turek, V.A., Paget, J. et al. (2013). *Nat. Mater.* 12: 165–171.
5 Urban, J.J., Talapin, D.V., Shevchenko, E.V. et al. (2007). *Nat. Mater.* 6: 115–121.
6 Zhang, Y.S., Zhu, C.L., and Xia, Y.N. (2017). *Adv. Mater.* 29: 1701115.
7 Michael Schnall-Levin, E.L. and Brenner, M.P. (2006). *Langmuir* 22: 5.
8 Malaquin, L., Kraus, T., Schmid, H. et al. (2007). *Langmuir* 23: 11513–11521.
9 Sun, Y.J. and Walker, G.C. (2002). *J. Phys. Chem. B* 106: 2217–2223.
10 Winkleman, A., Gates, B.D., McCarty, L.S., and Whitesides, G.M. (2005). *Adv. Mater.* 17: 1507–1511.
11 Sacanna, S., Rossi, L., and Pine, D.J. (2012). *J. Am. Chem. Soc.* 134: 6112–6115.
12 Tian, Y., Wang, T., Liu, W.Y. et al. (2015). *Nat. Nanotechnol.* 10: 637–644.
13 Guo, R., Wang, D.N., Wei, Y.Y. et al. (2020). *Microchim. Acta* 187: 194.
14 Demiroers, A.F., Pillai, P.P., Kowalczyk, B., and Grzybowski, B.A. (2013). *Nature* 503: 99–103.
15 Li, M.Z., Liao, Q., Liu, Y. et al. (2010). *Appl. Phys. A: Mater. Sci. Process.* 98: 85–90.

16 Huang, Y., Zhou, J.M., Su, B. et al. (2012). *J. Am. Chem. Soc.* 134: 17053–17058.
17 Ge, J.P. and Yin, Y.D. (2008). *J. Mater. Chem.* 18: 5041–5045.
18 Kim, J.Y., Pfeiffer, K., Voigt, A. et al. (2012). *J. Mater. Chem.* 22: 3053–3058.
19 Sirringhaus, H., Kawase, T., Friend, R.H. et al. (2000). *Science* 290: 2123–2126.
20 John, S. (1987). *Phys. Rev. Lett.* 58: 2486–2489.
21 Yablonovitch, E. (1987). *Phys. Rev. Lett.* 58: 2059–2062.
22 Cui, L.Y., Li, Y.F., Wang, J.X. et al. (2009). *J. Mater. Chem.* 19: 5499–5502.
23 Wang, J.X., Wen, Y.Q., Feng, X.J. et al. (2006). *Macromol. Rapid Commun.* 27: 188–192.
24 Wang, J.X., Wen, Y.Q., Ge, H.L. et al. (2006). *Macromol. Chem. Phys.* 207: 596–604.
25 Chen, M., Wu, L.M., Zhou, S.X., and You, B. (2006). *Adv. Mater.* 18: 801–806.
26 Qi, G.G., Wang, Y.B., Estevez, L. et al. (2010). *Chem. Mater.* 22: 2693–2695.
27 Pell, L.E., Schricker, A.D., Mikulec, F.V., and Korgel, B.A. (2004). *Langmuir* 20: 6546–6548.
28 Kim, D., Jeong, S., Park, B.K., and Moon, J. (2006). *Appl. Phys. Lett.* 89: 264101.
29 Deegan, R.D., Dupont, T.F., Huber, G. et al. (1997). *Nature* 389: 827–829.
30 Deegan, R.D. (2000). *Phys. Rev. E* 61: 475–485.
31 Deegan, R.D., Bakajin, O., Dupont, T.F. et al. (2000). *Phys. Rev. E* 62: 756–765.
32 Park, J. and Moon, J. (2006). *Langmuir* 22: 3506–3513.
33 de Gans, B.J. and Schubert, U.S. (2004). *Langmuir* 20: 7789–7793.
34 Tekin, E., de Gans, B.J., and Schubert, U.S. (2004). *J. Mater. Chem.* 14: 2627–2632.
35 Hu, H. and Larson, R.G. (2006). *J. Phys. Chem. B* 110: 7090–7094.
36 Still, T., Yunker, P.J., and Yodh, A.G. (2012). *Langmuir* 28: 4984–4988.
37 Sempels, W., De Dier, R., Mizuno, H. et al. (2013). *Nat. Commun.* 4: 1757.
38 Wang, L.B., Wang, J.X., Huang, Y. et al. (2012). *J. Mater. Chem.* 22: 21405–21411.
39 Zhou, J.M., Wang, J.X., Huang, Y. et al. (2012). *NPG Asia Mater.* 4: e21.
40 Liu, K.S., Tian, Y., and Jiang, L. (2013). *Prog. Mater. Sci.* 58: 503–564.
41 Sun, T.L., Feng, L., Gao, X.F., and Jiang, L. (2005). *Acc. Chem. Res.* 38: 644–652.
42 Liu, K.S., Yao, X., and Jiang, L. (2010). *Chem. Soc. Rev.* 39: 3240–3255.
43 Ko, H.Y., Park, J., Shin, H., and Moon, J. (2004). *Chem. Mater.* 16: 4212–4215.
44 Hou, J., Zhang, H., Yang, Q. et al. (2014). *Angew. Chem. Int. Ed.* 53: 5791–5795.
45 Kuang, M.X., Wang, J.X., Bao, B. et al. (2014). *Adv. Opt. Mater.* 2: 34–38.
46 Cui, L., Zhang, J., Zhang, X. et al. (2012). *Soft Matter* 8: 10448–10456.
47 Nguyen, T.A.H. and Nguyen, A.V. (2012). *Langmuir* 28: 1924–1930.
48 Shanahan, M.E.R., Sefiane, K., and Moffat, J.R. (2011). *Langmuir* 27: 4572–4577.
49 Masuda, T.I.Y. and Koumoto, K. (2005). *Langmuir* 21: 4478–4481.
50 Yang, S.M., MÍguez, H., and Ozin, G.A. (2002). *Adv. Funct. Mater.* 12: 425–431.
51 Guo, D., Zheng, X., Wang, X.H. et al. (2018). *Angew. Chem. Int. Ed.* 57: 16126–16130.
52 Guo, D., Wang, Y., Li, Y.N., and Song, Y.L. (2017). *Angew. Chem. Int. Ed.* 56: 15348–15352.

53 Guo, D., Li, Y.N., Zheng, X. et al. (2018). *J. Am. Chem. Soc.* 140: 18–21.

54 Wu, L., Dong, Z., Kuang, M. et al. (2015). *Adv. Funct. Mater.* 25: 2237–2242.

55 Kim, H., Ge, J.P., Kim, J. et al. (2009). *Nat. Photonics* 3: 534–540.

56 Lee, H.S., Shim, T.S., Hwang, H. et al. (2013). *Chem. Mater.* 25: 2684–2690.

57 Lee, S.Y., Kim, S.H., Hwang, H. et al. (2014). *Adv. Mater.* 26: 2391–2397.

58 Lee, H.S., Kim, J.H., Lee, J.S. et al. (2014). *Adv. Mater.* 26: 5801–5807.

59 Ge, J.P. and Yin, Y.D. (2011). *Angew. Chem. Int. Ed.* 50: 1492–1522.

60 Xuan, R.Y., Wu, Q.S., Yin, Y.D., and Ge, J.P. (2011). *J. Mater. Chem.* 21: 3672–3676.

61 Hu, Y.X., He, L., and Yin, Y.D. (2011). *Angew. Chem. Int. Ed.* 50: 3747–3750.

62 Xu, L., Wang, J.X., Song, Y.L., and Jiang, L. (2008). *Chem. Mater.* 20: 3554–3556.

63 Yin, S.N., Wang, C.F., Liu, S.S., and Chen, S. (2013). *J. Mater. Chem. C* 1: 4685–4690.

64 Hu, H.B., Chen, Q.W., Cheng, K., and Tang, J. (2012). *J. Mater. Chem.* 22: 1021–1027.

65 Richel, A., Johnson, N.P., and McComb, D.W. (2000). *Appl. Phys. Lett.* 76: 1816–1818.

66 Takeoka, Y. (2012). *J. Mater. Chem.* 22: 23299–23309.

67 Chung, K., Yu, S., Heo, C.J. et al. (2012). *Adv. Mater.* 24: 2375–2379.

68 Lee, I., Kim, D., Kal, J. et al. (2010). *Adv. Mater.* 22: 4973–4977.

69 Kim, S.H., Jeon, S.J., and Yang, S.M. (2008). *J. Am. Chem. Soc.* 130: 6040–6046.

70 Bao, B., Li, M.Z., Li, Y. et al. (2015). *Small* 11: 1649–1654.

71 Weissman, J.M., Sunkara, H.B., Tse, A.S., and Asher, S.A. (1996). *Science* 274: 959–960.

72 Debord, J.D. and Lyon, L.A. (2000). *J. Phys. Chem. B* 104: 6327–6331.

73 Tian, E.T., Wang, J.X., Zheng, Y.M. et al. (2008). *J. Mater. Chem.* 18: 1116–1122.

74 Lee, K. and Asher, S.A. (2000). *J. Am. Chem. Soc.* 122: 9534–9537.

75 Asher, S.A., Holtz, J., Liu, L., and Wu, Z.J. (1994). *J. Am. Chem. Soc.* 116: 4997–4998.

76 Jethmalani, J.M. and Ford, W.T. (1996). *Chem. Mater.* 8: 2138–2146.

77 Arsenault, A.C., Puzzo, D.P., Manners, I., and Ozin, G.A. (2007). *Nat. Photonics* 1: 468–472.

78 Goponenko, A.V. and Asher, S.A. (2005). *J. Am. Chem. Soc.* 127: 10753–10759.

79 Sharma, A.C., Jana, T., Kesavamoorthy, R. et al. (2004). *J. Am. Chem. Soc.* 126: 2971–2977.

80 Bai, L., Xie, Z.Y., Wang, W. et al. (2014). *ACS Nano* 8: 11094–11100.

81 Arya, S.K. and Bhansali, S. (2011). *Chem. Rev.* 111: 6783–6809.

82 Engel, Y., Elnathan, R., Pevzner, A. et al. (2010). *Angew. Chem. Int. Ed.* 49: 6830–6835.

83 Brettell, T.A., Butler, J.M., and Almirall, J.R. (2011). *Anal. Chem.* 83: 4539–4556.

84 Rodriguez, E., Navarro-Villoslada, F., Benito-Pena, E. et al. (2011). *Anal. Chem.* 83: 2046–2055.

85 Shen, W., Li, M., Ye, C. et al. (2012). *Lab Chip* 12: 3089–3095.

86 Hou, J., Zhang, H.C., Yang, Q. et al. (2015). *Small* 11: 2738–2742.

87 Anzenbacher, P., Lubal, P., Bucek, P. et al. (2010). *Chem. Soc. Rev.* 39: 3954–3979.
88 Bai, L., Xie, Z.Y., Cao, K.D. et al. (2014). *Nanoscale* 6: 5680–5685.
89 Huang, Y., Li, F.Y., Ye, C.Q. et al. (2015). *Sci. Rep.* 5: 9724.
90 Qin, M., Huang, Y., Li, Y.N. et al. (2016). *Angew. Chem. Int. Ed.* 55: 6911–6914.
91 Xie, Z.Y., Cao, K.D., Zhao, Y.J. et al. (2014). *Adv. Mater.* 26: 2413–2418.
92 Huang, Y., Li, F.Y., Qin, M. et al. (2013). *Angew. Chem. Int. Ed.* 52: 7296–7299.
93 Hardwick, B., Jackson, W., Wilson, G., and Mau, A.W.H. (2001). *Adv. Mater.* 13: 980–984.
94 Mullard, A. (2010). *Nat. Med.* 16: 361.
95 Prime, E.L. and Solomon, D.H. (2010). *Angew. Chem. Int. Ed.* 49: 3726–3736.
96 Chen, X., Wang, L.H., Wen, Y.Q. et al. (2008). *J. Mater. Chem.* 18: 2262–2267.
97 Ye, S.Y., Fu, Q.Q., and Ge, J.P. (2014). *Adv. Funct. Mater.* 24: 6430–6438.
98 Hu, H.B., Tang, J., Zhong, H. et al. (2013). *Sci. Rep.* 3: -1484.
99 Nam, H., Song, K., Ha, D., and Kim, T. (2016). *Sci. Rep.* 6: 30885.
100 Hou, J., Zhang, H.C., Su, B. et al. (2016). *Chem. Asian J.* 11: 2680–2685.
101 Xuan, R.Y. and Ge, J.P. (2012). *J. Mater. Chem.* 22: 367–372.
102 Ye, S.Y. and Ge, J.P. (2015). *J. Mater. Chem. C* 3: 8097–8103.

10

Light Extraction Efficiency Enhancement in GaN-Based LEDs by Colloidal Self-assembly

Tongbo Wei

Semiconductor Lighting Technology Research and Development Center, Institute of Semiconductors, Chinese Academy of Sciences, Beijing 100083, China

10.1 Introduction

III-nitride light-emitting diodes (LEDs) have become interestingly popular due to their various applications, such as traffic signals, full color display, backside lighting in liquid crystal display, and illumination lighting. In the past three decades, multiple quantum wells (MQWs) LEDs have been extensively investigated on GaN-based materials, novel device configurations, and extensions of working wavelengths from deep ultraviolet (DUV) to red. The luminous efficacy increases continuously, and efficiency record of 303 lm/W had been announced for white LEDs by Cree Ltd (http://www.cree.com/News-and-Events/Cree-News/Press-Releases/2014/March/300LPW-LED-barrier). The GaN LEDs have been regarded as the ultimate lighting solution in this century. However, further development of such devices is still needed to continuously improve the efficiency of GaN LEDs.

Efficiency of LEDs is best characterized by wall plug efficiency (WPE) [1] given by the ratio of radiated power P_{rad} to electrical power P_{elec} drawn by the device.

$$\text{WPE} = \frac{P_{rad}}{P_{elec}}$$

Here, $P_{rad} = \eta_{LEE} \cdot \hbar\omega \cdot \eta_{IQE} \cdot \frac{1}{q} \cdot \eta_{IE} \cdot I$. Electric power drawn by the device is $P_{elsc} = I\,V$; therefore, the WPE is given by

$$\text{WPE} = \eta_{LEE} \cdot \eta_{IQE} \cdot \eta_{IE} \cdot \left(\frac{\hbar\omega}{qV}\right)$$

In Figure 10.1, a summary of the reasons to reduce the efficiency of LEDs is shown. Currently, injection efficiency (η_{IE}) is relatively high, and thus the improvement of WPE is hindered by internal quantum efficiency (η_{IQE}) and light extraction efficiency (η_{LEE}). So far most of GaN-based LEDs are grown on sapphire substrate along (0001) orientation, which is susceptible to the strong internal electric field induced by the spontaneous and the piezoelectric polarization. First, high defect density in GaN due to large mismatch heteroepitaxy strongly impacts η_{IQE}. Second,

Functional Materials from Colloidal Self-assembly, First Edition. Edited by Qingfeng Yan and George Zhao.

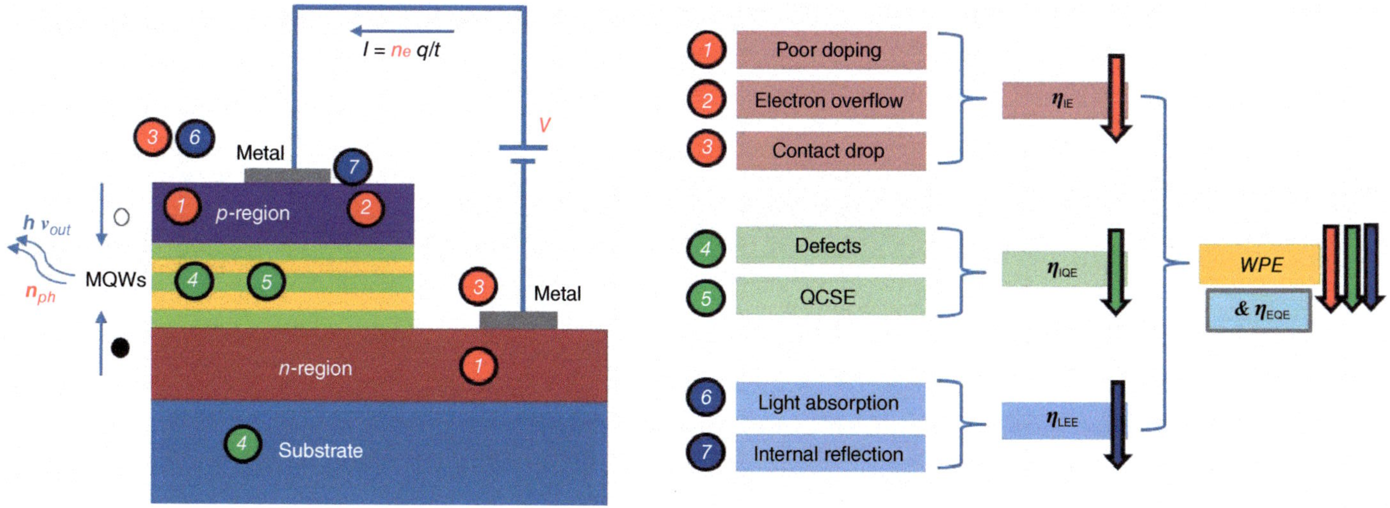

Figure 10.1 Schematic showing of efficiency loss mechanisms in GaN-based LEDs.

the spatial separation between electrons and holes in the MQWs due to tilting of the energy band also reduces η_{IQE} of carrier recombination, called quantum confined stark effect (QCSE). On the other hand, η_{LEE} remains low due to severe internal reflection occurring at the interface of semiconductor and air. The large difference in refractive index between GaN and air results in a finite escape cone of 23.5°, which limits a large portion of the emitted light to be trapped inside the LEDs as the guided modes. The light fails to emit from the active region and thus is absorbed by semiconductors and electrodes, further degrading the performance of the LEDs.

In order to solve the aforementioned problems, various micro- and nanoscale structures have been incorporated into GaN-based LEDs by diverse techniques, including conventional photolithography, e-beam lithography (EBL), holographic lithography, nanoimprint lithography (NIL), and nanosphere lithography (NSL). Such micro/nanostructures have been demonstrated to greatly enhance device performance of LEDs by improving the crystal quality of GaN epitaxial layer and reducing total internal reflections. Among these methods, conventional photolithography is hard to realize nanoscale features, while EBL and holographic lithography demonstrate high resolution but low throughput due to their processing nature [2, 3]. NIL permits high throughputs but is required to produce the expensive hard mold for imprinting [4]. On the contrast, NSL has proved to be a low-cost and high mass production technique to make features in nanoscale [5, 6].

In this chapter, we review the recent advances of efficiency improvement of GaN-based LEDs by using NSL-derived nanostructures. First, we introduce the representative fabrication process of nanostructure by NSL method. Then we provide an overview of the current application of nanostructure in the efficiency enhancement of GaN-based LEDs. A new application of highly ordered nanostructures on optoelectronic devices including LEDs and lasers is also presented. Finally, a perspective on the future development of GaN-based LEDs and other optoelectronic devices by using NSL is addressed.

10.2 Nanostructure Fabrication by MCC

A typical NSL process starts from the self-assembly of a monolayer colloidal crystal (MCC) on a flat substrate to realize the nanostructures. During the self-assembly, a balance of interactions including van der Waals forces, steric repulsions, and Coulombic repulsions is usually involved at the gas/liquid interface. Generally, self-assembly of colloidal spheres at the air/water interface requires that the surface of the colloidal spheres is hydrophobic, including polystyrene (PS) and silica [6]. Therefore, how to achieve large-area and crack-free MCCs is of great importance for the fabrication of ordered nanostructures.

Most of MCCs prepared by the self-assembly approach demonstrate a two-dimensional (2D) hexagonal close-packed (HCP) structure. In order to transfer the nano-pattern, the MCC is used as a template for the evaporation of an etch mask, and a material of choice is deposited onto the mask by using physical vapor deposition (Figure 10.2a) or wet chemical infiltration (Figure 10.2b). Subsequently,

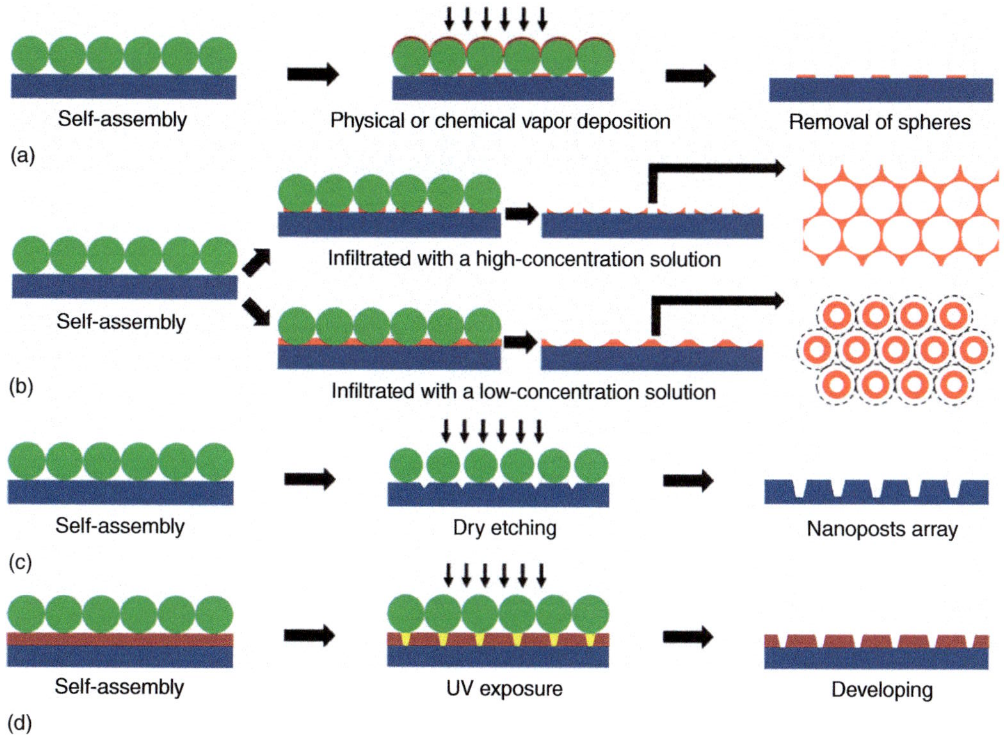

Figure 10.2 Schematic illustration of the nanostructures derived from a self-assembled MCC. MCC is used as a physical mask for selectively depositing materials by using (a) physical vapor deposition or (b) wet chemical infiltration. MCC is used as an etching mask to prepare an ordered array (c) and serves as an optical lens to generate sub-wavelength holes on PR layer. Source: Reproduced with permission from Geng et al. [7] Copyright 2014, Wiley-VCH Verlag GmbH & Co. KGaA.

metal or other mask materials can be partially or fully filling the interstices between the colloidal spheres. Finally, the MCC is removed by a liftoff process, and a periodic nanostructure of the deposited material is left on the substrate. The fast process allows for a choice of different nano-patterns with variable densities.

Alternatively, the MCC can also be used as an etching mask to prepare an ordered nanoposts array on the substrate (Figure 10.2c). First, MCC spheres may be shrunk by inductively coupled plasma (ICP). Then, MCC transfers the pattern to the substrate using dry etching method, and the pattern spacing may be controlled by changing the etch time [8]. Furthermore, MCC can be employed as an optical lenses to focus ultraviolet (UV) light and generate deep sub-wavelength regular patterns over large areas on photoresist (PR, Figure 10.2d). Here, it takes the role of phase mask in conventional contact photolithography. Depending on the fabrication routes, a variety of nanostructures including ordered nanoparticle (NP) arrays, nanoring arrays, nanopore network, nanopost arrays, nanohole arrays, and others can be derived from an MCC [9]. Over the past two decades, these nanostructures have been integrated into GaN-based LEDs and improved the device performance successfully by modification of spontaneous emission to increase internal quantum efficiency (IQE) and/or by molding the flow of light emitted from the active region to enhance LEE [7]. From top to bottom of device structures, we will discuss the influence of nanostructure on the efficiency enhancement of LEDs.

10.3 Applications of Nanostructures to LEDs

10.3.1 Surface Texturing

As the most direct and intuitive approach, surface texturing can effectively improve the light output power (LOP) of GaN-based LEDs by increasing the light scattering at the GaN/air interface. Typically, surface texturing of *p*-GaN layer is performed by maskless wet etching or dry etching using nanoscale metal clusters as an etch mask [10]. The surface texturing can also be applied to *n*-side GaN surface of LED with a combination of laser liftoff (LLO) and photo-electrochemical techniques [11]. However, the roughness of the surface textures obtained using these strategies is not uniform, and there is a big range of pattern size and thus leads to a variation in the improvement of LEE of the LED across the samples. To avoid the random roughened surface, EBL, NIL, and other semiconductor micromachining techniques have been used for surface texturing. Most of these surface-texturing approaches involve complicated fabrication steps and decrease production yield [12]. Therefore, developing a cost-effective and reproducible approach for surface texturization of LED chips with controllable roughness is urgent.

NSL, which is capable of producing well-ordered arrays on the substrate, has been proposed to be effective approach to construct periodic surface textures with nanoscale features. Nanospheres are commercially available with diameters from 20 nm to 3 μm, which makes it possible to vary the size and pitch of the patterns. Soh et al. [13] reported the fabrication of a nanocone structured *p*-GaN surface for enhanced light extraction using self-assembled PS nanospheres as an etch mask. A well-defined periodic nanocone showed an average cone diameter of 300 nm and height of 150 nm on the *p*-GaN surface. Besides the photoluminescence (PL) enhancement, the LOP from the LEDs with surface nanocones also showed significantly higher electroluminescence (EL) intensity at an injection current of 70 mA, due to the internal multiple scattering of light from the nanocone sidewalls. Furthermore, Sim et al. [14] fabricated a pillar-patterned *p*-GaN surface using the size-controllable NSL technique. In this process, the size and shape of the pillars could easily be controlled by changing the diameter of PS nanosphere using reactive ion etching (RIE) treatment. The LOP of *p*-GaN pillar-patterned LEDs was increased by up to 32.6% compared with the conventional LED at 20 mA.

For *p*-side-up LEDs, surface texturing directly on the *p*-GaN contact layer may induce plasma damage during dry etching, invariably degrading electrical characteristics of the devices. At the same time, the *p*-GaN layer is too thin to bear enough etching depth. To solve these problems, Zhang et al. [15] transferred the pattern shape on the indium-tin-oxide (ITO) electrode from self-assembled silica nanospheres by dry etching and realized ITO lens arrays, leaving *p*-GaN layer damage-free as shown in Figure 10.3a–c. As shown in Figure 10.3d, the current–voltage (I-V) curves substantiated that structuring of the ITO layer had not degraded the electrical properties of the devices. An enhancement of up to 63.5% on LOP from the LEDs had been observed in Figure 10.3e. Figure 10.3f showed a 26.8° reduction in emission divergence for the LED with 500 nm lenses compared with the bare LED, following the significant focusing effect. Kim et al. [16] demonstrated the increase of LEE of LEDs by pillar-patterned ITO contact layer and compared the effects of *p*-GaN and ITO surface texturing. A hole-patterned *p*-GaN layer and a pillar-patterned ITO layer were fabricated with size-tunable PS spheres. The LOPs of the LEDs with the hole-patterned *p*-GaN and the pillar-patterned ITO were enhanced 21% and 10% at 20 mA, respectively, compared with the bare LED due to the increase of the extraction probability of the internally reflected photons through the patterns. Recently, Chen et al. [17] also fabricated periodic ITO nanodisk (ND) arrays designed by the Bragg diffraction and maximum transmittance using PS NSL. The light output intensity of the LED with optimum periodic ND arrays was experimentally enhanced by a factor of 2.08 under the injection currents of 20 mA. The whole fabrication technique is fairly economical and efficient.

Vertical light-emitting diodes (VLEDs) have been suggested as a solution to overcome the limitations of conventional LEDs, with better current spreading, excellent heat dissipation, and simple packaging [18]. For VLED, thick *n*-GaN enabled a larger etching depth to increase the surface texture and the LEE. Even deeper surface textures can be acquired by using ordered nanostructures derive from MCC as an etching mask. An et al. [19] demonstrated the formation of ordered, cone-shaped deep-pillar surface textures on the *n*-GaN layer of VLED in Figure 10.4a. Both 1.0 and 1.5 μm deep nanostructures were fabricated at the *n*-GaN surface of the honeycomb TM-type VLEDs by using a double-layer PS nanosphere coating and an Ni deposition/liftoff technique to form a nanoscale dot array mask for deep etching in Figure 10.4b–c. The LOP was increased by 200%

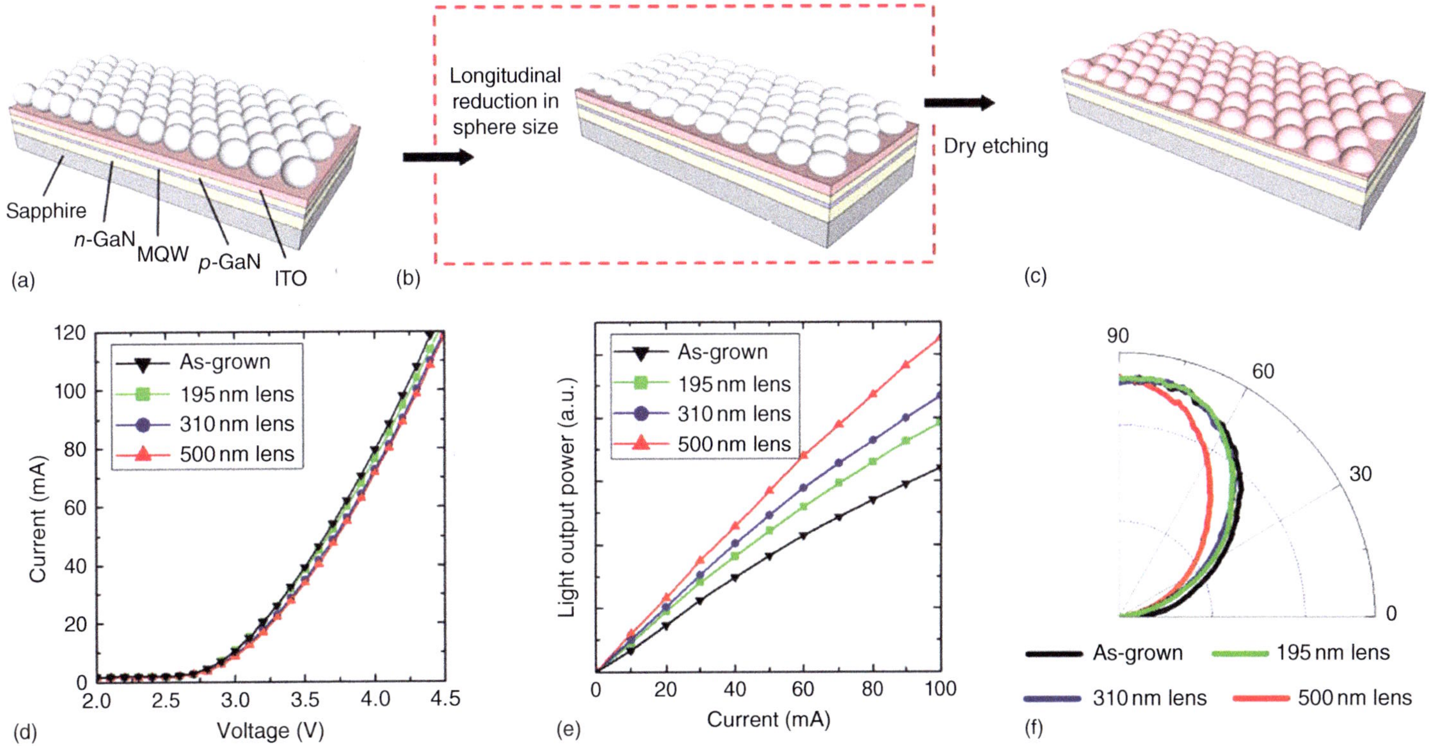

Figure 10.3 Schematic diagram showing the fabrication process flow: (a) silica spheres are coated onto the LED wafer. (b) RIE etching to reduce sphere size. (c) Pattern transfer to ITO using ICP etching. (d) I-V characteristics and (e) LOP as a function of injection current of ITO-roughened LEDs and as-grown LED. (f) Measured emission pattern of ITO-roughened LEDs and as-grown LED. Source: Reproduced with permission from Zhang et al. [15]. Copyright 2012, American Institute of Physics.

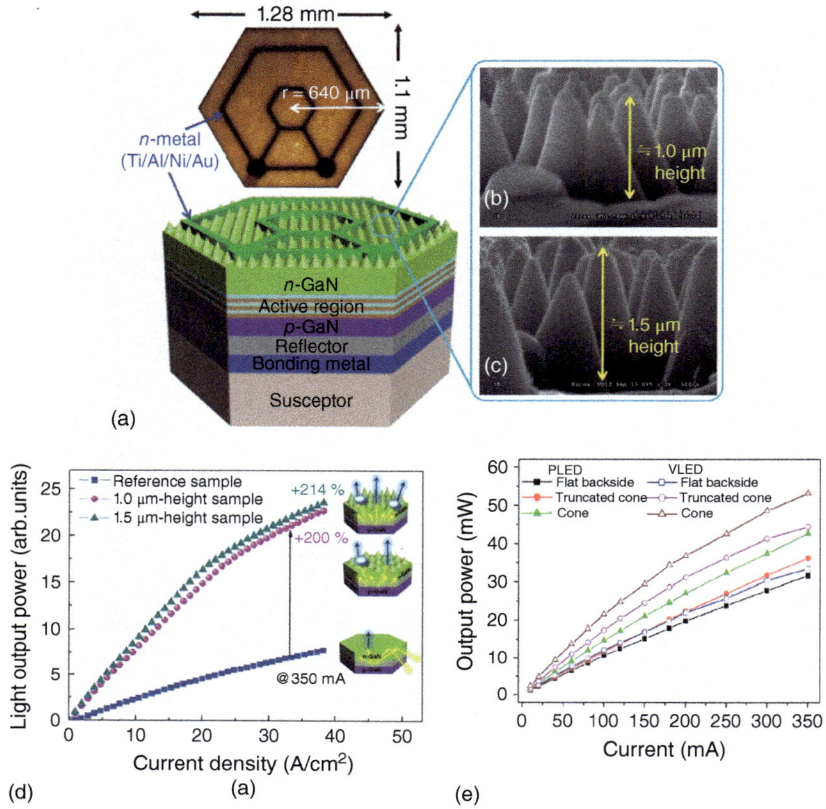

Figure 10.4 (a) Plane-view micrograph image (up) and the schematic diagram (down) of the honeycomb-type VLED chip used in this paper. Cross-sectional SEM images taken at the *n*-GaN surface of the VLED. (b) 1.0-μm height. (c) 1.5-μm-height cone-shaped nanostructures. (d) LOP versus injection current density characteristics. Source: An et al. [19]/with permission from IEEE. (e) LOP-current characteristics of PLEDs and various vertical-stand-type LEDs. Source: Reproduced with permission Wei et al. [20]. Copyright 2012, Institute of Electrical and Electronics Engineers.

and 214% for VLEDs with 1.0 and 1.5 μm deep cone-shaped nanostructures at 350 mA in Figure 10.4d, respectively, as compared with the reference VLED with flat surface. There was a slight sacrifice of operational voltage and leakage current. The enhancement in the LOP was attributed to the multiple scattering of light from the sidewall of the cone-shaped patterns and the increased surface dimension. Tsai et al. [21] also employed a similar method to fabricate large-scale biomimetic surface structures on VLEDs. The structures resulted in a modified heart-shaped radiation pattern, which is preferable for backlight applications in flat panel displays. Furthermore, we proposed a new vertical-stand-type LED with enlarged exposed surface and introduce conical array structures on the GaN substrate *n*-face for further enhancing light extraction [20]. The conical arrays were formed on the *n*-face surface of the GaN substrate using the size-controllable PS nanospheres as etch mask. In Figure 10.4e, the LOP of the vertical-stand-type LEDs with flat backside, truncated cone, and cone arrays improved by magnitudes of 16.5%, 66.8%, and 118.5% at 20 mA current, respectively, compared with that of conventional planar LEDs. These improvements could be attributed to the increased direct illumination surface and reduction in photon extraction path length.

10.3.2 Submicron Lenses

Compared to surface roughening by etching, the implementation of micron lenses on LED surface is a low-cost and straightforward approach for extracting out light of waveguide modes trapped in semiconductor layer. As the microlenses are deposited as the final step on the top emission area of the LEDs, the approach avoids any degradation on the electrical characteristics of the LEDs due to the absence of etching process. However, conventional microlens fabrication techniques, such as thermal resist reflow, just allow the formation of micron-scale microlenses [22]. Apart from feature dimensions, the packing density of the lens array is also limited since the reflow step requires sufficient spacing between PR posts.

MCC technique has been proved to be capable of directly serving as a submicron lens array at minimal cost and high efficiencies for the application in LEDs [23]. For example, hexagonal close-packed (CP) silica or PS sphere monolayer can directly serve as microlens arrays due to their convex lens shape. The diameter of the microspheres can be controlled accurately from 0.1 to 2 μm, resulting in good control and repeatability in the submicron lens structure and optical properties. This process also provides adequate adhesion to III-nitride materials at room temperature, and the materials are relatively transparent to visible light emission from the InGaN QWs. Furthermore, the refractive indices of silica and PS in the visible spectrum are 1.58 and 1.46, respectively. Hence, the SiO_2/PS MCC on the planar emitting surface of LEDs can provide graded refractive index transition between GaN and air. Thus, the LEE of LEDs can be increased due to the reduced Fresnel reflection, enlarged photon escape cone, and enhanced photon scattering.

Lee et al. [24] used a self-assembled monolayer (SAM) of silica submicron spheres to directly increase the LEE of GaN-based blue LEDs shown in Figure 10.5a. HCP silica sphere monolayer was formed onto the ITO layer of the LED by a spin-coating process. The size effect of silica spheres on the LEE of LEDs was theoretically studied, and their optimum size was determined. Figure 10.5b showed that the LOP of the LED with 650-nm-thick SAM of HCP silica spheres was experimentally enhanced by 1.28 and 1.23 times under the injection currents of 100 and 350 mA, respectively. By employing the SAM of HCP silica spheres, the directional emission pattern was relatively converged. Tansu and coworkers [25, 26] had systematically investigated the influence of a monolayer of silica colloidal spheres embedded in a planar PS layer on the LEE of InGaN LED. Fabrication of silica/PS submicron lens arrays was achieved by employing rapid convective deposition process. First, two monolayers of PS colloidal spheres and silica colloidal spheres were sequentially deposited on GaN-based top-emitting LED device. Then, heat treatment was performed using a hotplate at 140 °C to melt the PS microspheres. Thus, the silica

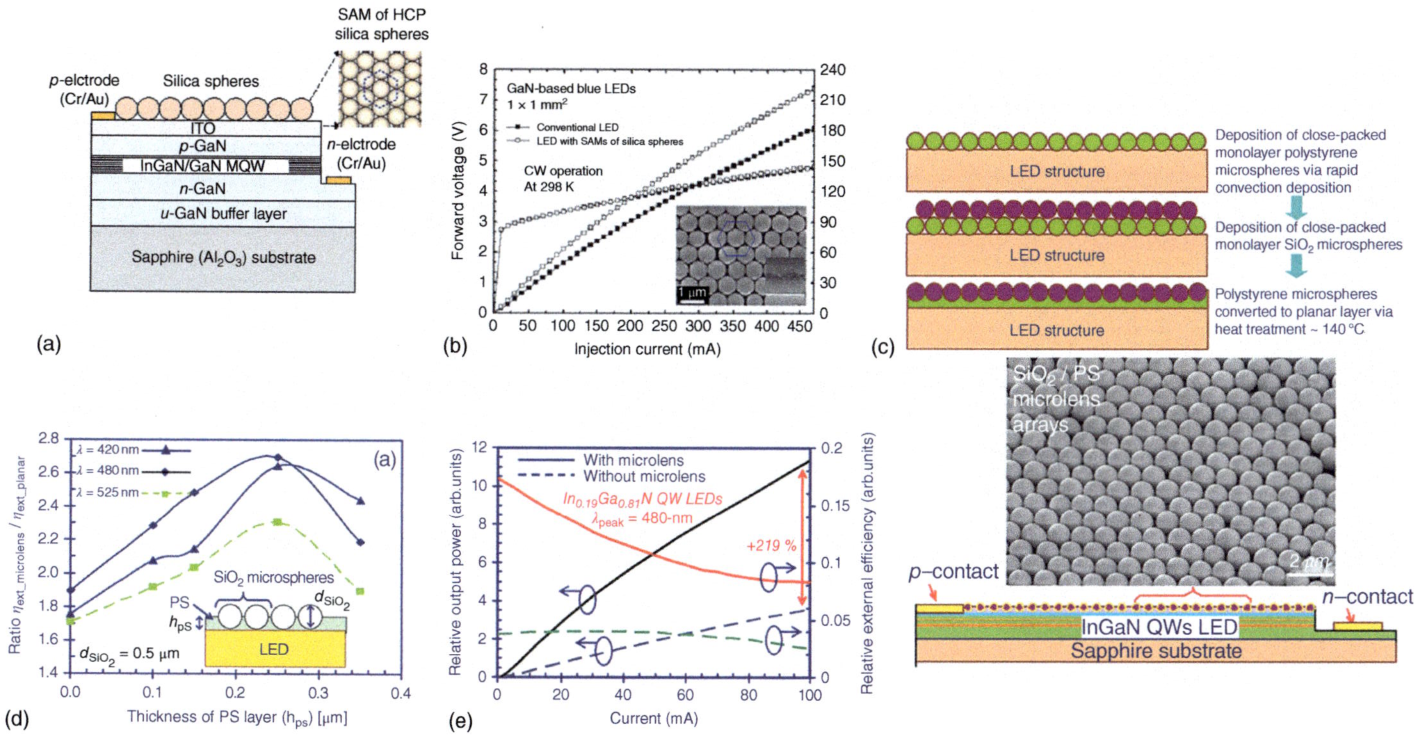

Figure 10.5 (a) Schematic diagrams of the GaN-based blue LEDs with SAM of silica spheres. (b) L-I-V curves of LEDs with and without SAMs of silica spheres. Source: Lee et al. [24]/with permission from The Optical Society (c) Process flow schematic of depositing SiO_2/PS submicron lens array on LEDs using rapid convective deposition and schematic of InGaN QWs LEDs utilizing SiO_2/PS submicron lens array. (d) Comparison of LEE ratios of InGaN with varying PS thickness and emission wavelength. Source: Ee et al. [25]/with permission from IEEE. (e) Comparison of relative LOP and EQE versus current for $In_{0.19}Ga_{0.81}N$ MQWs LEDs with and without SiO_2/PS submicron lens array. Source: Reproduced with permission from Ee et al. [26]. Copyright 2007, American Institute of Physics.

spheres were semi-buried in a planar PS layer without significant destroyment of the packing order of the silica spheres, forming HCP silica/PS submicron lens arrays shown in Figure 10.5c. Ray tracing simulations revealed that both the diameter of silica spheres and the thickness of PS layer affected enhancement of LEE. As shown in Figure 10.5d, the LEE for the LED with $d_{SiO_2} = 0.5\,\mu m$ submicron lens arrays was optimized for the case $h_{PS} = 0.25\,\mu m$, with improvement up to 2.7 times for the case of 480 nm emitting LED [26]. The LOP of submicron lens LEDs with various PS thicknesses exhibited 2.19 times enhancement over that of planar LEDs (Figure 10.5e) [25]. Both the normal direction and large angle far-field pattern for LEDs with submicron lens arrays exhibited significant enhancement over those of the planar LED, attributed to reduced Fresnel reflection and enlarged escape cone.

However, the refractive indices of SiO_2/PS submicron lens are still considerably smaller (n~1.5–1.6) than that of GaN (n~2.5). Thus, in order to optimize the coupling of the light from the GaN LEDs into the submicron lens arrays, the submicron lens are preferably formed by materials with refractive index matched to GaN. The optimized coupling of light from GaN into the submicron lens arrays with index matched to GaN will enable the optimization of the LEE of the nitride LEDs. Li et al. [27] reported detailed experimental studies on the use of self-assembled 2D TiO_2 monolayer microsphere arrays to enhance the LEE of InGaN QW LEDs. The refractive index of TiO_2 in the visible spectrum ranged from 1.5 to 2.6, depending on the crystalline phase and synthesis methods. Thus the implementation of TiO_2 microspheres provided higher indices that could match that of GaN, which resulted in larger light coupling from GaN to the TiO_2 microspheres arrays. The LEE of LEDs under electrical injection was enhanced by 1.8–1.9 times by utilizing 520-nm diameter amorphous and anatase TiO_2 microspheres. Further enhancement of extraction efficiency was expected with submicron lens structures by introducing planar materials (i.e. PS), which resulted in semi-buried TiO_2 submicron lens arrays, which was expected to result in higher light coupling from GaN LEDs into the TiO_2/PS submicron lens in comparison with the SiO_2/PS submicron lens.

10.3.3 Photonic Crystals (PhCs)

During the photon propagation in LEDs, random surface roughening offers random reflections of the internal light on roughened surface, and has no control over the direction of the emitted light. In contrast, a 2D PhC structure has strong control of the light emitted from the active region of the LED, such as directionally emitting. The PhCs are periodic optical materials or structures that are designed to affect the motion of photons in a way similar to that of the periodicity of semiconductor crystals affecting the motion of electrons. Thus, multiple modulating of photons by lattices of periodically varying refractive indices acts to form a photonic bandgap (PBG) in which propagation of certain wavelengths of the electromagnetic waves is prohibited, resulting in an enhanced light extraction in vertical direction from LEDs. The PBG can thus be exploited for suppressing lateral waveguiding and possibly redirect a significant proportion of trapped photons for extraction, overcoming one of the major limitations of nitride LEDs. As the characteristic length scales of

PhC structures are on the order of the wavelength, nano-patterning techniques are involved during the fabrication of visible light devices. Besides the nano-patterning using EBL or NIL techniques, MCCs have been extensively studied as routes to fabricate PhCs and PBG materials [5]. Typical MCCs can afford periodic, spatial variations in dielectric properties on the submicrometer length scale and overcome resolution issues arising from diffraction limit in optical lithography and even beam size limitations in EBL.

10.3.3.1 Surface PhCs

Athanasiou et al. [28] had fabricated a 2D PhC structure on an InGaN/GaN MQW samples by combining the liftoff process and the NSL technique. The lattice constant of the nanohole PhC structure was fixed to be 270 nm, which is actually the original diameter of the silica nanospheres used. Significant enhancement in PL intensity had been observed when the emission wavelength is within the PBG. Time-resolved PL had shown that the spontaneous emission rate was strongly reduced by a factor of ~4 due to the PhC effect. As a consequence, the emission intensity along 2D PhC slab plane directions was effectively suppressed and redistributed to the direction normal to the 2D PhC slab-plane. Similarly, HCP hemi-ellipsoidal PhC had been directly fabricated by ICP etching using silica nanosphere as hard mask. The existence of PBG is established whereby a transmission dip centered at 440 nm was identified, and threefold enhancement in light extraction was achieved [29]. Furthermore, Choi and coworkers [30] demonstrated the CP clover-shaped PhC on GaN by dual-step NSL. As shown in Figure 10.6a–e, by shrinkage of spheres prior to pattern transfer, a non-closed-packed (NCP) clover-shaped PBG structure was also realized. The PBG of the CP and NCP clover-shaped structures was verified through optical transmission spectroscopy, found to agree well with simulated results in Figure 10.6f–i. A threefold enhancement in PL intensity was observed from the optimized structure, when the PBG was tuned to overlap with the emission band of the InGaN/GaN multi-quantum wells.

In general, there are two methods to utilize PhC structures to improve the LEE of LEDs. One approach is to couple the trapped waveguide modes within the LED to the PBG, as mentioned previously the results. Another approach is to use the refractive index periodicity of PhC to diffract trapped waveguide modes above a certain cutoff frequency into externally propagating modes. Hou et al. [31] used the self-assembly method to fabricate a monolayer of PS microspheres, which were dip coated on LED wafer with PR on it, to serve as lenses to focus the UV light to locally expose the PR. After developing the wafer, the 2D periodic hole arrays were formed on the PR. Then a dry etching was performed to transfer the patterns onto LED surface. The LOP of LED with hole PhC was enhanced by 45% compared with the LED without PhC, due to the coupling between guided modes and Bragg modes. Furthermore, we obtained the single, twin, triple, and quadruple compound PCs on *p*-GaN by multiple-exposure nanosphere lens lithography (MENLL) process [32]. The tilt exposure produced the pattern combination of multiple ovals, and the optimized tilted angle was decided to be 26.6° through mathematic calculation to try to avoid the overlay of patterns. Compared with planar LED, the LOP of LEDs with single,

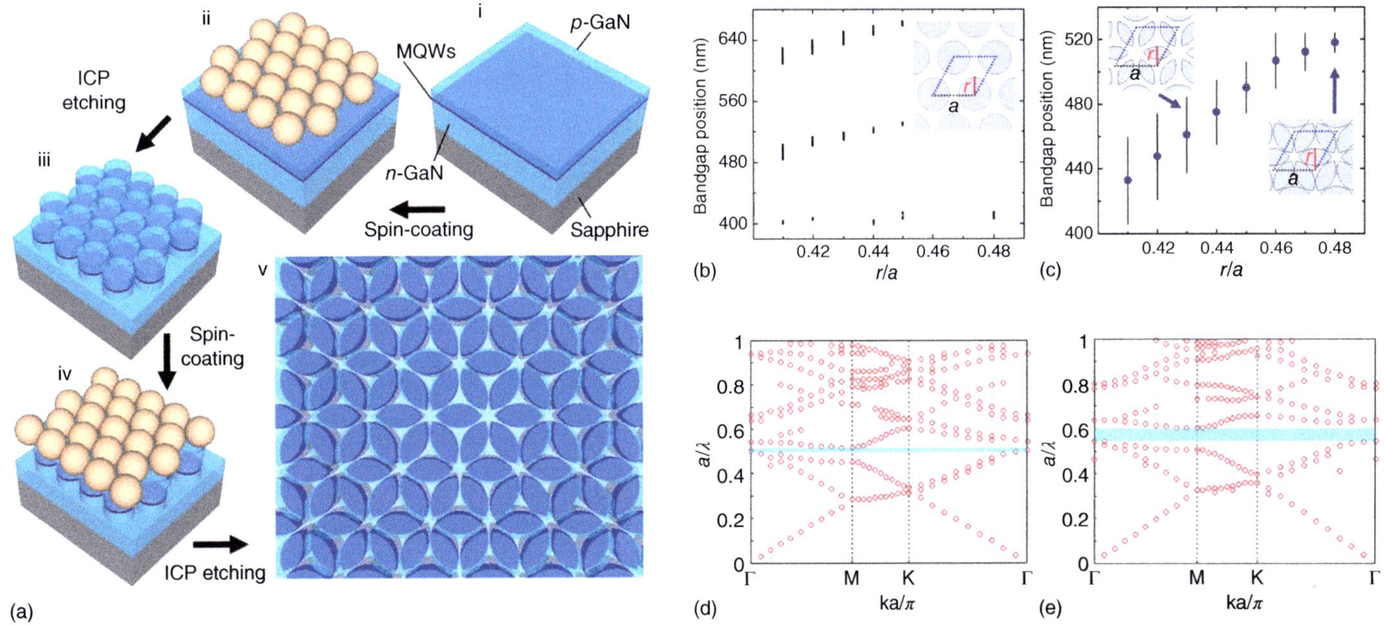

Figure 10.6 Schematic diagram of the process flow: (a) the starting LED wafer. (b) Silica spheres coated onto the wafer surface. (c) Pattern transfer to GaN by ICP etching. (d) Second monolayer of sphere array coated on top of pillar array. (e) Clover-shaped pattern formed after ICP etching. Calculated bandgap as a function of r/a for (f) pillar arrays and (g) clover-shaped PhCs structures. Simulated band diagram of (h) CP and (i) NCP clover-shaped structure, predicting PBG centered at ~517.72 and ~460.75 nm, respectively. Source: Li et al. [30]/with permission from AIP Publishing LLC.

twin, triple, and quadruple PCs was increased by 14.78%, 36.03%, 53.68%, and 44.85% under a drive current 350 mA, respectively. Such a technique is easy and economical for use in fabricating compound nano-patterns.

In the process of fabricating PhCs on the *p*-GaN, dry etching is usually performed to transfer the structure of colloidal sphere mask onto the *p*-GaN, which will induce the plasma damage and degrade electrical conduction in the device due to increased ohmic contact resistance and leakage currents. Instead of direct processing the GaN layer, fabricating PhCs on the ITO had been demonstrated as an alternative method without degradation of the electrical characteristics [33]. In addition, we provided a low-cost and high-throughput process for the fabrication of 2D SiO_2 PhCs by nanosphere lens lithography (NLL) method [34]. The PhC structures were realized by the selective area growth (SAG) of *p*-GaN using SiO_2 NDs, which were patterned utilizing a self-assembled nanosphere as an optical lens. Without prejudice to the electrical properties of LEDs, LOP of LEDs at 350 mA with the $SiO_2$and corresponding air-hole PhC was enhanced by 71.3% and 49.3%, respectively, compared to that without PhC. The LEDs with selectively grown PhC were also found to exhibit partial compression strain release and reduced emission divergence. Furthermore, we also fabricated size-controllable *p*-GaN hexagonal nanopyramid PhC by SAE on flat *p*-GaN layer [35]. The LED with nanopyramid PhC of 46.3% bottom fill factor (PhC lattice constant is 730 nm) showed an improved LOP by 99.9% at 350 mA compared to the LED with flat *p*-GaN. Especially, the effect on the LEE of LEDs is strongly affected by the shape of the PhC structure when it has large fill factor.

10.3.3.2 Bottom PhCs

Currently, the observed enhancement in the LEE of thick LEDs with surface-PhCs was much lower than the theoretical limit, due to the limited interaction of surface-PhCs with the low-order guided modes [36]. It has been proved that the embedded PhCs in LEDs have superior performance compared to surface PhCs, due to their better interaction with all of the guided modes. In addition, the planar top surface of the bottom PhC LED simplifies device fabrication, and eases issues related to making the best electrical contacts to the device. We fabricated large-scale 2D air-gap bottom PhCs, embedded in the *n*-type GaN layer to enhance the LEE of GaN-based LEDs [37]. The embedded PhCs were obtained through overgrowth on *n*-GaN templates, which had air holes array realized by NLL technique, as shown in Figure 10.7a–c. We compared the LOP of three kinds of PhC LEDs with a distance of 560, 1330, and 2370 nm to MQWs. Compared to conventional LED, the LOPs were improved by 58.6%, 77.1% and 80.8 on average at 350 mA without degradation of electrical performance in Figure 10.7d. Figure 10.7e demonstrated the full-width-at-half maximum (FWHM) of emission divergence for these PhCs LEDs were 146.9°, 147.8°, and 139.7°, respectively, which were a little smaller compared to that of 155.8° for conventional LEDs. This implied that the PhCs directional characteristic on the light emission was not strong because the period of embedded-PhC was large enough to have many diffraction orders in the blue light regime resulting in light leakage along many directions.

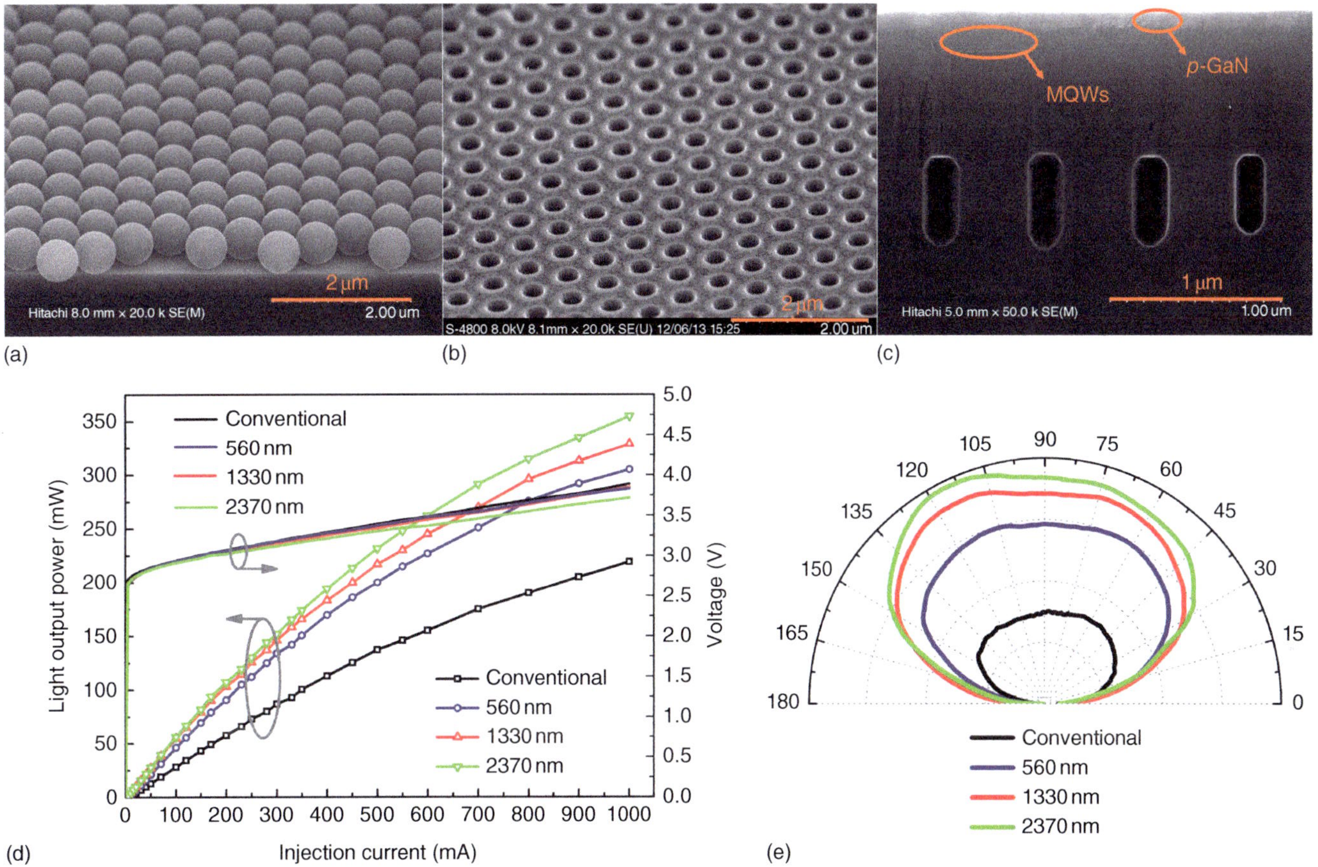

Figure 10.7 Tilted SEM images of (a) 600-nm-diameter PS spheres and (b) air holes array in *n*-type GaN layer. (c) The cross-section photograph of the embedded PhC with a distance of 560 nm to MQWs. (e) LOP-I-V curves (f) far-field radiation patterns of different LEDs at 350 mA. Source: Du et al. [37]/with permission from IEEE.

The periodically ordered bottom PhC structures not only promote LEE, but also partially suppress the piezoelectric field through strain relaxation of QWs and help to reduce the dislocations by epitaxial lateral overgrowth (ELO). Therefore, the IQE of bottom PhC LED can also be improved together. We reported the SiO_2 ND arrays by NLL method into *n*-GaN and *p*-GaN layer of LEDs to produce bottom and surface PhCs simultaneously [38]. The TDs, especially screw dislocations of *n*-GaN after inserting embedded PhC have clearly decreased due to ELO. The LOP of bottom, top, and double PC LEDs is enhanced by 74.5%, 60.1%, and 88.2% compared to that of a conventional LED at 350 mA, respectively. In addition, the PhC converging effect of embedded and top SiO_2 structures in double PhC LED was coupled and resulted in the smallest view angle. Therefore, double PhC structures provide a promising potential to increase both IQE and LEE of LEDs.

10.3.4 Localized Surface Plasmon (LSP)

Localized surface plasmon (LSP) is a kind of non-propagating electromagnetic mode, caused by the scattering of micro/nano structures or metallic NPs in the electromagnetic field. As long as the frequency of the incident electromagnetic field matches the frequency of the collective oscillation due to the excitation of the free electrons in the metal surface rough structure or the metal NPs, the plasmon resonance will be generated. When light-emitting materials are located within the penetration depth of LSP field, the emitted photons may couple with the LSP modes instead of free space. The coupling leads to the enhancement of radiative recombination rate and reduce PL decay time. Therefore, surface plasmon coupling is particularly useful for increasing the emission efficiency of LEDs of inherently low IQE [39].

In 2004, a pioneering work was reported by Okamoto et al. [40] that Ag LSPs-enhanced blue InGaN/GaN QWs showed 14-fold enhancement in the PL peak and 6.8-fold improvement in the IQE by thermal annealing of Ag film. The resonance frequency of the generated LSP is determined by the size, shape and spacing of Ag NP array, which is critical for the coupling. Currently, thermal annealing of metal films is the most general way to fabricate NP array. However, the NPs are subject to metal oxidation and diffusion in the GaN layer leading to performance degradation. On the other hand, the metal NPs generated by thermal annealing or chemical synthesis method show a large variety of size, shape and spacing, which leads to the difficulty in precisely controlling the resonance frequency for different LSP modes.

As an efficient and economical fabrication technique for nanostructure, NSL can produce ordered array metal NPs with good control over size, shape and spacing. It can be tailored by tuning the size of the nanospheres, annealing-induced deformation, and plasma-etching-driven non-closed packing. Kao et al. [41] fabricated Ag nanotriangle array (NTA) on a 40-nm-thick *p*-type GaN layer beneath the *p*-pad of the LED by means of NSL. The fabrication procedure was schematically shown in Figure 10.8a. The resonance frequency of a generated LSP can be precisely controlled by changing the size of the PS nanosphere and the Ag deposition thickness. Figure 10.8c showed the LOP as a function of injection current for the LED without

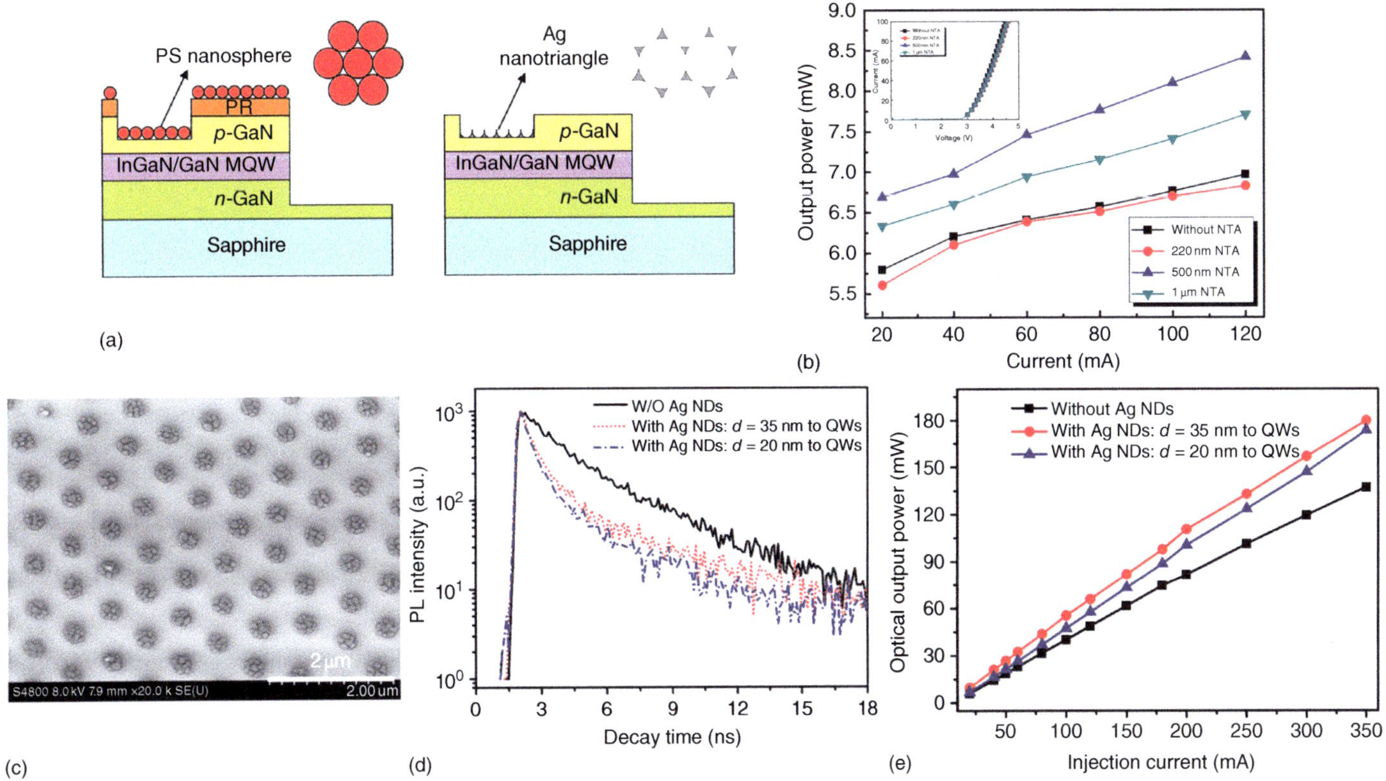

Figure 10.8 (a) Fabrication flowchart of the nitride-based LED with an Ag NTA. (b) LOP as a function of injection current for the LED without and with Ag NTA, respectively. Source: Reproduced with permission from Kao et al. [41]. Copyright 2010, Institute of Electrical and Electronics Engineers. (c) Ag NPs arrays after Ag deposition on ICP-etched PR and liftoff. (d) TRPL spectra of LSP-enhanced LEDs without and with Ag NDs. (e) LOP of LSP-enhanced LEDs as a function of injection current. Source: Wei et al. [42]/with permission from AIP Publishing LLC.

and with an Ag NTA. With a 20 mA current injection, the output powers of the LED without and with Ag NTA were 5.8, 5.6 (220 nm NTA), 6.7 (500 nm NTA), and 6.4 (1 μm NTA) mW, respectively. The resonance frequency of 500-nm Ag NTA with a side length of 115 nm was around 448 nm. Since the emitting wavelength of the LED was 450 nm, the resonance frequency of an 8-nm-thick 500-nm Ag NTA coincided with the emitting wavelength, with an enhancement of LOP (15.4%) for the LED. We also fabricated large-scale Ag NDs arrays using NLL [42], embedded in *p*-GaN layer of LED for generating LSP coupling with the radiating dipoles in QWs, as shown in Figure 10.8c. Based on the Ag NDs with the controlled surface coverage, LSP leaded to the improved crystalline quality of regrowth *p*-GaN epilayer and the increased PL intensity. From time resolved photoluminescence (TRPL) in Figure 10.8d, the LEDs with Ag NDs showed a pronounced decrease in the lifetime. Thus, fewer carriers were captured at the TDs through nonradiative recombination, and η_{IQE} was correspondingly increased. Figure 10.8e showed LOP of the LSP-enhanced LEDs with a distance of 20 and 35 nm to QWs was increased by 26.7% and 31.1% at 350 mA, respectively, compared to that without Ag NDs.

Up to now, most investigations on the enhancement of emission efficiency via LSP coupling have been performed for the blue or green LEDs. Due to the much lower IQE induced by the intrinsic defects in the AlGaN QW, it would be quite meaningful to LSP coupling to improve the emission efficiency of DUV LEDs. Taguchi et al. [43] reported size-controlled fabrication of Al nanostructure arrays providing direct control of the LSP at UV wavelengths. Plasmon resonances were obtained at wavelengths from near-UV down to 270 nm (4.6 eV) depending on the fabricated structure size. Such precise size control was realized by the NSL combined with additional microwave heating to shrink the spaces in a CP monolayer of colloidal nanosphere masks. Yin et al. [44] also achieved a significant DUV emission enhancement, as high as 3.2-fold, for AlGaN-based MQW by introducing the energy matched coupling of LSPs from Al NPs. The strongly enhanced, much DUV emission ranging from ~279 to ~286 nm in the MQWs was proposed to originate from the SP-enhanced emission from the high-order QW exciton state by enhancing the spontaneous radiative recombination rate through LSP-exciton coupling. The fabrication of size-tunable Al NP arrays in a wafer scale using NSL can potentially be applied to practical DUV LEDs with high IQEs.

10.3.5 Nano-patterned Sapphire Substrates (NPSS)

Patterned sapphire substrate (PSS), displaying microscale features obtained by photolithography and wet/dry etching, has been studied extensively to increase the output power of GaN-based LEDs. Both the IQE and LEE of LEDs are improved by utilizing the microscale PSS. Furthermore, the LEDs grown on nano-sized PSS (NPSS) show more improvement in EQE than those grown on microscale PSS [45]. NPSS can improve the LEE more efficiently because it has a higher density of patterns, thus increasing the chance of light scattering on the sapphire substrate. Furthermore, the growth of GaN epilayers on NPSS has also resulted in obvious reduction in threading dislocations (TDs) in GaN, which has been applied to achieve improved IQE in LEDs [46]. However, NPSS is not realized by conventional photolithography due to its diffraction-limited resolution, and generally requires EBL or NIL techniques, making it unfavorable for mass production.

NSL technique opens a door to fabricate ordered micro/nanostructures with large area and allows more flexible tuning of the feature size of the patterns. Chan et al. [47] reported the improved GaN-based LEDs grown on NPSS through etching an SAM of 750-nm-diameter SiO_2 nanospheres. Three-dimensional (3D) finite-difference time-domain calculation predicts a 40% enhancement in LEE of NPSS LEDs with various pattern depths in Figure 10.9a. The FWHM values of the ω-scan rocking curves for (102) planes are 370 arcsec and 620 arcsec for GaN film grown on NPSS and flat sapphire substrate (FSS), respectively. This means crystalline quality of the epitaxial LED film is also improved by the NPSS technique. Thus, LOP enhancement for the LED on NPSS reaches 76% relative to that on FSS in Figure 10.9b. Chen et al. compared the output power of LEDs grown NPSS, PSS, and conventional sapphire substrate (CSS) [48]. At the injection current of 20 mA, the LOPs of NPSS, PSS, and CSS LEDs were 10.27, 9.27, and 7.93 mW, respectively, as shown in Figure 10.9c. The LOP of the NPSS LED was 1.1 times higher than that of PSS LED. It indicates that the NPSS LED has superior capability for enhancing

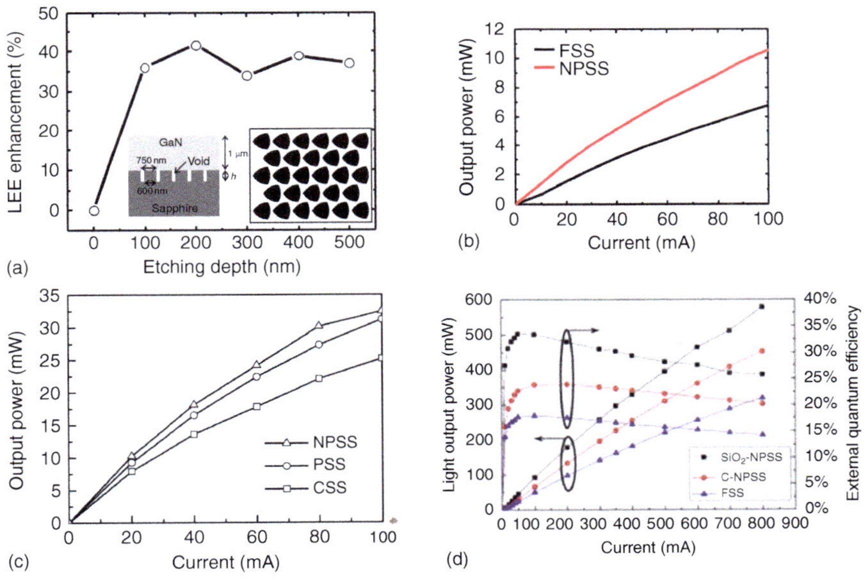

Figure 10.9 (a) Simulated values of the LEE enhancement. Left inset: schematic representation of the simulation scheme; right inset: profile of the simulated sapphire. (b) Light power versus current curves of the FSS and NPSS LEDs. Inset: side view of NPSS. Source: Reproduced with permission from Chan et al. [47]. Copyright 2009, American Institute of Physics. (c) LOP versus current curves of the LEDs grown NPSS, PSS, and CSS. Source: Reproduced with permission from Chen et al. [48]. Copyright 2008, Institute of Electrical and Electronics Engineers. (d) LOP and EQE as functions of current for SiO_2-NPSS, C-NPSS and FSS LEDs. Source: Reproduced with permission from Zhang et al. [49]. Copyright 2014, Elsevier B.V.

the LEE. We studied the pattern spacing size dependence of characteristics of GaN-based LED grown NPSS [8]. Under a driving current 350 mA, EQE of LEDs on NPSS with 0-, 50-, and 120-nm spacing increased by 43.3%, 50.6%, and 39.1, respectively, compared to that on FSS. However, the GaN growth on top of NPSS with 0-nm spacing showed the best crystalline quality because of laterally epitaxial overgrowth. The optimized pattern spacing is 50 nm for the NPSS with 600-nm pattern period.

It is reported that the growth of GaN on periodic SiO_2 nanosphere NPSS (SiO_2-NPSS) generated decreased dislocations by nanoscale epitaxial lateral overgrowth (NELOG). Here, the sapphire of SiO_2-NPSS does not suffer from etching, and is coated with self-assembled SiO_2 nanosphere monolayer. Ueda et al. [50] showed that the LOP of LED grown silica-sphere-coated sapphire was increased 2.5-fold compared with the LED constructed without silica spheres. We fabricated conventional NPSS (C-NPSS) and SiO_2-NPSS to grow LEDs [49]. Compared with the LEDs on C-NPSS and FSS, the EQE of LEDs on SiO_2-NPSS was increased to 30.7% and 81.9% at 350 mA current, respectively, as shown in Figure 10.9d. SiO_2-NPSS LED also showed more light in vertical direction and more uniform light distribution. The lower refractive index of SiO_2 nanosphere near the NPSS interface enabled more light to emit. Thus, the SiO_2-NPSS not only improved the crystalline quality of GaN but also enhanced the LEE of LEDs.

High-quality AlN and AlGaN materials are the basis for achieving high efficiency DUV LEDs. However, unlike the GaN, the growth of AlN usually follows a 3D Volmer–Weber growth mode due to the high surface adhesion coefficient of Al atom on sapphire [51]. Thus, the growth of AlN on PSS requires a coalescence thickness close to 10 μm and greatly increases the epitaxy time and cost. NPSS has been widely regarded as effective and necessary method to grow the DUV LEDs. We first reported AlGaN-based DUV LED [52] grown on NPSS prepared through an NSL technique and wet chemical etching in Figure 10.10a–c. Figure 10.10d showed that the AlN coalescence thickness on NPSS is only 3 μm due to NELOG of AlN, which leaded to low TDs in AlN and epilayers previously mentioned. On NPSS, the LOP of 282-nm UV-LED reaches 3.03 mW at 20 mA with EQE of 3.45, 98% higher than that on FSS in Figure 10.10e. Narrow X-ray diffraction (XRD) FWHMs and cross-section TEM images have both demonstrated high material quality of the AlN and its upper epilayers on NPSS. Recently, Lee et al. [53] demonstrated AlGaN-based DUV LEDs with periodic-air-voids-incorporated nanoscale pattern by NSL method. The NELOG improved the crystal quality of epitaxial layers at a relatively low growth temperature of 1050 °C and at small coalescence thickness less than 2 μm in Figure 10.10f. Figure 10.10g showed that the LOP of DUV LED on nano-pattern substrate was enhanced by 67% at current of 20 mA, attributed to the formation of embedded periodic air voids. They caused simultaneous improvements in the crystal quality and LEE enabled by breaking the predominant in-plane guided propagation of DUV photons.

Besides pattern size, shape of the pattern in NPSS also affects its efficiency of enhancing the LOP of GaN-based LEDs. Various shapes, such as conical, volcano, and pyramidal, have been employed in NPSS [51, 54]. Geng et al. have presented

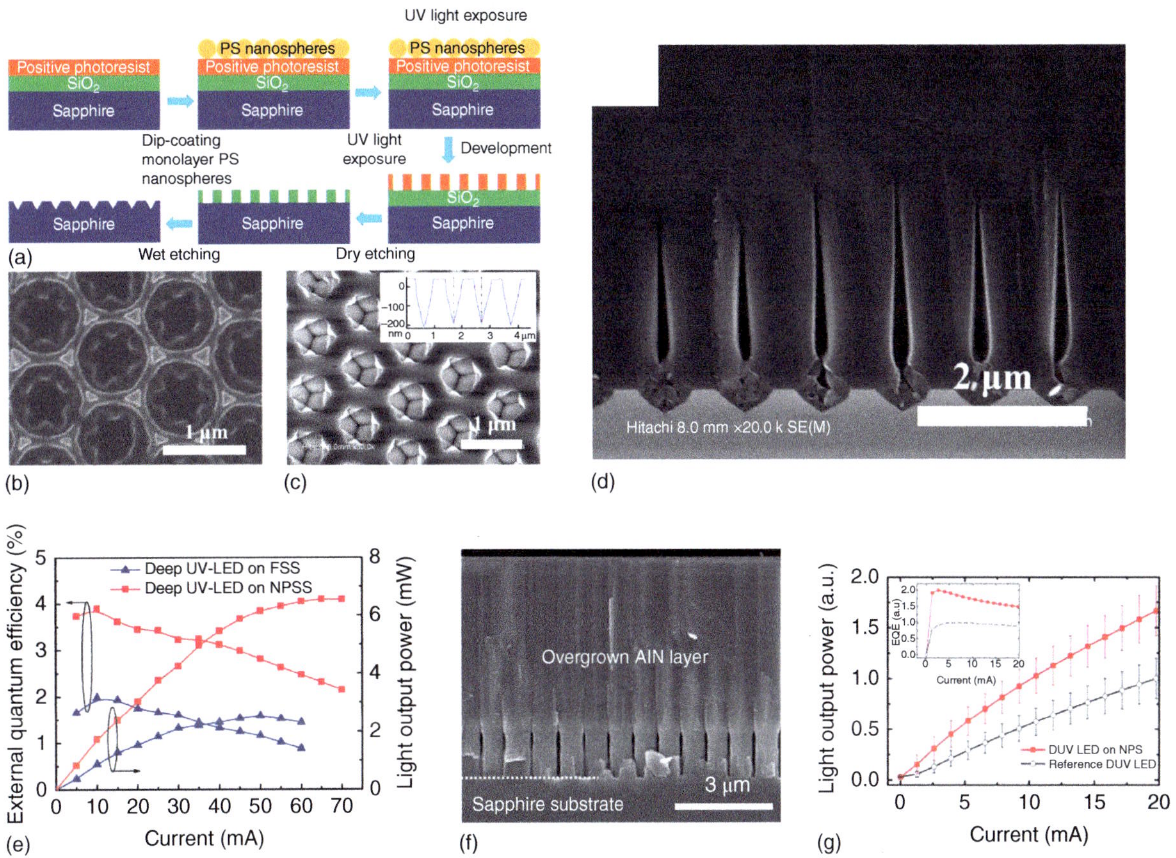

Figure 10.10 (a) Schematic of the fabrication flow to create NPSS. SEM images of the patterned PR (b) and wet etched NPSS (c). The inset in (c) shows the line profile of the patterns of NPSS by AFM. Source: Dong et al. [52]/with permission from AIP Publishing LLC. (d) Cross-sectional SEM image of AlN grown on NPSS. Source: Dong et al. [52]/with permission from AIP Publishing LLC. (e) LOP-I-EQE curves of DUV LED grown on NPSS and FSS. Source: Lee et al. [53]/with permission from AIP Publishing LLC.

the fabrication of nanoscale volcano-shaped NPSS using a wet chemical etching in combination with NSL [54]. Wet etching masks were fabricated by infiltrating the silica precursor solution into the NCP PS MCC. The formation of the volcano-shaped NPSS in nanoscale was ascribed to the different corrosion-resistant abilities of silica masks of different effective heights during wet chemical etching. Apart from patterning sapphire substrate, it is noteworthy that MCC can also be used to pattern other substrates, such as GaN and AlN materials. Kuo et al. [55] reported the fabrication of the nanoinverted pyramid (NIP) structures of a GaN template with SiO_2 nanospheres. The NIP GaN template can significantly reduce TDs density and thus improve the crystal quality of GaN epilayers. By using an NIP GaN, the LOP of LEDs can be increased by 32%.

10.3.6 Optical Reflector

Typically, photons generated from MQWs of LEDs can partially emit downward toward the substrate, leading to that a large fraction of light cannot be extracted and contribute to the effective light output. Thus, to alleviate this problem, the employment of a high-reflectivity reflector at the backside of sapphire substrate is one of effective methods to reflect downward photons into the escape cone on the front of the LEDs rather than be lost [56]. The commonly used reflectors include distributed Bragg reflectors (DBRs) and omnidirectional reflectors. However, the LEE of planar high-power GaN-based LEDs is still limited by the total internal reflection. Due to their 3D configurations, roughened reflectors have been proved more effective by randomizing the reflected light toward the top surface of LEDs compared with flat reflectors. Despite various methods to fabricate artificial nanostructures on substrate to realize broad band and omnidirectional reflective structures, the preparation of textured backside reflectors with nanoscale 3D configurations is still too complicated [57]. Instead, NSL technique provides an economic approach to fabricate 3D reflectors for improving LEE of LEDs.

In 2009, Kim et al. presented a 3D Ag reflector on GaN LEDs using SiO_2 nanospheres as the templates [58]. Hexagonal nano-lenses arrays were prepared through embedding SiO_2 nanosphere into a benzocyclobutene layer, while concave-shaped hemisphere arrays were obtained by etching embedded SiO_2 away. 3D Ag reflectors were fabricated through depositing Ag on concave and convex hemisphere arrays, which were found to enhance the LOP of LEDs by 29–33%. Subsequently, Liou et al. [59] inserted a self-assembled SiO_2 nanosphere monolayer between the substrate and backside reflectors. Due to the presence of concave surfaces and PhC-like air voids, downward photons emitted from MQWs toward 3D backside reflectors could be reflected, scattered, and redirected into arbitrary directions for light extraction. These textured 3D backside reflectors with an SiO_2 nanosphere monolayer could also extract the lateral light inside device into the normal direction and improve LEE, as shown in Figure 10.11a. As compared with a conventional LED without a backside reflector and LED with a planar hybrid backside reflector, at 350 mA, the studied device with a 3D hybrid backside reflector exhibited 136.4% (165%) and 23.6% (27.4%) enhancements in LOP without the

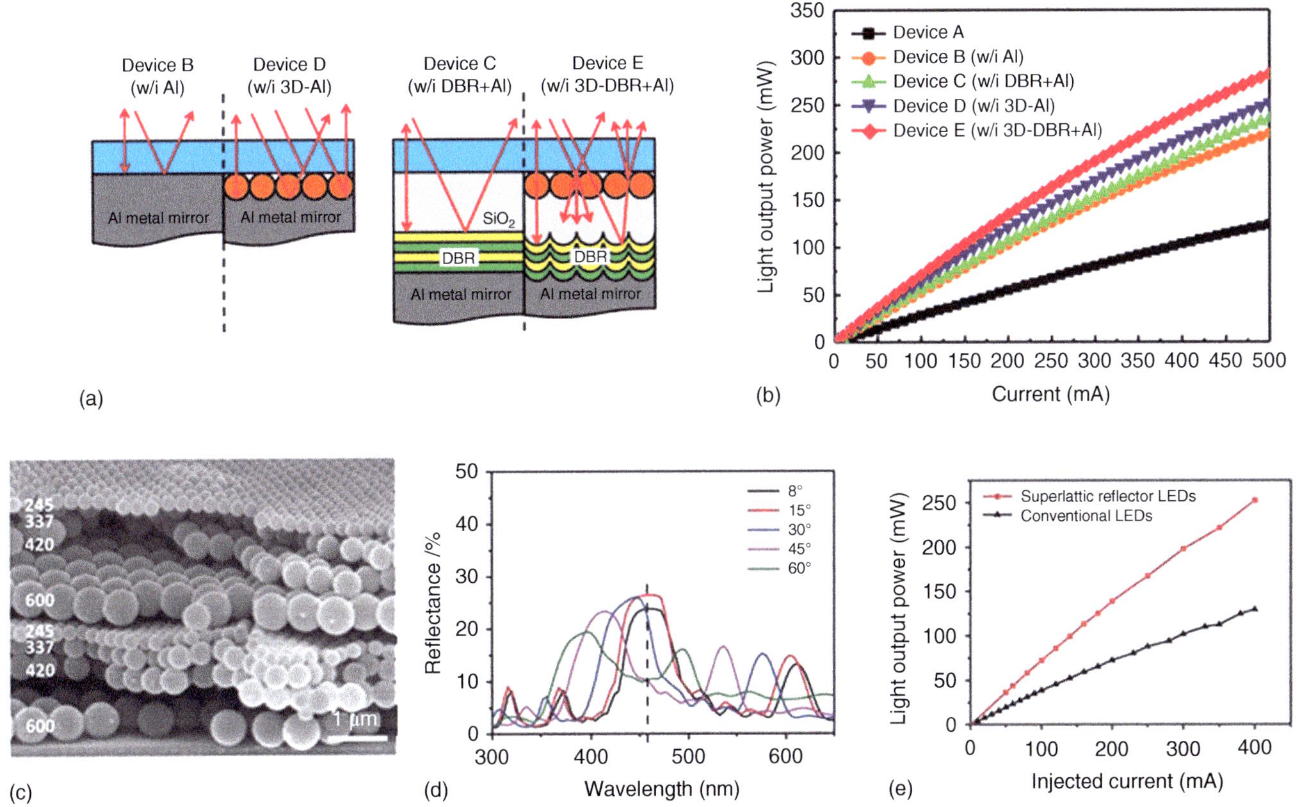

Figure 10.11 (a) Possible photon pathways in the LEDs with different hybrid structures. (b) LOP as a function of operation current of LEDs. Source: Reproduced with permission from Liou et al. [59]. Copyright 2014, Institute of Electrical and Electronics Engineers. (c) The (2 + 1)D colloidal PhC superlattice formed by alternately stacking colloidal monolayers in ABCDABCD arrangement. (d) The angle dependent specular reflection spectra of LBL $(ABCD)_6$ superlattice. (e) Relative LOP versus injection current (L-I) characteristic for traditional LED and LED with (2 + 1)D PhC superlattice reflector. Source: Zhang et al. [60]/with permission from John Wiley & Sons, Inc.

degradation of electrical properties in Figure 10.11b. Furthermore, to solve the weak adhesion between SiO_2 nanosphere and sapphire substrate, they fabricated a nanohemispherical hybrid backside reflector [61]. Using 100-nm SiO_2 nanosphere monolayer as a hard mask on the backside of the sapphire, nanohemispherical patterns could be transferred to the deposited hybrid backside reflector. As compared with a conventional LED without a backside reflector, at 350 mA, the studied device exhibits a 118.2% enhancement in LOP. Note the adhesion between an ICP-transferred sapphire and the hybrid backside reflector was better than that when directly inserting SiO_2 nanosphere monolayer in the device.

However, the PhC PBGs can only effectively reflect the vertical incident light. With the incident angle increasing, the intensity of reflectance declines dramatically. PhCs with a wide PBG may afford reflectivity with incident angle independence. Therefore, PhC superlattice featuring a wide PBG is desired for back reflector application in GaN-based LEDs. Huang et al. [62] directly fabricated multiple-layer 3D colloidal PhC bottom reflector on the backside of LED wafer using PS suspension with a diameter of 193 nm. The resulting LEDs exhibited significant improvement in light extraction (155%), slope efficiency, and EL spectra (219.6%), as compared with that of conventional LEDs. We presented a facile approach to the fabrication of (2 + 1)D PhC superlattice [60]. The colloidal monolayers of different sphere sizes were stacked layer by layer (LBL), resulting in a colloidal (2 + 1)D PhC superlattice with high mechanical strength and good crystalline integrity in Figure 10.11c. The reflection peak was centered at 457 nm, which well matched with the emitted wavelength of GaN-based LED. The intensity of the diffuse reflectance of LBL $(ABCD)_6$ could reach 88%. It could be seen from Figure 10.11d, for the wavelength of 457 nm, when the incident angle varied from 0° to 30°, the superlattice exhibited good specular reflectivity with an intensity over 23%. An LOP enhancement of 97.3% had been demonstrated employing such a (2 + 1)D PhC superlattice as a back reflector in Figure 10.11e.

10.4 New Applications for Optoelectronic Devices

10.4.1 Nanorod LEDs

In comparison with conventional thin film LEDs, GaN-based nanorod LED offers many potential advantages [63]. GaN nanorods with high aspect ratio and large surface-to-volume ratio can dramatically reduce the dislocation density in the upper part of the nanorods. Besides, nanorods may help to relieve the strain induced by thermal expansion mismatch and avoid cracks generation, which is a big problem for GaN growth on the substrates with large thermal expansion and lattice coefficient difference to GaN. LEE of nanorod LED is expected to increase owing to the nonplanar geometry of nanorods. Approaches for the fabrication of these functional nanostructures include bottom-up vapor–liquid–solid (VLS) process, metal organic chemical vapor deposition (MOCVD), thermal reaction, one-step self-masked top down dry etching, etc. [63] The critical issue of nanorod LED fabrication is to prevent shorting the *p*- and *n*-type semiconductors when depositing the top metal contact.

10.4.1.1 Top-Down Nanorod LEDs

Currently, the use of MCC-based colloidal lithography toward nanostructure LEDs has been extensively reported. The most straightforward method to define a nanorod LED is to etch through the p-GaN, the quantum well region, and the n-GaN by using a self-assembled MCC as an etching mask. Chen et al. [64] fabricated InGaN/GaN nanorod LED arrays using NSL for nanorod formation, plasma enhanced chemical vapor deposition (PECVD) grown SiO_2 layer for sidewall passivation, and chemical mechanical polishing for uniform nanorod contact. The nano-device demonstrated a reverse current 4.77 nA at −5 V, an ideality factor 7.35, and an optical output intensity 6807 mW/cm^2 at the injection current density 32 A/cm^2 (20 mA).

However, the plasma etching-induced sidewall damage has created a large amount of defects and leakage currents that limit the LOP. Li et al. [65] had developed a two-step top-down process for creating nanorod LED arrays from planar LED epitaxial structures. This top-down technique combined the plasma etch with a subsequent anisotropic wet etch, which allowed for removal of the plasma etch damage, enabling straight nanorod sidewalls and control over the nanorod geometry. The nanorod LEDs also showed reduced QCSE compared with the planar LEDs due to reduced piezoelectric strain. Subsequently, Latzel et al. [66] detailedly explored the performance of nanorod LEDs before and after potassium hydroxide (KOH) wet etch and alumina passivation. Figure 10.12a showed the LED fabrication process including silica nanospheres deposition, KOH etching, graphene transfer, and alumina passivation. Steady-state PL revealed that the luminescent performance of the nanorod LEDs was increased by about 50% after KOH and 80% after additional alumina passivation. The measured LOP increased linearly with injection current, indicating a negligible influence of droop in this low injection regime. The enhancement of EL after alumina passivation was about 30% compared to the KOH etched device in Figure 10.12b.

Especially, for AlGaN-based DUV LED, it is expected that transverse magnetic (TM) polarization becomes more dominant at relatively short wavelengths (<280 nm), which is hard to be extracted. Fortunately, the TM polarized light can be significantly enhanced in 3D nanostructure DUV LED by exploiting the lateral side emission [68]. Thus, 3D nanorod structures are crucial to address the LEE problems of DUV LEDs. Zhang et al. [67] reported the electrically injected 274-nm AlGaN nanorod array DUV LEDs fabricated by NSL and etching technique. Figure 10.12c showed the IQE of nanorod LEDs increased by 1.2 times due to the transformation of carriers from the exciton to the free electron-hole, possibly driven by the interface state effect of the nanorod sidewall surface. Nanorod DUV LED devices with good electrical properties were successfully realized. Compared with planar DUV LEDs, nanorod DUV LEDs presented >2.5 times improvement in LOP and EQE in Figure 10.12d. The nanorod arrays significantly facilitated photons escaping from the interior of LEDs along the vertical direction, via the waveguide effect, Bragg scattering effect, and reflection of the planar Al/Ti/Au stack layer on top of the nanorods.

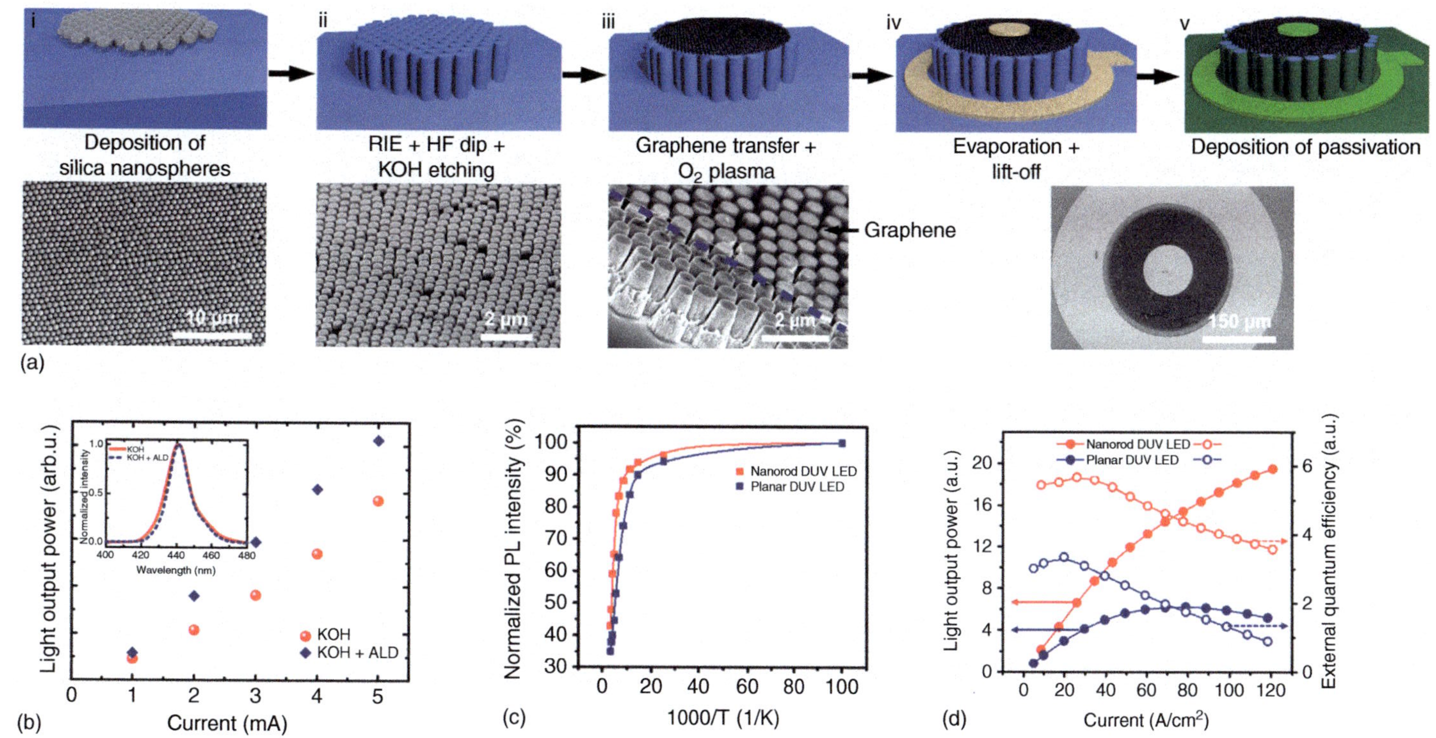

Figure 10.12 (a) LED fabrication process in chronological order: (i) deposition of silica nanospheres. (ii) RIE of nanorods, removal of silica nanospheres by HF and KOH etching. (iii) Transfer of graphene. (iv) Deposition of metal contacts. (v) Deposition of the passivating alumina layer. (b) Emitted power as function of injection current of the nanorod LEDs before and after passivation. Inset: normalized EL spectra of the nanorod LEDs before and after passivation. Source: Zhang et al. [66]/with permission from Institute of Physics. (c) Arrhenius plots of the normalized PL intensities as a function of temperature for nanorod LED and planar LED. (d) LOPs and EQEs of nanorod LED and planar LED under different current densities. Source: Zhang et al. [67].

10.4.1.2 Bottom-Up Nanorod LEDs

Through a direct top-down process, a large portion of III-nitrides is etched away, and cannot contribute to light emission, which will add additional cost during device fabrication. Besides, in addition to stress release and enhanced light extraction, other advantages of nanorod LEDs, for example, defect reduction and tuning the indium concentration by the nanorod size, will not be applicable. Thus, bottom-up method to grow nanorod/nanopyramid using NSL has been developed. We presented the phosphor-free nanopyramid white (NPW) LED grown on {1011} planes using NLL [69]. As Figure 10.13a showed, a silica layer was deposited on n-GaN and patterned by PS NLL and ICP etching. The nanopyramid LEDs were obtained through SAG using as-fabricated SiO_2/GaN template. Due to the dislocation filtering effect of SAG, Figure 10.13b–c showed that the nanopyramid LED was nearly perfect single crystalline. Blue emission was from the upper sidewalls of nanopyramid, and yellow lighting originated from the lower ridges with reference to the base of nanopyramid in Figure 10.13d. The EL spectrum was similar to white LEDs from blue LEDs with yellow phosphors as shown in Figure 10.13e, and NPW LEDs will imply great potential to form the basis of a low-cost technique for future white lighting.

Recently, we also reported optically and electrically pumped 280 nm DUV LEDs with ultrathin GaN/AlN quantum disks (QDs) inserted into AlGaN nanorods by selective epitaxial regrowth using molecular beam epitaxy (MBE) [70]. The nanospheres were utilized as masks in an ICP etching process to fabricate top-down AlGaN nanorods, followed by the selective regrowth of MBE. Due to high quantum confinement and the large band offset between AlN and GaN, emission wavelength of GaN/AlN QDs shifts to 280 nm. The IQE of the GaN/AlN QD nanorod DUV LED reaches 81.2% by comparing CL intensity at room temperature (RT) and 7 K. The LEE was calculated to be increased from 14.9% for the planar LED to 49.6% for the nanorod LED because most of the light is extracted from the top surface of the nanorod LED.

10.4.2 Microdisk Lasers

Nitride-based microcavity lasers have attracted much attention in past decades for both physical insight of lasing mechanisms in both the weak and strong coupling regimes and their widespread applications [71]. Among them, microdisk (MD) lasers are particularly standing out thanks to their high Q-factors and small mode volumes, as well as simple architectures. However, low-temperature operation, optical pumping, and high lasing thresholds are the existing obstacles lying ahead for the "state-of-art" nitride-based microcavity lasers stepping into real-life practical applications [72]. Because whispering-gallery (WG) modes are confined at the edge of MD cavities, sidewall surface roughness is dominantly responsible for the optical leakage. Therefore, to fabricate smooth sidewalls for GaN-based microcavities is crucial to achieve high Q-factors by suppressing the light scattering effect. Conventionally, mesa of the nitride-based MDs is defined by a dry etching process followed by a standard photolithography. However, due to the optical diffraction limit of the light source, the photolithography process can hardly reproduce the mask pattern

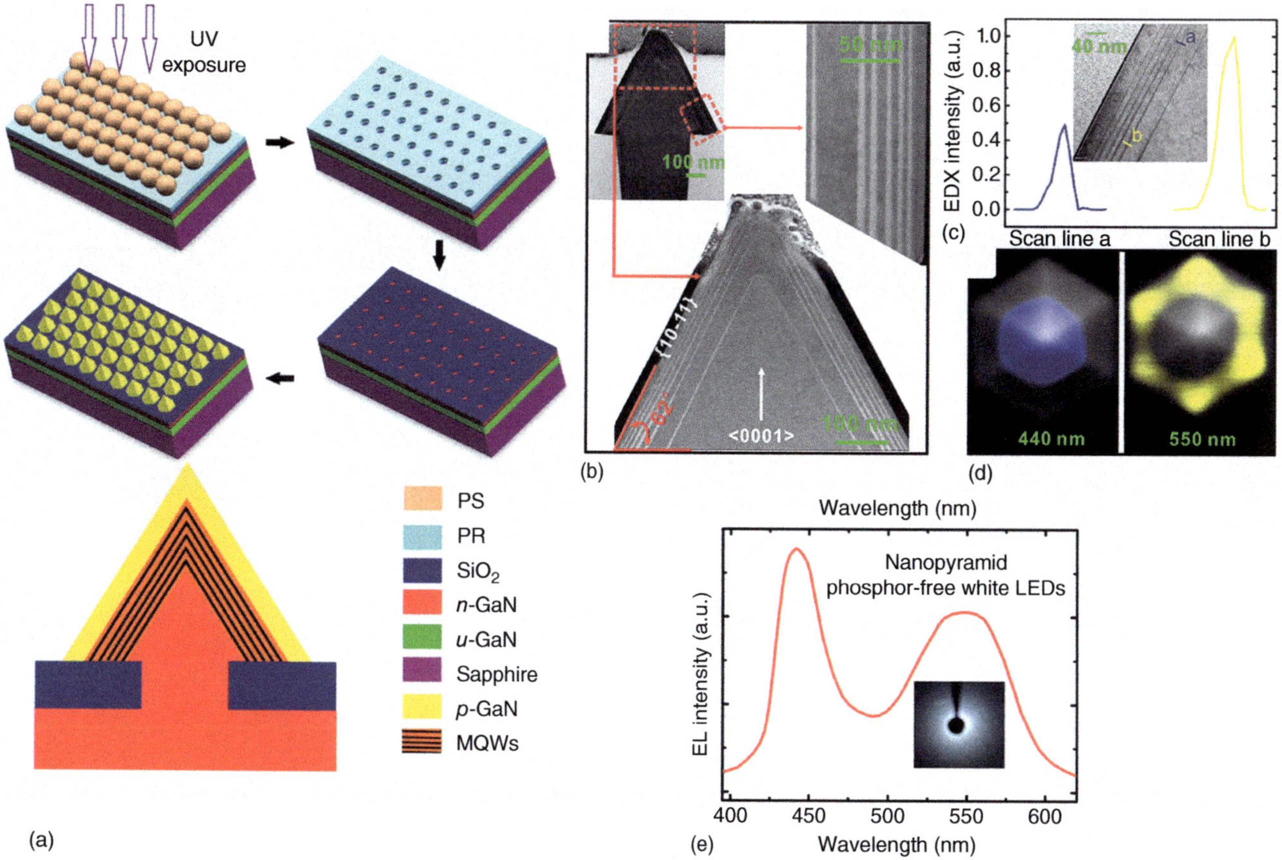

Figure 10.13 (a) Schematic illustrations of the process for fabricating phosphor-free NPW GaN LED by NLL. (b) TEM cross section of single NPW LED and the corresponding high magnification images. (c) For the cross-section cut along the ridges, the normalized TEM-EDX line scan analysis was carried out perpendicular to the QW. (d) Monochromatic CL images at the given wavelength. (e) EL spectrum of NPW LED at 20 mA current. Source: Wu et al. [69]/with permission from AIP Publishing LLC.

precisely at sub-wavelength scale. More precise patterning techniques, such as EBL and NIL techniques, have to be used to ensure more precise replication of the mask patterns, which are very expensive and always criticized for low production efficiency. As alternative, with a good etching selectivity and much less mask erosion, cost-effective microsphere lithography (MSL) has been developed to fabricate GaN-based microcavity lasers [73], in which silica microspheres are serving as hard masks in the subsequent dry etch process, whereby smooth sidewalls and near-perfect circularity can be obtained.

Zhang et al. [74] fabricated the InGaN/GaN MD lasers on Si platform using MSL, ion dry etching, and chemical wet etching, which leaded to the demonstration of optically pumped room-temperature WG lasing in the blue spectral range. As the excitation exceeding a threshold of 8.43 mJ/cm^2 (Eth), a sharp spectral peak at ~437.8 nm emerged in the PL spectra. Moreover, nonlinear increase in integrated PL intensities at ~437.8 nm could be observed as excitation increased beyond this threshold. Purcell factor (F_p) was evaluated to be 0.12, as the spontaneous emission coupling factor was about ~0.11. Furthermore, an optically pumped multicolor laser had been achieved by Athanasiou et al. using an InGaN/GaN-based MD with an undercut structure on a silicon substrate [75]. The MD laser was fabricated by means of a combination of a cost-effective MSL and subsequent dry/wet etching processes as shown in Figure 10.14a,b. The MD laser was approximately 1 μm in diameter. The structure was designed in such a way that the vertical components of the WG modes formed could be effectively suppressed. The optically pumped lasing was achieved at 443, 492, and 522 nm in continuous wave mode simultaneously in Figure 10.14c, forming multicolor lasing, where their corresponding thresholds for lasing are around 27, 15, and 5 kW/cm^2, respectively. This was the first report of a multicolor laser with a low threshold using III-nitride semiconductors.

Optically pumped single-mode DUV lasing actions operating at room temperature were also demonstrated by Zhang et al. [76] from 150-nm-thick undercut MDs with AlN/$Al_{0.35}Ga_{0.65}N$ (5.5 nm/2.5 nm) MQWs. These AlGaN-based MDs were grown on Si substrate by MOCVD. The lasing wavelength centered at ~300.1 nm with the linewidth of ~1.0 nm as the excitation exceeded the lasing threshold of ~24.2 mJ/cm^2. An emission coupling factor (β) of 9.2×10^{-2} was estimated based on the light output characteristics of the AlN/AlGaN MDs with increasing the pumping densities in Figure 10.14d. A Q-factor of ~300 at the transparency could be roughly estimated using the relation $\lambda/\Delta\lambda$. This value was relatively low perhaps due to the high dislocation density with the AlGaN MQWs grown on Si and increased intrinsic radiative losses with smaller disk radius. However, it is promising to serve as efficient DUV coherent light sources for the photonic integration of the nitride-based semiconductor devices on the Si platform.

10.5 Conclusions and Perspectives

The past 20 years have witnessed a number of encouraging advancements in LOP enhancement of the GaN-based LEDs. NSL-based MCCs have emerged as a

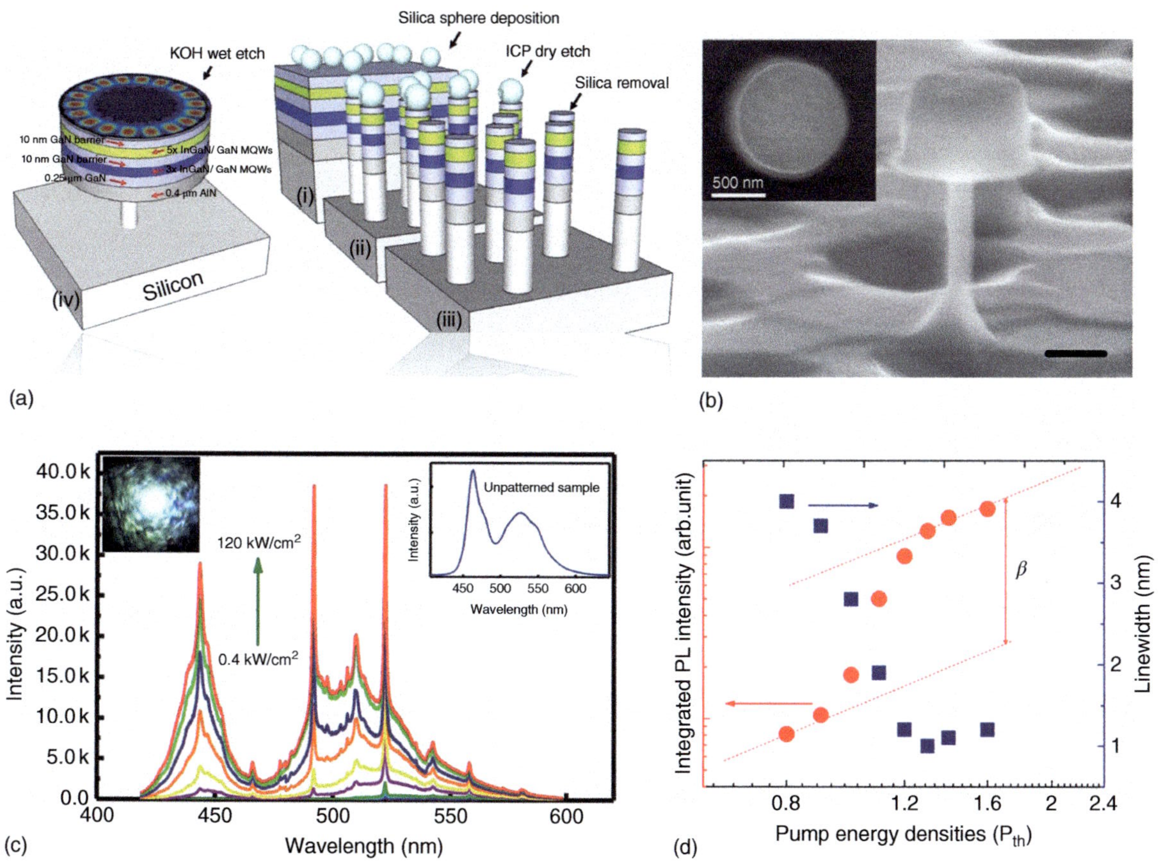

Figure 10.14 (a) Schematic illustration of multicolor MD laser, inset showing procedure of the fabrication of MD on silicon. (b) Side-view SEM image of typical microdisk with a diameter of ~1 μm. (c) Lasing spectra from MD laser as a function of optical pumping power density. Source: Athanasiou et al. [75]/Springer Nature/CC BY 4.0.. (d) Integrated PL intensities from an AlGaN/AlN MD and linewidth evolution under different pumping densities. Source: Reproduced with permission from Zhang et al. [76]. Copyright 2017, Institute of Electrical and Electronics Engineers.

promising technique for the creation of highly ordered nanostructures because it is cheap, fast, and simple, and does not require any sophisticated equipment. These nanostructures fabricated by NSL have been extensively used to improve the crystal quality, IQE, and LEE of nitride LEDs heteroepitaxially grown on lattice mismatch substrates. In this chapter, we have reviewed recent advancement of nanostructures derived from MCCs and their detailed applications in enhancing IQE and LEE of GaN-based LEDs. Particular attention is devoted to summarizing the representative principles of the enhancement of LOP of LEDs by using the nanostructures derived from these colloidal monolayers. A variety of approaches, including surface texturing, submicron lens, PhC, LSP, NPSS, optical reflector, have been adopted to improve LOP of LEDs. Based on these techniques, significant enhancement in LOP has been achieved from dozens of percent to several times. As an ongoing interesting topic, large surface area of 3D nanostructures induces new optical and electrical properties of LEDs and lasers, demonstrating a great potential in solid-state lighting.

For the future work regarding the further enhancement of LOP from GaN-based LEDs, especially DUV LEDs and green LEDs, by using the nanostructures derived from MCCs, the preparation of large-area highly ordered MCC without defects is still a big challenge. New techniques that may reduce the possibility of the appearance of defects (vacancies, dislocations, grain boundaries, cracks, etc.) during the self-assembly process of colloidal nanospheres are highly needed. One more important task is to gain a deeper understanding on the electro-optical transformation mechanism in nitride LEDs, which involves carrier injection, transportation and recombination, radiative transition, photon transportation, and so forth. Especially, for the DUV LED with the wavelength below 280 nm, the efficiency improvement is urgent. The mechanisms of these processes at interface or/and surface are of particular importance since incorporation of nanostructures generally leads to more interface or surface in GaN-based LEDs. Furthermore, the revolutionary ideas on processing of GaN nanoLEDs and MD lasers that could greatly simplify the device processing are required. It can be expected that high-quality ND lasers with higher coupling factor, reduced modal volume, and single-mode lasing can be achieved by using NSL in the future [77]. These advances in the fabrication of high-quality nitride MDs by MSL have also opened the way for the research in strong coupling regime and quantum optics, such as light–matter coupling for extremely low threshold polariton lasers and cavity-enhanced single photon emission for quantum information processing.

References

1 Schubert, E.F., Gessmann, T., and Kim, J.K. (2006). Inorganic semiconductors for light-emitting diodes. In: *Organic Light Emitting Devices. Synthesis, Properties and Applications* (eds. K. Müllen and U. Scherf). Weinheim: Wiley VCH.
2 Lai, C.F., Kuo, H.C., Chao, C.H. et al. (2007). *Appl. Phys. Lett.* 91: 123117.
3 Kwon, M.-K., Kim, J.-Y., Park, I.-K. et al. (2008). *Appl. Phys. Lett.* 92: 251110.

4 Inoue, S.-I., Tamari, N., and Taniguchi, M. (2017). *Appl. Phys. Lett.* 110: 141106.
5 Yadav, A., Kaushik, A., Mishra, Y.K. et al. (2020). *Mater. Today Chem.* 15: 100208.
6 Zhang, Z.M., Geng, C., Hao, Z.B. et al. (2016). *Adv. Colloid Interface Sci.* 228: 105–122.
7 Geng, C., Wei, T., Wang, X. et al. (2014). *Small* 10: 1668–1686.
8 Zhang, Y.H., Wei, T.B., Wang, J.X. et al. (2014). *AIP Adv.* 4: 027123.
9 Ye, X.Z. and Qi, L.M. (2011). *Nano Today* 6: 608–631.
10 Wei, T.B., Kong, Q.F., Wang, J.X. et al. (2011). *Opt. Express* 19: 1065–1071.
11 Fujii, T., Gao, Y., Sharma, R. et al. (2004). *Appl. Phys. Lett.* 84: 855–857.
12 Huang, H.W., Lin, C.H., Yu, C.C. et al. (2008). *Semicond. Sci. Technol.* 23: 045022.
13 Soh, C., Wang, B., Chua, S. et al. (2008). *Nanotechnology* 19: 405303.
14 Sim, J.I., Lee, B.G., Yang, J.W. et al. (2011). *Jpn. J. Appl. Phys.* 50: 102101.
15 Zhang, Q., Li, K.H., and Choi, H.W. (2012). *Appl. Phys. Lett.* 100: 061120.
16 Kim, T.S., Kim, S.-M., Jang, Y.H., and Jung, G.Y. (2007). *Appl. Phys. Lett.* 91: 171114.
17 Chen, Z.X., Liu, W.J., Wan, W. et al. (2016). *J. Display Technol.* 12: 1588–1593.
18 Li, W.J., Zhang, X., Meng, R.L. et al. (2019). *Micromachines* 10: 322.
19 An, H.-M., Sim, J.I., Shin, K.S. et al. (2012). *IEEE J. Quantum Electron.* 48: 891–896.
20 Wei, T.B., Wu, K., Chen, Y. et al. (2012). *IEEE Electron Device Lett.* 33: 857–859.
21 Tsai, M.-A., Yu, P.C., Chiu, C.H. et al. (2010). *IEEE Photon. Technol. Lett.* 22: 12–14.
22 Lin, C.-P., Yang, H., and Chao, C.-K. (2003). *J. Micromech. Microeng.* 13: 748–757.
23 Kim, B.J., Jung, H., Shin, J. et al. (2009). *Thin Solid Film* 517: 2742–2744.
24 Lee, H.K., Ko, Y.H., Raju, G.S.R., and Yu, J.S. (2012). *Opt. Express* 20: 25058–25063.
25 Ee, Y.-K., Kumnorkaew, P., Arif, R.A. et al. (2009). *IEEE J. Sel. Top. Quantum Electron.* 15: 1218–1225.
26 Ee, Y.-K., Arif, R.A., and Tansu, N. (2007). *Appl. Phys. Lett.* 91: 221107.
27 Li, X.H., Zhu, P.F., Liu, G.Y. et al. (2013). *J. Display Technol.* 9: 324–332.
28 Athanasiou, M., Kim, T.K., Liu, B. et al. (2013). *Appl. Phys. Lett.* 102: 191108.
29 Fu, W.Y., Wong, K.K.-Y., and Choi, H.W. (2009). *Appl. Phys. Lett.* 95: 133125.
30 Li, K.H., Ma, Z.T., and Choi, H.W. (2013). *Appl. Phys. Lett.* 100: 141101.
31 Hou, C.-H., Tseng, S.-Z., Chan, C.-H. et al. (2009). *Appl. Phys. Lett.* 95: 133105.
32 Zhang, Y.H., Wei, T.B., Xiong, Z. et al. (2014). *Appl. Phys. Lett.* 105: 013108.
33 Li, K. and Choi, H. (2011). *J. Appl. Phys.* 110: 053104.
34 Wei, T., Wu, K., Lan, D. et al. (2012). *Appl. Phys. Lett.* 101: 211111.
35 Du, C., Wei, T., Zheng, H. et al. (2013). *Opt. Express* 21: 25373–25380.
36 Wierer, J.J., David, A., and Megens, M.M. (2009). *Nat. Photon.* 3: 163–169.
37 Du, C.X., Zhang, W.R., Wei, T.B. et al. (2017). *IEEE Photon. J.* 9: 8201507.
38 Wei, T.B., Ji, X.L., Wu, K. et al. (2014). *Opt. Lett.* 39: 379–382.

39 Fan, X.C., Hao, Q., Qiu, T., and Chu, P.K. (2020). *J. Appl. Phys.* 127: 040901.
40 Okamoto, K., Niki, I., Shvartser, A. et al. (2004). *Nat. Mater.* 3: 601–605.
41 Kao, C.C., Su, Y.K., Lin, C.L., and Chen, J.J. (2010). *IEEE Photon. Technol. Lett.* 22: 984–986.
42 Wei, T., Wu, K., Lan, D. et al. (2014). *AIP Adv.* 6: 067119.
43 Taguchi, A., Saito, Y., Watanabe, K. et al. (2012). *Appl. Phys. Lett.* 101: 081110.
44 Yin, J., Li, Y., Chen, S. et al. (2014). *Adv. Opt. Mater.* 2: 451–458.
45 Gao, H., Yan, F., Zhang, Y. et al. (2008). *J. Appl. Phys.* 103: 014314.
46 Li, Y., You, S., Zhu, M. et al. (2011). *Appl. Phys. Lett.* 98: 151102.
47 Chan, C.H., Hou, C.H., Tseng, S.Z. et al. (2009). *Appl. Phys. Lett.* 95: 011110.
48 Chen, J.J., Su, Y.K., Lin, C.L. et al. (2008). *IEEE Photon. Technol. Lett.* 20: 1193–1195.
49 Zhang, Y.H., Wei, T.B., Wang, J.X. et al. (2014). *Growth* 394: 7–10.
50 Ueda, K., Tsuchida, Y., Hagura, N. et al. (2008). *Appl. Phys. Lett.* 92: 101101.
51 Chang, H.L., Chen, Z.L., Liu, B.Y. et al. (2020). *Adv. Sci.* 7: 2001272.
52 Dong, P., Yan, J.C., Wang, J.X. et al. (2013). *Appl. Phys. Lett.* 102: 241113.
53 Lee, D., Lee, J.W., Jang, J. et al. (2017). *Appl. Phys. Lett.* 110: 191103.
54 Geng, C., Zheng, L., Fang, H.J. et al. (2013). *Nanotechnology* 24: 335301.
55 Kuo, C.H., Chang, L.C., Kuo, C.W., and Tun, C.J. (2009). *J. Electrochem. Soc.* 156: H986–H989.
56 Chou, Y.-L., Lin, R.-M., Tung, M.-H. et al. (2011). *IEEE Photon. Technol. Lett.* 23: 393–395.
57 Son, J., Verma, L.K., Danner, A.J. et al. (2011). *Optics Exp.* 19: A35–A40.
58 Kim, B.J., Jung, H., Kim, S.H. et al. (2009). *IEEE Photon. Technol. Lett.* 21: 700–702.
59 Liou, J.-K., Chou, P.-C., Chen, C.-C. et al. (2014). *IEEE Trans. Electron Devices* 61: 831–837.
60 Zhang, L.J., Xiong, Z., Shan, L. et al. (2015). *Small* 11: 4910–4921.
61 Liou, J.-K., Chen, W.-C., Chang, C.-H. et al. (2015). *IEEE Trans. Electron. Devices* 62: 3296–3301.
62 Huang, K.-M., Chang, H.-J., Ho, C.-L., and Wu, M.-C. (2012). *IEEE Photon. Technol. Lett.* 24: 1298–1300.
63 Li, S.F. and Waag, A. (2012). *J. Appl. Phys.* 111: 071101.
64 Chen, L.-Y., Huang, Y.-Y., Chang, C.-H. et al. (2010). *Opt. Express* 18: 7664–7669.
65 Li, Q.M., Westlake, K.R., Crawford, M.H. et al. (2011). *Opt. Express* 19: 25528–25534.
66 Latzel, M., BÜttner, P., Sarau, G. et al. (2017). *Nanotechnology* 28: 055201.
67 Zhang, L., Guo, Y.N., Yan, J.C. et al. (2019). *Photon. Res.* 7: B66–B72.
68 Djavid, M. and Mi, Z.T. (2016). *Appl. Phys. Lett.* 108: 051102.
69 Wu, K., Wei, T.B., Zheng, H.Y. et al. (2014). *J. Appl. Phys.* 115: 123101.
70 Wei, T.B., Islam, S.M., Jahn, U. et al. (2020). *Opt. Lett.* 45: 121–124.
71 Christmann, G., Butté, R., Feltin, E. et al. (2008). *Phys. Rev. B* 77: 085310.
72 Zhang, Y., Feng, C., Wang, T., and Choi, H.W. (2016). *Appl. Phys. Lett.* 108: 031110.
73 Zhang, Y., Zhang, X., Li, K.H. et al. (2015). *Phys. Status Solidi A* 212: 960.

74 Zhang, Y., Ma, Z., Zhang, X. et al. (2014). *Appl. Phys. Lett.* 104: 221106.
75 Athanasiou, M., Smith, R.M., Pugh, J. et al. (2017). *Sci. Rep.* 7: 10086.
76 Zhang, Y., Li, H., Li, P. et al. (2017). *IEEE Photon. J.* 9: 2400508.
77 Aharonovich, I., Woolf, A., Russell, K.J. et al. (2013). *Appl. Phys. Lett.* 103: 021112.

11

Self-assembled Photonic Crystals for Solar Cells

Likui Wang

Jiangnan University, School of Chemical and Materials Engineering, Department of Materials Engineering, Lihu Avenue 1800, Wuxi 214122, P. R. China

11.1 Introduction

With the continuous development of world economy, global consumption of energy grows rapidly. The US Energy Information Administration [1] predicted that the total world energy consumption would increase from 627 quadrillion British thermal units (Btus) in 2019 to 911 quadrillion Btu by 2050. In 2019, 79.5% of the energy consumption depends on nonrenewable fossil fuels such as oil, coal, and natural gas. However, not only the use of fossil fuel causes the emission of carbon dioxide and other pollutants, leading to the deterioration of environment, but also the depletion of fossil fuel poses a serious threat to world economy in the near future. Renewable energies, such as hydroelectric, wind, marine tidal, biomass, and solar energy have long been considered as alternative energy resources for tackling the environmental problem and energy crisis. Renewable energy consumption is expected to increase to 252 quadrillion Btu by 2050.

Among all the renewable energy sources, solar energy is considered inexhaustible and pollution-free. Photovoltaic technologies, converting solar energy to electricity has attracted tremendous research interest. Great efforts have been devoted to develop advanced solar energy technologies with enhanced power conversion efficiency (PCE) [2–9]. Photon management has emerged as an attractive research topic in order to promote light harvesting efficiency in photovoltaics [10–15]. The development of nanostructured materials with feature size around visible light wavelength range [16–24], including photonic crystals (PCs) [25–30], light scattering materials [31–34], and plasmonic nanometals [35–37], has paved a way in manipulating the propagation and distribution of photons in solar cells, especially the third-generation thin film solar cells [14].

PC is a concept proposed to control the spontaneous emission of light and to realize localization of photons by John [38] and Yablonovitch [39] independently in 1987. Composed of periodically arranged two or more materials with different dielectric constant, PCs induce Bragg scattering of light when they propagate

Functional Materials from Colloidal Self-assembly, First Edition. Edited by Qingfeng Yan and George Zhao.

through, causing the formation of a photonic bandgap, a range of frequencies for which photons are forbidden to exist within the interior of the PC. Due to the existence of such a photonic bandgap, PCs are able to control light like semiconductors control electrons. Depending on the dimension of the arrangement, PCs can be classified into three categories, namely, one-dimensional (1D), two-dimensional (2D), and three-dimensional (3D) PCs, exhibiting different photonic band structures. 3D PCs possess a complete photonic bandgap, which is open for all directions and allows for controlling the light flow omnidirectionally. There are mainly four approaches to constructing 3D PCs, namely, top-down micromachining [40], holographic lithography [41, 42], laser direct writing [43], and bottom-up self-assembly [44–49]. Self-assembly of colloidal microspheres is an effective and inexpensive method of fabricating 3D PCs. Although colloidal crystals produced through self-assembly normally do not possess a complete photonic bandgap, it could be employed as template for fabricating inverse opals (IOs), which could have a complete bandgap with appropriate structure and index contrast. Additionally, the pseudo bandgaps of colloidal crystals and IOs are also applicable in many fields.

Self-assembled PCs [44, 45], benefiting from their photonic bandgap, have shown their great potential of controlling the behavior of light in solar cells in a new level. By coupling self-assembled PCs to solar cells, the absorption of light could be enhanced, leading to increased light harvesting efficiency. Self-assembled PCs could improve the light harvesting of solar cells [50], mainly due to the following four mechanisms. First, due to the existence of photonic bandgap, self-assembled PCs could work as a back reflector, reflecting the transmitted light to be reutilized by the light absorbing layer in solar cells [51]. Figure 11.1a,b illustrates the conventional metal mirrors and the PCs used in solar cells. The former uses the traditional geometric optics, while the latter employs wave optics to trap light. PCs could reflect light within the bandgap from any incident angles. Second, the light scattering effect effectively increases the path length of light inside the solar cells. Especially when the self-assembled PCs are in intimate contact with an absorbing layer, the interface of PC and light absorbing layer couples incident light into resonant modes [52]

Figure 11.1 Schematic of the four mechanisms of PCs to enhance the light harvesting efficiency of solar cells. (a, b) Comparison of the effect of a conventional metal mirror and a PC [51]. Source: Ref. 51. Adapted with permission of the Optical Society. (a) A silicon solar cell with a thin layer of aluminum as the back reflector and charge collector; (b) a silicon solar cell with PCs to enhance light absorption. (c) Illustration of a surface resonant mode at the interface of silicon film and PC [52]. Source: Ref. 52. Adapted with permission of John Wiley & Sons. (d, e) The mechanism of slow photons induced by a photonic bandgap [53]. Source: Ref. 53. Adapted with permission of American Chemical Society. (d) A simplified PC band structure. When a photon with energy below the bandgap approaching a photonic bandgap, the group velocity of light diminishes and a standing wave forms at the high index part of the structure. For a photon with energy above the bandgap, the standing wave forms at the low refractive part. (e) The effect of standing wave localization on dye absorbance. When the dye (light absorption materials) is located at the high index part of the PC (dotted line), the red absorption is enhanced, while the blue absorption is inhibited. (f) A scheme illustrating an IO sensitized with dye or quantum dots.

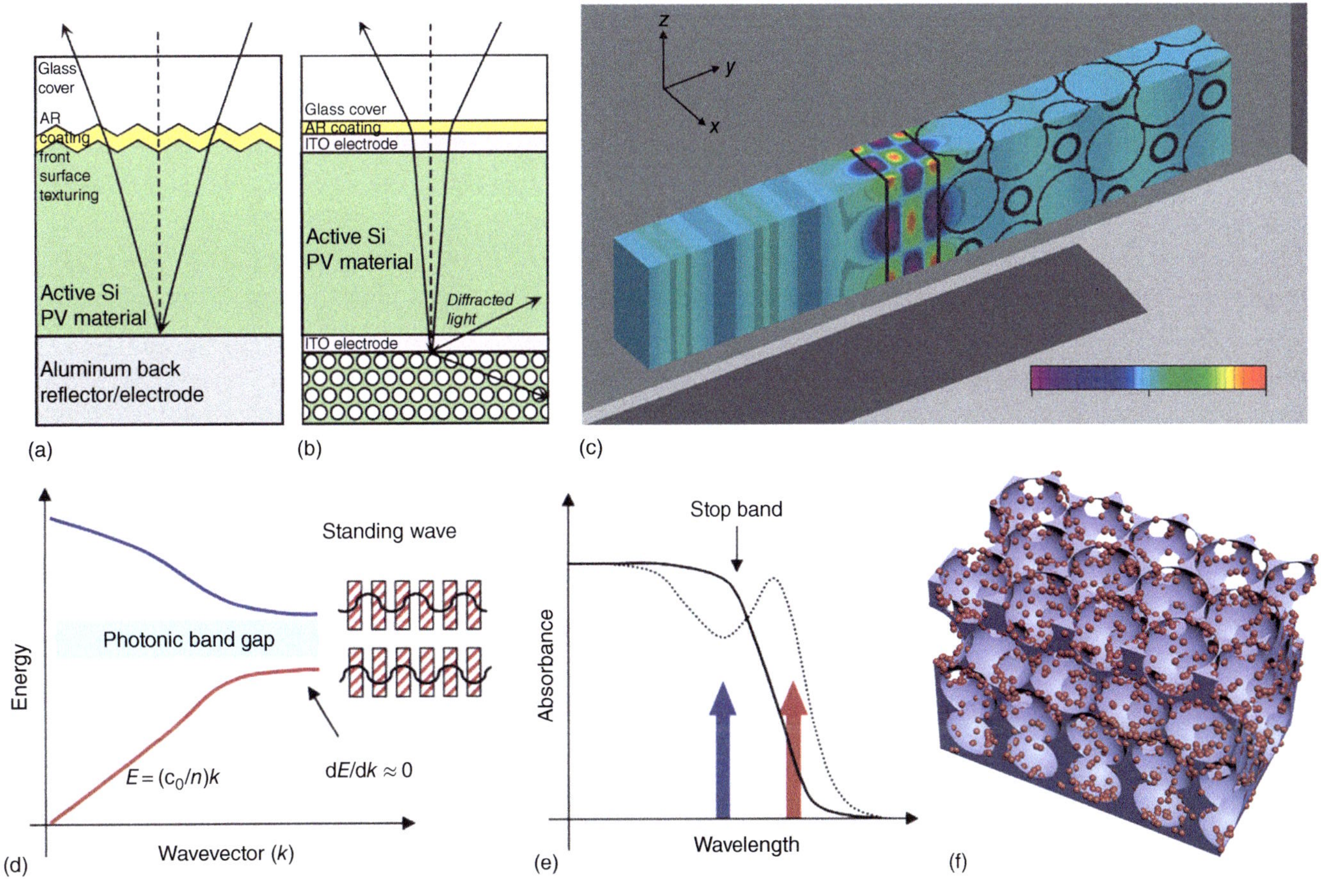
Glass cover
AR coating
front surface texturing
Active Si PV material
Aluminum back reflector/electrode
(a)
Glass cover
AR coating
ITO electrode
Active Si PV material
Diffracted light
ITO electrode
(b)
z
y
x
(c)
Energy
Photonic band gap
Standing wave
$E = (c_0/n)k$
$dE/dk \approx 0$
Wavevector (k)
(d)
Stop band
Absorbance
Wavelength
(e)
(f)

that propagate along the plane of the film, greatly increasing the distance that light could interact with the absorbing materials. As shown in Figure 11.1c, the solid rectangular cross section illustrates the electric field of the PC-coupled film. When incident light propagates along the *y*-axis, parallel to the [111] direction of a colloidal crystal, the electric field is the strongest at the surface of silicon film contacting the PC, outlined by two straight lines [52, 54–56]. Furthermore, on the red side of the photonic bandgap, photons have reduced group velocity, which are called slow light or slow photons [53]. The slow photons effect could effectively enhance the interaction between the photons and the light absorbing materials, as shown in Figure 11.1d,e. Finally, self-assembled PCs have uniform porous structures, serving as an additional merits for both dye-sensitized solar cells (DSSCs) and perovskite solar cells (PSCs). In a DSSC, self-assembled PCs could absorb sensitizers, including dye molecules or quantum dots, promoting the effective loading of sensitizer (Figure 11.1f) [57, 58]. In a PSC, the unique porous structure facilitates uniform infiltration of perovskite materials, inhibiting defects formation.

Due to their excellent photon management capabilities, PCs have been widely adopted in the research of DSSCs, PSCs, and silicon-based solar cells. DSSCs and PSCs are acknowledged as promising low-cost alternatives to traditional silicon or CdTe-based photovoltaic devices, due to their low cost, simple fabrication processes, and high PCEs. Because the porous structure and preparation methods of self-assembled PCs are highly compatible, tremendous research efforts have been put on the integration of self-assembled PCs into DSSCs and PSCs. The application of PCs in silicon-based solar cells has also been studied by various researchers. In this chapter, we review the theoretical and experimental research progresses on the application of self-assembled PCs in solar cells. The roles of self-assembled PCs in solar cells, their working mechanisms, and the different preparation methods of coupling PCs with the light absorption layer are discussed.

11.2 The Application of Self-assembled PCs in DSSCs

DSSCs are a type of low-cost photoelectrochemical solar cells that mimic natural photosynthesis process [14, 16, 59–61]. They differ from conventional silicon-based solar cells because the light absorption and charge transport are separated in the cells. The former is performed by the dye molecules adsorbed on oxide semiconductor (typically TiO_2), and the semiconductor transports the photoelectron to the current collector.

As shown in Figure 11.2, DSSCs are a photoelectrochemical solar cells composed of photoanode (also called working electrode), counter electrode, and electrolyte. The photoanode is normally fabricated by depositing nanocrystalline (NC) oxides, typically TiO_2, to form a mesoporous semiconductor film with a thickness of 10–15 μm on a transparent conducting oxide (TCO)-covered glass, followed by sensitizing the NC oxides with dye molecules. TiO_2 is the most widely adopted NC oxide in DSSCs due to its suitable bandgap and high stability. The counter electrode is normally Pt-covered TCO glasses. Between the photoanode and counter electrode, an electrolyte with redox couples I^-/I_3^- is filled. When the incident light

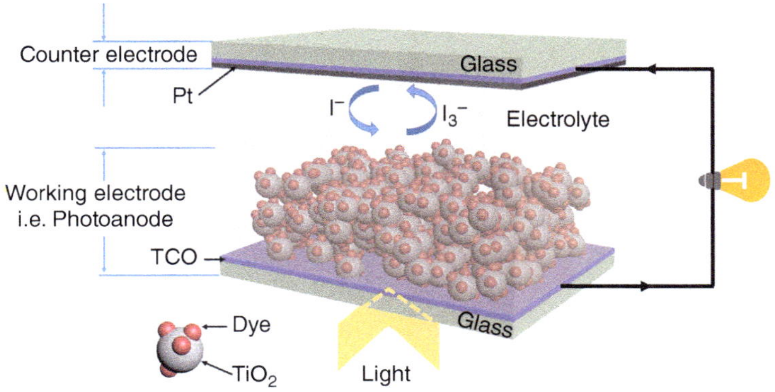

Figure 11.2 Schematic structure of a typical DSSC.

is captured by the dye molecules adsorbed on the NC oxide, photoexcited electrons are generated and injected to the conduction band of the semiconductor, accompanied by the oxidation of the dye molecules. The electrons from the semiconductor are collected by TCO layer and become available for electrical work in the external circuit. At the same time, the oxidized dye molecules are reduced to their original state, i.e. regenerated, by I^-, which is oxidized to I_3^- in return. The I_3^- ions diffuse to the counter electrode and are reduced back to I^- by the platinum layer after acceptance of a low energy electron from the external circuit.

Since Mallouk's group [53] first coupled a TiO_2 IO into DSSCs in 2003 and improved their short circuit photocurrent efficiency by 26%, great efforts have been devoted to investigate incorporation of PCs into DSSCs. Self-assembled PCs are introduced into DSSC mainly in two approaches. One is to directly apply TiO_2 PCs sensitized with dye as photoanodes. The other is to utilize the photonic bandgap of the PCs to enrich the photon absorption of a NC film in photoanodes. Depending on the positions of PCs, these DSSCs could be classified into five different configurations, with their schematic structures illustrated in Figure 11.3. NC-only is a simplified standard Grätzel cell structure with only the NC TiO_2 layer for comparison without the incorporation of IOs. IO-only represents a structure that dye-sensitized IO is directly employed as photoanode. In an IO-outside or IO-on-CE structure, IO is placed outside or inside on the counter electrode, respectively, to reflect the transmitted light back into the light absorbing layer. In both cases, IOs are normally not sensitized with dyes since they solely function as reflecting mirrors. Both NC-on-IO and IO-on-NC structures integrate IOs into photoanodes. IO is constructed between the NC TiO_2 film and TCO glasses for the former and on the NC film for the latter. Since IOs are integrated into photoanodes for both cases, IOs are normally sensitized to enlarge the effective dye loading of the DSSCs.

11.2.1 Application of Dye-Sensitized IOs as Photoanodes of DSSC

The interconnected porous structures of IOs make their internal surface readily available for sensitizer loading. TiO_2 IOs sensitized with dyes could be directly

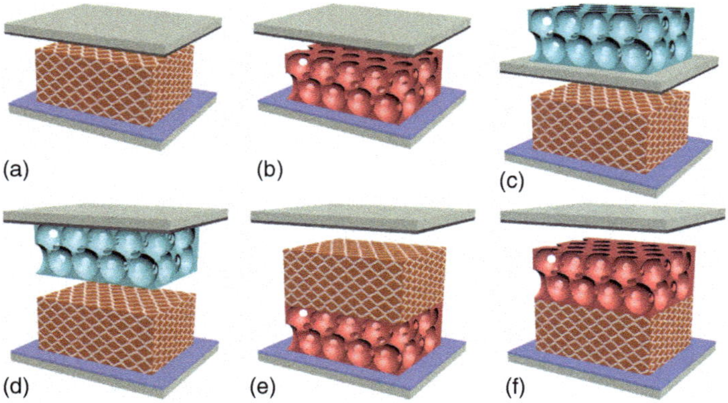

Figure 11.3 The structures of PC-incorporated DSSCs and their photovoltaic performances: (a) Grätzel cell (NC-only), a standard DSSC structure, (b) IO-only, (c) IO-outside, (d) IO-on-CE, (e) NC-on-IO, and (f) IO-on-NC.

employed as photoanodes of DSSCs [62]. Chen et al. prepared WO_3 IO as photoanodes and adjusted their stopbands conventionally by tuning the diameter of the colloidal sphere used for the CC templates [57]. The slow light wavelength (in the vicinity of 440 nm) of WO_3 IO with pore size of 260 nm overlaps with the electronic absorption edge of WO_3, which induces strong interaction of light with the dielectric matrix through multiple scattering and diffraction, thus generating much more photoinduced carriers in WO_3, resulting in increased light absorption and enhanced light utilization for photoelectric conversion (Figure 11.4a). The incident photon-to-current conversion efficiency (IPCE) in the spectral regime from 410 to 460 nm is enhanced much more significantly than other wavelength range, confirming that the slow light effect improves the light absorption around the absorption edge of WO_3. The photoanode composed of WO_3 IO with pore size of 260 nm produces twofold photocurrent density of that of a disordered porous WO_3 photoanode (Figure 11.4b).

Kim and Moon [58] coated TiO_2 IO with Nb_2O_5 through the oxidation of hydrated $NbCl_5$ precursor. The fully connected porous IO structure ensures the uniformity of the Nb_2O_5 coatings. The short circuit current density (J_{SC}) of Nb_2O_5-coated TiO_2 IO electrodes shows an increase of 60% from uncoated one, from 10.45 to 16.73 mA/cm^2, which could be ascribed to that Nb_2O_5 coating suppresses the reduction in J_{SC} under backside illumination.

Although IOs have interconnected pores, their large pore size, typically several hundred nanometers, limits their surface area available for dye adsorption. Consequently, their efficiency could be improved through enlarging their effective surface area either by directly incorporating mesoporous structure inside the IOs during preparation or by infiltrating the IOs with NC particles. Lee's group [63] developed a method of fabricating IO composed of TiO_2 nanoparticles by co-assembly of TiO_2 nanoparticles and polystyrene (PS) microspheres, which was employed as photoanode of DSSCs. Binary mixture of TiO_2 nanoparticles and PS microspheres were directly deposited on an fluorine-doped tin oxide (FTO)

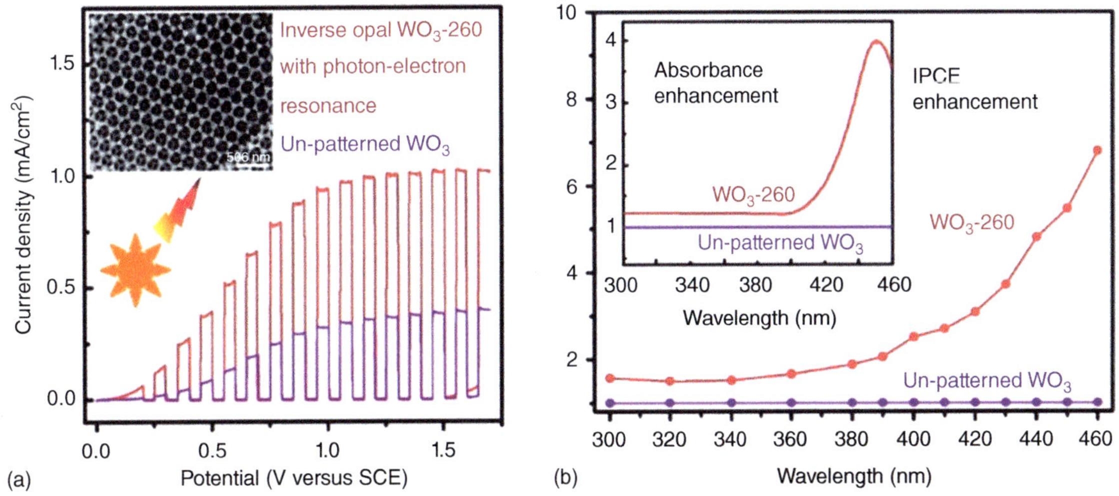

Figure 11.4 (a) Photocurrent-potential curves obtained under visible light illumination ($\lambda > 400$ nm) (inset: SEM image of a WO_3 IO, WO_3-260); (b) IPCE enhancement factor (the ratio of the IPCE of WO_3 IO over that of the unpatterned WO_3 photoanode) and absorbance enhancement spectra (absorbance of WO_3 IO over that of the unpatterned WO_3 photoanode, inset image) at different wavelength. Source: Chen et al. [57]/with permission from American Chemical Society.

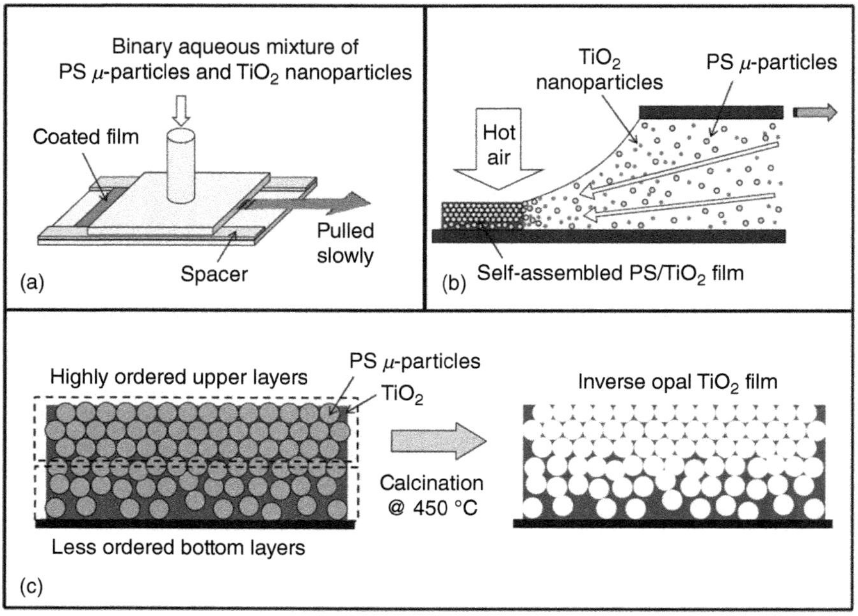

Figure 11.5 A schematic of fabricating IO composed of TiO_2 nanoparticles by co-assembly of TiO_2 nanoparticles and PS microspheres. (a) Experimental setup. The binary colloid is dispensed from the top. The movement of the top unit is at constant speed of 0.5–8 mm/min with bottom substrate fixed. (b) Side view of the colloidal assembly. (c) The crystallization of PS spheres is induced by water evaporation and capillary force at the surface and hindered by decreasing capillary force and increasing TiO_2 content near the substrate. After calcination, higher volume occupation of TiO_2 particles at the bottom and lower volume occupation at the surface are achieved [63]. Source: Ref. 63. Adapted with permission of John Wiley & Sons.

glass with two parallel spacers, followed by hot air blowing to accelerate the water evaporation (Figure 11.5). After formation of colloidal crystals and removal of PS spheres through calcination, IO with mesoporous structures were formed. After sensitized with N719 dye, the PCE of DSSC based on a 10 μm IO composed of TiO_2 nanoparticle was measured to be 4.2%, much higher than IO electrodes fabricated from TiO_2 precursor.

Kwak et al. [64] constructed IOs through filling polymer colloidal crystals with non-aggregated surface-modified TiO_2 nanoparticles, followed by removal of the colloidal crystal templates, which were further sensitized with N-719 dye and employed as photoanode of DSSC. This NC IO structure exhibits enlarged surface area and the corresponding solar cell shows a photo-conversion efficiency of 3.47% under air mass 1.5 illumination. Wu's group [65] fabricated hierarchical ordered porous (HOP) TiO_2 by directly casting a colloid of monodisperse polymer spheres, triblock copolymer P123, and titania precursors, followed by removal of the polymer templates through calcination (Figure 11.6). The spheres self-assemble into 3D colloidal crystals, serving as the template for macroporous IO structures, while P123 directs the formation

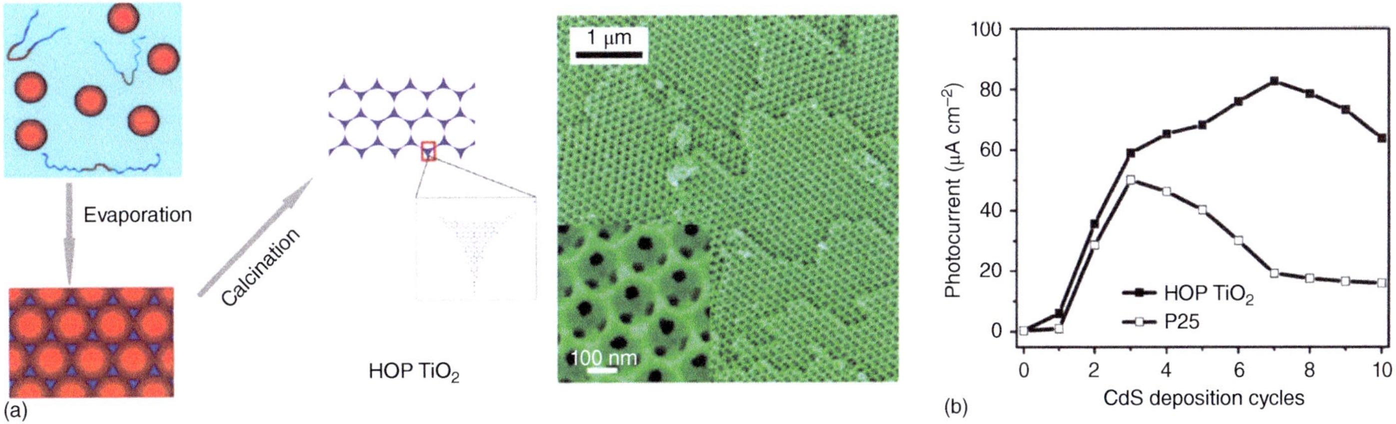

Figure 11.6 (a) Schematic of preparing HOP TiO_2. (b) Photocurrents of HOP TiO_2 and P-25 photoelectrodes sensitized with different CdS deposition cycles under the visible light irradiation. Source: Sun et al. [65]/with permission from American Chemical Society.

of mesoporous structures with a pore size of 3.5 nm. This HOP structure shows BET surface area of 135 m^2/g, almost three times as large as that of P-25 (50 m^2/g). After being sensitized with CdS nanoparticles, the highest photocurrent of the HOP TiO_2 film (seven CdS deposition cycles) is about 70% higher than that of the P-25 film (three CdS deposition cycles).

Bein's group [66] synthesized hierarchical porous TiO_2 structure with both macropores and mesopores through a confined synthesis approach. Titania IO was infiltrated with the mixture of titania precursor and a pluronic P123, triblock copolymer P123. After calcination, mesostructure formed in the macropores of IOs (Figure 11.7). They found that the confinement effect of IOs strongly influences the assembly of mesostructure and induced much larger structural parameters. This fully accessible porous structure has a total surface area of 154 m^2/g and an average

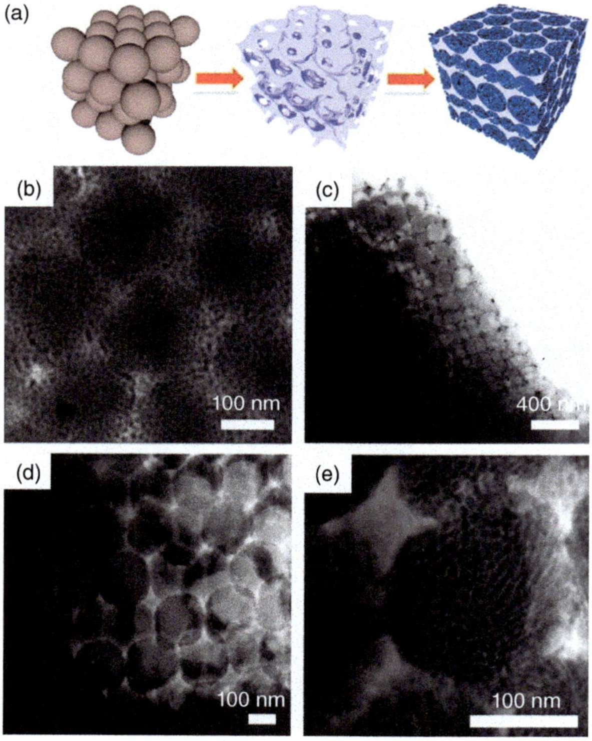

Figure 11.7 (a) An approach of fabricating hierarchical TiO_2 frameworks: IO is formed by infiltrating a CC with TiO_2 precursor solution and removal of the template. IO is further infiltrated with hydrolyzed TiO_2 containing surfactant and a hierarchical TiO_2 framework is obtained after calcination. Images of TiO_2 frameworks obtained with 200-nm PMMA spheres: top-view FESEM image (b), TEM image of the cross sections (c), and HAADF-STEM (d, e). The brighter areas in (b), (d), and (e) and the darker spots in (c) correspond to larger TiO_2 crystals in the dense walls of the IO. All films were prepared on silicon substrates and calcined at 450 °C. Source: Mandlmeier et al. [66]/with permission from American Chemical Society.

mesopore size of 6.1 nm, which is about 20% larger than the pore size of 5.1 nm for the mesoporous film fabricated on a flat substrate. After being sensitized with N719, this hierarchical structure is employed as photoanode of DSSC, exhibiting nearly 10 times higher conversion efficiency of 4.0% as compared with the unfilled macroporous IO with an efficiency around 0.4%.

Tetreault et al. [67] constructed a disordered IO of Al/ZnO as host structure for photoanode, which is infiltrated with TiO_2 nanoparticles (Figure 11.8). The host has much higher electron mobility and diffusion coefficient than that of anatase TiO_2 nanoparticles, leading to a strong decrease in dark current and increase in photovoltage up to 110 mV. Xu's group applied a fluorine-doped tin oxide IO, coated with TiO_2, as photoanode of DSSCs. This electrode shows enhanced charge transport and strong light scattering in the visible region [68].

11.2.2 Self-assembled Photonic Crystals for Light Harvesting Enhancement in DSSCs

A large portion of photons are not captured by the dye molecules in the photoanodes of DSSCs. To enhance the photon capture, two approaches could be employed: the use of geometric optics or the use of wave optics. The former approach employs traditional mirrors, while the latter utilizes PCs, both of which could reflect the transmitted light to reenter the photoanodes of DSSCs for reuse. Chang's group [69] compared the effects of a Ag mirror and a ZnO IO attached to the outside of counter electrode (Figure 11.9). They found that although there is no change in V_{oc}, fill factor, and the transport properties of the cells, the cells with a PC back reflector shows 13% higher cell efficiency than with Ag mirrors, due to the systematic increase of J_{sc} and reduction of internal cell resistance.

In 2003 Mallouk's group [53] first coupled a TiO_2 IO into DSSCs. They found that the photon absorbance was suppressed at wavelengths shorter than the stopband center and was enhanced at longer wavelengths. The IO PCs serve as a dielectric mirror for wavelengths around the stopband and as a medium for promoting light absorption on the long wavelength side of the stopband. The short circuit photocurrent efficiency is increased by about 26%. They found that the bilayer structure, instead of enhanced light harvesting within the IO structures, is responsible for the enhancement of IPCE [70]. They ascribed the light conversion efficiency in the red spectral range to the combined effect of localization of slow photons near the edges of the photonic bandgap, Bragg diffraction in the IOs, and multiple scattering at disordered regions.

Míguez's group [71] theoretically investigated the optical absorption of DSSCs with four different configurations, namely, NC-only (a standard Grätzel cell), NC-on-IO, IO-only, and IO-on-NC. Scalar wave approximation was employed to calculate the effect of different configurations on the optical absorption of these self-assembled PC-based DSSCs. They confirmed that the enhancement of light absorption over a wide spectral range demands the combination of PC and NC absorbing layer with illumination from the NC TiO_2 side, which induces partial localization of light within the latter, rather than in the former. The localization

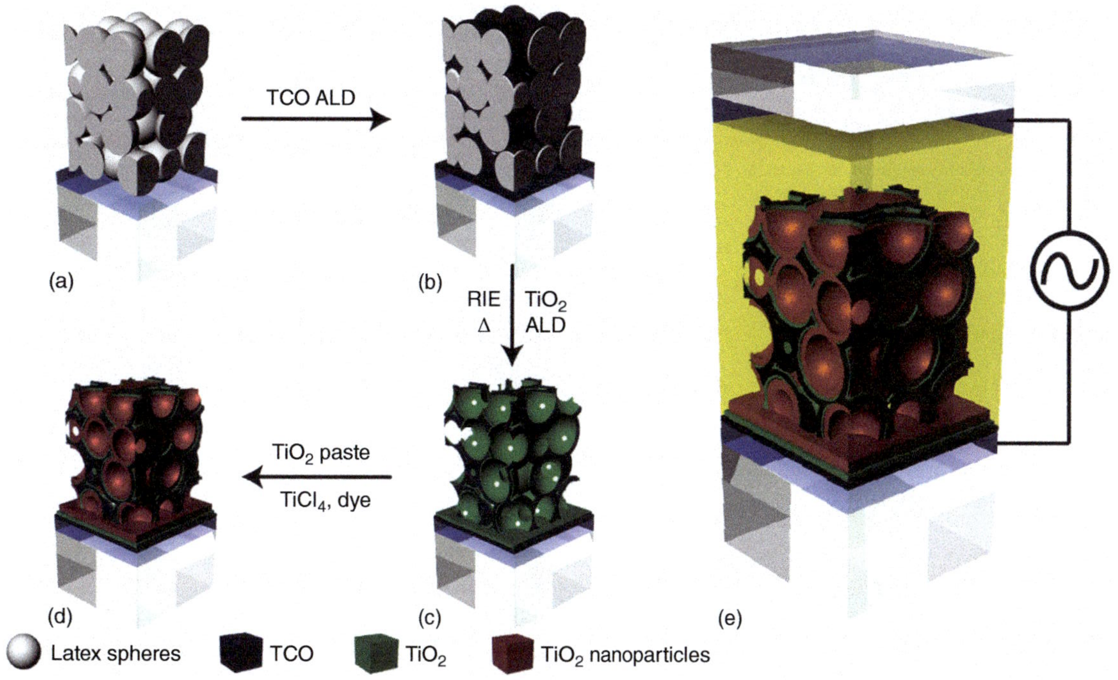

Figure 11.8 Schematic of fabricating a 3D host–guest DSSC. (a) Disordered template of 2.2-μm PS spheres prepared by doctor-blade method. (b) ALD deposition of the host material (Al/ZnO, TiO_2). (c) After reactive ion etching to open the top surface, the template is removed by annealing at 360 °C to obtain a disordered IO of host material, followed by ALD coating of 25 nm dense TiO_2. (d) Infiltrating the pores with nanocrystalline TiO_2 and dye adsorption. (e) Liquid electrolyte infiltration and assembly of the DSSC [67]. Source: Ref. 67. Reproduced with permission of American Chemical Society.

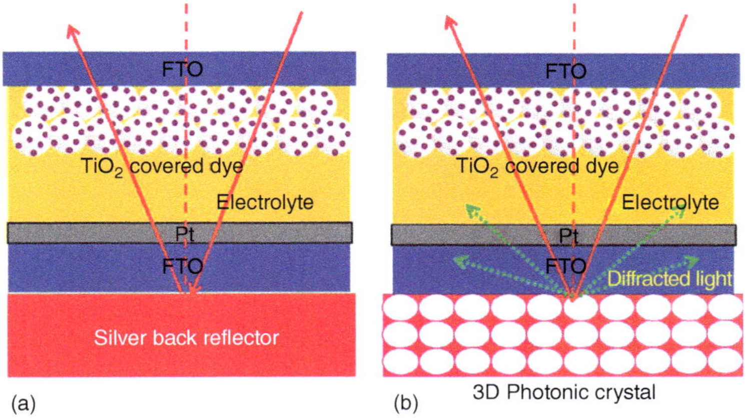

Figure 11.9 (a) Schematic of a DSSC with a conventional metal reflector. (b) DSSC with wave optics (3D PC) to trap light with reflection and diffraction [69]. Source: Ref. 69. Reproduced with permission of American Chemical Society.

of light originated from the multiple resonant modes produced by the interface between a PC and NC absorbing layer. Subsequently, they [72] experimentally analyzed the spectral dependence of the photogenerated current of IO-based solar cells (Figure 11.10). They suggested that DSSCs with a TiO_2 IO as photoanode caused a decrease of efficiency for the spectral region falling in a photonic stop bandgap. When an IO is coupled to a TiO_2 NC layer, IPCE at photonic bandgap frequencies was effectively increased. Through the introduction of double-layer IO with different pore sizes, IPCE enhancement in wider frequency range could be achieved, which leads to higher overall PCE.

Mallouk's group [73] quantitatively compared the IPCE and PCE of DSSCs with bilayers of NC TiO_2 and TiO_2 IOs of four different configurations, namely, NC-on-IO configurations with cathode-side illumination and anode-side illumination, split configuration (IO-on-CE), and IO-on-NC, as shown in Figure 11.11. The DSSC with NC-on-IO configuration and cathode side illumination shows remarkably enhanced photocurrent in the spectral region (600–750 nm) to the red of the stop-band maximum. However, the photocurrent declined in the blue spectral region due to the strong light absorption by the I^-/I^{3-} redox electrolyte. If the illumination is from anode side, there is a decrease in IPCE in the vicinity of the stopband and slight enhancement in the red spectral region. For the split bilayer configuration (IO-on-CE), the IPCE is elevated slightly across the whole spectrum.

Normally, to construct an IO-on-NC bilayer structure, colloidal crystals of polymer spheres are deposited on NC-TiO_2 layer first, followed by infiltration with TiO_2 and removal of the spheres to convert the CC to IO. However, during the conversion, infiltration of the voids of the CC easily causes the filling of the mesoporous NC-TiO_2 layers at the same time. Mallouk and coworkers prevented the infiltration of NC-TiO_2 by covering the NC-TiO_2 layer with a ~200 nm thick of 80 : 20 or 75 : 25 MMA/MA copolymer [73]. As shown in Figure 11.11g, the IPCE of the DSSC with

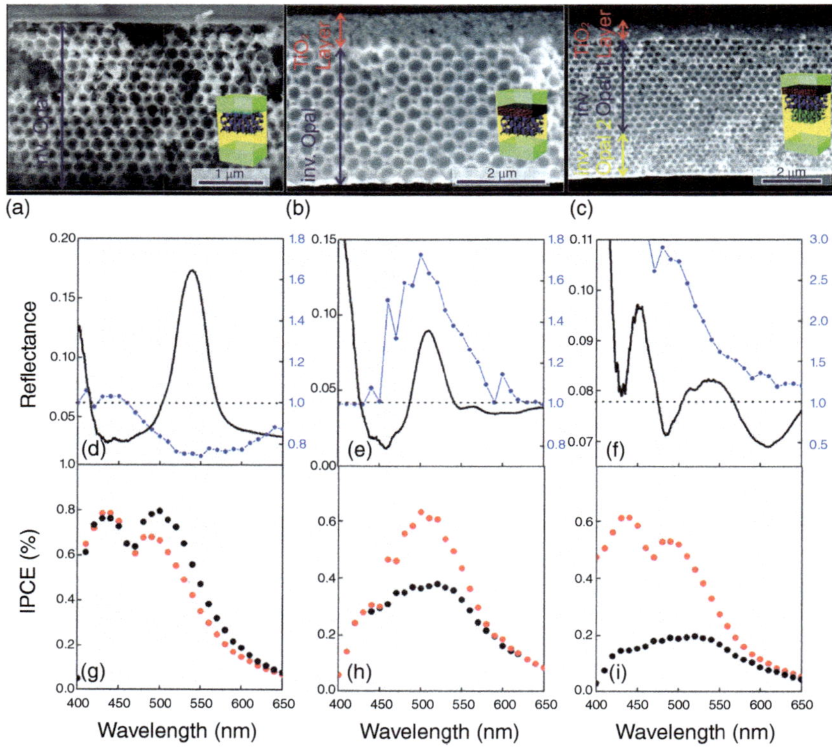

Figure 11.10 SEM image of cross-sectional view of a TiO_2 IO (a), a TiO_2 IO prepared on an NC-TiO_2 layer (b), and a TiO_2 IOs with two different lattice parameter prepared on a NC-TiO_2 film (c). Reflectance and IPCE spectra of the DSSCs (d–j). (d) Specular reflectance spectrum (in black) of a dye-sensitized TiO_2 IO ($\Phi = 300$ nm) and the enhancement factor γ (in blue). (e) Specular reflectance spectrum of a DSSC including a TiO_2 IO ($\Phi = 240$ nm) on an NC-TiO_2 layer (in black) and the corresponding enhancement factor. (f) Specular reflectance spectrum of a DSSC with two TiO_2 IO of different lattice parameter (spheres diameter $\Phi_1 = 240$ nm and $\Phi_2 = 300$ nm) on an NC-TiO_2 layer (in black) and the corresponding enhancement factor. In all cases, the enhancement factor γ refers to the ratio between the IPCE of the cell (represented in h–j in a red dotted curve) and the IPCE of a standard cell (subjected to similar silica deposition treatments for the purpose of comparison, black dotted curves). Source: Mihi et al. [72]/with permission from American Chemical Society.

Figure 11.11 Top right is the scheme of a standard Grätzel cell (NC-only); (a)–(d) are the schemes of three DSSC configurations and the arrows illustrate the illumination directions: NC-on-IO configurations with (a) cathode-side illumination and (b) anode-side illumination. (c) Split configuration (IO-on-CE). (d) IO-on-NC. The thicknesses of the PC and nc-TiO2 layers are typically 5–10 µm. (e) IPCE curves of a standard Grätzel cell (NC-only, open circles) and bilayer structures illustrated in Scheme a–c. Dashed line: NC-on-IO with cathode-side illumination (Scheme a). Solid line: NC-on-IO with anode-side illumination (Scheme b). Crosses: IO-on-CE, anode-side illumination (Scheme c); the corresponding enhancement factor is shown in (f). (g) IPCE spectra and (i) IV curves of DSSCs with IO-on-NC configuration prepared from spheres of two different diameters with anode-side illumination (Scheme d): (solid line) standard Grätzel cell, (open circle) IO-on-NC (sphere size: 291 nm), and (diamonds) IO-on-NC (sphere size: 327 nm). The dotted line in panel (g): the transmittance of the blank FTO-glass. IPCE was not corrected for reflectance or scattering losses of the FTO-glass. (h) enhancement factor (the ratio of the IPCE to that of the standard Grätzel cell) versus wavelength calculated from (g) [73]. Source: Ref. 73. Adapted with permission of American Chemical Society.

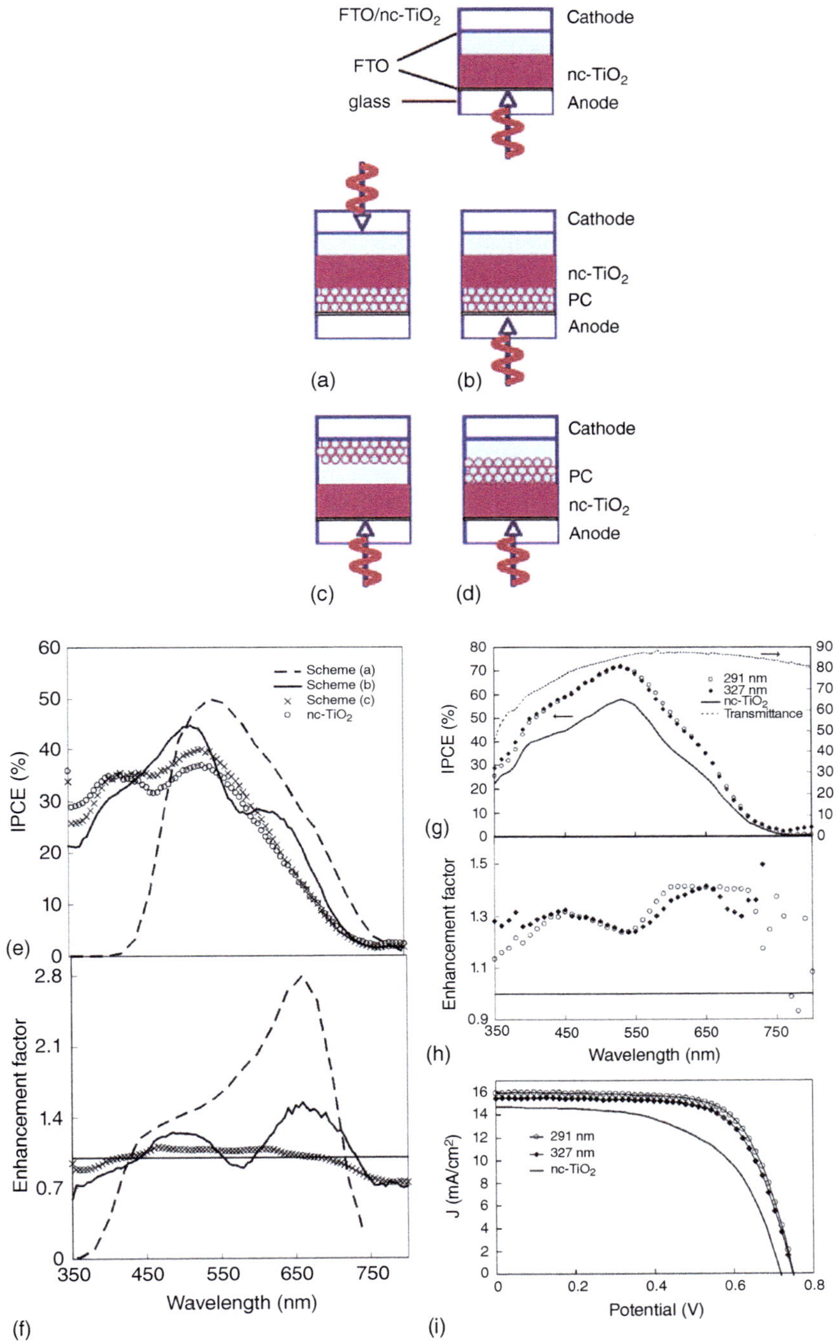
FTO/nc-TiO2
Cathode
FTO
nc-TiO2
glass
Anode
Cathode
nc-TiO2
PC
Anode
(a)
(b)
Cathode
PC
nc-TiO2
Anode
(c)
(d)
IPCE (%)
Scheme (a)
Scheme (b)
Scheme (c)
nc-TiO2
(e)
Enhancement factor
Wavelength (nm)
(f)
IPCE (%)
Transmittance (%)
291 nm
327 nm
nc-TiO2
Transmittance
(g)
Enhancement factor
Wavelength (nm)
(h)
J (mA/cm2)
291 nm
327 nm
nc-TiO2
Potential (V)
(i)

IO-on-NC configuration are increased in the whole spectral region, which could be attributed to the combination of the three effects: the dielectric mirror effect, the multidirectional scattering from defects in IO layer, and the resonant modes caused by the combination of NC and IO layer. The PCE of the IO-on-NC DSSC is increased to 8.3% as compared with 6.5% of a standard Grätzel cell (NC-only) at a constant N719 dye loading of 155–160 nmol/cm^2 (Figure 11.11i). The striking difference between the IPCE of DSSCs with IO-on-CE and IO-on-NC configuration further illustrates the intimate physical contact between NC and IO layers is needed to achieve strong red enhancement [71].

Tétreault's group [74] built a smooth composite film composed of the mixture of PI-b-PEO copolymer and TiO_2 nanoparticles on the FTO glass, which allowed subsequent self-assembling of colloidal crystals on its surface with a vertical deposition method (Figure 11.12). After infiltration of the CCs with TiO_2 and removal of the polymers, a structure of IO on NC TiO_2 layer is formed.

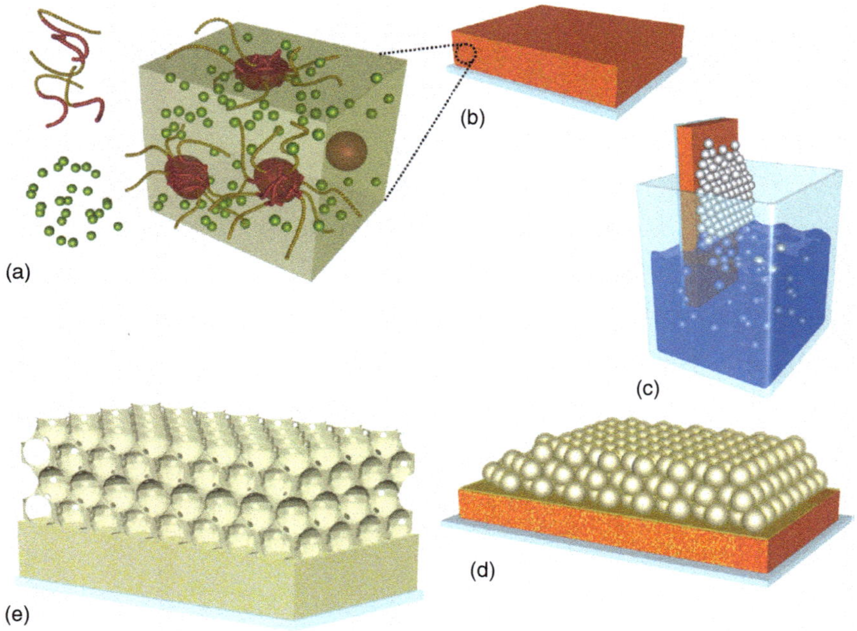

Figure 11.12 Schematic of the fabrication of an IO-on-NC structure. A colloid of PI-*b*-PEO copolymer and 1–4-nm TiO_2 nanoparticles (a) is spin-coated on an FTO glass, and a layer of NC-TiO_2 embedded in polymer (b) is formed after solvent evaporation. (c) The covered slide is placed vertically in an ethanol suspension of PS spheres to allow the self-assembly of the 3D colloidal crystal. (d) Infiltration of PS colloidal crystals with amorphous TiO_2 by APCVD. (e) Calcination for removal of polymer materials and obtaining TiO_2 IO-on-NC photoanode with full pore and electronic connectivity [74]. Source: Ref. 74. Reproduced with permission of American Chemical Society.

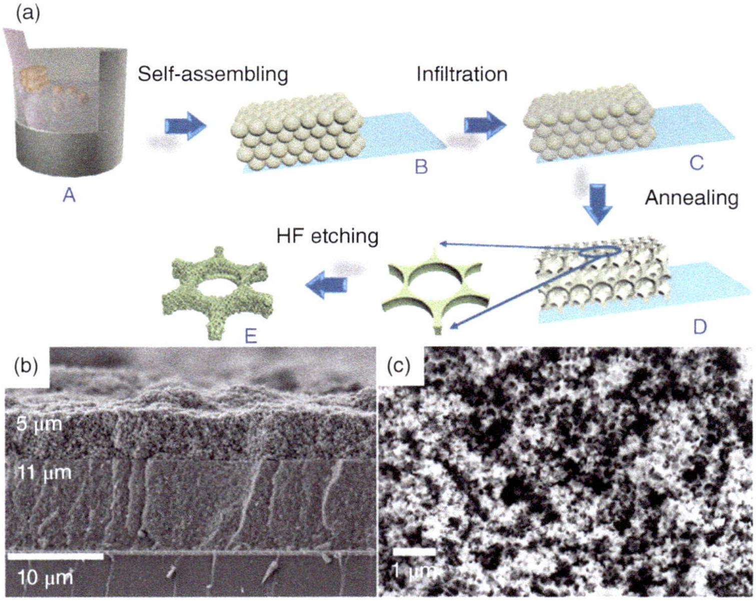

Figure 11.13 (a) Scheme of the fabrication of mesoporous TiO_2 3D-IO nanostructures. (b) SEM cross-sectional view of a photoanode containing ~11-μm thick layer of TiO_2 NPs and ~5-μm layer of mesoporous locally ordered 3D-IO TiO_2 (5 mol% SiO_2 removed). (c) Top view of the photoanode. Source: Xu et al. [75]/with permission from John Wiley & Sons.

Reiss's group [75] prepared TiO_2 IOs with mesopores through impregnating the colloidal crystals with a mixture of TiO_2 and SiO_2 precursor, followed by removal of the CC by calcination and SiO_2 by etching. These mesoporous IOs are scraped from the substrate and deposited on a NC TiO_2 layer to form photoanodes of DSSCs, as shown in Figure 11.13. These locally ordered IOs serve as the light scattering layer and at the same time enhance the dye-absorption due to their mesoporous structures. Consequently, the corresponding DSSC achieves an efficiency of 10.35% versus 9.26% for the same devices with conventional photoelectrodes.

Ravine's group [76] incorporated TiO_2 IOs into solid state DSSCs as light scattering layer behind the NC-TiO_2 layer. During the fabrication, the colloidal crystals were deposited on the NC-TiO_2 film in a layer-by-layer manner (Figure 11.14) [77], so that the exact layer number of IO could be controlled. The influence of the layer number and the pore size of the IOs on the PCEs were studied. They found that six layers of IO synthesized from 260 nm PS spheres promote the PCE from 2.73% up to 4.25%.

Jung's group [78] introduced quasi-inverse opal (QIO) layers with imperfect periodicity as scattering layers in DSSCs (Figure 11.15). The QIO layers are composed of crystalline anatase nanocrystals, which exhibits excellent dye-absorption. The PCE

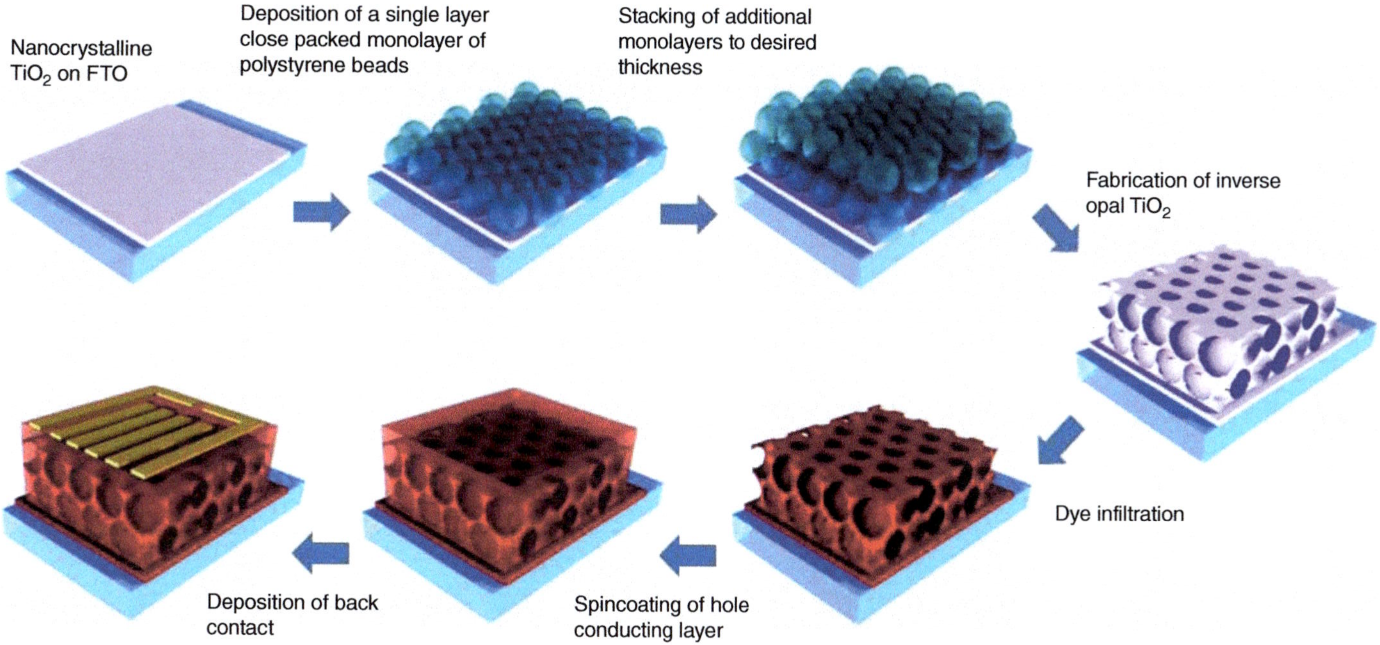

Figure 11.14 Schematic of the fabrication process of integrating TiO_2 IOs with an exact thickness into a solid state DSSC. Source: Zheng et al. [76]/with permission from Elsevier.

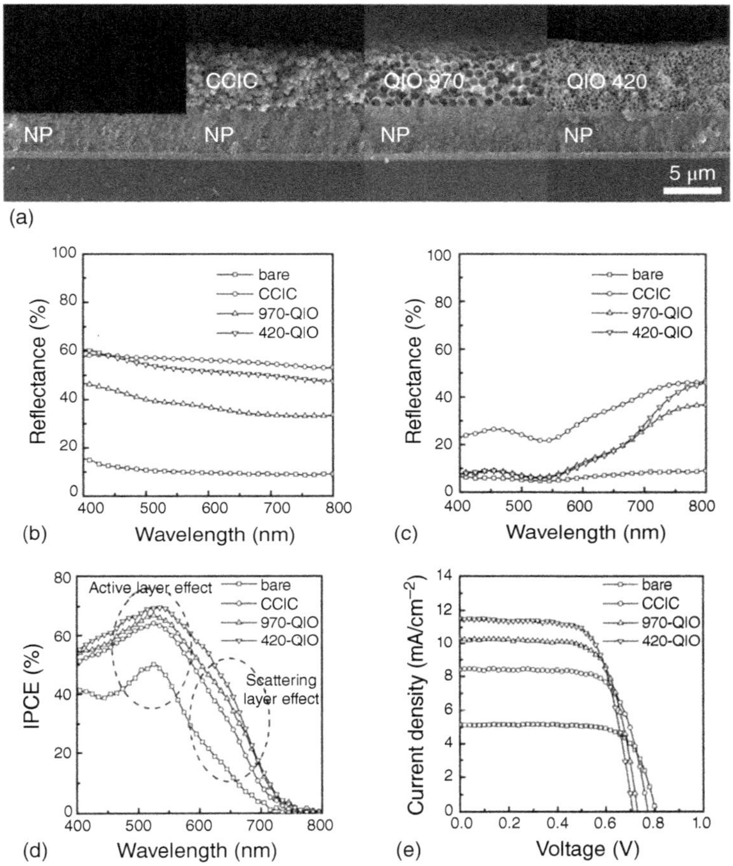

Figure 11.15 (a) Cross-sectional SEM images and UV–Vis reflectance spectra of the photoanodes composed of bare NC-TiO_2 layer, with commercial CCIC (Catalysts & Chemical Ind. Co., Japan) light scattering layers, with QIO of 970 nm, and with QIO of 420 nm as scattering layer, respectively. (b) Before and (c) after dye-loading and (d) IPCE spectrum and (e) J–V curves of the DSSCs employing each photoelectrode. Source: Han et al. [78]/with permission from John Wiley & Sons.

of DSSC based on 420 nm-QIO layer achieves 5.7%, significantly higher than the DSSC based on commercial scattering layers.

Additionally, 2D nanopillar array was fabricated on the surface of a TiO_2 photoelectrode by colloidal lithography with a PS sphere monolayer as the mask (Figure 11.16) [79]. The 430-nm thickness 2D PC reflector could reflect up to 40% of the light at around 400–500 nm wavelength. The DSSC exhibited a PCE of 8.23%, which is 18% higher than a bare DSSC with the same configuration. Conducting polymer IOs were also employed as counter electrodes for DSSCs and their photoelectric conversion efficiencies approach that with a Pt electrode [80, 81].

Figure 11.16 (a) Schematic diagram of colloidal lithography for fabricating 2D PCs of nanopillar array. Tilted view SEM images of the (b) PS monolayer colloidal mask and (c) nanopillar 2D PC. (d) Magnified SEM image of nanopillars. (e) Large-area nanopillar 2D PCs (inset: a digital camera image). Source: Baek et al. [79]/with permission from American Chemical Society.

11.3 Self-assembled Photonic Crystals for Perovskite Solar Cells

PSCs have emerged as one of the most efficient photovoltaic devices in the last decade, with their power conversion efficiencies rising remarkably from 3.8% in 2009 [82] to 25.5% in 2020 [83], rivaling established silicon and other thin film technologies [7, 84–86]. The organic–inorganic hybrid halide perovskite semiconductors have a structure of ABX_3, where A, B, and X are monovalent organic cation, metal cation, and halide anion, respectively (Figure 11.17a) [87]. They exhibit excellent light absorption efficiency and high charge-carrier mobilities [89]. PSCs are normally composed of a submicrometer layer of perovskite materials sandwiched by an electron transport layer (ETL) with a low work function and a hole transport layer (HTL) with a high function, as shown in Figure 11.17 [88]. Under light illumination, the perovskite materials capture photons and produce pairs of electrons and

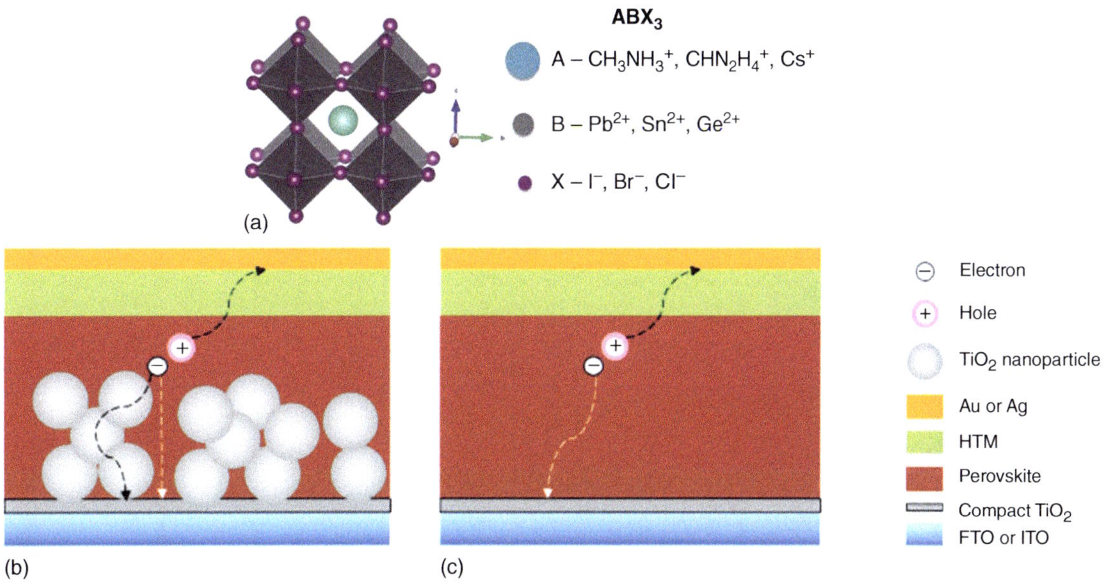

Figure 11.17 (a) Cubic crystal structure of an organic–inorganic hybrid ABX_3 perovskite with common substituents listed for the A, B, and X positions [87]. Source: Ref. 87. Adapted with permission of John Wiley & Sons. Schematic of the structure and electron transport mechanism of (b) a mesoporous structure and (c) a planar structure PSC. In the mesoporous structure, photogenerated electrons can be collected and transported by TiO_2 nanoparticles [88]. Source: Ref. 88. Reproduced with permission of from the Royal Society of Chemistry.

holes. The electrons diffuse through the ETL to the anode, while the holes move through the HTL to the counter electrode, generating an electrical current in the external circuit connected to the two electrodes. Typically, based on the morphology of the ETL, PSCs could be classified into two typical structures, mesoporous structures and planar structures [7, 90]. The former (Figure 11.17a) uses mesoporous TiO_2 scaffold as the ETL and perovskite materials are infiltrated into and cap the scaffold, while the planar devices (Figure 11.17b) employ smooth and mesopore-free ETLs. Huge progresses have been made in terms of materials design [89, 91–95], structure optimization [96–101], and interface engineering [102–113] in the field of PSCs.

The microstructures of the ETL have a significant influence on the infiltration of perovskite, light harvesting, charge injection, and transportations. Various self-assembled IO structures have been adopted in the ETL to facilitate the uniform filtration of perovskite materials, enhance the light harvesting through scattering effect or slow photon effect, and render the PSCs with engineerable colors.

Yang's group first introduced a TiO_2 IO to replace the conventional combination of mesoporous scaffold composed of nanoparticles and compact layer. The light manipulation ability of IOs contributed to enhanced PCEs and the bottom of the IO could replace the traditional compact hole blocking layer [114]. Moon and coworkers [115] employed a TiO_2 meso-IO as the scaffold in $MAPbI_3$ PSCs (Figure 11.18). Benefiting from the interconnected pores of the IOs, perovskite materials could be completely infiltrated in the scaffold. The thickness of the IO scaffold is correlated positively with J_{SC} but negatively with V_{OC} and FF. When an optimal thickness of 600 nm is selected, the PSC exhibits PCE of 17.1%, along with a low efficiency hysteresis between forward and reverse scans of 0.5%. The fewer defects of the mesoporous IO structure, compared with conventional mesoporous scaffold, make the infiltration of perovskite materials more uniform, producing better interfaces between ETL and perovskite and less pin holes defects in perovskite layers. Consequently, the combination of charge carriers are minimized, which contributes to the higher efficiency and lower hysteresis.

Yin and coworkers [116] demonstrated PSCs based on carbon quantum dot (CQD)-sensitized all-inorganic $CsPbBr_3$ perovskite with IO structure ($CQD/CsPbBr_3$ IO), as shown in Figure 11.19. CQDs facilitate the charge transfer process and enhances light absorption, while the slow photon effect induced by the PBG of IOs greatly strengthens the light utilization efficiency. Those PSCs exhibit excellent stability and PCE of 8.29%, higher than twice of the planar $CsPbBr_3$-based counterpart.

Additionally, although perovskite materials show excellent light absorption ability, due to the submicrometer thickness of perovskite layer, the performance of PSCs is limited by loss of light due to reflection and transmission. The introduction of light scattering effect is an effective way to increase light path of the solar cells. 2D PCs composed of mesoporous TiO_2 hemispheres were fabricated using a nano-imprinting method, with the aid of PS sphere array as the initial template [117]. The light scattering effect produced in the PSCs effectively enhances the light utilization efficiency, and the recombination of charge carriers is reduced.

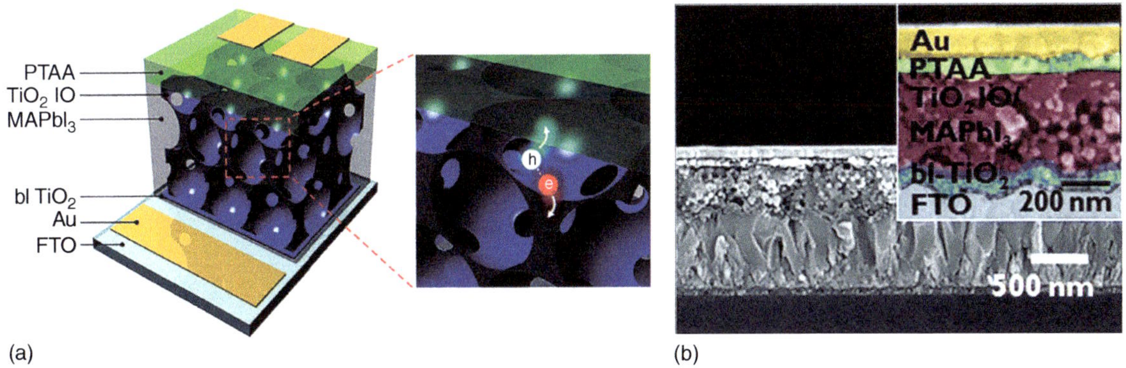

Figure 11.18 (a) Schematic of the structure of the PSC with TiO_2 IO as ETL scaffold. (b) Cross-sectional SEM image of the PSC employing TiO_2 IO with thicknesses of ~400 nm as ETL scaffold (inset: a magnified view). Source: Ha et al. [115]/with permission from Royal Society of Chemistry.

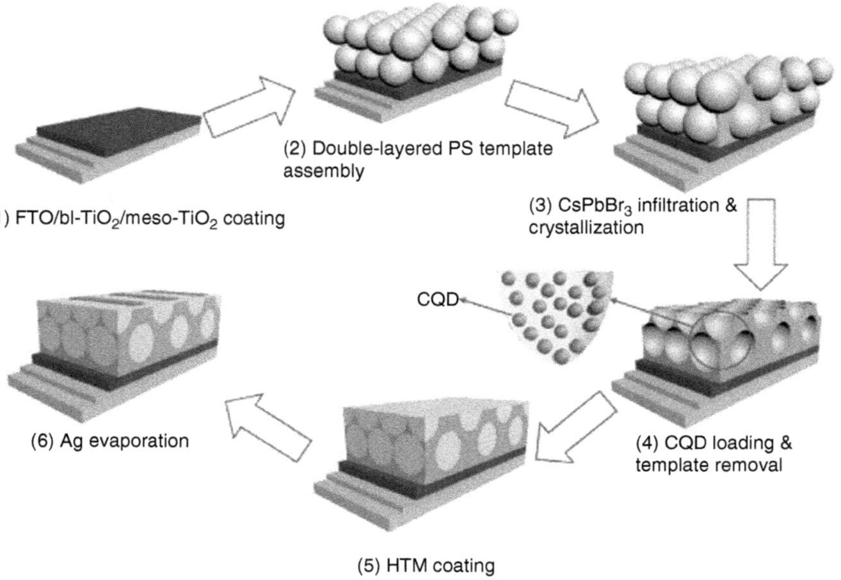

Figure 11.19 Preparation diagram of CQD/$CsPbBr_3$ IO-based perovskite solar cells [116]. Source: Ref. 116. Reproduced with permission of John Wiley & Sons.

The PCE of the corresponding PSC reaches 15.2%, as compared with 10.5% for the counterpart with a planar ETL.

Moreover, color tunability is one of the key advantages of PSCs facilitating their integration into urban context [13, 118, 119]. In contrast to the opaque, rigid, and heavy silicon-based SCs, PSCs are promising photovoltaic devices with high potential of being implemented in built environment. The tuning of the colors of the PSCs is normally achieved through modification of the chemical compositions [120] or introduction of structural color [25, 118, 119, 121–125]. Although modifying the chemical compositions or cell structure of perovskite materials is a straightforward method to endow PSCs with different hues, available colors that can be achieved are limited. The introduction of PC structures is one of the efficient approaches to achieving colorful PSCs. Snaith and coworkers [118] integrated porous alternating TiO_2–SiO_2 layers within photoactive layer, which form a 1D photonic structure. Through tuning the thickness of the corresponding layers, reflection colors covering the entire visible region could be generated. PCEs of 8.8% is achieved for the blue PSC and 4.5% for the red. Huang's group [126] reported a scalable doctor-blade coating method to fabricate perovskite films with a concentric ring photonics structure, generating vivid colors. The coffee-ring-like photonic structures are induced by the convection during precursor solution drying. The periodicity of the rings can be altered by the substrate temperature or the precursor solution concentration, with potential of producing different colors.

Chen's group [25] fabricated 2D IO structured SnO_2 and SnO_2–TiO_2 composite as ETLs for PSCs, using the template-assisted spin-coating method (Figure 11.20).

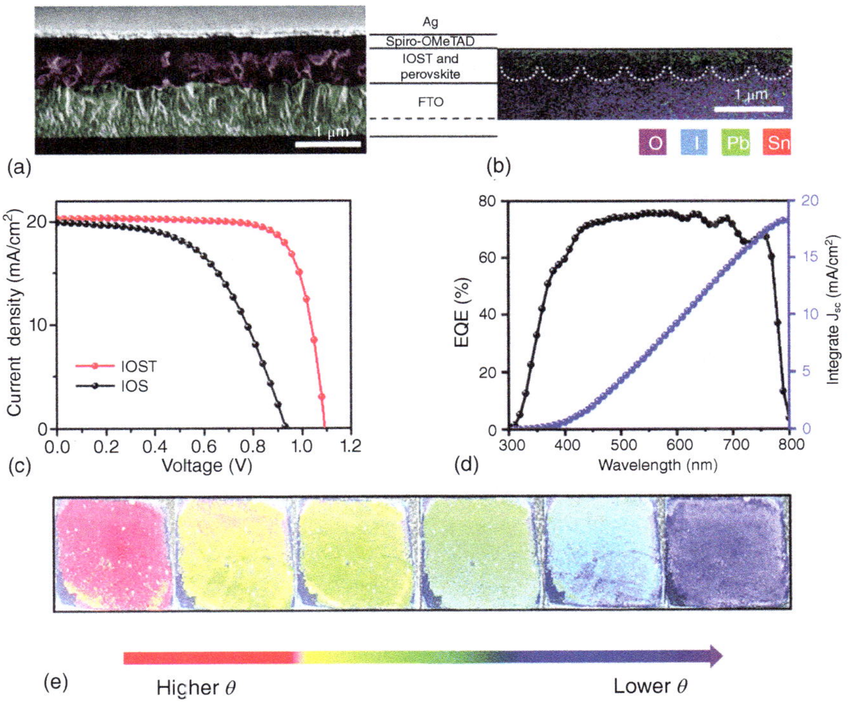

Figure 11.20 (a) The cross-sectional SEM image and (b) the EDS mapping of the PSC device using the SnO_2–TiO_2 IO (IOST) as ETL. (c) The *J*–*V* curves of the SnO_2 IO and IOST-based PSCs. (d) The external quantum efficiency spectrum and the integrated J_{sc} of the IOST-based PSC. (e) Digital photos (from left to right: higher to lower θ) of the IOST-based perovskite film (15 × 15 mm). Source: Liu et al. [25]/with permission from Royal Society of Chemistry.

The combination of 2D IO structure and material modification to the ETLs results in good electron transfer ability and hole-blocking properties and renders the PSCs with efficiency of 16.8% and vivid colors (Figure 11.20) [127]. Qi's group demonstrated PSCs with angle-dependent colors, using TiO_2 nanobowl arrays prepared by PS sphere lithography as ETL. They adopted a Lewis acid–base adduct approach using lead acetate to form a uniform overlayer of $CH_3NH_3PBI_3$ crystals on the inner surface of the nanobowls array [122].

11.4 The Application of Self-assembled PCs in Silicon-Based Solar Cells

Most photovoltaic solar cells made of crystalline silicon have an indirect bandgap and weak absorption in the near-infrared region, giving rise to weak absorption of only one-third of usable solar photons [27, 128]. Micromachining method are widely adopted to produce 2D PC structures to enhance the light absorption of silicon solar cells [27, 129–131]. However, there are limited number of literatures on the application of self-assembled PCs in silicon solar cells, probably because the top-down approaches are more compatible with the materials and device fabrications.

Bermel et al. [51] compared the effect of improving PCEs of thin film c-Si solar cells by geometric optics and wave optics. The latter is obviously superior within the range of 700–1000 nm, where thin film c-Si has a weak absorption. They found that the PCE could be improved by 24% by adding an optimized 1D grating to a six bilayer distributed Bragg reflector (DBR), 31.3% with addition of an optimized 2D grating, 26.3% by using six-period 2D triangular PC composed of air holes in silicon, and 26.5% by placing an eight-period 3D IO composed of air holes arranged in an fcc lattice in c-Si. Kherani and coworkers [52] illustrated that photoconductivity in a silicon thin film can be improved by preparing an opaline PC on the backside of the film. When the photonic bandgap of the PC falls in the range of frequencies where the silicon film has weak absorption, the greatest improvement can be achieved. It is demonstrated that when the surface resonant modes have wavelengths within the [111] bandgap of the PC, the photoconductivity could be enhanced by 130%.

Chen and coworkers designed a 2D inverted nanopyramid structure, which shows excellent antireflection and enhanced long-wavelength photon absorption [130]. Bhattacharya et al. employed wavelength-scale inverted-pyramid arrays for wave-interference-based light trapping, and an overall PCE of 30% was achieved [132]. Bermel and coworkers [133] prepared silicon solar cells with silicon IOs as back reflectors. In their strategy, free-standing silicon IO films are prepared and transferred to couple with the solar cells. An enhancement factor of 10% was achieved in short circuit current without obvious sacrifice of open-circuit voltage and fill factor.

11.5 Summary and Outlook

In this chapter, the recent development of self-assembled PCs for solar cells are reviewed. Self-assembled PCs have many advantages over the conventional geometric optic components, including metallic mirrors or particulate scattering layer, due to their unique optical properties endowed by the photonic bandgap and their uniform porous structure. Self-assembled PCs can be integrated into the electrodes of solar cells or implemented as back reflector at the opposite side of illumination. They can effectively enhance the PCE of the solar cells or render the solar cells with vivid colors across the visible region without sacrificing their efficiency.

Self-assembled PCs are most widely adopted in DSSCs because of their highly compatible structure and fabrication methods. The photon absorption could be enhanced by enlarged optical path because of refection, scattering, resonant mode induced by the close contact of PCs with TiO_2 NC layer, and the slow photon effect induced by the photonic bandgap. Additionally, when mesoporous IOs are introduced into the photoanodes, the dye loading could be enlarged. Self-assembled

PCs could be embedded into DSSCs in different positions. To maximize their potential, PCs are normally built on the surface of the NC TiO_2 layer, and close contact with the NC layer is believed to be the key to induce the resonant mode. Various fabrication methods have been explored to couple IOs of desired structures with the NC TiO_2 layer. 2D or 3D IOs structured TiO_2 or SnO_2 are often adopted as ETL for PSCs for better control of the infiltration of perovskite materials, enhance the power efficiency, or produce solar cells with different hues. For silicon solar cells, although 2D PC structure fabricated with top-down methods are widely adopted to enhance the photon absorption, there are only a few reports on the application of self-assembled PCs as back reflectors.

To harness the full potential of the self-assembled PCs in solar cells, the development of novel technologies for large-scale production of self-assembled PCs with different structures, compatible with relative solar cell fabrication approaches, is expected. A deeper understanding of the photon propagation and interaction with materials in the solar cells, especially at the interface of PCs and light absorbing layer will be essential for the future development. Additionally, the design of PC-coupled solar cells structures with the aid of theoretical simulations should also be further explored.

References

1 U.S. Energy Information Administration, International Energy Outlook (2019), 11 Dec, 2020. https://www.eia.gov/outlooks/archive/ieo19/.
2 Lee, C., Lee, S., Kim, G.-U. et al. (2019). *Chem. Rev.* 119: 8028.
3 Wadsworth, A., Moser, M., Marks, A. et al. (2019). *Chem. Soc. Rev.* 48: 1596.
4 Pan, Z., Rao, H., Mora-Sero, I. et al. (2018). *Chem. Soc. Rev.* 47: 7659.
5 Zhang, J., Tan, H.S., Guo, X. et al. (2018). *Nat. Energy* 3: 720.
6 Shi, Z. and Jayatissa, A.H. (2018). *Materials* 11: 729.
7 Correa-Baena, J.-P., Abate, A., Saliba, M. et al. (2017). *Energy Environ. Sci.* 10: 710.
8 Li, X., Lv, Z., and Zhu, H. (2015). *Adv. Mater.* 27: 6549.
9 Urbani, M., Graetzel, M., Nazeeruddin, M.K., and Torres, T. (2014). *Chem. Rev.* 114: 12330.
10 Wang, H.-P. and He, J.-H. (2017). *Adv. Energy Mater.* 7: 1602385.
11 Moakhar, R.S., Gholipour, S., Masudy-Panah, S. et al. (2020). *Adv. Sci.* 7: 1902448.
12 Hossain, M.I., Qarony, W., Ma, S. et al. (2019). *Nano-Micro Lett.* 11: 58.
13 Wang, W. and Qi, L. (2019). *Adv. Funct. Mater.* 29: 1807275.
14 Xie, K., Guo, M., and Huang, H. (2015). *J. Mater. Chem. C* 3: 10665.
15 Sandhu, S., Yu, Z., and Fan, S. (2015). *J. Phys. D Appl. Phys.* 48: 413001.
16 Aftabuzzaman, M., Lu, C., and Kim, H.K. (2020). *Nanoscale* 12: 17590.
17 Babar, F., Mehmood, U., Asghar, H. et al. (2020). *Renew. Sustain. Energy Rev.* 129: 109919.

18 Samantaray, M.R., Mondal, A.K., Murugadoss, G. et al. (2020). *Materials* 13: 2779.
19 Wan, T., Ramakrishna, S., and Liu, Y. (2018). *J. Appl. Polym. Sci.* 135: 45649.
20 Cavallo, C., Di Pascasio, F., Latini, A. et al. (2017). *J. Nanomater.* 2017: 5323164.
21 Ali, N., Hussain, A., Ahmed, R. et al. (2016). *Renew. Sustain. Energy Rev.* 59: 726.
22 Liu, X., Fang, J., Liu, Y., and Lin, T. (2016). *Front. Mater. Sci.* 10: 225.
23 Yun, J.-H., Wang, L., Amal, R., and Ng, Y.H. (2016). *Energies* 9: 1030.
24 Narasimhan, V.K. and Cui, Y. (2013). *Nanophotonics* 2: 187.
25 Liu, Z., Wu, L.L., Wang, X. et al. (2020). *Nanoscale* 12: 8425.
26 Zhao, C., Ma, Y., Wang, Y. et al. (2018). *Acta Chim. Sinica* 76: 9.
27 Ishizaki, K., De Zoysa, M., Tanaka, Y. et al. (2018). *Jpn. J. Appl. Phys.* 57: 060101.
28 Liu, W., Ma, H.L., and Walsh, A. (2019). *Renew. Sustain. Energy Rev.* 116: 19.
29 Wang, H.P., Lien, D.H., Tsai, M.L. et al. (2014). *J. Mater. Chem. C* 2: 3144.
30 Liu, J., Yao, M., and Shen, L. (2019). *J. Mater. Chem. C* 7: 3121.
31 Deepak, T.G., Anjusree, G.S., Thomas, S. et al. (2014). *RSC Adv.* 4: 17615.
32 Byranvand, M.M., Kharat, A.N., and Bazargan, M.H. (2012). *Nano-Micro Lett.* 4: 253.
33 Jiang, C., Liu, H., Zhang, F. et al. (2018). *J. Synth. Cryst.* 47: 292.
34 Mustafa, M.N. and Sulaiman, Y. (2021). *Sol. Energy* 215: 26.
35 Erwin, W.R., Zarick, H.F., Talbert, E.M., and Bardhan, R. (2016). *Energy Environ. Sci.* 9: 1577.
36 Rho, W.Y., Song, D.H., Yang, H.Y. et al. (2018). *J. Solid State Chem.* 258: 271.
37 Jang, Y.H., Jang, Y.J., Kim, S. et al. (2016). *Chem. Rev.* 116: 14982.
38 John, S. (1987). *Phys. Rev. Lett.* 58: 2486.
39 Yablonovitch, E. (1987). *Phys. Rev. Lett.* 58: 2059.
40 Takahashi, S., Suzuki, K., Okano, M. et al. (2009). *Nat. Mater.* 8: 721.
41 Zhang, R.Y., Ning, H.L., Krueger, N.A. et al. (2016). *Adv. Opt. Mater.* 4: 1533.
42 Jin, W.M., Shin, J.H., Cho, C.Y. et al. (2010). *ACS Appl. Mater. Interfaces* 2: 2970.
43 Arnold, C.B., Serra, P., and Pique, A. (2007). *MRS Bull.* 32: 23.
44 Xia, Y.N., Gates, B., and Li, Z.Y. (2001). *Adv. Mater.* 13: 409.
45 Norris, D.J., Arlinghaus, E.G., Meng, L.L. et al. (2004). *Adv. Mater.* 16: 1393.
46 Yan, Q., Wang, L., and Zhao, X.S. (2007). *Adv. Funct. Mater.* 17: 3695.
47 Wang, L. and Zhao, X.S. (2007). *J. Phys. Chem. C* 111: 8538.
48 Wang, L., Wan, Y., Li, Y. et al. (2009). *Langmuir* 25: 6753.
49 Wang, L., Yan, Q., and Zhao, X.S. (2006). *J. Mater. Chem.* 16: 4598.
50 Prasad, T., Rengarajan, R., Mittleman, D.M., and Colvin, V.L. (2005). *Opt. Mater.* 27: 1250.
51 Bermel, P., Luo, C., Zeng, L. et al. (2007). *Opt. Express* 15: 16986.
52 O'Brien, P.G., Kherani, N.P., Zukotynski, S. et al. (2007). *Adv. Mater.* 19: 4177.

53 Nishimura, S., Abrams, N., Lewis, B.A. et al. (2003). *J. Am. Chem. Soc.* 125: 6306.

54 Villeneuve, P.R., Fan, S.H., and Joannopoulos, J.D. (1996). *Phys. Rev. B* 54: 7837.

55 O'Brien, P.G., Kherani, N.P., Chutinan, A. et al. (2008). *Adv. Mater.* 20: 1577.

56 Liu, J., Zhao, H., Wu, M. et al. (2017). *Adv. Mater.* 29: 1605349.

57 Chen, X.Q., Ye, J.H., Ouyang, S.X. et al. (2011). *ACS Nano* 5: 4310.

58 Kim, H.-N. and Moon, J.H. (2012). *ACS Appl. Mater. Interfaces* 4: 5821.

59 Yun, S., Liu, Y., Zhang, T., and Ahmad, S. (2015). *Nanoscale* 7: 11877.

60 Bella, F., Gerbaldi, C., Barolo, C., and Graetzel, M. (2015). *Chem. Soc. Rev.* 44: 3431.

61 Hagfeldt, A., Boschloo, G., Sun, L. et al. (2010). *Chem. Rev.* 110: 6595.

62 Huisman, C.L., Schoonman, J., and Goossens, A. (2005). *Sol. Energy Mater. Sol. Cells* 85: 115.

63 Seo, Y.G., Woo, K., Kim, J. et al. (2011). *Adv. Funct. Mater.* 21: 3094.

64 Kwak, E.S., Lee, W., Park, N.-G. et al. (2009). *Adv. Funct. Mater.* 19: 1093.

65 Sun, W., Zhou, S., You, B., and Wu, L. (2012). *Chem. Mater.* 24: 3800.

66 Mandlmeier, B., Szeifert, J.M., Fattakhova-Rohlfing, D. et al. (2011). *J. Am. Chem. Soc.* 133: 17274.

67 Tetreault, N., Arsenault, E., Heiniger, L.P. et al. (2011). *Nano Lett.* 11: 4579.

68 Yang, Z.Z., Gao, S.M., Li, W. et al. (2011). *ACS Appl. Mater. Interfaces* 3: 1101.

69 Lee, B., Hwang, D.-K., Guo, P. et al. (2010). *J. Phys. Chem. B* 114: 14582.

70 Halaoui, L.I., Abrams, N.M., and Mallouk, T.E. (2005). *J. Phys. Chem. B* 109: 6334.

71 Mihi, A. and Miguez, H. (2005). *J. Phys. Chem. B* 109: 15968.

72 Mihi, A., Calvo, M.E., Anta, J.A., and Miguez, H. (2008). *J. Phys. Chem. C* 112: 13.

73 Lee, S.H.A., Abrams, N.M., Hoertz, P.G. et al. (2008). *J. Phys. Chem. B* 112: 14415.

74 Guldin, S., Huttner, S., Kolle, M. et al. (2010). *Nano Lett.* 10: 2303.

75 Xu, L., Aumaitre, C., Kervella, Y. et al. (2018). *Adv. Funct. Mater.* 28: 1706291.

76 Zheng, H.B., Shah, S.K., Abbas, M. et al. (2016). *Photonics Nanostruct. Fundam. Appl.* 21: 13.

77 Vogel, N., Goerres, S., Landfester, K., and Weiss, C.K. (2011). *Macromol. Chem. Phys.* 212: 1719.

78 Han, S.-H., Lee, S., Shin, H., and Jung, H.S. (2011). *Adv. Energy Mater.* 1: 546.

79 Baek, S., Ha, S.J., Lee, H. et al. (2017). *ACS Appl. Mater. Interfaces* 9: 37006.

80 Deng, L.D., Wang, L.K., Li, Y.X. et al. (2018). *Macromol. Chem. Phys.* 219: 6.

81 Wang, L., Jin, L., Shi, G. et al. (2018). *New J. Chem.* 42: 12960.

82 Kojima, A., Teshima, K., Shirai, Y., and Miyasaka, T. (2009). *J. Am. Chem. Soc.* 131: 6050.

83 U.S. Energy Information Administration, International Energy Outlook (2019), 24 Sept, 2019. https://www.nrel.gov/pv/cell-efficiency.html.

84 Niu, G., Guo, X., and Wang, L. (2015). *J. Mater. Chem. A* 3: 8970.

85 Rong, Y., Hu, Y., Mei, A. et al. (2018). *Science* 361: eaat8235.

86 Chen, B., Rudd, P.N., Yang, S. et al. (2019). *Chem. Soc. Rev.* 48: 3842.
87 Leijtens, T., Eperon, G.E., Noel, N.K. et al. (2015). *Adv. Energy Mater.* 5: 1500963.
88 Liu, H., Huang, Z., Wei, S. et al. (2016). *Nanoscale* 8: 6209.
89 Huang, F., Pascoe, A.R., Wu, W.-Q. et al. (2017). *Adv. Mater.* 29: 1601715.
90 Deng, K.M. and Li, L. (2020). *Small Methods* 4: 19.
91 Im, J.-H., Jang, I.-H., Pellet, N. et al. (2014). *Nat. Nanotechnol.* 9: 927.
92 Liu, T., Hu, Q., Wu, J. et al. (2016). *Adv. Energy Mater.* 6: 1501890.
93 Jeon, N.J., Noh, J.H., Yang, W.S. et al. (2015). *Nature* 517: 476.
94 Chen, H. (2017). *Adv. Funct. Mater.* 27: 1605654.
95 Ono, L.K., Juarez-Perez, E.J., and Qi, Y. (2017). *ACS Appl. Mater. Interfaces* 9: 30197.
96 Mali, S.S. and Hong, C.K. (2016). *Nanoscale* 8: 10528.
97 Teh, C.H., Daik, R., Lim, E.L. et al. (2016). *J. Mater. Chem. A* 4: 15788.
98 Li, J., Xia, R., Qi, W. et al. (2021). *Sources* 485: 229313.
99 Zhao, Z., Sun, W., Li, Y. et al. (2017). *J. Mater. Chem. A* 5: 4756.
100 Green, M.A. and Ho-Baillie, A. (2017). *ACS Energy Lett.* 2: 822.
101 Green, M.A., Ho-Baillie, A., and Snaith, H.J. (2014). *Nat. Photonics* 8: 506.
102 Bai, Y., Meng, X., and Yang, S. (2018). *Adv. Energy Mater.* 8: 1701883.
103 Ran, C., Xu, J., Gao, W. et al. (2018). *Chem. Soc. Rev.* 47: 4581.
104 Abdi-Jalebi, M., Andaji-Garmaroudi, Z., Cacovich, S. et al. (2018). *Nature* 555: 497.
105 Luo, D., Su, R., Zhang, W. et al. (2020). *Nat. Rev. Mater.* 5: 44.
106 Chen, J. and Park, N.-G. (2019). *Adv. Mater.* 31: 1803019.
107 Ke, W., Xiao, C., Wang, C. et al. (2016). *Adv. Mater.* 28: 5214.
108 Peng, J., Wu, Y., Ye, W. et al. (2017). *Energy Environ. Sci.* 10: 1792.
109 Wang, Q., Dong, Q., Li, T. et al. (2016). *Adv. Mater.* 28: 6734.
110 Zheng, X., Chen, B., Dai, J. et al. (2017). *Nat. Energy* 2: 17102.
111 Bu, T., Li, J., Zheng, F. et al. (2018). *Nat. Commun.* 9: 4609.
112 Bi, D., Yi, C., Luo, J. et al. (2016). *Nat. Energy* 1: 16142.
113 Correa-Baena, J.-P., Saliba, M., Buonassisi, T. et al. (2017). *Science* 358: 739.
114 Chen, X., Yang, S., Zheng, Y.C. et al. (2015). *Adv. Sci.* 2: 1500105.
115 Ha, S.-J., Heo, J.H., Im, S.H., and Moon, J.H. (2017). *J. Mater. Chem. A* 5: 1972.
116 Zhou, S., Tang, R., and Yin, L. (2017). *Adv. Mater.* 29: 1703682.
117 Kim, D.I., Lee, J.W., Jeong, R.H. et al. (2019). *Appl. Surf. Sci.* 487: 409.
118 Zhang, W., Anaya, M., Lozano, G. et al. (2015). *Nano Lett.* 15: 1698.
119 Wang, H., Li, J., Dewi, H.A. et al. (2021). *J. Phys. Chem. Lett.* 12: 1321.
120 Noh, J.H., Im, S.H., Heo, J.H. et al. (2013). *Nano Lett.* 13: 1764.
121 Du, J., An, Y.D., Wu, D.H. et al. (2020). *Opt. Lett.* 45: 6326.
122 Wang, W.H., He, Y.T., and Qi, L.M. (2020). *Sci. China Mater.* 63: 35.
123 Yoo, G.Y., Azmi, R., Kim, C. et al. (2019). *ACS Nano* 13: 10129.
124 Harwell, J., Burch, J., Fikouras, A. et al. (2019). *ACS Nano* 13: 3823.
125 Lu, J.H., Yu, Y.L., Chuang, S.R. et al. (2016). *J. Phys. Chem. C* 120: 4233.
126 Deng, Y., Wang, Q., Yuan, Y., and Huang, J. (2015). *Mater. Horiz.* 2: 578.
127 Meng, K., Gao, S., Wu, L. et al. (2016). *Nano Lett.* 16: 4166.

128 Goldschmidt, J.C. and Fischer, S. (2015). *Adv. Opt. Mater.* 3: 510.

129 Bhattacharya, S. and John, S. (2018). *Phys. Rev. Appl.* 9: 044009.

130 Branham, M.S., Hsu, W.-C., Yerci, S. et al. (2015). *Adv. Mater.* 27: 2182.

131 Mallick, S.B., Agrawal, M., and Peumans, P. (2010). *Opt. Express* 18: 5691.

132 Bhattacharya, S., Baydoun, I., Lin, M., and John, S. (2019). *Phys. Rev. Appl.* 11: 014005.

133 Varghese, L.T., Xuan, Y., Niu, B. et al. (2013). *Adv. Opt. Mater.* 1: 692.

12

Mesoporous Zeolites: Synthesis and Catalytic Applications

Feng Yu[1] *and Feng-Shou Xiao*[1,2]

[1] *Zhejiang University, Key Lab of Applied Chemistry of Zhejiang Province, Department of Chemistry, No148 Tianmushan Road, Xihu District, Hangzhou 310028, China*
[2] *Zhejiang University, Key Lab of Biomass Chemical Engineering of Ministry of Education, College of Chemical and Biological Engineering, No38 Zheda Road, Xihu District, Hangzhou 310027, China*

12.1 Introduction

Zeolites are low-density crystalline aluminosilicates with interconnected and ordered micropores, which render them extraordinary properties, such as physical and chemical stability and tunability of pore size, and surface properties like high surface area and acidity, making these materials as efficient heterogeneous catalysts widely used in a broad spectrum of industrial applications such as oil refining and petrochemistry [1–3]. The uniform microporous structures of zeolites offer them a unique shape selectivity in catalysis; the mere presence of the microporous network, however, also gives rise to significant diffusion limitation [4]. Bulky molecules that have larger dimensions than the microporous channels of the zeolites are not able to enter into the pore system and diffuse to the active sites placed in the zeolite crystals. Even in the cases that micropore size is larger than the molecule dimensions of the reactants and products, reduced mass transfer of these compounds in the pore system may appear. The catalytic reaction rate can be restrained due to the diffusion limitation that resulted from restricted access and low diffusion coefficients. It may even lead to the deactivation of catalysts owing to the blockage of the pores caused by large molecules or coke formation.

Considerable efforts have been devoted to develop new materials with improved pore architecture to overcome the diffusion limitation issue related to the conventional microporous zeolites [5]. Exploring zeolites with larger micropore size and ordered mesoporous aluminosilicates with pore size in the range of 2–50 nm represents the early attempts to this end [6–8]. Unfortunately, the unsatisfactory properties, such as low hydrothermal stability and low acidity, substantially hamper their practical application.

Recently, mesoporous zeolites, also referred to be hierarchical zeolites, possessing at least two levels of pore size including inherent microporosity and additional

Functional Materials from Colloidal Self-assembly, First Edition. Edited by Qingfeng Yan and George Zhao.

mesoporosity (or macroporosity) have been exploited [9–12]. Strategies via creating intracrystalline mesopores in the microporous zeolite crystals or inducing intercrystalline mesopores in between the intergrown nanosized zeolite crystals can be adopted to prepare this zeolite. The generation of additional mesopore structure could reduce transport path length and increase diffusivity for molecules involved in the reaction. The hierarchical porosity enables the zeolite to integrate both efficient mass transfer and shape selectivity, which are contributed by mesopore and micropore structure, respectively. Over the past two decades, there has been a great number of literature reports on the synthesis strategies for the tailoring of pore architecture of the mesoporous zeolite with the aim to maximize the functionality of zeolite in catalysis, which is expected to have a significant economic impact [13–16].

In this chapter, a summary on the progress of the synthesis strategies for fabricating mesoporous zeolites with emphasis on the pioneering examples will be given in Section 12.2. It should be noted that packed nanosized zeolites with transient mesoporosity will not be covered herein. Considering the fact that the benefits of a secondary mesopore system in the microporous zeolites are established today by an impressive number of publications covering diverse catalytic applications, we will present prominent recent examples of the use of mesoporous zeolites as the heterogeneous catalysts for different reaction types, especially in the last decade, in Section 12.3.

12.2 Synthesis of Mesoporous Zeolite

With the aim to form mesoporous zeolite materials, an additional mesopore needs to be introduced with an interconnectivity with the inherent micropore system dictated by the zeolitic framework structure. A wide variety of approaches to accomplish the synthesis of zeolites with additional mesoporosity have been developed in the last two decades. It can be generally classified into bottom-up, top-down, and mixed zeolite synthetic methods.

The bottom-up zeolite synthetic strategies referred to the ones that build the mesopore structure either during the zeolite crystallization period or via an aggregation process of pre-synthesized crystals. These methods can be mainly categorized into three routes, which are hard templating, soft templating, and non-templating. The former two routes are defined according to the physicochemical nature of the mesopore template, being of more or less rigid nature, while the latter one indicates the synthetic process without using any mesopore template. The top-down strategies, also known as demetallization methods, mainly include dealumination and desilication routes, in which the mesostructure is introduced through selective removal of aluminum or silicon atoms from the zeolite crystal framework via a posttreatment. The mixed synthetic method represents a combination of bottom-up and top-down methods, or in other words zeolite recrystallization approach, in which usually with the aid of surfactants mesoporosity of zeolite is produced through dissolution and recrystallization of the preformed zeolite crystals.

The pros and cons of each synthetic route for preparation of mesoporous zeolites with respect to the final zeolite properties as well as preparative challenges need to be considered seriously prior to selecting a certain strategy. A comparison of different main routes to achieve this objective is shown in Table 12.1.

12.2.1 Bottom-up Zeolite Synthetic Strategies

So far a number of preparation approaches to mesoporous zeolite denoting the bottom-up strategy have been developed. In addition to the conventional micropore templates or structure-directing agents (SDA) employed in the synthesis of zeolite, the use of mesopore templates for the creation of mesopore structure in zeolite takes a major proportion of these preparation approaches. Depending on rigidity of mesopore templates used, these synthetic approaches can be classified as hard and soft templating. The mesopore templates, including hard and soft ones, are usually encapsulated within the zeolite during the early synthetic process and leached afterward by calcination to generate mesoporosity. Besides the hard and soft templating, several non-templating synthetic methods for creating mesopore structure without using mesopore templates have also been studied. In this regard, the intergrowth of the zeolite nanocrystals without the participation of mesopore templates accounts for giving rise to mesoporosity.

12.2.1.1 Hard Templating Methods

Hard templating methods involves introducing porous or hollow solids, usually with a relatively rigid structure, into the zeolite gels to serve as mesoporous templates during the zeolite crystallization process [17]. The general procedures of the hard templating methods consist of the following main steps: (i) introducing the liquor of zeolite precursor into the pores/cavities of the template, (ii) performing a zeolite crystallization process in the hard template by hydrothermal treatment in a controlled atmosphere, and (iii) removing the template via combustion or dissolution processes. A wide range of solid materials, such as metal oxide nanoparticles, resin beads, or plant materials, have been explored as hard templates in the last two decades, among which carbonaceous materials are the ones that have been most extensively studied, showing superior operational characteristics such as chemical inertness, structural diversity, and easy removal by simple combustion [9]. A variety of zeolites with tailored mesopore architecture have been prepared by using different carbonaceous materials, which include carbon nanoparticles, nanotubes, nanofibers, aerogels, and mesoporous carbons.

The pioneering work of mesoporous zeolite synthesis with the aid of carbon nanoparticle was reported by the group of Jacobsen. They used commercial carbon black pearls (BP2000 or BP700) as a matrix to assist the synthesis of nanozeolites with controlled particle size [18, 19]. To this end, zeolites were crystallized in the voids of the porous carbons or in the confined space between the carbon particles. When the amount of zeolite synthesis gel was equal to or lower than the total pore volume of the carbon matrix, zeolites were only confined to grow in the voids of the matrix and finally provided as nanozeolite with very small nanoparticle sizes

Table 12.1 Pros and cons of different preparative routes toward mesoporous zeolites.

	Bottom-up strategy			Top-down strategy		Mixed strategy
Method	**Hard templating**	**Soft templating**	**Non-templating**	**Dealumination**	**Desilication**	**Dissolution and recrystallization**
Pros	• High zeolitic character • A wide Si/Al ratio scope • Applicable to a broad variety of zeolites • High degree of additional porosity	• Controllable mesoporosity • Variation of Si/Al possible • Application to a broad variety of zeolites • High degree of additional porosity	• Eco-friendly • Cost-effective • Medium zeolitic character	• High zeolitic character • Applicable to different zeolites • Cost-effective • Scale-up possible	• High zeolitic character • Applicable to a broad variety of zeolites with a wide Si/Al ratio range • Cost-effective • High pore connectivity possible • Scale-up possible	• High degree of additional porosity
Cons	• Low pore interconnectivity • High production costs • A challenge to tune the additional porosity	• High production costs • Low to medium zeolitic character • Most templates are not available commercially	• Rigorous synthesis conditions and time-consuming process involved • Controlling of the zeolite topology and additional porosity is still a great challenge • Applicable to few zeolites	• Low pore interconnectivity • Applicable to Al-rich zeolites only • May alter the Si/Al ratio of the parent zeolite • Difficult to tune additional porosity	• May alter the Si/Al ratio of the parent zeolite • Difficult to tune additional porosity	• High production costs • Partial dissolution of silica is possible • Challenging to scale-up • Limited to the certain Si/Al ratio zeolites

(ZSM-5, ~20 nm). In contrast, when an excess of the zeolite precursor gel was used in the impregnating process, carbon nanoparticles was encapsulated in zeolite crystals during the crystallization, which resulted in mesoporous zeolite single crystals with a wide mesopore size distribution in the range of 5–50 nm instead (Figure 12.1a,f) [20]. This method has been extensively used for the synthesis of different structures and compositions of zeolites and zeotypes, including MFI-type zeolites (ZSM-5, TS-1, and Silicalite-1) [25], MEL-type zeolites (ZSM-11, TS-2, and Silicalite-2) [26], and MTW (ZSM-12) [27]. In addition, other zeolite types and aluminophosphates, such as BEA (beta), AFI ($AlPO_4$-5), and CHA ($AlPO_4$-34), were also synthesized as mesoporous single crystals by the hydrothermal fluoride approach, in which the fluoride ion instead of the more common OH^- acted as mineralizer [28]. Due to the formation of extra mesoporosity in the structure, mesoporous zeolites demonstrated a significant improvement in the diffusion property and accordingly promoted the catalytic performance in comparation with the traditional microporous counterpart. The resulting mesopores, however, generally present a broad size distribution, and some formed cavity-like mesopores are only accessible via passing through the neighboring micropores, making them not contribute much to the enhancement of mass transfer, which leads to a less obvious improvement in catalysis.

In view of the broad size distribution resulted from carbon nanoparticles, other carbon sources were therefore investigated to produce more defined mesoporous imprints and to better control the crystallization conditions. Alternative materials include carbon aerogels, carbon nanotubes, carbon fibers, and ordered mesoporous carbons.

The use of carbon aerogels as hard templates can fabricate hierarchical zeolite monoliths with interconnected mesoporous channels. Mesoporous zeolite ZSM-5 (MFI), A (LTA), and Y (FAU) monoliths were successfully prepared by Tao et al. by using the carbon aerogels, which were made via CO_2 supercritical drying and subsequent thermal pyrolysis of resorcinol–formaldehyde gels (Figure 12.1b,g and g) [21, 29, 30]. The mesopore sizes from the porous aerogels can be tuned through controlling the reaction conditions. The resultant mesoporous zeolites show a narrow pore size distribution in accordance with the thickness of the pore wall of carbon aerogels and a good interconnection of channels. In 2005, this strategy was further improved by Schüth and coworkers through introducing an additional step of impregnation to ensure the Silicalite-1 monolith product with a high mechanical stability [31]. The monolith is composed of well-defined nanosized crystals aggregated in such a way that micropores, mesopores, and macropores are generated. With a similar synthetic mechanism to that of carbon aerogels, sustainable biomass-derived carbon materials, such as sucrose and glucose, were also used as the carbonaceous templates for synthesis of mesoporous zeolites in some studies [22, 32, 33]. For instance, Christensen's group developed a new method that applied *in situ* generation of the required carbon template via decomposition of sucrose directly onto the silica raw material employed for the preparation of mesoporous zeolites (Figure 12.1c,h) [22]. The mesoporous zeolite crystals contain interconnected micropores and mesopores inside each individual single crystal with

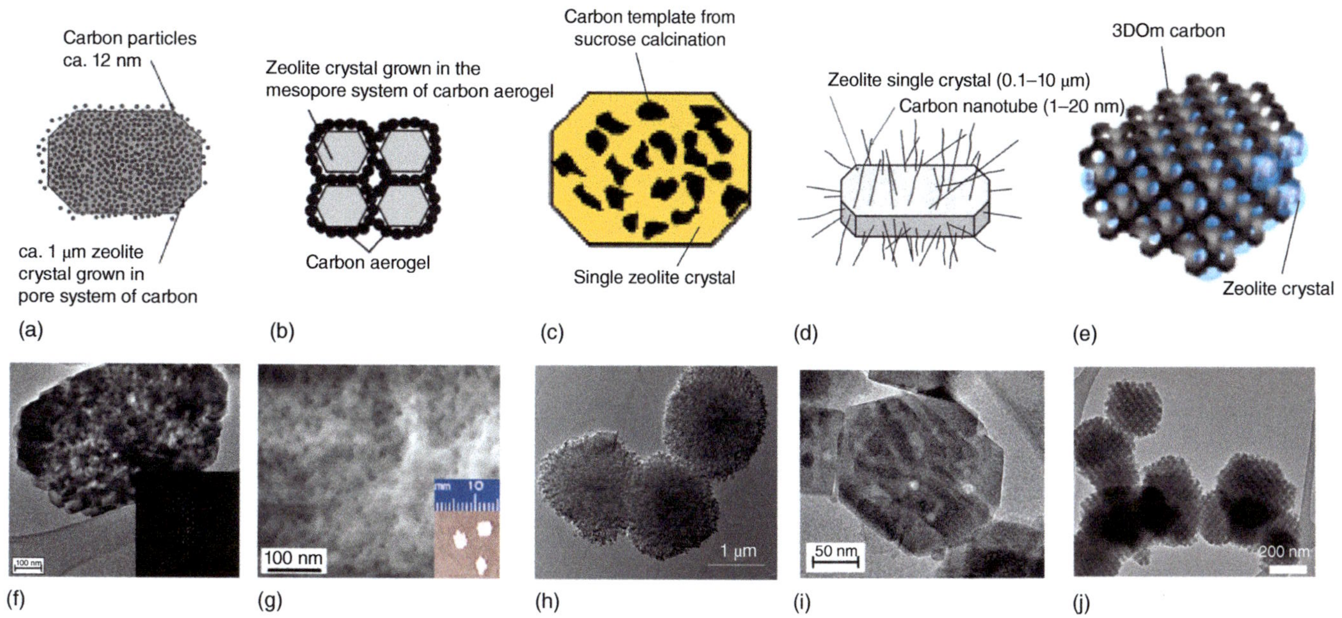

Figure 12.1 Overview of hard templating strategies using different forms of carbon materials. (a–e) Schematic diagram showing mesoporous zeolite templated by different carbon. (a) carbon nanoparticles, (b) carbon aerogel, (c) carbon from sugar calcination, (d) carbon nanotubes, and (e) 3D ordered mesoporous (3DOm) carbon. (f–j) Electron micrographs of final mesoporous zeolite products of (a)–(e), respectively. Source: Adapted from Refs. [20–24] / with permission from American Chemical Society.

a disordered character. It should be noted that this approach allows fine control of the porosity of the mesoporous zeolite in a very simple manner through just tuning the molar carbon/silica ratio. In addition, a monolithic N-doped carbonaceous template, prepared by the hydrothermal treatment of glucose and ovalbumin, was also employed by White and Thomas and coworkers for the synthesis of relatively small single-crystal ZSM-5 particles ($\leq$200 nm) featuring intracrystalline mesopores of 12–16 nm diameter [34]. The texture, dimensions, and chemistry of the N-doped carbonaceous monoliths, which act as mesoporous templates, can be tuned via adjusting precursor ratio, solvent volume, and post-synthesis thermal annealing, thus offering a cost-effective and highly flexible synthetic manner.

Alternatively, even more structure-defined but higher priced single- or multi-walled carbon nanotubes or more cost-effective carbon nanofibers were used to improve control over the size and shape of the resultant mesopores [23, 35, 36]. For instance, in 2001 Jacobsen's group first reported the use of commercial multiwall carbon nanotube templates, with the diameter of 12 nm and the length of several micrometers, for synthesis of mesoporous silicalite-1, giving rise to the formation of straight and uniform mesoporous channels penetrating the zeolite single crystals (Figure 12.1d,i) [23]. Additionally, de Jong and coworkers used carbon nanofibers as the template for the preparation of silicalite-1 crystals, which demonstrates cylindrical mesopores with low tortuosity [35].

Mesopore design is not only for the creation of evenly sized pores but also for arranging these pores into an ordered secondary pore system. For this purpose, highly ordered mesoporous carbon scaffolds were prepared by nanocasting from ordered mesoporous silicates or imprinting from colloid silica and further employed for the synthesis of zeolites with ordered mesoporous channels [24, 37–43]. In 2004, ordered mesoporous carbons CMK-1 and CMK-3, replicated from ordered mesoporous silicates MCM-48 and SBA-15, were used by Mokaya and Liu and coworkers, respectively, to obtain mesoporous zeolites [37, 38]. For example, ZSM-5 with pore sizes of around 2 nm was prepared via the synthesis procedure, which consisted of impregnating the CMK-3 carbon by the incipient wetness approach with the zeolite precursor gel, subsequent hydrothermal crystallization, and final combustion of the carbon template. Hu and coworker presented a success in the synthesis of an ordered mesoporous aluminosilicate with a completely crystalline wall structure, denoted as OMZ-1, which was prepared by the recrystallization of SBA-15 using *in situ* formed CMK-5 [39]. The key to the success in this case lies in the use of CMK-5, which not only serves as a hard template but also kinetically controls the crystallization process to produce large crystals. In addition, alternative colloidal mesoporous silica was used to imprint ordered mesoporous carbon for the preparation of desired mesoporous zeolites (Figure 12.1e,j) [24, 41–43]. Tsapatsis' group realized a wide range of crystal morphologies through confined growth within three-dimensionally ordered mesoporous (3DOm) carbon, which synthesized by replication of colloidal crystals composed of size-tunable silica nanoparticles (10–40 nm) [43]. Confined crystal growth within these templates gave rise to uniformly shaped silicalite-1 nanocrystals with size tunability as well as 3DOm-imprinted single-crystal zeolite particles. Steam-assisted crystallization

(SAC) was employed in this method due to the fact that zeolite precursors tended to migrate from mesopores to external surfaces in crystallization process. With the use of 3Dom carbon template, the preparation of several zeolites including BEA, FAU, LTA, and LTL were achieved afterward by conventional hydrothermal treatment, featuring these zeolites with highly ordered and tunable mesopores in 3–7-nm range [24]. Due to time-consuming and costly synthesis issues, the strategy of ordered mesoporous carbon is rather limited in the practical application.

Other hard templates include urea-formaldehyde resin [44, 45], $CaCO_3$ [46], bread [47], and raw biological materials [48–52]. Besides sugars mentioned previously, several types of raw biological materials were investigated as templates, such as wood [48, 49], sugarcane bagasse [50], leaves [51, 52], and bacteria [53]. Most of these hard templates generate secondary porosity with a broad range of pore size, usually resulting from the existence of nanosized polycrystalline ensembles.

12.2.1.2 Soft Templating Methods

In comparison with the hard templating method, soft templating route uses relatively flexible species, such as surfactants and polymers, as the mesopore templates to achieve better control of the mesopore size and pore connectivity. Simply mixing ordinary organic surfactant as the mesoporous template with the micropore template, in most cases, only gave a mixture of zeolite crystals with amorphous materials as a result of phase segregation, due to the interfacial incompatibility between surfactant and zeolite gel. Delightedly, a lot improvement has been achieved by researchers from the beginning of this century to overcome this limitation via searching for appropriate soft templates or kinetic control of zeolite seed formation.

Due to high stabilities under alkaline crystallization conditions (up to 200 °C) and efficient interaction with negatively charged inorganic silica gel, cationic polymers were corroborated as a type of effective soft templates for mesoporous zeolite preparation. In 2006, Xiao et al. firstly reported a strategy that combined small organic ammonium salts and cationic polymers as micropore and mesopore templates, respectively, for the synthesis of mesoporous zeolites [54]. More specifically, they used tetraethylammonium hydroxide (TEAOH) and polydiallyldimethylammonium chloride (PDADMAC) templates and successfully synthesize a beta zeolite with a mesopore size distribution of 5–40 nm. When a mixture of tetrapropylamine hydroxide and dimethyldiallyl ammonium chloride acrylamide copolymer was employed, mesoporous ZSM-5 was formed via this approach. It was further found that the mesoporosity of beta and ZSM-5 zeolites could be adjusted by simply altering the amounts of cationic polymers added in the starting gels [55]. With the use of PDADMAC as the mesoscale template, mesoporous EU-1 zeolites with controllable mesoporosity could also be achieved [56]. Additionally, Möller and coworkers found that cationic polymers PDADMAC can act as flocculating agents to produce interparticle mesopores between the zeolite beta nanoparticles [57]. The beta nanocrystals are highly aggregated and therefore significantly superior in recoverability when compared with the colloidal nanosized beta crystals achieved without adding PDADMAC. The interstitial pore size between 40 and 400 nm can be effectively tuned by controlling the growth of the zeolite particles

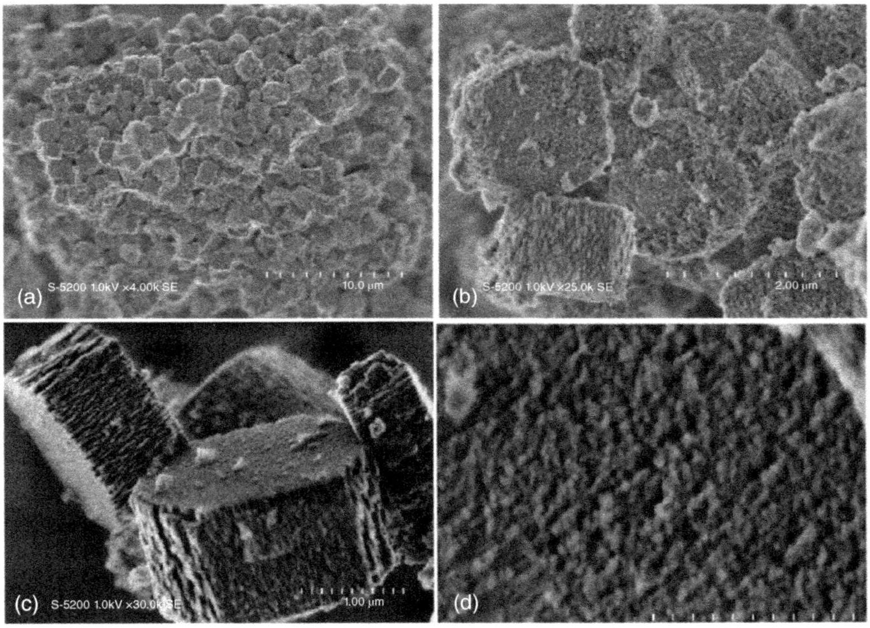

Figure 12.2 SEM images of ZSM-5 synthesized in the presence of C-PSt-*co*-P4VP template exhibiting oriented mesopores. Source: Liu et al. [58] / with permission from American Chemical Society.

through changing the polymer concentration. Moreover, Xiao's group successfully synthesized *b*-axis-aligned mesoporous ZSM-5 with a size range of 10–50 nm by using the designed amphiphilic copolymer (C-PSt-*co*-P4VP) as the mesoporous soft template (Figure 12.2) [58]. C-PSt-*co*-P4VP was obtained from copolymer polystryrene-*co*-4-polyvinylpyridine after positively charging by methyl iodide treatment. The particular orientation of the acquired mesopores was probably due to the copolymer template being energetically prone to occupy the (010) face in the self-assembly according to density-functional theory (DFT) calculations.

Polymers can also act as dual-function soft templates for synthesis of mesoporous zeolites. The cationic polymer PDADMAC was found its use as a dual-function template to prepare single beta crystals with mesoporosity, which was reported by Xiao's group [59]. This study indicates that the quaternary ammonium groups on the flexible polymer PDADMAC serve as an SDA for the structure of beta zeolite. In the meantime, the polymer itself acted as a "porogen" to give rise to disordered mesopores rather than self-assemble to form mesoscale micelle structure resulting in ordered mesopores. The mesopore size of the synthesized beta zeolite can be tuned in the range of 4–10 nm by simply adjusting the molecular weight of PDADMAC. In addition, amorphous linear-chain polymers functionalized with a random distribution of side groups, which can participate in inorganic crystallization, were employed by Ryoo and coworkers for preparation of mesostructured zeolites and their aluminophosphate analog [60]. It was found that tight binding of the multi-ammonium side groups on the polymers with negatively charged inorganic species enforced strain in the polymer backbones, giving rise to the crystallization to an ultrathin micellar scale of only a few nanometers (Figure 12.3). The mesopore size can be tailored by varying functionalization degree. Recently, Hensen and coworkers showed that cetyltrimethylammonium hydroxide can be used as both mesoporogen and SDA to synthesize mesoporous ZSM-5 zeolite [61]. In this synthetic process, KOH was employed instead of NaOH in order to improve silica dissolution, thereby significantly promoting the desired zeolite nucleation. Furthermore, Ryoo and coworkers also reported a novel method, namely, pseudomorphic crystallization, to prepare nanocrystalline zeolites with intercrystalline mesopores [62, 63]. The method relies on the use of cyclic diquaternary ammoniums (CDA) as an SDA and simultaneously restraining the mobility of silicates during crystallization. Besides mesoporous BEA structure zeolites, MFI and MTW nanocrystallites with intercrystalline mesopores can be also obtained by properly tuning the structure of CDA.

In addition to the previously mentioned cationic polymer templates, quaternary ammonium surfactants including single and dual-function ones, were also designed for the preparation of mesoporous zeolites. In 2009, a whole new idea was introduced by the group of Ryoo to synthesize ultrathin MFI zeolite nanosheets with the use of an elegantly designed dual-function surfactant $C_{22}H_{45}\text{–}N^+(CH_3)_2\text{–}C_6H_{12}\text{–}N^+(CH_3)_2\text{–}C_6H_{13}(Br^-)_2$, which is abbreviated as $C_{22\text{-}6\text{-}6}$ [64]. The study demonstrates that the hydrophobic long-chain C_{22} group causes the formation of mesoscale micellar structure and restrains the excessive growth of zeolite; meanwhile two quaternary ammonium sites on the surfactant $C_{22\text{-}6\text{-}6}$ spaced by a C_6 alkyl chain direct the microporous MFI structure generation, as shown in Figure 12.4. Multilamellar and unilamellar nanosheets were obtained via this synthetic strategy. The multilamellar nanosheets exhibited 20–40-nm thickness, which consist of alternate layers of MFI zeolite framework and surfactant micelles with 2.0 and 2.8 nm thickness, respectively. They showed a high mesoporosity with a rather broad mesopore size distribution, probably resulting from the pillar effect by the crystal intergrowth and the slight deviation of the crystal orientation hampering the complete condensation of zeolite layers. The unilamellar nanosheets possessed higher surface area than the multilamellar ones, reaching 720 m^2/g. Through prolonging hydrothermal aging time, the unilamellar structure can be transformed into the ordered multilamellar mesostructured via a dissolution–recrystallization process, which was reported in a subsequent study [65]. On the basis of their modular character, it is possible to tailor the dual-function surfactants by changing the number of cationic centers, the length, and the structure of the spacer between these centers and the hydrophobic tails, for constructing different meso- and microporous structures. The first examples of truly hexagonally ordered mesoporous zeolites were demonstrated later on when a symmetric bifunctional surfactant with multiple ammonium centers $C_{18}H_{37}\text{–}N^+(CH_3)_2\text{–}C_6H_{12}\text{–}N^+(CH_3)_2\text{–}C_6H_{12}\text{–}N^+(CH_3)_2\text{–}C_{18}H_{37}(Br^-)_3$ (abbreviated as $C_{18}\text{–}N_3\text{–}C_{18}$) was employed [66]. Relying on this type of templates, ultrathin MFI zeolite sheets (1.7-nm thickness) were shaped into cylinders, generating

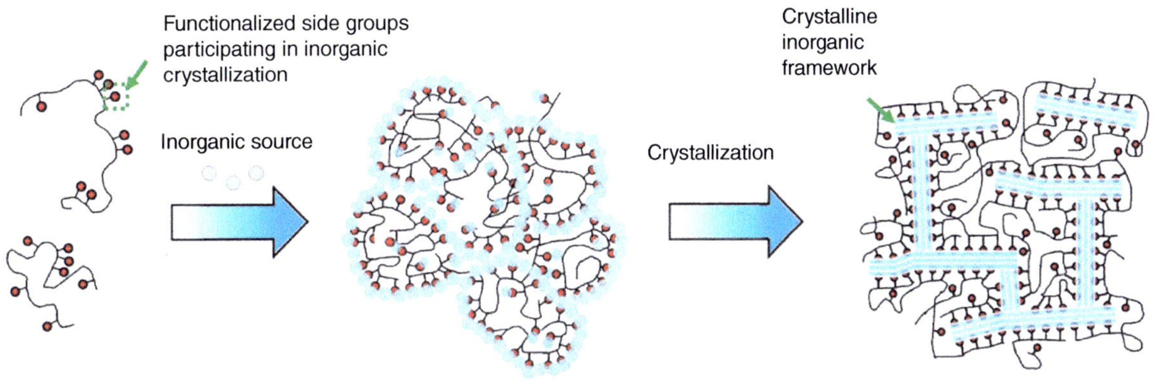

Figure 12.3 Description of random-graft polymer-directed inorganic crystallization. Source: Jo et al. [60]. Reproduced with permission of John Wiley & Sons, Ltd.

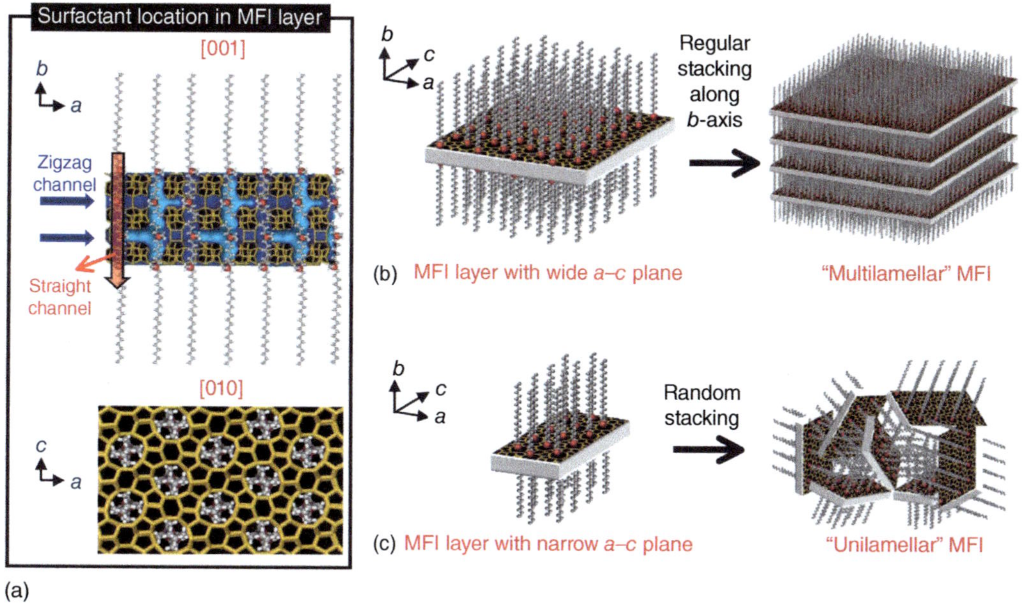

Figure 12.4 Crystallization of MFI nanosheets. (a) Proposed structure model for the single MFI nanosheet. Surfactant molecules are aligned along the straight channel of MFI framework. Two quaternary ammonium groups (indicated as a red sphere) are located at the channel intersections; one is inside the framework, and the other is at the pore mouth of the external surface. (b) Many MFI nanosheets form either multilamellar stacking along the *b*-axis. (c) A random assembly of unilamellar structure. Source: Choi et al. [64]. Reproduced with permission of Springer Nature.

mesopores with a diameter of around 3.5 nm and possessing an extremely large surface of over 1000 m^2/g for potential catalytic applications. It is worth noting that when the surfactant template was modified by introducing two or three bridging phenyl groups to the center part of the template, beta zeolite, a different zeolite type, was prepared accordingly with this designed bifunctional surfactant. Although these products did not feature ordered mesoporosity, the mesopore wall thickness and the mesopore size were uniform, giving rise to a high surface area up to 940 m/g. Che and coworkers developed a new type of dual-function amphiphilic surfactants, which possesses aromatic groups in the hydrophobic tails and thereby exhibits strong ordered self-assembling ability owing to pi-stacking [67]. Single-crystalline mesopore MFI zeolite nanosheets with a lamellar structure were prepared with the use of the template with biphenyl and naphthyl at the alkyl tail and a single quaternary ammonium at the head. In the presence of bolaform amphiphilic templates with bi-quaternary ammonium head groups and biphenyl groups, mesostructured MFI nanosheets joined with a 90° rotational boundary were synthesized. It was found that house-of-cards-like morphologies can be obtained from the bolaform templates with proper hydrophobic chain lengths, resulting in a well-defined micro–meso–macroporous architecture featuring a high external surface area of 658 m^2/g owing to the mutual pillaring between ultrathin plates. A combination of surfactant $C_{22}H_{45}–N^+(CH_3)_2–C_4H_8–N^+(CH_3)_2–C_4H_9(Br^-)_2$ ($C_{22\text{-}4\text{-}4}$) and a conventional micropore template (trimethyl adamantanammonium hydroxide) was also employed for the preparation of mesoporous SSZ-13 with CHA topology [68]. The less amount of costly dual-function surfactant templates used in this approach made it more cost-effective though the resultant mesopores show a broad size distribution. Additionally, multiamines with amphiphilic structure can be served as simultaneous SDAs in micro- and mesostructural levels for the synthesis of aluminophosphate materials (AlPOs) and their analogs including silicoaluminophosphate, cobalt aluminophosphate, and gallium phosphate [69].

In 2006, the crystallization of silanized zeolitic seeds or protozeolitic units was firstly reported by Serrano et al. as a novel approach developed for the preparation of mesopore zeolites [70]. This approach generally consists of the following four steps: (i) precrystallization of the zeolite precursor gel to acquire the formation of zeolitic seeds or protozeolitic nanounits; (ii) functionalization of the zeolitic seeds by refluxing with organosilane, such as phenylaminopropyl-trimethxysilane (PHAPTMS), in order to form a protective layer (Figure 12.5); (iii) crystallization of the functionalized zeolite gel under hydrothermal conditions; and (iv) removal of the zeolite SDA and the seed silanization agent by calcination in air. The addition of the silanization agent in the initial stages of zeolite crystallization can result in its anchoring on the external surface of the zeolitic seeds, which could hinder the aggregation of zeolitic seeds and therefore avoid their growth into large zeolite crystals. In the case of ZSM-5, the size of the synthesized mesoporous zeolite particles was in the range of around 300–400 nm, resulting from the aggregation of nanocrystals below 10 nm with a significant intergrowth. The mesopores in between the nanocrystals were revealed to be relatively uniform with a size of about 4.5 nm. Through varying the synthesis parameters such as precrystallization temperature,

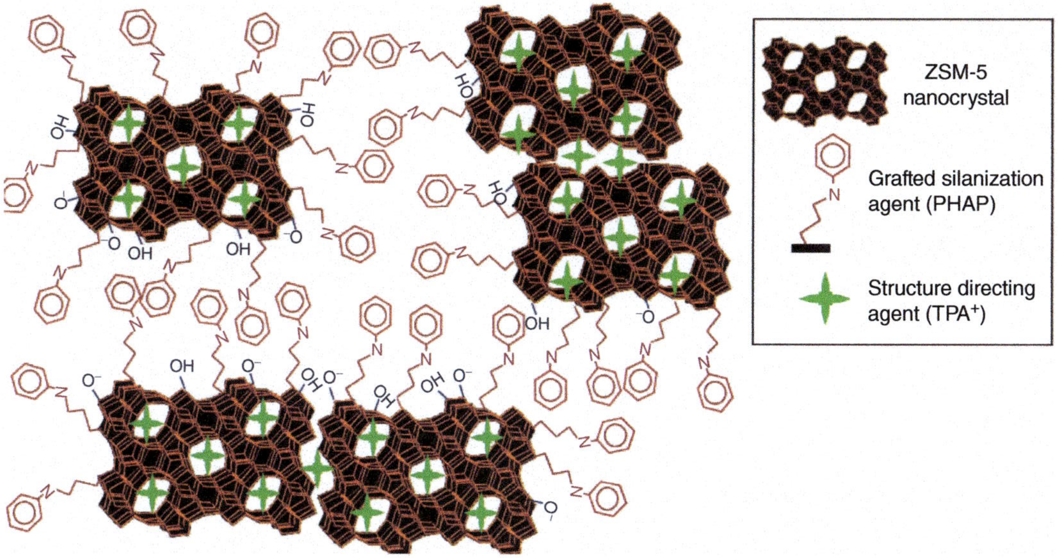

Figure 12.5 Schematic diagram showing the distributions of the TPA^+ zeolite-structure-directing and PHAPTMS surface-silanization species with respect to the interior nanopores and exterior surfaces of ZSM-5 zeolite nanocrystals. Source: Serrano et al. [71]. Reproduced with permission of ACS.

the concentration, and the organic moiety nature of the silanization agent, it is possible to tune to some extent the size of the nanocrystals and the intercrystalline mesopores [71, 72]. The addition of alcohols, such as methanol or 2-propanol in this approach could improve the zeolite textural properties, which was resulted from the strong interaction between the alkoxylated alcohol and the organosilane tail, leading to an enhancement in the hydrophobicity and stability of the protective layer [73].

Besides the seed-silanization method, a one-step approach for synthesis of mesoporous zeolites with the use of an amphiphilic organosilane as the mesopore template was also developed by Ryoo's group [74]. The required amphiphilic organosilanes are comprised of a hydrolyzable methoxysilyl moiety, a long-chain hydrophobic group, and a quaternary ammonium moiety functioning as the zeolite SDA. Taking 3-(trimethoxysilyl)propyl hexadecyl dimethyl ammonium chloride (TPHAC) as an example, the presence of silanol group, generated from the hydrolysis of methoxysilyl moiety, is favorable for a strong interaction with growing crystal domains through forming covalent bonds with other SiO_2 and Al_2O_3 sources, and the long-chain hydrophobic group could induce forming mesoscale micelles. The synthesized mesoporous ZSM-5 exhibited globular morphologies with rugged surfaces and a uniform mesopore size, which can also be finely tuned in the range of 2–8 nm by changing the chain length of the organosilane and the hydrothermal synthesis temperature. Moreover, the use of this synthetic strategy was also extended to synthesize other types of mesoporous zeolites, such as LTA, SOD, aluminophosphate, and silicoaluminophosphate zeolite analogs, via employing either amphiphilic organosilanes or alkylphosphonic acid [75–78]. All the previously mentioned mesoporous zeolites, obtained via this strategy, demonstrated compact morphology with uniform mesopores located inside the crystallites. However, hierarchical FAU-type zeolite X with house-of-cards-like nanosheet assemblies and hierarchical FAU/EMT (effective medium theory) zeolite with a small amount of EMT structure in the FAU-type nanosheets were also prepared by Schwieger and coworkers and Tsapatsis and coworkers, respectively, with the use of amphiphilic organosilane templates [79, 80]. In addition, the group of Pinnavaia used a silane-functionalized polyethylenimine as the mesoporogen for constructing intracrystalline mesopores within the ZSM-5 zeolite [81]. The addition of the silylated polymer to the zeolite precursor gel allowed it to be grafted on the zeolite surface through covalent Si–O–Si linkages in the nucleation stage. The grafted polymer became phase-segregated from the zeolite matrix as the zeolite crystal grew, constructing an intracrystal polymer network linked to the zeolite framework in the crystal. The ZSM-5 zeolite with uniform intracrystal mesopores in the range of 2–3 nm was obtained after calcination. It is noteworthy that mesoporous FAU zeolites can also be prepared via this strategy.

12.2.1.3 Non-templating Methods

Non-templating methods enable the fabrication of additional mesoporosity in zeolites without the use of mesopore templates. Several synthesis routes have been reported typifying these methods in the past decade, mainly being based on the

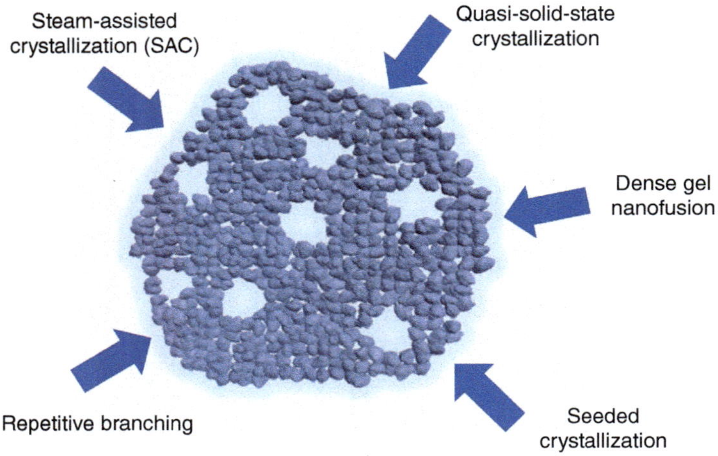

Figure 12.6 Synthetic routes to prepare mesoporous zeolites through nanozeolite assembly. Source: Adapted from Möller and Bein [10] with the permission of RSC.

aggregation of nanocrystals to generate intercrystalline mesopores and the controlled crystallization of amorphous gels into zeolite crystals with intracrystalline mesopores (see Figure 12.6). In these examples, only conventional zeolite micropore SDA was employed or even not required in some cases.

SAC can be used to synthesize mesoporous zeolites without using any mesopore templates [82–84]. For example, in the presence of TEAOH, a common zeolite SDA, mesopore zeolite beta was synthesized by Bein and coworkers through employing the SAC procedure [84]. This was achieved by a dense-gel synthesis to induce a burst of nucleation. The final self-sustaining zeolite beta aggregates were assembled from 20-nm crystalline domains, resulting in a mesoporous structure with dominant pore sizes of around 13 nm. The amount of water used for the steam treatment was revealed to be crucial to the zeolite growth kinetics.

A quasi-solid-state reaction, which was related to the SAC procedure, was used to synthesize hierarchically micro–meso–macroporous zeolites by Yang and coworkers, affording TS-1 aggregates with pores at three length scales [85]. Specifically, a macroporous amorphous titanosilicate, obtained from the spontaneous reaction of tetramethyl orthosilicate and titanium isopropoxide, was impregnated in a silica suspension containing the micropore SDA. The gel was dried and finally converted in a glycerol medium at 130 °C. The amorphous walls with around 1-μm macropores were transformed into 200 nm TS-1 nanoparticles with the interparticle mesopore size of 4.7 nm, providing a hierarchical TS-1 with almost twice the surface area (580 m^2/g) as TS-1 nanoparticles prepared conventionally.

A new concept of the synthetic strategy named "nanofusion" was raised by Möller et al. toward the preparation of mesopore zeolite beta [86]. A concentrated precursor gel containing TEAOH as the SDA was firstly hydrothermally converted to nano-sized beta crystals with sizes in the range of 20–40 nm, which was then instantly fused into stable zeolite aggregates after drying and calcination, showing tunable mesopore sizes from 15 to 35 nm. It was found that the dissolved aluminosilicate species in the gel enabled the process of nanozeolites fusion. The fused zeolite beta product exhibited a high micro- and mesopore volume as well as high surface area.

A strategy named "Repetitive Branching" was also used during one-step hydrothermal crystal growth by Tsapatsis's group to generate a new mesoporous MFI-type zeolite made of orthogonally connected microporous nanosheets [87]. The house-of-cards arrangement of the nanosheets formed a permanent network of 2–7-nm mesopores, along with the high external surface area and reduced micropore diffusion length, accounting for high reaction rates for bulky molecules. In addition, through a sequential intergrowth route, Okubo and coworkers constructed hierarchically organized MFI zeolites in the presence of an SDA, that is, $(C_3H_7)_3N^+$–$(CH_2)_5$–$N^+(C_3H_7)_3$ [88]. The use of this SDA, through causing framework distortion, was thought to be responsible for the formation of very thin zeolite plates with enhanced 90° rotational intergrowth, thereby creating the hierarchical structure with three classes of porosity.

Seed-directed synthesis (SDS) strategy was recently adopted by Xiao's group for the synthesis of mesoporous Al-rich MFI zeolite in the absence of organic template [89]. The MFI crystals were found to be assembled with rodlike nanocrystals showing almost the same aligned direction, and mesoporosity was generated from the space between nanocrystals. The analysis results indicate that the synthesized zeolite possessed abundant tetrahedral Al species in the zeolite framework, showing the Si/Al ratio as low as 9.5, which provided it with enormous potential for catalytic applications.

Some other examples were also reported by employing non-templating methods [90–93]. For instance, Fan and coworkers directly synthesized mesoporous ZSM-5 by controlling the nucleation, growth, and template-free self-assembly of zeolite precursors formed in the initial stage of the zeolite crystallization process [90]. MFI nanocrystal assembly coupled with a possible dissolution–crystallization mechanism accounts for the resulted structure with broadly distributed mesopores. Additionally, mesoporous SAPO-34 and FAU-type zeolites with intergrown nanosheet structure have been successfully prepared by utilizing the natural layered material kaolin and inorganic salts respectively [91, 92].

Besides nanozeolite assembly routes, mesoporous zeolite can be also synthesized in a template-free way with the assistance of high-pressure gas, such as CO_2 [94, 95]. In the presence of compressed CO_2, fast crystallization of mesoporous MFI zeolites without using any mesoscale organotemplate can be achieved, which was reported by Xiao and coworkers in 2010 [94]. This synthesis route required much shorter time for the crystallization of MFI zeolites compared with the conventional one. It was revealed that the presence of CO_2-in-water emulsions and foams accounted for the generation of mesoporosity in zeolite and the mesopore volume can be tuned by adjusting CO_2 pressure. The synthesized ZSM-5 zeolite was found to be uniform chainlike crystal morphology, while silicalite-1 showed monodispersed crystals. The route was further extended to the preparation of other zeolites including TS-1, beta, and Y.

Non-templating routes, in which mesoporous zeolites are prepared without using any mesoporogens, are therefore a favorable approach with respect to the cost. However, compared with the hard or soft templating methods, it is still considered as a less general approach due to the following factors: very limited choice in zeolite topologies, rigorous crystallization conditions involved, and a relatively low control on the mesopore size.

12.2.2 Top-down Approaches via Demetallization

A simple and direct approach to develop mesoporosity in zeolite may be through demetallization, namely, removing framework T atoms of the zeolite. Specifically, special dealumination or desilication leaching procedures can be carried out to generate amorphous areas in the zeolite framework, which could be further transformed to the mesopores upon extraction of the amorphous debris. Demetallization presents a destructive process intended to sacrifice part of the original micron-sized zeolite crystals for generating larger external surfaces in the form of mesopores.

12.2.2.1 Dealumination

Dealumination emerged in the beginning as a solution to prepare zeolites with high Si/Al ratios that were not viable by direct synthesis and was realized later to be a feasible way to generate mesoporosity. Steaming at high temperature and acid leaching are the most widespread techniques for dealumination. The extraction of aluminum atom is caused by the cleavage of the Si—O—Al bonds in zeolite through hydrolysis under certain conditions, leading to the formation of voids and a partial collapse of the zeolite structure. The voids constitute an additional porosity to the original microporous zeolite structure, being usually in the mesopore range. However, relying on the dealumination treatment used, severe changes in the acidic properties of the zeolite may occur during the process of expelling aluminum atoms, which should be taken into account in studies.

Steaming, a hydrothermal treatment, is generally carried out at temperature above 500 °C in the presence of steam. The fabrication of mesopores via steaming is widely considered to be highly dependent on the Al content and the stability of Al sites. Thus, zeolites with low pristine Si/Al are usually chosen as the substrate for steaming. Taking Y zeolite as the example, steaming process has been used for many years for dealuminating zeolite Y [96, 97]. The ultrastable zeolite Y with a higher Si/Al ratio (USY) generated after condensation of the defect sites is predominantly utilized as the cracking catalyst in fluidized catalytic cracking process. To gain a deeper understanding of the dealumination process of Y zeolite via steaming, Lamberti and coworkers investigated it with *in situ*, time-dependent, synchrotron radiation X-ray powder diffraction (XRPD) and *in situ* Al K-edge X-ray absorption spectroscopy (XAS) [98]. It was found that considerable structural collapse caused by steaming did not occur at the highest temperature; however, defects formed at much lower temperature when water was able to enter the pores again, leading to significant migration of framework Al^{3+} to extraframework positions. In order to avoid obstruction of the micropores, amorphous debris deposited on the mesopore surface or on the external surface of the treated zeolite crystals might be necessarily removed through the treatment of mild acids, such as diluted nitric acid, diluted hydrochloric acid, and oxalic acid [99].

In addition to zeolite Y, dealumination has also been applied to other zeolites, such as beta [100], mordenite [101], and ferrierite [102], most of which are leached by using more concentrated acids. In these cases, the desired mesopores are produced by direct attack of the acid rather than by hydrolysis from steam. Koningsberger and coworkers dealuminated the mordenite zeolite by using 1 M HCl, affording an acid-leached Pt/mordenite catalyst with mesopores, which showed an improvement in the activity of hydroisomerization, attributing to the better access to the acid sites and the shorter diffusion path length in zeolite [103].

There are still some other reports on the dealumination methods, involving in calcination and the use of silicon tetrachloride [104] or ammonium hexafluorosilicate [102]. Due to the extraction of aluminum atoms from the framework, the number of acid sites in the zeolite declines during the dealumination process. It should also be noted that the mesopore volume of the zeolite prepared by dealumination is largely based on isolated cavities rather than on interconnected mesopores, which cannot be beneficial to substantially enhance diffusion of large molecules for reactions.

12.2.2.2 Desilication

Desilication has become a versatile route to tailor zeolites with a mesoporous framework since the beginning of this century. Since the preferential removal of silica is the objective in this method, bases are utilized for the post-synthetic treatment of conventional micron-sized zeolite crystals. The procedure usually involves in treating zeolite sample with a 0.2 M NaOH solution under controlled conditions. Silica is preferentially leached from the crystal framework in the process, giving rise to the occurrence of mesopores at the expense of a decline in the Si/Al ratio of the zeolite.

Matsukata and coworkers were the first authors who realized that mesopores in ZSM-5 zeolite can be acquired from the alkaline treatments [105]. It was found that the alkaline treatment dissolved roughly 40% of the ZSM-5 crystals and led to a distinct increment of the mesopore volume in the zeolite, whereas the crystallinity of the zeolite was preserved well. It is worth noting that scanning electron microscope (SEM) micrographs provided an explicit evidence for the mesopore formation in ZSM-5 crystals. Additionally, the impact of desilication time and NaOH concentration on the ZSM-5 desilication was also studied [106]. In order to optimize the mesopore formation in MFI zeolites, a deep investigation of base treatment conditions was conducted by Groen et al. [107–109]. It was found that a molar Si/Al ratio in the range of 25–50 was identified as the optimal window for the pristine zeolite, leading to an optimal mesoporosity centered around 10 nm and mesopore surface areas of up to 235 m^2/g while preserving the intrinsic crystalline and acidic properties, as shown in Figure 12.7. Due to the repulsion between the negatively charged lattice and OH^- in ZSM-5 zeolite with lower Si/Al ratios, minor extraction of framework silicon and limited mesopore formation occurs, whereas at higher Si/Al non-selective and excessive extraction appears, affording substantial mesopore formation at the expense of a severe crystallinity loss. In addition to Si/Al ratio, the

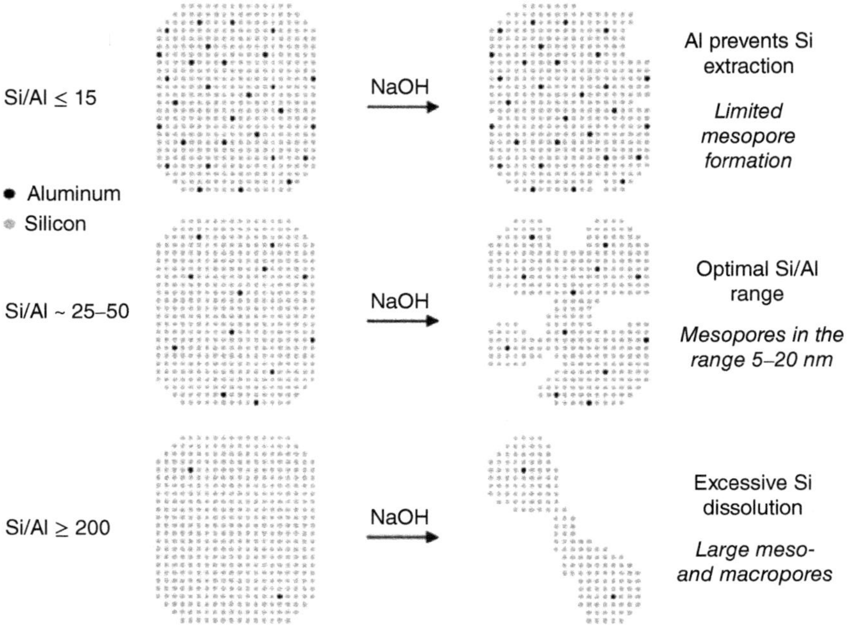

Figure 12.7 Simplified schematic representation of the influence of the Al content on the desilication treatment of MFI zeolites in NaOH solution and the associated mechanism of pore formation. Source: Groen et al. [107]. Reproduced with permission of ACS.

morphology of the parent zeolites, comprised of either small intergrown particles or large single crystals with a substantial external surface area, also gives a significant impact on the dissolution process in desilication [110]. Boundaries, intergrowths, and defects within each zeolite particle are much more susceptible to dissolution or etching.

Besides Al^{3+}, other trivalent cations, such as B^{3+}, Fe^{3+}, and Ga^{3+}, have also shown the possibility to control the pore size during the desilication procedure [111]. Thus, the term pore directing agent (PDA) was introduced for indicating to the role played by these trivalent cations. With the development of desilication methods, organic additives, such as tetrapropyl ammonium and tetrabutyl ammonium, were also found their function acting as the external PDA to regulate the intracrystalline mesoporosity during the base etching process. The specific interaction of the organic additives with the zeolite surface under alkaline conditions was found by the group of Pérez-Ramírez [112–114]. This interaction can provide an adjustable protection against zeolite dissolution and enable the creation of similar mesopore surface areas of the zeolite with a smaller mesopore size of around 5 nm and less damage to the original micropore volumes when compared with the one treated solely by an alkaline. The PDA was considered to play a key role in the pore formation process through its affinity to the zeolite surface. In addition, molecular criteria for the selection of organic PDAs in NaOH leaching were studied on USY and beta zeolites with Si/Al ratio in a broad range of 15–385 [115]. It was found that an efficient PDA tended to be positively charged and contain ca. 10–20 carbon atoms. The addition of PDAs restrained the realumination during desilication, yielding the desired mesoporous zeolite with Si/Al ratios similar to that of the parent zeolite. Moreover, controlled mesopore formation of all-silica zeolites can also be achieved with the use of external PDA. Besides PDAs, the organic template residues in zeolite also revealed their positive role in tailoring the mesoporosity for zeolite. For example, Pérez-Ramírez et al. demonstrated that hierarchical beta zeolites with tailored degree of mesoporosity can be achieved through desilication of partial detemplated zeolites in alkaline medium [116]. This method is based on the fact that the template-containing zeolite is virtually inert to silicon extraction upon treatment in aqueous NaOH solution. Partial removal of the template (SDA) provides regions in the crystal susceptible to the formation of mesopores by subsequent desilication, while template-containing regions are protected from silicon leaching. Additionally, de Jongh and coworkers reported that a base leaching treatment to organic template-containing zeolites gave rise to intercrystalline mesopore formation without loss of micropore volume and acidity [117].

The application of the strategy of base leaching could also be expanded to Al-rich zeolites. To achieve this goal, the Si/Al ratio of the Al-rich zeolite firstly needs to be increased to an optimal range through a dealumination step, which is beneficial to the mesopore introduction upon the subsequent base leaching process. For instance, crystalline mesoporous mordenite with Si/Al ratio of around 15 was synthesized by van Bokhoven and coworkers and de Jong and coworkers, respectively, via a sequential acid and base leaching route [118, 119]. Pérez-Ramírez's group reported the preparation of mesoporous clinoptilolite and L zeolites by using optimized tandem dealumination–desilication treatments [120]. The prior dealumination steps using aqueous HCl solution and ammonium hexafluorosilicate were applied to natural clinoptilolite and L zeolite, respectively. The subsequent desilication by NaOH treatment afforded both mesoporous zeolites with surface of up to fourfold relative to the starting zeolites. de Jong et al. conducted a base leaching treatment to a commercially available zeolite H–Y that had been previously steamed and acid-leached, leading to an unique intracrystalline trimodal porosity including micropores (~1 nm) from the zeolite synthesis, smaller spongelike mesopores (~3 nm) from desilication, and large mesopores (~30 nm) from steaming and acid leaching [121].

Like dealumination, desilication unavoidably modulates the Si/Al ratio of the zeolite framework. Nevertheless, the desilication results in a decrease in Si/Al ratio. Also, due to the existence of realumination, it is most likely to observe some extra-framework aluminum species after the base leaching, and introduction of an additional acid treatment or ion-exchange step thus seems necessary to get rid of these species for unblocking the micropores and mesopores.

12.2.3 Mixed Synthetic Approaches

To date, bottom-up and top-down synthetic methods are both mainstream strategies to fabricate mesoporous zeolites. However, there are still several reported works by using approaches that cannot be easily considered as either of the two main strategies mentioned earlier. These approaches usually involve the combination of both bottom-up and top-down methods in order to achieve a fine control of the mesoporosity and other influential properties in mesoporous zeolite preparation. The most representative example is the method for mesoporous zeolite preparation following a dissolution–recrystallization route, in which dissolution achieved through base treatment of zeolites is most often carried out in the presence of surfactants.

Local rearrangement of the zeolite framework during the process of zeolite recrystallization could enable forming zeolite crystals with ordered mesopores, whereas the composites of microporous zeolite and mesoporous oxide are usually obtained when the conditions for the dissolution process are not elaborately optimized [122]. Ivanova et al. carried out extensive research works in this subject [123–127]. Several conventional zeolites, including mordenite [123, 125, 127], beta [124], and ferrierite [126], were investigated to prepare mesoporous materials via the dissolution–recrystallization route in the presence of NaOH and cetyltrimethylammonium bromide. It was found that with the increase of base concentration, the volume of mesopores rose at the expense of the vanishment of zeolite micropores. Thus, it is possible to completely transform the microporous zeolite into ordered mesoporous MCM-41 material when an appropriately high NaOH concentration is employed in the process. In addition, it should be noted that the recrystallization of microporous mordenites in the presence of NaOH with a relatively low concentration could also yield a truly hierarchical mordenites featuring both micropores and mesopores in the structure.

Tsapatsis and coworkers synthesized ZSM-5 zeolite crystals with secondary mesopores by alkaline desilication and surfactant-induced reassembly of dissolved species originating from the pristine ZSM-5 crystals [128]. The synthesized mesoporous ZSM-5 product exhibited a dual-mesopore size distribution, well-preserved crystallinity and acidity, and enhanced external surface areas. The textures of the products revealed that desilication and reassembly occurred simultaneously to generate a dual-pore system. The desilication process created large mesopores (10–30 nm) in the zeolite crystals, while reassembly of the dissolved species, consisting of silicates, aluminosilicates, and fragments of the ZSM-5 crystals, by micellization of a surfactant produced smaller mesopores (~3 nm). In addition, mesoporous zeolite Y with uniformly distributed intracrystalline mesopores of around 4 nm was prepared by García-Martínez et al. through a hydrothermal treatment of the parent zeolite Y with a mixture of diluted NH_4OH and cetyltrimethylammonium bromide surfactant at 150 °C [129]. The mesopores can be tuned in the range of 2.5–4.5 nm by varying the chain length of the surfactants. The introduction of controlled mesoporosity into crystalline zeolites proceeded presumably via a surfactant-assisted crystal rearrangement mechanism, as shown in Figure 12.8. Some of the Si–O–Si linkages were first cleaved in basic reaction conditions to provide some flexibility in the crystalline structure and resulted in the creation of negatively charged sites in the zeolite framework that attracted cationic surfactants. Electrostatic interactions between the negatively charged sites and the positively charged surfactants, and the formation of micelles within the zeolite

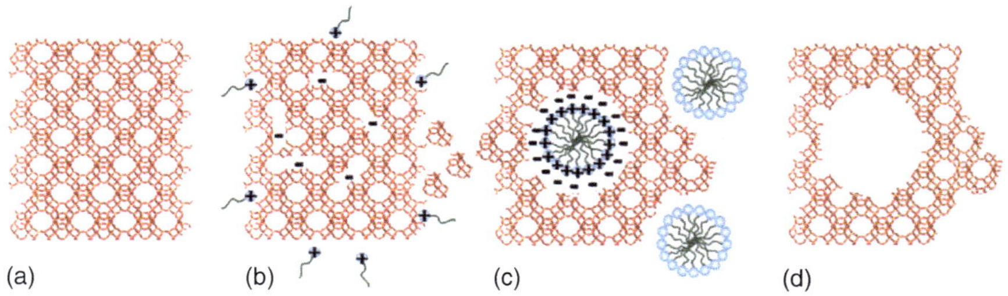

Figure 12.8 Schematic of the speculated zeolite mesopore formation process: (a) original zeolite Y, (b) Si—O—Si bond opening/reconstruction in basic media, (c) crystal rearrangement to accommodate the surfactant micelles, and (d) removal of the template to expose the mesoporosity introduced. Source: García-Martínez et al. [129]. Reproduced with permission of RSC.

crystals through self-assembly of surfactant cations, led to the rearrangement of the crystal structure to generate mesopores around the micelles. This process allowed an almost complete recovery of the zeolite Y through effectively preventing the dissolution of the zeolite crystals.

12.3 Catalytic Application of Mesoporous Zeolites

In catalytic reactions, the presence of additional mesopores in zeolites, as a consequence of the employed synthetic strategy mentioned earlier, typically accounts for the often observed enhancement in activity, stability, and longer catalyst lifetime. Over the years, mesoporous zeolites have been employed in various classes of reactions, such as hydrogenation, oxidation, alkylation, hydrodeoxygenation (HDO), isomerization, and combustion. These zeolites have also shown their superior catalytic performance in both fuel chemistry and biomass valorization fields.

12.3.1 Fuel Chemistry

Zeolite catalysis has long played a crucial role in petroleum refining and petrochemical industry, with reaction processes such as catalytic cracking, Fischer–Tropsch (FT) chemistry, and methanol to fuels production. Owing to the current strong demands to improve the utilization of petroleum and fossil-derived resources and environmental concerns for cleaner transportation fuels, mesoporous zeolite-based materials have recently emerged as a type of efficient catalysts to address these significant challenges.

12.3.1.1 Catalytic Cracking

Catalytic cracking of crude oil plays a vital role in the large-scale production of commodity compounds. Zeolites are the catalyst of choice although they are far from ideal catalysts with respect to their microporous nature suppressing mass transfer of bulky compounds, and problems resulting from pore blockage and coking.

Introduction of a secondary mesopore network to microporous zeolites can improve catalyst performance in cracking reactions. Conspicuous examples of such zeolite materials are presented herein.

Mesoporous zeolite Y have been widely applied in several catalytic cracking processes, which include *n*-paraffins [130], aromatics [131–133], heavy oils [134–136], vacuum gas oil [137], and so on. For instance, Shen and coworkers used mesoporous Y zeolites in the cracking reaction of cumene and 1,3,5-triisopropylbenzene [131]. In this study, mesoporous zeolite Y prepared by sequential NaOH desilication and ammonium hexafluorosilicate dealumination was compared with the one prepared via ammonium hexafluorosilicate dealumination only. It was found that Y zeolite obtained by sequential desilication and dealumination exhibited significantly better dealumination uniformity and aluminum–silicon distribution, affording higher activity and stability in the cracking cumene and 1,3,5-triisopropylbenzene, which resulted from the improved accessibility of the active acid sites and the reduced diffusional path lengths of the zeolite. Hierarchical Y zeolites featuring microporous zeolite in center and mesoporous silica–alumina in outer shell were prepared by Ishihara et al. and employed in the catalytic cracking of *n*-dodecane [130]. The Y zeolite with mesopores showed higher activity and selectivity for gasoline production compared with the microporous Y zeolite. It was revealed that the existence of mesopore and appropriate amount of acid sites in the matrix contributed to the efficient generation of the branched alkane products (Scheme 12.1). Hensen and coworkers used mesoporous Y zeolites as face-centered cubic (FCC) catalysts in the catalytic cracking of vacuum gas oil, providing an excellent catalytic

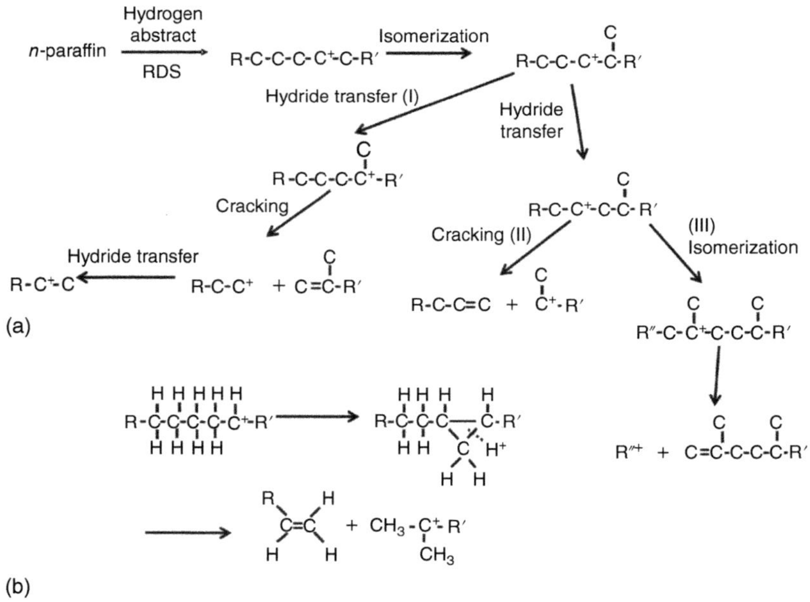

Scheme 12.1 Reaction routes of carbenium ion in *n*-paraffin catalytic cracking. Source: Ishihara et al. [130]. Reproduced with permission of Elsevier.

performance [137]. The catalytic data indicate that the relatively weak acid sites were responsible for the FCC activity and the well-embedded mesoporosity in the pristine zeolite crystals resulted in higher diesel and lower gasoline yield.

In addition to Y zeolites, mesoporous ZSM-5 zeolites have also received substantial attention in the application of catalytic cracking reactions [138–148]. For example, the group of Wu synthesized a type of mesoporous ZSM-5 zeolites with unique hydrocarbon cracking properties using a small molecular organic amine-assisted desilication technique [141]. Introduction of controllable mesopores into as-synthesized ZSM-5 crystals was achieved by alkaline-derived desilication with the addition of piperidine or hexamethyleneimine due to the fact that the organic amines could protect the zeolite crystals from severe dissolving by NaOH attacking. The mesoporous ZSM-5 zeolites showed much superior catalytic performances in the cracking of *n*-hexane to the parent ZSM-5, providing a higher propylene yield and a longer lifetime. Hierarchical ZSM-5 zeolite aggregates with controllable intercrystalline mesopores were also employed in the catalytic cracking of hexane, exhibiting similar light olefin selectivity and remarkably improved lifetime compared with the highly dispersed ZSM-5 zeolite nanocrystals [142]. It was also reported that a phosphorus modification to mesoporous ZSM-5 zeolites resulted in a catalyst with a great property of substantially restraining carbon deposition, thereby maintaining the catalytic activity and stability at a high level in *n*-hexane cracking to light alkenes [145]. In addition, hierarchical ZSM-5 zeolite fibers with macro–meso–microporosity were fabricated by coaxial electrospinning [139]. The synergistic integration of the suitable acidity and the hierarchical porosity of the prepared ZSM-5 hollow fibers gave rise to a high yield of propylene and excellent anti-coking stability in the catalytic cracking process of iso-butane. Hierarchical ZSM-5 composites with inner ZSM-5 micropores and large outer mesopores, such as ZSM-5/SBA-15 and ZSM-5/MCM-41, were also investigated in the catalytic cracking of *n*-dodecane, exhibiting superior activity and selectivity for gasoline fraction [149, 150]. In the case of ZSM-5/MCM-41, it should be noted that a proper catalyst treatment with NaCl solution under ultrasonic conditions is capable of removing undesirable microporous and mesoporous acid sites, which could avoid annoying coke deposition on these sites and therefore significantly enhance the stability of catalyst [150].

Other mesoporous zeolites such as USY [151], ZSM-48 [152], MTT [153], BEA [154], and IM-5 [155] also found their roles in a variety of catalytic cracking reactions. Taking USY as the example, Corma and coworkers reported a strategy of post-synthetic modifications to USY zeolites aimed at introducing intracrystalline mesopores and stabilizing zeolite framework [151]. When compared with the conventional commercial USY-based catalyst, the stabilized mesoporous USY zeolites showed improved hydrothermal stability, increased bottoms conversion capacity, and improved product selectivity, e.g. higher diesels yield, less gases generated, and higher olefinicity within the liquefied petroleum gas (LPG) fraction. It was found that the combination of appropriate Brønsted acidity and shorted diffusion pathway in the zeolite crystals accounted for the obtained selectivity of products. Recently, the incorporation of metals into the mesoporous zeolite framework, such as Ni/Co-BEA [154] and Ni-MFI [156], has also emerged as an new strategy to improve the catalytic activity in the cracking reactions. The metal introduction is considered to give rise to the acidity variation of the zeolite, which accordingly leads to the enhancement of catalytic activity.

12.3.1.2 Fischer–Tropsch Reaction

FT synthesis, referring to the conversion of syngas (H_2 and CO) to hydrocarbons, is considered to be a recognized route to produce liquid transportation fuels [157]. A range of products can be produced during the FT reaction. The product selectivity is usually reported on the basis of hydrocarbon chain length, which is an important indicator for evaluating the efficiency of the catalysts. In addition, a secondary process, that is, the water gas shift reaction (WGSR), can occur and yield CO_2 and H_2 from the reaction of *in situ* generated water and CO.

To date, mesoporous zeolites such as ZSM-5, beta, and Y-supported Co or Ru, have been considerably reported to be efficient FT catalysts due to the unique mesoporous structure and acidity of the zeolites [158–169]. Ru/meso-ZSM-5 catalyst was prepared and first employed in the FT reaction by Wang and coworkers [158]. This catalyst exhibited a high selectivity for the production of C_5–C_{11} isoparaffins from syngas. The selectivity reached around 80% with a ratio of isoparaffins to *n*-paraffins of about 2.7, which was significantly higher than the maximum expected from Anderson–Schulz–Flory distribution (~45%). It was further revealed that the hierarchical porous structure and the relatively weak acidity, mainly Lewis acidity, contributed to the enhancement of selectivity to C_5–C_{11} and the decrease in selectivity to light hydrocarbons (CH_4 and C_5–C_{11} alkanes). Later on the authors further extended the mesoporous zeolite support to beta and Y zeolites, which led to Ru/meso-beta and Co/Na-meso-Y catalyst, respectively, affording excellent selectivity to the specific length of alkanes in FT reactions [159, 166]. Ryoo and coworkers used mesoporous MFI zeolite nanosponge supporting Co nanoparticles, Co/NS-MFI, as the FT catalyst in the syngas conversion [162]. The mesoporous MFI nanosponge was synthesized by a seed-assisted hydrothermal preparation route with the use of $C_{22}H_{45}$–$N^+(CH_3)_2$–C_6H_{12}–$N^+(CH_3)_2$–C_6H_{13} as the SDA, possessing disordered network of 2.5-nm-thick MFI zeolite nanolayers with a narrow distribution of mesopore sizes of around 4 nm. The Co/NS-MFI zeolite presented high catalytic conversion of CO to hydrocarbons and long catalytic lifetime in the FT synthesis due to the excellent stability of the Co nanoparticles against sintering (Figure 12.9a). When compared with conventional alumina-based catalysts, the Co/NS-MFI catalyst demonstrated higher selectivity for branched hydrocarbons in the gasoline range, achieving 74% gasoline selectivity at 82% CO conversion (Figure 12.9b), which was ascribed to the short diffusion path lengths in the extremely thin zeolite frameworks for branched hydrocarbons during hydroisomerization reactions. Recently, *in situ* crystal Co/SiO_2 transformation to a mesoporous Co@NaZSM-5 catalyst with Co particles encapsulated in the ZSM-5 via a solvent-free process was reported [168]. The catalyst afforded an efficient catalysis performance in the FT synthesis for the production of gasoline-range products, showing a selectivity of 68%. Besides Ru- and Co-based FT catalysts, Fe-based catalyst with mesoporous

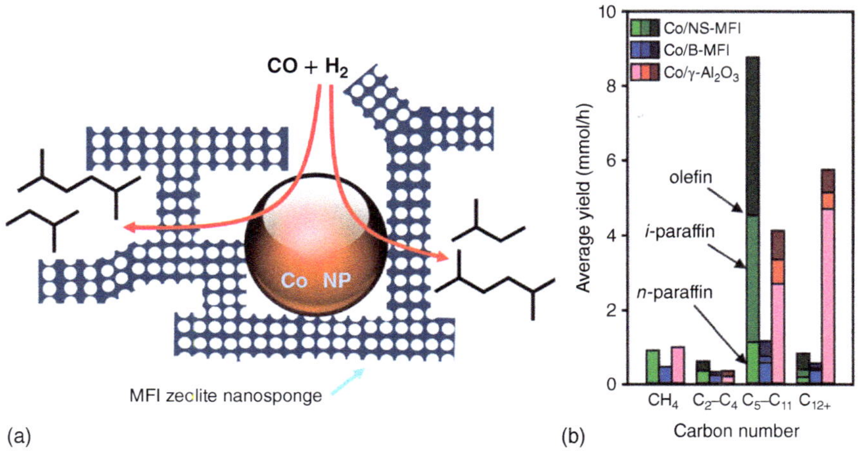

Figure 12.9 (a) Schematic diagram presenting FT reaction over Co/NS-MFI; (b) product selectivity of Co/NS-MFI, Co/B-MFI, and Co/γ-Al_2O_3 catalysts, which were averaged over 100 h of the FT reaction time. Source: Kim et al. [162]. Reproduced with permission of ACS.

ZSM-5 as the support, denoted as Fe-meso-ZSM-5, showed its potential in syngas conversion to aromatics [170]. It was found that the mesoporous zeolites with high acidity could improve the selectivity to aromatics in the hydrocarbon products. The maximum aromatics selectivity of 23% in hydrocarbon could be obtained by using a low Si/Al ratio zeolite catalyst under low H_2/CO syngas ($H_2/CO = 1$).

12.3.1.3 Methanol to Gasoline or Olefins

The processes of converting methanol to either gasoline, olefins, or aromatics are considered to afford promising alternative routes for the generation of hydrocarbons, such as gasoline, ethylene, propylene, and aromatics, from nonpetroleum sources (natural gas, coal, and biomass) via intermediate products including methanol and syngas. A vast number of products with a sequence from light olefins–paraffins–higher olefins–aromatics and naphthenes to final coke can be resulted from these complex reactions. The mechanism to explain deactivation and coke generation during the reaction is still under debate although there have been some intriguing observations. For example, the formation of naphthene-derived compounds was observed during the reaction of methanol and various methyl benzenes in the presence of beta zeolite [171]. The use of different zeolites can lead to varied rate of coke formation, in which the topology, acid density, and strength are believed to play important roles. Considering the fact that rapid deactivation of the catalyst may occur due to the coke formation, the use of mesoporous zeolites in methanol conversion has been attempted in order to suppress the catalyst deactivation.

For the conversion of methanol to gasoline, ZSM-5 zeolites with mesopore channels as the decent catalysts have been extensively studied [172–178]. Early investigation by Svelle and coworkers in 2008 was conducted on the alkaline leached ZSM-5 zeolite for converting methanol to gasoline [172]. After treatment by 0.2 M NaOH, the total conversion capacity of the zeolite, given in grams of methanol converted per gram catalyst until complete deactivation, increased by a factor of 3.3, while selectivity toward the gasoline fraction improved by a factor of 1.7. The Si/Al ratio of the catalyst decreased as a result of the alkaline treatment, the total acidity determined by NH_3-TPD (temperature programmed desorption), however, was found to be unchanged. Zhang and coworkers synthesized mesoporous ZSM-5 zeolites with intercrystalline mesopores and evaluated their performance in catalytic transformation of methanol to gasoline by using a fixed-bed reactor, which was operated at 350 °C with ca. 16.5% methanol in nitrogen at a total pressure of 1.1 MPa [173]. The mesoporous ZSM-5 zeolites demonstrated almost 100% conversion of methanol and 59% selectivity of the hydrocarbons within the gasoline range. The study further indicated that after 24 h reaction the coke formation in the case of using mesoporous ZSM-5 as the catalyst reduced by nearly 92% compared with the case with the conventional microporous ZSM-5 zeolites. Recently, extensive studies with respect to the effect of synthesis conditions for mesoporous ZSM-5 zeolites on the catalytic performance in methanol-to-hydrocarbons conversion were carried out by Hensen and coworkers [174, 175]. It was found that the catalytic performance of the mesoporous zeolites in the reaction can be altered by using different bases, such as LiOH, NaOH, KOH, RbOH, and CsOH, for the alkaline posttreatment or by employing different organic amines, such as diethylamine, *n*-propylamine, 1,4-diaminobutane, and 1,6-diaminohexane, as the SDA for zeolite preparation. The obtained results further revealed that ZSM-5 zeolites with a large mesopore volume or few silanol groups or few external Brønsted acid sites can lead to a low rate of deactivation in methanol conversion.

In order to efficiently produce olefins, mainly propylene, from methanol, a variety of zeolites with mesopore structure, including SSZ-13, SAPO-34, beta, ZSM-5, ZSM-11, and ZSM-58, have been widely investigated as the catalysts in the reaction [179–192]. For instance, hierarchical SSZ-13 and SAPO-34 zeolites were prepared with the use of a surfactant mesoporogen and studied by Hensen and coworkers in the methanol-to-olefins (MTO) reaction [179, 180]. The synthesized SSZ-13 zeolites possessed trimodal porosity consisting of native micropores of the CHA framework, larger micropores, and mesopores, which attributed to a substantial improvement of mass transport inside the zeolite crystals. The mesoporous SSZ-13 zeolites exhibited a lower rate of deactivation in the MTO reaction when compared with the bulk SSZ-13 zeolite. Confocal fluorescence microscopy indicated that aromatic species that gave rise to deactivation were evenly deposited throughout the crystals of such mesoporous SSZ-13 zeolites, implying the fact that micropore space was well utilized. In contrast to SSZ-13, SAPO-34 zeolites, however, showed a different catalytic performance in this reaction. The addition of intracrystalline mesopores to SAPO-34 could not lower the deactivation rate. The capacity of mesoporous and microporous SAPO-34 for methanol conversion was found to be comparable. Mesoporous SAPO-34 zeolites have been successfully prepared by Yu's group using a facile and green route via recycling of waste mother liquors [181]. After three recycles of mother liquors, the prepared catalysts remained a high MTO catalytic performance, which was comparable to the initial catalyst, showing the lifetime of up to eight hours and the combined selectivities to ethylene and propylene of above 83%. In addition, Xiao and coworkers recently investigated the catalytic application of mesoporous ZSM-5 zeolites with large crystalline particles, which were prepared via a solvent-free and mesoporogen-free route, in the MTO reaction [182]. Catalytic tests showed that these S-ZSM-5 zeolites provided superior catalytic performances, which included extraordinarily long lifetime and excellent propylene selectivity in the reaction. For instance, in the case of employing the crystallization time of 30 h and the Si/Al ratio of 150 in the starting solid mixture, the resultant HS-ZSM-5-30h-150 catalyst exhibited large particle sizes in the range of 10–20 μm and afforded propylene selectivity as high as 50% as well as catalyst lifetime up to nine hours (Figure 12.10a,b). It is worth noting that the mesopore sizes in the zeolites can be tuned by controlling the crystallization time. With the increase of crystallization time from 6 to 72 h, the mesopore size gradually largened from 5 to 30 nm. The existence of mesopores in these zeolites was also verified by high-resolution transmission electron microscope (TEM) image, as shown in Figure 12.10c. The tunable features of mesoporosity and the sustainability of the synthesis processes could make this type of zeolites potentially important in the catalytic applications in the future.

Various aromatic compounds can also be obtained in the conversion of methanol. ZSM-5 zeolites seem to be the most studied catalysts for this process [193–198]. For instance, mesoporous ZSM-5 zeolites, synthesized by using natural kaolin clay as the silica and aluminum resource and saccharide source as the secondary mesopore agent, was studied in the conversion of methanol to aromatic hydrocarbons [195]. The reaction was carried out in a fixed bed reactor, which provided superior methanol conversion and aromatics selectivity showing 90% over 100 h and 85%, respectively. Mesoporous ZSM-5 zeolites modified by ZnO were also tested by Wang and coworkers for the methanol conversion to aromatics [196]. It was clarified that the mesopore structure, acidity, and ZnO dispersion played crucial roles in the formation of aromatics, which mainly constituted with benzene, toluene, and xylenes (BTX). The study indicated that the higher density of Brønsted acidity favored the production of BTX at the initial stage whereas seemed unfriendly to the stability of the catalyst. The enhancement in pore hierarchy improved the coke tolerance through restraining coke deposition in the micropores, thereby increased the catalyst stability in the reaction. ZnO, in particular the highly dispersed ZnO clusters, was supposed to improve the selectivity for aromatics via facilitating the dehydrogenation pathway.

12.3.2 Selective Organic Reactions

Selective conversion of organic compounds to the desired molecule products are critical to both the fine chemical and pharmaceutical industries. Implementation of zeolites possessing hierarchical micro- and mesoporous networks is able to improve internal mass and heat transfer as well as accessibility of the active sites in a number of catalytic organic transformations.

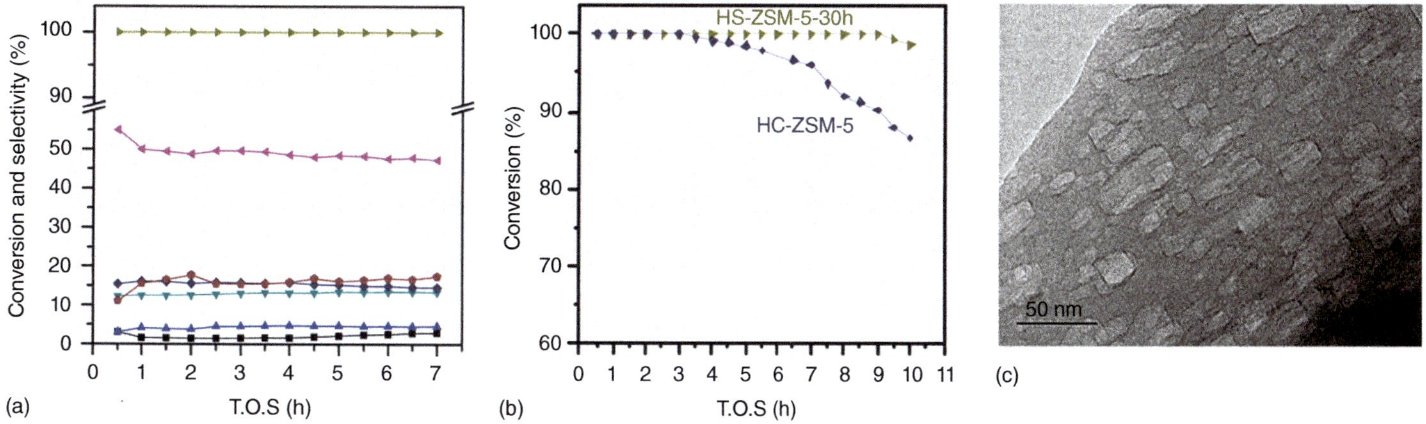

Figure 12.10 (a) Dependences of catalytic conversion and product selectivities on reaction time in MTO over HS-ZSM-5-30h; (b) lifetime of HC-ZSM-5 and HS-ZSM-5-30h in MTO reaction; (c) HR-TEM image of the S-ZSM-5 zeolite with crystallization time of 30 h. Source: Zhang et al. [182] / with permission from American Chemical Society.

12.3.2.1 Hydrogenation

In the presence of metal decorated mesoporous zeolites, a variety of aromatics, such as naphthalene [199–201], phenanthrene [202, 203], pyrene [199, 204], dibenzothiophene [203, 205], and nitroarenes [206], have been chosen for the selective hydrogenation reactions. Xiao and Tang and coworkers carried out catalytic hydrogenation studies on aromatics, including naphthalene, pyrene, and phenanthrene, with the use of mesoporous zeolite-supported metal catalysts [199, 202–204]. Pd/beta-H, a catalyst with Pd metal supported on the mesoporous beta zeolite, was found to exhibit better sulfur tolerance in the hydrogenation of naphthalene and pyrene when compared with Pd/Al-MCM-41 catalyst. The study further revealed that the differences in the catalytic performance from these two catalysts were attributed to the discrepancy of the acidity of the two zeolite supports. Moreover, metal sulfide (NiMoS) or metal phosphide (Ni_2P) supported on mesoporous ZSM-5 zeolites also showed decent catalytic performance for converting phenanthrene to the deep hydrogenated products [202, 203]. These tested catalysts also exhibited good catalytic performance in the hydrodesulfurization of 4,6-dimethyldibenzothiophene. It is noteworthy that mesoporous Y zeolite-supported Pd catalyst, Pd/HY-M, provided much higher activity, achieving over 97% conversion, in the hydrodesulfurization reaction as compared with Al_2O_3, mesoporous beta, and ZSM-5-supported Pd catalysts [205]. The superior activity of Pd/HY-M was ascribed to the larger micropore size of Y zeolite than that of beta and ZSM-5.

Shape-selective hydrogenation of nitroarenes was investigated by Chen and coworkers by using Pd@mnc-S1 catalyst [206]. The catalyst was elaborately designed with Pd nanoparticles encapsulated inside mesoporous MFI zeolite nanocrystals. The catalyst presented high stability and general shape selectivity in the hydrogenation of nitroarenes due to its unique nanostructure and porosity. For example, in the presence of Pd@mnc-S1 catalyst, the hydrogenation of nitrobenzene proceeded smoothly to yield aniline with 94% yield in five minutes, while only a negligible amount of 1-nitronaphthalene, a relatively bulky nitroarene, was transformed into 1-naphthylamine (Figure 12.11). The study further revealed that

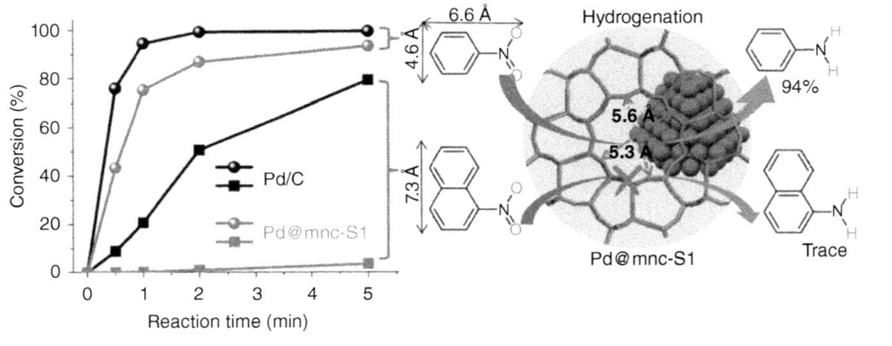

Figure 12.11 Hydrogenation reactions; reaction conditions: 0.1 mmol nitrobenzene (spheres)/1-nitronaphthalene (cubes), 20 mg of Pd@mnc-S1 (gray) or 7 mg of 5 wt.% Pd/C (black), 0.2 mmol $NaBH_4$, 5 ml of H_2O, room temperature, five minutes. Source: Cui et al. [206]. Reproduced with permission of John Wiley & Sons, Ltd.

the mesoporous structure and nanosized crystals of the catalyst support contributed to facilitating mass transfer to Pd nanoparticles and therefore improved its catalytic performance.

12.3.2.2 Oxidation

Mesoporous zeolites with or without metal modification can be utilized as the catalysts for the catalytic oxidation of various industrial important chemicals, such as alcohols [206, 207], cyclic olefins [207–210], alkylbenzenes [211, 212], thiols [213], and so on.

Mesoporous ZSM-5 zeolites were directly employed by García Martínez and coworkers for the selective oxidation of benzyl alcohol to benzaldehyde [207]. In the presence of hydrogen peroxide (H_2O_2), these metal-free zeolites substantially promoted the process of benzyl alcohol oxidation under microwave irradiation, affording benzaldehyde product with a yield of up to 45% after five minutes of reaction. A plausible mechanism for this oxidation was proposed, as shown in Scheme 12.2. Terminal and well-exposed Al–OH group in the zeolite framework could react with H_2O_2 to form highly reactive aluminum hydroperoxide surface species, which can then oxidize benzyl alcohol to form benzaldehyde. Moreover, the study found that cyclohexene can be also selectively oxidized to the corresponding cyclohexanediol with the use of these mesoporous ZSM-5 zeolites. Shape-selective oxidation of bulky cyclic olefins like cyclododecene were also investigated by Park and coworkers using mesoporous TS-1 zeolites, which were synthesized via an approach using microwave-assisted posttreatment of conventional TS-1 with H_2O_2 [208]. Mesopores generated through H_2O_2 treatment coupled with microwave irradiation were considered to account for this shape-selective performance. In addition, a catalyst with Pd nanoparticles encapsulated inside mesoporous S-1 zeolite was reported to provide an excellent catalytic performance for the aerobic oxidation of benzyl alcohol, affording product benzaldehyde with a yield of up to 94% [206]. Alkylbenzenes including toluene and ethyl benzene can also be selectively oxidized to the corresponding benzaldehyde and acetophone by using mesoporous MFI-based zeolites [211, 212]. A self-assembled mesoporous Mn-ZSM-5 catalyst was found its potential for the conversion of benzenethiol to diphenyldithiol [213]. The zeolitic catalyst possessed good pore-wall stability, high Brunauer-Emmett-Teller (BET) surface areas, and reactive Mn^{4+} sites, which were

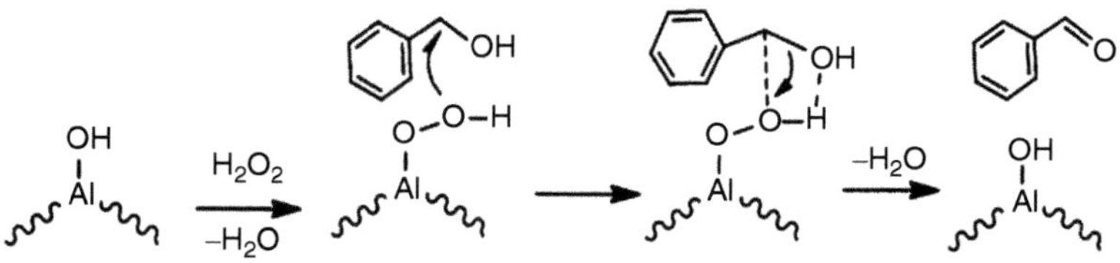

Scheme 12.2 Proposed peroxide species from Al–OH terminal groups involved in selective benzyl alcohol oxidation. Source: Ojeda et al. [207]. Reproduced with permission of John Wiley & Sons, Ltd.

believed to facilitate the cleavage of sulfur–hydrogen bond and form the disulfide bond from thiols in air.

12.3.2.3 Friedel–Crafts Alkylation

The Friedel–Crafts alkylation is an important class of reactions in fine chemical industries and petrochemistry. Alkylation of aromatics by zeolite-based catalysts is a frequently studied reaction in the liquid phase. The introduction of a secondary (meso-)porosity into the zeolites is expected to improve the catalytic performance considering the presence of bulky reaction intermediates and alkylated products. Compared with the parent zeolites, hierarchical zeolites usually exhibit enhanced catalytic performance, which is attributed to the existence of mesopores, thereby facilitating the transport of substrate to the active sites.

The alkylation of benzene with olefin was evaluated with the use of mesoporous zeolites. For example, mesoporous MCM-49 zeolites were used by Xu and coworkers in the catalytic alkylation of benzene with ethylene [214]. The zeolites were prepared by a posttreatment of conventional MCM-49 zeolite in CTAB and NaOH mixed solution. In the alkylation reaction, the zeolite with proper intracrystalline mesopores demonstrated better reaction stability than the conventional MCM-49 zeolite while providing comparable selectivity of ethylbenzene. The alkylation of benzene with 1-dodecene was also investigated in the presence of mesoporous beta zeolites [215]. The study revealed that the synergistic interaction of the regular mesopore structure and the relatively mild acidity of beta zeolites played a crucial role in promoting the alkylation reaction.

The benzylation of aromatic compounds is of importance for the production of diphenylmethane and substituted diphenylmethanes, which are key industrial building blocks for production of pharmaceutical and fine chemicals. A variety of mesoporous zeolites, such as ZSM-5 [216–218], Y [219], beta [220, 221], and FAU zeolite [222], have been applied in the benzylation of aromatics. For example, a series of mesoporous ZSM-5 zeolites synthesized using polyethylene glycol as the mesoporous soft template were employed by Xu et al. in the alkylation of phenol with *t*-butanol [218]. When compared with the pure microporous ZSM-5 zeolite, the prepared mesoporous ZSM-5 zeolites with high overall surface area and large pore size as well as strong acidity exhibited higher activity and selectivity in the alkylation of phenol with *t*-butanol, achieving up to 97% phenol conversion and 54% selectivity to 2,4-di-*tert*-butylphenol. Ma and coworkers synthesized mesoporous beta zeolites using organic functionalized fumed silica as the silicon source and applied them in the catalytic benzylation of naphthalene with benzyl chloride [221]. An infrared spectroscopy study revealed that increasing the mesoporosity of beta zeolites could not only enhance the diffusion ability of the reactants and products but also improve the accessibility of acidic sites, which could substantially increase the activity for the benzylation reaction. In addition, mesoporous beta zeolites with bulky particle size and strong acidity were employed by Tang and coworkers in the direct *N*-alkylation of sulfonamide with alcohols recently [220]. Owing to the variation in acidity, the mesoporous beta zeolite with stronger acidity gave an initial turnover frequency of 11×10^{-2}/s, which was higher than that from

H-form mordenite nanosheets and montmorillonite, showing 3.3×10^{-2}/s and 4.0×10^{-2}/s, respectively. A number of substituted benzhydrols and sulfonamides with different substituent were examined in the catalytic *N*-alkylation reactions over the mesoporous beta catalyst, giving the desired *N*-substituted sulfonamide products with overall yield above 90%.

12.3.2.4 Isomerization

Catalytic isomerization of hydrocarbon substrates into value-added chemicals is always highly desirable in the petrochemical industry. Zeolites with mesopore structure are the catalysts of choice for this transformation. Somorjai and coworkers investigated several types of aluminosilicate catalyst for the isomerization of *n*-hexane, both with and without Pt nanoparticles [223]. The catalysts studied included MCF-17 mesoporous silica, an aluminum-modified MCF-17 (Al-MCF-17/Pt), mesoporous MFI, and BEA-type zeolites. Al-MCF-17 catalyst was found to provide the best catalytic performance, showing the highest isomer production, which was due to the mild Lewis acid and Brønsted acid sites supplied by the aluminum modification. The study indicated that Al-MCF-17/Pt catalyst exhibited higher activity and selectivity in the reaction than the other zeolitic counterparts. The isomer selectivity for each catalyst with the same Si/Al ratio is shown in Figure 12.12a,b. For the Al-MCF-17/Pt catalyst, the Pt loading played an important role in activity and isomer selectivity (Figure 12.12c,d). If the Pt loading was too low, the activity was too low to react to the adsorbed reaction intermediates. If the Pt loading was too high, catalysis dominated the reaction process and selectivity came to harm. Al-MCF-17 with a Si/Al ratio of 8 as well as loaded with 0.5 wt.% Pt was found to be the optimum catalyst for isomerization of *n*-hexane. Nano-ferrierite zeolite aggregates with mesopores were studied in the 1-butene skeletal isomerization reaction by Xu and coworkers [224]. It was found that low crystallization temperature and high alkalinity were beneficial to generate nano-ferrierite zeolite aggregates with high hierarchical porosity. Compared to the commercial ferrierite zeolite, mesoporous nano-ferrierite aggregates afforded better catalytic stability and selectivity for the desired isobutene product in the 1-butene isomerization reaction. Additionally, mesoporous ZSM-5 single zeolite crystals modified with Ni-nanoparticles were also reported for the isomerization of *n*-octane, showing higher activity than the catalyst based on the conventional ZSM-5 zeolite [156].

The isomerization of aromatic hydrocarbons like alkylbenzenes over mesoporous zeolite catalysts has also been extensively studied [225–230]. Xiao and coworkers prepared chainlike mesoporous ZSM-5 crystals with tailorable *b*-axis length and tested them in *m*-xylene isomerization [225]. The catalytic results indicated that ZSM-5 crystals with long *b*-axis length and high volume of mesopores exhibited relatively high *m*-xylene conversion and *p*-xylene selectivity. Ryoo and coworkers developed a mesoporous EU-1 zeolite with a large volume of intercrystalline mesopores for the industrial hydroisomerization of ethylbenzene to *p*-xylene [226]. The catalytic results indicated that the impact of a large exposed surface area and a high amount of readily accessible Brønsted acid sites in the EU-1 nanosponge

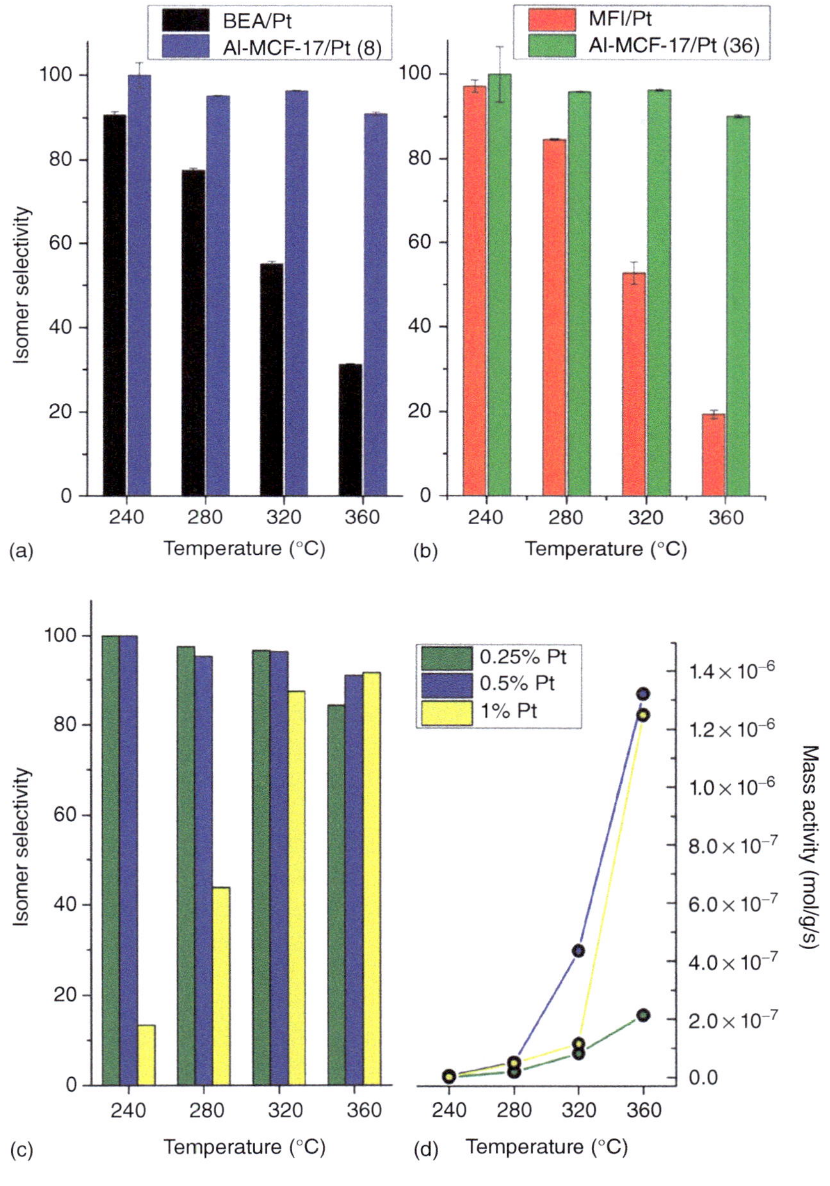

Figure 12.12 (a) Isomer selectivities for BEA/Pt zeolite (black) and Al-MCF-17/Pt (blue), both catalysts shown contain a Si/Al ratio of 8 : 1; (b) isomer selectivities for MFI/Pt (red) and Al-MCF-17/Pt (green), both catalysts shown contain a Si/Al ratio of 36 : 1; (c) effect on Pt nanoparticle loading on the isomer selectivity; and (d) the overall hexane conversion mass activity for the Al-MCF-17 (8) supported material. Source: Adapted from Musselwhite et al. [223] with the permission of ACS.

catalyst on the product yield reached up to 55% and 29%, respectively. A threefold improvement in the conversion of ethylbenzene was achieved by using the best EU-1 catalyst. Mesoporous MCM-68 zeolite was recently compared with the microporous MCM-68 in the isomerization of *m*-xylene [227]. The mesoporous zeolite apparently improved resistance ability to coke formation as compared with the microporous one.

12.3.2.5 HDO Phenolics

Bio-oils obtained from biomass, containing up to 35 wt.% oxygen, require significant deoxygenation to increase the H/C atomic ratio for converting into conventional transport fuels. Phenolic compounds are generally considered as important model compounds for bio-oils. HDO of these phenolics using developed mesoporous zeolite catalysts has gained considerable attention in the last few years.

In the HDO tests, phenol is usually used as a model compound [231–233]. Mesoporous ZSM-5 and beta zeolites supported Ni or Co phosphide catalysts were evaluated by Pizarro and coworkers in the HDO of phenol reaction [232]. The study revealed that metal phosphide deposition affected the textural and acidic properties of the zeolite supports, especially for beta zeolite. Ni_2P/h-ZSM-5 catalyst was found to exhibit outstanding catalytic results, achieving almost full conversion of phenol and up to 95% selectivity to cyclohexane product. This excellent catalytic performance of Ni_2P/h-ZSM-5 catalyst was attributed mainly to the preservation of the mesopores in zeolite, which afforded a good Ni_2P dispersion and an easy accessibility to the acid sites. Hermans and coworkers also tested mesoporous aluminosilicate nanofibers with Pd nanoparticles modified in the HDO process of phenol [233]. A full phenol conversion with a high selectivity toward cyclohexane (up to 95%) was obtained in the study.

Besides phenol, bulky phenolics, which are more like bio-oils, have also been employed as the model compounds for the HDO process [234, 235]. Xiao and coworkers reported a study on the application of mesoporous ZSM-5 zeolite supported Ru nanoparticles as the HDO catalyst for upgrading a variety of phenolic biomolecules into the corresponding alkanes [234]. Ru/HZSM-5-OM, a new type of mesoporous zeolite with *b*-axis-aligned mesopore-supported Ru nanoparticles was found to be highly active for the production of alkane from phenolics. The study further revealed that the higher catalytic performance in both activity and selectivity over Ru/HZSM-5-OM catalyst compared with the conventional ZSM-5 zeolite-supported Ru catalyst was due to its abundant exposed acidic sites with easy accessibility.

12.3.3 Catalytic Combustion

Motor vehicles and industrial processes daily release a huge amount of volatile organic compounds (VOCs), which are the major components of air pollutants and cause a great threat to the environment and human health due to their high toxicity. Elimination of VOCs has recently attracted much attention and therefore become a prevalent research topic. Catalytic combustion is considered to be one of most efficient ways to address this issue. Mesoporous zeolite-supported noble metal catalysts have recently found their potentials in the catalytic combustion of VOCs, especially aromatics [236–240]. Beta zeolite-supported Pt catalysts were utilized by Xiao's group in 2013 as the catalysts for the catalytic combustion of toluene [236]. A comparison of the catalysts with mesoporous and microporous beta zeolites as the respective supports was conducted in the study. It was found that the catalyst based on mesoporous beta (Pt-R/beta-H) showed superior performance in both activity and catalyst lifetime as compared with the conventional beta catalyst, giving complete combustion of toluene and affording CO_2 and H_2O as the only products. These excellent catalytic results were attributed to the presence of mesoporosity of the zeolite, which facilitated mass transfer and dispersion of Pt nanoparticles. Additionally, mesoporous ZSM-5 zeolites with Pd- and Pt-modified catalysts were subsequently developed by the same group for benzene and toluene combustion, both exhibiting excellent catalytic results as well [237, 238]. Very recently, mesoporous ZSM-5 zeolite-supported Ru nanoparticle was verified by Zhou and coworkers as the efficient catalyst for eliminating aromatic VOCs, including toluene, *o*-xylene, and 1,3,5-trimethylbenzene (TMB) [239]. Compared with the conventional Ru/ZSM-5 catalyst, the activity improvement of mesoporous ZSM-5 zeolite-supported Ru catalyst appeared more significant when bulkier aromatic substrate was employed in the catalytic combustion, implying the importance of the mesoporosity in the catalyst.

12.3.4 Biomass Valorization via Catalytic Fast Pyrolysis

The catalytic fast pyrolysis (CFP) of biomass is an interesting approach to produce bio-aromatics in consideration of its relatively simple process arrangement without using any solvent. During the CFP process, the oxygenated volatiles generated by biomass pyrolysis diffuse within the catalyst particles and contact the catalytic acid sites to produce the desired aromatics and also unavoidable coke and gas, mainly CO and CO_2. Zeolites are the main catalysts used for biomass CFP due to their acidic property, which enables the degradation of polymeric biomass into small molecules via the cleavage of C—O bonds. However, the small micropore channels of the zeolite significantly restrain the diffusion of the reactive intermediates and the aromatic products and can be easily blocked by coke deposition. Mesoporosity is thereby introduced into the zeolite crystals to overcome the limitations of the micropores.

Several mesoporous zeolites, mainly ZSM-5, have been tested and tailored for biomass CFP with the aim to enhance the product selectivity to mono-aromatics and reduce coke formation. Introducing a metal (Mg, Zn, Fe, etc.) [241, 242] and tuning the acidity as well as controlling the size and shape of pores are the main investigated parameters [243–251]. Van Bokhoven and coworkers systemically studied the effect of porosity and acidity on the lignin CFP over several mesoporous zeolite catalysts, which included ZSM-5, Al-MCM-15, Al-MSU-J, and Al-SBA-15 [244]. The study demonstrated that both mild Bronsted and strong Lewis acid sites could catalyze pyrolysis intermediates into fewer oxygenated phenols and aromatic hydrocarbons (Figure 12.13a). The fact that nanosized MCM-41 with 0.1 mmol

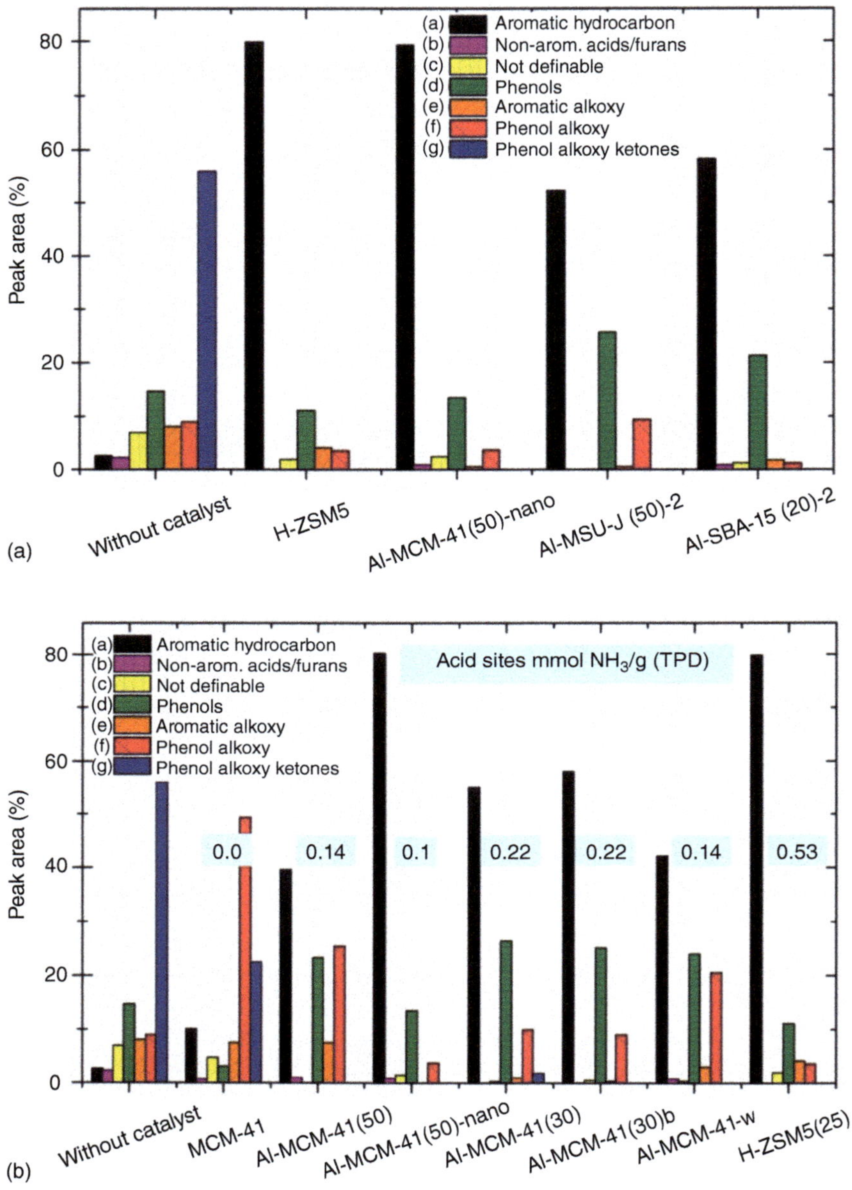

Figure 12.13 (a) Selectivity of CFP at 650 °C by the mesoporous catalysts yielding the highest content of aromatics; (b) product selectivity of the condensable fraction of CFP of the differently synthesized catalysts and the number of acid sites determined by NH_3-TPD. Source: Custodis et al. [244]. Reproduced with permission of John Wiley & Sons, Ltd.

NH_3/g (TPD) acid sites provided a high liquid yield and aromatic hydrocarbon selectivity, which was comparable to the results from ZSM-5 with higher acidity (0.53 mmol NH_3/g), indicated that zeolitic strong acid sties were not necessarily required for achieving high conversion and selectivity (Figure 12.13b). It was found that the diffusion and active sites were the two most essential parameters in lignin CFP reaction. The diffusion can be influenced by grain and pore size, and the active sites could be dominated by Lewis acid sites, which showed mildly acidic. Mild acidity and nanosized grains made the catalyst efficient in the lignin conversion. Dufour and coworkers investigated two ZSM-5 zeolites, microporous and mesoporous ones, for the CFP of oak in a microfluidized bed reactor at 500 °C [247]. The mesoporous ZSM-5 zeolite was prepared from the microporous ZSM-5 through a desilication process, showing a doubled selectivity to the targeted mono-aromatic compounds as compared with the parent zeolite. The stability of the zeolite was also found to be significantly improved after the desilication treatment. Very recently, Zn or Fe-modified mesoporous ZSM-5 zeolite catalysts were used by Sadrameli and coworker for the CFP of beech wood [242]. The introduction of metal was found to alter textural properties and acid site distribution of the mesoporous ZSM-5 zeolite. The catalytic results revealed a rather significant effect of metal, particularly in 2 wt.% loading, on the deoxygenation activity, product selectivity, and coke formation. Zn-modified ZSM-5 catalyst demonstrated a high selectivity toward oxygenates, such as ketones, and a decrease in coke formation.

12.4 Conclusions and Perspectives

Mesoporous zeolites are considered as a fascinating upgrade of conventional zeolites in various catalytic reactions, in particular with bulky molecules. In the recent years, the ever-increasing interest in mesoporous zeolites has boosted the development and progress in both synthesis approaches and practical catalytic applications.

In this chapter, we have firstly presented an overview of the so far developed synthesis strategies (Section 12.2), which include bottom-up (hard, soft, and non-templating routes), top-down (dealumination and desilication routes), and mixed (sequential dissolution and recrystallization) approaches. The bottom-up strategies mainly depend on the hard or soft templates used for generating mesopores. A high control of the additional porosity has already been achieved through exploiting efficient templates. Nevertheless, most of these templates are costly and unrecoverable, which accentuates the burden of these approaches on economic and environmental costs. Although the use of non-templating methods has been explored, it is still difficult to be carried out on a large-scale owing to the stringent preparation conditions and time-consuming processes. The top-down strategies through the removal of framework Al or Si atoms have been widely applied in industry for the preparation of zeolite catalysts. However, a major drawback of these methods is that a great part of the zeolite mass can be dissolved and lost during the process (up to 40% weight loss). These methods may also cause a significant change in the acidic properties of zeolite and the creation of isolated mesoporous cavities, which in fact hardly facilitate the diffusion. In the presence of conventional surfactants, zeolite recrystallization via the mixed approaches may provide a way at the moment closest to achieve industrial realization due to the high synthesis yield and relatively low environmental cost. Although substantial progresses have been made so far, it is still of great interest and challenge to develop new synthesis strategies for fabricating mesoporous zeolites with significant amounts of controllable mesopores, high zeolitic character, a wide range of zeolite types, and Si/Al ratios via a low economic and environmental cost route.

Section 12.3 is focused on the catalytic applications of mesoporous zeolites. Most catalytic reactions in which mesoporous zeolites participated so far have been summarized herein, which mainly include hydrogenation, oxidation, alkylation, isomerization, HDO, combustion, and the reactions involved in fuel chemistry and biomass conversion. Compared with the conventional zeolites, the use of mesoporous zeolites indeed affords obvious improvements in the catalytic performance in many reactions due to their diffusion strength, exhibiting higher catalytic activity and selectivity as well as longer catalyst lifetime. However, caution should be taken in comparison to a conventional zeolite catalyst. Thorough characterization of both catalysts, to ensure that they possess similar particle size, Si/Al ratio, quantity, and strength of active sites, is mandatory for a valid evaluation of the benefit from the mesoporosity. Big challenges within this field lie in the assessment of the level and type of mesoporosity in zeolite crystals and in comparing the mesoporous zeolite to a parent zeolite with exactly same physical and chemical properties except the absence of mesopores, due to the fact that it is difficult to acquire the zeolites with these features.

References

1 Corma, A. (1995). Inorganic solid acids and their use in acid-catalyzed hydrocarbon reactions. *Chem. Rev.* 95 (3): 559–614.

2 Xu, R., Pang, W., Yu, J. et al. (2010). Structural chemistry of microporous materials. In: *Chemistry of Zeolites and Related Porous Materials*, 19–116. Chichester, UK: Wiley.

3 Meng, X. and Xiao, F.S. (2014). Green routes for synthesis of zeolites. *Chem. Rev.* 114 (2): 1521–1543.

4 Smit, B. and Maesen, T.L.M. (2008). Towards a molecular understanding of shape selectivity. *Nature* 451 (7179): 671–678.

5 Davis, M.E. (2002). Ordered porous materials for emerging applications. *Nature* 417 (6891): 813–821.

6 Kresge, C.T., Leonowicz, M.E., Roth, W.J. et al. (1992). Ordered mesoporous molecular sieves synthesized by a liquid-crystal template mechanism. *Nature* 359 (6397): 710–712.

7 Zhao, D., Feng, J., Huo, Q. et al. (1998). Triblock copolymer syntheses of mesoporous silica with periodic 50 to 300 angstrom pores. *Science* 279 (5350): 548–552.

8 Jiang, J., Yu, J., and Corma, A. (2010). Extra-large-pore zeolites: bridging the gap between micro and mesoporous structures. *Angew. Chem. Int. Ed.* 49 (18): 3120–3145.

9 Lopez-Orozco, S., Inayat, A., Schwab, A. et al. (2011). Zeolitic materials with hierarchical porous structures. *Adv. Mater.* 23 (22–23): 2602–2615.

10 Möller, K. and Bein, T. (2013). Mesoporosity – a new dimension for zeolites. *Chem. Soc. Rev.* 42 (9): 3689–3707.

11 Serrano, D.P., Escola, J.M., and Pizarro, P. (2013). Synthesis strategies in the search for hierarchical zeolites. *Chem. Soc. Rev.* 42 (9): 4004–4035.

12 Schwieger, W., Machoke, A.G., Weissenberger, T. et al. (2016). Hierarchy concepts: classification and preparation strategies for zeolite containing mat erials with hierarchical porosity. *Chem. Soc. Rev.* 45 (12): 3353–3376.

13 Chen, L.H., Li, X.Y., Rooke, J.C. et al. (2012). Hierarchically structured zeolites: synthesis, mass transport properties and applications. *J. Mater. Chem.* 22 (34): 17381–17403.

14 Perego, C. and Millini, R. (2013). Porous materials in catalysis: challenges for mesoporous materials. *Chem. Soc. Rev.* 42 (9): 3956–3976.

15 Zhang, J., Wang, L., Ji, Y. et al. (2018). Mesoporous zeolites for biofuel upgrading and glycerol conversion. *Front. Chem. Sci. Eng.* 12 (1): 132–144.

16 Hartmann, M., Machoke, A.G., and Schwieger, W. (2016). Catalytic test reactions for the evaluation of hierarchical zeolites. *Chem. Soc. Rev.* 45 (12): 3313–3330.

17 Li, K., Valla, J., and Garcia-Martinez, J. (2014). Realizing the commercial potential of hierarchical zeolites: new opportunities in catalytic cracking. *ChemCatChem* 6 (1): 46–66.

18 Madsen, C. and Jacobsen, C.J.H. (1999). Nanosized zeolite crystals - convenient control of crystal size distribution by confined space synthesis. *Chem. Commun.* 1999 (8): 673–674. https://doi.org/10.1039/a901228a.

19 Schmidt, I., Madsen, C., and Jacobsen, C.J.H. (2000). Confined space synthesis. A novel route to nanosized zeolites. *Inorg. Chem.* 39 (11): 2279–2283.

20 Jacobsen, C.J.H., Madsen, C., Houzvicka, J. et al. (2000). Mesoporous zeolite single crystals. *J. Am. Chem. Soc.* 122 (29): 7116–7117.

21 Tao, Y., Kanoh, H., and Kaneko, K. (2003). ZSM-5 monolith of uniform mesoporous channels. *J. Am. Chem. Soc.* 125 (20): 6044–6045.

22 Kustova, M., Egeblad, K., Zhu, K., and Christensen, C.H. (2007). Versatile route to zeolite single crystals with controlled mesoporosity: *in situ* sugar decomposition for templating of hierarchical zeolites. *Chem. Mater.* 19 (12): 2915–2917.

23 Schmidt, I., Boisen, A., Gustavsson, E. et al. (2001). Carbon nanotube templated growth of mesoporous zeolite single crystals. *Chem. Mater.* 13 (12): 4416–4418.

24 Chen, H., Wydra, J., Zhang, X. et al. (2011). Hydrothermal synthesis of zeolites with three-dimensionally ordered mesoporous-imprinted structure. *J. Am. Chem. Soc.* 133 (32): 12390–12393.

25 Schmidt, I., Krogh, A., Wienberg, K. et al. (2000). Catalytic epoxidation of alkenes with hydrogen peroxide over first mesoporous titanium-containing zeolite. *Chem. Commun.* 2000 (21): 2157–2158.

26 Kustova, M.Y., Hasselriis, P., and Christensen, C.H. (2004). Mesoporous MEL - type zeolite single crystal catalysts. *Catal. Lett.* 96 (3–4): 205–211.

27 Wei, X. and Smirniotis, P.G. (2006). Synthesis and characterization of mesoporous ZSM-12 by using carbon particles. *Microporous Mesoporous Mater.* 89 (1–3): 170–178.

28 Egeblad, K., Kustova, M., Klitgaard, S.K. et al. (2007). Mesoporous zeolite and zeotype single crystals synthesized in fluoride media. *Microporous Mesoporous Mater.* 101 (1–2 SPEC. ISS): 214–223.

29 Tao, Y., Kanoh, H., and Kaneko, K. (2003). Uniform mesopore-donated zeolite Y using carbon aerogel templating. *J. Phys. Chem. B* 107 (40): 10974–10976.

30 Tao, Y., Kanoh, H., and Kaneko, K. (2005). Synthesis of mesoporous zeolite a by resorcinol-formaldehyde aerogel templating. *Langmuir* 21 (2): 504–507.

31 Li, W.C., Lu, A.H., Palkovits, R. et al. (2005). Hierarchically structured monolithic silicalite-1 consisting of crystallized nanoparticles and its performance in the Beckmann rearrangement of cyclohexanone oxime. *J. Am. Chem. Soc.* 127 (36): 12595–12600.

32 Zhu, K., Egeblad, K., and Christensen, C.H. (2007). Mesoporous carbon prepared from carbohydrate as hard template for hierarchical zeolites. *Eur. J. Inorg. Chem.* 2007 (25): 3955–3960.

33 Wang, X., Li, G., Wang, W. et al. (2011). Synthesis, characterization and catalytic performance of hierarchical TS-1 with carbon template from sucrose carbonization. *Microporous Mesoporous Mater.* 142 (2–3): 494–502.

34 White, R.J., Fischer, A., Goebel, C., and Thomas, A. (2014). A sustainable template for mesoporous zeolite synthesis. *J. Am. Chem. Soc.* 136 (7): 2715–2718.

35 Janssen, A.H., Schmidt, I., Jacobsen, C.J.H. et al. (2003). Exploratory study of mesopore templating with carbon during zeolite synthesis. *Microporous Mesoporous Mater.* 65 (1): 59–75.

36 Schmidt, F., Paasch, S., Brunner, E., and Kaskel, S. (2012). Carbon templated SAPO-34 with improved adsorption kinetics and catalytic performance in the MTO-reaction. *Microporous Mesoporous Mater.* 164: 214–221.

37 Yang, Z.X., Xia, Y.D., and Mokaya, R. (2004). Zeolite ZSM-5 with unique supermicropores synthesized using mesoporous carbon as a template. *Adv. Mater.* 16 (8): 727–732.

38 Sakthivel, A., Huang, S.J., Chen, W.H. et al. (2004). Replication of mesoporous aluminosilicate molecular sieves (RMMs) with zeolite framework from mesoporous carbons (CMKs). *Chem. Mater.* 16 (16): 3168–3175.

39 Fang, Y. and Hu, H. (2006). An ordered mesoporous aluminosilicate with completely crystalline zeolite wall structure. *J. Am. Chem. Soc.* 128 (33): 10636–10637.

40 Cho, H.S. and Ryoo, R. (2012). Synthesis of ordered mesoporous MFI zeolite using CMK carbon templates. *Microporous Mesoporous Mater.* 151: 107–112.

41 Kim, S.S., Shah, J., and Pinnavaia, T.J. (2003). Colloid-imprinted carbons as templates for the nanocasting synthesis of mesoporous ZSM-5 zeolite. *Chem. Mater.* 15 (8): 1664–1668.

42 Li, H., Sakamoto, Y., Liu, Z. et al. (2007). Mesoporous silicalite-1 zeolite crystals with unique pore shapes analogous to the morphology. *Microporous Mesoporous Mater.* 106 (1–3): 174–179.

43 Fan, W., Snyder, M.A., Kumar, S. et al. (2008). Hierarchical nanofabrication of microporous crystals with ordered mesoporosity. *Nat. Mater.* 7 (12): 984–991.

44 Kang, Y., Shan, W., Wu, J. et al. (2006). Uniform nanozeolite microspheres with large secondary pore architecture. *Chem. Mater.* 18 (7): 1861–1866.

45 Shi, Y., Li, X., Hu, J. et al. (2011). Zeolite microspheres with hierarchical structures: formation, mechanism and catalytic performance. *J. Mater. Chem.* 21 (40): 16223–16230.

46 Zhu, H., Liu, Z., Wang, Y. et al. (2008). Nanosized $CaCO_3$ as hard template for creation of intracrystal pores within silicalite-1 crystal. *Chem. Mater.* 20 (3): 1134–1139.

47 Wang, L., Yin, C., Shan, Z. et al. (2009). Bread-template synthesis of hierarchical mesoporous ZSM-5 zeolite with hydrothermally stable mesoporosity. *Colloids Surf. A Physicochem. Eng. Asp.* 340 (1–3): 126–130.

48 Dong, A., Wang, Y., Tang, Y. et al. (2002). Zeolitic tissue through wood cell templating. *Adv. Mater.* 14 (12): 926–929.

49 De La Iglesia, Ó., Sánchez, J.L., and Coronas, J. (2011). Hierarchical silicalite-1 structures based on pyrolized materials. *Mater. Lett.* 65 (19–20): 3124–3127.

50 Ocampo, F., Cunha, J.A., De Lima Santos, M.R. et al. (2010). Synthesis of zeolite crystals with unusual morphology: application in acid catalysis. *Appl. Catal. A Gen.* 390 (1–2): 102–109.

51 Valtchev, V., Smaihi, M., Faust, A.C., and Vidal, L. (2003). Biomineral-silica-induced zeolitization of equisetum arvense. *Angew. Chem. Int. Ed.* 42 (24): 2782–2785.

52 Valtchev, V.P., Smaihi, M., Faust, A.C., and Vidal, L. (2004). Equisetum arvense templating of zeolite beta macrostructures with hierarchical porosity. *Chem. Mater.* 16 (7): 1350–1355.

53 Zhang, B., Davis, S.A., Mendelson, N.H., and Mann, S. (2000). Bacterial templating of zeolite fibres with hierarchical structure. *Chem. Commun.* 2 (9): 781–782.

54 Xiao, F.S., Wang, L., Yin, C. et al. (2006). Catalytic properties of hierarchical mesoporous zeolites templated with a mixture of small organic ammonium salts and mesoscale cationic polymers. *Angew. Chem. Int. Ed.* 45 (19): 3090–3093.

55 Wang, L., Zhang, Z., Yin, C. et al. (2010). Hierarchical mesoporous zeolites with controllable mesoporosity templated from cationic polymers. *Microporous Mesoporous Mater.* 131 (1–3): 58–67.

56 Pan, S., Wu, Q., Wang, X. et al. (2016). Mesoporous EU-1 zeolite synthesized in the presence of cationic polymer. *Microporous Mesoporous Mater.* 235: 246–252.

57 Möller, K., Yilmaz, B., Müller, U., and Bein, T. (2011). Hierarchical zeolite beta via nanoparticle assembly with a cationic polymer. *Chem. Mater.* 23 (19): 4301–4310.

58 Liu, F., Willhammar, T., Wang, L. et al. (2012). ZSM-5 zeolite single crystals with *b*-axis-aligned mesoporous channels as an efficient catalyst for conversion of bulky organic molecules. *J. Am. Chem. Soc.* 134 (10): 4557–4560.

59 Zhu, J., Zhu, Y., Zhu, L. et al. (2014). Highly mesoporous single-crystalline zeolite beta synthesized using a nonsurfactant cationic polymer as a dual-function template. *J. Am. Chem. Soc.* 136 (6): 2503–2510.

60 Jo, C., Seo, Y., Cho, K. et al. (2014). Random-graft polymer-directed synthesis of inorganic mesostructures with ultrathin frameworks. *Angew. Chem. Int. Ed.* 53 (20): 5117–5121.

61 Meng, L., Mezari, B., Goesten, M.G., and Hensen, E.J.M. (2017). One-step synthesis of hierarchical ZSM-5 using cetyltrimethylammonium as mesoporogen and structure-directing agent. *Chem. Mater.* 29 (9): 4091–4096.

62 Choi, M., Na, K., and Ryoo, R. (2009). The synthesis of a hierarchically porous BEA zeolite via pseudomorphic crystallization. *Chem. Commun.* 2009 (20): 2845–2847.

63 Na, K., Choi, M., and Ryoo, R. (2009). Cyclic diquaternary ammoniums for nanocrystalline BEA, MTW and MFI zeolites with intercrystalline mesoporosity. *J. Mater. Chem.* 19 (37): 6713–6719.

64 Choi, M., Na, K., Kim, J. et al. (2009). Stable single-unit-cell nanosheets of zeolite MFI as active and long-lived catalysts. *Nature* 461 (7261): 246–249.

65 Na, K., Park, W., Seo, Y., and Ryoo, R. (2011). Disordered assembly of MFI zeolite nanosheets with a large volume of intersheet mesopores. *Chem. Mater.* 23 (5): 1273–1279.

66 Na, K., Jo, C., Kim, J. et al. (2011). Directing zeolite structures into hierarchically nanoporous architectures. *Science* 333 (6040): 328–332.

67 Xu, D., Ma, Y., Jing, Z. et al. (2014). Π-Π interaction of aromatic groups in amphiphilic molecules directing for single-crystalline mesostructured zeolite nanosheets. *Nat. Commun.* 5 (May): 1–9.

68 Wu, L., Degirmenci, V., Magusin, P.C.M.M. et al. (2012). Dual template synthesis of a highly mesoporous SSZ-13 zeolite with improved stability in the methanol-to-olefins reaction. *Chem. Commun.* 48 (76): 9492–9494.

69 Seo, Y., Lee, S., Jo, C., and Ryoo, R. (2013). Microporous aluminophosphate nanosheets and their nanomorphic zeolite analogues tailored by hierarchical structure-directing amines. *J. Am. Chem. Soc.* 135 (24): 8806–8809.

70 Serrano, D.P., Aguado, J., Escola, J.M. et al. (2006). Hierarchical zeolites with enhanced textural and catalytic properties synthesized from organofunctionalized seeds. *Chem. Mater.* 18 (10): 2462–2464.

71 Serrano, D.P., Aguado, J., Morales, G. et al. (2009). Molecular and meso- and macroscopic properties of hierarchical nanocrystalline ZSM-5 zeolite prepared by seed silanization. *Chem. Mater.* 21 (4): 641–654.

72 Serrano, D.P., Aguado, J., Escola, J.M. et al. (2008). Effect of the organic moiety nature on the synthesis of hierarchical ZSM-5 from silanized protozeolitic units. *J. Mater. Chem.* 18 (35): 4210–4218.

73 Serrano, D.P., Aguado, J., Escola, J.M. et al. (2011). Synthesis of hierarchical ZSM-5 by silanization and alkoxylation of protozeolitic units. *Catal. Today* 168 (1): 86–95.

74 Choi, M., Cho, H.S., Srivastava, R. et al. (2006). Amphiphilic organosilane-directed synthesis of crystalline zeolite with tunable mesoporosity. *Nat. Mater.* 5 (9): 718–723.

75 Choi, M., Srivastava, R., and Ryoo, R. (2006). Organosilane surfactant-directed synthesis of mesoporous aluminophosphates constructed with crystalline microporous frameworks. *Chem. Commun.* 42 (42): 4380–4382.

76 Shanbhag, G.V., Choi, M., Kim, J., and Ryoo, R. (2009). Mesoporous sodalite: a novel, stable solid catalyst for base-catalyzed organic transformations. *J. Catal.* 264 (1): 88–92.

77 Cho, K., Cho, H.S., De Ménorval, L.C., and Ryoo, R. (2009). Generation of mesoporosity in LTA zeolites by organosilane surfactant for rapid molecular transport in catalytic application. *Chem. Mater.* 21 (23): 5664–5673.

78 Fan, Y., Xiao, H., Shi, G. et al. (2012). Alkylphosphonic acid- and small amine-templated synthesis of hierarchical silicoaluminophosphate molecular sieves with high isomerization selectivity to di-branched paraffins. *J. Catal.* 285 (1): 251–259.

79 Inayat, A., Knoke, I., Spiecker, E., and Schwieger, W. (2012). Assemblies of mesoporous FAU-type zeolite nanosheets. *Angew. Chem. Int. Ed.* 51 (8): 1962–1965.

80 Khaleel, M., Wagner, A.J., Mkhoyan, K.A., and Tsapatsis, M. (2014). On the rotational intergrowth of hierarchical FAU/EMT zeolites. *Angew. Chem. Int. Ed.* 53 (36): 9456–9461.

81 Wang, H. and Pinnavaia, T.J. (2006). MFI zeolite with small and uniform intracrystal mesopores. *Angew. Chem. Int. Ed.* 45 (45): 7603–7606.

82 Majano, G., Mintova, S., Ovsitser, O. et al. (2005). Zeolite beta nanosized assemblies. *Microporous Mesoporous Mater.* 80 (1–3): 227–235.

83 Li, C., Wang, Y., Shi, B. et al. (2009). Synthesis of hierarchical MFI zeolite microspheres with stacking nanocrystals. *Microporous Mesoporous Mater.* 117 (1–2): 104–110.

84 Möller, K., Yilmaz, B., Jacubinas, R.M. et al. (2011). One-step synthesis of hierarchical zeolite beta via network formation of uniform nanocrystals. *J. Am. Chem. Soc.* 133 (14): 5284–5295.

85 Chen, L.H., Li, X.Y., Tian, G. et al. (2011). Highly stable and reusable multimodal zeolite TS-1 based catalysts with hierarchically interconnected three-level micro-meso-macroporous structure. *Angew. Chem. Int. Ed.* 50 (47): 11156–11161.

86 Möller, K., Yilmaz, B., Müller, U., and Bein, T. (2012). Nanofusion: mesoporous zeolites made easy. *Chemistry* 18 (25): 7671–7674.

87 Zhang, X., Liu, D., Xu, D. et al. (2012). Synthesis of self-pillared zeolite nanosheets by repetitive branching. *Science* 336 (6089): 1684–1687.

88 Chaikittisilp, W., Suzuki, Y., Mukti, R.R. et al. (2013). Formation of hierarchically organized zeolites by sequential intergrowth. *Angew. Chem. Int. Ed.* 52 (12): 3355–3359.

89 Zhang, H., Wang, L., Zhang, D. et al. (2016). Mesoporous and Al-rich MFI crystals assembled with aligned nanorods in the absence of organic templates. *Microporous Mesoporous Mater.* 233: 133–139.

90 Wang, Z., Li, C., Cho, H.J. et al. (2015). Direct, single-step synthesis of hierarchical zeolites without secondary templating. *J. Mater. Chem. A* 3 (3): 1298–1305.

91 Zhu, J., Cui, Y., Wang, Y., and Wei, F. (2009). Direct synthesis of hierarchical zeolite from a natural layered material. *Chem. Commun.* 45 (22): 3282–3284.

92 Inayat, A., Schneider, C., and Schwieger, W. (2015). Organic-free synthesis of layer-like FAU-type zeolites. *Chem. Commun.* 51 (2): 279–281.

93 Shan, Z., Lu, Z., Wang, L. et al. (2010). Stable bulky particles formed by TS-1 zeolite nanocrystals in the presence of H_2O_2. *ChemCatChem* 2 (4): 407–412.

94 Sun, J., Wang, C., Wang, L. et al. (2010). Synthesis of MFI zeolites with improved crystallization rate and mesoporosity in the presence of CO_2-in-water emulsions. *Catal. Today* 158 (3–4): 273–278.

95 Zhu, G.W., Chen, X.B., Jiang, H. et al. (2013). Synthesis and characterization of TS-1 with the aid of supercritical CO_2. *Chem. Res. Chin. Univ.* 29 (6): 1036–1039.

96 Zukal, A., Patzelová, V., and Lohse, U. (1986). Secondary porous structure of dealuminated Y zeolites. *Zeolites* 6 (2): 133–136.

97 Sasaki, Y., Suzuki, T., Takamura, Y. et al. (1998). Structure analysis of the mesopore in dealuminated zeolite Y by high resolution TEM observation with slow scan CCD camera. *J. Catal.* 178 (1): 94–100.

98 Agostini, G., Lamberti, C., Palin, L. et al. (2010). *In situ* XAS and XRPD parametric rietveld refinement to understand dealumination of Y zeolite catalyst. *J. Am. Chem. Soc.* 132 (2): 667–678.

99 Van Donk, S., Janssen, A.H., Bitter, J.H., and De Jong, K.P. (2003). Generation, characterization, and impact of mesopores in zeolite catalysts. *Catal. Rev. Sci. Eng.* 45 (2): 297–319.

100 Roberge, D.M., Hausmann, H., and Hölderich, W.F. (2002). Dealumination of zeolite beta by acid leaching: a new insight with two-dimensional multi-quantum and cross polarization 27 Al MAS NMR. *Phys. Chem. Chem. Phys.* 4 (13): 3128–3135.

101 Chung, K.H. (2008). Dealumination of mordenites with acetic acid and their catalytic activity in the alkylation of cumene. *Microporous Mesoporous Mater.* 111 (1–3): 544–550.

102 Peixoto, D.P.B., Cabral de Menezes, S.M., and Pais da Silva, M.I. (2003). Influence of different processes of dealumination on acid properties of an H-ferrierite zeolite. *Mater. Lett.* 57 (24–25): 3933–3942.

103 Tromp, M., Van Bokhoven, J.A., Garriga Oostenbrink, M.T. et al. (2000). Influence of the generation of mesopores on the hydroisomerization activity and selectivity of *n*-hexane over Pt/mordenite. *J. Catal.* 190 (2): 209–214.

104 Müller, M., Harvey, G., and Prins, R. (2000). Comparison of the dealumination of zeolites beta, mordenite, ZSM-5 and ferrierite by thermal treatment, leaching with oxalic acid and treatment with SiCl4 by 1H, 29Si and 27A1 MAS NMR. *Microporous Mesoporous Mater.* 34 (2): 135–147.

105 Ogura, M., Shinomiya, S.Y., Tateno, J. et al. (2000). Formation of uniform mesopores in ZSM-5 zeolite through treatment in alkaline solution. *Chem. Lett.* 29 (8): 882–883.

106 Ogura, M., Shinomiya, S.Y., Tateno, J. et al. (2001). Alkali-treatment technique - new method for modification of structural and acid-catalytic properties of ZSM-5 zeolites. *Appl. Catal. A Gen.* 219 (1–2): 33–43.

107 Groen, J.C., Jansen, J.C., Moulijn, J.A., and Pérez-Ramírez, J. (2004). Optimal aluminum-assisted mesoporosity development in MFI zeolites by desilication. *J. Phys. Chem. B* 108 (35): 13062–13065.

108 Groen, J.C., Peffer, L.A.A., Moulijn, J.A., and Pérez-Ramírez, J. (2005). Mechanism of hierarchical porosity development in MFI zeolites by desilication: the role of aluminium as a pore-directing agent. *Chemistry* 11 (17): 4983–4994.

109 Groen, J.C., Moulijn, J.A., and Pérez-Ramírez, J. (2006). Desilication: on the controlled generation of mesoporosity in MFI zeolites. *J. Mater. Chem.* 16 (22): 2121–2131.

110 Svelle, S., Sommer, L., Barbera, K. et al. (2011). How defects and crystal morphology control the effects of desilication. *Catal. Today* 168 (1): 38–47.

111 Groen, J.C., Caicedo-Realpe, R., Abelló, S., and Pérez-Ramírez, J. (2009). Mesoporous metallosilicate zeolites by desilication: on the generic pore-inducing role of framework trivalent heteroatoms. *Mater. Lett.* 63 (12): 1037–1040.

112 Abelló, S., Bonilla, A., and Pérez-Ramírez, J. (2009). Mesoporous ZSM-5 zeolite catalysts prepared by desilication with organic hydroxides and comparison with NaOH leaching. *Appl. Catal. A Gen.* 364 (1–2): 191–198.

113 Peréz-Ramírez, J., Verboekend, D., Bonilla, A., and Abelló, S. (2009). Zeolite catalysts with tunable hierarchy factor by pore-growth moderators. *Adv. Funct. Mater.* 19 (24): 3972–3979.

114 Verboekend, D. and Pérez-Ramírez, J. (2011). Desilication mechanism revisited: highly mesoporous all-silica zeolites enabled through pore-directing agents. *Chemistry* 17 (4): 1137–1147.

115 Verboekend, D., Vilé, G., and Pérez-Ramírez, J. (2012). Mesopore formation in USY and beta zeolites by base leaching: selection criteria and optimization of pore-directing agents. *Cryst. Growth Des.* 12 (6): 3123–3132.

116 Pérez-Ramírez, J., Abelló, S., Bonilla, A., and Groen, J.C. (2009). Tailored mesoporosity development in zeolite crystals by partial detemplation and desilication. *Adv. Funct. Mater.* 19 (1): 164–172.

117 Van Laak, A.N.C., Zhang, L., Parvulescu, A.N. et al. (2011). Alkaline treatment of template containing zeolites: introducing mesoporosity while preserving acidity. *Catal. Today* 168 (1): 48–56.

118 Li, X.F., Prins, R., and van Bokhoven, J.A. (2009). Synthesis and characterization of mesoporous mordenite. *J. Catal.* 262 (2): 257–265.

119 Van Laak, A.N.C., Sagala, S.L., Zečević, J. et al. (2010). Mesoporous mordenites obtained by sequential acid and alkaline treatments - catalysts for cumene production with enhanced accessibility. *J. Catal.* 276 (1): 170–180.

120 Verboekend, D., Keller, T.C., Milina, M. et al. (2013). Hierarchy brings function: mesoporous clinoptilolite and L zeolite catalysts synthesized by tandem acid-base treatments. *Chem. Mater.* 25 (9): 1947–1959.

121 De Jong, K.P., Zečević, J., Friedrich, H. et al. (2010). Zeolite Y crystals with trimodal porosity as ideal hydrocracking catalysts. *Angew. Chem. Int. Ed.* 49 (52): 10074–10078.

122 Wang, S., Dou, T., Li, Y. et al. (2005). A novel method for the preparation of MOR/MCM-41 composite molecular sieve. *Catal. Commun.* 6 (1): 87–91.

123 Ivanova, I.I., Kuznetsov, A.S., Yuschenko, V.V., and Knyazeva, E.E. (2004). Design of composite micro/mesoporous molecular sieve catalysts. *Pure Appl. Chem.* 76 (9): 1647–1658.

124 Ordomsky, V.V., Murzin, V.Y., Monakhova, Y.V. et al. (2007). Nature, strength and accessibility of acid sites in micro/mesoporous catalysts obtained by recrystallization of zeolite BEA. *Microporous Mesoporous Mater.* 105 (1–2): 101–110.

125 Ivanova, I.I., Kuznetsov, A.S., Knyazeva, E.E. et al. (2011). Design of hierarchically structured catalysts by mordenites recrystallization: application in naphthalene alkylation. *Catal. Today* 168 (1): 133–139.

126 Khitev, Y.P., Kolyagin, Y.G., Ivanova, I.I. et al. (2011). Synthesis and catalytic properties of hierarchical micro/mesoporous materials based on FER zeolite. *Microporous Mesoporous Mater.* 146 (1–3): 201–207.

127 Ivanova, I.I., Kasyanov, I.A., Maerle, A.A., and Zaikovskii, V.I. (2014). Mechanistic study of zeolites recrystallization into micro-mesoporous materials. *Microporous Mesoporous Mater.* 189: 163–172.

128 Yoo, W.C., Zhang, X., Tsapatsis, M., and Stein, A. (2012). Synthesis of mesoporous ZSM-5 zeolites through desilication and re-assembly processes. *Microporous Mesoporous Mater.* 149 (1): 147–157.

129 García-Martínez, J., Johnson, M., Valla, J. et al. (2012). Mesostructured zeolite Y - high hydrothermal stability and superior FCC catalytic performance. *Catal. Sci. Technol.* 2 (5): 987–994.

130 Ishihara, A., Inui, K., Hashimoto, T., and Nasu, H. (2012). Preparation of hierarchical β and Y zeolite-containing mesoporous silica-aluminas and their properties for catalytic cracking of *n*-dodecane. *J. Catal.* 295: 81–90.

131 Qin, Z., Shen, B., Gao, X. et al. (2011). Mesoporous Y zeolite with homogeneous aluminum distribution obtained by sequential desilication-dealumination and its performance in the catalytic cracking of cumene and 1,3,5-triisopropylbenzene. *J. Catal.* 278 (2): 266–275.

132 Qin, Z., Shen, B., Yu, Z. et al. (2013). A defect-based strategy for the preparation of mesoporous zeolite Y for high-performance catalytic cracking. *J. Catal.* 298: 102–111.

133 Li, W., Zheng, J., Luo, Y. et al. (2017). The influence of TPOACl on the post-synthesis of hierarchical zeolite Y and the catalytic cracking application. *ChemistrySelect* 2 (13): 3872–3877.

134 Jin, J., Peng, C., Wang, J. et al. (2014). Facile synthesis of mesoporous zeolite Y with improved catalytic performance for heavy oil fluid catalytic cracking. *Ind. Eng. Chem. Res.* 53 (8): 3406–3411.

135 Li, W., Zheng, J., Luo, Y. et al. (2017). Hierarchical zeolite Y with full crystallinity: formation mechanism and catalytic cracking performance. *Energy Fuel* 31 (4): 3804–3811.

136 Lv, P., Yan, L., Liu, Y. et al. (2019). Catalytic upgrading of coal pyrolysis gaseous tar over hierarchical Y-type zeolites synthesized using a microwave hydrothermal method. *Ind. Eng. Chem. Res.* 58 (47): 21817–21826.

137 Tempelman, C.H.L., Zhu, X., Gudun, K. et al. (2015). Texture, acidity and fluid catalytic cracking performance of hierarchical faujasite zeolite prepared by an amphiphilic organosilane. *Fuel Process. Technol.* 139: 248–258.

138 Ishihara, A., Kimura, K., Owaki, A. et al. (2012). Catalytic cracking of VGO by hierarchical ZSM-5 zeolite containing mesoporous silica-aluminas using a Curie point pyrolyzer. *Catal. Commun.* 28: 163–167.

139 Liu, J., Jiang, G., Liu, Y. et al. (2014). Hierarchical macro-meso-microporous ZSM-5 zeolite hollow fibers with highly efficient catalytic cracking capability. *Sci. Rep.* 4: 1–6.

140 Zhao, T., Li, F., Yu, H. et al. (2019). Synthesis of mesoporous ZSM-5 zeolites and catalytic cracking of ethanol and oleic acid into light olefins. *Appl. Catal. A Gen.* 575 (October 2018): 101–110.

141 Wang, D., Zhang, L., Chen, L. et al. (2015). Postsynthesis of mesoporous ZSM-5 zeolite by piperidine-assisted desilication and its superior catalytic properties in hydrocarbon cracking. *J. Mater. Chem. A* 3 (7): 3511–3521.

142 Li, W., Ma, T., Zhang, Y. et al. (2015). Facile control of inter-crystalline porosity in the synthesis of size-controlled mesoporous MFI zeolites via in situ conversion of silica gel into zeolite nanocrystals for catalytic cracking. *CrystEngComm* 17 (30): 5680–5689.

143 Zhang, X., Cheng, D.G., Chen, F., and Zhan, X. (2017). *n*-Heptane catalytic cracking on hierarchical ZSM-5 zeolite: the effect of mesopores. *Chem. Eng. Sci.* 168: 352–359.

144 Liu, H., Diao, Z., Liu, G. et al. (2017). Catalytic cracking of *n*-dodecane over hierarchical HZSM-5 zeolites synthesized by double-template method. *J. Porous. Mater.* 24 (6): 1679–1687.

145 Ji, Y., Yang, H., and Yan, W. (2018). Catalytic cracking of n-hexane to light alkene over ZSM-5 zeolite: influence of hierarchical porosity and acid property. *Mol. Catal.* 448 (January): 91–99.

146 Wang, Y.H., Sun, H.M., Peng, P. et al. (2018). Synthesis of hierarchical ZSM-5 zeolites via two stage varying temperature crystallization with enhanced catalytic cracking performance. *Chinese J. Inorg. Chem.* 34 (5): 989–996.

147 Zhang, X., Cheng, D.G., Chen, F., and Zhan, X. (2018). The role of external acidity of hierarchical ZSM-5 zeolites in *n*-heptane catalytic cracking. *ChemCatChem* 10 (12): 2655–2663.

148 Xu, S., Zhang, X., Cheng, D.G. et al. (2018). Effect of hierarchical ZSM-5 zeolite crystal size on diffusion and catalytic performance of n-heptane cracking. *Front. Chem. Sci. Eng.* 12 (4): 780–789.

149 Ishihara, A., Mori, K., Mori, K. et al. (2019). Preparation of hierarchical catalysts with the simultaneous generation of microporous zeolite using a template and large mesoporous silica by gel skeletal reinforcement and their reactivity in the catalytic cracking of: *N*-dodecane. *Catal. Sci. Technol.* 9 (14): 3614–3618.

150 Xian, X., Ran, C., Yang, P. et al. (2018). Effect of the acidity of HZSM-5/MCM-41 hierarchical zeolite on its catalytic performance in supercritical catalytic cracking of *n*-dodecane: experiments and mechanism. *Catal. Sci. Technol.* 8 (16): 4241–4256.

151 Martínez, C., Verboekend, D., Pérez-Ramírez, J., and Corma, A. (2013). Stabilized hierarchical USY zeolite catalysts for simultaneous increase in diesel and LPG olefinicity during catalytic cracking. *Catal. Sci. Technol.* 3 (4): 972–981.

152 Ahmed, M.H.M., Muraza, O., Nakaoka, S. et al. (2017). Stability assessment of regenerated hierarchical ZSM-48 zeolite designed by post-synthesis treatment for catalytic cracking of light naphtha. *Energy Fuel* 31 (12): 14097–14103.

153 Muraza, O., Bakare, I.A., Tago, T. et al. (2014). Selective catalytic cracking of *n*-hexane to propylene over hierarchical MTT zeolite. *Fuel* 135: 105–111.

154 Ahmed, M.H.M., Muraza, O., Jamil, A.K. et al. (2017). Steam catalytic cracking of *n*-dodecane over Ni and Ni/Co bimetallic catalyst supported on hierarchical BEA zeolite. *Energy Fuel* 31 (5): 5482–5490.

155 Yu, Q., Sun, H., Sun, H. et al. (2019). Highly mesoporous IM-5 zeolite prepared by alkaline treatment and its catalytic cracking performance. *Microporous Mesoporous Mater.* 273 (August 2018): 297–306.

156 Abildstrøm, J.O., Kegnæs, M., Hytoft, G. et al. (2016). Synthesis of mesoporous zeolite catalysts by *in situ* formation of carbon template over nickel nanoparticles. *Microporous Mesoporous Mater.* 225: 232–237.

157 Khodakov, A.Y., Chu, W., and Fongarland, P. (2007). Advances in the development of novel cobalt Fischer-Tropsch catalysts for synthesis of long-chain hydrocarbons and clean fuels. *Chem. Rev.* 107 (5): 1692–1744.

158 Kang, J., Cheng, K., Zhang, L. et al. (2011). Mesoporous zeolite-supported ruthenium nanoparticles as highly selective Fischer-Tropsch catalysts for the production of C_5-C_{11} isoparaffins. *Angew. Chem. Int. Ed.* 50 (22): 5200–5203.

159 Cheng, K., Kang, J., Huang, S. et al. (2012). Mesoporous beta zeolite-supported ruthenium nanoparticles for selective conversion of synthesis gas to C_5-C_{11} isoparaffins. *ACS Catal.* 2 (3): 441–449.

160 Sartipi, S., Alberts, M., Meijerink, M.J. et al. (2013). Towards liquid fuels from biosyngas: effect of zeolite structure in hierarchical-zeolite-supported cobalt catalysts. *ChemSusChem* 6 (9): 1646–1650.

161 Xing, C., Shen, W., Yang, G. et al. (2014). Completed encapsulation of cobalt particles in mesoporous H-ZSM-5 zeolite catalyst for direct synthesis of middle isoparaffin from syngas. *Catal. Commun.* 55: 53–56.

162 Kim, J.C., Lee, S., Cho, K. et al. (2014). Mesoporous MFI zeolite nanosponge supporting cobalt nanoparticles as a Fischer-Tropsch catalyst with high yield of branched hydrocarbons in the gasoline range. *ACS Catal.* 4 (11): 3919–3927.

163 Lin, Q., Yang, G., Chen, Q. et al. (2015). Design of a hierarchical meso/macroporous zeolite-supported cobalt catalyst for the enhanced direct synthesis of isoparaffins from syngas. *ChemCatChem* 7 (4): 682–689.

164 Xing, C., Yang, G., Wu, M. et al. (2015). Hierarchical zeolite Y supported cobalt bifunctional catalyst for facilely tuning the product distribution of Fischer-Tropsch synthesis. *Fuel* 148: 48–57.

165 Zhao, W., Wang, Y., Li, Z. et al. (2015). The influence of hierarchical zeolite composition and pore structure on behavior of cobalt-based Fischer–Tropsch synthesis catalysts. *J. Porous. Mater.* 22 (4): 1097–1104.

166 Kang, J., Wang, X., Peng, X. et al. (2016). Mesoporous zeolite Y-supported Co nanoparticles as efficient Fischer-Tropsch catalysts for selective synthesis of diesel fuel. *Ind. Eng. Chem. Res.* 55 (51): 13008–13019.

167 Wang, Y., Cao, X., Jiang, Y. et al. (2017). The influence of alkali treatment for synthesizing hierarchical zeolite on behavior of cobalt Fischer–Tropsch synthesis catalysts. *Catal. Surv. Asia* 21 (1): 28–36.

168 Mazonde, B., Cheng, S., Zhang, G. et al. (2018). A solvent-free *in situ* synthesis of a hierarchical co-based zeolite catalyst and its application to tuning Fischer-Tropsch product selectivity. *Catal. Sci. Technol.* 8 (11): 2802–2808.

169 Li, H., Hou, B., Wang, J. et al. (2018). Direct conversion of syngas to isoparaffins over hierarchical beta zeolite supported cobalt catalyst for Fischer-Tropsch synthesis. *Mol. Catal.* 459 (May): 106–112.

170 Wen, C., Wang, C., Chen, L. et al. (2019). Effect of hierarchical ZSM-5 zeolite support on direct transformation from syngas to aromatics over the iron-based catalyst. *Fuel* 244 (January): 492–498.

171 Bjørgen, M., Olsbye, U., and Kolboe, S. (2003). Coke precursor formation and zeolite deactivation: mechanistic insights from hexamethylbenzene conversion. *J. Catal.* 215 (1): 30–44.

172 Bjørgen, M., Joensen, F., Spangsberg Holm, M. et al. (2008). Methanol to gasoline over zeolite H-ZSM-5: improved catalyst performance by treatment with NaOH. *Appl. Catal. A Gen.* 345 (1): 43–50.

173 Wan, Z., Wu, W., Chen, W. et al. (2014). Direct synthesis of hierarchical ZSM-5 zeolite and its performance in catalyzing methanol to gasoline conversion. *Ind. Eng. Chem. Res.* 53 (50): 19471–19478.

174 Meng, L., Mezari, B., Goesten, M.G. et al. (2017). Direct synthesis of hierarchical ZSM-5 zeolite using cetyltrimethylammonium as structure directing agent for methanol-to-hydrocarbons conversion. *Catal. Sci. Technol.* 7 (19): 4520–4533.

175 Meng, L., Zhu, X., Wannapakdee, W. et al. (2018). A dual-templating synthesis strategy to hierarchical ZSM-5 zeolites as efficient catalysts for the methanol-to-hydrocarbons reaction. *J. Catal.* 361: 135–142.

176 Wang, Y.H., Wang, X.D., Xu, J.W. et al. (2018). Hierarchical ZSM-5 zeolite: preparation by sequential desilication-dealumination and catalytic performance in methanol to gasoline reaction. *Wuji Cailiao Xuebao/J. Inorg. Mater.* 33 (11): 1193–1200.

177 Sadrara, M., Khorrami, M.K., Garmarudi, A.B. et al. (2019). Optimization of desilication parameters in fabrication of mesoporous ZSM-48 zeolite employed as excellent catalyst in methanol to gasoline conversion. *Mater. Chem. Phys.* 237 (July): 121817.

178 Sadrara, M., Khanmohammadi Khorrami, M., and Bagheri Garmarudi, A. (2019). Improved catalytic performance of mesoporous ZSM-5 nanocrystalline zeolite prepared by the cationic surfactant-ammonium salt mixed agent method in the methanol to gasoline reaction. *React. Kinet. Mech. Catal.* 128 (2): 1111–1126.

179 Wu, L. and Hensen, E.J.M. (2014). Comparison of mesoporous SSZ-13 and SAPO-34 zeolite catalysts for the methanol-to-olefins reaction. *Catal. Today* 235: 160–168.

180 Zhu, X., Hofmann, J.P., Mezari, B. et al. (2016). Trimodal porous hierarchical SSZ-13 zeolite with improved catalytic performance in the methanol-to-olefins reaction. *ACS Catal.* 6 (4): 2163–2177.

181 Xi, D., Sun, Q., Chen, X. et al. (2015). The recyclable synthesis of hierarchical zeolite SAPO-34 with excellent MTO catalytic performance. *Chem. Commun.* 51 (60): 11987–11989.

182 Zhang, C., Wu, Q., Lei, C. et al. (2017). Solvent-free and mesoporogen-free synthesis of mesoporous aluminosilicate ZSM-5 zeolites with superior catalytic properties in the methanol-to-olefins reaction. *Ind. Eng. Chem. Res.* 56 (6): 1450–1460.

183 Chen, H., Wang, Y., Sun, C. et al. (2018). Synthesis of hierarchical ZSM-5 zeolites with CTAB-containing seed silicalite-1 and its catalytic performance in methanol to propylene. *Catal. Commun.* 112 (April): 10–14.

184 Chen, H., Yang, M., Shang, W. et al. (2018). Organosilane surfactant-directed synthesis of hierarchical ZSM-5 zeolites with improved catalytic performance in methanol-to-propylene reaction. *Ind. Eng. Chem. Res.* 57 (32): 10956–10966.

185 Song, G., Chen, W., Dang, P. et al. (2018). Synthesis of an excellent MTP catalyst: hierarchical ZSM-5 zeolites with great mesoporosity. *R. Soc. Open Sci.* 5 (12).

186 Ahmed, M.H.M., Muraza, O., Galadima, A. et al. (2019). Choreographing boron-aluminum acidity and hierarchical porosity in *BEA zeolite by in-situ hydrothermal synthesis for a highly selective methanol to propylene catalyst. *Microporous Mesoporous Mater.* 273 (June 2018): 249–255.

187 Ahmadpour, J. and Taghizadeh, M. (2015). Catalytic conversion of methanol to propylene over high-silica mesoporous ZSM-5 zeolites prepared by different combinations of mesogenous templates. *J. Nat. Gas Sci. Eng.* 23: 184–194.

188 Zhou, J., Teng, J., Ren, L. et al. (2016). Full-crystalline hierarchical monolithic ZSM-5 zeolites as superiorly active and long-lived practical catalysts in methanol-to-hydrocarbons reaction. *J. Catal.* 340: 166–176.

189 Wang, X., Meng, F., Chen, H. et al. (2017). Synthesis of a hierarchical ZSM-11/5 composite zeolite of high SiO_2/Al_2O_3 ratio and catalytic performance in the methanol-to-olefins reaction. *Comptes Rendus Chim.* 20 (11–12): 1083–1092.

190 Li, H., Wang, Y., Fan, C. et al. (2018). Facile synthesis of a superior MTP catalyst: hierarchical micro-meso-macroporous ZSM-5 zeolites. *Appl. Catal. A Gen.* 551 (December 2017): 34–48.

191 Selzer, C., Biemelt, T., Werner, A., and Kaskel, S. (2018). Hierarchical zeolite ZSM-58 as shape selective catalyst for methanol-to-olefins reaction. *Microporous Mesoporous Mater.* 261 (September 2017): 51–57.

192 Zhao, X., Wang, L., Guo, P. et al. (2018). Synthesis of high-Si hierarchical beta zeolites without mesoporogen and their catalytic application in the methanol to propene reaction. *Catal. Sci. Technol.* 8 (11): 2966–2974.

193 Wang, X., Zhang, J., Zhang, T. et al. (2016). Mesoporous ZnZSM-5 zeolites synthesized by one-step desilication and reassembly: a durable catalyst for methanol aromatization. *RSC Adv.* 6 (28): 23428–23437.

194 Zhou, M., Wang, F., Xiao, W. et al. (2016). The comparison of mesoporous HZSM-5 zeolite catalysts prepared by different mesoporous templates and their catalytic performance in the methanol to aromatics reaction. *React. Kinet. Mech. Catal.* 119 (2): 699–713.

195 Asghari, A., Khorrami, M.K., and Kazemi, S.H. (2019). Hierarchical H-ZSM5 zeolites based on natural kaolinite as a high-performance catalyst for methanol to aromatic hydrocarbons conversion. *Sci. Rep.* 9 (1): 1–9.

196 Shen, X., Kang, J., Niu, W. et al. (2017). Impact of hierarchical pore structure on the catalytic performances of MFI zeolites modified by ZnO for the conversion of methanol to aromatics. *Catal. Sci. Technol.* 7 (16): 3598–3612.

197 Jia, Y., Wang, J., Zhang, K. et al. (2018). Hierarchical ZSM-5 zeolite synthesized via dry gel conversion-steam assisted crystallization process and its application in aromatization of methanol. *Powder Technol.* 328: 415–429.

198 Goodarzi, F., Herrero, I.P., Kalantzopoulos, G.N. et al. (2020). Synthesis of mesoporous ZSM-5 zeolite encapsulated in an ultrathin protective shell of silicalite-1 for MTH conversion. *Microporous Mesoporous Mater.* 292 (July 2019): 109730.

199 Tang, T., Yin, C., Wang, L. et al. (2008). Good sulfur tolerance of a mesoporous beta zeolite-supported palladium catalyst in the deep hydrogenation of aromatics. *J. Catal.* 257 (1): 125–133.

200 Ju, C., Wang, Y., Huang, Y., and Fang, Y. (2015). Design of mesoporous KA zeolite supported sulfur-tolerant noble metal catalyst for naphthalene hydrogenation. *Fuel* 154: 80–87.

201 Lysenko, S.V., Baranova, S.V., Maksimov, A.L. et al. (2013). Hydrogenation of aromatic compounds in the presence of dibenzothiophene over bimetallic catalysts containing mesoporous aluminosilicates. *Pet. Chem.* 53 (2): 97–101.

202 Fu, W., Zhang, L., Wu, D. et al. (2015). Mesoporous zeolite-supported metal sulfide catalysts with high activities in the deep hydrogenation of phenanthrene. *J. Catal.* 330: 423–433.

203 Tang, T., Fu, W., Zhang, L. et al. (2016). Mesoporous zeolite ZSM-5 supported Ni2P catalysts with high activity in the hydrogenation of phenanthrene and 4,6-dimethyldibenzothiophene. *Ind. Eng. Chem. Res.* 55 (26): 7085–7095.

204 Tang, T., Yin, C., Wang, L. et al. (2007). Superior performance in deep saturation of bulky aromatic pyrene over acidic mesoporous beta zeolite-supported palladium catalyst. *J. Catal.* 249 (1): 111–115.

205 Fu, W., Zhang, L., Tang, T. et al. (2011). Extraordinarily high activity in the hydrodesulfurization of 4,6-dimethyldibenzothiophene over Pd supported on mesoporous zeolite Y. *J. Am. Chem. Soc.* 133 (39): 15346–15349.

206 Cui, T.L., Ke, W.Y., Zhang, W.B. et al. (2016). Encapsulating palladium nanoparticles inside mesoporous MFI zeolite nanocrystals for shape-selective catalysis. *Angew. Chem. Int. Ed.* 55 (32): 9178–9182.

207 Ojeda, M., Grau-Atienza, A., Campos, R. et al. (2015). Hierarchical zeolites and their catalytic performance in selective oxidative processes. *ChemSusChem* 8 (8): 1328–1333.

208 Kim, S.K., Reddy, B.M., and Park, S.E. (2018). High-performance microwave synthesized mesoporous TS-1 zeolite for catalytic oxidation of cyclic olefins. *Ind. Eng. Chem. Res.* 57 (10): 3567–3574.

209 Chen, Y., Wang, X., and Zhang, L. (2020). Synthesis and catalytic activity of hierarchical NbS-1 zeolites for cyclohexene oxidation. *Microporous Mesoporous Mater.* 295 (December 2019): 109963.

210 Cheng, W., Jiang, Y., Xu, X. et al. (2016). Easily recoverable titanosilicate zeolite beads with hierarchical porosity: preparation and application as oxidation catalysts. *J. Catal.* 333: 139–148.

211 Jesudoss, S.K., Vijaya, J.J., Sivachidambaram, M. et al. (2018). Liquid phase catalytic oxidation of toluene over rich silica and alumina composition of hierarchical ordered ZSM-5 zeolites prepared without organic templates. *J. Nanosci. Nanotechnol.* 18 (8): 5367–5379.

212 Nandi, M. and Talukdar, A.K. (2019). Ceria–zirconia solid solution loaded hierarchical MFI zeolite: an efficient catalyst for solvent free oxidation of ethyl benzene. *Arab. J. Chem.* 12 (8): 3753–3763.

213 Patra, A.K., Dutta, A., Pramanik, M. et al. (2014). Synthesis of hierarchical mesoporous Mn-MFi zeolite nanoparticles: a unique architecture of heterogeneous catalyst for the aerobic oxidation of thiols to disulfides. *ChemCatChem* 6 (1): 220–229.

214 Gao, N., Xie, S., Liu, S. et al. (2015). Development of hierarchical MCM-49 zeolite with intracrystalline mesopores and improved catalytic performance in liquid alkylation of benzene with ethylene. *Microporous Mesoporous Mater.* 212: 1–7.

215 Jin, Y., Zhang, L., Liu, J. et al. (2017). Mesopore modification of beta zeolites by sequential alkali and acid treatments: composition-dependent T-atoms removal behavior back donating to hierarchical structure and catalytic activity in benzene alkylation. *Microporous Mesoporous Mater.* 248: 7–17.

216 Miao, H.X., Ma, L., Ma, J.H. et al. (2014). Benzylation of naphthalene over a mesoporous ZSM-5 zeolite microsphere catalyst. *Wuli Huaxue Xuebao/Acta Phys. Chim. Sin.* 30 (8): 1518–1526.

217 Zhao, L., Yu, D., Dai, W. et al. (2017). Catalytic performance of micro–/micro-mesoporous H-ZSM-5 zeolites for alkylation of toluene with methanol. *Gaodeng Xuexiao Huaxue Xuebao/Chem. J. Chin. Univ.* 38 (10): 1834–1840.

218 Xu, L., Wang, F., Feng, Z. et al. (2019). Hierarchical ZSM-5 zeolite with enhanced catalytic activity for alkylation of phenol with tert-butanol. *Catalysts* 9 (2): 1–12.

219 Fu, X., Sheng, X., Zhou, Y. et al. (2016). Design of micro-mesoporous zeolite catalysts for alkylation. *RSC Adv.* 6 (56): 50630–50639.

220 Fu, W., Shen, R., Bai, E. et al. (2018). Reaction route and mechanism of the direct *N*-alkylation of sulfonamides on acidic mesoporous zeolite β-catalyst. *ACS Catal.* 8 (10): 9043–9055.

221 Hao, W., Zhang, W., Guo, Z. et al. (2018). Mesoporous beta zeolite catalysts for benzylation of naphthalene: effect of pore structure and acidity. *Catalysts* 8 (11).

222 Yutthalekha, T., Wattanakit, C., Warakulwit, C. et al. (2017). Hierarchical FAU-type zeolite nanosheets as green and sustainable catalysts for benzylation of toluene. *J. Clean. Prod.* 142: 1244–1251.

223 Musselwhite, N., Na, K., Sabyrov, K. et al. (2015). Mesoporous aluminosilicate catalysts for the selective isomerization of *n*-hexane: the roles of surface acidity and platinum metal. *J. Am. Chem. Soc.* 137 (32): 10231–10237.

224 Chu, W., Li, X., Zhu, X. et al. (2017). Size-controlled synthesis of hierarchical ferrierite zeolite and its catalytic application in 1-butene skeletal isomerization. *Microporous Mesoporous Mater.* 240: 189–196.

225 Liu, Y., Zhou, X., Pang, X. et al. (2013). Improved para-xylene selectivity in meta-xylene isomerization over ZSM-5 crystals with relatively long *b*-axis length. *ChemCatChem* 5 (6): 1517–1523.

226 Mota, F.M., Eliášová, P., Jung, J., and Ryoo, R. (2016). Mesoporous EU-1 zeolite as a highly active catalyst for ethylbenzene hydroisomerization. *Catal. Sci. Technol.* 6 (8): 2735–2741.

227 Hao, H., Chang, Y., Yu, W. et al. (2018). Microporous and mesoporous materials hierarchical porous MCM-68 zeolites: synthesis, characterization and catalytic performance in m-xylene isomerization. *Microporous Mesoporous Mater.* 263 (December 2017): 135–141.

228 Zhou, J., Liu, Z., Li, L. et al. (2013). Hierarchical mesoporous ZSM-5 zeolite with increased external surface acid sites and high catalytic performance in o-xylene isomerization. *Cuihua Xuebao/Chin. J. Catal.* 34 (7): 1429–1433.

229 Chaida-Chenni, F.Z., Belhadj, F., Grande Casas, M.S. et al. (2018). Synthesis of mesoporous-zeolite materials using beta zeolite nanoparticles as precursors and their catalytic performance in m-xylene isomerization and disproportionation. *Appl. Catal. A Gen.* 568 (October): 148–156.

230 Sun, H., Wang, A., Sun, K. et al. (2019). Effect of acidity and porosity of hierarchical HBEA zeolite on catalytic stability of α-methylnaphthalene isomerization. *J. Porous. Mater.* 26 (4): 961–970.

231 Liu, J., Xiang, M., and Wu, D. (2017). Enhanced phenol hydrodeoxygenation over a Ni catalyst supported on a mixed mesoporous ZSM-5 zeolite and Al_2O_3. *Catal. Lett.* 147 (10): 2498–2507.

232 Berenguer, A., Gutiérrez-Rubio, S., Linares, M. et al. (2019). On the feasibility of using hierarchical ZSM-5 and beta zeolites as supports of metal phosphides for catalytic hydrodeoxygenation of phenol. *Energy Technol.* 7 (6): 1–13.

233 Haynes, T., D'hondt, T., Morritt, A.L. et al. (2019). Mesoporous aluminosilicate nanofibers with a low Si/Al ratio as acidic catalyst for hydrodeoxygenation of phenol. *ChemCatChem* 11 (16): 4054–4063.

234 Wang, L., Zhang, J., Yi, X. et al. (2015). Mesoporous ZSM-5 zeolite-supported Ru nanoparticles as highly efficient catalysts for upgrading phenolic biomolecules. *ACS Catal.* 5 (5): 2727–2734.

235 Lee, H.W., Jun, B.R., Kim, H. et al. (2015). Catalytic hydrodeoxygenation of 2-methoxy phenol and dibenzofuran over Pt/mesoporous zeolites. *Energy* 81: 33–40.

236 Chen, C., Zhu, J., Chen, F. et al. (2013). Enhanced performance in catalytic combustion of toluene over mesoporous beta zeolite-supported platinum catalyst. *Appl. Catal. B Environ.* 140–141: 199–205.

237 Liu, F., Zuo, S., Wang, C. et al. (2014). Pd/transition metal oxides functionalized ZSM-5 single crystals with *b*-axis aligned mesopores: efficient and long-lived catalysts for benzene combustion. *Appl. Catal. B Environ.* 148–149: 106–113.

238 Chen, C., Wang, X., Zhang, J. et al. (2015). Superior performance in catalytic combustion of toluene over mesoporous ZSM-5 zeolite supported platinum catalyst. *Catal. Today* 258: 190–195.

239 Wang, Y., Yang, D., Li, S. et al. (2018). Ru/hierarchical HZSM-5 zeolite as efficient bi-functional adsorbent/catalyst for bulky aromatic VOCs elimination. *Microporous Mesoporous Mater.* 258: 17–25.

240 Chen, M.Q., Wang, Y., Yang, D.Y. et al. (2019). Pt supported hierarchical ZSM-5 zeolite as adsorbent/catalytic combustion catalyst for o-xylene elimination. *Wuji Cailiao Xuebao/J. Inorg. Mater.* 34 (2): 173–178.

241 Hernando, H., Moreno, I., Fermoso, J. et al. (2017). Biomass catalytic fast pyrolysis over hierarchical ZSM-5 and beta zeolites modified with Mg and Zn oxides. *Biomass Convers. Biorefin.* 7 (3): 289–304.

242 Palizdar, A. and Sadrameli, S.M. (2020). Catalytic upgrading of beech wood pyrolysis oil over iron- and zinc-promoted hierarchical MFI zeolites. *Fuel* 264 (December 2019): 116813.

243 Li, J., Li, X., Zhou, G. et al. (2014). Catalytic fast pyrolysis of biomass with mesoporous ZSM-5 zeolites prepared by desilication with NaOH solutions. *Appl. Catal. A Gen.* 470: 115–122.

244 Custodis, V.B.F., Karakoulia, S.A., Triantafyllidis, K.S., and Van Bokhoven, J.A. (2016). Catalytic fast pyrolysis of lignin over high-surface-area mesoporous aluminosilicates: effect of porosity and acidity. *ChemSusChem* 9 (10): 1134–1145.

245 Gamliel, D.P., Cho, H.J., Fan, W., and Valla, J.A. (2016). On the effectiveness of tailored mesoporous MFI zeolites for biomass catalytic fast pyrolysis. *Appl. Catal. A Gen.* 522: 109–119.

246 Lazaridis, P.A., Fotopoulos, A.P., Karakoulia, S.A., and Triantafyllidis, K.S. (2018). Catalytic fast pyrolysis of Kraft lignin with conventional, mesoporous and nanosized ZSM-5 zeolite for the production of alkyl-phenols and aromatics. *Front. Chem.* 6 (JUL): 295.

247 Jia, L.Y., Raad, M., Hamieh, S. et al. (2017). Catalytic fast pyrolysis of biomass: superior selectivity of hierarchical zeolites to aromatics. *Green Chem.* 19 (22): 5442–5459.

248 Bi, Y., Lei, X., Xu, G. et al. (2018). Catalytic fast pyrolysis of Kraft lignin over hierarchical HZSM-5 and Hβ zeolites. *Catalysts* 8 (2): 82.

249 Hertzog, J., Carré, V., Jia, L. et al. (2018). Catalytic fast pyrolysis of biomass over microporous and hierarchical zeolites: characterization of heavy products. *ACS Sustain. Chem. Eng.* 6 (4): 4717–4728.

250 Tang, S., Zhang, C., Xue, X. et al. (2019). Catalytic pyrolysis of lignin over hierarchical HZSM-5 zeolites prepared by post-treatment with alkaline solutions. *J. Anal. Appl. Pyrolysis* 137: 86–95.

251 Kim, Y.M., Jeong, J., Ryu, S. et al. (2019). Catalytic pyrolysis of wood polymer composites over hierarchical mesoporous zeolites. *Energy Convers. Manag.* 195: 727–737.

13

Colloidal Self-assembly of Block Copolymers for Drug Loading and Controlled Release

Guangze Yang, Yun Liu*, and Chun-Xia Zhao*

The University of Queensland, Australian Institute for Bioengineering and Nanotechnology, Corner College and Cooper Road, St. Lucia, QLD 4072, Australia

13.1 Introduction

Block copolymers contain two or more chemically distinct monomer clusters that form blocks of repeating units. The phase separation between two (or more) blocks takes place only within a size range of tens of nanometers because of chain connectivity and thermodynamic inconsistency [1, 2]. Various block copolymers have been developed with different structures including AB, ABA, BAB, and ABC (Figure 13.1). These block copolymers have demonstrated great variety in morphology and characteristics depending on their molecular architectures and the constitution of the block copolymers [1].

Self-assembly of block copolymers has attracted great attention in the past decades for fabricating a vast variety of functional nanomaterials for drug delivery and controlled release. It has been regarded as one of the most commonly used bottom-up methods in nanotechnology. Block copolymers have been explored for producing various well-organized structures such as spherical and cylindrical micelles, nanoparticles, vesicles, and other morphologies [1, 3, 4], depending on the mass fraction of each block, molar mass, and polymer composition. By virtue of their attractive properties such as biocompatibility, degradability, and tunability, polymeric colloidal nanoparticles of block copolymers have been applied to deliver drugs and imaging agents for various biomedical applications [5, 6]. Since the development of poly(ethylene glycol) (PEG) for prolonging the circulation time of nanoparticles *in vivo* in 1994 [7], it has become the most extensively used hydrophilic block of copolymers due to its high water solubility, low toxicity, excellent biocompatibility, and unique anti-fouling performance [8]. In combination with other hydrophobic or hydrophilic blocks, a wide variety of block copolymers have been developed for various applications [9, 10].

This chapter focuses on colloidal self-assembly block copolymers for drug loading and controlled release. Different types of block copolymers, including

* Guangze Yang and Yun Liu are contributed equally.

Functional Materials from Colloidal Self-assembly, First Edition. Edited by Qingfeng Yan and George Zhao.

Figure 13.1 Various block copolymers.

di- and triblock copolymers are introduced, and their self-assembly mechanisms are discussed. In addition, various polymeric nanostructures such as micelle, nanoparticle, and vesicle and their applications are presented and compared. Several strategies are elaborated for achieving single-, dual-, and even multiple-drug loading in polymeric nanoparticles. Finally, various stimuli-responsive polymeric nanoparticles are discussed for triggered release applications.

13.2 Block Copolymers

13.2.1 Diblock Copolymers

Diblock copolymers are the most widely used block copolymers. They include two blocks of homopolymers in a linear sequence and usually contain a hydrophobic block and a hydrophilic block, so they can form nanocarriers by self-assembly [11]. Recently, various synthetic diblock copolymers with different combinations of polymer blocks have been developed to produce polymeric nanoparticles having attractive properties such as excellent biocompatibility and stability, controllable degradability, high drug loading, and controlled release [12–24].

Biodegradable polymers are defined as polymers that can be degraded hydrolytically or oxidatively by enzymes of microorganisms [25]. Aliphatic polyesters from bacterial or chemical synthesis have been developed as a major type of biocompatible and biodegradable polymers [26]. Biodegradable polyesters, such as poly(lactic-*co*-glycolic acid) (PLGA), polycaprolactone (PCL), and poly(lactic acid) (PLA), have attracted significant interests for various biomedical applications due to their excellent biocompatibility and biodegradability by non-enzymatic or enzymatic hydrolysis [27]. Therefore, they have great potential in addressing not only environmental concerns about nondegradable polymer pollutions but also urgent biomedical needs for biodegradable polymers for human clinical applications [26]. Polyester-based amphiphilic block copolymers have been extensively used for the production of nanoparticles for a wide range of drug delivery applications.

PLA as a biodegradable and biocompatible material has attracted significant interest in biomedical applications due to their successful clinical translation [28]. It has many different structures, such as poly(L-lactide) (PLLA) and poly(D-lactide) (PDLA) [29]. As the polyester-based amphiphilic diblock copolymers with a hydrophilic portion of PEG, the enantiomeric PLLA-PEG and PDLA-PEG are widely used for producing drug-loaded polymeric NPs by self-assembly. Two approaches have been developed. One approach is to form nanoparticles via self-assembly

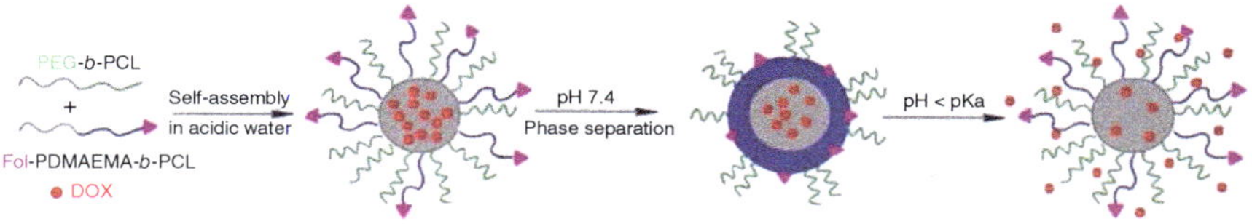

Figure 13.2 Synthesis of PCL-based micelles for controlled DOX release. Source: Zhao et al. [35]/with permission from John Wiley & Sons, Inc.

with a drug incorporated in the nanoparticles through the noncovalent interaction between the drug and the amphiphilic copolymer. Using this approach, a chemotherapeutic drug paclitaxel (PTX)-loaded PDLA-PEG NPs were synthesized with a drug loading of 4.73 wt% and an encapsulation efficiency of 99.38% [30]. On the other hand, drug molecules can be conjugated to a copolymer, and then the self-assembly of this drug–polymer conjugates induces the formation of drug-loaded polymeric nanoparticles. For example, a drug cabazitaxel was conjugated with PDLA-PEG to form a polymeric prodrug [31]. As both cabazitaxel and PDLA are hydrophobic and PEG is hydrophilic, this amphiphilic polymer–drug conjugate can self-assemble into stable colloidal nanoparticles. The resultant NPs had a size of around 30 nm and narrow size distribution with drug loadings of 4.9–10.6%. Unlike pure cabazitaxel NPs having an initial burst release, the polymeric NPs showed a slower and sustained release. *In vivo* studies showed that the polymeric NPs had higher antitumor efficacy. Similarly, the polyester-based amphiphilic diblock copolymer poly(ester amide)–poly(ethylene oxide) (PEA-PEO) was conjugated with PTX via ester linkages and assembled into micelles by nanoprecipitation. The drug loading of PTX was around 11% and the covalently conjugated PTX demonstrated a prolonged release [32].

PLGA-PEG is another widely used polyester-based amphiphilic diblock copolymer, owing to its excellent biocompatibility and biodegradability. PLGA and PEG have already been approved by the US Food and Drug Administration (FDA) for drug delivery applications [33]. The polyester-based amphiphilic diblock copolymer poly(epsilon-caprolactone)–poly(ethylene glycol) (PCL-PEG) can self-assemble into spherical or wormlike micelles [34]. By changing the molecular weight of the PCL block, different drug release can be achieved. Shorter PCL blocks exhibited faster release than those with longer PCL chains because of the higher hydrophobicity of the longer ones resulting in their stronger hydrophobic interactions with drugs. Additionally, the introduction of glycolide (GA) moieties (P(CL/GA)-PEG) can enhance drug release because the chain structures of the PCL blocks are disrupted, resulting in much faster chain cleavage [10]. Moreover, combining two polyesters (e.g. PCL-PEG and folate-PDMAEMA-PCL) offers additional functions for micelles such as pH-responsive drug release, which is attractive for cancer targeted drug delivery (Figure 13.2) [35].

Other amphiphilic diblock copolymers, for instance, poly(oxyethylene)–poly(oxybutylene)(PEO-PBO), have also been synthesized with a lower critical micelle concentration (CMC) than that of the polyester-based amphiphilic block copolymer PEG-PDLA [30]. Therefore, the micellization of PEO-PBO happened at a lower concentration resulting in higher stability when diluted in plasma. Similar to the PEG-PDLA, the PEO-PBO can self-assemble to form NPs, and hydrophobic drugs such as PTX can be loaded into the hydrophobic portions of PEO-PBO NPs through hydrophobic interactions with a drug loading of 4.62% and encapsulation efficiency of 97.03%.

13.2.2 Triblock Copolymers

Triblock copolymers are composed of ABA and BAB types with hydrophobic A-blocks (usually biodegradable polyester) and hydrophilic B-blocks, as well as ABC type from three different blocks. The ABA triblock copolymers have been used to deliver hydrophilic macromolecular drugs, such as proteins and peptides for more than 20 years [36]. With the smart design, ABA triblock copolymers can swell fast when contacting with water, followed by forming a biodegradable, physically cross-linked hydrogel [37]. For instance, when the content of A-block is high for the triblock copolymers, the copolymers become water-insoluble but can swell in water. Moreover, the ABA- or BAB-type copolymers having low molecular weight are thermosensitive, for instance, PLGA-PEG-PLGA and PEG-PLGA-PEG. Polymer–polymer attraction is growing with the increase of temperature, facilitating micelle formation [38]. These thermosensitive triblock copolymers have been applied for *in situ* gelling applications [39].

Thermo-responsive copolymers have been widely used for controlled release in the past two decades. Thermosensitive hydrogels prepared by low molecular weights ABA- or BAB-type copolymers are attractive as injectable biomaterials, as they can switch quickly from fluid having low viscosity to gel with high viscosity under certain temperature [40]. Therefore, these hydrogels are very attractive for topical or parenteral injection for a site-specific action. Moreover, the drug release rate can be controlled easily by altering material properties.

Poly(ethylene glycol)-*b*-poly(propylene glycol)-*b*-poly(ethylene glycol) (PEG-PPG-PEG) is a BAB-type triblock copolymer. It has been commercialized as a non-ionic surfactant (Pluronic) [41]. The Pluronics have diverse states from liquid to paste to solid based on their molar weight and the ratio of A-block to B-block. The PPG blocks are hydrophobic and PEG blocks are hydrophilic under room temperature. When the concentration of these amphiphilic copolymers is higher than their CMC, they self-assemble into micelles in water. When the concentration of copolymers is higher than their critical gelation concentration (CGC), some Pluronic solutions can form thermosensitive hydrogels. However, Pluronic hydrogels are not biodegradable, which hinders their applications as drug carriers.

The triblock copolymers with PLGA as A-block and PEG as B-block are the most generally used among the thermo-responsive polymers. The solutions of these polymers experience a revocable sol–gel change from room temperature to a body temperature, resulting in the formation of a highly viscous gel [42]. The ABA-type triblock copolymers PLGA-PEG-PLGA are especially appealing because they are not only biodegradable but also have a good safety profile. Contrary to Pluronics, the PLGA-PEG-PLGA copolymers are biodegradable and biocompatible with no systemic toxicity, and both *in vitro* and *in vivo* experiments confirmed that the gel state of PLGA-PEG-PLGA copolymers can persist for a longer time [41]. Compared to the diblock copolymers (such as the PLGA-PEG), the thermosensitive triblock copolymers (such as the PLGA-PEG-PLGA) have some extra advantages. Firstly, the thermosensitive triblock copolymers do not require any organic solvent for drug loading [40, 43]. Secondly, they can deliver both hydrophobic and hydrophilic drugs, such as paclitaxel [44], spironolactone, and ketoprofen [45], due to their hydrophilicity at sol state, and they can also stabilize and solubilize biologic drugs, such as insulin [46], calcitonin [47], 5-fluorouracil [48], porcine growth hormone [49], testosterone, etc. [50] Bioactive molecules can be incorporated by simply mixing with sols, and the mixture can be sterilized by filtration. Moreover, the sols are injectable, and if the body temperature is higher than the gelling temperature, *in situ* gelling occurs rapidly after injection [41]. Besides the PLGA-PEG-PLGA copolymers, other ABA-type triblock copolymers such as the PCL-PEG-PCL can form micelles for drug loading using a pH-induced self-assembly method. Also, these micelles demonstrated a considerably improved thermodynamic stability against dilution due to their low CMC [51].

ABC-type triblock copolymers comprise of three different blocks A, B, and C. As a result, their structural variety is significantly increased [52]. When using ABC triblock copolymers, solvents are often selected for either dissolving the two consecutive blocks (A and B or B and C) or one terminal block (A or C) [53, 54] to form core–shell–corona spheres [55] or cylinders [56]. For instance, an ABC-type triblock copolymer consisting of poly(tert-butyl acrylate)-block-poly(2-cinnamoyloxyethyl methacrylate)-block-poly(glyceryl monomethacrylate) (PtBA-*b*-PCEMA-*b*-PGMA) can be dissolved in a solvent mixture of methanol and pyridine at different volume ratios [57]. With the increase of methanol volume ratio from 80 to 95%, particle morphology varied from spherical, cylindrical micelles, to vesicles. When using 100% of methanol, tubular and vesicular micelle-like aggregates coexisted. ABC-type triblock copolymers can also be used to load drugs. An amphiphilic triblock copolymer consisting of a hydrophobic poly(methyl methacrylate) (PMMA) block, a middle poly(acrylic acid) (PAA) block, and a hydrophilic PEG block can self-assemble in aqueous solutions. Hydrophobic molecules such as an anticancer drug cabazitaxel and Nile Red dye were co-loaded during the formation of micelles [58]. The PEG-PAA-PMMA micelles demonstrated a pH-responsive drug release.

13.2.3 Self-assembly Mechanism

Amphiphilic block copolymers tend to self-assembly in solutions forming core–shell micelles or nanoparticles with the insoluble hydrophobic blocks at the core covered with hydrophilic blocks at the shell [59–61]. This interfacial activity of amphiphilic block copolymers has been exploited as colloidal dispersants, surface modifiers, and drug carriers [62]. Moreover, hydrophobic NPs can be encapsulated in micelles or nanoparticles by an interfacial instability method. Amphiphilic block copolymers and hydrophobic NPs are dissolved in a water-immiscible volatile solvent followed by emulsification [63]. The self-assembly of block copolymers and NP encapsulation were induced by transient interfacial instabilities at the surface of an emulsion droplet [64].

For thermo-responsive triblock copolymers such as PLGA-PEG-PLGA, a stepwise mechanism explains their physical gelation process in water. Amphiphilic triblock copolymers self-assemble into micelles firstly with the hydrophobic PLGA blocks forming the core and the hydrophilic PEG blocks as the coronas [41]. With the increase of temperature, the formed micelles in the solution are further associated to form a gel with an integral micellar structure retained during the phase transition. Further increasing temperature could lead to micelle-network coarsening. Consequently, the PLGA-PEG-PLGA precipitates with the increase of the molecular motion of hydrophobic PLGA blocks and the collapse of the micellar structure due to the high hydrophobicity of the block copolymers at high temperatures [65]. Hydrophobic interaction is the thermodynamic driving force for the physical gelation of thermo-responsive triblock copolymers [66]. Therefore, the balance of hydrophilic and hydrophobic blocks is critical to induce a thermos-reversible phase transition. Normally, longer hydrophobic blocks result in a lower sol–gel transition temperature, a lower CGC, and a broader gel window with the same hydrophilic block length. Moreover, ABA-type copolymers often exhibit a relative lower sol–gel transition temperature as well as a lower CGC compared with BAB-type copolymers.

13.3 Nanoscale Structures

Three different nanostructures are summarized in this section including micelles, nanoparticles, and vesicles or polymersomes. Despite the ambiguity and discrepancy in defining micelles and nanoparticles [67], this chapter differentiates micelles, nanoparticles, and vesicles (polymersomes) based on the fraction of the hydrophilic portion in amphiphilic copolymers. Generally, when the hydrophilic fraction of block copolymers is higher than 45–50%, micelles are formed in most cases. When the hydrophilic fraction is less than 25%, solid nanoparticles can be obtained. When the hydrophilic fraction is around $35 \pm 10\%$, vesicles are often observed [68]. Although this simple and basic empirical definition is applicable to most of the systems, the exact nanostructure is subject to synthesis conditions as well as the properties of block copolymers. Many factors of the block copolymers, such as hydrophobicity, length of the hydrophobic blocks, fraction of the hydrophilic block, the chain length, the ratio of the hydrophilic and hydrophobic blocks, etc., play critical roles in affecting the formation of various nanostructures as well as their properties and their practical applications [59].

13.3.1 Micelles

Polymeric micelles have spherical inner cores and an outer shells via self-assembly of synthetic block copolymers or graft copolymers. Generally, the size of polymeric micelles ranges from 10 to 100 nm with narrow size distributions [69].

Compared to low-molecular-weight surfactants, most polymeric micelles have lower CMC values from 1 to 10 μg/ml. However, the CMC of nonionic surfactants, such as tween-80 is up to 3500 μg/ml [70]. Although polymeric micelles and low-molecular-weight-surfactant micelles are both micelles, they have different physicochemical properties. The higher structural stability of polymeric micelles is important for their drug delivery applications to avoid undesirable premature disassembly.

When the concentration of an amphiphilic polymer is above the critical aggregation concentration (CAC), hydrophobic interactions between the hydrophobic chains induce the micelles formation. Polymeric micelles have many advantages such as good biodegradability, biocompatibility, and thermodynamical stability. With their small size (10–100 nm) in combination with PEGylation for a long circulation effect and the loading of insoluble drugs into the hydrophobic cores, these systems are desirable for intravenous drug delivery [71].

Diblock copolymers have been widely used for making micelles. PLGA-PEG copolymer forms micelles when the PEG fraction is higher than 45% [72]. At concentrations below the CAC, PLGA-PEG copolymer exists unimers. While at concentrations above CAC, it self-assembles to micelles. It is important to have extremely low CAC; otherwise the formed micelles would disassemble upon dilution [71]. In addition to PEG, other hydrophilic polymers have also been used as the hydrophilic block. The self-assembly of a poly(epsilon-caprolactone)-*b*-poly(methacrylic acid) (PCL-*b*-PMAA) copolymer also creates polymer micelles with the PCL block forming the core and the PMAA at the shell [73].

Micelle systems have been utilized to deliver both various drugs and hydrophobic metal nanoparticles. PLGA-PEG micelles were synthesized to encapsulate an anticancer drug PTX and iron oxide nanoparticles [74]. Similarly, PEG–poly(amino acid) block copolymer micelles were produced to encapsulating hydrophobic drugs indomethacin and resveratrol [75]. To achieve effective drug delivery, micelles are capable of achieving drug loading higher than 10 wt%. PAA-PEG micelles were created with a kinase inhibitor dactolisib conjugated to the PAA block. A high dactolisib loading of 17% was obtained. More importantly, these drug-loaded micelles demonstrated good colloidal stability and a sustained drug release over 96 hours [76].

13.3.2 Nanoparticles

Polymeric nanoparticles are often synthesized using copolymers having low hydrophilic fractions. For instance, when the PEG fraction is under 25%, the PLGA-PEG copolymers are likely to form solid-like nanoparticles [72, 77], nanospheres [78], or nanocapsules [79] due to high hydrophobicity. Generally, such nanoparticles are fabricated using two approaches. The traditional one is nanoprecipitation, in which a solvent solution containing polymers and drugs is mixed with an antisolvent solution (water or buffer) inducing the self-assembly of polymers to form drug-loaded polymeric nanoparticles. Particle size and drug loading of these polymeric nanoparticles are dependent on many factors, such as temperature, mixing, the ratio of solvent to water, and the type of solvents [11, 71]. Liu et al.

fabricated a combinatorial library of PLGA-PEG nanoparticles with various ligand density, ratios, and lengths using a one-step microfluidic nanoprecipitation method [77]. Several review articles have discussed this traditional nanoprecipitation in details [11, 80, 81]. Another method is the emulsion/solvent evaporation approach. Typically, a polymer and a drug are dissolved in a solvent, which is then mixed with an aqueous phase through an emulsification process using high-pressure homogenizer. Then, evaporation of the solvent leads to the formation of solid nanoparticles containing the polymer and drug [71]. Although the O/W single emulsion method is usually utilized to encapsulate hydrophobic drugs, hydrophilic drugs such as proteins can be incorporated using W/O/W double emulsion methods.

13.3.3 Vesicles or Polymersomes

Polymeric vesicles, or polymersomes, are polymeric capsules with a bilayer membrane of amphiphilic block copolymers. Their particle sizes range from nano- to micrometer [68]. The name of vesicles was firstly coined in 1999 to differentiate themselves from liposomes [82]. The formation of vesicles is dependent on the hydrophilic fraction of the polymers. For instance, when the PEG fraction of the PLGA-PEG copolymers is between 25% and 45%, the copolymers intend to self-assemble to form vesicles [72]. Polymeric vesicles or polymersomes have been exploited for drug delivery and imaging applications.

Vesicles and liposomes have a similar structure composed of a bilayer membrane, but they also have many differences. In most cases, the building blocks of liposomes are naturally occurring phospholipids with a molecular weight below 1 kDa [83, 84]. In contrast, vesicles are assembled by amphiphilic block copolymers with a molecular weight up to 100 kDa, which is much bigger than phospholipids. Compared to those phospholipids for liposomes, the polymer chains for forming vesicles are longer in size and more flexible in conformation, offering improved toughness, decreased permeability, and less fluid [85, 86]. The wide variety of block copolymers with different properties, such as different molecular weights, block fraction, and block architecture, offers polymeric vesicles with tunable membrane properties [68, 82].

Vesicles have been prepared by various methods. One of the mostly used methods is nanoprecipitation, also called "solvent switch" [87, 88]. This nanoprecipitation is a simple, quick, and cheap method for forming vesicles, and the quality of the formed vesicles depends not only on the nature of the block copolymers but also on the process parameters of the nanoprecipitation. An alternative method is a technique based on rehydration. Typically, a polymer film is casted onto a substrate and an aqueous solution is utilized to rehydrate the polymer to produce vesicles. Additionally, approaches such as electroformation, bulk rehydration, and rehydration from pretreated substrates have also been utilized [80]. The selection of synthesis method has a great impact on the properties of the resulting vesicles. Due to the coexistence of the aqueous lumen and bilayer of the vesicles, they can be used to deliver both hydrophobic and hydrophilic drugs similar to the liposomes. Furthermore, polymeric vesicles have a more stable and less leaky membrane, which is particularly beneficial for eliminating the uncontrolled drug release thus improving their stability during blood circulation.

13.4 Drug Loading

Most commercial hydrophobic anticancer drugs are formulated with surfactants or oil to solubilize water-insoluble drugs, for example, Cremophor EL (Taxol®) and tween-80 (Taxotere®) [89]. However, the solvent Cremophor EL causes severe side effects, including neurotoxicity, hypersensitivity reactions, and hyperlipidemia, and the presence of surfactant tween-80 leads to hemolysis [71, 90]. Besides, they are unstable upon administration, because drugs precipitate rapidly when these formulations are injected into blood stream due to the large volume dilution, low water solubility of these hydrophobic drugs, and high CMC concentration [71]. Compared to surfactant micelles, polymer nanostructures are more biocompatible and biodegradable, so less toxic. More importantly, they are more stable and have better PK and biodistribution. Therefore, polymeric nanosystems represent an attractive and alternative nanocarriers for drug delivery and controlled release [91].

13.4.1 Single Drug Encapsulation

Nanoprecipitation has been utilized to produce a broad range of polymeric nanocarriers for encapsulating both hydrophobic and hydrophobic therapeutics. It has been reported that more than 90% of drugs under current development and about 40% of approved clinical drugs have low water solubility [92–95]. It is important to formulate these hydrophobic drugs into nanoparticle formulations. Generally, single drug-loaded polymeric nanocarriers can be manufactured using various methods such as adsorption of the drug on particle surfaces, co-precipitation of the drug and polymer by microfluidic nanoprecipitation [96–101], flash nanoprecipitation [102–104], covalent conjugating the drug to the polymer [105], or sequential precipitation [106, 107].

Drugs can be incorporated into polymeric nanocarriers by the co-precipitation of a polymer and a drug. Amphiphilic block copolymers such as PCL-PEG [108], PLA-PEG [109], and PLGA-PEG [110] have been used extensively in formulating hydrophobic drugs. For example, the amphiphilic block copolymers PCL-PEG self-assembled into micelles with doxorubicin (DOX) incorporated using nanoprecipitation [111]. The micelles exhibits a quicker release of DOX at pH 5.0 but slower at pH 7.4 under 37 °C. Interestingly, the size of micelles raised from 22.9 to 104.9 nm and the drug loading of DOX increased from 3.1% to 4.4% with the increase of the PCL length (hydrophobic block) from 2.5 to 24.7 kDa. In contrast, the length of the PEG (hydrophilic block) showed no obvious effect on drug loading.

In addition to the composition of the block copolymer, the solubility of the amphiphilic polymers and drugs in an organic solvent is vital for controlling particle morphology and drug loading. Adjusting the solubility of polymers and drugs by using different solvent mixtures [106] or via chemical modification [112]

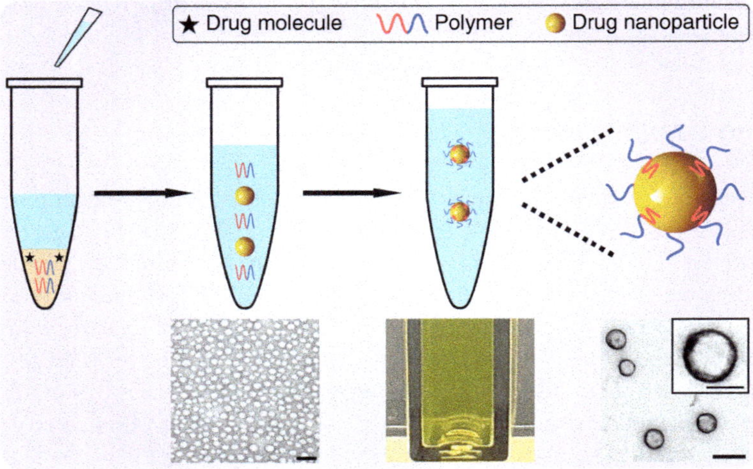

Figure 13.3 A sequential nanoprecipitation method for making polymeric nanoparticles with high drug loading (up to 58.5%). Source: Liu et al. [106]. Reproduced with permission of Wiley-VCH Verlag GmbH & Co. KGaA, Weinheim.

offers better control over particle morphology and size as well as improving drug loading [113]. Traditional nanoprecipitation approaches often lead to nanoparticles with low drug loading (<10 wt%) due to the intrinsic difference in the precipitation times of the polymer and drug. Therefore, it is critical to develop new method for improving drug loading [114]. Liu et al. recently developed a robust, rapid, and simple sequential nanoprecipitation method to prepare high drug loading (up to 58.5%) nanoparticles with drug-core and polymer-shell from an extensive range of polymers (PLGA-PEG, PLGA, PLA-PEG, and shellac), and drugs (docetaxel, paclitaxel, ketamine, curcumin, amphotericin B, ibuprofen, bulleyaconitine A, and scutellarin) [106, 107]. Their strategy is to tune the solubility of drug and polymer using a solvent mixture containing two or three solvents instead of a single solvent, so that the drug can precipitate first followed by immediate precipitation of the polymer, thus forming drug-core polymer-shell nanoparticles with tunable drug loading (Figure 13.3). Their polymeric nanoparticles with high drug loading exhibited programmed release and good long-term stability.

13.4.2 Dual Drug Encapsulation

Co-delivery is a promising strategy to overcome the restrictions of a single drug as well as to achieve the synergistic effect of two drugs having different functions in clinics. Block copolymers are capable of loading dual drugs with either similar or opposite hydrophobicity [115].

13.4.2.1 Similar Hydrophobicity

Polymeric nanoparticles and micelles are often used to co-deliver two hydrophobic drugs. Dual-drug-loaded nanoparticles are generally prepared using nano-precipitation with two drugs encapsulated via hydrophobic interactions between the two drugs as well as between the hydrophobic drugs and the hydrophobic blocks of the block copolymers. A reduction-sensitive lipid–polymer hybrid nanoparticles composed of monomethoxy-poly(ethylene glycol)-*S*-*S*-hexadecyl, soybean lecithin, and PLGA were synthesized by self-assembly [116]. The DOX and a Chinese herb extract triptolide-loaded lipid–polymer hybrid nanoparticles exhibited a reduction sensitive release profile. Also, they showed a high level of synergistic effects *in vitro* and *in vivo* with enhanced cell uptake and anticancer efficacy. Similarly, Chen et al. reported a dual–drug-loaded polymeric micellar system with two types of hydrophobic anticancer drugs, DOX and methotrexate, by a one-pot strategy [117]. Basically, micelles formation and drug encapsulation took place simultaneously under simple dialysis. The micelles were pH-responsive with a quicker drug release at acidic pH. This dual–drug-loaded system demonstrated a synergistic effect against cancer cells with significantly reduced half-maximum inhibition concentrations.

13.4.2.2 Opposite Hydrophobicity

Polymeric vesicles or polymersomes are often used for co-delivery of a hydrophilic and a hydrophobic drug because of the hydrophobic biolayer and hydrophilic vesicle core structure. So, hydrophobic drugs can be loaded in the bilayer, while hydrophilic drugs can be encapsulated in the core. For instance, vesicles made of the carboxylic-functionalized polycaprolactone block copolymer (CPCL) were reported for dual-drug delivery and pH-responsive release for oral delivery [118]. The amphiphilic diblock polymers PEG-CPCL were self-assembled into 100–250 nm vesicles in water. These vesicles showed the ability to encapsulate both hydrophobic drugs such as ibuprofen and camptothecin in their bilayer and hydrophilic molecules (Rhodamine B) in the core. These pH-responsive PCL vesicles were stable in strongly acidic conditions similar to the stomach but disassembled quickly under neutral or basic pH conditions, mimicking the small intestine to release the loaded cargoes. The drug release kinetics of the dual–drug-loaded vesicles mainly followed the diffusion-controlled process under the simulated gastrointestinal tract. This type of pH stimuli-responsive polymer vehicles opens new opportunities for oral delivery of multiple drugs with different hydrophobicity.

Core–shell NPs also provide a valuable system for loading dual drugs by virtue of their core–shell structures [119–125]. For instance, polymeric nanoparticles composed of a PLGA core and a monomethoxy polyethylene glycol graft-chitosan (mPEG-g-CS) shell were developed to co-deliver a hydrophilic chemotherapeutic drug epirubicin and a hydrophobic antiangiogenic drug PTX [126]. As chemotherapeutic drugs are commonly used in killing tumor cells, and antiangiogenic drugs are effective in limiting the growth of tumor blood vessels, thus inhibiting the growth of solid tumor [127, 128], the combination of these two types of drugs could lead to synergic effect in cancer treatment [129]. Similar to polymeric vesicles, this double-layer core–shell structure NP had the hydrophilic drug loaded inside the core and the hydrophobic drug in the outer layer. The dual–drug-loaded core–shell NPs showed enhanced antitumor and antiangiogenic effects, prolonged drug retention, improved distribution, and notably reduced microvessel density

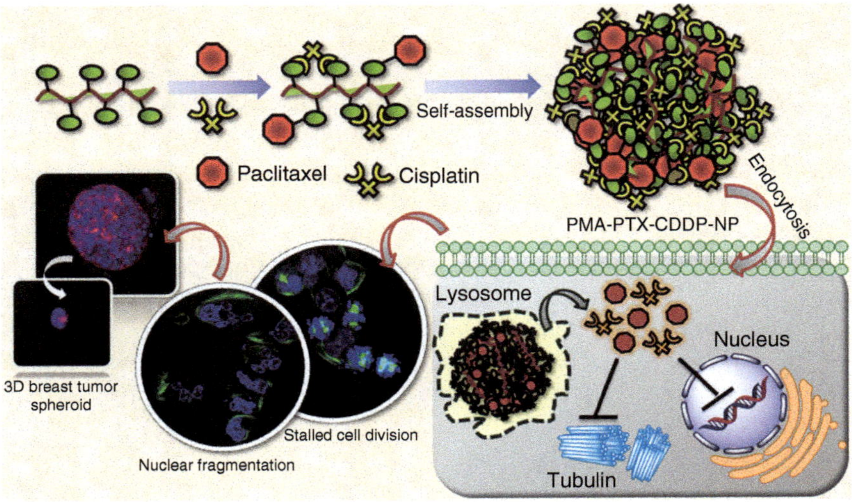

Figure 13.4 Polymeric nanoparticles for co-delivery of a hydrophobic drug PTX and a hydrophilic drug cisplatin. Source: Palvai et al. [130]/with permission from American Chemical Society.

in tumor tissues. Therefore, these approaches represent new strategies for the combination of chemotherapy and antiangiogenesis therapy for controlled drug delivery applications.

Polymeric nanoparticles are also able to deliver two chemotherapeutic drugs with different hydrophobicity. For example, Palvai et al. developed a polymeric nanoparticle containing a hydrophobic antiangiogenic drug PTX and hydrophilic chemotherapeutic drug cisplatin (Figure 13.4) [130]. The polymers were conjugated with the PTX by ester linkage followed by another conjugation with the hydrophilic cisplatin. Upon self-assembly, the drug-modified polymers formed spherical nanoparticles. The dual-drug-loaded nanoparticles exhibited a significant enhanced cancer cell death.

13.4.3 Multiple Drugs

Multiple-drug delivery offers great values in improving anti-cancer efficacy through a synergistic effect, reducing adverse effects, and delaying the onset of drug resistance [131]. Co-delivery of various antitumor drugs in one nanocarrier is promising for addressing some of challenges in cancer treatment. A successful combinational therapy depends on three important parameters: ratiometric dose control, similar pharmacokinetic profile, and suitable drug grouping. Among these parameters, the ratiometric dose control is critical, but remains challenging in combinational nanomedicine [132].

The easiest method to load multiple drugs in single polymeric nanocarriers was to use polymer–drug co-assembly. Shin et al. prepared three-drug-loaded PLA-PEG micelles via self-assembly [133]. Three anticancer drugs PTX, etoposide, and 17-allylamino-17-demethyoxygeldanamycin (17-AAG) were dissolved in acetonitrile along with the PLA-PEG copolymers followed by evaporation and rehydration to obtain the drug-loaded PLA-PEG micelles. The drug loading of the three-drug-loaded micelles was about 26.2%, which was much higher than single-drug-loaded micelles produced by the same material and method. The three-drug-loaded micelles exhibited the best stability and slowest release rate compared to single- and two-drug-loaded micelles because of the stabilization ability of 17-AAG. However, the multiple-drug-loaded polymeric nanocarriers prepared using this method were unable to achieve ratiometric dose control.

Drug-loaded polymeric nanocarriers via conjugation have demonstrated their potentials in achieving ratiometric dose control. Generally, drugs are conjugated to polymers at a predefined molar ratio followed by self-assembly, thus forming drug-loaded polymeric nanocarriers with a precise control over the ratiometric dose of multiple drugs. The drug–polymer conjugation can be implemented in two different ways: one is to link one polymer with multiple drugs at a certain ratio, and another is to link one drug to a polymer, then mix several different drug–polymer conjugates together for self-assembly. Cyclodextrin-containing amphiphilic polymer micelles have been reported for loading three hydrophobic drugs, i.e. DOX, camptothecin, and curcumin in a single nanocarrier (Figure 13.5) [132]. Beta-cyclodextrin was firstly conjugated to an amphiphilic copolymer methoxyl poly(ethylene glycol)–poly(glutamic acid) (mPEG-PGlu). As cyclodextrins are easy to form inclusion complexes with hydrophobic composites, each drug can be physically trapped in the beta-cyclodextrin-modified polymer through guest–host complexation. The dose- and ratio-controlled multiple-drug-loaded micelles were prepared by the self-assembly of the three drug-loaded polymers with a predetermined ratio. These multiple-drug-loaded micelles showed high drug

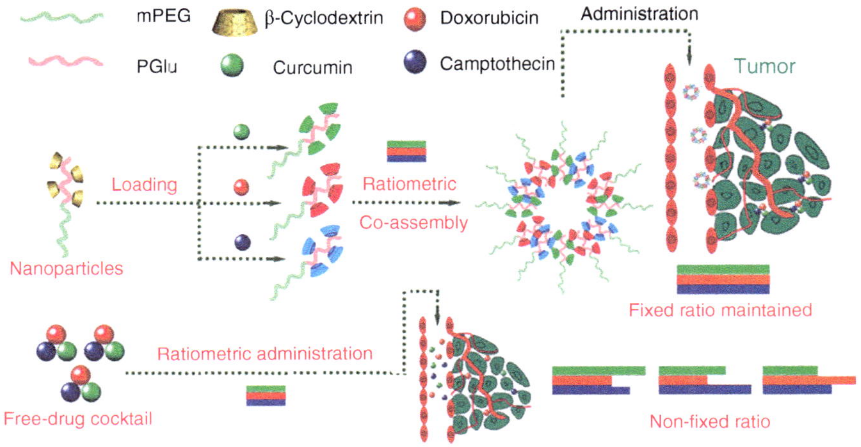

Figure 13.5 Co-assembled mixed micelles for ratiometric loading and delivery of three model drugs. Source: Chen et al. [132]. Reproduced with permission of Elsevier B.V.

loading of 22.5% for all three drugs with 6.0% for curcumin, 7.5% for camptothecin, and 9.0 % for DOX, respectively.

13.4.4 Drug and Imaging Reagent

The term "theranostics" emerging in 2002 was defined as a material that combines the modalities of both therapy and diagnostic imaging [134]. So theranostics can deliver therapeutics and diagnostic agents simultaneously. As a combined nanotherapeutic system, theranostic nanomedicine can be exploited to diagnose, deliver therapy, and monitor the response to therapy [135]. Many interesting nanostructures have been developed for bioimaging including fluorescent proteins [136], organic dyes [137–139], metal complexes [140, 141], semiconductor nanocrystals [142], and upconversion nanophosphors [143]. Upconversion is a nonlinear optical process that converts two or more low-energy excitation photons to produce an anti-Stokes higher-energy luminescence [144, 145]. Lanthanide-doped upconversion nanoparticles in combination with therapeutic drugs are promising for theranostic applications. For instance, a one-step self-assembly method was developed to formulate multifunctional PLGA nanoparticles to encapsulate the antitumor drug DOX and upconversion nanoparticles (NaYF4:Yb, Er@NaGdF4) simultaneously for drug delivery and imaging (Figure 13.6) [146]. In addition to fluorescence imaging, lanthanide-doped upconversion nanoparticles having paramagnetic gadolinium ions (Gd^{3+}) have also been exploited as contrast agents for magnetic resonance imaging (MRI).

Semiconducting polymer nanoparticles have become a promising material for theranostic applications. Near-infrared-absorbing organic semiconducting nanoparticles have been explored for *in vivo* fluorescence and photoacoustic imaging [147]. Amphiphilic semiconducting oligomer having a hydrophobic semiconducting oligomer backbone conjugated to hydrophilic PEG chains self-assemble to form nanoparticles. They showed high structural stability and bright photoacoustic signals. Their small size and the PEGylation facilitated their accumulation in tumor, thus giving idea for tumor imaging with high signal-to-background ratios.

Magnetic nanoparticles represent one of the most promising candidates for theranostic applications because of their success in clinical applications. Polymeric nanocarriers with magnetic iron oxide and anticancer drug encapsulated are promising for drug delivery and MRI imaging [148]. Kim et al. prepared magnetic

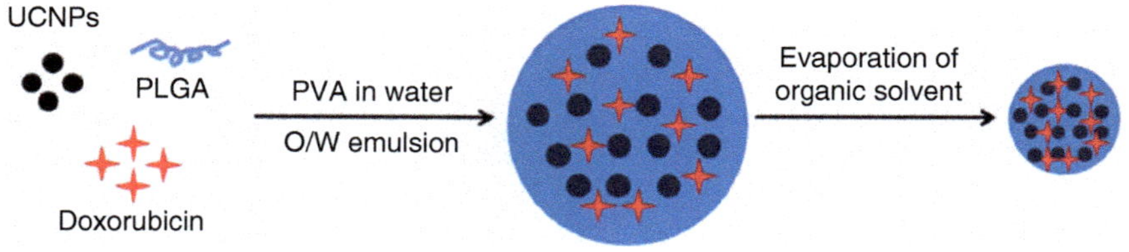

Figure 13.6 Synthesis of PLGA(UCNPs/DOX) nanocapsules. Source: Zhao et al. [146]. Reproduced with permission of Springer Nature.

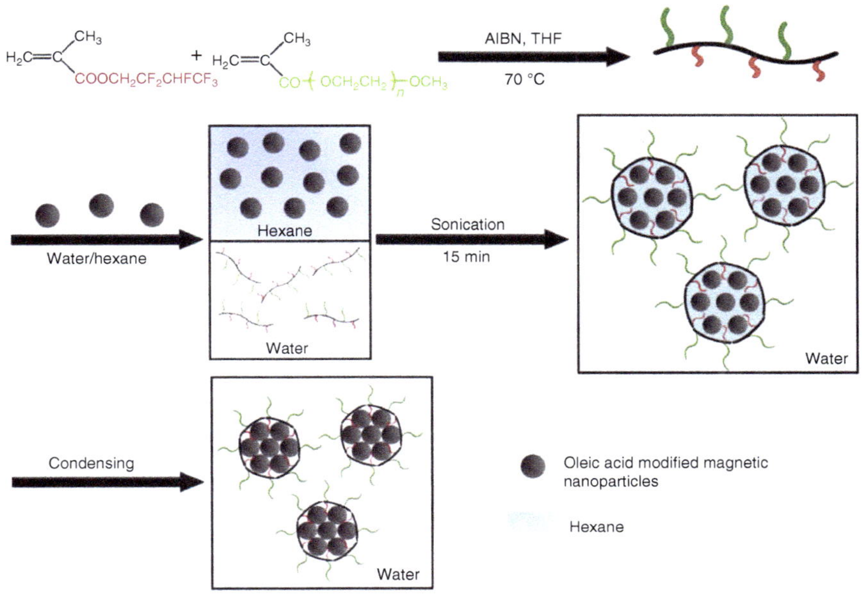

Figure 13.7 Synthesis of magnetic polymeric nanoparticles by self-assembly. Source: Li et al. [150]. Reproduced with permission of Elsevier Ltd.

PLGA polymeric nanoparticles for theranostic applications [149]. The polymeric nanoparticles comprised of an anti-cancer drug DOX and magnetite (Fe_3O_4) nanoparticles that were functionalized with PEGylation and a cancer targeting ligand folic acid for synergetic tumor targeting and imaging. Similarly, Li et al. produced stable magnetite-loaded polymeric nanoparticles using fluorine-containing amphiphilic copolymers and oleic acid-modified Fe_3O_4 nanoparticles (Figure 13.7) [150]. The Fe_3O_4 nanoparticles formed small clusters inside the polymeric nanocarriers and a hydrophobic drug 5-fluorouracil was also loaded with a high drug loading of 20.94%.

Apart from polymeric nanoparticles and micelles, polymeric vesicles have also been commonly used as theranostic nanocarriers. Fluorescent vesicles are promising for bioimaging and monitoring drug distribution and delivery *in vivo,* while organic dyes trapped in the fluorescent vesicles often suffer from aggregation-caused quenching (ACQ), which presents a challenge for their applications [151]. In contrast, a new type of dye molecules, aggregation-induced emission (AIE) dyes, has drawn increasing attention. The AIE dyes show opposite property in contrast with ACQ dyes. They have almost no emission in dilute solution but become very luminescent when aggregated [152]. These vesicles can be applied as a stable fluorescent tool for bioimage as well as drug delivery vehicles for both hydrophobic and hydrophilic drugs [153]. Several review articles have summarized ACQ-based nanoparticles for applications in drug delivery [154, 155].

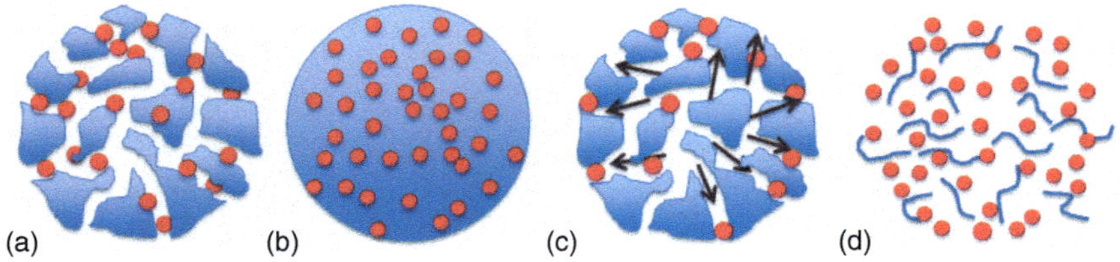

Figure 13.8 Drug release mechanisms for polymeric NPs: (a) diffusion through water-filled pores, (b) diffusion through the polymer matrix, (c) osmotic pumping, and (d) erosion. Source: Kamaly et al. [157]. Reproduced with permission of American Chemical Society.

13.5 Drug Release

13.5.1 Drug Release Mechanism

Drug release of drug-loaded polymeric systems refers to the transport of drug molecules from the polymeric matrix toward the surface of the particles and ultimately to the surrounding environment [156]. In general, classical mechanisms for drug release from polymeric matrix include drug diffusion through water-filled pores, diffusion through polymer matrix, osmotic pumping, and degradation/erosion (Figure 13.8) [157, 158].

Drug diffusion through water-filled pores describes a release process of some degradable polymer matrixes capable of adsorbing water rapidly, thus forming water-filled networks. Over time, these water-filled pores become larger in size and more in number, thus facilitating the transport of drug molecules as a result of chemical potential gradient [159]. Diffusion through the polymer matrix refers to the process that drug molecule diffuses through the polymer matrix directly, especially for nondegradable polymers. This nondegradable polymer matrix enables a constant release (zero-order) kinetics that is not affected by the concentration gradient of drugs but is affected by the property of the polymers such as permeability and thickness [160]. Osmotic pumping is another mechanism that determines the release behavior of encapsulated drugs. The osmotic pumping drug delivery system (DDS) usually contains a core containing drugs and osmogens as well as a semipermeable polymeric shell. The osmotic pressure drives external water influx into the polymer matrix; consequently, the drugs are pumping out simultaneously [161]. Degradation of polymer structure is a common driving force for the release of drugs, especially for biodegradable polymers. It can be further divided into surface degradation and bulk degradation. The surface degradation describes those systems that the erosion rate of surface polymer is faster than the water penetration rate in the bulk polymer, resulting in an overall size reduction [162]. Whereas the bulk degradation illustrates those polymers that water penetrates the bulk polymer matrix faster than their erosion rate, which causes an overall size expansion and a homogeneous degradation [163, 164]. Surface degradation is generally preferable than bulk degradation because it is more controllable and predictable. It should be noted that a polymeric drug delivery system may have other mechanisms and more than one classical mechanism that controls its drug release, resulting in a case-by-case release kinetic.

Drug release solely relying on slow diffusion and degradation is usually insufficient in drug delivery applications. As the biological environment is dynamic and different, DDSs may only remain at target sites for a limited time. Sufficient amount of drug need to be released quickly at targeting sites; otherwise the DDSs are pumped out. The ideal release profile of DDS is minimal release rate at undesired sites but maximal release rate at targeting sites. Based on the facts that human body has various microenvironments, e.g. different pH in stomach and colon, different temperature in healthy and inflammable tissues, different glutathione (GSH) concentrations in normal and cancer cells, etc., various stimuli-responsive polymeric nanoparticles have been developed for trigger-release, e.g. pH (acidic or basic), elevated temperature, oxidation, reduction, etc. [165].

13.5.2 pH-Triggered Release

Naturally, the human body has different biological environments with different pH. For example, gastric acid has a pH of 1.5–3.5, whereas the pH of the common duct bile ranges from 7.5 to 8.05. Even within a single cell, cytosol has a slightly basic pH of 7.2, whereas lysosome has a pH of 4.5–5.0. The very different pH from organs to compartments in a cell provides the basis for the design of pH-responsive polymer-based DDSs. One strategy to design pH-responsive polymeric DDSs is to include a pH-sensitive linker in the polymer backbone or polymer–drug conjugating bond. These pH-sensitive linkers include hydrazone, amide, ester, imine, and acetal [166–169]. When the pH changes to specific values, the pH-sensitive linkers can be cleaved, leading to the disintegration of the polymer structure thus drug release. Another strategy is to incorporate polyelectrolytes with ionic functional groups, e.g. acidic groups such as carboxylic and sulfonic groups and basic groups such as amines, imidazole, and pyridine groups [170]. These groups are capable of accepting or donating H^+ ions from external environments. The protonation or deprotonation changes the net charge of the polymer. Increased net charge can lead to elevated electrostatic repulsion resulting in the expansion of the polymer matrix and *vice versa* [171, 172]. Also, higher net charge of the polymer enhances its water solubility, which leads to the dissolution of the polymer, thus resulting in quicker drug release. Dong et al. designed a multifunctional nanocomplex composed of PEI-hydrazone-DOX, PEI-PEG-folate, and siRNA [166]. The hydrazone linkage was stable at physiological pH but could be cleaved at lysosomal pH, releasing the drug DOX rapidly in cancer cells. Song et al. developed a pH-sensitive polymeric tadpole nano-assemblies for programmed tumor therapy [173]. Amine groups were incorporated inside the polymer backbone. At basic conditions, these single-chain polymers self-assembled to a micelle-like structure. However, when turning the pH to slightly acidic, the amine groups can be protonated, and the nanoparticles disassembled rapidly to single chains with a significant size reduction (~70 to ~10 nm) and charge reverse as well as increased net charge (zeta potential from −5 to 17 mV). These pH-sensitive DDSs facilitated targeted delivery of drugs at specific sites.

13.5.3 Thermo-Triggered Release

Temperature-responsive polymeric DDS is suitable for numerous applications because they are easy to fabricate and control [174]. Temperature change in biological environments exists in various ways. For example, inflammation is normally associated with heat where the temperature of inflammable tissues is higher than that of healthy tissues. In addition, the difference between room temperature and body temperature enables self-assembly of polymers *in situ* upon temperature change [175]. An important concept for designing temperature-responsive polymers is the lower critical solution temperature (LCST). Below the LCST, the polar groups on the polymer chains are favorable to form hydrogen bonds with water molecules, resulting in the "swollen state" of the polymer matrix. At higher temperatures than the LCST, the increase in Gibbs free energy (ΔG) makes the favorable interactions transit from polymer–water to polymer–polymer and water–water, leading to the "collapsed state" of the polymer matrix [176, 177]. Commonly used thermosensitive polymers include acrylamide-based hydrogels, especially poly[*N*-isopropylacrylamide] (PNIPAAm), elastin-like polypeptides, PEG analogs, thermogelling systems, and thermosensitive liposomes [175, 178–180]. Zhang and Misra developed a magnetic-targeting drug nanocarrier encapsulated with a thermosensitive polymer dextran-g-poly(*N*-isopropylacrylamide-*co*-*N*,*N*-dimethylacrylamide) [181]. The thermosensitive polymer presented an LCST of ~38 °C. At the temperatures below its LCST, it showed a very slow release. However, at a higher temperature of 40 °C, the drug release became significantly faster. This smart nanocarrier had great potentials in reducing side effects and controlling drug release in response to external temperature change. Employing a more biocompatible material, Wang et al. synthesized a temperature-responsive poly(ether-urethane) particles based on PEG-diisocyanates and aliphatic diols [182]. Upon the temperature rising, the size of nanoparticle significantly grew because of the increased hydrophobicity of the polymer. Interestingly, the phase transition temperature (T_p) could be well-tuned by controlling the length of hydrophobic and hydrophilic segments of the polymer. The increase in PEG length enhances the intermolecular forces between the polymers and water molecules, resulting in higher T_p, whereas the increase in aliphatic diol segments leads to lower T_p. Moreover, Moktan and Raucher used another thermosensitive polymers elastin-like polypeptides to deliver the KLAKLAKKLAKLAK peptide, which can induce apoptosis via disrupting the mitochondria [183]. Under mild hyperthermia, selectively cell killing function was observed.

13.5.4 Redox-Triggered Release

Redox-responsive polymeric DDS is to use the distinctive differences in redox potentials between different tissues. The most common redox couple in animal cells is the GSH/glutathione disulfide (GSSG), which has been exploited to develop reduction-responsive polymeric DDSs containing disulfide linkage for many years [165]. The GSH concentration in the cytosol and nuclei inside cells (about 10 mM) is much higher than the GSH concentration outside the cell (about 2–20 μM) [184]. Also, tumor tissues have more than fourfold higher concentrations of GSH than normal tissue in mice [185]. The disulfide linker acting as the end group can be introduced into olefins on the side chains or be introduced to the end of the polymer chain by post-polymerization [186–189]. Other reduction-responsive bonds include diselenide linkage [190, 191], cisplatin derivatives–polymer conjugate [192], and trimethyl-locked benzoquinone [193]. De Vries et al. developed self-assembled reduction-responsive polymeric DDSs comprised of a cyclodextrin vesicle core and a thin reductively cleavable polymer containing disulfide linkage on the vesicle surface (Figure 13.9a) [194]. The polymeric DDS showed highly efficient host–guest recognition as well as controlled release of cargo by a reductive trigger. In addition to the disulfide linkage, Ma et al. used amphiphilic copolymers comprised of diselenide-containing polyurethane hydrophobic blocks and hydrophilic PEG blocks to form micelles by self-assembly [195]. The drug encapsulated inside the micelles can be released in a reductive environment.

Oxidation-responsive systems rely on reactive oxygen species (ROS) such as hydrogen peroxide (H_2O_2) and hydroxyl radicals from the by-products of aerobic metabolism. ROS is abundant in the body. For example, the concentration of ROS rises considerably during surgery or suffering from diseases such as cancer, heart injury, and arteriosclerosis [196, 197]. Various oxidation-responsive polymeric DDSs have been developed for ROS-triggered drug release applications [198, 199]. Oxidation-responsive bonds in polymers include boronic ester [200], tetrathiafulvalene [201], sulfide-contained systems [202], selenium-contained systems [203], and ferrocene-contained systems [204]. The ROS-responsive drug release is often achieved by an external stimulus such as ultrasound [205] or light [206]. For instance, a ROS-responsive polymeric DDS (IR780/PTL) was prepared via self-assembly. It is comprised of a ROS-cleavable thioketal linker (TL) for stimuli-responsive release of PTX during sonodynamic therapy (Figure 13.9b) [205]. Upon ultrasound stimulation, IR780/PTL NPs produced ROS, inducing cell apoptosis and simultaneously enhancing PTX release by cleaving the ROS-sensitive TL. This novel sono-chemotherapy polymeric DDS has great potential for future clinical applications.

13.5.5 Other Triggered Release

In addition to the triggers mentioned earlier, other stimuli include light [207, 208], glucose [209, 210], enzymes [211, 212], ultrasound [213, 214], ATP [215, 216], voltage [217, 218], CO_2 [219, 220], etc. [221, 222] For example, photo-responsive biocompatible micelles were produced through self-assembly of a photo-responsive molecule (2-diazo-1,2-naphthoquinone-containing amphiphilic hyperbranched polyphosphate polymers) to control coumarin release [207]. Under ultraviolet (UV) irradiation, photochemical reaction of the polymer destabilized the micelles, thus resulting in the release of coumarin. Yang et al. developed glucose-sensitive polymeric vesicles using a α-CD/PEG inclusion complex-templated method, which was

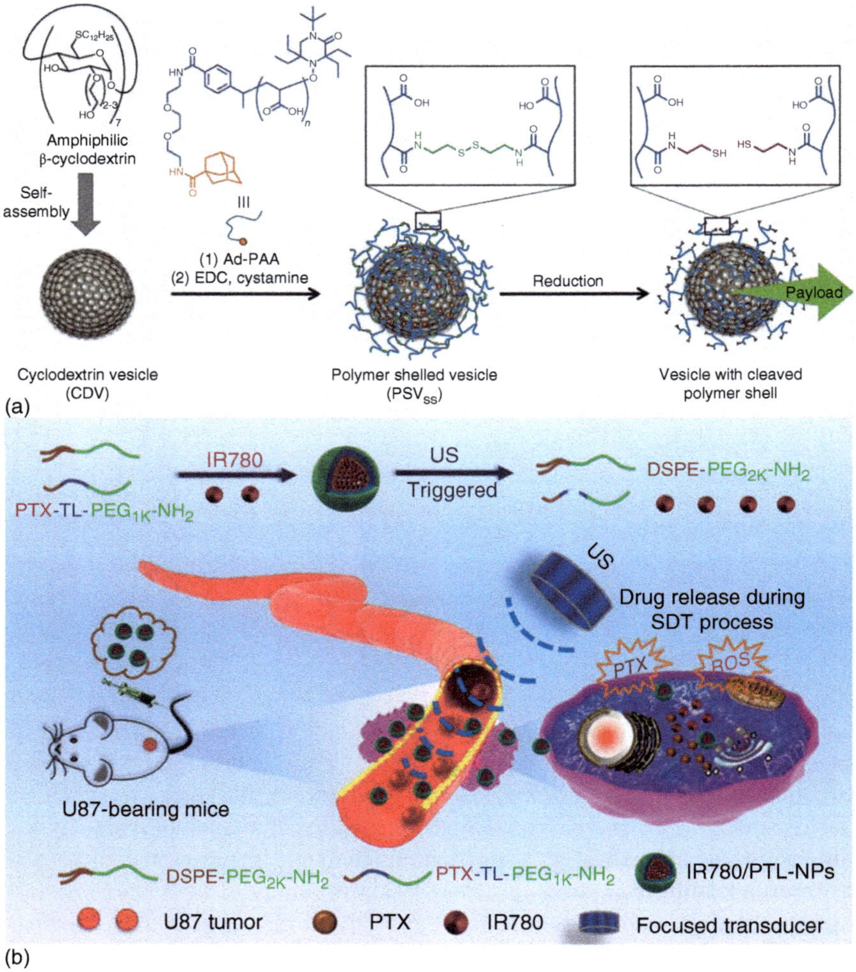

Figure 13.9 Redox-responsive polymeric nanoparticles. (a) Stepwise preparation of a glutathione-responsive nanoparticle and glutathione-triggered release of a hydrophilic payload. Source: De Vries et al. [194]. Reproduced with permission of Wiley-VCH Verlag GmbH & Co. KGaA, Weinheim. (b) Synthesis of IR780/PTL-NPs with ultrasound-induced ROS-triggered cascade chemotherapy to synergistically clear cancer cells. Source: Wu et al. [205]. Reproduced with permission of WILEY-VCH Verlag GmbH & Co. KGaA, Weinheim.

based on the complexation between a glucosamine (GA)-contained block copolymer PEG45-*b*-P(Asp-*co*-AspGA) and a phenylboronic acid (PBA)-contained block copolymer PEG114-*b*-P(Asp-*co*-AspPBA). These vesicles rapidly disassembled in the presence glucose. Li et al. synthesized penicillin G amidase (PGA) and β-lactamase (Bla)-responsive polymeric vesicles using amphiphilic block copolymers [211]. They exhibited strain-selective delivery of potent antibiotics, enhanced structural stability, and improved burn wound healing as well as reduced side effects.

13.5.6 Multiple-Triggered Release

Multiple-triggered release systems often consist of dual-responsive and triple-responsive DDSs that respond to two or more different stimulations [221]. Different combinations include magnetic/thermos [223], magnetic/ultrasonic [224], pH/ionic strength [225], and pH/thermos responsive [226]. Various functional groups, cross-linkers, or linkage groups are available for these different responsive functions. Formulating multiple-triggered polymeric DDS needs the rational design of synthesis strategies and proper selection of molecular building blocks to improve the tunability of controlled drug release [227]. Li et al. developed a redox and pH dual-responsive polymeric micelle [228]. The DOX-loaded dual-responsive polymer micelles containing many disulfide bonds and tertiary amines remained stable in circulation but quickly disassociated inside tumor cells due to the reductive and acidic cellular environment. Chen et al. fabricated photo-, pH-, and thermo-responsive polymeric nanoparticles using spiropyran-based amphiphilic random copolymer via self-assembly [229]. The copolymer contains pH- and photo-responsive spiropyran and temperature-responsive *N*-isopropyl acrylamide in the amphiphilic polymer. The self-assembly of this amphiphilic copolymer in water resulted in the formation of spiropyran core and the *N*-isopropyl acrylamide shell. The change of temperature or pH with UV irradiation promoted faster drug. Multiple-triggered release offers various mechanism to control drug release, but it is important to recognize that the increase in complexity of these systems will make the reproducible fabrication of these nanocarriers in large scales challenging.

13.6 Conclusions

In this chapter, we have discussed the colloidal self-assembly of polymeric nanoparticles for drug loading and controlled release. Compared to other materials such as inorganic materials, polymers offer better biocompatibility and biodegradability. Block copolymers hold great promise for encapsulating various therapeutic compounds due to their advantages including easy to synthesize by self-assembly, good colloidal stability, flexibility in choosing different blocks with various properties, and tenability in particle structure, morphology, and stimuli-responsive properties. Diblock copolymers generally consists of a portion of hydrophobic block that can attract hydrophobic drugs via hydrophobic or π–π interactions and a fraction of hydrophilic block for good water dispersibility, stability, and biocompatibility. Triblock copolymers offer great flexibilities in tuning their self-assembly and the resultant particle property.

Three typical structures of polymeric nanostructures have been introduced and discussed in this chapter, including micelles, nanoparticles, and vesicles or polymersomes. Micelle often has relatively low fraction of hydrophobic sections and high CMC, therefore relatively unstable compared with the other two nanostructures. Whereas nanoparticle normally has high fraction of hydrophobic portions and low CMC, so they often perform better in particle stability and controlled drug release.

When hydrophobic segments of the polymer are within a specific range, vesicle can be formed. Vesicle has a liposome-like structure; therefore they are attractive for loading both hydrophobic and hydrophilic drugs. In terms of drug loading, micelles and nanoparticles are both suitable for loading hydrophobic drugs, but due to the structure of micelles, drug loading might be limited. In contrast, for loading two or more drugs having same hydrophobicity, self-assembly of different drug–polymer conjugates offers a better solution as it offers a better ratiometric control of the two encapsulated drugs. On the other hand, vesicle provides an excellent system for loading two drugs with opposite hydrophobicity. Similarly, drug and imaging agents can be loaded to polymeric nanostructure for theranostic applications. Polymeric nanostructures are versatile in release behaviors. Apart from classic release routes by diffusion or degradation in a sustained and slow mode, numerous functional groups can be incorporated to the polymer blocks, endowing different triggered-release properties. A wide variety of polymers have been designed to be stimuli-responsive, including temperature, pH, redox, light, enzyme, ultrasound, etc. These stimuli-responsive nanostructures are able to achieve minimal release at healthy tissues but maximum release at targeting disease sites, which improves therapeutic efficacy and drug delivering efficiency while reducing undesirable toxicity to healthy tissues.

Despite the tremendous progress in this field, the translation of polymer-based drug delivery system remains challenging. First, safety is the most important factor in affecting their biomedical applications. The safety evaluation of newly designed polymer is often time-consuming and expensive; therefore it takes very long time to develop polymeric nanomedicines of new polymers from laboratory research to clinical applications. Then, the synthesis of polymer nanomedicines often involves solvents for dissolving polymers and drugs, so the possible residual of toxic solvents could also pose health risks. Furthermore, extensive research has been directed to the development of very complex systems such as multi-drug-loaded nanomedicines, multi-stimuli-responsive systems for controlled release, and multifunctional nanomedicines for theranostic applications. However, there has been a trade-off between the complexity and real clinical applicability. The increase in complexity of nanomedicines often leads to difficulties in fabricating these materials with reproducible properties and minimum batch-to-batch variation in a large-scale. To accelerate the clinical translation of polymeric nanomedicines, clinically approved polymers (such as PLGA-PEG, PCL-PEG, PLA-PEG, etc.) have been often adopted for formulating various drugs. However, how to deliver drugs specifically to targeting disease sites while spare healthy tissues remains a long-lasting challenge especially for systematic intravenous administration. More collaborative efforts from various disciplines including material engineering, nanotechnology, pharmacology, clinical oncology, etc. are essential for the successful development of polymeric nanomedicines for practical clinical applications.

References

1 Kim, J.K., Yang, S.Y., and Lee, Y. (2010). Functional nanomaterials based on block copolymer self-assembly. *Prog. Polym. Sci.* 35: 1325–1349.

2 Kim, H.-C., Park, S.-M., Hinsberg, W.D. et al. (2010). Block copolymer based nanostructures: materials, processes, and applications to electronics. *Chem. Rev.* 110: 146–177.

3 Albert, J.N. and Epps, T.H. III, (2010). Self-assembly of block copolymer thin films. *Mater. Today* 13: 24–33.

4 Wang, H.-F., Ran, R., Liu, Y. et al. (2018). Tumor-vasculature-on-a-chip for investigating nanoparticle extravasation and tumor accumulation. *ACS Nano* 12: 11600–11609.

5 Wilhelm, S., Tavares, A.J., Dai, Q. et al. (2016). Analysis of nanoparticle delivery to tumours. *Nat. Rev. Mater.* 1: 1–12.

6 Nicolas, J., Mura, S., Brambilla, D. et al. (2013). Design, functionalization strategies and biomedical applications of targeted biodegradable/biocompatible polymer-based nanocarriers for drug delivery. *Chem. Soc. Rev.* 42: 1147–1235.

7 Gref, R., Minamitake, Y., Peracchia, M.T. et al. (1994). Biodegradable long-circulating polymeric nanospheres. *Science* 263: 1600–1603.

8 Adams, M.L., Lavasanifar, A., and Kwon, G.S. (2003). Amphiphilic block copolymers for drug delivery. *J. Pharm. Sci.* 92: 1343–1355.

9 Gaucher, G., Dufresne, M.H., Sant, V.P. et al. (2005). Block copolymer micelles: preparation, characterization and application in drug delivery. *J. Control. Release* 109: 169–188.

10 Sun, X.K., Liu, X., Li, C.L. et al. (2018). Self-assembled micelles prepared from poly(epsilon-caprolactone)-poly(ethylene glycol) and poly(epsilon-caprolactone/glycolide)-poly(ethylene glycol) block copolymers for sustained drug delivery. *J. Appl. Polym. Sci.* 135: 45732.

11 Liu, Y., Yang, G., Zou, D. et al. (2019). Formulation of nanoparticles using mixing-induced nanoprecipitation for drug delivery. *Ind. Eng. Chem. Res.* 59: 4134–4149.

12 Szymusiak, M., Kalkowski, J., Luo, H. et al. (2017). Core-shell structure and aggregation number of micelles composed of amphiphilic block copolymers and amphiphilic heterografted polymer brushes determined by small-angle X-ray scattering. *ACS Macro Lett.* 6: 1005–1012.

13 Cai, K., He, X., Song, Z. et al. (2015). Dimeric drug polymeric nanoparticles with exceptionally high drug loading and quantitative loading efficiency. *J. Am. Chem. Soc.* 137: 3458–3461.

14 Perevyazko, I.Y., Delaney, J.T., Vollrath, A. et al. (2011). Examination and optimization of the self-assembly of biocompatible, polymeric nanoparticles by high-throughput nanoprecipitation. *Soft Matter.* 7: 5030–5035.

15 Perevyazko, I.Y., Vollrath, A., Pietsch, C. et al. (2012). Nanoprecipitation of poly(methyl methacrylate)-based nanoparticles: effect of the molar mass and polymer behavior. *J. Polym. Sci., Part A: Polym. Chem.* 50: 2906–2913.

16 Salvage, J.P., Thom, C., Lewis, A.L. et al. (2015). Nanoprecipitation of polymeric nanoparticle micelles based on 2-methacryloyloxyethyl phosphorylcholine (MPC) with 2-(diisopropylamino)ethyl methacrylate (DPA), for intracellular delivery applications. *J. Mater. Sci. Mater. Med.* 26: 1–14.

17 Reisch, A., Runser, A., Arntz, Y. et al. (2015). Charge-controlled nanoprecipitation as a modular approach to ultrasmall polymer nanocarriers: making bright and stable nanoparticles. *ACS Nano* 9: 5104–5116.

18 Panyam, J., Zhou, W.Z., Prabha, S. et al. (2002). Rapid endo-lysosomal escape of poly(DL-lactide-co-glycolide) nanoparticles: implications for drug and gene delivery. *FASEB J.* 16: 1217–1226.

19 Sahin, A., Esendagli, G., Yerlikaya, F. et al. (2017). A small variation in average particle size of PLGA nanoparticles prepared by nanoprecipitation leads to considerable change in nanoparticles' characteristics and efficacy of intracellular delivery. *Artif. Cells Nanomed. Biotechnol.* 45: 1657–1664.

20 Lee, J.S., Hwang, S.J., Lee, D.S. et al. (2009). Formation of poly(ethylene glycol)-poly(epsilon-caprolactone) nanoparticles via nanoprecipitation. *Macromol. Res.* 17: 72–78.

21 Farrag, Y., Montero, B., Rico, M. et al. (2018). Preparation and characterization of nano and micro particles of poly(3-hydroxybutyrate-*co*-3-hydroxyvalerate) (PHBV) via emulsification/solvent evaporation and nanoprecipitation techniques. *J. Nanopart. Res.* 20: 1–17.

22 Gavory, C., Durand, A., Six, J.L. et al. (2011). Polysaccharide-covered nanoparticles prepared by nanoprecipitation. *Carbohydr. Polym.* 84: 133–140.

23 Pooch, F., Sliepen, M., Knudsen, K.D. et al. (2019). Poly(2-isopropyl-2-oxazoline)-*b*-poly(lactide) (PiPOx-*b*-PLA) nanoparticles in water: interblock van der waals attraction opposes amphiphilic phase separation. *Macromolecules* 52: 1317–1326.

24 Liu, Y., Yang, G., Jin, S. et al. (2020). Development of high-drug-loading nanoparticles. *ChemPlusChem* 85: 2143–2157.

25 Ikada, Y. and Tsuji, H. (2000). Biodegradable polyesters for medical and ecological applications. *Macromol. Rapid Commun.* 21: 117–132.

26 Pan, P. and Inoue, Y. (2009). Polymorphism and isomorphism in biodegradable polyesters. *Prog. Polym. Sci.* 34: 605–640.

27 Raycraft, B.M., MacDonald, J.P., McIntosh, J.T. et al. (2017). Post-polymerization functionalization of poly(ethylene oxide)-poly(beta-6-heptenolactone) diblock copolymers to tune properties and self-assembly. *Polym. Chem.* 8: 557–567.

28 Athanasiou, K.A., Niederauer, G.G., and Agrawal, C.M. (1996). Sterilization, toxicity, biocompatibility and clinical applications of polylactic acid/polyglycolic acid copolymers. *Biomaterials* 17: 93–102.

29 Chen, L., Xie, Z., Hu, J. et al. (2007). Enantiomeric PLA–PEG block copolymers and their stereocomplex micelles used as rifampin delivery. *J. Nanopart. Res.* 9: 777–785.

30 Wang, L.J., Yao, J., Zhang, X.M. et al. (2018). Delivery of paclitaxel using nanoparticles composed of poly(ethylene oxide)-*b*-poly(butylene oxide) (PEO-PBO). *Colloids Surf. B* 161: 464–470.

31 Shuai, Q., Zhao, G.K., Lian, X.F. et al. (2020). Self-assembling poly(ethylene glycol)-block-polylactide-cabazitaxel conjugate nanoparticles for anticancer therapy with high efficacy and low in vivo toxicity. *Int. J. Pharm.* 574: 118879.

32 Soleimani, A., Moustafa, M.M.A., Borecki, A., and Gillies, E.R. (2015). A comparison of covalent and noncovalent strategies for paclitaxel release using poly(ester amide) graft copolymer micelles. *Can. J. Chem.* 93: 399–405.

33 Cheng, J., Teply, B.A., Sherifi, I. et al. (2007). Formulation of functionalized PLGA–PEG nanoparticles for in vivo targeted drug delivery. *Biomaterials* 28: 869–876.

34 Kang, G.Y., Sun, L., Liu, Y.P. et al. (2019). Micelles with cyclic poly(epsilon-caprolactone) moieties: greater stability, larger drug loading capacity, and slower degradation property for controlled drug release. *Langmuir* 35: 12509–12517.

35 Zhao, L.Z., Wu, C.L., Wang, F. et al. (2014). Fabrication of biofunctional complex micelles with tunable structure for application in controlled drug release. *Colloid Polym. Sci.* 292: 1675–1683.

36 Youxin, L., Volland, C., and Kissel, T. (1994). In-vitro degradation and bovine serum albumin release of the ABA triblock copolymers consisting of poly (L (+) lactic acid), or poly (L (+) lactic acid-co-glycolic acid) A-blocks attached to central polyoxyethylene B-blocks. *J. Control. Release* 32: 121–128.

37 Kissel, T., Li, Y., and Unger, F. (2002). ABA-triblock copolymers from biodegradable polyester A-blocks and hydrophilic poly (ethylene oxide) B-blocks as a candidate for in situ forming hydrogel delivery systems for proteins. *Adv. Drug Deliv. Rev.* 54: 99–134.

38 Jeong, B., Bae, Y.H., and Kim, S.W. (1999). Biodegradable thermosensitive micelles of PEG-PLGA-PEG triblock copolymers. *Colloids Surf. B* 16: 185–193.

39 Ghahremankhani, A.A., Dorkoosh, F., and Dinarvand, R. (2008). PLGA-PEG-PLGA tri-block copolymers as in situ gel-forming peptide delivery system: effect of formulation properties on peptide release. *Pharm. Dev. Technol.* 13: 49–55.

40 Eeckman, F., Moës, A.J., and Amighi, K. (2004). Synthesis and characterization of thermosensitive copolymers for oral controlled drug delivery. *Eur. Polym. J.* 40: 873–881.

41 Yu, L. and Ding, J. (2008). Injectable hydrogels as unique biomedical materials. *Chem. Soc. Rev.* 37: 1473–1481.

42 Kim, Y.J., Choi, S., Koh, J.J., and Lee, M. (2001). Controlled release of insulin from injectable biodegradable triblock copolymer. *Pharm. Res.* 18: 548.

43 Khodaverdi, E., Tekie, F.S.M., Mohajeri, S.A. et al. (2012). Preparation and investigation of sustained drug delivery systems using an injectable, thermosensitive, in situ forming hydrogel composed of PLGA–PEG–PLGA. *AAPS PharmSciTech* 13: 590–600.

44 Wei, J., Wang, H., Zhu, M. et al. (2013). Janus nanogels of PEGylated taxol and PLGA–PEG–PLGA copolymer for cancer therapy. *Nanoscale* 5: 9902–9907.

45 Jeong, B., Bae, Y.H., and Kim, S.W. (2000). Drug release from biodegradable injectable thermosensitive hydrogel of PEG–PLGA–PEG triblock copolymers. *J. Control. Release* 63: 155–163.

46 Kwon, Y.M. and Kim, S.W. (2003). New biodegradable polymers for delivery of bioactive agents. In: Macromol. Symp., vol. 201, 179–186. Weinheim: WILEY-VCH Verlag.

47 Ghahremankhani, A.A., Dorkoosh, F., and Dinarvand, R. (2007). PLGA-PEG-PLGA tri-block copolymers as an in-situ gel forming system for calcitonin delivery. *Polym. Bull.* 59: 637–646.

48 Qiao, M., Chen, D., Ma, X., and Liu, Y. (2005). Injectable biodegradable temperature-responsive PLGA–PEG–PLGA copolymers: synthesis and effect of copolymer composition on the drug release from the copolymer-based hydrogels. *Int. J. Pharm.* 294: 103–112.

49 Chen, S. and Singh, J. (2008). Controlled release of growth hormone from thermosensitive triblock copolymer systems: in vitro and in vivo evaluation. *Int. J. Pharm.* 352: 58–65.

50 Chen, S. and Singh, J. (2005). Controlled delivery of testosterone from smart polymer solution based systems: in vitro evaluation. *Int. J. Pharm.* 295: 183–190.

51 Zhang, J., Men, K., Gu, Y.C. et al. (2011). Preparation of core cross-linked PCL-PEG-PCL micelles for doxorubicin delivery in vitro. *J. Nanosci. Nanotechnol.* 11: 5054–5061.

52 Ludwigs, S., Böker, A., Voronov, A. et al. (2003). Self-assembly of functional nanostructures from ABC triblock copolymers. *Nat. Mater.* 2: 744–747.

53 Talingting, M.R., Munk, P., Webber, S., and Tuzar, Z. (1999). Onion-type micelles from polystyrene-block-poly (2-vinylpyridine) and poly (2-vinylpyridine)-block-poly (ethylene oxide). *Macromolecules* 32: 1593–1601.

54 Lei, L., Gohy, J.-F., Willet, N. et al. (2004). Tuning of the morphology of core–shell– corona micelles in water. I. Transition from sphere to cylinder. *Macromolecules* 37: 1089–1094.

55 Njikang, G., Liu, G., and Gao, J. (2007). Preparation and quencher diffusion study of pyrene-tagged water-dispersible abc triblock nanospheres. *Macromolecules* 40: 9174–9180.

56 Yan, X., Liu, G., and Li, Z. (2004). Preparation and phase segregation of block copolymer nanotube multiblocks. *J. Am. Chem. Soc.* 126: 10059–10066.

57 Njikang, G., Han, D., Wang, J., and Liu, G. (2008). ABC triblock copolymer micelle-like aggregates in selective solvents for A and C. *Macromolecules* 41: 9727–9735.

58 Aydin, O., Youssef, I., Durmaz, Y.Y. et al. (2016). Formulation of acid-sensitive micelles for delivery of cabazitaxel into prostate cancer cells. *Mol. Pharm.* 13: 1413–1429.

59 Allen, C., Maysinger, D., and Eisenberg, A. (1999). Nano-engineering block copolymer aggregates for drug delivery. *Colloids Surf. B* 16: 3–27.

60 Tsukruk, V.V. (1997). Assembly of supramolecular polymers in ultrathin films. *Prog. Polym. Sci.* 22: 247–311.

61 Tuzar, Z. and Kratochvil, P. (1976). Block and graft copolymer micelles in solution. *Adv. Colloid Interf. Sci.* 6: 201–232.

62 Otsuka, H., Nagasaki, Y., and Kataoka, K. (2001). Self-assembly of poly (ethylene glycol)-based block copolymers for biomedical applications. *Curr. Opin. Colloid Interface Sci.* 6: 3–10.

63 Bae, J., Lawrence, J., Miesch, C. et al. (2012). Multifunctional nanoparticle-loaded spherical and wormlike micelles formed by interfacial instabilities. *Adv. Mater.* 24: 2735–2741.

64 Nabar, G.M., Winter, J.O., and Wyslouzil, B.E. (2018). Nanoparticle packing within block copolymer micelles prepared by the interfacial instability method. *Soft Matter.* 14: 3324–3335.

65 Yu, L., Zhang, H., and Ding, J. (2006). A subtle end-group effect on macroscopic physical gelation of triblock copolymer aqueous solutions. *Angew. Chem. Int. Ed.* 45: 2232–2235.

66 Yu, L., Chang, G., Zhang, H. et al. (2007). *J. Polym. Sci., Part A: Polym. Chem.* 45: 1122–1133.

67 Husseini, G.A. and Pitt, W.G. (2008). Micelles and nanoparticles for ultrasonic drug and gene delivery. *Adv. Drug Deliv. Rev.* 60: 1137–1152.

68 Brinkhuis, R.P., Rutjes, F.P., and van Hest, J.C. (2011). Polymeric vesicles in biomedical applications. *Polym. Chem.* 2: 1449–1462.

69 Yokoyama, M. (2011). Clinical applications of polymeric micelle carrier systems in chemotherapy and image diagnosis of solid tumors. *J. Exp. Clin. Med.* 3: 151–158.

70 Feng, J., Zeng, Y., Ma, C. et al. (2006). The surfactant tween 80 enhances biodesulfurization. *Appl. Environ. Microbiol.* 72: 7390–7393.

71 Zhang, K., Tang, X., Zhang, J. et al. (2014). PEG–PLGA copolymers: their structure and structure-influenced drug delivery applications. *J. Control. Release* 183: 77–86.

72 Discher, D.E. and Eisenberg, A. (2002). Polymer vesicles. *Science* 297: 967–973.

73 Lee, H.J. and Lee, S.C. (2010). Fabrication of core-shell hybrid nanoparticles by mineralization on poly(epsilon-caprolactone)-*b*-poly(methacrylic acid) copolymer micelles. *Polym. Bull.* 65: 743–752.

74 Zhang, K.L., Zhou, J., Zhou, H. et al. (2017). Bioinspired "active" stealth magneto-nanomicelles for theranostics combining efficient MRI and enhanced drug delivery. *ACS Appl. Mater. Interfaces* 9: 30502–30509.

75 Yotsumoto, K., Ishii, K., Kokubo, M., and Yasuoka, S. (2018). Improvement of the skin penetration of hydrophobic drugs by polymeric micelles. *Int. J. Pharm.* 553: 132–140.

76 Kelly, E.A., Houston, J.E., and Evans, R.C. (2019). Probing the dynamic self-assembly behaviour of photoswitchable wormlike micelles in real-time. *Soft Matter* 15: 1253–1259.

77 Liu, Y., Hui, Y., Ran, R. et al. (2018). Synergetic combinations of dual-targeting ligands for enhanced in vitro and in vivo tumor targeting. *Adv. Healthc. Mater.* 7: 1800106.

78 Suh, H., Jeong, B., Rathi, R., and Kim, S.W. (1998). Regulation of smooth muscle cell proliferation using paclitaxel-loaded poly (ethylene oxide)–poly (lactide/glycolide) nanospheres. *J. Biomed. Mater. Res.* 42: 331–338.

79 Arunkumar, R., Prashanth, K.V.H., Manabe, Y. et al. (2015). Biodegradable poly (lactic-*co*-glycolic acid)-polyethylene glycol nanocapsules: an efficient carrier for improved solubility, bioavailability, and anticancer property of lutein. *J. Pharm. Sci.* 104: 2085–2093.

80 Lepeltier, E., Bourgaux, C., and Couvreur, P. (2014). Nanoprecipitation and the "ouzo effect": application to drug delivery devices. *Adv. Drug Deliv. Rev.* 71: 86–97.

81 Schubert, S., Delaney, J.T. Jr., and Schubert, U.S. (2011). Nanoprecipitation and nanoformulation of polymers: from history to powerful possibilities beyond poly (lactic acid). *Soft Matter.* 7: 1581–1588.

82 Discher, B.M., Won, Y.-Y., Ege, D.S. et al. (1999). Polymersomes: tough vesicles made from diblock copolymers. *Science* 284: 1143–1146.

83 Sercombe, L., Veerati, T., Moheimani, F. et al. (2015). Advances and challenges of liposome assisted drug delivery. *Front. Pharmacol.* 6: 286.

84 Ran, R., Wang, H.F., Hou, F. et al. (2019). A microfluidic tumor-on-a-chip for assessing multifunctional liposomes' tumor targeting and anticancer efficacy. *Adv. Healthc. Mater.* 8: 1900015.

85 Ran, R., Wang, H., Liu, Y. et al. (2018). Microfluidic self-assembly of a combinatorial library of single-and dual-ligand liposomes for in vitro and in vivo tumor targeting. *Eur. J. Pharm. Biopharm.* 130: 1–10.

86 Anajafi, T. and Mallik, S. (2015). Polymersome-based drug-delivery strategies for cancer therapeutics. *Ther. Deliv.* 6: 521–534.

87 Lim Soo, P. and Eisenberg, A. (2004). Preparation of block copolymer vesicles in solution. *J. Polym. Sci. B Polym. Phys.* 42: 923–938.

88 Kita-Tokarczyk, K., Grumelard, J., Haefele, T., and Meier, W. (2005). Block copolymer vesicles—using concepts from polymer chemistry to mimic biomembranes. *Polymer* 46: 3540–3563.

89 Pazdur, R., Kudelka, A.P., Kavanagh, J.J. et al. (1993). The taxoids: paclitaxel (Taxol®) and docetaxel (Taxotere®). *Cancer Treat. Rev.* 19: 351–386.

90 Gelderblom, H., Verweij, J., Nooter, K., and Sparreboom, A. (2001). Cremophor EL: the drawbacks and advantages of vehicle selection for drug formulation. *Eur. J. Cancer* 37: 1590–1598.

91 Kakkar, A., Traverso, G., Farokhzad, O.C. et al. (2017). Evolution of macromolecular complexity in drug delivery systems. *Nat. Rev. Chem.* 1: 1–17.

92 Van Ngo, H., Nguyen, P.K., Van Vo, T. et al. (2016). Hydrophilic-hydrophobic polymer blend for modulation of crystalline changes and molecular interactions in solid dispersion. *Int. J. Pharm.* 513: 148–152.

93 Tran, T.T.-D., Tran, P.H.-L., Lim, J. et al. (2010). Physicochemical principles of controlled release solid dispersion containing a poorly water-soluble drug. *Ther. Deliv.* 1: 51–62.

94 Nguyen, T.N.G., Tran, V.T., Duan, W. et al. (2017). Nanoprecipitation for poorly water-soluble drugs. *Curr. Drug Metab.* 18: 1000–1015.

95 Fahr, A. and Liu, X. (2007). Drug delivery strategies for poorly water-soluble drugs. *Expert Opin. Drug Deliv.* 4: 403–416.

96 Liu, Y., Cheng, C., Prud'homme, R.K., and Fox, R.O. (2008). Mixing in a multi-inlet vortex mixer (MIVM) for flash nano-precipitation. *Chem. Eng. Sci.* 63: 2829–2842.

97 Johnson, B.K. and Prud'homme, R.K. (2003). Chemical processing and micromixing in confined impinging jets. *AIChE J.* 49: 2264–2282.

98 Baby, T., Liu, Y., Middelberg, A.P., and Zhao, C.-X. (2017). Fundamental studies on throughput capacities of hydrodynamic flow-focusing microfluidics for producing monodisperse polymer nanoparticles. *Chem. Eng. Sci.* 169: 128–139.

99 Baby, T., Liu, Y., Yang, G. et al. (2021). Microfluidic synthesis of curcumin loaded polymer nanoparticles with tunable drug loading and pH-triggered release. *J. Colloid Interface Sci.* 594: 474–484.

100 Li, W., Chen, Q., Baby, T. et al. (2021). Insight into drug encapsulation in polymeric nanoparticles using microfluidic nanoprecipitation. *Chem. Eng. Sci.* 235: 116468.

101 Yang, M., Gao, Y., Liu, Y. et al. (2021). Integration of microfluidic systems with external fields for multiphase process intensification. *Chem. Eng. Sci.* 234: 116450.

102 Tao, J.S., Chow, S.F., and Zheng, Y. (2019). Application of flash nanoprecipitation to fabricate poorly water-soluble drug nanoparticles. *Acta Pharm. Sin. B* 9: 4–18.

103 Pagels, R.F., Edelstein, J., Tang, C., and Prud'homme, R.K. (2018). Controlling and predicting nanoparticle formation by block copolymer directed rapid precipitations. *Nano Lett.* 18: 1139–1144.

104 Pustulka, K.M., Wohl, A.R., Lee, H.S. et al. (2013). Flash nanoprecipitation: particle structure and stability. *Mol. Pharm.* 10: 4367–4377.

105 Dreis, S., Rothweiler, F., Michaelis, M. et al. (2007). Preparation, characterisation and maintenance of drug efficacy of doxorubicin-loaded human serum albumin (HSA) nanoparticles. *Int. J. Pharm.* 341: 207–214.

106 Liu, Y., Yang, G., Baby, T. et al. (2020). Stable polymer nanoparticles with exceptionally high drug loading by sequential nanoprecipitation. *Angew. Chem. Int. Ed.* 59: 4720–4728.

107 Han, F.Y., Liu, Y., Kumar, V. et al. (2020). Sustained-release ketamine-loaded nanoparticles fabricated by sequential nanoprecipitation. *Int. J. Pharm.* 581: 119291.

108 Gou, M., Wei, X., Men, K. et al. (2011). PCL/PEG copolymeric nanoparticles: potential nanoplatforms for anticancer agent delivery. *Curr. Drug Targets* 12: 1131–1150.

109 Avgoustakis, K. (2004). Pegylated poly (lactide) and poly (lactide-*co*-glycolide) nanoparticles: preparation, properties and possible applications in drug delivery. *Curr. Drug Deliv.* 1: 321–333.

110 Kumari, A., Yadav, S.K., and Yadav, S.C. (2010). Biodegradable polymeric nanoparticles based drug delivery systems. *Colloids Surf. B* 75: 1–18.

111 Shuai, X., Ai, H., Nasongkla, N. et al. (2004). Micellar carriers based on block copolymers of poly (ε-caprolactone) and poly (ethylene glycol) for doxorubicin delivery. *J. Control. Release* 98: 415–426.

112 Clay, N.E., Whittenberg, J.J., Leong, J. et al. (2017). Chemical and mechanical modulation of polymeric micelle assembly. *Nanoscale* 9: 5194–5204.

113 Buzea, C., Pacheco, I.I., and Robbie, K. (2007). Nanomaterials and nanoparticles: sources and toxicity. *Biointerphases* 2: MR17–MR71.

114 Yang, G., Liu, Y., Wang, H. et al. (2019). Bioinspired core-shell nanoparticles for hydrophobic drug delivery. *Angew. Chem. Int. Ed.* 131: 14495–14502.

115 He, C., Tang, Z., Tian, H., and Chen, X. (2016). Co-delivery of chemotherapeutics and proteins for synergistic therapy. *Adv. Drug Deliv. Rev.* 98: 64–76.

116 Wu, B., Lu, S.-T., Zhang, L.-J. et al. (2017). Codelivery of doxorubicin and triptolide with reduction-sensitive lipid–polymer hybrid nanoparticles for in vitro and in vivo synergistic cancer treatment. *Int. J. Nanomed.* 12: 1853.

117 Chen, J.-X., Shi, Y., Zhang, Y.-R. et al. (2016). One-pot construction of boronate ester based pH-responsive micelle for combined cancer therapy. *Colloids Surf. B* 143: 285–292.

118 Surnar, B. and Jayakannan, M. (2013). Stimuli-responsive poly (caprolactone) vesicles for dual drug delivery under the gastrointestinal tract. *Biomacromolecules* 14: 4377–4387.

119 Perez, R.A. and Kim, H.-W. (2015). Core–shell designed scaffolds for drug delivery and tissue engineering. *Acta Biomater.* 21: 2–19.

120 Wibowo, D., Yang, G.Z., Middelberg, A.P., and Zhao, C.X. (2017). Non-chromatographic bioprocess engineering of a recombinant mineralizing protein for the synthesis of silica nanocapsules. *Biotechnol. Bioeng.* 114: 335–343.

121 Yang, G.-Z., Wibowo, D., Yun, J.-H. et al. (2017). Biomimetic silica nanocapsules for tunable sustained release and cargo protection. *Langmuir* 33: 5777–5785.

122 Hui, Y., Wibowo, D., Liu, Y. et al. (2018). Understanding the effects of nanocapsular mechanical property on passive and active tumor targeting. *ACS Nano* 12: 2846–2857.

123 Peters, B.C., Wibowo, D., Yang, G.-Z. et al. (2019). Evaluation of baiting fipronil-loaded silica nanocapsules against termite colonies in fields. *Heliyon* 5: e02277.

124 Hui, Y., Yi, X., Wibowo, D. et al. (2020). Nanoparticle elasticity regulates phagocytosis and cancer cell uptake. *Sci. Adv.* 6: eaaz4316.

125 Yang, G., Liu, Y., Jin, S., and Zhao, C.X. (2020). Development of core-shell nanoparticle drug delivery systems based on biomimetic mineralization. *Chembiochem* 21: 2871–2879.

126 Qin, J., Wei, X., Chen, H. et al. (2018). mPEG-g-CS-modified PLGA nanoparticle carrier for the codelivery of paclitaxel and epirubicin for breast cancer synergistic therapy. *ACS Biomater. Sci. Eng.* 4: 1651–1660.

127 Li, Y., Cai, B., Shen, L. et al. (2017). MiRNA-29b suppresses tumor growth through simultaneously inhibiting angiogenesis and tumorigenesis by targeting Akt3. *Cancer Lett.* 397: 111–119.

128 Feng, Q., Zhang, C., Lum, D. et al. (2017). A class of extracellular vesicles from breast cancer cells activates VEGF receptors and tumour angiogenesis. *Nat. Commun.* 8: 1–17.

129 Lin, J.-T., Liu, Z.-K., Zhu, Q.-L. et al. (2017). Redox-responsive nanocarriers for drug and gene co-delivery based on chitosan derivatives modified mesoporous silica nanoparticles. *Colloids Surf. B* 155: 41–50.

130 Palvai, S., Anandi, L., Sarkar, S. et al. (2017). Drug-triggered self-assembly of linear polymer into nanoparticles for simultaneous delivery of hydrophobic and hydrophilic drugs in breast cancer cells. *ACS Omega* 2: 8730–8740.

131 Li, H., Li, M., Chen, C. et al. (2015). On-demand combinational delivery of curcumin and doxorubicin via a pH-labile micellar nanocarrier. *Int. J. Pharm.* 495: 572–578.

132 Chen, C., Tao, R., Ding, D. et al. (2017). Ratiometric co-delivery of multiple chemodrugs in a single nanocarrier. *Eur. J. Pharm. Sci.* 107: 16–23.

133 Shin, H.-C., Alani, A.W., Rao, D.A. et al. (2009). Multi-drug loaded polymeric micelles for simultaneous delivery of poorly soluble anticancer drugs. *J. Control. Release* 140: 294–300.

134 Funkhouser, J. (2002). Reinventing pharma: the theranostic revolution. *Curr. Drug Discov.* 2: 17–19.

135 Sumer, B. and Gao, J. (2008). Theranostic nanomedicine for cancer. *Nanomedicine* 3: 137–140.

136 Betzig, E., Patterson, G.H., Sougrat, R. et al. (2006). Imaging intracellular fluorescent proteins at nanometer resolution. *Science* 313: 1642–1645.

137 Li, L., Chen, Y., Chen, W. et al. (2019). Photodynamic therapy based on organic small molecular fluorescent dyes. *Chin. Chem. Lett.* 30: 1689–1703.

138 Liu, Y., Yang, G., Jin, S. et al. (2020). J-aggregate-based FRET monitoring of drug release from polymer nanoparticles with high drug loading. *Angew. Chem. Int. Ed.* 59: 20065–20074.

139 Yang, G., Liu, Y., Hui, Y. et al. (2021). Implications of quenching-to-dequenching switch in quantitative cell uptake and biodistribution of dye-labeled nanoparticles. *Angew. Chem. Int. Ed.* 60: 15426–15435.

140 Fernández-Moreira, V., Thorp-Greenwood, F.L., and Coogan, M.P. (2010). Application of d6 transition metal complexes in fluorescence cell imaging. *Chem. Commun.* 46: 186–202.

141 Zou, D., Yu, L., Sun, Q. et al. (2020). A general approach for biomimetic mineralization of MOF particles using biomolecules. *Colloids Surf. B*: 111108.

142 Erwin, S.C., Zu, L., Haftel, M.I. et al. (2005). Doping semiconductor nanocrystals. *Nature* 436: 91–94.

143 Zhou, J., Liu, Z., and Li, F. (2012). Upconversion nanophosphors for small-animal imaging. *Chem. Soc. Rev.* 41: 1323–1349.
144 Chen, G., Qiu, H., Prasad, P.N., and Chen, X. (2014). Upconversion nanoparticles: design, nanochemistry, and applications in theranostics. *Chem. Rev.* 114: 5161–5214.
145 Wang, F., Banerjee, D., Liu, Y. et al. (2010). Upconversion nanoparticles in biological labeling, imaging, and therapy. *Analyst* 135: 1839–1854.
146 Zhao, J.W., Yang, H., Li, J.L. et al. (2017). Fabrication of pH-responsive PLGA(UCNPs/DOX) nanocapsules with upconversion luminescence for drug delivery. *Sci. Rep.* 7: 1–11.
147 Yin, C., Zhen, X., Zhao, H. et al. (2017). Amphiphilic semiconducting oligomer for near-infrared photoacoustic and fluorescence imaging. *ACS Appl. Mater. Interfaces* 9: 12332–12339.
148 Howes, P., Green, M., Bowers, A. et al. (2010). Magnetic conjugated polymer nanoparticles as bimodal imaging agents. *J. Am. Chem. Soc.* 132: 9833–9842.
149 Kim, J., Lee, J.E., Lee, S.H. et al. (2008). Designed fabrication of a multifunctional polymer nanomedical platform for simultaneous cancer-targeted imaging and magnetically guided drug delivery. *Adv. Mater.* 20: 478–483.
150 Li, X.L., Li, H., Liu, G.Q. et al. (2012). Magnetite-loaded fluorine-containing polymeric micelles for magnetic resonance imaging and drug delivery. *Biomaterials* 33: 3013–3024.
151 Chen, M. and Yin, M. (2014). Design and development of fluorescent nanostructures for bioimaging. *Prog. Polym. Sci.* 39: 365–395.
152 Hong, Y., Lam, J.W., and Tang, B.Z. (2011). Aggregation-induced emission. *Chem. Soc. Rev.* 40: 5361–5388.
153 Zhang, N., Chen, H., Fan, Y. et al. (2018). Fluorescent polymersomes with aggregation-induced emission. *ACS Nano* 12: 4025–4035.
154 Qin, S.-Y., Zhang, A.-Q., Cheng, S.-X. et al. (2017). Drug self-delivery systems for cancer therapy. *Biomaterials* 112: 234–247.
155 Qi, J., Hu, X., Dong, X. et al. (2019). Towards more accurate bioimaging of drug nanocarriers: turning aggregation-caused quenching into a useful tool. *Adv. Drug Deliv. Rev.* 143: 206–225.
156 Langer, R. (1990). New methods of drug delivery. *Science* 249: 1527–1533.
157 Kamaly, N., Yameen, B., Wu, J., and Farokhzad, O.C. (2016). Degradable controlled-release polymers and polymeric nanoparticles: mechanisms of controlling drug release. *Chem. Rev.* 116: 2602–2663.
158 Blasi, P., D'Souza, S.S., Selmin, F., and DeLuca, P.P. (2005). Plasticizing effect of water on poly (lactide-*co*-glycolide). *J. Control. Release* 108: 1–9.
159 Webber, W.L., Lago, F., Thanos, C., and Mathiowitz, E. (1998). Characterization of soluble, salt-loaded, degradable PLGA films and their release of tetracycline. *J. Biomed. Mater. Res.* 41: 18–29.
160 Fu, Y. and Kao, W.J. (2010). Drug release kinetics and transport mechanisms of non-degradable and degradable polymeric delivery systems. *Expert Opin. Drug Deliv.* 7: 429–444.

161 Keraliya, R.A., Patel, C., Patel, P. et al. (2012). Osmotic drug delivery system as a part of modified release dosage form. *ISRN Pharm.* 2012: 1–9.

162 Marin, E., Briceño, M.I., and Caballero-George, C. (2013). Critical evaluation of biodegradable polymers used in nanodrugs. *Int. J. Nanomedicine* 8: 3071.

163 Uhrich, K.E., Cannizzaro, S.M., Langer, R.S., and Shakesheff, K.M. (1999). Polymeric systems for controlled drug release. *Chem. Rev.* 99: 3181–3198.

164 Park, H., Park, K., and Shalaby, W.S. (1993). *Biodegradable Hydrogels for Drug Delivery*. CRC Press.

165 Huo, M., Yuan, J., Tao, L., and Wei, Y. (2014). Redox-responsive polymers for drug delivery: from molecular design to applications. *Polym. Chem.* 5: 1519–1528.

166 Dong, D.-W., Xiang, B., Gao, W. et al. (2013). pH-responsive complexes using prefunctionalized polymers for synchronous delivery of doxorubicin and siRNA to cancer cells. *Biomaterials* 34: 4849–4859.

167 Kakinoki, A., Kaneo, Y., Ikeda, Y. et al. (2008). Synthesis of poly (vinyl alcohol)–doxorubicin conjugates containing cis-aconityl acid-cleavable bond and its isomer dependent doxorubicin release. *Biol. Pharm. Bull.* 31: 103–110.

168 Li, C., Yu, D.-F., Newman, R.A. et al. (1998). Complete regression of well-established tumors using a novel water-soluble poly (L-glutamic acid)-paclitaxel conjugate. *Cancer Res.* 58: 2404–2409.

169 Gillies, E.R., Goodwin, A.P., and Fréchet, J.M. (2004). Acetals as pH-sensitive linkages for drug delivery. *Bioconjug. Chem.* 15: 1254–1263.

170 Qiu, Y. and Park, K. (2001). Environment-sensitive hydrogels for drug delivery. *Adv. Drug Deliv. Rev.* 53: 321–339.

171 Bajpai, A.K., Shukla, S.K., Bhanu, S., and Kankane, S. (2008). Responsive polymers in controlled drug delivery. *Prog. Polym. Sci.* 33: 1088–1118.

172 Bawa, P., Pillay, V., Choonara, Y.E., and Du Toit, L.C. (2009). Stimuli-responsive polymers and their applications in drug delivery. *Biomed. Mater.* 4: 022001.

173 Song, C., Lin, T., Zhang, Q. et al. (2019). pH-sensitive morphological transitions in polymeric tadpole assemblies for programmed tumor therapy. *J. Control. Release* 293: 1–9.

174 Strong, L.E. and West, J.L. (2011). Thermally responsive polymer–nanoparticle composites for biomedical applications. *Wiley Interdiscip. Rev. Nanomed. Nanobiotechnol.* 3: 307–317.

175 Bikram, M. and West, J.L. (2008). Thermo-responsive systems for controlled drug delivery. *Expert Opin. Drug Deliv.* 5: 1077–1091.

176 Schild, H.G. (1992). Poly (*N*-isopropylacrylamide): experiment, theory and application. *Prog. Polym. Sci.* 17: 163–249.

177 Sasaki, S., Kawasaki, H., and Maeda, H. (1997). Volume phase transition behavior of *N*-isopropylacrylamide gels as a function of the chemical potential of water molecules. *Macromolecules* 30: 1847–1848.

178 Chilkoti, A., Dreher, M.R., Meyer, D.E., and Raucher, D. (2002). Targeted drug delivery by thermally responsive polymers. *Adv. Drug Deliv. Rev.* 54: 613–630.

179 Galaev, I.Y. and Mattiasson, B. (1999). 'Smart'polymers and what they could do in biotechnology and medicine. *Trends Biotechnol.* 17: 335–340.

180 Meyer, D.E., Shin, B., Kong, G. et al. (2001). Drug targeting using thermally responsive polymers and local hyperthermia. *J. Control. Release* 74: 213–224.

181 Zhang, J. and Misra, R. (2007). Magnetic drug-targeting carrier encapsulated with thermosensitive smart polymer: core–shell nanoparticle carrier and drug release response. *Acta Biomater.* 3: 838–850.

182 Wang, Y., Wu, G., Li, X. et al. (2013). Synthesis, characterization and controlled drug release from temperature-responsive poly (ether-urethane) particles based on PEG-diisocyanates and aliphatic diols. *J. Biomater. Sci. Polym. Ed.* 24: 1676–1691.

183 Moktan, S. and Raucher, D. (2012). Anticancer activity of proapoptotic peptides is highly improved by thermal targeting using elastin-like polypeptides. *Int. J. Pept. Res. Ther.* 18: 227–237.

184 Cheng, R., Feng, F., Meng, F. et al. (2011). Glutathione-responsive nano-vehicles as a promising platform for targeted intracellular drug and gene delivery. *J. Control. Release* 152: 2–12.

185 Kuppusamy, P., Li, H., Ilangovan, G. et al. (2002). Noninvasive imaging of tumor redox status and its modification by tissue glutathione levels. *Cancer Res.* 62: 307–312.

186 Siegwart, D.J., Oh, J.K., and Matyjaszewski, K. (2012). ATRP in the design of functional materials for biomedical applications. *Prog. Polym. Sci.* 37: 18–37.

187 Boyer, C., Bulmus, V., Davis, T.P. et al. (2009). Bioapplications of RAFT polymerization. *Chem. Rev.* 109: 5402–5436.

188 Willcock, H. and O'Reilly, R.K. (2010). End group removal and modification of RAFT polymers. *Polym. Chem.* 1: 149–157.

189 Gauthier, M.A., Gibson, M.I., and Klok, H.A. (2009). Synthesis of functional polymers by post-polymerization modification. *Angew. Chem. Int. Ed.* 48: 48–58.

190 Xu, H., Cao, W., and Zhang, X. (2013). Selenium-containing polymers: promising biomaterials for controlled release and enzyme mimics. *Acc. Chem. Res.* 46: 1647–1658.

191 Liu, J., Pang, Y., Chen, J. et al. (2012). Hyperbranched polydiselenide as a self assembling broad spectrum anticancer agent. *Biomaterials* 33: 7765–7774.

192 Yang, J., Liu, W., Sui, M. et al. (2011). Platinum (IV)-coordinate polymers as intracellular reduction-responsive backbone-type conjugates for cancer drug delivery. *Biomaterials* 32: 9136–9143.

193 Cho, H., Bae, J., Garripelli, V.K. et al. (2012). Redox-sensitive polymeric nanoparticles for drug delivery. *Chem. Commun.* 48: 6043–6045.

194 De Vries, W.C., Grill, D., Tesch, M. et al. (2017). Reversible stabilization of vesicles: redox-responsive polymer nanocontainers for intracellular delivery. *Angew. Chem. Int. Ed.* 56: 9603–9607.

195 Ma, N., Li, Y., Xu, H. et al. (2010). Dual redox responsive assemblies formed from diselenide block copolymers. *J. Am. Chem. Soc.* 132: 442–443.

196 Lee, S.H., Gupta, M.K., Bang, J.B. et al. (2013). Current progress in reactive oxygen species (ROS)-responsive materials for biomedical applications. *Adv. Healthc. Mater.* 2: 908–915.

197 Colson, Y.L. and Grinstaff, M.W. (2012). Biologically responsive polymeric nanoparticles for drug delivery. *Adv. Mater.* 24: 3878–3886.

198 D'Autréaux, B. and Toledano, M.B. (2007). ROS as signalling molecules: mechanisms that generate specificity in ROS homeostasis. *Nat. Rev. Mol. Cell Biol.* 8: 813–824.

199 Lallana, E. and Tirelli, N. (2013). Oxidation-responsive polymers: which groups to use, how to make them, what to expect from them (biomedical applications). *Macromol. Chem. Phys.* 214: 143–158.

200 de Gracia Lux, C., Joshi-Barr, S., Nguyen, T. et al. (2012). Biocompatible polymeric nanoparticles degrade and release cargo in response to biologically relevant levels of hydrogen peroxide. *J. Am. Chem. Soc.* 134: 15758–15764.

201 Bigot, J., Charleux, B., Cooke, G. et al. (2010). Tetrathiafulvalene end-functionalized poly (*N*-isopropylacrylamide): a new class of amphiphilic polymer for the creation of multistimuli responsive micelles. *J. Am. Chem. Soc.* 132: 10796–10801.

202 Hu, P. and Tirelli, N. (2012). Scavenging ROS: superoxide dismutase/catalase mimetics by the use of an oxidation-sensitive nanocarrier/enzyme conjugate. *Bioconjug. Chem.* 23: 438–449.

203 Ren, H., Wu, Y., Ma, N. et al. (2012). Side-chain selenium-containing amphiphilic block copolymers: redox-controlled self-assembly and disassembly. *Soft Matter.* 8: 1460–1466.

204 Staff, R.H., Gallei, M., Mazurowski, M. et al. (2012). Patchy nanocapsules of poly (vinylferrocene)-based block copolymers for redox-responsive release. *ACS Nano* 6: 9042–9049.

205 Wu, P., Dong, W., Guo, X. et al. (2019). ROS-responsive blended nanoparticles: cascade-amplifying synergistic effects of sonochemotherapy with on-demand boosted drug release during SDT process. *Adv. Healthc. Mater.* 8: 1900720.

206 Dai, L., Yu, Y., Luo, Z. et al. (2016). Photosensitizer enhanced disassembly of amphiphilic micelle for ROS-response targeted tumor therapy in vivo. *Biomaterials* 104: 1–17.

207 Chen, C., Liu, G., Liu, X. et al. (2011). Photo-responsive, biocompatible polymeric micelles self-assembled from hyperbranched polyphosphate-based polymers. *Polym. Chem.* 2: 1389–1397.

208 Fomina, N., Sankaranarayanan, J., and Almutairi, A. (2012). Photochemical mechanisms of light-triggered release from nanocarriers. *Adv. Drug Deliv. Rev.* 64: 1005–1020.

209 Yang, H., Zhang, C., Li, C. et al. (2015). Glucose-responsive polymer vesicles templated by α-CD/PEG inclusion complex. *Biomacromolecules* 16: 1372–1381.

210 Ma, R. and Shi, L. (2014). Phenylboronic acid-based glucose-responsive polymeric nanoparticles: synthesis and applications in drug delivery. *Polym. Chem.* 5: 1503–1518.

211 Li, Y., Liu, G., Wang, X. et al. (2016). Enzyme-responsive polymeric vesicles for bacterial-strain-selective delivery of antimicrobial agents. *Angew. Chem. Int. Ed.* 55: 1760–1764.

212 De La Rica, R., Aili, D., and Stevens, M.M. (2012). Enzyme-responsive nanoparticles for drug release and diagnostics. *Adv. Drug Deliv. Rev.* 64: 967–978.

213 Sirsi, S.R. and Borden, M.A. (2014). State-of-the-art materials for ultrasound-triggered drug delivery. *Adv. Drug Deliv. Rev.* 72: 3–14.

214 Xia, H., Zhao, Y., and Tong, R. (2016). *Therapeutic Ultrasound*, 365–384. Springer.

215 Mo, R., Jiang, T., DiSanto, R. et al. (2014). ATP-triggered anticancer drug delivery. *Nat. Commun.* 5: 1–10.

216 Wang, G.-H., Huang, G.-L., Zhao, Y. et al. (2016). ATP triggered drug release and DNA co-delivery systems based on ATP responsive aptamers and polyethylenimine complexes. *J. Mater. Chem. B* 4: 3832–3841.

217 Szunerits, S., Teodorescu, F., and Boukherroub, R. (2016). Electrochemically triggered release of drugs. *Eur. Polym. J.* 83: 467–477.

218 Ge, J., Neofytou, E., Cahill, T.J. III et al. (2012). Drug release from electric-field-responsive nanoparticles. *ACS Nano* 6: 227–233.

219 Liu, H., Lin, S., Feng, Y., and Theato, P. (2017). CO_2-responsive polymer materials. *Polym. Chem.* 8: 12–23.

220 Yan, Q., Zhou, R., Fu, C. et al. (2011). CO_2-responsive polymeric vesicles that breathe. *Angew. Chem. Int. Ed.* 50: 4923–4927.

221 Fleige, E., Quadir, M.A., and Haag, R. (2012). Stimuli-responsive polymeric nanocarriers for the controlled transport of active compounds: concepts and applications. *Adv. Drug Deliv. Rev.* 64: 866–884.

222 Cheng, R., Meng, F., Deng, C. et al. (2013). Dual and multi-stimuli responsive polymeric nanoparticles for programmed site-specific drug delivery. *Biomaterials* 34: 3647–3657.

223 Müller-Schulte, D. and Schmitz-Rode, T. (2006). Thermosensitive magnetic polymer particles as contactless controllable drug carriers. *J. Magn. Magn. Mater.* 302: 267–271.

224 Asfaram, A., Ghaedi, M., Hajati, S., and Goudarzi, A. (2016). Synthesis of magnetic γ-Fe_2O_3-based nanomaterial for ultrasonic assisted dyes adsorption: modeling and optimization. *Ultrason. Sonochem.* 32: 418–431.

225 Arizaga, A., Ibarz, G., and Piñol, R. (2010). Stimuli-responsive poly (4-vinyl pyridine) hydrogel nanoparticles: synthesis by nanoprecipitation and swelling behavior. *J. Colloid Interface Sci.* 348: 668–672.

226 Chatterjee, S., Hui, P., Kan, C.W., and Wang, W. (2019). Dual-responsive (pH/temperature) Pluronic F-127 hydrogel drug delivery system for textile-based transdermal therapy. *Sci. Rep.* 9: 1–13.

227 Kelley, E.G., Albert, J.N., Sullivan, M.O., and Epps, T.H. III (2013). Stimuli-responsive copolymer solution and surface assemblies for biomedical applications. *Chem. Soc. Rev.* 42: 7057–7071.

228 Li, D., Bu, Y., Zhang, L. et al. (2016). Facile construction of pH-and redox-responsive micelles from a biodegradable poly (β-hydroxyl amine) for drug delivery. *Biomacromolecules* 17: 291–300.

229 Chen, S., Jiang, F., Cao, Z. et al. (2015). Photo, pH, and thermo triple-responsive spiropyran-based copolymer nanoparticles for controlled release. *Chem. Commun.* 51: 12633–12636.

14

Heat Management by Colloidal Self-assembly

Qimeng Song, Marius Schöttle, Pia Ruckdeschel, Fabian Nutz, and Markus Retsch

University of Bayreuth, Department of Chemistry, Physical Chemistry I, Universitätsstr. 30, Bayreuth 95447, Germany

14.1 Introduction

Thermal transport properties of nanostructured materials have increasingly attracted the interest of many research groups for the past two decades [1–5]. This surge in research activities on various aspects of thermal transport characterization is based on two major developments. Firstly, our globally limited energy resources require better management of heat and waste heat in a way to increase the overall energy efficiency. Therefore, novel strategies to control and to use the flow of heat are necessary. Secondly, the ongoing miniaturization in the broad field of nanotechnology requires a detailed understanding of heat management on these small length scales. An effective way of managing the involved heat and temperature is paramount since many of the intended device properties strongly depend on the effective device temperature. These requirements coincide with new possibilities along the entire research chain of the nanostructured matter; the major improvements in materials fabrication, characterization, and modeling allow now to specifically access, investigate, and understand nanoscale transport phenomena, which have been elusive before.

Consequently, thermal materials start to complement the traditionally well-investigated field of electronics and photonics. Quite often, the term phononics is used to describe this emerging area of research [6, 7]. One should note that the term phononics does not only relate to the transport of thermal energy (high energy) but also to lower energy regimes (hypersonic sound transport) [8]. Novel interesting material properties evolving around the concepts of phononic materials are thermal cloaks [9], thermal rectification [10, 11], thermal memory [12], dynamic insulation materials [13–15], or improved thermoelectric devices [16–18], to name a few. Whereas, historically, a strong focus has been laid on crystalline inorganic materials, recent endeavors point toward the great potential and many surprises to be found in polymer-based and consequently amorphous materials. Prominent

Functional Materials from Colloidal Self-assembly, First Edition. Edited by Qingfeng Yan and George Zhao.

examples are routes toward good thermal conductance in traditionally considered insulating polymers [19, 20].

Considering the tremendous potential and development of the field of nanoscale thermal transport, it is quite surprising that, up to now, the well-developed field of colloidal engineering has been barely in the focus of such studies. The intent of this book chapter is to raise attention to the many ways that colloidal assembly structures can contribute to this growing field. The reasons to look at colloidal materials from a thermal transport perspective are quite obvious: (i) the accessible feature size addresses the entire submicron range without any specialized nano- or microfabrication technology; (ii) there is a great selection of materials to choose from; and (iii) the interfacial properties of colloidal particles can be tuned. Combining all that, we are convinced that colloidal assembly structure will experience a renaissance to complement the established mechanical and photonic properties with heat transport applications.

Since the field of colloidal assembly structures is quite broadly defined, we will mainly restrict ourselves to colloidal materials based on simple spherical and sheet-like objects and the nonelectronic thermal transport along these structures. Consequently, we will not touch upon important related fields, where colloidal particles also play a paramount role, but refer the interested reader to the respective literature. Topics not included are near-field thermal radiation [21, 22], nanofluids [23–25], and surface phonon polaritons [26, 27]. We briefly introduce the wide parameter space of colloidal assembly structures and point the reader toward ways to synthetically realize them (Section 14.2). We then outline the fundamentals of thermal transport, beginning with bulk dielectric materials and then progressing toward structured devices (Section 14.3). Since such mesostructures are quite often only accessible in small quantities or as thin film materials, specialized analytical tools are required to determine the thermal conductivity or thermal diffusivity experimentally (Section 14.4). Having laid out these fundamental properties and techniques, we will then review intriguing examples of thermal transport in colloidal assemblies and heat management applications in Section 14.5.

14.2 Fabrication Methods

Colloidal crystal fabrication via self-assembly is well established and has been widely researched by many groups for the past 30 years. Here, we will give a brief introduction to the range of possibilities and the methods of fabricating colloidal assembly structures.

14.2.1 Categories of Colloidal Assemblies

There are several possible classifications for the rather broad term of colloidal assembly structures. A widespread distinction is based on dimensionality: two-dimensional (2D) and three-dimensional (3D). 2D colloidal assemblies, typically referred to as colloidal monolayers, comprise only a single layer of particles.

These can be close-packed or non-close-packed on a support substrate. Due to the 2D nature of the monolayers, these are mainly interesting regarding e.g. wetting or optical properties of interfaces. For thermal transport applications, only the 3D equivalents are of interest. We refer to such 3D assemblies as colloidal crystals or glasses. The vast majority of colloidal assembly structures are based on spherical monodisperse nano- and mesoparticles. These typically self-assemble into hexagonally close-packed structures or colloidal glasses. The shape and symmetry of the constituting colloidal particles have a strong influence on the resulting colloidal ensemble. Owing to unique anisotropic interparticle interactions, colloidal assemblies from structured colloids have also been explored [28]. A manifold of particle morphologies has been realized, ranging from simple spheres to stars to intricately structured patchy particles [28–33]. Such objects can be further modified by the introduction of functional groups, doping, or interfacial engineering via specific ligand exchange [34]. The resulting assembly structures reach from a preferential orientation in the case of ellipsoids [35] to quasicrystals based on soft core–shell systems [36]. Colloidal objects based on 2D sheets such as clays or graphene oxide (GO) assemble in textured materials with one or several axes of preferential orientation [37, 38].

Overall, a vast toolbox of appropriate colloidal building blocks has been established over the past decades. Nowadays, researchers have access to tiny nanocrystals (e.g. quantum dots and plasmonic nanoparticles); polymer beads, which span the entire colloidal range; and metal oxide nanoparticles of various compositions. Consequently, an almost endless spectrum of functional and structural properties can be accessed once the colloidal objects are transferred into a particular superstructure.

14.2.2 Self-assembly Strategies

To fully exploit the potential of colloidally structured materials, one needs to be able to control the specific interparticle interaction forces during the self-assembly process. The most prominent types of particle interactions in suspension and at interfaces are attractive ones, like van der Waals and capillary forces, and repulsive interactions, like electrostatic and steric forces (Figure 14.1a). The interplay of these forces, with respect to the particle–particle distance, the colloid composition, surface charge, and the colloid size, is the decisive contribution to access highly defined or randomly aggregated colloidal superstructures [40].

A variety of assembly techniques have been developed for colloidal assemblies in the last few decades [39, 41]. We will only focus on 3D self-assembly strategies.

Generally, a colloidally stable dispersion is the starting point for any self-assembly process. This means that the repulsive colloidal forces dominate at the timescale of interest and that the colloidal objects are dispersed at least in a metastable and single-object state. During the self-assembly process, one or several attractive interaction potentials start to dominate, which direct the colloidal particles toward each other. The most common self-assembly strategies for 3D colloidal crystals are summarized in Figure 14.1b. These methods comprise sedimentation, centrifugation, spray coating, convective assembly, doctor blading, filtering, or dispersion

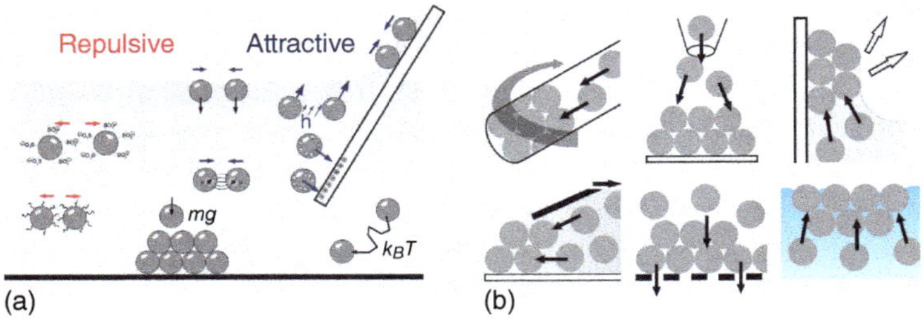

Figure 14.1 (a) Overview of attractive and repulsive colloidal interaction forces, which are relevant for the fabrication of colloidal superstructures. (b) Overview of common self-assembly strategies to fabricate colloidal crystals and ensembles. Source: Adapted from [39] with permission of American Chemical Society.

drying. The main distinction between these methods is, on the one hand, the accessible colloidal ensemble thickness, which can range from a few particle layers up to centimeters. On the other hand, the colloidal ensembles can be self-sustaining and hence free-standing, or they can be immobilized on a support structure. Further considerations for choosing one or the other fabrication method can be based on the time, scalability, or the intended ensemble quality. For instance, one can assemble highly ordered colloidal crystals from monodisperse latex particles by simple dispersion drying. This process, however, can take up to several days. Filtering the same dispersion through a nanoporous filter takes only a couple of minutes but quite commonly results in less ordered colloidal glasses.

In addition, colloidal ensembles can be used to template other materials in a way to access their porous counterparts. Such materials are commonly referred to as inverse opals. Inverse opals can comprise metals [42], polymers [43], or metal oxides [44]. For more information on the fabrication and properties of inverse opals, we refer the reader to these review articles [45–47].

Apart from colloidal crystals solely fabricated by (functionalized) colloidal objects, nanocomposites have emerged as a versatile material in the application of heat management in recent years. We will, therefore, introduce the fundamental fabrication aspects of this materials class.

Nanocomposites are multicomponent materials that combine colloidal particles with a matrix. (Nano)composite materials are widely used to synergistically benefit from the individual properties of both components to improve e.g. optical, thermal, electric, or mechanical properties. Compared to materials with macroscopic fillers, nanocomposites can considerably enhance the properties of a material due to the enormous surface-to-volume ratio of nanofillers [48]. However, phase separation and demixing have to be rigorously controlled by proper surface functionalization. A major challenge in material science is the combination of soft polymer-based adhesion with high thermal conductivity, in particular for thermal interface materials. Consequently, the thermal conductivity of dielectric materials such as polymers shall be increased by the incorporation of distinct organic and inorganic nanofillers with an intrinsically high thermal conductivity. Promising filler particles are ceramic particles [49], carbon nanotubes [50], boron nitride nanotubes, or nanosheets [51, 52], to name a few. Various methods were revealed for nanocomposite fabrication; most of them require the utilization of a common solvent for both components (matrix and filler). Freeze casting (also known as ice-templating) is a versatile technique that has been extensively applied in the fabrication of hierarchically structured nanocomposites. In a two-step process, a stable dispersed mixture is solidified via freezing, and, subsequently, the solvent, mostly water, is removed via sublimation. In this way, composites with a tailored micro- or macrostructure can be constructed with a large range of materials [53]. Electrospinning has been widely used to generate nanocomposite fiber materials. With this technique, the orientation of the nanofillers can be widely controlled, which circumvents suboptimal properties of composites without a defined filler orientation [52, 54]. Spray-casting of composite solutions containing clay sheets and polymer has been demonstrated to result in exceptionally periodically ordered hybrid materials [55]. Additionally, other techniques, such as the vacuum-assisted self-assembly technique [56–58], bidirectional freezing technique [59], and *in situ* polymerization [60], have also been reported for nanocomposite fabrication.

14.3 Fundamentals of Thermal Transport in Dielectric Materials

Colloidal materials are commonly composed of either crystalline or amorphous matter. Therefore, we will first contrast the bulk properties with respect to the thermal transport of these two fundamental material classes. Understanding the bulk properties is essential for a better understanding of the additional influence provided by the colloidal nano- and mesostructuring. Thermal transport in metals, where electronic contributions dominate, will not be discussed in this section.

14.3.1 Thermal Transport in Crystalline Matter

The thermal transport mechanisms are well characterized in classical crystalline materials, such as silicon. The thermal energy is transported by propagating delocalized lattice vibration waves called phonons.

Treating phonons as quasi-particles, the thermal conductivity can be expressed by the application of the kinetic theory using the Debye equation:

$$\kappa = \frac{1}{3} C_{\mathrm{v}} v_{\mathrm{g}} \Lambda \tag{14.1}$$

where C_{v} is the volumetric heat capacity, v_{g} is the phonon group velocity, and Λ is the average phonon mean free path (MFP), which is the characteristic length of phonons traveling between two scattering events.

The transport of phonons, however, can be easily obstructed by a range of processes, each leading to a reduction of the effective thermal conductivity. Scattering processes in dielectric crystals are the phonon–phonon scattering and the scattering at boundaries, impurities, imperfections, and grain boundaries [61].

Phonon–phonon scattering results as a consequence of the anharmonicity of the lattice potential and is dominant in pure crystals. It is also responsible for the common T^{-1} temperature dependency beyond the Debye temperature. Below this temperature the thermal transport is limited by boundary scattering, and the thermal conductivity follows the T^3 dependence of the heat capacity [62, 63]. The most common scattering process is three-phonon scattering, which can be classified into a normal process (N-process) and an Umklapp process (U-process) [64]. The Umklapp scattering generates a thermal resistance and leads to a finite thermal conductivity. However, in crystalline colloidal materials, other scattering processes caused by boundaries, grain boundaries, imperfections, or impurities are likely to dominate, owing to their colloidal nanostructure. Therefore, one would typically expect a reduced thermal conductivity for a colloidal ensemble compared to its pure bulk counterpart. The effect of confinement to nanostructured length scales was shown in silicon thin films. A decreasing film thickness resulted in increased boundary scattering [63, 65–67]. Additionally, the introduction of surface roughness reduces the thermal transport properties [68].

A very challenging approach to manipulate heat transport is the creation of forbidden bandgaps for phonons by wave interference at periodic nanostructures [69]. This requires preserving wave coherence and can only be realized when diffusive phonon scattering (incoherent scattering) is strictly prevented. This interference-based manipulation is analogously known in photonic materials leading to manipulation of light by photonic bandgaps [70]. The major issue for many synthetic systems is provided by surface roughness effects or changes to the crystal symmetry and orientation close to interfaces [71]. These often evoke diffusive scattering, which destroys any bandgap engineering [72]. For more information about phonon wave interference and the devices accessible through thermal bandgap materials, we want to refer to a review of Maldovan [69].

Very recently, Tambo et al. [73] demonstrated that even in amorphous materials, phononic bandgap engineering is possible. The periodic sub-100-nm hole array was structured on amorphous silicon nitride thin films and was shown to reduce the thermal transport effectively.

14.3.2 Thermal Transport in Amorphous Matter

In contrast to crystalline matter, amorphous materials have no translational symmetry and periodicity. This implies that heat transport mechanisms are intrinsically distinct. In these disordered systems, it is not possible to clearly define the dispersion and thus the group velocity due to the lack of periodicity. In an early theoretical work by Einstein, heat conduction in disordered solids is described by a "random walk" of independent oscillators with a characteristic frequency, i.e. the Einstein frequency [74]. Subsequently, Cahill and Pohl further extended this concept and proposed the minimum thermal conductivity model in disordered materials, referred to as the "amorphous limit" [75, 76]. However, these independent oscillators, known as diffusion modes (diffusons), contribute only partly to the thermal conductivity of amorphous materials, as discovered numerically and theoretically by Allen and

Feldman [77, 78]. They claimed that the diffusons could only be assigned with a diffusivity, and thus the concept of MFP and wave-vector is useless due to the MFP being shorter than the interatomic distance.

Besides diffusons, Allen and Feldman defined two other vibrational modes being present in amorphous materials. These are locons and propagons. Locons are localized, non-propagating vibrational modes that do not contribute to thermal transport in the harmonic approximation. Propagons are propagating and delocalized phonon-like modes that typically possess long wavelengths compared to the interatomic spacing. Thus, ballistic propagation takes place, which means that no scattering occurs during transport. The description of thermal transport by means of a particle-like or wavelike behavior was recently unified in a theoretical approach described by Simoncelli et al. [79]. Based on the Boltzmann transport equation, they derived a unifying theory that allowed for a correct description of crystalline and amorphous materials over a wide temperature range.

The fundamental difference in the transport mechanism between crystalline and amorphous matter often results in a much lower thermal conductivity and also a different temperature dependency in the latter [80]. Bulk crystalline silicon, for example, has a thermal conductivity of 148 W/m/K at 300 K [81], which reduces to approximately 2 W/m/K for its amorphous counterpart [82]. For crystalline materials, the thermal conductivity decreases after reaching the Debye temperature, whereas the thermal conductivity in amorphous materials increases monotonically with increasing temperature, which is mainly driven by the increase in heat capacity [83].

As polymers play an important role in colloid chemistry, it makes sense to look at the unique thermal properties of this amorphous material class. In most cases, amorphous or semicrystalline polymers are considered to be thermal insulators [84, 85]. Scattering of vibrational modes at interfaces between amorphous and crystalline domains as well as along the curvature of randomly coiled chains and at chain ends typically results in a low thermal conductivity. Several factors can, however, influence the thermal properties significantly. For instance, increasing the molecular weight minimizes scattering at chain ends and can increase the thermal conductivity [86]. Stiffening of the polymer backbone, e.g. in aromatic polymers, prevents rotation of C–C segments and also facilitates propagation of vibrational modes [87]. Interchain interactions can be adjusted by including side chains capable of forming hydrogen bonds. The bulk thermal conductivity of these polymers was shown to increase [88–90]. Orientation of chains has a very pronounced effect on linear polymers. Fully amorphous yet oriented polymer fibers can show a significant increase in the thermal conductivity [19], and highly crystalline polyethylene (PE) fibers could be prepared with incredible values up to 100 W/m/K [91]. The preferred propagation along the oriented polymer chain naturally leads to significant anisotropy of thermal properties [92, 93].

14.3.3 Thermal Transport in Composite and Porous Materials

While the previous two sections dealt with homogeneous materials, we now want to introduce the basics of thermal transport in multiphase materials. Colloidal materials comprise, by definition, two distinct phases. For the case of thermal transport, two scenarios are relevant: (i) porous colloidal materials and (ii) composite colloidal materials.

Consequently, several distinct pathways contribute to the overall thermal performance, which we want to introduce briefly. In porous materials, the main contributions to the total thermal conductivity (κ_{tot}) are the thermal conductivity through the solid material (κ_{solid}) and the gaseous phase (κ_{gas}). Additionally, radiation (κ_{rad}) and convection (κ_{conv}) can play a role [94].

$$\kappa_{tot} = \kappa_{solid} + \kappa_{gas} + \kappa_{rad} + \kappa_{conv} \tag{14.2}$$

It is important to note the difference between gaseous heat conduction and gaseous convection. Convection only applies to macroscopic systems, which allow for convective heat exchange. Convection may be relevant for the heat exchange at the outer interface of a colloidal material but is not a significant thermal transport path for the heat transfer inside porous/structured materials. Convective gas transport typically emerges at pore diameters larger than 4 mm [95]. Radiative heat transfer (κ_{rad}) increases with temperature (T^4-dependence) and strongly depends on the network connectivity, which is directly linked to the density [96]. It can be suppressed by the addition of opacifiers, which absorb light in the mid-IR range. For most systems under consideration here, the thermal transport is only characterized at relatively low temperatures (<500 °C), where radiative effects are small.

For pore sizes in the range of the MFP of the gas phase, the heat conduction of the gas phase (e.g. air) equals no longer that of a free (bulk) gas but follows the Knudsen formula. This results in reduced gas conduction:

$$\kappa_{gas} = \frac{\kappa_{gas,0}\Phi}{(1 + 2\beta Kn)} = \frac{\kappa_{gas,0}\Phi}{\left(1 + \frac{p_{1/2}}{p}\right)} \tag{14.3}$$

where $\kappa_{gas,0}$ is the thermal conductivity of the free gas, β is a constant depending on the gas, Kn is the Knudsen number, Φ is the porosity of the sample, and $p_{1/2}$ is the pressure p, at which the value of the gaseous thermal conductivity is half of the maximum value for the free gas. The Knudsen number is defined as the ratio of the MFP to a representative physical length scale of the porous structure. Thus, Knudsen diffusion takes place if the MFP is larger than the pore diameter. Gas MFPs range in the order of few tens of nanometer, which is commensurate with length scale and pore sizes in colloidal materials.

In the case of porous and dense composite materials, it is common to describe their effective thermal conductivity with binary mixing models. The most important ones are the series, the parallel, the Maxwell–Eucken-1 and 2, and the effective medium theory (EMT) model [97, 98]. These mixing models capture the effective thermal transport properties of even nanostructured hybrid materials remarkably well [55]. Depending on the thermal conductivity contrast between the two components, percolation theory is capable of capturing the composition-dependent properties. Deviations from such models hint toward a significant contribution of interfaces or other intramolecular interactions [20].

To conclude this introductory section, we want to stress several key aspects. Thermal transport comprises phonons or vibrational modes over a wide range of excitation frequencies and MFPs. Thermal transport is very sensitive to defects, imperfections, impurities, boundaries, and grain boundaries of a given material. In inorganic hard materials, crystallinity, doping, size, or surface roughness play an important role. In soft polymer materials, the polymer backbone and side groups translate into a potentially anisotropic thermal transport behavior, where interchain and intrachain conduction dominate. The accessible range of thermal conductivities depends on the heat capacity, the density, and speed of sound of the material under consideration. The temperature-dependent thermal conductivity is vastly different between crystalline and amorphous materials. Thermal transport comprises several transport mechanisms, with conduction being the most relevant one for colloidal materials. Binary mixtures of two material phases are typically described by mixing models.

14.4 Characterization Methods

Measuring thermal conductivity on small length scales is a challenge in itself. The ongoing miniaturization calls for ever better and more accurate methods to determine this important material parameter. Therefore, we introduce a range of techniques that, by our experience, are suitable for colloidal assemblies (Table 14.1). We also refer to a review, which provides more detailed information on non-contact characterization techniques [99].

14.4.1 Laser Flash Analysis (LFA)/Xenon Flash Analysis (XFA)

The laser flash analysis (LFA) is a transient method to measure the thermal diffusivity α [m^2/s] of a sample. The thermal diffusivity is a material-specific property for characterizing unsteady temperature conduction. It describes how quickly a material reacts to a temperature change. The thermal conductivity κ can be calculated using the density ρ and the specific heat capacity c_p:

$$\kappa = \alpha \cdot \rho \cdot c_p \tag{14.4}$$

The principle of the LFA is shown in Figure 14.2a.

The time-dependent temperature rise (V) at the top face of the specimen is measured by a contactless IR detector after a short light flash has heated the front face (Figure 14.2b). A numerical fit of the measured signal allows the determination of the thermal diffusivity. For ideal, adiabatic conduction, the thermal conductivity is described by the following equation:

$$\alpha = 0.1388 \cdot \frac{d^2}{t_{1/2}} \tag{14.5}$$

where d is the thickness of the sample and $t_{1/2}$ is the time of the half-maximum [101]. This equation is only valid for adiabatic conditions with no heat loss and an infinitely short light pulse compared to the measuring time. Refined numerical models exist to fully fit the measured temperature rise, taking into account thermal

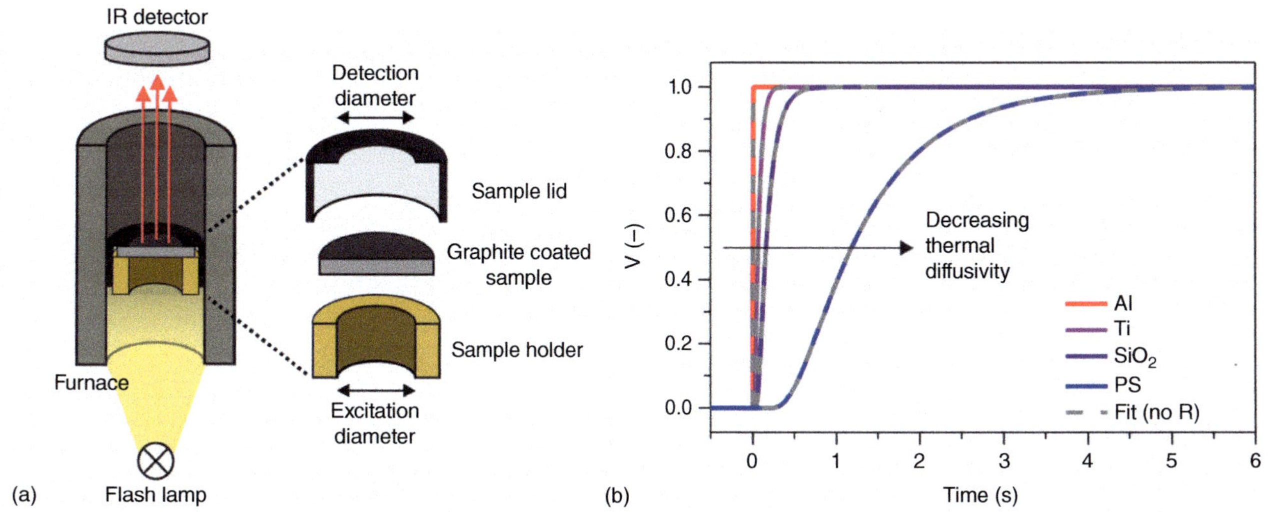

Figure 14.2 (a) LFA setup and the corresponding measurement signal with radiation fit model. (b) Normalized temperature-rise-versus-time curves of materials with different thermal diffusivity. Source: Adapted from [100] with permission of Springer Nature.

Table 14.1 Thermal transport characterization methods for nanostructured materials.

Method	Absolute/ relative	Measured quantity	Lateral sample size	Sample thickness	Complexity	Commercially available
Laser flash analysis/xenon flash analysis (LFA/XFA)	Abs	α	Several mm	>100 μm	Medium	Yes
Transient plane source/modified plans source (TPS/MPS)	Abs	κ	cm	>mm	Low	Yes
Time-domain thermore-flectance/ frequency-domain thermoreflectance (TDTR/FDTR)	Abs	κ	Few μm	Bulk to <100 nm	High	Yes
Guarded hot plate/heat flow meter (GHP/HFM)	Abs	κ	Several cm	mm to cm	Medium	Yes
Photoacoustic (PA)	Rel	κ	Several mm	<μm	Medium	No
Lock-in infrared thermography (LIT)	Abs	α	Several mm	Freestanding, μm–mm	Medium	No

losses and a finite pulse function [102, 103]. An advantage of the LFA is the high sample throughput due to the short measurement time. However, the thickness of the sample must be well defined and cannot drop below a lower limit (typically hundreds of micrometer) based on the time resolution of the IR detector. Furthermore, infrared absorbing/emitting coatings of the sample surface are necessary to receive a sufficient signal/noise ratio. Optically clear samples or porous materials need to be treated such that the primary light flash cannot penetrate into the material. The sample's density and heat capacity have to be measured separately to receive the thermal conductivity of the sample. Whereas this method has been mainly used on freestanding, single-layer samples, analytical models are also available to characterize multilayer samples. The use of specialized sample holders allows characterizing liquids and powder samples as well.

14.4.2 Transient Plane Source (TPS)/Modified Plane Source (MPS) Technique

The transient plane source (TPS) technique is an absolute method for measuring thermal conductivity and thermal diffusivity and is widely used on different materials, e.g. solid, powder, and liquid [104, 105]. Generally, a TPS hot disk sensor consists of a planar double metal spiral covered on both sides with insulation film, e.g. Teflon and Kapton [105]. During the measurement, the TPS sensor simultaneously acts as a heat source and temperature sensor, which is sandwiched by the test samples (Figure 14.3a). By applying an optimized power into the TPS sensor, the TPS sensor and the test material will be heated up. The time-dependent temperature increase of the sensor directly depends on the thermal conductivity of the test material. However, it is important to note that due to the thermal contact resistance between the insulation film and the sample, as well as the heat capacity of the TPS sensor, the initial data points from the measurement should be ignored [106].

The modified plans source (MPS) technique was established based on TPS, which simplified the measurement environment [107, 108] (Figure 14.3b). Instead of sandwiching the TPS sensor between two specimens, the sample is placed on top of the MPS sensor for measuring the thermal properties. As the plot in Figure 14.3b shows, with applying identical power to the MPS metal tracks, the rates of temperature increase for material A and B are different, which implies that material A possesses higher thermal diffusivity in comparison to material B.

Owing to the simple sample preparation and automated data analysis, TPS or MPS have been widely applied to investigate the thermal conductivity and thermal diffusivity of solid [106], liquid [107], powders [109], food items [110]. TPS was also used to monitor the moisture content in textiles [111] and on human skin [112]. However, the accuracy of the TPS technique is limited when approaching thin films or materials with a low thermal conductivity [113]. In addition, heating power, probe size, and sensor thickness may also cause uncertainties during the measurement [114].

14.4.3 Time-Domain Thermoreflectance (TDTR) and Frequency-Domain Thermoreflectance (FDTR) Methods

Time-domain thermoreflectance (TDTR) [115, 116] and frequency-domain thermoreflectance (FDTR) [62, 117, 118] are optical pump–probe techniques to determine the thermal properties of materials with a smooth surface. The methods are based on the ultrafast detection of changes in the reflectivity ΔR of the sample after the modulated heating of a transducer layer. The temperature-dependent change of the reflectivity is used to monitor the temperature decay at the sample surface after a heating pulse.

A schematic illustration of a TDTR setup is shown in Figure 14.4a. The experimental setup consists of a pulsed laser source split into a "pump" and "probe" beam. An electro-optic modulator (EOM) modulates the pulses of the pump beam (Figure 14.4b, red lines). The transducer layer absorbs the pump beam, inducing a surface temperature rise within the sample (Figure 14.4b, blue lines). The temperature decays into the sample until the next laser pulse arrives. The probe beam can be time delayed (picosecond resolution) compared with the pump beam using a mechanical delay-stage within the setup. By incrementally scanning the time decay of the probe beam (Figure 14.4b, black dots), it is possible to monitor the entire

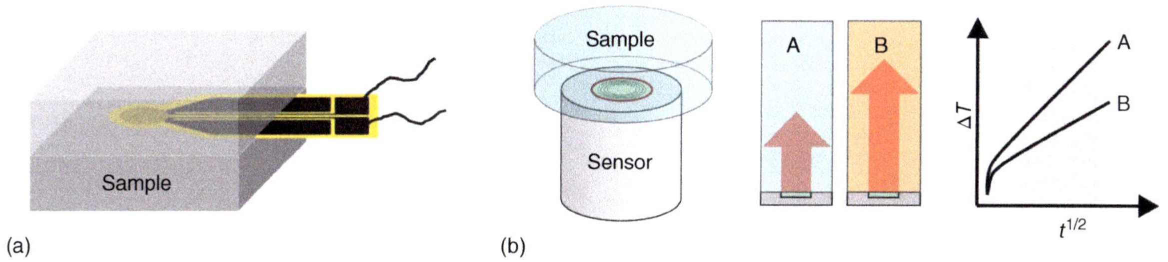

Figure 14.3 (a) Schematic of TPS measurements with a hot disk sensor. (b) Schematic of MPS measurement on two homogeneous materials with different thermal diffusivity and the corresponding temperature change versus time.

temperature decay. The thermal properties of the sample are extracted from a theoretical fit model of the experimental decay curve and adjusting the free parameters.

A nano-/picosecond resolution is achieved by this method. This enables to measure the thermal properties of thin film samples possessing a submicron thickness. Since radial heat flow is negligible, the heat dissipation is treated as one-dimensional (1D) heat conduction into the sample.

For FDTR, the frequency of the modulated pump beam is swept as the experimentally independent parameter instead of the delay time between both beamlines in TDTR. FDTR offers a less expensive and simpler laser setup since no delay stage is necessary, and continuous-wave lasers are sufficient for the experiment. A lock-in amplifier is referenced to the pump beam frequency. The changes in the amplitude and phase of the reflected probe beam are compared with the original signal. The thermal properties of the sample are extracted from the differences between amplitude/phase of pump and probe beam.

Both techniques enable measurements of thin samples possessing a thickness of only several hundred nanometers. Additionally, owing to the high temporal resolution and the high heating frequencies (up to 200 MHz), in-depth insight into the contribution of specific phonons and vibrational modes to the overall thermal conductivity can be extracted. Still, both methods are experimentally challenging as the optical setups need to be aligned and calibrated accurately. They are only applicable to samples with a smooth surface, which feature specular (thermo)reflections. Furthermore, the layer thickness of the metal transducer and the pump and probe spot size need to be accurately determined to yield quantitative results. Modifications of these primary techniques include lateral scanning of heterogeneous samples for local thermal conductivity measurements [120], pump and probe beam offsets for a distinction of in-plane and cross-plane thermal conductivity [121, 122], and changes to the beam spot size [123] and shape for accessing anisotropic thermal transport [124].

14.4.4 Guarded Hot Plate (GHP) and Heat Flow Meter (HFM)

Hot plate techniques such as guarded hot plate (GHP) or heat flow meter (HFM) represent common methods to determine the thermal conductivity of bulk samples. GHP/HFM are absolute methods to measure thermal conductivity. Both methods differ in their experimental setup and can typically be used for low conducting samples in the range between 0.01 W/m/K and 6 W/m/K [125]. The schematic setups of these methods are illustrated in Figure 14.5.

Both methods are steady-state techniques, where a constant temperature gradient ΔT is applied to the sample. In steady-state methods, the measurement is conducted in equilibrium over a long period of time. The sample is located between a heated and cooled plate, providing a defined temperature gradient. For GHP, the power applied to keep the hot side at a constant temperature is used for data evaluation. In HFM experiments, heat flow transducer layers between the heated/cooled plate and the sample surface record the direct heat flow into/out of the sample. By knowing the geometry of the specimen, it is possible to calculate its thermal conductivity by 1D

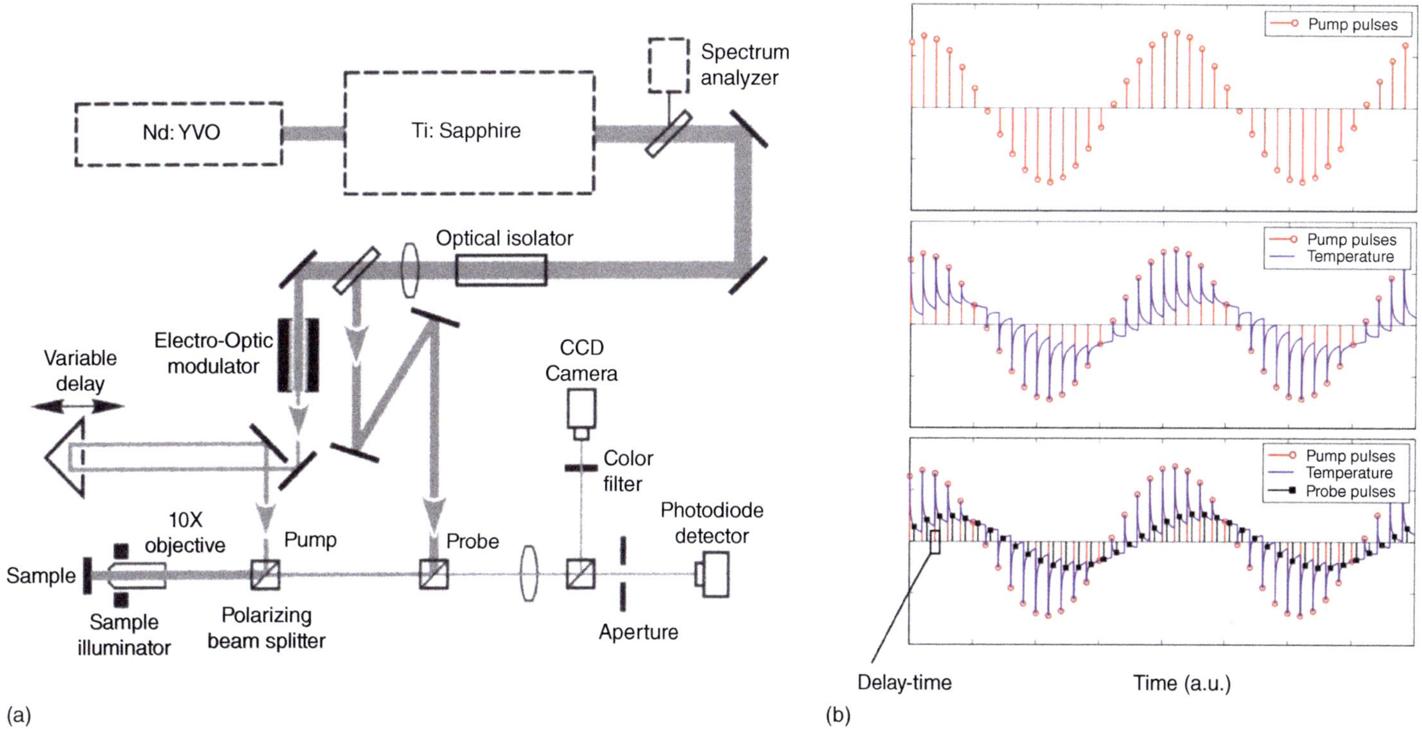

Figure 14.4 (a) Schematic TDTR laser setup. Source: Adapted from [1] with permission of AIP Publishing. (b) Modulated pump laser pulses (red), temperature decay between the pulses (blue), probe pulse at one specific delay time (black). Source: Adapted from [119] with permission from AIP Publishing.

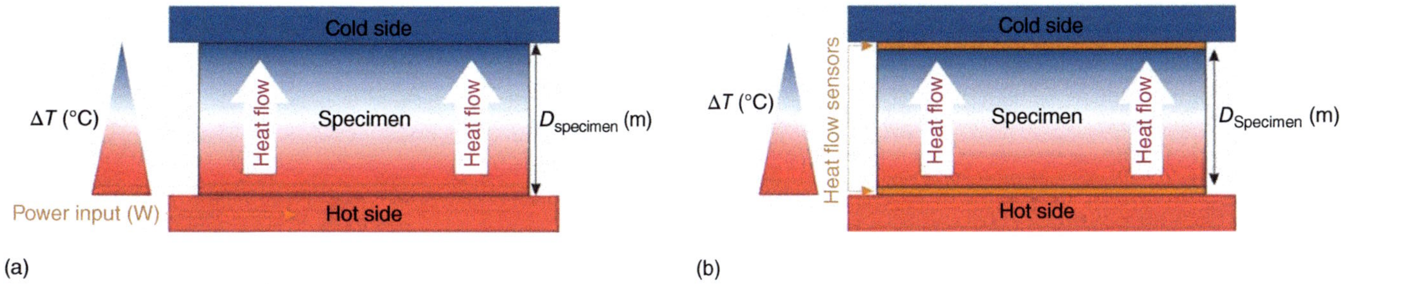

Figure 14.5 Schematic setup of the (a) GHP and (b) HFM technique. Whereas for GHP, the power input is used for the evaluation, the direct heat flow is extracted by the HFM method.

Fourier's law:

$$\text{GHP}: \ \kappa = \frac{Q \cdot D}{A \cdot \Delta T} \ \text{HFM}: \ \kappa = \vec{q}\,\frac{D}{\Delta T} \tag{14.6}$$

where κ is the thermal conductivity of the sample, Q represents the electric power applied to the hot plate, A is the cross-section area, and D is the thickness of the sample. $\vec{q}$ represents the heat flux, and ΔT is the temperature difference between the hot and the cool plate. Since the measurements are performed in equilibrium, accurate data acquisition is possible, however, at the expense of long measurement time. Both methods are limited to macroscopically sized, low conducting samples, and heat losses to the environment via convection or radiation need to be strictly excluded.

14.4.5 Photoacoustic (PA) Method

The photoacoustic method determines the thermal properties of a sample based on the photoacoustic effect. The setup is shown in Figure 14.6.

A modulated laser beam periodically heats a thin metal transducer layer coated onto the top surface of a sample. The temperature increase at the sample surface translates into a pressure increase in the surrounding (enclosed) He-gas atmosphere. These modulated pressure waves can be recorded by a microphone. Owing to the dynamic range of typical microphones, the frequency range of interest ranges from 400 to 20.000 Hz, which is considerably lower compared with FDTR discuss earlier. The acoustic signal is analyzed by its phase shift and amplitude component relative to a bulk reference substrate at each modulated frequency. A (multi)layer model is used to extract the thermal properties of the sample from the phase and amplitude signals [127, 128]. The amplitude fitting can be used for both bulk and thin film samples, whereas the evaluation of the phase signal is only appropriate for thin films. Due to the necessity of a gaseous sealing of the measurement chamber, only samples immobilized on a solid support can be investigated. Acoustic interferences from the sample chamber itself need to be strictly excluded as well as any laser absorption inside the sample.

14.4.6 Lock-In Infrared Thermography (LIT)

Lock-in infrared thermography (LIT) is a versatile technique to image the temperature distribution of a surface [129]. Knowledge about the temperature distribution can be used to determine the lateral thermal diffusivity of a sample, especially for freestanding thin films [130–132]. A typical lock-in measurement setup is shown in Figure 14.7a.

An intensity-modulated laser beam at a known frequency heats the sample. In equilibrium, the temperature of the sample oscillates with the same frequency. The phase and amplitude of the temperature oscillations are measured by a position-sensitive infrared detector. The position-dependent decay of the phase and amplitude component is then used to determine the in-plane thermal diffusivity of the material. As shown in Figure 14.7b, the temperature distribution around the

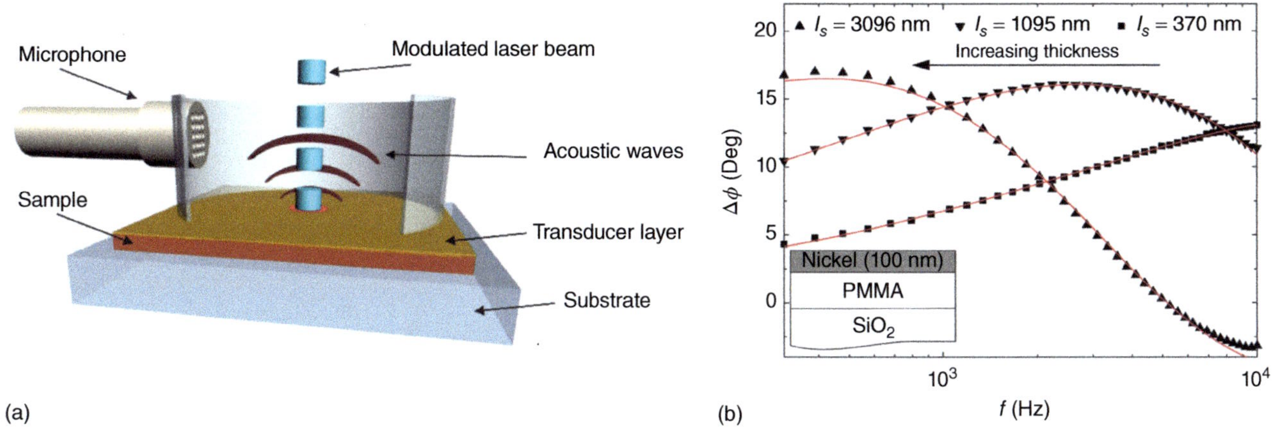

Figure 14.6 (a) Photoacoustic setup. A modulated laser beam is absorbed in a metallic transducer layer. The sound wave caused by the heated transducer layer is recorded by a microphone. (b) Phase shift data of a photoacoustic measurement: PMMA thin films with different thicknesses on quartz substrates. Source: Based on Herrmann et al. [126].

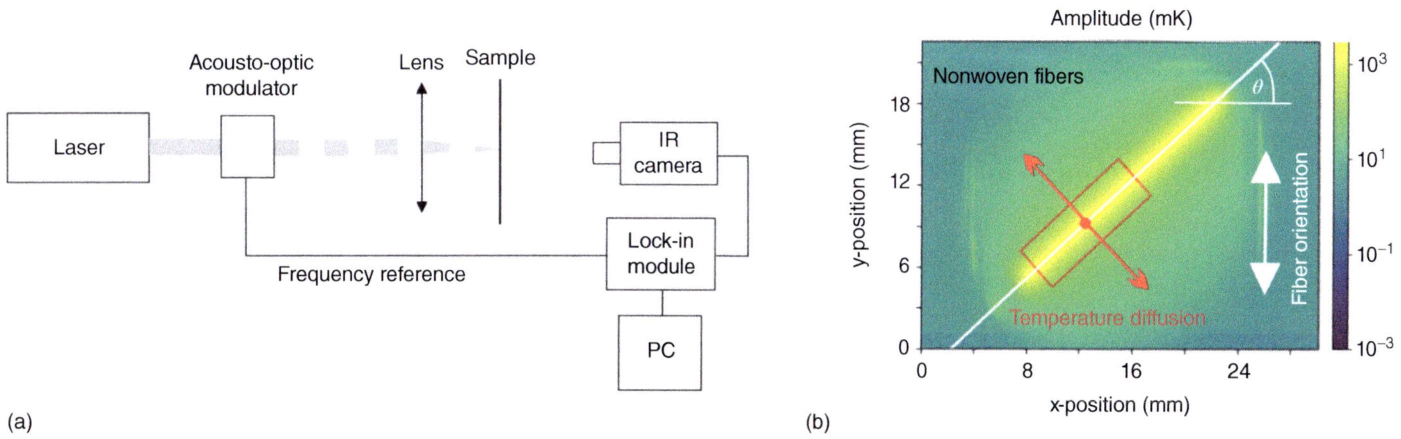

Figure 14.7 (a) Setup of lock-in infrared thermography. A periodically modulated laser beam is focused on a freestanding thin film sample. Source: Adapted from [131] with permission of AIP Publishing. (b) The temperature distribution around the focal laser line is measured by a high-resolution IR camera. The lateral change in phase and amplitude allows calculating the in-plane thermal diffusivity. Source: Gao et al. [133]/John Wiley & Sons, Inc./CC BY 4.0.

focal laser line of a thermoplastic polyurethane (TPU) membrane was determined with LIT. By measuring the thermal diffusivity in different directions, the thermal anisotropy in-plane can be thus determined. LIT offers a flexible, nondestructive platform to measure the in-plane thermal diffusivity. Ideally, the sample itself absorbs some of the laser light energy to serve as the local heat source. Freestanding samples are required, which need to be larger than the (lateral) thermal diffusion length at the respective lock-in frequency. To obtain highly accurate data, measurements at different frequencies are necessary, which extends the measurement time. Additionally, the thermal conductivity can only be determined by knowing the density and the heat capacity of the sample.

14.5 Heat Transport in Colloidal Structures

14.5.1 Porous Structures

Colloidal assemblies, especially if they are porous and exhibit gaseous inclusions in a solid scaffold, are predestined to exhibit exceptional thermal insulation properties. A high concentration of interfaces and, consequentially, a high degree of phonon boundary scattering can be expected to reduce the thermal conductivity beyond previously assumed lower limits.

Colloidal crystals consisting of solid spheres represent the most classic colloidal material. The interstitial pores occupy 26 vol.% of the entire structure, and the pore sizes are fully controlled by the diameter of the constituting spheres. These interparticle pores range from 15% to 41% of the constituting spheres [134]. Thus, colloidal crystals can be considered nanoporous, open-cell structures, with pores easily reaching the sub-100-nm range. Consequently, only a minor sensitivity to the surrounding atmosphere was observed by Nutz et al. [135], measuring 366-nm polystyrene latex colloidal crystals in a vacuum, air, and helium – the latter being a particularly good gaseous thermal conductor. Even more interesting was the observed absolute thermal conductivity. Applying the Maxwell–Eucken model to this porous material (see Section 14.3.3), a 30% higher effective thermal conductivity would have been expected. This deviation can be explained by the additional constriction provided by the small interfaces between adjacent spheres. The combination of high interface density, small contact points, and the inherently low conducting polymer results in a very low conductivity of 0.051 W/m/K at a comparatively high density of 750 kg/m^3 (Figure 14.8a). This strong reduction in thermal conductivity is corroborated by a theoretical continuum approach, where a critical reduction is reached close to the maximum porosity achievable in colloidal crystals [137].

Yet, the most striking property was revealed during temperature-dependent measurements. An almost threefold increase of the thermal conductivity results upon heating the polymer crystal beyond the T_g of the polystyrene particles (105 °C) (Figure 14.8a). Comparison between this and the general behavior of soft amorphous materials (Section 14.3.2) highlights the unique behavior of colloidal matter. Observing such a large and sudden increase in thermal transport is rather

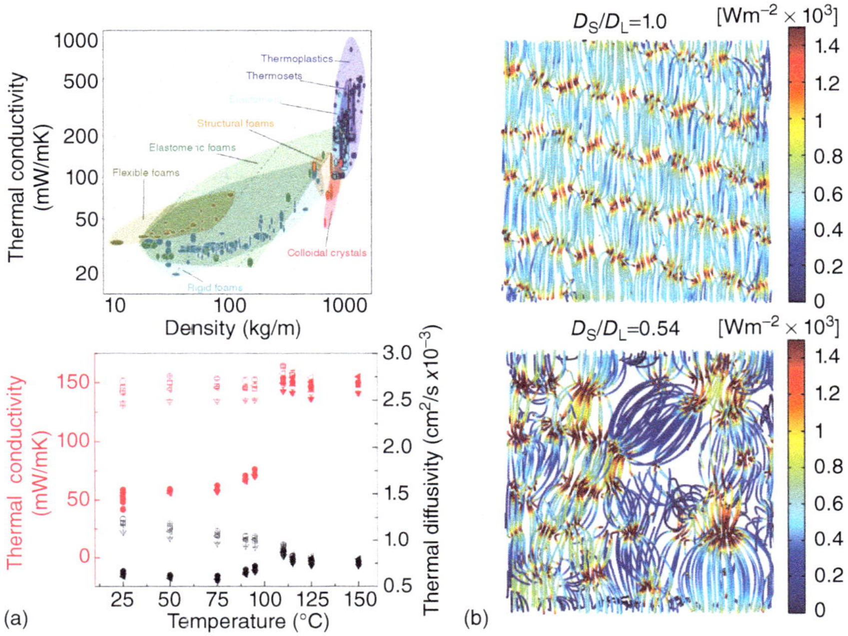

Figure 14.8 (a) Thermal conductivity of polymer opals. Top row: comparison to polymer foams; bottom row: temperature dependence of thermal diffusivity (black) and thermal conductivity (red). The increase at the glass transition temperature is caused by the film formation. Source: Adapted from [135] with permission of Elsevier. (b) Influence of particle ordering in a colloidal ensemble on the thermal transport pathways. The heat conduction pathway is more tortuous in the case of disordered samples (bottom) compared with ordered ones (top). Source: Nutz et al. [136]/with permission from John Wiley & Sons, Inc.

unusual – the concomitant changes to the shape of the colloidal crystal monolith are small with a shrinkage of about 12% in thickness. The reason for this increase can be found in the softening of the particles. The polymer chains gain mobility, which in turn leads to an enlargement of the contact points and a loss of porosity within the structure. A detailed kinetic investigation of this dry sintering process revealed two distinct timescales for the film formation process [138]: a fast process being dominated by interfacial changes between adjacent spheres, which dominates thermal diffusivity properties, and a slow process being dominated by volume filling of the interstitial voids, which dominates the structural color. To further boost the thermal insulation properties, the mesoscopic order of the colloidal ensemble can be specifically suppressed by defined binary mixtures. Not only does disorder decrease the effective density of the colloidal ensemble, but it also increases the thermal transport pathway from the hot to the cold side, which further reduces the thermal conductivity (Figure 14.8b) [136]. The unique temperature-dependent changes to the thermal conductivity can be further controlled by a mesoscopic heterostructure of the colloidal ensemble. Binary mixtures or layered structures provide access to gradual and stepwise changes of the thermal conductivity, respectively [139].

Even stronger insulating materials can be realized by changing the colloidal structure itself. Ruckdeschel et al. [140] demonstrated a low thermal conductivity of 0.071 W/m/K in vacuum at 25 °C for a hollow silica nanosphere colloidal crystal (Figure 14.9a,b) at a relatively high density of 1040 kg/m^3. The absolute value of this thermal conductivity is higher compared to silica aerogels, owing to the higher effective density. However, the major advantage of this colloid-based approach is the tight control on the nanostructure in these ensembles [142]. The relevant length scales are *a priori* defined by the morphology of the colloidal object. The hollow spheres used in this case entail two distinct length scales (few 10-nm shell thickness and few hundred nanometer particle diameter). They can be synthesized in large quantities and demonstrate a remarkably high degree of monodispersity [143]. Only this stringent control on the hierarchical mesostructure allowed the authors to deduce the influence of the microstructure within the silica shell as well as the interfacial contact strength, separately. This was achieved by thermal conductivity measurements after calcination at various temperatures. The calcination of a silica colloidal crystal leads to an enhancement of the interfacial bonding between neighboring hollow silica spheres, a stiffening of the silica network, and closure of remaining microspores within the silica shell. A schematic sketch of these effects is shown in the inset in Figure 14.9b.

Ruckdeschel et al. found that the interfacial bonding between the particles has a higher contribution to the overall thermal transport than the amorphous silica shell material. The temperature-dependency of these hollow sphere colloidal crystals demonstrated a purely diffusive thermal transport behavior with a $T^{0.4}$ scaling law. Thus, although being fully crystalline on the few 100-nm length scale, the amorphous structure on the microscopic scale fully dictates the thermal transport. Further adjustments to the colloid geometry, order of the particle assembly structure, and interfacial bonding between adjacent spheres allowed accessing ultralow thermal transport properties; in vacuum <10 mW/m/K could be realized (Figure 14.9c,d) [141].

Colloidal crystals can also be used as a template to produce a class of porous materials known as inverse opals. The matrix material can be readily varied. Polymeric matrices are applicable via *in situ* polymerization, inorganic oxides are applicable by simple sol–gel processes, and even metallic inverse opals can be prepared via electrodeposition or physical vapor deposition. These inverse opals are constituted of interconnected spherical pores, where the matrix skeleton comprises several length scales, which are determined by the templating particles size and the interstitial space. In analogy to the size of the open pores in colloidal crystals, the length scale of the matrix can easily reach to sub-100-nm range.

This concept was exploited to understand thermal transport in three sorts of crystalline materials, i.e. silicon, copper, and nickel. Silicon inverse opals with periodicities and shell thicknesses in the range of 420–900 nm and 18–38 nm, respectively, were studied by Ma et al. [144] (Figure 14.10a). They feature low thermal conductivities of about 0.6–1.4 W/m/K at 300 K, owing to the high porosity of the inverse opal. Most interesting is the temperature dependence of the thermal transport in such inverse silicon opals. It follows a $\kappa \sim T^{1.8}$ dependency instead

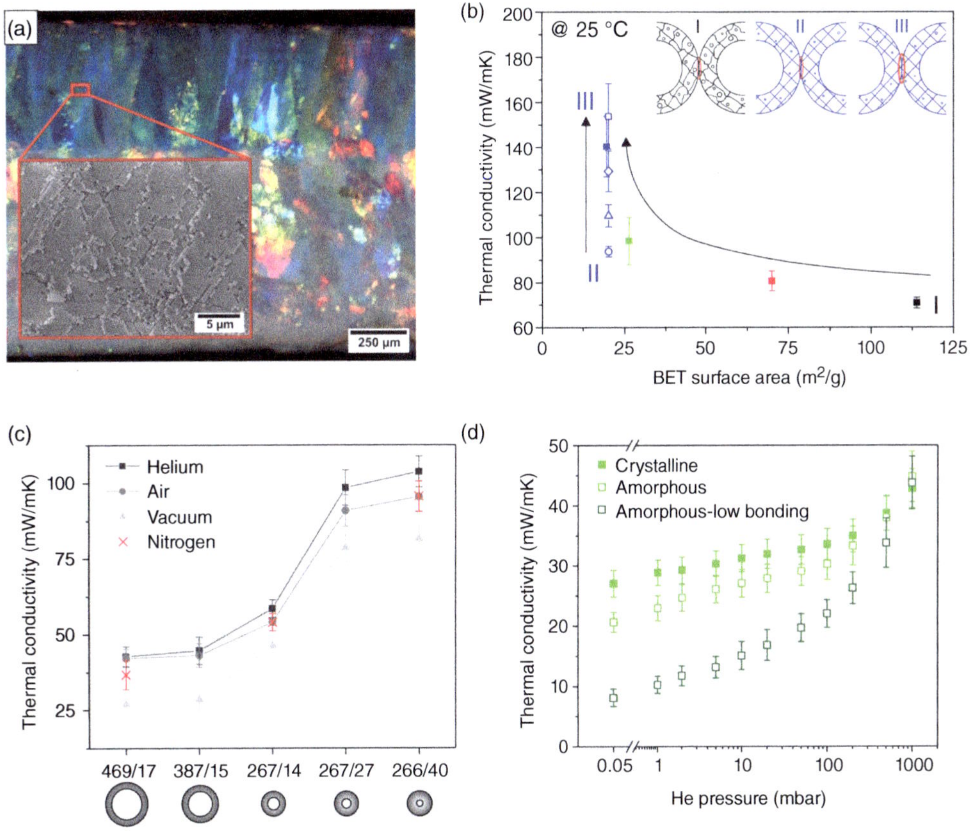

Figure 14.9 (a) Optical side-view image of a hollow silica nanoparticle (HSNP) colloidal crystal; the strong opalescence indicates a high order among the particles within the monolith. Inset: SEM image proves a crystalline particle structure. (b) Thermal conductivity in dependence of the BET surface area, indicating changes in the microstructure of the silica shell. The blue data points show the thermal conductivity of HSNP colloidal crystals, with the particles being calcined before the assembly. This reveals the strong influence of interfacial bonding. Source: Ruckdeschel et al. [140]/Royal Society of Chemistry/CC BY 3.0. (c) Influence of the particle morphology and the environment on the thermal conductivity of the colloidal ensemble. (d) Helium-atmosphere dependence of the thermal conductivity for three types of colloidal ensembles. Amorphous structures with low interfacial adhesion demonstrate the strongest thermal insulation. Source: Adapted from [141] with permission of John Wiley and Sons.

of the expected $\kappa \sim T^3$ scaling law in crystalline materials. The T^3 dependency was confirmed for low-pressure chemical vapor deposited polysilicon films with ~1–4 μm film thickness and ~190–550 nm grain sizes. The different temperature dependency was shown to originate from coherent phonon grain boundary scattering, and it could explain the thermal conductivity down to temperatures of 30 K. The coherent phonon grain boundary scattering furthermore depends on the phonon frequency and was best described by a ω^2 dependency. This is an impressive example for the elucidation of very fundamental thermal transport questions using the structural control and small length scales accessible by colloidal materials.

Barako et al. [42] investigated inverse opals based on copper and nickel (Figure 14.10b). Here, the electron thermal conductivity plays an important role, too. Consequently, the effective thermal conductivity of the metallic inverse opals remained at a high level in the tens of W/m/K. An important difference between electrons and phonons is the smaller MFP of the electronic energy carriers (~10 nm). By reducing the pore size (1000–100 nm), and thereby the feature size down to ~23 nm, the thermal transport pathway is strongly confined. Thus, electron conduction is reduced due to an increased surface scattering. If the size of the conduction pathway becomes comparable to the MFP of the thermal energy carriers (electrons), a transition from diffusive to quasi-ballistic conduction occurs. This is observable for copper inverse opals (Figure 14.10b, right). For large structures, the thermal conductivity is independent of the pore size and converges to the diffusive limit (dashed line, Figure 14.10b). When reducing the pore size to a comparable length as the diffusive MFP of the electrons (39 nm in copper at room temperature), an increased surface scattering takes place, and the effective thermal conductivity is reduced by more than ~57%. In nickel inverse opals, the electron MFP is much shorter (~6 nm). Thus, the thermal conduction is diffusive and constant over the entire size range. This study once more underlines the advantage of sub-μm and sub-100-nm control provided by templating with colloidal ensembles. Yet, these results are not limited to the particular class of inverse opals but can also be used to better understand heat conduction in other porous metals.

Inverse opals possess a well-defined order and homogeneously distributed pores. Even stronger thermal insulation has been reported for disordered systems with very low densities. These are referred to as aerogels and comprise a nanoporous, low-density network of inorganic or organic moieties. Typical inorganic aerogels consist of SiO_2 or Al_2O_3. They are lightweight and thermally stable and exhibit an exceptionally low thermal conductivity as low as 0.013 W/m/K [96]. Aerogels are, however, most commonly fabricated through a sol–gel process and successive supercritical drying. Colloidal assembly forces are barely relevant for these lightweight materials.

Freeze-casting of colloidal dispersions provides alternative access to porous materials. In contrast to inverse opals or aerogels discussed earlier, the pore sizes in freeze cast materials range in the few micrometer and tens of micrometer range. The freeze casting procedure has the added benefit of yielding anisotropic pores and hence the possibility to control the orientation of heat spreading.

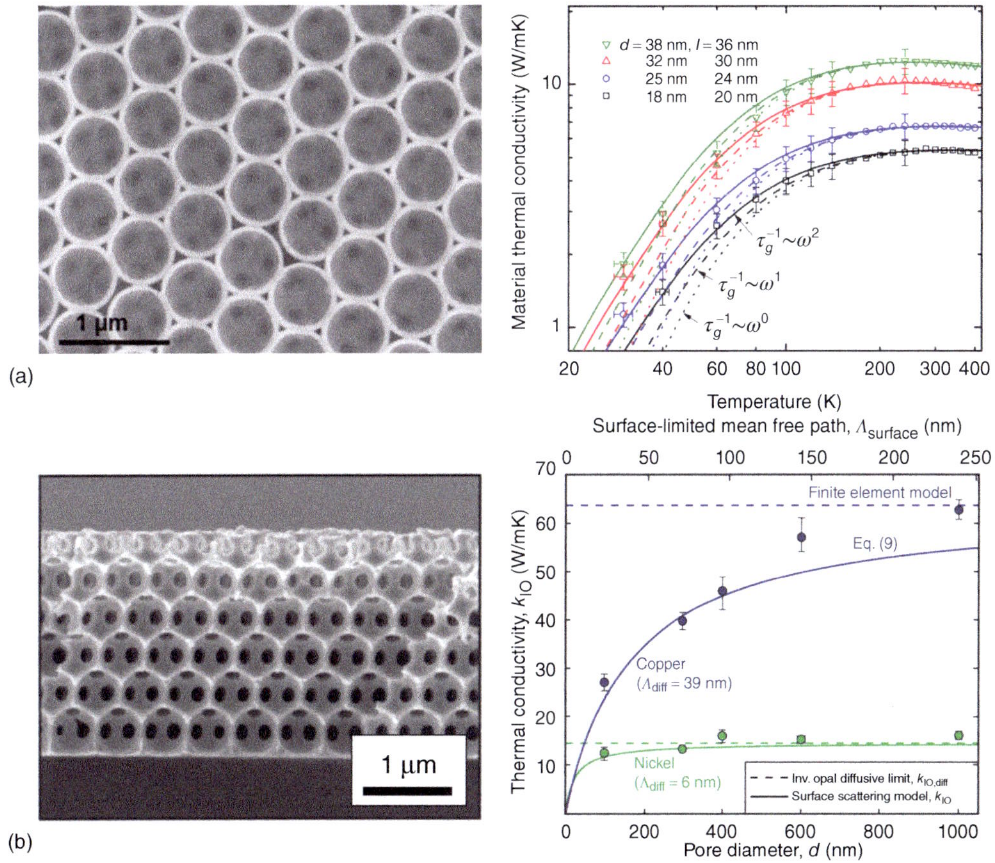

Figure 14.10 Thermal transport in inverse opals. (a) Silicon inverse opals. Scanning electron microscopy (SEM) image of the top surface of a silicon inverse opal (left side) and thermal conductivity of the silicon inverse opals with different shell thicknesses d and average grain sizes l. Source: Ma et al. [144]/with permission from American Chemical Society. (b) Metal inverse opals. SEM image of a nickel inverse opal with a pore diameter of about 600 nm (left side) and thermal conductivity of copper and nickel inverse opals with different pore diameters (right side). Source: Barako et al. [42]/with permission from American Chemical Society.

The anisotropy originates from the higher density of interfaces/pores orthogonal to the ice crystal growth direction [53]. Xie et al. [145] used this method to prepare a type of organic aerogel using aramide fibers. The aromatic rings and a partial double-bond character of the amide moiety ensure a pronounced rigidity in this type of polymer. Ice templating, therefore, resulted in a "lasagna-like" microstructure with parallel lamellae exhibiting a long-range order. The thermal conductivity, however, only had a slight anisotropy with a value of 0.037 W/m/K orthogonal and 0.042 W/m/K parallel to the freezing direction. Wicklein et al. [146] prepared lightweight, thermally insulating foams by freeze casting of a ternary dispersion. A combination of cellulose nanofibers (CNFs), GO, and sepiolite nanorods resulted in foams with a parallel arrangement of cylindrical pores about 20 μm in diameter and in the millimeter-range lengthwise. Pore walls were shown to be mesoporous at approximately 3-nm pore size, which is far below the MFP of gas molecules, thereby making use of the Knudsen effect. The addition of carbonaceous GO increases the mass attenuation coefficient in the mid-IR range significantly and thereby reduces the effect of radiative heat transfer. The cell walls, made up of three different types of colloidal particles, exhibit a rather complex nanostructure with many interfaces. This, in turn, reduces the conduction through the solid part of the interconnected cellular scaffold. The interplay of all these factors explains the obtained ultralow axial thermal conductivity of 0.015 W/m/K and a pronounced anisotropy ratio ($\lambda_{\parallel}/\lambda_{\perp}$) larger than 10. Applying appropriate building blocks with high elastic moduli and cross-linking these with boric acid ensured a pronounced mechanical stability. At 77 kNm/kg the specific Young's modulus is in the range of commercial polymeric foams. In contrast to conventional foams, e.g. expanded polystyrene, the nanocomposite foams prepared via freeze casting are inherently flame retardant and could be applied as a much safer class of highly insulating material. Chau et al. [15] presented a more functional and responsive type of material prepared via freeze casting of a polyurethane nanoparticle dispersion. Also showing a lamellar microstructure, this material showed a pronounced anisotropic thermal conductivity under vacuum conditions with 59 and 11 mW/m/K parallel and perpendicular to the lamellae, respectively. The novelty of this report is a reversible transition between anisotropic and fully isotropic thermal conduction. The latter is realized by measuring the effective thermal conductivity in a helium atmosphere. The high thermal conductivity of helium outweighs the low conductivity in the vertical direction caused by the high pore density and results in thermal conductivity of 115 mW/m/K in both directions. Han et al. [59] went a step further and infiltrated freeze-cast boron nitride nanosheet (BNNS) networks with epoxy resin. The freeze-drying procedure was controlled such that randomly distributed, uniaxially aligned, or long-range lamellae BNNS were obtained, respectively (Figure 14.11). This gradually increasing degree of order has a clear effect on the axial thermal conductivity, which increases from 1.0 to 1.9 to 6.1 W/m/K in the long-range lamellar case at only 15 vol% filler content. The latter also shows a high thermal anisotropy of 12.

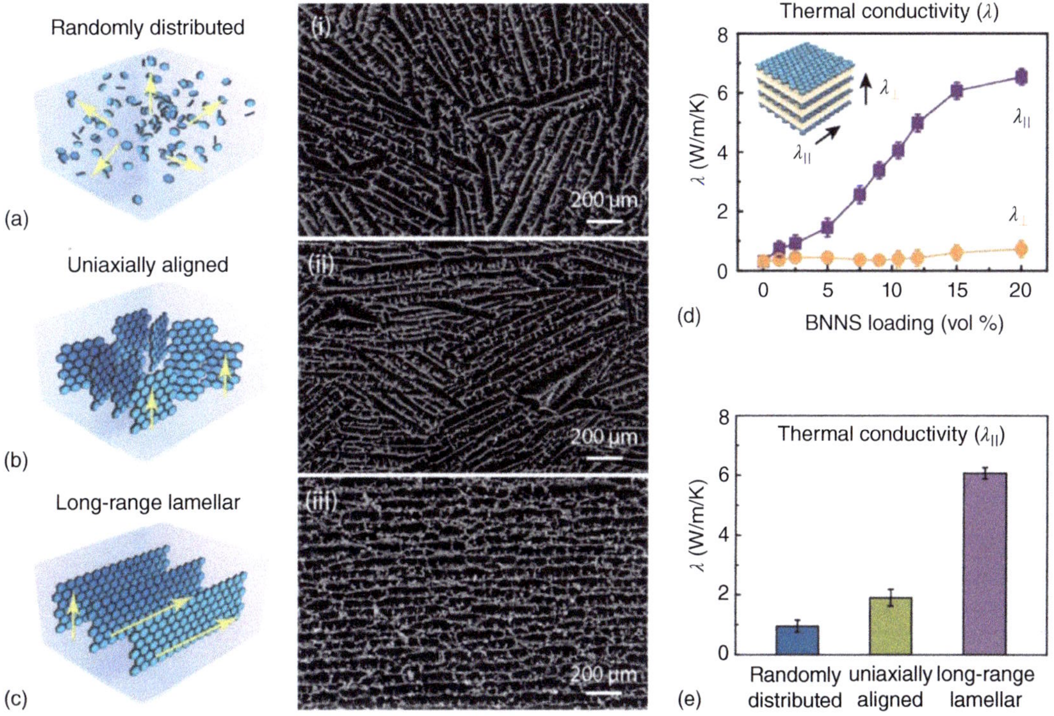

Figure 14.11 Microstructure and thermal conductivity of BNNS networks with different filler loadings. Schematics and SEM images of BNNS/epoxy composites with randomly distributed (a), uniaxially aligned (b), and long-range lamellar nanosheets (c). (d) Thermal conductivity of composites as a function of BNNS loading. (e) Comparison of thermal conductivity between different orientated BNNS in composites. Source: Han et al. [59]/with permission from John Wiley & Sons, Inc.

14.5.2 Dense Structures

Next, we want to focus on thermal transport in dense, i.e. non-porous, colloidal structures called nanocrystal assemblies (NCAs) [147–149]. Here, the colloidal objects possess diameters <10 nm, and we will lay a particular focus on interfacial chemistry. These 3D materials possess two length scales: (i) the nanoparticle in the size range of a few nanometers and (ii) the array of periodically ordered nanoparticles separated by ligand monolayers. Thus, a vast number of inorganic–organic and organic–organic interfaces exist in these assembled materials. One should note that ligand layers >0.5 nm effectively prevent electronic transport through such systems, even though the core particles may be semiconductors or even conductors.

Ong et al. [149] studied the thermal transport properties of such nanocrystal arrays using different core diameters, core materials, and ligand groups. The structure of these nanocrystal arrays is shown in Figure 14.12a.

They obtained the following decisive parameters on the heat transport: (i) the volume fraction of nanoparticles and surface ligands and (ii) the density and chemistry of the organic/inorganic interface. They measured an increasing thermal conductivity from 0.1 to 0.3 W/m/K with increasing core diameter (Figure 14.12b). This is explainable by the reduced interface density and the increased core volume fraction with increasing nanoparticle size. The core material composition has a minor influence on thermal conductivity. This is shown by the insensitivity of the NCA thermal conductivity on the thermal conductivity of the bulk core material (see inset of Figure 14.12b). Furthermore, this work highlights the importance of the interfaces on thermal properties. The thermal transport is reduced by increasing the mass ratio of the core species and by exchanging the ligands with longer ligands of a smaller mass. Quite remarkably, the small length scale and good structural control even allowed to compare the measured data to molecular dynamics simulations. This, of course, provides an exquisite insight into the fundamental physics governing thermal transport through such densely packed nanocomposite structures. A good overlap of the vibrational density of states between the core and the ligand shell is necessary for a low thermal interface conductance. Further effects may influence the interfacial thermal conductance, such as the grafting density, which in turn can also depend on the size of the core [148].

Liu et al. [147] confirmed the measurements of Ong et al. discussed earlier. Using PbS nanoparticles as a model system, they varied (i) the ligand length, (ii) the ligand-binding group, and (iii) the nanoparticle diameter (Figure 14.12c, top). They concluded that (i) reducing the ligand length increases the thermal conductivity, (ii) the influence of the ligand-binding group on the thermal conductivity is negligible, and (iii) increasing the nanoparticle diameter increases the thermal conductivity. Overall, the thermal conductivity of nanocrystals was varied by a factor of four from 0.1 to 0.4 W/m/K. As it was also concluded by Ong et al., the ligand predominantly dictates the thermal properties. It varies the thermal conductivity by up to a factor of 2.5. The different parameters influencing the thermal conductivity are shown in Figure 14.12c, bottom. The surface chemistry itself was found to play only a minor role based on the results from Liu et al. This is puzzling based on the findings

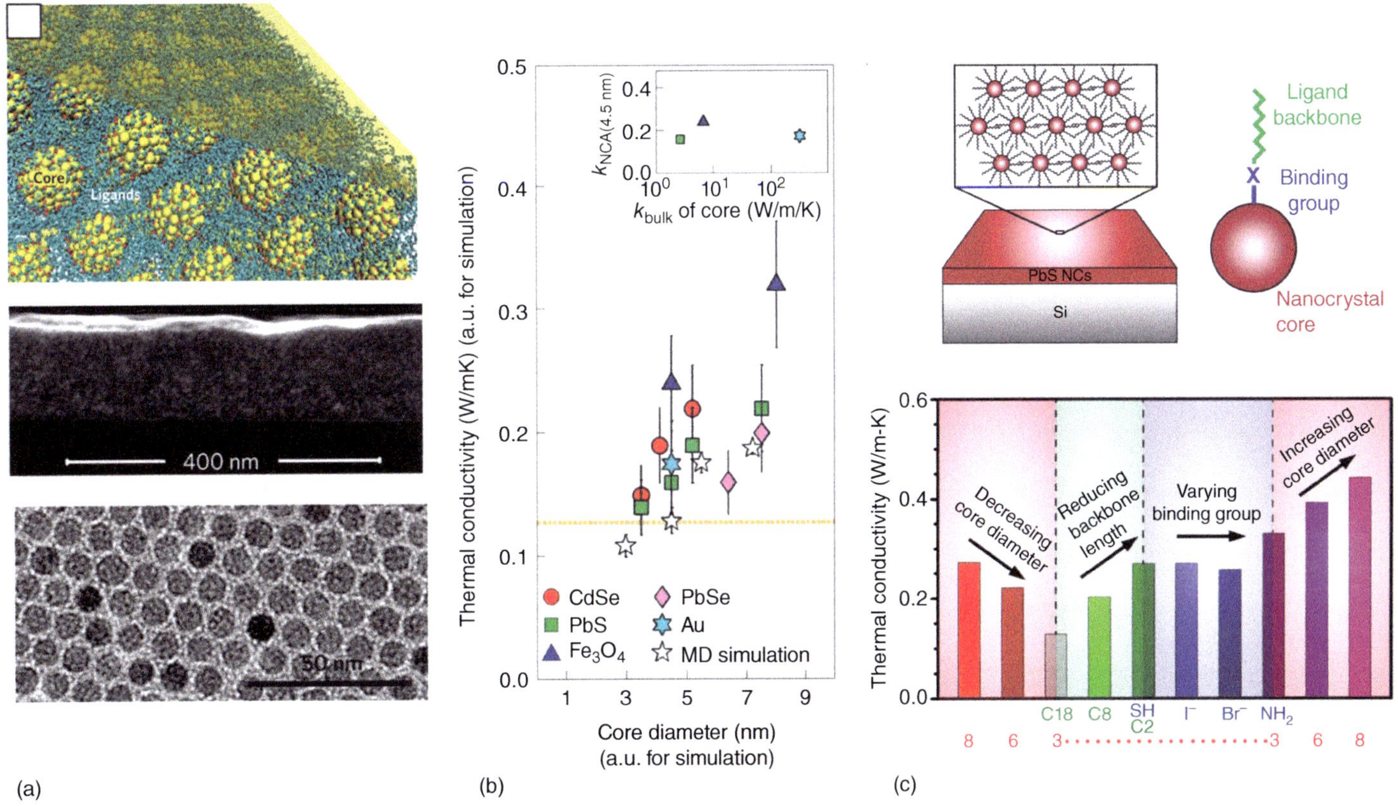

Figure 14.12 Thermal transport in nanocrystal arrays. (a) The structure of the nanocrystal array. Schematic illustration, side-view scanning electron microscopy (SEM) image, and transmission electron microscopy (TEM) image (from top to bottom). Source: Ong et al. [149]/with permission from Springer Nature. (b) Effects of the core diameter on the thermal conductivity for various nanocrystal arrays. Source: Adapted from [149] with permission of Springer Nature. (c) Influence of the nanocrystal core diameter, the binding group, and the ligand backbone on the thermal conductivity in a nanocrystal array. Source: Adapted from [147] with permission of American Chemical Society.

of Losego et al. where the interfacial thermal conductance between a single gold interface and a self-assembly monolayer was determined [150]. The interfacial conductance was shown to vary between 36 $MW/m^2/K$ for van der Waals interaction and 65 $MW/m^2/K$ for covalently bonded thiol ligands. Such strong variations in the interfacial conductance, however, will be hard to realize in a colloidal material. Due to the necessity to sufficiently stabilize the colloidal dispersion, typically, only ligands with a high bonding strength to the colloidal core can be used.

One common goal of many dense composite materials is an increase in thermal conductivity while electric insulation is still maintained. 2D materials have been widely investigated in this regard owing to their benign intrinsically high thermal conductivity along the basal plane and low percolation limit. One of the most widely examined inorganic fillers for such nanocomposites is hexagonal boron nitride nanosheets (BNNS). These combine a high thermal conductivity with a large bandgap of 5.9 eV and hence a low electronic conductivity. An intrinsic 2D morphology, as well as the possibility of facile mechanical exfoliation, makes these excellent candidates for nanocomposite thin films. Chen et al. [52] incorporated BNNS in a polyvinylidene fluoride matrix via electrospinning and subsequent folding and pressing. Electron microscopy images revealed interconnected chains of overlapping BNNS. These percolation pathways are essential for a noticeable increase in thermal conductivity. Randomly dispersed BNNS would at best show a slight improvement due to phonon scattering at the organic/inorganic interfaces being the predominant effect. In the case of the electrospun samples, a high thermal conductivity of 16.3 W/m/K was achieved at 33 wt.% filler content. These electrically insulating and flexible films were applied as a thermal interface material in metal-oxide-semiconductor field effect transistors (MOSFETs).

The underwhelming thermal conductivity of nanocomposites is a general problem in this field of research. Fillers with an enormous theoretical thermal conductivity such as carbon nanotubes (κT_{heo} = 3500 W/m/K) or graphene sheets (κ_{Theo} = 2000–4000 W/m/K) have only resulted in moderately increased values in composites, usually below 10 W/m/K. BNNS also seem to have much greater potential with a bulk value of κ_{BN} = 390 W/m/K and a theoretical value in single layer BNNS of κ_{BNNS} = 600 W/m/K. Similar to the graphite/graphene system, this effect of exfoliation can be explained by suppression of out-of-plane acoustic phonons as a result of interlayer interactions [151]. Nevertheless, the low thermal conductivity of respective composite structures is a direct result of weak interfacial adhesion and a resulting high interfacial thermal resistance (ITR). For low filler loadings, the most relevant contribution is the resistance at the filler/polymer interface, and for larger loadings, i.e. overlapping filler particles, this is dominated by the ITR of the filler/filler interface. Several recent reports have concentrated on ways to reduce the ITR in nanocomposites and to achieve better performing materials. Wang et al. [152] used a process similar to doctor blading to prepare BNNS/epoxy composites. At a constant BNNS content of 25 vol.%, they additionally prepared samples with a small amount of silver nanoparticles (AgNPs), formed *in situ* on the BNNS. This increased the thermal conductivity from 1.63 to 3.06 W/m/K and is explained by bridging between filler platelets that reduces the ITR. Shortly after, Zeng et al. [51] published a report on BNNS/CNF composites prepared by a simple filtration method. At only 25 wt.% filler content, this material showed a comparatively high thermal conductivity of 21.4 W/m/K. The strong van der Waals interactions between BNNS and CNF are presented as the main factor influencing the ITR and causing this effective thermal transport. The two concepts are combined in the work by Sun et al. [153] on BNNS/nanocellulose composites with the addition of AgNPs, resulting in a fivefold increase of the thermal conductivity to 65.7 W/m/K. Similar to what was presented in the previous studies, three contributions to the thermal conductivity can be summarized as follows. (i) BNNS at a high loading above the percolation threshold show a parallel alignment as well as sufficient overlap to enable the formation of an interconnected network. (ii) Strong van der Waals interaction between filler and matrix reduces the ITR between the two components. (iii) AgNPs added to the system act as thermal bridges between the filler platelets. The resulting reduction of the ITR at the filler/filler interfaces has a dramatic effect on the thermal transport through the ternary nanocomposite structure. It is stressed, however, that the size of the silver colloids should not exceed a diameter of 12 nm as this can have an adverse effect by increasing the gap size between adjoining BNNS particles. Taking into account the degree of orientation and applying both the EMT as well as a nonlinear model for filler/filler interactions, the effect of AgNP addition could be examined in more detail. Both the resistance at the filler/matrix and the filler/filler interface are reduced, the latter by one full magnitude. Lastly, the effect on the electric properties had to be examined. As the content of silver was held very low (0.16 wt.%), the AgNPs are sufficiently well dispersed and far from forming interconnected, metallic pathways. The increase of the electronic conductivity is, therefore, only marginal and makes this a promising approach to easily enhance properties of other established heat spreader materials based on nanocomposites.

Clay minerals, both of natural and synthetic origin, show a prominent aspect ratio as well as good processability, especially in polar solvents. These inorganic colloids have only been moderately characterized regarding their thermal transport. The high interface density due to a rather small thickness in the range of about 1 nm and a concomitantly high anisotropy, however, make them promising 2D materials as well. Philipp et al. [154] examined a highly optimized synthetic clay mineral, i.e. sodium fluorohectorite (Hec), processed via spray coating from aqueous dispersions (Figure 14.13a). Their anisotropic thermal conductivity was compared to analogously prepared GO samples. Both single-phase materials show a high degree of order and alignment of the 2D colloidal components. The in-plane thermal conductivity, measured with LIT, was high at 5.7 W/m/K, whereas photoacoustic measurements of the cross-plane thermal conductivity gave a rather low value of 0.21 W/m/K. The resulting anisotropy ratio of ~30 is higher than comparable samples consisting of natural clay minerals by at least a factor of 3 (Figure 14.13b). This is attributed to the superior order in the spray-coated films caused by the full delamination prior to the coating process, which is corroborated by electron microscopy and X-ray scattering techniques. A significant reduction of the platelet diameter via sonication from 20 to ~0.4 μm had no effect on these values. It is concluded that the order in the samples is not impaired due to a remaining high aspect ratio and that the phonon MFP must be far below the lateral extension of the Hec particles. Scattering at platelet edges, therefore, has little effect in both cases. Films prepared analogously with GO showed a lower in-plane thermal conductivity and, consequently, a lower anisotropy ratio. This is attributed to the defect-rich structure of this undefined material class. Thermal transport properties depend strongly on the exact synthesis parameters, which in some cases can even lead to large holes in the 2D sheets as a result of overoxidation. In summary, this report showed how clay minerals, especially of synthetic origin, are a promising material both for application in heat spreaders and as model systems for the examination of thermal transport in 2D nanostructures. Wang et al. [55] used the same starting material to examine the interplay of mechanical and thermal transport properties in 2D hybrid nanostructures. Spray coating an aqueous nematic solution of delaminated Hec and dissolved polyvinylpyrrolidone produced highly ordered, nacre-mimetic hybrid materials. In-plane conductivity measurements of samples with varying compositions stood in compliance with the EMT. This indicates that even though the interlayer spacing is far smaller than the radius of gyration and strong polymer confinement must be the case, no nanoconfinement effect regarding thermal transport through the organic phase existed. The cross-plane thermal conductivity was shown to be lower than that of both pure Hec and polymer and confirms the important role of polymer/inorganic interfaces and the corresponding acoustic mismatch. The highest anisotropy ratio for these samples was found to be 38 at 51 vol.% Hec (Figure 14.13c). The processability via spray coating and the resulting large-scale, highly ordered, and transparent films allowed the characterization of direction-dependent mechanical properties via Brillouin light scattering. A determination of Young's modulus both in in-plane and cross-plane direction showed an increase of the respective ratio by a factor of 7 when increasing the Hec concentration from 0% to 100%. Furthermore, as the phonon MFP is much lower than the diameter of the Hec platelets, a correlation of the in-plane conductivity and the respective Young's and shear moduli could be found, whereas the cross-plane thermal transport is governed by interfaces. This study further elucidates the importance of high-quality Hec as a starting material to produce well-ordered nanostructures, which enable examination and correlation of material properties not feasible with inferior samples. Besides the fabrication of nanocomposites, interfaces in clays can be tuned by exchanging the inorganic interlayer cations for alkylammonium ions. Losego et al. [155] investigated the thermal transport behavior of these fully dense organoclay nanocomposites fabricated from modified montmorillonite by TDTR. They were able to demonstrate how this high amount of interfaces on the nanometer scale will increase the insulation properties of fully dense matter. The anisotropic 2D nanoclay stacking represents serially connected thermal resistances, which impede phonons from traveling through the material. Consequently, they measured a cross-plane thermal conductivity of only 0.10 W/m/K, which is a fivefold decrease compared with unmodified clay (Figure 14.13d).

As introduced earlier, GO is a widely used type of 2D colloid that can easily be processed from aqueous suspensions. Subsequent reduction recovers some of the

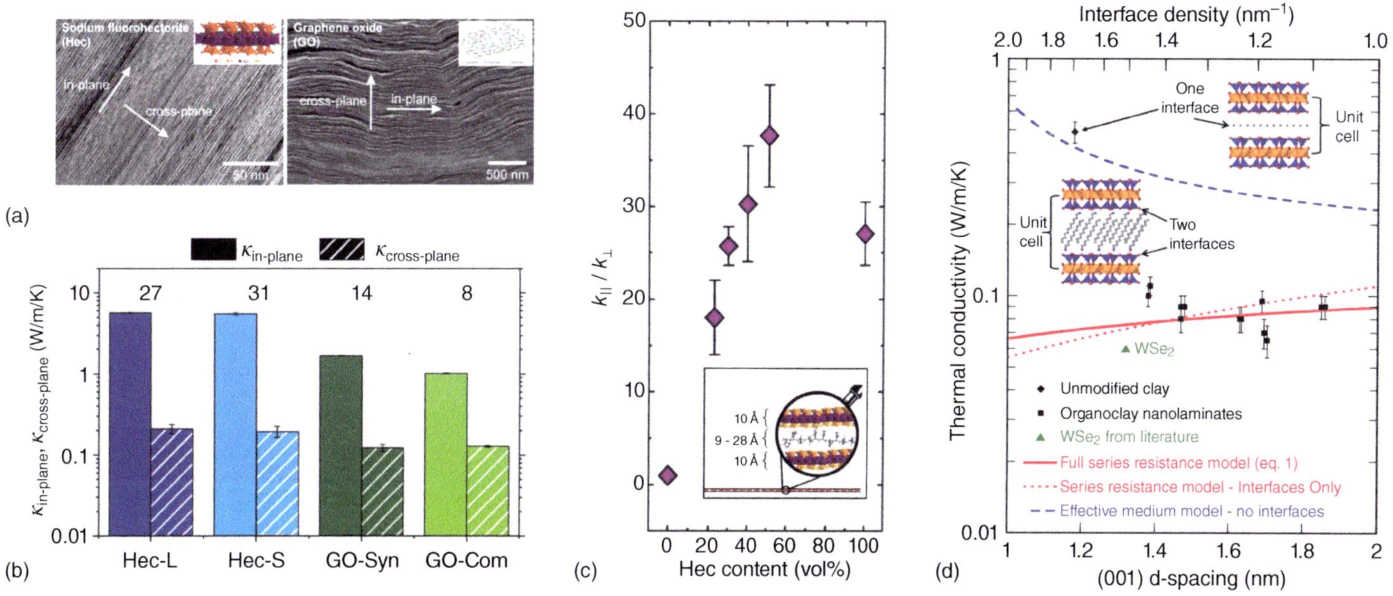

Figure 14.13 (a) TEM image and SEM image of an aligned smectic Hec film and GO film, respectively. (b) In-plane and cross-plane thermal conductivity of Bragg stacks of GO and Hec. Source: Philipp et al. [154]/with permission from American Chemical Society. (c) The ratio of the in-plane to cross-plane thermal conductivities of Hec/PVP hybrid Bragg stacks against the Hec volume fraction. Source: Adapted from [55] with permission of John Wiley and Sons. (d) Cross-plane thermal conductivity of the nanocomposite against the d-spacing/interface density. The thermal conductivity follows a full series resistance model. Source: Adapted from [155] with permission of American Chemical Society.

outstanding graphene-like properties in the form of reduced graphene oxide (rGO). While it may seem odd to chemically modify graphite and, in the end, convert it back, this step is essential to enable any type of processability that is not feasible with pristine graphitic carbon. The oxidation usually proceeds chemically according to the Hummers or Brodie method [156]. Polar oxygen-containing moieties can thereby be introduced into the sp^2 carbon lattice, which enables swelling and exfoliation of the produced graphite oxide to generate suspensions of monolayer GO ready for further processing. Song et al. [157], for example, prepared hybrid films from suspensions of GO and nanofibrillated cellulose via simple vacuum filtration. Via a chemical reduction with hydrazine and a resulting conversion to rGO, the films were imparted with a highly anisotropic thermal conductivity with an in-plane conductivity of 6.2 W/m/K and an anisotropy factor of 86. The latter is a direct result of the inherently anisotropic bonding properties in graphite. Strong covalent bonds in the sp^2 layers stand in contrast to the weak interlayer van der Waals interactions. GO films prior to any type of treatment usually show a rather low thermal conductivity of 0.5–1.0 W/m/K due to the great number of defects, heteroatoms, and holes in the undefined, heterogeneous material. A more prevalent posttreatment is thermal annealing in an inert atmosphere, as shown by Renteria et al. [158]. Here, a still comparatively moderate temperature of 1000 °C was used to prepare thermally reduced GO films with an in-plane conductivity of 61 W/m/K. The anisotropy ratio is extremely high (675), which is caused by the formation of gas voids between layers and leads to a reduction of the cross-plane conductivity. Compositional analysis with X-ray photoelectron spectroscopy (XPS) showed a strong reduction of impurity heteroatoms (mainly oxygen) and a gradually increasing amount of sp^2 hybridized carbon atoms. Comparison with a theoretical approach showed that these two factors, as well as an increase of the sp^2-grain size from several nanometer to several 100 nm, all influence the thermal transport via acoustic phonons and explain the observed increase of the thermal conductivity. Another interesting effect is the temperature dependency. Samples annealed at temperatures up to 600 °C showed little change and only a slight increase of the thermal conductivity with increasing temperature, which is typical for highly disordered structures. When applying an annealing temperature of 1000 °C, however, a clear decrease in thermal conductivity with increasing temperature can be observed. This is a sign of Umklapp-scattering limited thermal transport and is an indicator of crystalline-like behavior. The method of thermal reduction was further examined by Peng et al. [159] with an ultra-high annealing temperature of 3000 °C (Figure 14.14). In order to obtain a higher quality starting material, debris-free GO was prepared via physical separation. GO thin films were then subjected to 3000 °C in an inert atmosphere, causing thermal reduction and consequentially the formation of pyrolysis gases. Compositional and structural analysis showed an extremely high quality, crystalline carbonaceous material. The formed pyrolysis gases were captured in the interlayer region and resulted in a porous, foamlike microstructure. These microgasbags, or semifullerenes, were subsequently removed via isostatic compression at high pressures and the structure was returned to its previous state of parallel aligned 2D sheets. The excellent crystallinity and alignment on a mesoscale result in an extremely high in-plane thermal conductivity of 1940 W/m/K, outperforming other high-end materials such as copper or even assemblies of graphene sheets. In addition, the microfolds generated by geometric constriction during the compression of microgasbags have a significant effect on the mechanical properties. Compensation of local stress by unfolding increases the bending capability of the prepared films and makes this an interesting example of a high κ, flexible material.

14.6 Heat Management Applications

Going beyond the thermal transport properties of colloidal assembly structures, we want to outline emerging areas of applications where colloidal materials are utilized in an integrated and dynamic fashion. Consequently, colloidal assemblies can contribute to heat management in various functional systems. In this section, we will show how tailored nanostructures can develop interesting new features ranging from thermal recorders, switches, or even rectifiers to applications such as thermoelectrics and passive cooling.

14.6.1 Thermal Recorders

Knowledge of the thermal history of a given material or process is of high importance to many applications. Recording of temperatures electronically can be conducted with a minimum of effort but require a continuous power supply. Colloidal superstructures have recently been shown to allow an off-grid recording of time and temperature. In such applications, the sensitivity of a polymer colloidal crystal or inverse opal to film formation and concomitant loss of its stopband is exploited [160]. When properly designing the film formation kinetic behavior, e.g. by varying the degree of cross-linking, multiple measurement spots can be used to work out time and temperature information just by measuring the changes of the photonic stopband [161] (Figure 14.15a). Such recorders can certainly not provide the detailed time/temperature evolution as electronic sensors can. However, such colloidal structures are much harder to manipulate and operate autonomously without any external power supply. The photonic stopband is, furthermore, a very simple and robust optical readout signal.

14.6.2 Thermal Switches

Thermal switches based on dispersions of graphitic colloids dispersed in an organic medium were presented by Sun et al. [162]. Percolation of the (semi)conducting graphite flakes or carbon nanotubes by a simple first-order phase transition, namely, the freezing of the liquid majority phase, opens highly conductive transport pathways. A schematic sketch of the switching process during the phase transition of a carbon nanotube/hexadecane mixture and the thermal conductivity are illustrated in Figure 14.15b.

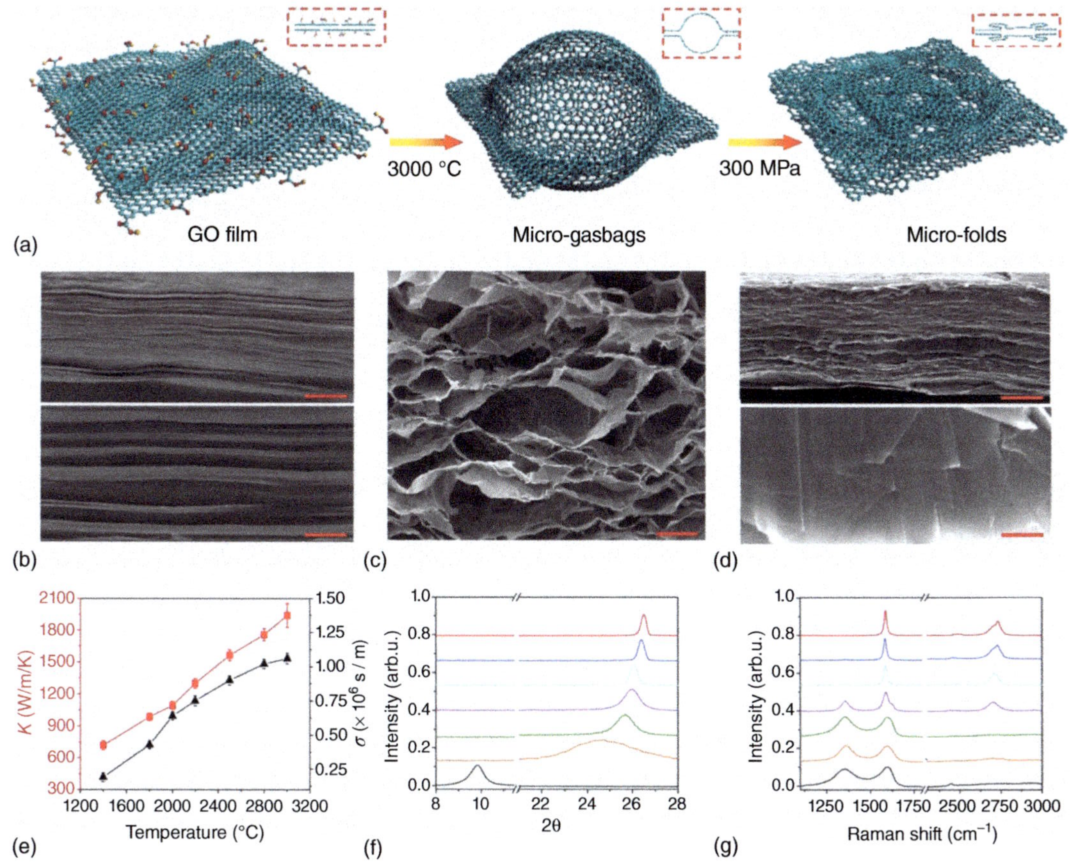

Figure 14.14 (a) Schematic of the thermal reduction of GO films resulting in gas-filled semifullerenes and further conversion to fully aligned films with incorporated microfolds via isostatic compression. SEM images of cross sections of the dfGF (b), foamed dfGF (c), and dense dfGF (d). Scale bars: 4 μm in (b, up), (c), and (d, up); 300 nm in (b, down) and (d, down). (e) The thermal and electrical conductivity of dfGFs as a function of annealing temperatures. XRD (f) and Raman (g) spectra of dfGFs with different annealing temperatures, i.e. red (3000 °C), blue (2500 °C), cyan (2000 °C), magenta (1500 °C), olive (1300 °C), orange (500 °C), and black (20 °C). Source: Peng et al. [159]/with permission from John Wiley & Sons, Inc.

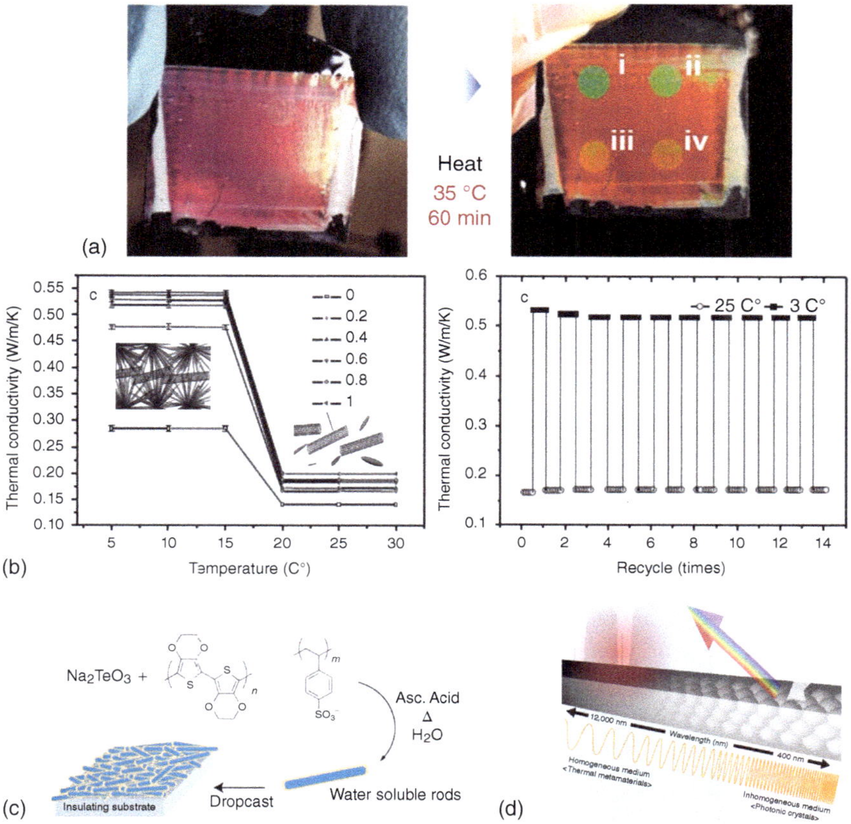

Figure 14.15 (a) Inverse opal structure-based thermal recorder, reported by Lee et al. Source: Lee et al. [161]/with permission from John Wiley & Sons, Inc. (b) Thermal conductivity of graphitic colloids dispersions based on thermal switches, prepared by Sun et al. Source: Adapted from [162] with permission of John Wiley and Sons. (c) Thermoelectrics, with high thermoelectrical performance, synthesized by See et al., based on solution-processable hybrid material. Source: Adapted from [163] with permission of American Chemical Society. (d) Colloidal photonic assembles applied in daytime passive cooling, reported by Kim et al. Source: Adapted from [164] with permission of American Chemical Society.

The thermal conductivity changes are reversible (Figure 14.15b) since melting of the majority phase will redisperse the colloidal objects and destroy the percolating network. Owing to the anisotropy of the colloidal sheets or fibers, this percolation pathway is created even for low concentrations of the colloidal material during the phase separation that accompanies the crystallization of the water or wax phase upon cooling. This leads to an intimate contact between the colloidal particles, demonstrating the possibilities provided by controlling the interfaces. Notably, the strong increase in thermal conductivity is driven by the onset of electrical conductance. The macrostructure of the resulting composite is controlled by the nucleation and crystallization process of the liquid majority phase and is typically phase-separated on a tens of micrometer scale. The reversibility of the melting and freezing process results in a switchable thermal conductivity behavior. Transitions between contacting and isolated colloidal structures have also been introduced to increase the inherent safety of Li-ion batteries [165]. Here, the thermal conductivity of the composite structure played a minor role, but the battery temperature influenced the thermal expansion of a polymer matrix comprising spiky graphene-coated Ni nanoparticles. Upon excessive heating, e.g. during the onset of thermal runaway, the contact and, consequently, the electric conductivity between the spiky particles in this layer are shut down.

14.6.3 Thermal Rectification

While the fabrication and measurement of thermal switches can be challenging, the principle in itself is fairly straightforward. Regulating the direction of a heat flow becomes more complicated when attempting to prepare the thermal equivalent of an electric diode. Rectification of heat flow, i.e. enabling a larger heat flux in the forward compared to the reverse direction, is far from trivial. Thermal energy being much more ubiquitous than electrical energy, nevertheless, makes the fabrication of phononic devices a promising research topic. A common approach toward rectification is the serial combination of two materials with different thermal dependencies of the thermal conductivity $\kappa(T)$. The rectification coefficient R resulting from this setup is defined as the ratio of forward (+) and backward (−) heat flux: $r = J^+/J^- - 1$. The interface between the two materials must, however, be carefully controlled. The ITR should be kept as low as possible; otherwise, this will be the dominant effect and outweigh any possible thermal rectification. A few studies have examined this problem and postulate solutions based on nanostructured materials. Some of these methods allow using the same material on both sides, thereby eliminating the problem of a large ITR. Cartoixa et al. [166] presented a computational study on telescopic silicon nanowires, utilizing the change of $\kappa(T)$ due to confinement effects. Si nanowires are known to show different $\kappa(T)$ dependencies depending on the thickness. In wires with a diameter of 22 nm, boundary scattering becomes a more relevant factor, and the thermal transport properties differ distinctly from wires with a diameter of 56 nm, which show behavior consistent with bulk silicon. Verification of the validity of the applied heat equation was done by comparing the resulting temperature profile along a telescopic wire to a profile determined by molecular dynamics simulations. The good agreement of the two indicates that provided sufficient experimentally or computationally determined values for $\kappa(T)$ are available, the rectification capability of telescopic nanowires can be determined by simply solving the heat equation. In this work, samples with a length of 60 nm exhibited a preferential heat transfer from the thick to the thin side and a maximum rectification coefficient of $r = 0.5$. Zhang and Luo [167] reported a different approach. In this computational study, crystalline PE nanofibers were examined. This material shows a morphology change at 397 K, at which segmental rotations cause a sharp decrease of the thermal conductivity. To achieve a directional dependency, cross-linking sites were introduced on one side of the sample. This inhibits the mentioned morphology change and thereby results in a significantly different temperature dependency of the thermal conductivity. As the entire sample still consists of PE, covalent bonds are present at the interface of the cross-linked and non-cross-linked portion, thereby minimizing the effect of ITR. All this results in a fairly high calculated rectification coefficient of 1.2. Additionally, this material shows rectification behavior at a small temperature gradient of only 20 K, which is much lower than e.g. combinations of bulk materials that require gradients >100 K. This is an important point to take into account when determining the practical applicability of such materials.

14.6.4 Thermoelectrics

Besides the control of the heat flow direction, the conversion of heat to electrical energy with materials showing thermoelectric properties is an important aspect of heat management. The underlying mechanism can be described by the Seebeck effect: in a circuit consisting of two different conductive materials, a temperature gradient is applied between the two contact points. Due to thermodiffusion of charge carriers, this results in a measurable electric voltage. The maximum efficiency of such a thermoelectric material is best characterized by the thermoelectric figure of merit: $zT = \sigma S^2 T/\kappa$ (σ, electric conductivity; S, material-dependent Seebeck constant). A high zT indicates a high thermoelectric energy conversion. This, in turn, implies that an ideal material requires a low thermal yet a high electric conductivity at a retained high Seebeck coefficient. Nanocomposite and hybrid structures are promising candidates to meet the requirements, owing to boundary phonon scattering and high mobility of charge carriers. Nanostructures exhibiting a high interface density and consisting of selected semiconductors naturally show great potential in this regard [168, 169]. Ibanez et al. [170] used a simple method, drying and subsequent hot-pressing of a binary PbS/AgNP dispersion, to produce a consolidated nanocrystalline assembly of the two materials. Due to both diffusive and acoustic mismatch at grain boundaries, a low thermal conductivity is maintained while the electric conductivity is kept fairly high by the adjustment of the fermi-level of the metal component to the conduction band of the semiconductor. This results in a high zT of 1.7 at 850 K. A few years later the same group used a stepwise colloidal synthesis to prepare a thermoelectric material [171]. First, AgNPs were prepared and then converted to Ag_2S-tipped PbS rods. These were functionalized with K_2Te, after which the resulting suspension was dried and hot-pressed. The result, a Pb–Ag–K–S–Te composite, shows how colloidal assembly can enable the fabrication of highly complex multicomponent systems. A large number of interfaces between the different crystal phases existing in this material minimize the thermal conductivity. Doping with K-ions ensures a high charge carrier concentration and, therefore, a high zT of 1.0 at 620 K. In addition, without using conventional energy-intensive and high-temperature processing technique, a solution-processable hybrid material, with high thermoelectrical performance, was synthesized by See et al. [163] (Figure 14.15c). By synthesizing Tellurium (Te) nanorods, which inherently possesses high thermopower, *in situ* with water-soluble electrical conducting polymer poly(3,4-ethylene dioxythiophene):poly(styrene–sulfonate) (PEDOT:PSS), a polymer–inorganic hybrid material with a combination of high thermopower, low thermal conductivity, and high electrical conductivity was obtained. As a result, this composite material was demonstrated to possess a remarkable thermoelectric transport property, i.e. a room temperature $zT \sim 0.1$.

14.6.5 Passive Cooling

Besides thermal conduction, thermal radiation is also an important transport mechanism to transfer heat. In this regard, colloidal entities have made substantial contributions during the past decade to the emerging field of passive daytime cooling [172]. In passive cooling applications, the heat from hot terrestrial surfaces is transferred to cold outer space through the "atmospheric window" (8–13 μm) without additional energy consumption. Most important is the fulfillment of stringent optical properties spanning from the ultraviolet to the mid-infrared range. SiO_2, SiC particles, and composites derived from them are widely used for novel material designs, owing to their inherent adsorption/emission properties in the atmospheric window range. For instance, daytime passive cooling devices based on 2D and 3D SiO_2 colloidal crystals were revealed by Jaramillo-Fernandez et al. [173], Kim et al. [164], and Liu et al. [174], respectively. In the case of colored passive cooling structures, colloidal crystals are particularly interesting owing to their opalescence, which evokes color without the absorption of light (Figure 14.15d).

Additionally, composite materials of a polymer matrix and embedded colloidal particles have been investigated by various groups. Typical polymers being used for such composite materials are PE, polymethylpentene (TPX), polymethylmethacrylate (PMMA), or polydimethylsiloxane (PDMS) in combination with particles comprising SiO_2, SiC, or Al_2O_3, to name a few [175–178]. In all of these composite cases, the particles are rather used as random scatterers with little to no control on their interparticle distance or superlattice order.

14.7 Conclusion and Summary

Thermal transport on small length scales is of high importance for many nano- and mesostructured materials and applications. Colloidally assembled structures provide a convenient and readily available way to access suitable model systems, which allow to develop an in-depth understanding of the underlying physics and to study the governing structure–property relationships. This chapter presented an overview on the exciting topic of nanoscale thermal transport, with a particular focus on materials accessible via colloidal self-assembly strategies. We, therefore, provided a fundamental introduction to the physics of thermal transport in hard and soft bulk materials. These constitute the most relevant material classes for common dispersion colloids. Working with nanostructured materials entails challenges for their thermal characterization. Consequently, specialized and often home-built methods are required to measure the thermal transport properties of self-assembled colloidal objects. We introduced the most important methods and provided a brief overview of their working principle and respective limitations. The combination of colloid synthesis, self-assembly, and post-processing provides an almost endless range of nanostructured materials with unique properties. With select examples from recent literature, we demonstrated the parameter space ranging from ultralow to extremely high thermal conductivities.

Finally, we surveyed the recent literature that readily showcases the vast potential provided by colloidal assemblies. The reviewed examples allowed insights into the validity of existing mixing models, size dependencies, or the contributions of diffusive and ballistic thermal transport. The well-defined structures also allow distinguishing between contributions from the material's composition itself and the interfacial thermal conductances. In composite and soft materials, the phase transitions, e.g. glass transition temperatures, can lead to structural changes, ensuing a drastic change to the thermal conductance. The massive degree of nanostructuring provided by colloidal objects, e.g. stacked layers of 2D clays, is furthermore an attractive route toward highly insulating yet mechanically robust materials.

Colloidal materials further mature from their use as a model structure and the investigation of their material properties. They can play a significant role in emerging applications, where heat management is of pivotal importance. First examples demonstrated their capability to function as thermal recorders, thermal switches, or thermal rectifiers. They can also contribute to sustainable energy management, for example, via hybrid thermoelectric materials or via photonic structures for passive daytime cooling.

References

1 Cahill, D.G., Ford, W.K., Goodson, K.E. et al. (2003). *J. Appl. Phys.* 93: 793.
2 Cahill, D.G., Braun, P.V., Chen, G. et al. (2014). *Appl. Phys. Rev.* 1: 011305.
3 Chernatynskiy, A., Clarke, D., and Phillpot, S. (2012). *Thermal Transport in Nanostructured Materials. In Handbook of Nanoscience, Engineering and Technology*, 3e. CRC Press.
4 Xu, Z. (2016). *Theor. Appl. Mech. Lett.* 6: 113.
5 Yang, N., Xu, X., Zhang, G., and Li, B. (2012). *AIP Adv.* 2: 041410.
6 Maldovan, M. (2013). *Nature* 503: 209.
7 Li, N., Ren, J., Wang, L. et al. (2012). *Rev. Mod. Phys.* 84: 1045.
8 Cheng, W., Wang, J., Jonas, U. et al. (2006). *Nat. Mater.* 5: 830.
9 Farhat, M., Chen, P.Y., Bagci, H. et al. (2015). *Sci. Rep.* 5: 9876.
10 Wu, G. and Li, B. (2007). *Phys. Rev. B* 76: 085424.
11 Zhu, J., Hippalgaonkar, K., Shen, S. et al. (2014). *Nano Lett.* 14: 4867.
12 Kubytskyi, V., Biehs, S.-A., and Ben-Abdallah, P. (2014). *Phys. Rev. Lett.* 113: 074301.
13 Liu, M., Ma, Y., Wu, H., and Wang, R.Y. (2015). *ACS Nano* 9: 1341.
14 Risk, W.P., Rettner, C.T., and Raoux, S. (2009). *Appl. Phys. Lett.* 94: 101906.

15 Chau, M., Kopera, B.A.F., Machado, V.R. et al. (2017). *Mater. Horiz.* 4: 236.
16 Mingo, N., Hauser, D., Kobayashi, N.P. et al. (2009). *Nano Lett.* 9: 711.
17 Dresselhaus, M.S., Chen, G., Tang, M.Y. et al. (2007). *Adv. Mater.* 19: 1043.
18 Poudel, B., Hao, Q., Ma, Y. et al. (2008). *Science* 320: 634.
19 Singh, V., Bougher, T.L., Weathers, A. et al. (2014). *Nat. Nanotechnol.* 9: 384.
20 Kim, G.-H., Lee, D., Shanker, A. et al. (2015). *Nat. Mater.* 14: 295.
21 Volokitin, A.I. and Persson, B.N.J. (2007). *Rev. Mod. Phys.* 79: 1291.
22 Basu, S., Zhang, Z.M., and Fu, C.J. (2009). *Int. J. Energy Res.* 33: 1203.
23 Keblinski, P., Prasher, R., and Eapen, J. (2008). *J. Nanopart. Res.* 10: 1089.
24 Özerinç, S., Kakaç, S., and Yazıcıoğlu, A.G. (2010). *Microfluid. Nanofluid.* 8: 145.
25 Cardellini, A., Fasano, M., Bigdeli, M.B. et al. (2016). *J. Phys.: Condens. Matter* 28: 483003.
26 Tervo, E.J., Adewuyi, O.S., Hammonds, J.S., and Cola, B.A. (2016). *Mater. Horiz.* 3: 434.
27 Ordonez-Miranda, J., Tranchant, L., Gluchko, S., and Volz, S. (2015). *Phys. Rev. B* 92: 115409.
28 Yang, S.-M., Kim, S.-H., Lim, J.-M., and Yi, G.-R. (2008). *J. Mater. Chem.* 18: 2177.
29 Sau, T.K. and Rogach, A.L. (2010). *Adv. Mater.* 22: 1781.
30 Matijevic, E. (1993). *Chem. Mater.* 5: 412.
31 Sacanna, S. and Pine, D.J. (2011). *Curr. Opin. Colloid* 16: 96.
32 Murray, C.B., Kagan, C.R., and Bawendi, M.G. (2000). *Annu. Rev. Mater. Sci.* 30: 545.
33 Groschel, A.H., Walther, A., Lobling, T.I. et al. (2013). *Nature* 503: 247.
34 Ehlert, S., Taheri, S.M., Pirner, D. et al. (2014). *ACS Nano* 8: 6114.
35 Hopkins, P.E., Mittal, M., Phinney, L.M. et al. (2011). *Appl. Phys. Lett.* 99: 133106.
36 Fischer, S., Exner, A., Zielske, K. et al. (2011). *Proc. Natl. Acad. Sci. U. S. A.* 108: 1810.
37 Möller, M.W., Kunz, D.A., Lunkenbein, T. et al. (2012). *Adv. Mater.* 24: 2142.
38 Walther, A., Bjurhager, I., Malho, J.-M. et al. (2010). *Nano Lett.* 10: 2742.
39 Vogel, N., Retsch, M., Fustin, C.A. et al. (2015). *Chem. Rev.* 115: 6265.
40 Israelachvili, J.N. (2011). *Intermolecular and Surface Forces*. Elsevier Science.
41 Lotito, V. and Zambelli, T. (2017). *Adv. Colloid Interface Sci.* 246: 217.
42 Barako, M.T., Sood, A., Zhang, C. et al. (2016). *Nano Lett.* 16: 2754.
43 Stachowiak, A.N., Bershteyn, A., Tzatzalos, E., and Irvine, D.J. (2005). *Adv. Mater.* 17: 399.
44 Retsch, M. and Jonas, U. (2013). *Adv. Funct. Mater.* 23: 5381.
45 Stein, A. and Schroden, R.C. (2001). *Curr. Opin. Solid State Mater. Sci.* 5: 553.
46 Stein, A., Li, F., and Denny, N.R. (2008). *Chem. Mater.* 20: 649.
47 Phillips, K.R., England, G.T., Sunny, S. et al. (2016). *Chem. Soc. Rev.* 45: 281.
48 Mishra, R. and Militky, J. (2019). *Nanotechnology in Textiles*, 263. Woodhead Publishing.
49 Yao, Y., Zeng, X., Guo, K. et al. (2015). *Compos. Part A Appl. Sci. Manuf.* 69: 49.

50 Wang, H., Tazebay, A.S., Yang, G. et al. (2016). *Carbon* 106: 152.
51 Zeng, X., Sun, J., Yao, Y. et al. (2017). *ACS Nano* 11: 5167.
52 Chen, J., Huang, X., Sun, B., and Jiang, P. (2019). *ACS Nano* 13: 337.
53 Shao, G., Hanaor, D.A.H., Shen, X., and Gurlo, A. (2020). *Adv. Mater.*: e1907176.
54 Terao, T., Zhi, C., Bando, Y. et al. (2010). *J. Phys. Chem. C* 114: 4340.
55 Wang, Z., Rolle, K., Schilling, T. et al. (2020). *Angew. Chem. Int. Ed.* 59: 1286.
56 Zeng, X., Ye, L., Yu, S. et al. (2015). *Nanoscale* 7: 6774.
57 Zhu, H., Li, Y., Fang, Z. et al. (2014). *ACS Nano* 8: 3606.
58 Yao, Y., Zeng, X., Sun, R. et al. (2016). *ACS Appl. Mater. Interfaces* 8: 15645.
59 Han, J., Du, G., Gao, W., and Bai, H. (2019). *Adv. Funct. Mater.* 29: 1900412.
60 Li, Q., Guo, Y., Li, W. et al. (2014). *Chem. Mater.* 26: 4459.
61 Asheghi, M., Kurabayashi, K., Kasnavi, R., and Goodson, K.E. (2002). *J. Appl. Phys.* 91: 5079.
62 Regner, K.T., Majumdar, S., and Malen, J.A. (2013). *Rev. Sci. Instrum.* 84: 064901.
63 Marconnet, A.M., Asheghi, M., and Goodson, K.E. (2013). *J. Heat Transfer* 135: 061601.
64 Peierls, R. (1929). *Ann. Phys. (Berl.)* 395: 1055.
65 Chavez-Angel, E., Reparaz, J.S., Gomis-Bresco, J. et al. (2014). *APL Mater.* 2: 012113.
66 Ju, Y.S. and Goodson, K.E. (1999). *Appl. Phys. Lett.* 74: 3005.
67 Turney, J.E., McGaughey, A.J.H., and Amon, C.H. (2010). *J. Appl. Phys.* 107: 024317.
68 Neogi, S., Reparaz, J.S., Pereira, L.F.C. et al. (2015). *ACS Nano* 9: 3820.
69 Maldovan, M. (2015). *Nat. Mater.* 14: 667.
70 Joannopoulos, J.D., Villeneuve, P.R., and Fan, S. (1997). *Nature* 386: 143.
71 Wagner, M.R., Graczykowski, B., Reparaz, J.S. et al. (2016). *Nano Lett.* 16: 5661.
72 Vega-Flick, A., Duncan, R.A., Eliason, J.K. et al. (2016). *AIP Adv.* 6: 121903.
73 Tambo, N., Liao, Y., Zhou, C. et al. (2020). *Sci. Adv.* 6: eabc0075.
74 Einstein, A. (1911). *Ann. Phys. (Berl.)* 340: 679.
75 Cahill, D.G., Watson, S.K., and Pohl, R.O. (1992). *Phys. Rev. B* 46: 6131.
76 Cahill, D.G. and Pohl, R.O. (1989). *Solid State Commun.* 70: 927.
77 Allen, P.B. and Feldman, J.L. (1993). *Phys. Rev. B* 48: 12581.
78 Allen, P.B., Feldman, J.L., Fabian, J., and Wooten, F. (1999). *Philos. Mag. B* 79: 1715.
79 Simoncelli, M., Marzari, N., and Mauri, F. (2019). *Nat. Phys.* 15: 809.
80 Shenogin, S., Bodapati, A., Keblinski, P., and McGaughey, A.J.H. (2009). *J. Appl. Phys.* 105: 034906.
81 Ho, C.Y., Powell, R.W., and Liley, P.E. (1972). *J. Phys. Chem. Ref. Data* 1: 279.
82 Wada, H. and Kamijoh, T. (1996). *Jpn. J. Appl. Phys., Part 2* 35: L648.
83 Zeller, R.C. and Pohl, R.O. (1971). *Phys. Rev. B* 4: 2029.
84 A. Henry (2013) Thermal Transport in Polymers, Annual Review of Heat Transfer, Vol 17, Chapter 13, p. 485-520, Begell House.
85 Choy, C.L. (1977). *Polymer* 18: 984.

86 Hansen, D. and Ho, C.C. (1965). *J. Polym. Sci., Part A: General Papers* 3: 659.
87 Liu, J. and Yang, R. (2012). *Phys. Rev. B* 86: 104307.
88 Zhang, L., Ruesch, M., Zhang, X. et al. (2015). *RSC Adv.* 5: 87981.
89 Zhang, L., Chen, T., Ban, H., and Liu, L. (2014). *Nanoscale* 6: 7786.
90 Xie, X., Li, D., Tsai, T.-H. et al. (2016). *Macromolecules* 49: 972.
91 Shen, S., Henry, A., Tong, J. et al. (2010). *Nat. Nanotechnol.* 5: 251.
92 Choy, C.L., Chen, F.C., and Luk, W.H. (1980). *J. Polym. Sci., Part B: Polym. Phys.* 18: 1187.
93 Kurabayashi, K., Asheghi, M., Touzelbaev, M., and Goodson, K.E. (1999). *J. Microelectromech. Syst.* 8: 180.
94 Reichenauer, G., Heinemann, U., and Ebert, H.P. (2007). *Colloids Surf. Physicochem. Eng. Aspects* 300: 204.
95 Hrubesh, L.W. and Pekala, R.W. (1994). *J. Mater. Res.* 9: 731.
96 Koebel, M., Rigacci, A., and Achard, P. (2012). *J. Sol-Gel Sci. Technol.* 63: 315.
97 Carson, J.K., Lovatt, S.J., Tanner, D.J., and Cleland, A.C. (2005). *Int. J. Heat Mass Transfer* 48: 2150.
98 Wang, J., Carson, J.K., North, M.F., and Cleland, D.J. (2006). *Int. J. Heat Mass Transfer* 49: 3075.
99 Abad, B., Borca-Tasciuc, D.A., and Martin-Gonzalez, M.S. (2017). *Renew. Sustain. Energy Rev.* 76: 1348.
100 Philipp, A., Eichinger, J.F., Aydin, R.C. et al. (2019). *Heat Mass Transfer* 56: 811.
101 Parker, W.J., Jenkins, R.J., Butler, C.P., and Abbott, G.L. (1961). *J. Appl. Phys.* 32: 1679.
102 Dusza, L. (1995). *High Temp.-High Press* 27: 467.
103 Cape, J.A. and Lehman, G.W. (1963). *J. Appl. Phys.* 34: 1909.
104 Gustafsson, S.E., Karawacki, E., and Khan, M.N. (1979). *J. Phys. D: Appl. Phys.* 12: 1411.
105 Gustavsson, M., Karawacki, E., and Gustafsson, S.E. (1994). *Rev. Sci. Instrum.* 65: 3856.
106 Gustafsson, S.E. (1991). *Rev. Sci. Instrum.* 62: 797.
107 Harris, A., Kazachenko, S., Bateman, R. et al. (2014). *J. Therm. Anal. Calorim.* 116: 1309.
108 Oudebrouckx, G., Vandenryt, T., Nivelle, P. et al. (2021). *IEEE Trans. Instrum. Meas.* 70: 1.
109 Liu, B., Wang, H., and Qin, Q.H. (2018). *Materials* 11: 133.
110 Gustavsson, M. and Gustafsson, S.E. (2006). *Thermochim. Acta* 442: 1.
111 Schönfisch, D., Göddel, M., Blinn, J. et al. (2019). *Phys. Status Solidi A* 216: 1800765.
112 Webb, R.C., Bonifas, A.P., Behnaz, A. et al. (2013). *Nat. Mater.* 12: 938.
113 Ahadi, M., Andisheh-Tadbir, M., Tam, M., and Bahrami, M. (2016). *Int. J. Heat Mass Transfer* 96: 371.
114 Zhang, H., Li, Y.-M., and Tao, W.-Q. (2017). *Int. J. Heat Mass Transfer* 108: 1634.
115 Hurley, D.H., Khafizov, M., and Shinde, S.L. (2011). *J. Appl. Phys.* 109: 083504.

116 Paddock, C.A. and Eesley, G.L. (1986). *J. Appl. Phys.* 60: 285.
117 Regner, K.T., Sellan, D.P., Su, Z. et al. (2013). *Nat. Commun.* 4: 1640.
118 Schmidt, A.J., Cheaito, R., and Chiesa, M. (2009). *Rev. Sci. Instrum.* 80: 094901.
119 Schmidt, A.J., Chen, X., and Chen, G. (2008). *Rev. Sci. Instrum.* 79: 114902.
120 Yang, J., Maragliano, C., and Schmidt, A.J. (2013). *Rev. Sci. Instrum.* 84: 104904.
121 Feser, J.P. and Cahill, D.G. (2012). *Rev. Sci. Instrum.* 83: 104901.
122 Rodin, D. and Yee, S.K. (2017). *Rev. Sci. Instrum.* 88: 014902.
123 Donovan, B.F., Warzoha, R.J., Cosby, T. et al. (2020). *Macromolecules* 53: 11089.
124 Li, M., Kang, J.S., and Hu, Y. (2018). *Rev. Sci. Instrum.* 89: 084901.
125 Hammerschmidt, U. (2002). *Int. J. Thermophys.* 23: 1551.
126 Herrmann, K., Pech-May, N.W., and Retsch, M. (2021). *Photoacoustics* 22: 100246.
127 Hu, H., Wang, X., and Xu, X. (1999). *J. Appl. Phys.* 86: 3953.
128 Wang, X., Hu, H., and Xu, X. (2000). *J. Heat Transfer* 123: 138.
129 Breitstein, O., Warta, W., and Langenkamp, M. (2010). *Lock-in Thermography*. Springer.
130 Wolf, A., Pohl, P., and Brendel, R. (2004). *J. Appl. Phys.* 96: 6306.
131 Mendioroz, A., Fuente-Dacal, R., Apiñaniz, E., and Salazar, A. (2009). *Rev. Sci. Instrum.* 80: 074904.
132 Salazar, A., Mendioroz, A., and Fuente, R. (2009). *Appl. Phys. Lett.* 95: 121905.
133 Gao, Q., Kopera, B.A.F., Zhu, J. et al. (2020). *Adv. Funct. Mater.* 30: 1907555.
134 Wang, J., Ahl, S., Li, Q. et al. (2008). *J. Mater. Chem.* 18: 981.
135 Nutz, F.A., Ruckdeschel, P., and Retsch, M. (2015). *J. Colloid Interface Sci.* 457: 96.
136 Nutz, F.A., Philipp, A., Kopera, B.A.F. et al. (2018). *Adv. Mater.* 30: e1704910.
137 Bogomolov, V.N., Kurdyukov, D.A., Parfen'eva, L.S. et al. (1997). *Phys. Solid State* 39: 341.
138 Nutz, F.A. and Retsch, M. (2017). *Phys. Chem. Chem. Phys.* 19: 16124.
139 Nutz, F.A. and Retsch, M. (2017). *Sci. Adv.* 3: eaao5238.
140 Ruckdeschel, P., Kemnitzer, T.W., Nutz, F.A. et al. (2015). *Nanoscale* 7: 10059.
141 Ruckdeschel, P., Philipp, A., and Retsch, M. (2017). *Adv. Funct. Mater.* 27: 1702256.
142 Jia, Z., Wang, Z., Hwang, D., and Wang, L. (2018). *ACS Appl. Energy Mater.* 1: 1146.
143 Ruckdeschel, P., Dulle, M., Honold, T. et al. (2016). *Nano Res.* 9: 1366.
144 Ma, J., Parajuli, B.R., Ghossoub, M.G. et al. (2013). *Nano Lett.* 13: 618.
145 Xie, C., He, L., Shi, Y. et al. (2019). *ACS Nano* 13: 7811.
146 Wicklein, B., Kocjan, A., Salazar-Alvarez, G. et al. (2015). *Nat. Nanotechnol.* 10: 277.
147 Liu, M., Ma, Y., and Wang, R.Y. (2015). *ACS Nano* 9: 12079.
148 Ong, W.-L., Majumdar, S., Malen, J.A., and McGaughey, A.J.H. (2014). *J. Phys. Chem. C.* 118: 7288.
149 Ong, W.-L., Rupich, S.M., Talapin, D.V. et al. (2013). *Nat. Mater.* 12: 410.
150 Losego, M.D., Grady, M.E., Sottos, N.R. et al. (2012). *Nat. Mater.* 11: 502.
151 Wang, C., Guo, J., Dong, L. et al. (2016). *Sci. Rep.* 6: 25334.

152 Wang, F., Zeng, X., Yao, Y. et al. (2016). *Sci. Rep.* 6: 19394.
153 Sun, J., Yao, Y., Zeng, X. et al. (2017). *Adv. Mater. Interfaces* 4: 1700563.
154 Philipp, A., Hummel, P., Schilling, T. et al. (2020). *ACS Appl. Mater. Interfaces* 12: 18785.
155 Losego, M.D., Blitz, I.P., Vaia, R.A. et al. (2013). *Nano Lett.* 13: 2215.
156 Feicht, P., Biskupek, J., Gorelik, T.E. et al. (2019). *Chem. Eur. J.* 25: 8955.
157 Song, N., Jiao, D., Ding, P. et al. (2016). *J. Mater. Chem. C* 4: 305.
158 Renteria, J.D., Ramirez, S., Malekpour, H. et al. (2015). *Adv. Funct. Mater.* 25: 4664.
159 Peng, L., Xu, Z., Liu, Z. et al. (2017). *Adv. Mater.* 29: 1700589.
160 Jurewicz, I., King, A.A.K., Shanker, R. et al. (2020). *Adv. Funct. Mater.* 30: 2002473.
161 Lee, S.Y., Lee, J.S., and Kim, S.H. (2019). *Adv. Mater.* 31: e1901398.
162 Sun, P.C., Wu, Y.L., Gao, J.W. et al. (2013). *Adv. Mater.* 25: 4938.
163 See, K.C., Feser, J.P., Chen, C.E. et al. (2010). *Nano Lett.* 10: 4664.
164 Kim, H.H., Im, E., and Lee, S. (2020). *Langmuir* 36: 6589.
165 Chen, Z., Hsu, P.-C., Lopez, J. et al. (2016). *Nat. Energy* 1: 1.
166 Cartoixa, X., Colombo, L., and Rurali, R. (2015). *Nano Lett.* 15: 8255.
167 Zhang, T. and Luo, T. (2015). *Small* 11: 4657.
168 Toshima, N. (2017). *Synth. Met.* 225: 3.
169 He, M., Qiu, F., and Lin, Z. (2013). *Energy Environ. Sci.* 6: 1352.
170 Ibanez, M., Luo, Z., Genc, A. et al. (2016). *Nat. Commun.* 7: 10766.
171 Ibanez, M., Genc, A., Hasler, R. et al. (2019). *ACS Nano* 13: 6572.
172 Li, Z., Chen, Q., Song, Y. et al. (2020). *Adv. Mater. Technol.* 5: 1901007.
173 Jaramillo-Fernandez, J., Whitworth, G.L., Pariente, J.A. et al. (2019). *Small* 15: e1905290.
174 Liu, D., Xia, Z., Shi, K. et al. (2019). *Opt. Mater.* 92: 330.
175 Gentle, A.R. and Smith, G.B. (2010). *Nano Lett.* 10: 373.
176 Zhai, Y., Ma, Y., David, S.N. et al. (2017). *Science* 355: 1062.
177 Suichi, T., Ishikawa, A., Tanaka, T. et al. (2020). *Sci. Rep.* 10: 6486.
178 Zhang, H., Ly, K.C.S., Liu, X. et al. (2020). *Proc. Natl. Acad. Sci. U.S.A.* 117: 14657.

Index

a

ABA tri-block copolymers 460–461
acetylcholinesterase (AChE) 65, 256, 257
4-acryloylaminobenzo-18-crown-6 (AAB18C6) 252
Advanced Industrial Science and Technology (AIST) 88
aggregation induced emission (AIE)-doped polyion liquid (PIL) PhC sensor 266, 471
aggregation-caused quenching (ACQ) 471
air-liquid interface self-assembly 197
  gas flow induced self-assembly 144–146, 147
  Langmuir–Blodgett (LB) method 144
  needle tip flow method 144, 145
AlGaN-based DUV LED 352, 357
alternating adsorption method 97, 98, 100, 105
amino acids 258, 266–267
amorphous matter thermal transport 500–501
amorphous photonic structures (APSs) coatings 126, 209–211, 215, 216
amphiphilic block copolymers 27, 458, 461, 464, 465, 476
amphiphilic di-block copolymers 458, 459, 467
anisotropic magnetic particles 21
anodic aluminum oxide (AAO) film 123, 285, 286
anti-Counterfeiting 129, 130, 179, 227, 318, 321, 326–327
aptamer-based hydrogel PhC sensor 269
artificial cell membrane model 98
artificial liquid-liquid interfaces 285
as-prepared superparamagnetic colloidal particles 113
atmospheric window 534
atomic layer deposition (ALD) 193, 222
attractive force gradient method 146
AX_nE_m-type CMs 3

b

band gap engineering 171–173
bending-induced oscillatory shear (BIOS) 214, 215
bioactive molecules 461
biomass valorization via catalytic fast pyrolysis 435–437
biomolecules 63, 65–66, 68, 88, 90, 254, 258, 284, 434
biphasic droplet emulsions 202
block copolymers
  definition 457
  di-block copolymers 458–460
  drug loading
    drug and imaging reagent 470–471
    dual drug encapsulation 466–468
    multiple drug delivery 468–470
    single drug encapsulation 465–466
  drug release
    mechanism 472–473

Functional Materials from Colloidal Self-assembly, First Edition. Edited by Qingfeng Yan and George Zhao.

block copolymers (*contd.*)
multiple-triggered release systems 477
pH-triggered release 473
photo-responsive biocompatible micelles 475–477
redox-triggered release 474–475
thermo-triggered release 474
morphology and characteristics 457
nanoscale structures
polymeric micelles 462–463
polymeric nanoparticles 463–464
vesicles or polymersomes 464–465
self-assembly mechanism 461–462
self-assembly of 457
structures 457
tri-block copolymers 460–461
bolaform amphiphilic templates 411
Boltzmann transport equation 501
boron nitride nanosheet (BNNS) networks 520, 521, 524, 525
bottom PhCs 346–348
bottom-up nanorod LEDs 359
Bragg diffraction 150
effects 193
Bragg stacks 244, 251, 527
Bragg's law 150, 241, 243
Bragg–Snell law of diffraction 241
Braun research group 261
Brownian motion 110, 193
brush block copolymers (BBCPs) 204, 206, 207

c

cabazitaxel 459, 461
cadmium sulfide (CdS) quantum dots 321
carbamide bonds 3
carbohydrates 258, 259–264, 269
carbon aerogels 403, 404
carbon fiber yarns 222, 223
carbon microparticles 94
carbon nanoparticle 401, 403, 404
carbon-black powder 214
Casimir forces 29
cellulose nanofiber (CNF) 520, 525
cerasome 98, 99, 100
cetyl trimethylammonium bromide (CTAB) 26, 27, 258, 431
chain stopper 27
chemical colors 185
chemical self-assembly 90
chemical vapor deposition (CVD) 291, 356, 357
chemically modified graphenes (CMGs) 280, 283–285, 288, 293, 294, 302
chromatic patchy particles 52–53
close-packed planar CMs 6
cluster-encapsulation method 19
C_{70} molecular assemblies 94, 96
"coffee-ring" effect 112, 197, 310, 319
coherent scattering 186, 190, 207
colloidal assembly 109, 111, 115, 130, 131, 133, 374, 496, 497, 518, 529, 533
colloidal atoms (CAs)
definition 1
colloidal crystal (CCs)
fabricate
cylinder assembly from capillary tube 128–130
electrospun 130–131
microfluidics 131–133
Langmuir–Blodgett (LB) approach 115
large area
convective assembly 121–123
doctor blade technique 118
LB approach 116–118
spin coating 118–120
spray coating 120–121
vertical deposition 115–116
patterned colloidal 123–128
colloidal materials 1, 496, 499, 500, 502, 503, 518, 535
colloidal molecules (CM) 41
classification for 1–2
definition 1
organic and inorganic molecules

isotropic colloids 3–8
patchy particles clustering 8–13
colloidal particles
patterning assembly 307–328
colloidal PC-based sensors 321, 322
colloidal photonic crystal hydrogels (CPCHs) 254
colloidal photonic crystals (CPCs)
biological species
amino acids 266–267
biomarkers 267–269
carbohydrates 259–262
microorganisms 269–270
proteins 262–265
chemicals responsive
heavy metal ions 252
organic solvents 249
organophosphorus (OP) compounds 255
surfactants 258
VOCs 251–252
(2+1)D 160–179
(2+1)D colloidal photonic crystal 139–179
2D colloidal photonic crystal 139, 140–159
definition 139
fundamentals of 241–244
physical stimuli
humidity 245–247
mechanically responsive 247–249
colloidal polymers (CP)
organic macromolecules
dipole-directed formation 14–15
field-directed formation 17–22
ligand-directed formation 22–30
nanowelding 15
physical or chemical confinement 15–17
colloidal structures, heat transport
dense structures 522–529
porous structures 514–521
color tunability 207, 385
Commission Internationale de L'Eclairage (CIE) 129
composite materials thermal transport 501–503
Concanavalin A (Con A) 263
concave Au nanorods (cAuNRs) 217, 218
controlled micellization self-assembly 193, 204–207, 231
convective assembly 63, 66, 121–123, 141, 146, 497
conventional methods 195, 269
conventional responsive materials 244
copolymer poly(ester amide)-poly (ethylene oxide) (PEA-PEO) 459
covalent bonding 13, 23–25
crack-free single-crystalline colloidal PC 112, 113
critical aggregation concentration (CAC) 463
critical gelation concentration (CGC) 460, 462
crystalline colloidal arrays (CCAs) 244, 252, 262, 265
crystalline matter, thermal transport 499–500
cuprous oxide (Cu_2O) nanoparticles 211
cyclic diquaternary ammoniums (CDA) 408
β-cyclodextrin (β-CD) 266
cylinder assembly from capillary tube 110, 115, 128–130

d

3D Ag reflector 354
2D colloidal photonic crystal
application
detection and sensing 150, 157–158
templates 152–157
close-packed hexagonal colloidal particles 140
optical properties
Debye diffraction ring 151–152
diffraction 150–151
preparation methods

2D colloidal photonic crystal (*contd.*)
air/liquid interface self-assembly 140, 142–146, 179
attractive force gradient method 146–150
direct-writing method 146
electrophoretic deposition 142
roll-to-roll technique (R2R) 146–150
solvent-evaporation-induced self-assembly 141–142
spin-coating 142
(2+1)D colloidal photonic crystal 139
application
FL enhancement substrate 174–179
high performance back reflector 174
macroporous materials 174
growth monolayer 161
optical properties
band gap engineering 171–173
LB film and CE film 167, 171
preparation method
in-situ annealing combined layer-by-layer transfer technique 164–170
Langmuir-Blodgett method 162–164
layer-by-layer scooping transfer technique 164
PDMS sheet assisted layer-by-layer transfer technique 164, 165
SEM images 160
2–D CPCs 140–142, 144, 146, 150, 157, 158, 179, 239, 241, 258, 262, 263
2–D CPC sensors 262, 271
dealumination 400, 416–417, 419, 422, 437
Debye diffraction ring 151–152, 157, 158, 262
Debye equation 499
Debye ring diameter 152, 243
deep ultraviolet (DUV) LEDs 333, 350, 352, 353, 357, 359, 361, 363
deionized water and diethyl zinc (DEZ) 222
dense structures, heat transport 522–529
desilication 400, 416–420, 422, 423, 437
1D-Fourier's law 511
di-block copolymers 458–461, 477
dielectric materials, thermal transport
amorphous matter 500–501
composite and porous materials 501–503
crystalline matter 499–500
dielectric patchy particles 28
diffusion modes 500
dipole-directed colloidal polymerization
electric dipoles 14
magnetic dipoles 14–15
direct dropping 109, 123, 125–126, 133
direct writing 110, 123, 126–127, 146, 178, 207, 368
direct-functionalization 263
direct-writing method 146
disodium chromoglycate (DSCG) 26
dispersion polymerization 72, 309
distributed Bragg reflector (DBR) 354, 391
3D macroporous "bubble" graphene film (MGF) 296
DNA-based assembly methods 41–42
2D nanopillar array 381
doctor blade technique 118
dodecyltrimethylammonium bromide (DTAB) 258
Donnan potential 252, 256, 258, 259, 263
double-line chains 24, 25
DOX-loaded dual-responsive polymer micelles 477
2–D PhCs diffraction 243
3–D PhC hydrogel biosensor 263
2–D PhC hydrogel sensor 269
(2+1)D PhC superlattice 355, 356
drug delivery system (DDS) 472–475, 477, 478
drug diffusion 472
drug release mechanism 472–473

drug-loaded polymeric nanocarriers via conjugation 469
drug-loaded polymeric systems 472
dual drug encapsulation
  hydrophobic drugs 466, 467
  opposite hydrophobicity 467–468
dual-bandgap (2+1)D PC heterostructures 174, 178
dual-core ethane-like CMs 7
dual-drug loaded nanoparticles 466, 468
dual-drug loaded system 467
dye-sensitized solar cells (DSSCs)
  inverse opals photoanodes 371–377
  light harvesting enhancement in 377–381
  structures of 371, 372

e

e-beam lithography (EBL) 335, 337, 344, 351, 361
edge-induced rotational shearing self-assembly 193, 212–215, 231
effective medium theory (EMT) model 413, 502, 525, 526
egg-like molecular assemblies 96, 97
electro-optic modulator (EOM) 506
electrochemical energy storage (EES) devices 279, 280, 282, 295, 296, 301–302
electrode materials 60, 68–71, 296
electron transport layer (ETL) 383, 384, 387, 390, 392, 393
electrospinning 293, 423, 499, 524
electrospun 110, 128, 130–131, 524
electrostatic interactions 6, 22–24, 44, 45, 68, 97, 420
electrostatic repulsion 8, 22, 42, 47, 110, 193, 283, 284, 473
emulsion polymerization 40, 139, 309
entropic depletion interactions 44, 47–52
ethoxylated trimethylolpropane triacrylate (ETPTA) 118, 119
ethylene glycol 249, 309–312, 319
evaporation self-assembly 193–195
external stimuli induced assembly 318

f

fabricate dumbbell-shaped colloids 40
fabrication methods
  colloidal assemblies 496–497
  self-assembly strategies 497–499
fabrication of anisotropic colloids
  bottom-up routes
    cluster-based methods 41
    DNA-based assembly methods 41–42
    surface modification methods 42
    template(seed)-assisted synthesis 38–41
  natural materials 38, 39, 40, 44
  top-down routes
    photolithographic fabrication methods 42–43
    physical methods 43
Fabry–Perot interference fringes 160
field-directed formation of colloidal polymers
  electric field 17–19
  magnetic fields 19–22
finite-difference time-domain (FDTD) method 209–210
Fischer–Tropsch reaction 424–425
flow-directed self-assembly 280, 284
fluorine-containing amphiphilic copolymers 471
fluorosilane functionalized silica (F-SiO_2) nanoparticles 209
formamide 311, 312
Fourier's law 511
frequency-domain thermo-reflectance (FDTR) 506–508, 511
Friedel-Crafts alkylation 431–432
fuel chemistry
  catalytic cracking 421–424
  Fischer-Tropsch reaction 424–425
  methanol to gasoline or olefins 425–428
fullerene nanoarchitectonics 91

fullerene nanoarchitectonics (*contd.*)
  carbon microparticles 94
  C_{70} molecular assemblies 94, 96
  liquid-liquid interfacial precipitation processes 91–93
  rhombic and polygonal nanosheets 92
  supramolecular assembly 93–95, 97
Functional Engineered Nano Architectonics (FENA) 88
functional monomers 311

g
GaN-based LEDs
  efficiency loss mechanisms in 333–334
  MCC 335–337
  techniques 335
gas flow induced self-assembly 142, 144–146, 147
gem opals 139, 140, 240
Gibbs free energy 474
glucose oxidase (GOx) 259
glycolide (GA) moieties 459
gold nanorods 25–28, 68
Gox-functionalized hydrogel PCCA sensor 259
graphene
  CMGs 280
  composite materials
    graphene/carbon material assembly 293–294
    graphene/metal or metal oxide assembly 294–295
    graphene/polymer assembly 294
  definition 279
  EES 280, 282
  EES devices 279
  GO or rGO sheet 280, 281
  Li-S battery 299–301
  LIBs 298–299
  SC 295–298
  self-assembly strategy
    flow-directed 284
    interface-induced self-assembly 285–286
    Langmuir-Blodgett (LB) technique 283
    layer-by-layer (LbL) assembly approach 283–284
    spinning and space-confinement 288
    template-directed self-assembly and hydrothermal process 286–288
  SSA 279
  structural tuning and surface modification
    compact assembly via capillary and mechanical compression 289
    electrostatic and magnetic field 293
    tunable porous structure and surface chemistry 291–293
graphene aerogel mesoporous carbons (GA-MC) 298
graphene gold rush 279
graphene macroform (GF) 289, 297
graphene membrane 296
graphene nanosheets 69, 103, 120, 121, 279, 284, 289, 298, 301
graphene oxides (GOs) 280, 526
graphene paper 296, 299
graphene quantum dots 120, 121
graphene/metal or metal oxide assembly 294–295
gravity 3, 110, 111, 115, 193
guarded hot plate (GHP) 508–511

h
hard templating methods 401–406
heat flow meter (HFM) 508–511
heat management
  passive cooling 534
  thermal rectification 532–533
  thermal switches 529–532
  thermal recorders 529
  thermoelectrics 533–534
heat transport, colloidal structures
  dense structures 522–529
  porous structures 514–521
helical nanofilament (HNF) liquid crystal 220

hexagonal close-packed (HCP) structure 141, 335
Hibiscus trionum flower 186–187
high-resolution transmission electron microscopy (HRTEM) image 299, 300, 427
hole transport layer (HTL) 383
holey graphene framework (HGF) film 291, 292
hollow silica nanoparticle (HSNP) 517
holographic lithography 335, 368
hybridization chain reaction (HCR) 267
hydrodeoxygenation (HDO) 421, 434, 438
hydrogen bond-guided template polymerization method 217
hydrogenation 228, 421, 429, 430, 438
hydrophobic interactions 44–47, 459, 460, 462, 463, 466
hydrophobic molecules 461
hydrothermal process 280, 286–288
hydrothermal synthesis 309, 413

i

immunoglobulin G (IgG) 265
in-situ annealing combined layer-by-layer transfer technique 164–167, 170
incident photon to current conversion efficiency (IPCE) 372, 373, 377, 379, 380, 383, 387
incoherent scattering 190, 500
inductively coupled plasma (ICP) 63, 337
injection efficiency (η_{IE}) 333
ink formulations 308–312, 314
inkjet printing 115, 123, 127–128, 130, 131, 133, 207, 227, 228, 231, 245, 308–316, 319, 321, 323
interface-induced self-assembly 280, 285–286
interfacial thermal resistance (ITR) 524, 525, 532, 533
inverse opals photoanodes 371–377
inverse patchy colloid (IPC) 44, 45
inward Marangoni flow 311
isomerization 421, 432–434, 438
isotropic colloids clustering
  external fields 3–4
  geometrical confinement 4–6
  physical or chemical interactions 6–8

j

Janus particles 8–10, 18, 19, 22, 29, 40, 42, 45, 47, 52–54, 59, 73, 74

k

Knudsen effect 520
Knudsen number 502

l

Langmuir–Blodgett (LB) approach 97, 115–118, 144, 162, 280, 283
Langmuir–Blodgett (LB) effect 199
lanthanide-doped upconversion nanoparticles 470
laser flash analysis (LFA) 503–505
laser lift-off (LLO) 337
layer-by-layer assembly (LbL) method 283
  adsorption process 97
  cerasome 98–100
  graphene 103–104
  graphene nanosheet and ionic liquid 103–104
  mesoporous silica capsules 100–102
  sensor film 102, 103
layer-by-layer scooping transfer technique 164
layered hydrogels self assembly 207–209
Lepidothrix coronata 186–187
Li-S battery 299–301
ligand-directed formation of colloidal polymers
  covalent bonding 23–25
  electrostatic interactions 22–24
  isotropically capped particles, monomers 25–26
  patchy particles as monomers 26–30
light emitting diodes (LEDs)
  efficiency of 333

light emitting diodes (LEDs) (*contd.*)
GaN-based 335
III-nitride 333
nanostructures applications
LSP 348–350
NPSS 350–354
optical reflector 354–356
photonic crystals (PhCs) 343–348
sub-micron lenses 341–343
surface texturing 337–341
optoelectronic devices 356–361
lipopolysaccharides (LPS) 66, 270
liquid crystals 60–61, 288
liquid-liquid interfacial precipitation method 91–93, 95, 96
lithium-ion batteries (LIBs) 68, 69, 71, 279, 280, 298–299
Littrow configuration 151, 158, 243, 264
InGaN/GaN MD lasers 361
local surface plasmon (LSP) 348–350, 363
localized surface plasmon resonance (LSPR) 67, 68, 249
"lock-and-key" assembly 47
lock-in infrared thermography (LIT) 511–514, 525

m

magnetic induced self-assembly 193, 217–220, 231
magnetic nanoparticles 114, 220, 231, 326, 470
magnetic-plasmonic nanorods 217
magnetic/electric field 113–115, 133
Marangoni eddies 311
Marangoni effect 142, 144
Marangoni flow 310, 311
Maxwell-Eucken equation 502, 514
mean free path (MFP) 499, 501, 502, 518, 526
membrane separation-assisted assembly 193, 196–197
meniscus effects 307
mesoporous silica capsules 100–102
mesoporous zeolites
bottom-up
hard templating methods 401–406
non-templating methods 413–416
soft templating methods 406–413
catalytic application of
biomass valorization via catalytic fast pyrolysis 435–437
catalytic combustion 434–435
fuel chemistry 421–427
selective organic reactions 427–434
defintion 399
mixed synthetic approaches 419–421
pros and cons of 401, 402
top-down approaches
dealumination 416–417
desilication 417–419
mesoporous ZSM-5 zeolite 408, 423, 426, 427, 429–431, 435, 437
metal organic chemical vapor deposition (MOCVD) 356, 361
methanol to gasoline or olefins 425–427, 428
methoxyl poly(ethylene glycol)-poly (glutamic acid) (mPEG-PGlu) 469
N-methyl-*N*-dodecyldithiocarbamate (MeDTC) 65
3-methylpyridine (3MP) 29
Micky Mouse-shaped colloids 50
microbead 324
microdisk Lasers 359–361
microfluidics 10, 41, 43, 110, 115, 128, 131–133
microhorns 94, 95
micrometer-sized fullerene cubes with nanorods 95–96
microorganisms 269–270, 458
Mie scattering 130, 189–191
M-Ink 113
modified plans source (MPS) technique 505–506, 507
MOFs-based 1–D PhCs 249, 251
molecular dynamics (MD) simulations 45

molecular imprinted photonic hydrogel (MIPH) 266
monodisperse 1, 6, 15, 20, 21, 27, 45, 61, 109, 113, 118, 121, 123, 131, 132, 139, 162, 209, 212, 220, 308, 309, 314, 318, 374, 498
  colloidal particles 139, 309
  core-shell colloids 113
  oil-in-water emulsion droplets 131
monolayer colloidal crystal (MCC) 164, 253, 311, 335
Monte Carlo (MC) simulations 45
multilayer reflector 188, 189
multiple drug delivery 468–470
multiple heterostructure photonic crystal (MHPC) 178, 179
multiple strand crossover structures 42
multiple-quantum-wells (MQWs) 333, 335, 342, 346, 347, 350, 354, 361
multiple-triggered release systems 477
myristoylcholine chloride (Myr) 65

n
NANO 87
nano-centipedes 14
nano-imprint lithography (NIL) 335, 337, 344, 351, 361
nano-patterned sapphire substrates (NPSS) 350–354, 363
nanoarchitectonics
  concept 89–91
  concept of 88
  fullerene 91–97
  historical backgrounds of 89
  layer-by-later 97–104
nanoarchitecture 88, 90
nanocomposites 66, 69, 118, 119, 498, 499, 520, 522, 524, 525, 527
nanocrystal assemblies (NCAs) 522
nanoeconomics strategy 89
nanopyramid white (NPW) LED 359, 360
nanorod LEDs
  bottom-up 359
  top-down 357–359
nanoscale epitaxial lateral overgrowth (NELOG) 352
nanosphere lithography (NSL) 123, 155, 335
  derived nanostructures 335
National Institute for Materials Science (NIMS) 88
National Nanotechnology Initiative 87
needle tip flow method 142, 144, 145
nematic liquid crystal structures 37, 49, 50, 61
nickel nanoparticles (NiNSs) 65
non-templating methods 413–416
Norland Optical Adhesive 71(NOA 71) 204

o
oil-oil interface self-assembly 193, 202
oil-water interface self-assembly 193, 202–204, 231
one-dimensional (1–D) PhCs 2, 38, 239, 244, 283, 368, 508
opposite hydrophobicity 466–468, 478
optical reflector 354–356
optoelectronic devices
  microdisk lasers 359–361
  nanorod LEDs
    bottom-up 359
    top-down 357–358
organic-inorganic hybrid vesicle 98, 99
organophosphorus (OP) compounds 255
organophosphorus hydrolase (OPH) 66, 256, 257
orientational-dependent inter-particle interactions 54
osmotic pumping 472
oxygen-containing functional groups 280, 287

p
passive cooling heat management 534
patchy metallodielectric particles 18
patchy particles clustering
  colloidal satellites 10–13
  Janus particles 8–10

patterned colloidal PC
direct dropping 125–126
direct writing 126–127
inkjet printing 127–128
physical confinement 123–124
wettability induced template assembly 124–125
patterned PC microchips 323, 324
patterned sapphire substrate (PSS) 350–354
patterned-substrate induced assembly 316–317
patterning assembly of colloidal particles
definition 307
external stimuli induced assembly 318
inkjet printing technology
ink formulations 308–312
surface wettability 312–316
patterned-substrate induced assembly 316–317
photonic crystal (PC)
anti-Counterfeiting 326–327
displays 318–321
sensors 321–325
practical application 307
PDGI/PAAm hydrogel 207
peacock spiders 186
perovskite solar cells (PSCs) 370, 381–388, 391
perovskite, ionic liquids 179
PGM 289
pH-triggered release 473
phenylaminopropyl-trimethxysilane (PHAPTMS) 411, 412
phononic materials 495
photo-responsive biocompatible micelles 475
photoacoustic (PA) method 511
photoinduced self-assembly 193, 220–222, 231
photolithographic fabrication methods 42–43
photonic band gaps (PBGs) 61, 167, 193, 204, 239, 343, 368, 370, 371, 377, 379, 391, 392, 500
photonic crystal (PC\PhCs) 61, 63
colloidal crystal 115–133
color change mechanism 241
definition 109, 239
example of 109
fabricated via both top-down microfabrication approaches 239
latex particles assembly of
general process 110–111
magnetic/electric field 113–114
substrate wettability 111–113
LED
bottom PhCs 346–348
surface PhCs 344–346
PCCA sensors 241
primitive fabrication approach 109
responsive materials and novel photonic structures 244–245
physical colors 185
physical confinement 15, 115, 123–124
physical vapor deposition self-assembly 193, 222–225
Pickering emulsions 10, 71–74
poly (methacrylic acid) (PMAA) hydrogels 269, 463
poly (styrene-methyl methacrylate-acrylic acid) (P(St-MMA-AAc)) 245
poly (vinyl alcohol) (PVA) 43, 121, 130, 164, 165, 207, 249, 261, 296
poly(2-cinnamoyloxyethyl methacrylate)-block-poly(glyceryl monomethacrylate) (PtBA-b-PCEMA-b-PGMA) 461
poly(diallyldimethylammonium chloride) (PDDA) 66, 98–100
poly(dimethylsiloxane) (PDMS) 164, 165, 190, 204, 212, 224, 226–228, 231, 247, 249, 314, 324, 326, 534
poly(dodecyl glyceryl itaconate) (PDGI) amphiphilic bilayers 207–209
poly(ethylene glycol)-*b*-poly(propylene glycol)-*b*-poly(ethylene glycol) (PEG-PPG-PEG) 460
poly(ionic liquids) (PILs) 254

poly(L-lactide) (PLLA) 458
poly(lactic acid) (PLA) 458
poly(N-isopropylacrylamide) (PNIPAM) 4, 27, 254, 258
poly(oxyethylene)- poly(oxybutylene) (PEO-PBO) 459–460
poly(styrene-butadiene) (PSB) colloids 247
polyacrylonitrile (PAN) 299
polydiallyldimethylammonium chloride (PDADMAC) templates 406, 407
polydimethylsiloxane (PDMS) 164, 165, 190, 204, 212, 217, 224, 226–228, 231, 247, 249, 314, 324, 534
 elastomer 204
polyethylene (PE) 207, 224, 294, 431, 467, 501, 532–534
polymer-polymer attraction 460
polymerase chain reaction (PCR) 24, 25, 269
polymeric micelles 462, 463
polymeric nanoparticles 458, 459, 463–464, 466–468, 471, 473, 476, 477
polymerized crystalline crystal arrays (PCCA) sensors 241, 252, 259, 260
polymersomes 462, 464–465, 467, 477
polymethyl methacrylate (PMMA) 17–18, 52, 116, 120–121, 139, 150–151, 195, 249, 266, 296, 319, 376, 461, 534
polymethylpentene (TPX) 534
polymorphic meniscus convergence (PMC) method 16
polystyrene (PS) 3, 23, 25, 40, 50, 114, 159, 199, 207, 239, 254, 319, 335, 372, 514
polystyrene-block-poly(acrylic acid) (PS-b-PAA) 25
polysulfide microspheres (PSFMs) 211
polyurethane (PPU) 215, 254, 514
polyvinyl alcohol (PVA) 43, 121, 130, 164, 207, 249, 261, 296
pore directing agent (PDA) 418
porogen 407
porous materials thermal transport 501
programmable DNA hybridization 30
Pyjama Squid 189

q

quantum confined stark effect (QCSE) 335, 357
quartz crystal microbalance (QCM) 100, 102–103
quasi-inverse opal (QIO) layers 381
quasi-solid-state reaction 414

r

Raman scattering 189
random walk 500
Rayleigh scattering 189–190
reactive ion etching (RIE) treatment 153, *155*, 338
redox-triggered release 474–475
reduced graphene oxide (rGO) 66, 68–69, 196, 280, 283–287, 528
reduced graphene oxide-sulfopropyl methacrylate potassium salt (rGO-SPM) 196–197
reduced graphene oxide/Nafion (RGON) 66
Repetitive Branching 415
residual stress 113
resonant mode 190, 368, 379, 380, 391–393
roll-to-roll technique (R2R) 117–118, 146–150, 215
roughness controlled depletion attractions (RCDA) 50–52

s

sandwich-type cells 220
scattering 189
 coherent scattering 190
 incoherent scattering 190
screen printing self-assembly 193, 215–217, 231

seed-directed synthesis (SDS) 144, 258, 415
seed-silanization method 413
selective organic reactions
  Friedel–Crafts alkylation 431–432
  HDO phenolics 434
  hydrogenation 429–430
  isomerization 432–434
  oxidation 430–431
self-aggregation of one-patch particles 8–10
self-assembled monolayer (SAM) 42, 73, 227, 311, 341, 351
self-assembled photonic crystals
  conventional metal mirrors 368
  DSSCs
    light harvesting enhancement 377–381
    structures of 372
  light scattering effect 368
  PSCs 381–388
  silicon-based solar cells 388–392
self-assembly mechanisms of anisotropic colloids
  DNA mediated interactions 52–54
  electric-field-assisted 54–56
  electrode materials 68, 69, 71
  electrostatic interactions 44
  entropic depletion interactions 47–52
  functional materials 71–74
  hydrophobic interactions 45–47
  liquid crystals 60–61
  magnetic-field-assisted 56–58
  photonic crystals 61–63
  self-propelled colloidal motors 58–60
  sensors
    biomolecules 65–66
    electrode materials 69–71
    environmental cues 68
    gases 66–68
    gold nanorods 68
    metal ions 63–65
self-freestanding holy film 291
self-propelled colloidal motors 58–60
semiconducting polymer nanoparticles 470
shape-selective hydrogenation of nitroarenes 429
shuttle effect 299
signal transduction scheme 241
silicon-based solar cells 370, 388–392
silver nanoparticles (AgNPs) 25, 27, 310, 524–525, 533
single drug encapsulation 465–466
single-walled CNT(SWCNT) 293, 301
slow photons effect 370
soft photonic materials 207, 209
soft templating methods 406–408, 411, 413
specific surface area (SSA) 131, 251, 279–280, 288–289, 291, 293, 296, 322
spherical colloidal crystal beads (SCCBs) 244–247, 254–255, 266–267, 269
spin coating 98, 118–120, 140–142, 193, 240, 249, 341, 387
spinning and space-confinement self-assembly 288
spray coating 120, 121
  colloidal crystal 120, 121
  self assembly 193, 209–211, 231
ssDNA-functionalized SCCB hydrogel sensors 254
Stöber process 38
Stöber synthesis 309
steam-assisted crystallization (SAC) 406, 414
Stober method 139
structural color materials (SCMs)
  colloid dispersions 230
  diffraction 190–193
  dynamic color printing scheme 228
  in nature 185–188
  interference 188–189
  quantum dots film 228
  scattering 189–190
  self assembly method
    air-liquid interface 197–201